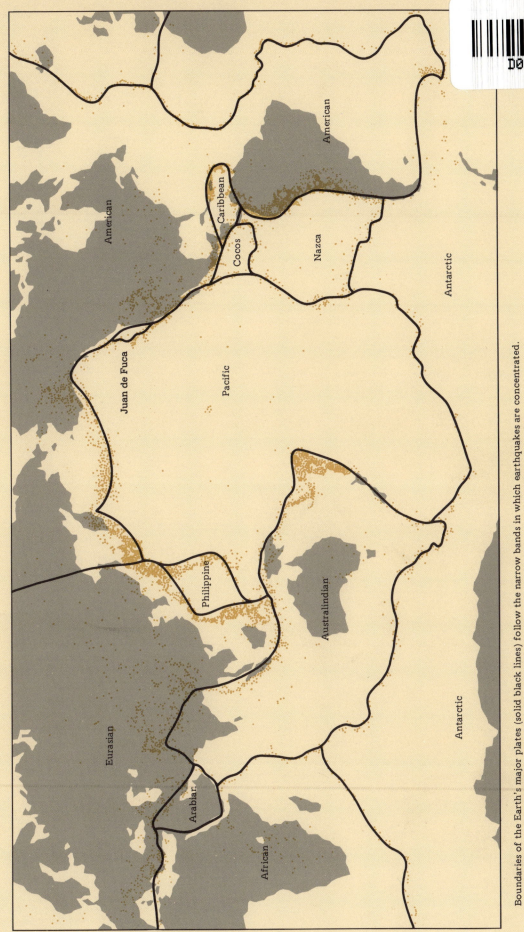

Boundaries of the Earth's major plates (solid black lines) follow the narrow bands in which earthquakes are concentrated.

PHYSICAL GEOLOGY

EIGHTH EDITION

PHYSICAL GEOLOGY

SHELDON JUDSON

Princeton University

MARVIN E. KAUFFMAN

American Geological Institute

PRENTICE HALL

Englewood Cliffs, New Jersey 07632

Library of Congress Cataloging-in-Publication Data

JUDSON, SHELDON.
  Physical geology / Sheldon Judson, Marvin E. Kauffman. — 8th ed. p.   cm.

  Includes bibliographical references.
  ISBN 0-13-666405-9 — ISBN 0-13-666421-0 (instructor's ed.)
  1. Physical geology.   I. Kauffman, Marvin Earl, date.
  II. Title.
  QE28.2.J82 1990   89-23065
  551 — dc20   CIP

Editorial/production supervision: Virginia Huebner
Interior design: Peggy Kenselaar and Christine Gehring-Wolf
Cover design: Peggy Kenselaar
Manufacturing buyer: Paula Massenaro
Page layout: Gail Cocker
Photo editor: Lori Morris-Nantz
Photo research: Tobi Zausner
Cover photograph: Yosemite Falls [Charlie Blecker]

Illustrations Copyright by William L. Chesser	
Figure 5.17, page 87	Figures B.13.2.1a & b, page 288
Figure 6.17, page 115	Figures B.13.2.1c & d, page 288
Figure 13.1, page 273	Figure 14.18, page 318

© 1990, 1987, 1982, 1978, 1971, 1965, 1958, 1954 by Prentice-Hall, Inc.
A Division of Simon & Schuster
Englewood Cliffs, New Jersey 07632

Printed in the United States of America
10 9 8 7 6 5 4 3 2 1

ISBN 0-13-666405-9

Prentice-Hall International (UK) Limited, *London*
Prentice-Hall of Australia Pty. Limited, *Sydney*
Prentice-Hall Canada Inc., *Toronto*
Prentice-Hall Hispanoamericana, S. A., *Mexico*
Prentice-Hall of India Private Limited, *New Delhi*
Prentice-Hall of Japan, Inc., *Tokyo*
Simon & Schuster Asia Ptc. Ltd., *Singapore*
Editora Prentice-Hall do Brasil, Ltda., *Rio de Janeiro*

CONTENTS

PREFACE

In the 35 years since the first edition of this book appeared our knowledge about our Earth has grown tremendously. In 1954 plate tectonics and sea-floor spreading were yet to be discovered and continental drift was generally discounted. The Russians had yet to launch the first Earth-orbiting satellite and usher in a period of spectacular discoveries as we explored our planetary system. Since 1954 we have greatly augmented our ability to "see" deep into the Earth. New field explorations and laboratory techniques have extended our vision back into time and shown the Earth to be 1.5 billion years older than we thought it was. These advances have indeed been significant. Just as important, however, has been our belated recognition that human activity can affect geologic processes on both a local and a global scale and not always to the advantage of living things.

The developments of the last three and a half decades have certainly changed the ways we view our Earth. Successive editions of this textbook have been modified to take these developments and understandings into account. Revolutionary as many of these have been, however, they have found a congenial place in the framework of geologic knowledge that began to build 200 years ago toward the end of the 18th century. This edition of *Physical Geology,* as did each of its predecessors, combines the new with the old in a continuing effort to present, interestingly and understandably, the accumulating body of knowledge that is physical geology.

The basic organization of this edition remains unchanged from the previous edition. Comments from students and colleagues, however, have resulted in extensive revisions in several places. In Chapter 3, *Origin and Occurrence of Igneous Rocks,* we have clarified the treatment of the Bowen reaction series and the generation of different types of magma. In Chapter 9, *Deformation,* we have simplified somewhat the treatment of stress and strain in rocks. We have integrated a section on the process of mountain building into Chapter 11, *Plate Tectonics and Mountain Building.* Chapter 12, formerly *Mass Movement of Surface Material,* has been recast and expanded, and retitled *Hillslopes: Form and Process* in order to emphasize the importance of slopes in the area of the surface processes. In Chapter 20, *Planets, Moons, and Meteorites,* we have added a section on meteorite impacts on Earth. We have included a new Appendix, E, which lists the names and addresses of Earth Science organizations and government surveys in Canada, Mexico and the United States. We have added a few new Boxes on subjects of general interest. Additional changes, although less extensive, have been made throughout the book.

Over a hundred photographs, most of them in color,

appear for the first time in this edition. An almost equal number of diagrams are either new or redesigned and a new color system has been adopted for all color diagrams.

Each chapter now opens with a color photograph appropriate to its subject. We have retained the opening quotation and short overview for each chapter. Chapters end with concise summaries. We have added some new questions, and have again worked out the answers for those that are quantitative. A few new titles appear among the Supplementary Readings and again the nature of each is indicated with a short comment.

In preparing this edition we have had the benefit of comments from users of the seventh edition. In addition we have had very effective reviews from colleagues as this edition developed. We are pleased to acknowledge the time and expertise of Gary C. Allen, University of New Orleans; William E. Bonini, Princeton University; Philip E. Brown, University of Wisconsin-Madison; John H. Cleveland, Indiana State University; Grenville Draper, Florida International University; John B. Droste, Indiana University, Bloomington; Steven B. Esling, Southern Illinois University; Stewart S. Farrar, Eastern Kentucky University; Gordon Frey, University of New Orleans; Robert B. Hargraves, Princeton University; Pamela Hemphill, West Chester State University; Kenneth Johnson, Skidmore College; Albert M. Kudo, University of New Mexico; Lawrence L. Malinconico, Jr., Southern Illinois University; Robert L. Nusbaum, College of Charleston; Steven M. Richardson, Iowa State University; Lisa Rossbacher, California State Polytechnic University; Douglas B. Sherman, College of Lake County; and Robert A. Wiebe, Franklin and Marshall College.

We are deeply indebted to colleagues and organizations for the use of materials in the illustrations for this book. We acknowledge their contributions specifically where they appear. Illustrations without attribution have been provided by the authors. At Prentice Hall, Holly Hodder again served efficiently as our editor and did so with reassuring confidence and ésprit. Virginia Huebner guided this volume, as she did its predecessor, through the various stages of production with unusual patience and accuracy. New and redesigned diagrams testify to the talents of Lorraine Abramson and John Smith of Network Graphics, and of William Chesser, Littleton, Colorado.

Our debts to the people, named and unnamed, are many. The errors and shortcomings of this volume, however, are ours.

SHELDON JUDSON

MARVIN E. KAUFFMAN

Eruption of the cone of Pu'u O'o along east rift at Hawaii's Kilauea Volcano, July 1984. [Katharine Cashman.]

CHAPTER ▪ 1

TIME AND A CHANGING EARTH

"O earth, what changes hast thou seen!"

Alfred, Lord Tennyson, from "In Memoriam"

This initial chapter provides a first look at *physical geology*. We introduce a few important concepts underlying much of what we shall discuss in some detail in later chapters.

First, Earth time is long—about 4.6 billion years—and like human history, it can be considered in either *relative* or *absolute* terms. This vast expanse of time has made it possible for very slow processes to effect very large changes in the Earth.

Second, we find that understanding present-day Earth processes provides the key to interpreting the *rock record*. This simple but powerful concept allows us to extrapolate from the present deep into the past. Geologic history is recorded in the rocks: *igneous, sedimentary,* and *metamorphic.*

A third major concept is that of the *rock cycle.* This provides a useful model by which to arrange the different rock groups, the changes they undergo, and the processes that connect the various parts of the rock record. In turn, the rock cycle leads logically to a preliminary description of *plate tectonics, sea-floor spreading,* and *continental drift.* These last three processes have unified much of our thinking about the Earth as a dynamic, changing body and have become an effective elaboration of the rock cycle.

Fourth, we make the perhaps overly obvious point that geology has had, and will continue to have, practical applications to a wide range of human activities.

1.1

SOME INTRODUCTORY OBSERVATIONS

Geology is the science of the Earth, an organized body of knowledge about the globe on which we live—about the mountains, plains, and ocean deeps, about the history of life from amoeba to humanity, and about the succession of physical events that accompanied this orderly development of life.

Geology helps us unlock the mysteries of our environment. Geologists explore the Earth from the ocean floors to the mountain peaks to discover the origins of our continents and the encircling seas. They try to explain an immensely complex landscape shaped, for example, by the sudden violence of earthquake or volcano (chapter-opening figure) and, as importantly, by the gentler and slower processes of water in streams (Figure 1.1) or underground or by the almost imperceptible changes of firm rock to pulverulent soil. They probe the action of glaciers that crawled over the land and then melted away hundreds of millions of years ago, and of some that even today cling to high valleys and cover most of Greenland and Antarctica, the remnants of a recent but presently receding glaciation.

Geologists search for the record of life from the earliest one-celled organisms of ancient seas to the complex plants and animals of the present. This story, from simple algae to seed-bearing trees and from primitive protozoa to highly organized mammals, is told against the ever-changing physical environment of the Earth.

Geologists are by no means earthbound, however. They have already applied their knowledge to the study of

1.1 The Clarks Fork, Wyoming, flows through its canyon out of the Beartooth Mountains. It carries the debris or weathered rock down to the Yellowstone River, then on to the Missouri and Mississippi Rivers, and finally to the Gulf of Mexico. [Tim Palmer, courtesy of American Rivers Conservation Council.]

other bodies in our planetary system (see Chapter 20). One can confidently predict that when observations of the planets and moons of other solar systems are possible, geologic principles will apply to their study as well (see Box 1.1).

The Earth has not always been as we see it today, and it is changing (but slowly) before our eyes. The highest mountains are built of materials that once lay beneath the oceans. Fossil remains of animals that swarmed the seas millions of years ago are now dug from lofty crags. Every continent is partially covered with sediments that were once laid down on the ocean floor, evidence of an intermittent rising and settling of the Earth's surface.

In this chapter we take a preliminary look at some of the important concepts in the study of our changing Earth. In subsequent chapters we shall discuss at greater length the subjects touched on here. Therefore you must read this chapter with the understanding that the assertions we make will be more fully explained at appropriate later points. This chapter is intended to provide a framework within which to organize your thinking about much of that more detailed material: It serves as a kind of map by which to chart our exploration of physical geology.

1.2
TIME

We now know that the Earth is about 4.6 billion years old—nearly a million times the age attributed to it in the seventeenth century. We are less certain about the age of the universe, of which our Earth is a part. But most current evidence suggests that the universe is more than three times the age of the Earth.

In the following paragraphs we briefly discuss some of the ways in which geologic time is divided and measured, deferring until Chapter 8 a more detailed consideration of the subject.

ABSOLUTE AND RELATIVE TIME

An initial and casual reaction to the notion of time is that we can mark it off without much difficulty, even though we recognize, as Thomas Mann wrote in *The Magic Mountain,* that

> time has no division to mark its passage, there is never a thunder-storm or blare of trumpets to announce the beginning of a new month or year. Even when a new century begins it is only we mortals who ring bells and fire off guns.[1]

Nevertheless, our modes—seconds, years, millennia, and the rest—are all we have to work with; consequently we can consider geologic time from two points of view: as relative or as absolute. **Relative time**—that is, whether one event in Earth history came before or after another event—disregards years (Figure 1.2). On the other hand, if we can determine how many years before the present an event took place—whether it was 10,000 years or 60 million years—we deal in **absolute time** (Figure 1.3).

Relative and absolute time in Earth history have their counterparts in human history. In tracing the history of the Earth we may wish to know whether some event, such as a volcanic eruption, occurred before or after another event,

---

[1] Thomas Mann, *The Magic Mountain,* trans. H. T. Lowe-Porter, p. 225, Modern Library, New York, 1955.

BOX 1.1   What's in a Name?

The word *geology* derives from the Greek *geo,* "Earth," and *logos,* "discourse," and comes to us through the Latin. The present meaning of geology, "science of the Earth," came into use in the late eighteenth and early nineteenth centuries.

The medieval Latin word *geologia* was apparently used for the first time by Richard de Bury in the fourteenth century. To him, it meant the study of the law. If you were reading the Italian F. Sessa's 1687 volume *Geologia,* you would discover that the author was trying to demonstrate that the "influences" ascribed by the astrologers to the stars actually came from the Earth.

Students of the Earth 150 to 200 years ago might have known their subject either as geology or as *geognosy.* They were essentially the same subjects. Geognosy didn't catch on as a word; nor did *geonomy,* introduced a century ago as a synonym for geology.

Today we pretty well know what we mean when we use the term *geology.* True, some colleges and universities call the study of the Earth *geoscience* and others maintain that *Earth science* is a better term. Indeed, there are some differences in meaning among these terms. For our purposes here, we prefer the term *geology,* and follow a general definition coined by one of our colleagues:

ge·ol·o·gy, n. The study of the Earth and other solid bodies in space. Geology applies the techniques originally devised for Earth problems to deciphering the present attributes, history, and origin of any natural solid body.

such as a rise in sea level, and how these two events are related in time to a third event, perhaps a mountain-building episode. In human history, too, we try to determine the relative position of events in time. In studying United States history we find it important to know that the Revolutionary War preceded the Civil War and that the Canadian-American boundary was fixed sometime between these two events.

Sometimes events in both Earth history and human history can be established only in relative terms. Yet our record becomes increasingly precise as we fit more and more events into an actual chronological calendar: If we did not know the date of the United States–Canadian boundary treaty—if we knew only that it was signed between the two wars—we would place it between 1783 and 1861. (Recorded history, of course, provides us with the actual date, 1846.)

Naturally, we should like to date geologic events with precision. But so far this has been impossible, and the accuracy we have achieved in dating human history—that is, written human history—will likely never be duplicated in geologic dating. Still, for many geologic events we can determine approximate dates that are probably of the correct order of magnitude. We can say that dinosaurs became extinct about 66 million years ago and that about 11,000 years ago the last continental glacier began to recede from New England and the area bordering the Great Lakes.

Radioactivity   Radioactive elements (those whose nuclei spontaneously emit particles to produce new elements, as discussed in Section 8.3) have provided the most effec-

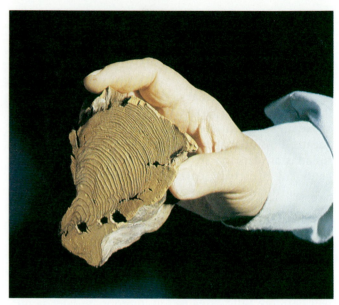

1.3   This fragment of wood has been cut from a spruce log found in a buried forest bed at Two Creeks, Wisconsin. It clearly shows the annual rings decreasing in width out to the bark layer when the tree died, presumably as the result of an advancing ice sheet. The tree's rings show that it lived for 79 years. Radiometric measurements (Carbon 14) give the time of death of the trees as about 11,350 radiocarbon years ago. [John Simpson.]

tive means of measuring absolute time. The rate at which a given radioactive element decays is (so far as we have been able to determine) unaffected by changes in physical conditions or by time. So if we know the amount of original radioactive material (the **parent**) that remains, the rate of radioactive decay, and the amount of new elements (the **daughters**) that has formed, then we can calculate the time elapsed since radioactive decay began. Of course, the calculation is not quite so simple, as we discuss in Section 8.3. Nevertheless, from the time that the first radioactive-age determinations were made (in 1906) to now, we have learned enough about the techniques and pitfalls of radioactive dating to be confident about the thousands of dates now available, particularly those determined during the last two decades. A variety of elements has proved useful, with a time range from that of carbon 14, which can be used to date events that occurred a few hundred to a few tens of thousands of years ago, to that of an element such as uranium 238, which has the potential of dating events several times greater than the age of the Earth.

1.2   These layers of rock in the walls of the Grand Canyon of the Colorado River in Arizona record relative time. Each layer is younger relative to the layers beneath and older relative to the layers above. [Neil Lundberg.]

DIVISIONS OF GEOLOGIC TIME

The application of radioactive elements to the measurement of geologic time became useful only after geologists had already constructed a calendar of geologic events. This calendar, still in use, was based on the ages of rock units relative to each other. Such relative ages were determined by conclusions drawn from a number of phenomena, including the superposition of younger rocks on older rocks, the cutting of older rocks by more recently formed rocks, and the progressive evolutionary stages of plant and animal life, as represented by remains in some rocks. The methods are discussed in detail in Section 8.1. It suffices to say here that the arrangement of rock units and the Earth events they record in the geologic calendar, as determined before the twentieth century, have been confirmed by the absolute dates of later radioactive dating.

The rock units in their proper chronological order make up the **geologic column,** which is reproduced on the front endpaper of this book.

UNIFORMITARIANISM

Modern geology was born in 1795, when James Hutton (1726–1797), a Scottish medical man, gentleman farmer, and geologist, formulated the principle now known to geologists as the **doctrine of uniformitarianism** (Figure 1.4). This principle simply means that the physical processes operating in the present to modify the Earth's surface also operated in the geologic past—which is another way of saying that the laws of nature are unchanging. We are reasoning by analogy when we say that the record of the past was created by processes that are still operating today. True, the intensity of any process may—and does—change with time, but the basic process remains the same.

Here is an example: We know from observations that modern glaciers deposit a distinctive type of debris made up of rock fragments that range in size from submicroscopic particles to boulders weighing several tons. This debris is jumbled, and many of the large fragments are scratched and broken. We know of no agent other than glacier ice that produces such a deposit. Now suppose that in the New England hills or across the plains of Ohio or in the deep valleys of the Rocky Mountains we find deposits that in every way resemble glacial debris but find no glacier in the area. We can still assume that the debris was deposited by now-vanished glaciers. On the basis of evidence like this, geologists have worked out the concept of the great Ice Age (Chapter 15).

Such an example can be multiplied many times. For instance, most Earth features and rocks exposed at the Earth's surface today are explained as the result of past processes similar to those of the present. We shall find that many conclusions of physical geology are based on the conviction that modern processes also operated in the past.

Armed with Hutton's concept of uniformitarianism, nineteenth-century geologists were able to explain Earth features on a logical basis. But the very logic of the explanation gave rise to a new concept for students of the Earth. Presumably, past processes operated at the same slow pace as do those of today. Consequently very long periods of time must have been available for those processes to accomplish their tasks. It was apparent that a great deal of

THEORY

OF THE

EARTH,

WITH

PROOFS AND ILLUSTRATIONS.

IN FOUR PARTS.

Br JAMES HUTTON, M. D. & F. R. S. E.

VOL. I.

EDINBURGH:
PRINTED FOR MESSRS CADELL, JUNIOR, AND DAVIES,
LONDON; AND WILLIAM CREECH, EDINBURGH.

1795.

1.4 Title page of James Hutton's *Theory of the Earth*, which was published in Edinburgh in 1795. This work, which appeared in two volumes, is widely credited with marking the birth of modern geology.

time was needed for a river to cut its valley, or for hundreds or thousands of feet of mud and sand to be deposited on an ocean bottom, and then harden into solid rock and rise far above the level of the sea.

The concept of almost unlimited time in Earth history is thus a necessary outgrowth of the application of the principle that **the present is the key to the past.** For example, geologists know that mountains as high as the modern Rockies once towered over what are now the low uplands of northern Wisconsin, Michigan, and Minnesota. But only the roots of these mountains are left, the great peaks having long since disappeared. Geologists explain that the ancient mountains were destroyed by rain and running water, creeping glaciers and wind, and landslides and slowly moving rubble, and that these processes acted essentially as they do in our present-day world.

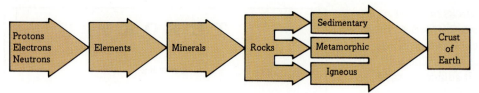

1.5 Relation of particles, elements, minerals, rocks, and the crust of the Earth.

Now think of what this explanation means. We know from firsthand observation that streams, glaciers, and winds have some effect on the surface of the Earth. But how can such feeble forces level whole mountain ranges? Instinct and common sense tell us that they cannot. This is where the factor of time comes into the picture. True, the small, almost immeasurable amount of erosion that takes place in a human lifetime has little effect; yet when the erosion during one lifetime is multiplied by millions of lifetimes, mountains can be worn away. Time makes possible what seems impossible.

Before ending these observations on uniformitarianism, we should raise a warning flag already hinted at in the final sentence of the opening paragraph of our discussion. No process is "uniform" all over the world at any one time. Its rate and intensity can be—and usually are—different from place to place. Some areas erode faster than others; some are less prone to earthquakes than others. By the same token, rates and intensities may change with time. For example, a meteoritic impact is relatively rare today compared to during the earliest history of the Earth, when impacts were not only extremely numerous but also (at least some of them) enormously large.

A final word: Some events are "catastrophic." These are rare and unlikely to be experienced in the average human lifetime. But we should view catastrophic events not as the result of a unique process, but rather as extreme end members of a familiar process. Thus a gigantic flood that happens once every 100 years or so appears unique when compared with an average annual flood. But the process of flooding obeys the same basic principles whether the event is large or small, catastrophic or benign.

1.3
EARTH MATERIALS AND THE ROCK CYCLE

Geology is based on the study of rocks. We seek to know their composition, their distribution, how they are formed and destroyed, and why they are lifted up into continental masses and depressed into ocean basins.

Rock is the most common material on Earth. We may recognize it as the gravel in a driveway, the boulders in a stream, or the cliffs along a ridge. It is the stuff that forms the crust of our earth. Close examination demonstrates that rock can be divided into three major groups based on mode of origin: **igneous, sedimentary,** and **metamorphic.**

When we look deeper, we find that rock is merely an assemblage of minerals. These minerals are chemical compounds with definite compositional and physical characteristics. Over 2,000 different minerals have been described, but only a handful (a dozen, more or less) make up the great bulk of the rocks of the Earth's crust. Digging farther, as we do in Chapter 2, we recognize that minerals are composed of chemical elements that in turn are made up of differing arrangements of protons, electrons, and neutrons. Figure 1.5 is a convenient way of relating these various levels of matter.

THE THREE ROCK FAMILIES

In later chapters we will examine the three rock families—igneous, sedimentary, and metamorphic—in some detail. Here we will merely take an introductory look at them.

Igneous rocks, the ancestors of all other rocks, take their name from the Latin *ignis,* meaning "fire." They form when a hot molten mass cools. When this material lies beneath the surface, we call it a **magma.** Igneous rocks that form from magma are hidden from our view until erosion strips away the overlying rock. When the magma works its way to the surface it may explode violently to form layers of **ash and cinders,** or it may erupt more quietly as a stream of molten material that we call **lava** (Figure 1.6). Both of these may combine to build a volanic cone. The lava cools rapidly—in days or weeks—into firm, solid rock. By contrast, the molten mass at depth cools and solidifies very slowly over hundreds of thousands, even millions, of years.

Most sedimentary (Latin *sedimentum,* "settling") rocks are made up of particles derived from the breakdown of preexisting rocks. Usually these particles are transported by gravity, water, wind, or ice to new locations, where they are deposited in new arrangements. For example, waves beating against a rocky shore may provide the sand grains and pebbles for a nearby beach. If these beach deposits were to harden, we should have sedimentary rock. As we will see later, however, a few sedimentary rocks form from the direct chemical precipitation of minerals. One of the most characteristic features of sedimentary rocks is the layering of the depositis that go to make them up (Figure 1.7).

1.6 A river of molten lava flows down Mt. Etna, Sicily, during a May 1983 eruption. This has cooled to form the extrusive igneous rock basalt. [John P. Lockwood, U.S. Geological Survey.]

1.7 Layering or stratification is one of the most characteristic features of sedimentary rocks. Here, at Trenton Falls, New York, the waters of the West Canada Creek tumble over the horizontally layered beds of limestone.

1.8 These twisted bands of metamorphic rock in Aberdeenshire, Scotland, began as flat-lying layers of limey muds. They were then turned into the sedimentary rock limestone. Later they were subjected to high pressure and temperature and the resulting contorted beds are the metamorphic rock marble. [Geological Survey of Great Britain.]

Metamorphic rocks compose the third large family of rocks. Metamorphic (from the Greek words *meta,* "change," and *morphē,* "form") refers to the fact that the original rock has changed from its primary form to a new form. Earth pressures, heat, and chemically active fluids beneath the surface may all be involved in changing an originally sedimentary or igneous rock into a metamorphic rock (Figure 1.8).

THE ROCK CYCLE

We have suggested that there are definite relationships among sedimentary, igneous, and metamorphic rocks. With time and changing conditions, any one of the rock types may be changed into some other form. These relationships form a cycle, as shown in Figure 1.9, which is simply a way of tracing out the various paths that earth materials follow. The outer circle represents the complete cycle; the arrows within the circle represent shortcuts in the system that can be, and often are, taken. Notice that the igenous rocks are shown as having been formed from a magma and as providing one link in a continous chain. From these parent rocks, through a variety of processes, all other rocks can be derived.

First, weathering attacks the solid rock, which either has been formed by the cooling of a lava flow at the surface or is an igneous rock that was formed deep beneath the Earth's surface and then was exposed by erosion. The products of weathering are the materials that will eventually go into the creation of new rocks—sedimentary, metamorphic, and even igneous. Landslides, running water, wind, and glacier ice all help to move the materials from one place to another. In the ideal cycle this material seeks the ocean floors, where layers of soft mud, sand, and gravel are consolidated into sedimentary rocks. If the cycle continues without interruption, these new rocks may in turn be deeply buried and subjected to pressures caused by overlying

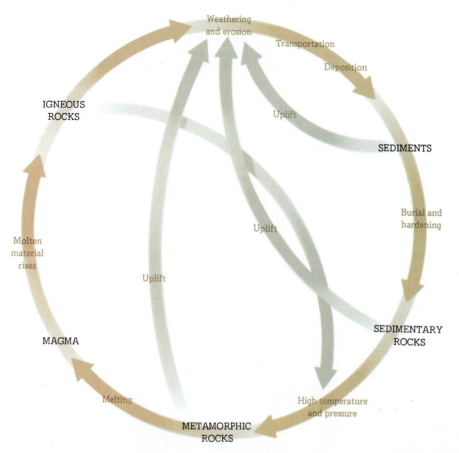

1.9 The rock cycle shown diagrammatically. If uninterrupted, the cycle will continue clockwise around the outer margin of the diagram from magma to igneous rocks to sediments to sedimentary rocks to metamorphic rocks and back to magma. The path may be interrupted, however, and follow one of the arrows through the interior of the diagram.

rocks, to heat, and to forces developed by Earth movements. The sedimentary rocks may then change in response to these new conditions and become metamorphic rocks. If these metamorphic rocks undergo continued and increased pressure and heat, they may eventually lose their identity and melt into a magma. When this magma cools, we have an igneous rock again; we have come full circle.

But notice too that the complete rock cycle may be **interrupted.** An igneous rock, for example, may never be exposed at the surface and hence may never be converted to sediments by weathering. Instead, it may be subjected to pressure and heat and converted directly into a metamorphic rock without passing through the intermediate sedimentary stage. Other interruptions may take place if sediments or sedimentary rock or metamorphic rock are attacked by weathering before they continue to the next stage in the larger, complete cycle.

This concept of the rock cycle was probably first stated in the late eighteenth century by James Hutton, whom we have already mentioned.

> We are thus led to see a circulation in the matter of this globe, and a system of beautiful economy in the works of nature. This earth, like the body of an animal, is wasted at the same time that it is repaired. It has a state of growth and augmentation; it has another state, which is that of diminution and decay. This world is thus destroyed in one part, but it is renewed in another; and the operations by which this world is thus constantly renewed are as evident to the scientific eye, as are those in which it is necessarily destroyed.[2]

We can consider the rock cycle to be a kind of outline of physical geology, as a comparison of Figure 1.9 with the Table of Contents of this book will show.

1.4

PLATE TECTONICS, SEA-FLOOR SPREADING, AND CONTINENTAL DRIFT

To generations of geologists it has been clear that the Earth is a dynamic, changing body. As we have pointed out, new rocks — sedimentary rocks — are made from the weathered debris of older rocks, and they can be crumpled, metamorphosed, and lifted into high mountain chains; or old rocks can be melted and the resulting magma cooled to form igneous rocks. Indeed, these changes can be traced in nature and pictured as we have in the rock cycle in Figure 1.9. But only recently have we been able to fit the rocks and their

alterations into a worldwide, integrated system and to explain in a general way the origin of continents and ocean basins and of mountain ranges and continental plains, as well as the location of volcanoes and earthquake belts. Two of the processes involved are referred to as **plate tectonics** and **sea-floor spreading.** Although we shall later extensively discuss these processes — particularly in Chapter 11 — we take a preliminary look at them here. They include the movement of several large plates that, fitted together, form the rigid rind of the Earth. This movement causes the growth as well as the closure of ocean basins and the creation of earthquakes, volcanoes, and mountain building along the plate boundaries. The movement also accounts for the shifting positions of continents over the last several hundred million years. The processes focus on the outer 200 km (kilometers) of the Earth, a subject to which we shall also return later. Meanwhile a brief sketch of what we know about this zone is presented below.

LITHOSPHERE, ASTHENOSPHERE, CRUST, AND MANTLE

The outer 50 to 100 km of the Earth is a rigid shell of rock called the **lithosphere** (from the Greek *lithos,* "rock," and "sphere"). Yet as observations from deep mines tell us, the temperature of the Earth increases by around 15°C (Celsius) with each kilometer of depth. So at a depth of about 70 km the temperature averages 1000°C, at which rock will slowly flow if pressure is applied. This "soft" zone, which is encountered at a depth from 70 to 100 km, is called the **asthenosphere** (Greek *asthenēs,* "weak"). As we shall see, its existence helps explain some of the Earth's major movements, both vertical and horizontal.

The lithosphere and asthenosphere are distinguished by temperature, but we can also divide the outer part of the Earth into shells on the basis of composition. We speak, therefore, of the skin of the Earth as the **crust** and of the bulk of the Earth beneath the crust as the **mantle.** A wide variety of rock types constitutes the Earth's crust, and we can make direct observations of most of the types. A discontinuous cover of sedimentary rocks, a few meters to a few kilometers thick, overlies igneous and metamorphic rocks. The crust beneath the continents is thicker (averaging about 35 km) than that beneath the oceans (about 5 km). A dark-colored, relatively heavy igneous rock called **basalt** dominates the crust beneath the oceans, and we quite naturally call it **oceanic crust.** Its density, about 3 g/cm^3 (grams per cubic centimeter), contrasts with the lighter-weight **continental crust,** with a density of about 2.8 g/cm^3. A large portion of the continental crust is composed of the igneous rock called **granite,** which not only is less dense than basalt but also is light gray to pink in color. Beneath the rocks of the crust at a depth of 5 to 50 km lies the mantle, made of rocks with a density of about 3.2 g/cm^3; the asthenosphere lies within the upper mantle (Figure 1.10).

[2] James Hutton, *Theory of the Earth,* vol. 2, p. 562. Edinburgh, 1795. Hutton's theory of the earth was first presented as a series of lectures before the Royal Society of Edinburgh in 1785. These lectures were published in book form in 1795. Seven years later Hutton's concepts were given greater circulation in a more readable treatment, called *Illustrations of the Huttonian Theory,* by John Playfair.

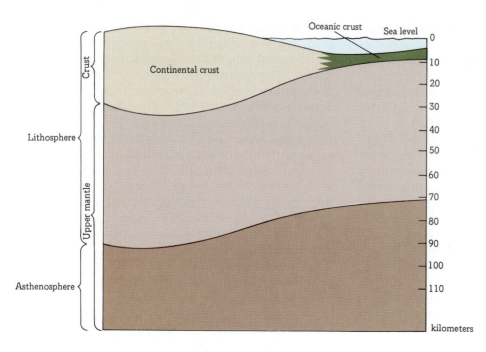

1.10 The relationships between the upper mantle and crust continental and oceanic) and the lithosphere and asthenosphere. (See also Figure 10.29.)

VERTICAL MOVEMENTS AND ISOSTASY

Precise surveying shows that when a sufficiently large lake forms behind a dam, it will depress the Earth's crust slightly. We also know that when glaciers expanded during the Ice Age, their weight depressed large areas of the crust. Conversely, when the glaciers waned and disappeared, the land rose, recovering its earlier elevation; and even today some areas of Scandinavia and Canada are still rising in response to the melting of the last glaciers. Again, we know that surface elevations can be affected by changes in temperature in the zones beneath. For instance, the Midatlantic Ridge stands high in the Atlantic Ocean basin because of increased heat flow from below along this axis of rifting. As we will see in the following paragraphs, the ocean floor moves slowly away from this zone of high heat, and as it does, it cools and subsides to lower and lower elevations below sea level. The ancient geologic record provides other examples. Studies of thick sequences of sedimentary rocks show numerous situations in which thousands of meters of sediments accumulated in shallow marine environments. There is no way to explain a continuous pile of shallow-water sediments unless the basin in which they had accumulated was shallow from the beginning and kept sinking slowly as additional sediments were added.

In all these examples of **vertical movement** portions of the Earth's crust behave as if they were floating in a soft, slowly flowing zone. And that is what we believe happens. The crust and the uppermost mantle that lie above the asthenosphere have some strength. But they cannot resist the pull of gravity, and they respond to the addition or removal of a load. So if material such as water, ice, or sediments is added at the surface, that overloaded area (and the column of rigid rock beneath) sinks slightly into the asthenosphere; conversely, if a load is removed, an area of

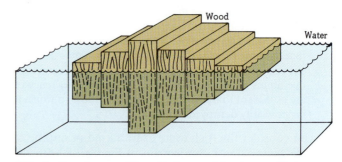

1.11 Isostasy is a condition of balance, of floating equilibrium, illustrated here by blocks of wood of varying height floating in water. Each block sinks to a different depth in the water, but the amount above the surface is proportionally the same for each — in this example about one-third. For each block the amount above the water is balanced by the part below the water. Large blocks of the Earth's lithosphere behave similarly, floating in the asthenosphere.

the crust floats upward. This floating balance of the crust is called **isostasy** (meaning "equal standing"; Figure 1.11). It explains why the thick, lightweight continental crust stands high in relation to the ocean basins underlain by thinner, heavier oceanic crust.

HORIZONTAL MOVEMENTS

The geologic record also reveals that large portions of the Earth's crust have moved horizontally. And if we accept the idea that continents shift position, then we can say that the entire surface of the Earth is subject to lateral movements, measured in thousands of kilometers. This notion brings us back to our preliminary look at plate tectonics and sea-floor spreading.

Plates of the Earth   If we plot earthquakes on a world map, we find them concentrated in narrow, well-defined belts (as in Figure 1.12). These belts are also marked by extensive volcanic activity. Between the belts are large areas, both on the continents and in the ocean basins, where earthquakes and volcanoes, although not entirely absent, are infrequent. We now know that the zones of intense activity mark the boundaries between rigid **plates** of the Earth's lithosphere. These plates form a jigsaw puzzle whose pieces constantly jostle one another, and it is this movement that is associated with volcanoes and earthquakes.

We are now able to demonstrate (as will be discussed in Section 11.4) that plates move at rates between 1 and 9 cm/year (centimeters per year). The relative motion along plate margins can be any of three types: Two plates may slide parallel to each other but in opposite directions; two plates may diverge from each other; or two plates may converge. In fact, each of these movements is found along the margin of any given plate.

What goes on along the boundary of diverging or converging plates?

If two plates are pulling apart, it must mean that something is being added to their separating margins. For example, the broad rise that runs down the center of the Atlantic Ocean — the Midatlantic Ridge — marks the axis of separation between the Old and New Worlds, with the American plate moving generally westward as the Eurasian and African plates drift eastward. The crest of the Midatlantic Ridge is marked by a precipitously walled **rift valley** along its entire length. Along the valley new volcanic material (basalt) wells upward from the mantle. The newly formed oceanic crust is then slowly transported laterally away from the ridge (sea-floor spreading), allowing room for still younger material to be added along the axial valley.

If the plates grow along a diverging boundary, they must be destroyed elsewhere to allow space for the new material. This happens along the converging boundaries of plates. Look, for example, at the western boundary of the American plate, where that plate, which includes both the southwestern Atlantic Ocean basin and the South American continent, collides along most of its western margin with a large Pacific Ocean plate, the Nazca plate. The convergence of the two plates is marked by the Andes Mountains, intense earthquakes, and volcanic activity; in addition, a deep, narrow trench characterizes the ocean floor just west of the South American shore (see Figure 1.13).

The leading edge of the American plate is riding over the leading edge of the Nazca plate. The horizontal movements of the crust are piling up an excess of rock material where the two plates collide, and isostasy operates to ensure that this thickened column of the Earth's crust not only sinks lower into the asthenosphere but also stands higher above sea level.

Now look at the trench just off South America. The ocean floor here is depressed much below the average deep

1.12 Earthquakes are concentrated in narrow bands around the Earth. The boundaries of the Earth's major plates follow these bands. (See also Figure 11.16.)

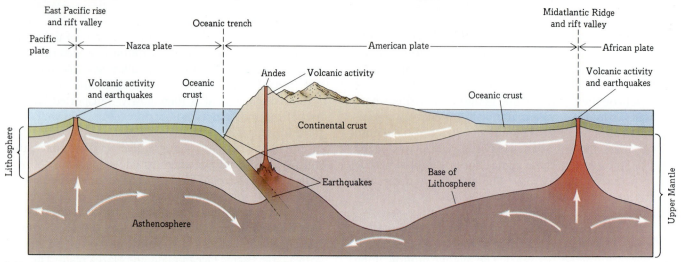

1.13   The relationships among the African, American, and Nazca plates. See text for discussion.

ocean floor. The leading edge of the Nazca plate is being shoved downward and eastward beneath the American plate, pulling part of the oceanic crust and floor with it. Along the dipping zone of contact between the two plates, called a **subduction zone,** earthquakes occur as the plates slip by each other. The bulk of the Nazca plate continues downward under the leading edge of the American plate until the temperature is so great that, along its front edge, it begins to lose its rigidity and is reincorporated into the mantle in the asthenosphere. During this process igneous activity goes on. The thermal energy needed to generate molten material comes from the friction generated between the plates and from the heat of radioactivity in the thickened rock pile.

The hypothesis of **continental drift** was first seriously introduced in 1912 by Alfred Wegener, a German meteorologist, but it received little acceptance until the mid-1960s, when the reality of plate tectonics and sea-floor spreading became apparent. Wegener had pictured the continents as moving about as individual units. Indeed, geology has shown that they do move, but not as isolated units. We now know that they move as portions of plates that include both landmasses and ocean floors. It is the plate that moves, and any Earth feature — whether continent or ocean — that the plate includes moves with it. So the meaning of "continental drift" is somewhat different for us than it was for Wegener.

THE ROCK CYCLE AGAIN

In later sections we shall return to the subject of plate tectonics and sea-floor spreading, expanding the bare outline just presented and laying out the data that have led to acceptance of this model of crustal movement. For the moment, let's assume the reality of plate tectonics and apply it to the concept of the rock cycle that we introduced in Section 1.3 and Figure 1.9.

In describing the rock cycle we pictured it without reference to world geography and without providing a general mechanism for the cycle. The plate-tectonic model furnishes some of this detail. We can visualize sediments moved by water, wind, and ice from the continents to the oceans, only to be brought back to the continents by the slowly moving ocean floors. Again, plate movements, particularly along plate boundaries, create temperatures and pressures that lead to metamorphism and igneous activity; and, especially along converging boundaries, continents grow in thickness and elevation, allowing erosion to eat deeper and deeper into the thickened wedge of continental rocks, thus creating more sediments for transfer back into the oceans, as shown in Figure 1.14.

1.5
SOME PRACTICAL CONSIDERATIONS

In this opening chapter we have touched on some major concepts in geology and, in doing so, have sought to provide a general guide to the details in the chapters to follow. You may want to come back to this chapter from time to time for the perspective it provides.

But before beginning more detailed discussion, we need to emphasize an additional important point. It is this: Geology has had and will continue to have practical applications for a wide range of human activities. The processes that have gone on within the Earth and at its surface have created the natural resources of fuels and metallic and nonmetallic deposits. The very search for these resources has provided us with the data and techniques that allow us to understand much about the operation of the Earth. And this new knowledge, directly and indirectly, will help us to find additional resources and to better understand those already discovered.

But the practical aspects of the subject are by no

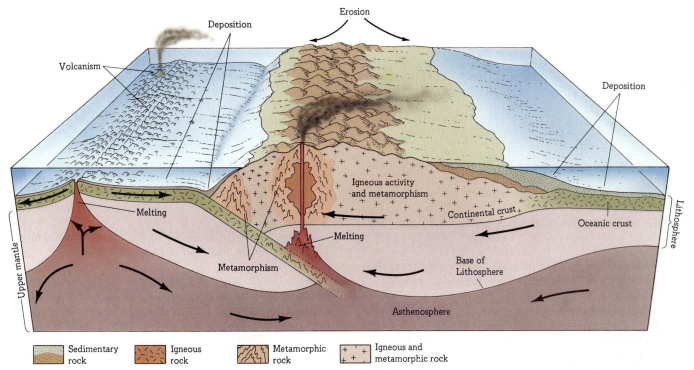

1.14 A cross section through a growing oceanic plate, which plunges beneath the edge of a continent, shows that the igneous, sedimentary, and metamorphic components of the rock cycle are present in both continental and oceanic regions.

means limited to the search for natural resources, important as this may be. By the end of the 1960s people began to realize that human activity could have startling, unexpected, and often disastrous effects on the Earth's environment. The next decade saw the beginning of the search for answers to the problems that our society has generated by interfering with natural systems. In the 1980s recognition that our natural resources were finite became generally widespread. We began to realize as well that the earthly environment produces events, from floods to earthquakes, that can be disastrous in human terms and that cannot be forestalled.

Because of the burgeoning human population, increasing demands on resources and the environment have produced a myriad of social, political, and economic stresses. Amelioration of these stresses depends on many disciplines as different as astrophysics and zoology, philoso-phy and economics, sociology and chemistry. Here we point out that geology has some unique information to help solve many of these problems. The search for new resources is an obvious area for geologic expertise. In the area of environmental protection the geologist brings certain skills cultivated over years of study of natural systems to such pressing questions as: What is the probability of a flood of a given magnitude? What dangers are involved in the underground disposal of wastes? What are the factors that lead to erosion of the shoreline? Can earthquakes or volcanic eruptions be predicted? Can they be controlled?

Thus from time to time we draw attention to certain geological factors that bear directly on human activity. And we devote Chapters 18 and 19 to the subjects of energy and useful materials. There are many other places in this book where the reader will make the connection between geological information and the world of human activity.

Summary

Geology is the science of the Earth and deals with the materials that compose it, the history of life, and the physical events of Earth history.

Geologic time encompasses the age of the Earth, about 4.6 billion years.

Absolute time and relative time are used to divide geologic time. Relative time—whether an event in Earth history came before or after another event—disregards years. Absolute time measures the age of a geologic event in years.

Radioactivity provides most of the absolute dates in

geology and depends on the breakdown, at a constant rate, of unstable elements to stable ones.

Divisions of geologic time are listed in a **geologic-time scale,** and the rock units representing these time divisions are shown in a **geologic column,** as reproduced on the front endpaper of this book.

Uniformitarianism is the principle that holds that the Earth processes now going on also operated in the past.

Earth materials are composed of minerals, which in turn make up rocks.

The three rock families are **igneous, sedimentary,** and **metamorphic.**

The rock cycle traces the changes as rocks are transformed from one type into another.

Plate tectonics, sea-floor spreading, and **continental drift** describe the motion of large pieces of the outer 50 to 100 km of the earth.

Lithosphere, asthenosphere, crust, and **mantle** are terms applied to certain portions of the outer 200 km of the Earth.

Vertical movements of the crust often reflect the floating balance reached by portions of the Earth's lithosphere as it sinks or rises in the asthenosphere, a process called **isostasy.**

Horizontal movements are often the result of plate tectonics and sea-floor spreading.

Plates of the Earth's lithosphere are defined by narrow bands of earthquakes and volcanic activity. Plates move relative to each other in converging, diverging, and parallel directions. Plates grow at diverging boundaries and are destroyed at converging boundaries.

The rock cycle can be related to the events accompanying plate tectonics and sea-floor spreading.

Practical applications of geology include the search for natural resources and provide basic data on the natural systems of our environment.

Questions

1. Geologic events may be dated in relative or absolute terms. Give one or more examples of each type.

2. Who was James Hutton? Why is he considered important in the development of the science of geology?

3. What is uniformitarianism? Give one or more examples.

4. List the three major rock families, and in a brief sentence describe each.

5. Describe, with the help of a diagram, how the rock families are interrelated by the rock cycle.

6. The outer layer of the Earth is called the **crust.** What are the two types of crust and what are the differences between them?

7. Distinguish between crust and mantle.

8. Distinguish between lithosphere and asthenosphere. How do they relate to crust and mantle?

9. What is sea-floor spreading? Plate tectonics? Draw a cross section of the upper 150 km of the Earth to show the relation and motions of two or more plates.

10. If the basin of the South Atlantic Ocean is about 200 million years old and was formed by sea-floor spreading, what has been the average annual separation of South America from Africa? (Hint: You need to know the present distance between the two continents.)

(We got 25 mm per year. Using 5,000 km as the present distance between the two continents, we set up the problem as follows):

$$\frac{5 \times 10^9 \text{ mm (distance)}}{2 \times 10^8 \text{ yr (time)}} = 25 \text{ mm/year (distance/year)}$$

11. If you answered Question 10, then you should be able to predict how far apart Africa and South America will be in 60 million years. What is your prediction? What assumption have you made in arriving at this prediction?

(We got 6,500 km. By rearranging the formula in Question 10, another 60 million years would produce another 1,500 km of separation. Adding this to the present distance between the two continents gives 6,500 km. You can also get the answer of 1,500 km by multiplying the present distance by the ratio 6/20, the ratio between 60 and 200 million years.)

12. How does the concept of the rock cycle as shown in Figure 1.9 fit into the model for plate tectonics? A neat, well-labeled diagram would be helpful in answering this question.

13. Here's a project for the list maker. There are a great many subdisciplines in geology with the prefix *geo-. Geophysics,* the application of physics to the study of the Earth, is an example. Try making a list of other such *geo-* subdisciplines. What do they cover? You will probably wish to resort to a dictionary, an encyclopedia, or other reference book.

If you want to go further, you can assemble a list of *paleo-* words such as *paleontology,* which is the study of past (old) life in the geological record and is based on the examination of fossils.

If you insist on further lists, there are a number of subdisciplines that do not have either *geo-* or *paleo-* as prefixes. An example is *mineralogy,* the study of minerals.

Supplementary Readings

ALLÈGRE, CLAUDE *The Behavior of the Earth: Continental and Sea-Floor Mobility.* Trans. by Deborah Kurmes Van Dam. Harvard University Press, Cambridge, Mass. 1988.
This volume on the revolution in the Earth Sciences can be useful as a preview for the newcomer to geology, as companion reading to this text, or as a final overview. Highly recommended.

EICHER, DONALD L. *Geologic Time,* Prentice Hall, Englewood Cliffs, N.J., 1968.
A short book that describes the various ways by which we measure geologic time.

GLEN, WILLIAM *The Road to Jaramillo: Critical Years of the Revolution in Earth Science,* Stanford University Press, Stanford, Cal., 1982
A volume for the general audience on the development of potassium-argon dating and magnetic reversals of the Earth's field and how the two in combination have been significant in proving the existence of sea-floor spreading.

MARVIN, URSULA B. *Continent Drift,* Smithsonian Institution Press, Washington, D.C., 1973.
An easy-to-read but authoritative account of the history of the idea of continental drift.

MCPHEE, JOHN *Basin and Range,* Farrar, Straus & Giroux, New York, 1981.
An account of the geology along portions of Interstate Highway 80, with emphasis on the Basin and Range, that desert land that extends from Salt Lake City to the Sierra Nevada and from southern Oregon to Mexico. McPhee discusses geology, its modern practitioners, and the history of science. Read it as an introduction to, or summary of, much of what is covered in this volume.

_____ *In Suspect Terrain,* Farrar, Straus & Giroux, New York, 1983.
A sequel to Basin and Range, *cited above. In this volume McPhee travels from the New Jersey Highlands to the Indiana Dunes with geologist Anita Harris, who offers guidance to finding oil, to understanding the last continental glaciation in the area, and to the history of eastern North America—all replete with cautionary tales on excessive enthusiasm for plate tectonics.*

MOORES, ELDRIDGE M. AND F. MICHAEL WAHL, (EDS.) *The Art of Geology,* Special Paper 225, Geological Society of America, Boulder, Colo., 1988.
This is a unique book of handsome photographs. It differs, however, from the familiar "coffee table" and "art" books in that each picture, or group of pictures, tells a story seen through the eyes of the geologist. The text, really a series of captions, is at once informative and unobtrusive. A book for the general reader and the professional.

SULLIVAN, WALTER *Continents in Motion,* McGraw-Hill, New York, 1974.
The science editor of The New York Times *tells the story of ideas about a changing Earth, which led to the present concepts of plate tectonics, sea-floor spreading, and continental drift.*

_____ *Landprints,* New York Times Books, New York, 1984.
The North American continent from the air. The author interprets the landscape and provides the air traveler with itineraries between major cities.

Rhombs of the rhombic cleavage fragments of calcite, each approximately 6 cm across. [John Simpson.]

CHAPTER ▪2

MINERALS AND MATTER

"A casual glance at crystals may lead to the idea that they were sports of nature, but this is simply an elegant way of declaring our ignorance. With a thoughtful examination of them, we discover laws of arrangement. . . . How variable, and at the same time how precise and regular are these laws! How simple they are ordinarily, without losing anything of their significance."

Abbé René Just Hauy, Traité de Mineralogie, *Vol. 1, p. xiii, 1801*

In this chapter we examine the matter from which the Earth is made. We review the subatomic particles, the *electrons, protons,* and *neutrons,* and how they combine to form *elements,* a few of which make up 99 percent of the Earth's crust.

The elements combine to make *minerals.* "Mineral" means different things to different people. We use the term to refer to a natural element or compound formed by various processes (see Figure 2.1).

2.1 Crystal of potassium feldspar showing several well-defined faces on the sides and top. Maximum dimension 16 cm. [John Simpson.]

We examine the various types of *bonding* found in minerals, and then take a look at the most *common minerals,* particularly the rock-forming ones. We identify these minerals megascopically by their *physical characteristics.*

We look at the general *distribution of elements* in nature, then at the *organization and association of minerals,* and end the chapter with a brief preview of the *reshuffling of the elements and minerals.*

2.1
PARTICLES OF MATTER

The physical universe is composed of what we call **matter;** yet one of the most elusive problems in science is to define matter precisely. It has long been customary to refer to the states of matter as solid, liquid, and gas. We say that matter has physical properties, such as color or hardness, as well as chemical properties, which govern its ability to change or to react with other bits of matter. But all this simply tells us about matter, not what matter is.

ORGANIZATION OF MATTER SUMMARIZED

Matter is the substance of the physical universe. It is composed of atoms. **Atoms** are combinations of the fundamental particles protons, neutrons, and electrons (except for the common form of hydrogen, which is a combination of just one proton and one electron). An atom is the smallest unit of an element. There are 103 elements: 88 have been found in nature; 4 others have been identified as short-lived transients; and 11 others have been artificially prepared.

Atoms are rarely found alone in nature. Rather, they are found in combination with other atoms of the same elements or with atoms of different elements. Atoms of different elements combine by exchanging or sharing electrons to form compounds. The 88 naturally occurring elements form numerous compounds, both in nature and in the laboratory.

Water is the only substance to occur naturally on Earth in the three states of solid, liquid, and gas.

ATOMS: FUNDAMENTAL PARTICLES

Protons, neutrons, and electrons are the fundamental particles that combine to form atoms (Table 2.1).

TABLE 2.1
Fundamental Particles

	Electric charge	Mass, u[a]
Electron	−1	0.00055
Proton	+1	1.00760
Neutron	0	1.00890

[a] Atomic mass unit.

Atomic Structure   All information that we have about **atomic structure** has been established through physicists' experiments dealing with the bombardment of atoms by particles accelerated to high velocity. For example, if these particles are shot at a target, such as a piece of metal, made up of billions of atoms, no more than 1 particle in 10,000 hits anything inside the target. It has been concluded, therefore, that the target must be largely open space.

Repeated tests have shown that an atom contains a nucleus of protons that behave as though they are positively charged and electrically neutral neutrons surrounded by a cloud of electrons that behave as though they are negatively charged. Tests have also revealed that a normal atom has as many electrons as it has protons. The number of protons plus the number of neutrons in an atom constitutes its mass number (the neutrons contribute to the mass of the atom, but they do not affect its electrical charge). The electrons occur in orbitals around the nucleus. They do not significantly contribute to the mass of an atom.

ELEMENTS

Each **element** is a special combination of protons, neutrons, and electrons. An element has an atomic number corresponding to the number of protons in its nucleus.

Element 1 is a combination of one proton and one electron (Figure 2.2). Long before its atomic structure was known, this element was named **hydrogen,** or "water former" (from the Greek roots *hydor,* "water," and *-genes,* "born"), because water is formed when hydrogen burns in air. The symbol of hydrogen is H. Because it has a nucleus with only one proton, hydrogen assumes place 1 in the table of elements.

Element 2 consists of two protons (plus two neutrons in the most common form) and two electrons (Figure 2.3). It was named **helium,** with the symbol He, from the Greek *helios,* "the Sun," because it was identified in the solar spectrum before being isolated on the Earth. Because of the two protons in its nucleus, helium takes place 2 in the table of elements.

Each addition of a proton, with a matching electron to maintain electrical balance, produces another element. Neutrons seem to be included more or less indiscrimi-

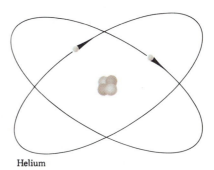

Helium

2.3 Diagrammatic representation of an atom of helium. The nucleus consists of two protons and two neutrons, and accordingly has a mass number of 4. There are two electrons (negative charges) to balance the positive charges of the two protons.

Since there are two protons in the nucleus, this atom is number 2 in the table of elements. The nucleus of helium (two protons + two neutrons) without any accompanying electrons is an alpha particle.

nately, although there are about as many neutrons as protons in the common form of many of the first 20 or so elements.

Of the total of 103 elements, 77 are classed as metals, 17 as nonmetals, and 9 as metalloids. These are keyed in the Periodic Chart (Appendix B). A metal typically shows a peculiar luster called **metallic luster,** is a good conductor of electricity or heat, is opaque, and may be fused, drawn into wire, or hammered into sheets. A nonmetal does not exhibit metallic luster, conductivity, or other features of metal. A metalloid has some metallic and some nonmetallic properties.

Isotopes   Elements have forms that, although somewhat similar in chemical and physical properties, have different masses. Such forms are called **isotopes**—from the Greek *isos,* "equal" or "the same," and *topos,* "place"—because each form has the same number of protons in its nucleus and occupies the same position in the table of elements. Isotopes show differences in mass as a result of differences in the number of neutrons in their nuclei. For example, hydrogen with one proton and no neutrons in its nucleus has a mass number of 1. When a neutron is present, however, the atom is an isotope of hydrogen with a mass number of 2 and is called **deuterium** (Figure 2.4). Two

2.2 Diagram of a hydrogen atom, which consists of one proton and one electron. This is the simplest atom.

Hydrogen

2.4 Schematic of deuterium, an isotope of hydrogen formed by the addition of a neutron to the nucleus. It has a mass number of 2.

neutrons in the nucleus forms the isotope **tritium,** with a mass number of 3. All elements have the capacity to form different isotopes.

Many elements as they occur in nature are mixtures of isotopes. Thus chlorine, whether occurring naturally or prepared in the laboratory, has an atomic weight of 35.5, which represents the average weight of a mixture of about three parts of Cl-35 to one part of Cl-37.

Isotopes are commonly found in varying proportions in different geologic settings. By calculating the proportions of certain isotopes, we can determine the source of many igneous and metamorphic rocks. Thus stable isotopes are important to geologists because they are not identical, but behave in subtly different ways. They are therefore separated by various geological processes, including thermal processes like evaporation.

Ions An atom is electrically neutral. If it gains or loses an electron from its outermost shell, it has a negative or positive charge, respectively. This unit is then known as an **ion.** A positively charged ion is a **cation.** A negatively charged ion is an **anion.** As we shall see later, more than one electron may be lost or gained, leading to the formation of ions with two or more units of electrical charge.

The electrons form a protective shield around the nucleus and define the size of the atom. In describing atomic dimensions we use a special unit of length, the **angstrom** (Å, or sometimes A), which is a hundred-millionth of a centimeter (usually written 1×10^{-8} cm). Atoms of the most common elements have diameters of about 2Å. Ions have sizes approximating atoms, but there are significant differences between the much larger anions and the relatively smaller cations.

2.2
BONDING

Combinations of atoms of different elements occur as **compounds.** The method by which the atoms are held together, or bonded, varies from one compound to another.

IONIC BOND

The number of electrons in an atom determines the manner and ease with which the atom can join with other atoms to form compounds. Compounds may be formed by different atoms losing or gaining electrons. For example, an atom of sodium has one "extra" electron, which it readily gives up, whereas chlorine readily accepts one electron. When a sodium atom and a chlorine atom join to form halite (NaCl), the sodium's electron is lost by the sodium and is taken up

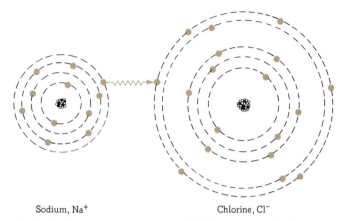

Sodium, Na⁺   Chlorine, Cl⁻

2.5   In the formation of the compound halite an atom of sodium, which has only one electron in its outermost shell, joins an atom of chlorine, which needs one electron to fill its outermost shell. The sodium's lone outermost electron slips into the vacant place to fill the chlorine's outermost shell.

by the chlorine. Because chlorine has gained an electron, it becomes a negative ion or anion, Cl⁻ (there is a surplus of one electron over the number of protons). By losing an electron, sodium becomes a positive ion, or cation, Na⁺ (Figure 2.5). Opposite charges attract; therefore a bond results, forming an **ionic compound.** The mineral halite, or common table salt, is a very abundant substance. Its symbol, NaCl, shows that the number of sodium atoms and chlorine atoms are equal. The smallest unit of a compound that displays all the properties of the compound is referred to as a **molecule.**

COVALENT BOND

Atoms can combine in various ways to form compounds. For example, in the formation of the mineral diamond, each atom of carbon is bonded to four neighboring carbon atoms. Each carbon atom has four electrons in its outermost shell. It is able to "share" one electron from each of its four neighboring atoms. In this way it fills its outermost shell to the full complement of eight electrons, even though it does not "gain" or "lose" any outright. This sharing of electrons produces a **covalent bond.**

Covalent bonding in the water molecule produces the unusually effective dissolving character of water. By sharing electrons from each of two hydrogen atoms (Figure 2.6), the oxygen part of a water molecule (H_2O) takes on a slightly negative aspect, whereas the hydrogen end becomes slightly positive, even though the entire molecule is neutral. As a result, a molecule of water acts like a magnet, one end attracting negative ions and the other positive ions. These ends are referred to as a **positive pole** and a **negative pole,** respectively, because of the molecule's similarity in behavior to a bar magnet. The molecule, then, is a **dipole** ("two-pole"); and water is known as a **dipolar compound.** This

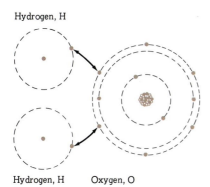

Hydrogen, H

Hydrogen, H   Oxygen, O

2.6 Two hydrogen atoms and one oxygen join to form water, H₂O, by a covalent bond. In this bond the hydrogen electrons do double duty in a sense, filling the two empty places in the outer shell of oxygen, yet remaining at normal distance from their nuclei. The result is the formation of a molecule of water, the smallest unit that displays the properties of that compound.

polarity gives water special properties that make it an extremely important agent in geological processes. The mechanism by which water dissolves salt is an illustration of the ease with which water dissolves various substances and participates in weathering and other geological processes. (See Figure 2.7.)

METALLIC BOND

The atoms of metallic elements have a special kind of bonding, which is responsible for some materials becoming good conductors of heat and electricity. In a piece of metal the atoms of a single element are packed closely together; their outermost electrons are not shared or exchanged but are free to move around and connect with any atoms in the solid. These wandering electrons, shared by any of the atoms in the metal, create the **metallic bond.**

The relative freedom of movement of the electrons in this relationship accounts for the malleability and ductility of metals, and especially for their high level of electrical conductivity. When there is no current, electrons jump randomly from atom to atom. When an electrical potential is applied to a good conductor, such as a copper wire, electrons flow through the wire without stopping to attach themselves to any particular atoms. They thus convey electricity (or heat) throughout the whole wire.

VAN DER WAALS' BOND

There are relatively weak attractive forces operating between neutral atoms and molecules in addition to those stronger forces described above. **Van der Waals' forces** commonly occur in all materials because of electrical polarization momentarily caused by the distribution of electrons

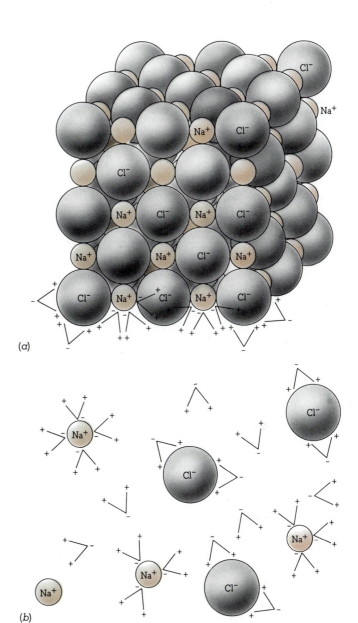

(a)

(b)

2.7 How water dissolves salt. (a) The water dipoles (the V-shaped symbols with plus and minus charges) attach themselves to the Na⁺ and Cl⁻ ions, some of which are labeled in the diagram. The dipoles of water overcome the ionic attraction that holds the sodium and chlorine together to form solid salt and then convey the sodium and chlorine ions into the liquid as suggested in (b).

in their orbits. When several electrons are temporarily on the same side of an atom or molecule, that side becomes negatively charged. Similarly, when electrons are momentarily clustered on the opposite side, an atom or a molecule momentarily becomes positively charged. Again, as with the water molecule, we get the equivalent of a bar magnet, with the negative end attracting cations, and the positive end, anions, if only for an instant. A few minerals such as talc and graphite exhibit this type of bonding more than others, although most substances are bonded to some degree by this weak force.

2.3

DISTRIBUTION OF ELEMENTS IN NATURE

Most astronomers today believe that only the lightest elements originated at the moment the universe began and that the heavier elements have accumulated since that time. Apparently the universe began 10 to 20 billion years ago as an expanding sea of high-energy photons—the so-called **Big Bang.** As the expansion of the universe cooled this sea of energy, the photons decayed into particles such as protons, electrons, and neutrons. A proton combined with an electron is a hydrogen atom, so the early universe was filled with hydrogen. But the temperature was so high that nuclear reactions could fuse particles to create helium, and by the time the temperature had fallen too low for these reactions to occur (roughly 3 minutes), about 20 percent (by mass) of the hydrogen had become helium.

Thus the gas that formed the first galaxies and stars was nearly all hydrogen and helium. Elements from atomic number 3 (lithium) through atomic number 26 (iron) could not have formed during the Big Bang. Instead, they have been formed by nuclear reactions inside successive generations of massive stars. The heaviest elements, from atomic number 27 (cobalt) on, are thought to form in supernovae explosions. Thus the elements of which planets are made are the by-products of the birth and death of stars.

ELEMENT ABUNDANCE

The relative abundance of elements found on the Earth contrasts markedly with the abundance of elements detected in the universe as a whole. About 90 percent by weight of the total mass of the Earth consists of four elements: iron (35 to 40 percent), oxygen (25 to 30 percent), silicon (13 to 15 percent), and magnesium (about 10 percent). Four other elements are present, together amounting to about 1 percent: nickel, calcium, aluminum, and sulfur. Another seven elements amount to from 0.1 to 1 percent: sodium, potassium, chromium, cobalt, phosphorus, manganese, and titanium. The total mass of the Earth is made up almost entirely of these 15 elements, with all the remaining elements contributing to less than 0.1 percent of its weight.

Within the crust of the Earth, on the other hand, only 10 elements make up over 99 percent by weight. These elements, with their approximate percentages, are: oxygen (the only anion in the top 10), 47; silicon, 28; aluminum, 8; iron, 5; calcium, 4; sodium, 3; potassium, 3; magnesium, 2; titanium, 0.4; and hydrogen, 0.1.

Let us compare these with abundances in the universe. Hydrogen and helium make up around 76 and 23 percent, respectively, of the weight of the universe. Thus all the remaining elements constitute approximately 1 percent of the weight of the universe. The Sun, like the universe as a whole, is mostly hydrogen and helium.

Apparently during the formation of the Earth and similar terrestrial planets of the solar system, the lighter gases—hydrogen and helium especially—were lost very quickly. This process left the Earth with a residue of unusual relative abundances of elements compared with the universe as a whole. Examination of the Jovian (or outer) planets (Jupiter, Saturn, Uranus, Neptune, and Pluto), in part by artificial satellites, together with earlier deductions from consideration of their masses, suggests that their composition differs markedly from that of the terrestrial planets (see Chapter 20).

2.4

CHARACTERISTICS OF MINERALS

At the beginning of this chapter we defined a mineral as a natural element or compound formed by various processes. Now we shall see specifically what this definition means. We shall also see that though a particular mineral may vary from one sample to another, the internal atomic arrangement of its component elements is always identical. So we find it necessary to include in our definition of a mineral not only that it is a naturally occurring element or compound with a diagnostic chemical composition, but also that it has a unique orderly internal arrangement of its elements.

Earlier in this chapter we showed that a mineral's composition can vary slightly: An occasional substitution by other elements whose atoms or ions are of similar size may occur. We said that every mineral is composed of elements in definite or slightly varying proportions.

PROPERTIES OF MINERALS

All the properties of minerals are determined by the composition and internal atomic arrangement of their elements. Although we can identify minerals on the basis of their chemical properties, physical properties are the ones we most often use. Such physical properties include crystal habit, cleavage, twinning, striations, hardness, specific gravity, color, streak, and luster.

Crystal Habit   When a mineral grows without interference or obstacle, it will be bounded by plane surfaces symmetrically arranged and will acquire a characteristic **crystal habit,** which is the external expression of its internal crystalline structure. Every mineral has a characteristic crystal habit.

The faces of crystals are defined by surface layers of atoms. These faces lie at angles to one another, angles that have definite characteristic values, with the same value for all specimens of the same mineral. The size of the faces may

2.8 An atomic view of a copper crystal, showing small octahedral faces and large cubic faces. [After Linus Pauling, *General Chemistry*, p. 23, W. H. Freeman and Co., San Francisco, copyright © 1953.]

vary from specimen to specimen, but the angles between the faces remain constant.

Every crystal consists of atoms arranged in a three-dimensional pattern that repeats itself regularly. The smallest repeat unit in this pattern is called its **unit cell.** In a crystal of copper all the atoms are alike and are arranged in a cube. This arrangement is limited by the size of the copper atoms. By repetition the entire crystal is built up (see Figure 2.8). The principal surface layers correspond to the faces of a cube; the smaller surface, obtained by cutting off a corner of a cube, is called an **octahedral face.** Native copper discovered in deposits of copper ore is often found in the form of crystals with cubic and octahedral faces. Remember that the atoms in the figure are greatly enlarged; if the crystal had sides only 1 mm (millimeter) long, there would be about 4 million atoms in a row along each edge.

In crystals of halite there are ions of two different kinds arranged in the regular pattern shown in Figure 2.9.

2.9 Arrangement of smaller sodium ions with positive electrical charge, Na^+, and larger chlorine ions with negative electrical charge, Cl^-, to form the ionic compound NaCl, halite, or common salt. Na^+ has a radius of 0.97 Å, and Cl^- a radius of 1.81 Å.

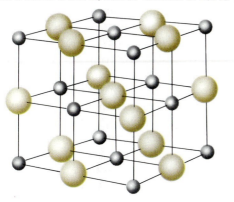

The smaller ones are those of sodium, and the larger ones are those of chlorine. The atomic arrangement is limited by the size of the ions. Halite is a cube made up of six chlorine ions surrounding one sodium ion, and each chlorine ion is surrounded by six sodium ions, all being equidistant. Repetition of the unit cell in three dimensions forms the cubic halite crystal.

The mineral quartz occurs in many rocks as irregular grains because the development of true crystal faces requires unrestricted growth. Even in these irregular grains, however, the atoms are arranged according to their typical crystalline structure. And where conditions have permitted the mineral to develop freely, quartz forms crystals that are always six-sided. (Note that only the external arrangement results in what we call crystals; the internal arrangement produces crystalline structures.)

Whether an individual crystal or quartz is only 1 mm long or 1 meter long, the faces of the prism always meet at the same angle (see Figure 2.10). This is an example of the **constancy of interfacial angles.** The fact that all minerals, when allowed to form in unrestricted space, produce solids bounded by plane surfaces is very powerful evidence to support the concept of a regular, predictable arrangement of the constituent atoms.

There are six different systems of crystals, classified according to the angles between their faces (see Table A.1 in Appendix A).

Some elements or compounds may develop in several different crystal systems, producing different minerals. This assumption of two or more crystal systems by the same substance is called **polymorphism.** Each of these crystal forms results from the conditions under which it was made. For example, carbon occurs as both graphite and diamond: The crystal shape of diamond is an eight-sided solid, an octahedron; the crystal shape of graphite is a flat crystal with six sides. Although both minerals are composed of

2.10 Quartz crystals from Crystal Mountain, Hot Springs, Arkansas. Note six-sided crystals with striations on side faces. Largest crystal is 12 cm long. [Smithsonian.]

carbon, the difference in their crystal shapes comes from the arrangement of their carbon atoms—one pattern in diamond, another in graphite (Figures 2.11 and 2.12). The crystal shape of pyrite is a cube, and that of marcasite is a flattened tabular shape. Both are composed of FeS_2. Here

2.11 Different arrangements of atoms of carbon produce the crystalline structure of (a) diamond and (b) graphite. In the diamond, each carbon atom is bonded with four other carbons. Each of these in turn is bonded to three other carbon atoms as well as with the original one in a three-dimensional framework. This bonding continues throughout the crystal and gives diamond its characteristic hardness. Graphite has a layered structure. Each carbon atom forms one double and two single bonds with its nearest neighbor. The resulting sheets lie parallel to each other in a loosely assembled pile. The layers are easily separated and account for graphite's softness. [After Linus Pauling, *General Chemistry*, W. H. Freeman and Co., San Francisco, copyright © 1953.]

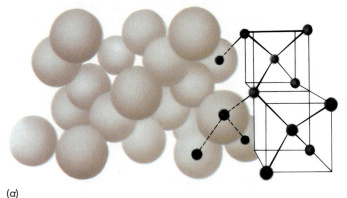

(a)

2.12 Graphite on left, and diamond, owe their differences to different arrangements of their atoms. Graphite crystal is 8 cm long. [Smithsonian.]

again, the reason for the difference in crystal system lies in the internal arrangement of the atoms (Figure 2.13). The crystal form of minerals can be examined in great detail with the help of a scanning electron microscope, as illustrated in Box 2.1.

Cleavage   Another manifestation of the orderly crystalline arrangement of atoms in a mineral is cleavage. **Cleavage** is the tendency of a mineral to break in certain preferred directions along smooth plane surfaces. Cleavage planes are governed by the internal arrangement of the atoms. Because cleavage is the breaking of a crystal between atomic planes, it is a directional property, and any plane throughout a crystal parallel to a cleavage surface is a potential cleavage plane. Moreover, the cleavage is always parallel to crystal faces or possible crystal faces because faces and cleavage both reflect the same crystalline structure.

Cleavage manifests a direction of weakness, and a mineral sample tends to break along planes parallel to this direction. This weakness may be due to a weaker type of atomic bond, to greater atomic spacing, or to a combination of the two. Graphite has a platy cleavage because of relatively weak bonds between the carbon layers but fairly strong bonds within the layers. Diamond has but one type

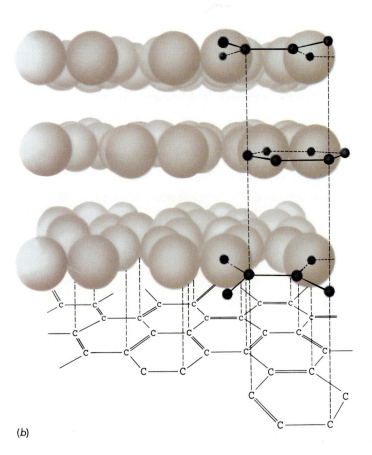

(b)

BOX 2.1 Scanning Electron Microscopes

The crystalline form of even submicroscopic crystals can now be observed by use of a scanning electron microscope, which focuses beams of electrons across a specimen. These electrons are electrically or magnetically moved sequentially back and forth until the reflected and emitted electrons produce an image of the sample. A great depth of field permits examination of three-dimensional opaque objects, with enormous magnifications (as much as 100,000 times). (See Figures B2.1.1–B2.1.3.)

B2.1.1 Scanning electron micrograph of books of chamosite, an iron alumino-silicate, and rosettes of hematite, Fe_2O_3., in oolitic ironstone from Aswan, Egypt. [Deba Bhattacharyya.]

B2.1.2 A scanning electron microscope photograph of iron crystals in a small cavity in recrystallized breccia (fragmented rock) from the *Apollo 15* lunar landing site. The largest crystal is 3 microns across (a micron being one-millionth of a meter). Well-shaped crystals like this indicate slow crystallization from a hot vapor. The crystals rest on a lattice of pyroxene (calcium-magnesium-iron silicate). [NASA.]

B2.1.3 Uranium crystals (carnotite) enlarged several hundred times, showing a flowerlike appearance, from Garland County, Arkansas. [U.S. Geological Survey.]

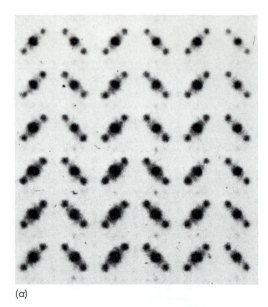

(a)

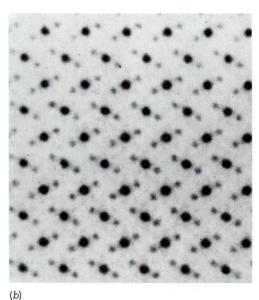

(b)

2.13 Representation of X-ray pattern of atoms. (a) Pyrite, showing orderly arrangement characteristic of crystalline structure. The mineral is composed of iron and sulfur, FeS_2. Large spots are atoms of iron; small ones, atoms of sulfur. Each atom of iron is bonded to two atoms of sulfur, and spacing of iron atoms is the same in both directions of the plane of the photograph. Magnification, approximately 2.2 million ×. (b) Marcasite, FeS_2. Note difference between horizontal and vertical spacing of iron atoms. Compare with pyrite. Magnification, about 2.8 million ×. [Martin J. Buerger.]

of bond joining all its carbon atoms; its cleavage takes place along planes having the largest atomic spacing.

Cleavage may be perfect, as in the micas; more or less obscure, as in beryl and apatite; or entirely lacking, as in quartz, in which case we say the mineral has only a fracture, sometimes with specific shape or form.

Twinning   The intergrowth of two or more single crystals of the same mineral may occur with different geometric

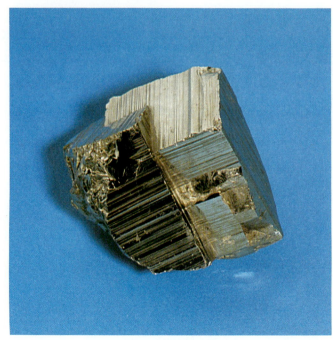

2.14   Intergrown cubes of pyrite make up this specimen. Note the striated crystal faces. Maximum dimension is 6 cm. [John Simpson.]

orientations. For example, alternate crystal layers may be in reversed position or even at right angles to each other. Such intergrowth is called **twinning.** Certain minerals have specific characteristic twinning patterns.

Striations   A few common minerals have parallel, threadlike lines or narrow bands, called **striations,** running across their crystal faces or cleavage surfaces. These lines can be clearly seen on crystal faces of quartz and pyrite (see Figure 2.14).

Once again, this property is a reflection of the internal arrangement of the atoms and the conditions of growth of the crystals.

Hardness   Another physical property governed by the internal atomic arrangement of the mineral elements is **hardness** (the stronger the binding forces between the atoms, the harder the mineral). The degree of hardness is determined by the relative ease or difficulty with which one mineral is scratched by another or by a fingernail or knife. It might be called the mineral's "scratchability." Hardness (H) ranges from 1 through 10 on what is known as Mohs' scale (see Box 2.2).

To illustrate, if you pick up a piece of granite and try to scratch one of its light-colored grains with a steel knife blade, this mineral simply refuses to be scratched. But if you drag one of these grains across a piece of glass, the glass is easily scratched. Clearly, then, these particular mineral grains in granite are harder than either steel or glass. But if you have a piece of topaz handy, you can reveal the vulnera-

BOX 2.2   Hardness or "Scratchability" of Minerals

Mohs' hardness scale for minerals			Common items for testing	
Softest	1	Talc		
	2	Gypsum	2.5	Most fingernails
	3	Calcite	3.0	Copper penny
	4	Fluorite		
	5	Apatite	5.5	Penknife or plate glass
	6	Orthoclase		
	7	Quartz		
	8	Topaz		
	9	Corundum		
Hardest	10	Diamond		

When we speak of the "hardness" of a mineral, we do *not* mean how easily it can be smashed, for even a diamond can readily be smashed. We mean, rather, how easily it can be scratched. If our fingernail can scratch the surface of a mineral, it must be rather soft; in fact, most fingernails have a hardness of about 2.5, so that particular mineral must be less than 2.5 in hardness. Similarly, a copper penny has a relative hardness of about 3 on Mohs' scale (named after Friedrich Mohs, an early mineralogist). We can determine the hardness of any mineral by scratching its smooth face with the sharp corner of another. We must be sure that the mineral tested is actually scratched. Sometimes particles simply rub off the specimen, fooling us into thinking it has been scratched.

The 10 minerals arranged as examples of the degrees of Mohs' scale simply show relative hardness. Each mineral listed will scratch those minerals lower in number on the scale and will be scratched by all those higher. In terms of absolute hardness, however, the steps are only approximate. Number 7 may be about 7 times harder than number 1, and number 9 may be 9 times harder, but number 10, diamond, is about 40 times as hard as number 1, talc. See the listing of many minerals in Table A.3 in Appendix A.

bility of these same granite grains, for although they are harder than steel or glass, they are not as hard as topaz.

Minerals differ widely in hardness (see Appendix A). Some are so soft that they can be scratched with a fingernail; some are so hard that a steel knife is required to scratch them; diamond, the hardest mineral known, cannot be scratched by any other substance.

Specific Gravity   Every mineral has an average mass per unit volume. This characteristic mass is usually described in comparison with the mass of the same volume of water. The number that represents this ratio is known as the **specific gravity** of the mineral. **Density** is given in grams per cubic centimeter (g/cm^3) or tonnes per cubic meter (t/m^3), and in numerical value is equal to specific gravity.

The specific gravity of a mineral increases roughly with the mass of its constituent elements and with the closeness with which these elements are packed together in their crystalline structure. Most rock-forming minerals have a specific gravity of around 2.7, although the average specific gravity of metallic minerals is 5. Pure gold has a specific gravity of 19.3.

It is not difficult to acquire a sense of relative weight by which to compare specific gravities. Just as we learn to distinguish between two bags of equal size, one filled with feathers and one filled with lead, experience in hefting stones gives most of us a sense of the "normal" weight of rocks.

Color   Although **color** is not a reliable property for identifying most minerals, it is strikingly characteristic for a few. These include the intense azure blue of azurite, the bright green of malachite, and the pale yellow of sulfur. Magnetite is iron black, and galena is lead gray. The color of other minerals, such as quartz, can be quite variable because of slight impurities.

Minerals containing iron are usually "dark-colored." In geologic usage **dark** includes dark gray, dark green, and black. Minerals that contain aluminum as a predominant element are usually "light-colored," a term that includes purples, deep red, and some browns among "normal" lights.

Streak   The **streak** of a mineral is the color it displays in finely powdered form. The streak may be different from the color of the hand specimen. Although the color of a mineral may vary between wide limits, the streak is usually constant.

One of the simplest ways of determining the streak of a mineral is to rub a specimen across a piece of unglazed porcelain known as a **streak plate**. The color of the powder left behind on the streak plate helps to identify some minerals. Because the streak plate has a hardness of 7, it cannot be used to identify minerals with greater hardness.

Hematite, Fe_2O_3, may be reddish brown to black; its streak, however, is light to dark blood-red, which becomes black on heating. Limonite, $FeO(OH) \cdot nH_2O$, sometimes known as "brown hematite" or "bog-iron ore," has a color that is dark brown to black but a streak that is yellowish brown. Cassiterite, SnO_2 (tinstone), is usually brown or black, but it has a white streak.

Luster   The reflection of light from the surface of a mineral, described by its quality and intensity, can be diagnostic for certain minerals. This appearance of a mineral in reflected light is known as **luster** and is one of the most obvious physical properties of a mineral. There are several kinds of luster: metallic, adamantine (diamondlike), resinous, pearly, and silky. Luster is commonly described simply as either metallic or nonmetallic.

Other Physical Properties   Minerals have other physical properties that may be helpful in identification: magnetism, electrical properties, fluorescence, fusibility, solubility, fracture, and tenacity (see Appendix A).

2.5

THE MOST COMMON MINERALS

Although more than 2,000 minerals are known, only a limited number compose most of the rocks of the Earth's crust. These can be grouped into closely related families of minerals. Minerals are homogeneous crystalline materials but not necessarily pure substances. Most **rock-forming minerals,** or minerals found most abundantly in the rocks of the Earth's crust, have variable compositions caused by ionic substitution of some elements for other elements. These substitutions are distributed throughout the crystalline structure. Such replacement is called **solid solution.** The resultant mineral has a single crystalline phase that may vary in composition within finite limits without the appearance of an additional phase.

The general tendency for two or more ions to substitute for one another is known as **isomorphism.** When the ions substitute in a continuous series (like magnesium and iron in olivine), the result is called a **solid-solution mineral series.**

We will look at the different mineral groups. The minerals are classified basically according to the principal anion, or negative ion, in the structure.

NATIVE ELEMENTS

A few minerals occur in nature as single elements, uncombined with any others. The more commonly occurring forms of these **native elements** are copper, carbon (as diamond and as graphite), gold, silver, and sulfur. Notice that all of these minerals are important to our economy.

HALIDES

A variety of minerals called **halides** results from the combination of certain cations with the halogen elements (chlorine, iodine, bromine, and fluorine): the very common rock salt halite, $NaCl$; its potassium partner, sylvite, KCl; and many other chlorides, iodides, bromides, and fluorides. Halides occur as precipitates from evaporating ponds, salt flats, and natural brines.

OXIDE MINERALS

Oxide minerals are formed by the direct union of an element with oxygen. These have relatively simple formulas compared to the complicated silicates (discussed below). The oxide minerals are usually harder than any other class except the silicates, and they are heavier than others except the sulfides. Within the oxide class are the chief ores of aluminum, iron, tin, chromium, and manganese. So we have the common oxide minerals ice, H_2O; corundum, Al_2O_3; hematite, Fe_2O_3; magnetite, Fe_3O_4; and cassiterite, SnO_2.

SULFIDE MINERALS

Sulfide minerals are formed by the union of an element with sulfur. The elements that occur most commonly in combination with sulfur are iron, copper, lead, zinc, silver, and mercury. Some of these sulfide minerals occur as valuable ores, such as chalcocite, Cu_2S; galena, PbS; and sphalerite, ZnS.

CARBONATE AND SULFATE MINERALS

A slightly more complex combination occurs in what are called complex ions. Several groups are of great importance in geology. One of these, the complex ion $(CO_3)^{2-}$, consists of a single carbon ion with three oxygen ions packed around it. Compounds in which this ion appears are called **carbonates.** For example, the combination of a calcium ion with a carbon-oxygen ion produces calcium carbonate, $CaCO_3$, known in one mineral form as **calcite.** Another common carbonate is **dolomite,** $CaMg(CO_3)_2$. Calcite and dolomite occur in enormous quantities in many sedimentary sequences as the rocks limestone and dolostone, respectively.

A combination of one sulfur ion and four oxygen ions produces the complex ion $(SO_4)^{2-}$, which combines with cations to form **sulfates;** for example, $CaSO_4$, the mineral **anhydrite;** and its hydrous form, $CaSO_4 \cdot 2H_2O$, the mineral **gypsum.**

SILICATE MINERALS

More than 90 percent of the rock-forming minerals are **silicates,** compounds containing silicon, oxygen, and one or more metals. Each silicate mineral has as its fundamental unit the **silicon-oxygen tetrahedron** (see Figure 2.15). This is a combination of one "small" silicon ion, with a radius of 0.42 Å, surrounded as closely as geometrically possible by four "large" oxygen ions, each with a radius of 1.40 Å, forming a tetrahedron. The oxygen ions contribute an electric charge of 8 − to the tetrahedron, and the silicon ion contributes 4 +. Therefore the silicon-oxygen tetrahedron is a complex ion with a net charge of 4 −. Its symbol is $(SiO_4)^{4-}$ and it has strong covalent bonds.

The most common rock-forming silicate minerals are olivine, pyroxene, amphibole, biotite, muscovite, feldspars, and quartz. They are listed and classified in Table 2.2.

Ferromagnesians In the first four of these rock-forming silicates — olivine, pyroxene, amphibole, and biotite — the silicon-oxygen tetrahedra are joined primarily by ions of iron and magnesium. These silicate minerals are known as **mafic** or **ferromagnesian** minerals, from the joining of the Latin *ferrum,* "iron," and "magnesium." All four ferromagnesian minerals are very dark or black and have a higher specific gravity than do the other rock-forming minerals. Except for olivine, which is made entirely of iron and magnesium silicates, the ferromagnesian minerals have other cations that help to bind the silica tetrahedra together (for example, Na^+, Ca^{++}, Al^{+++}).

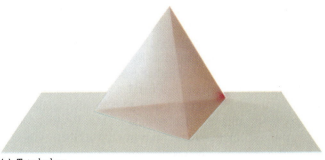

(a) Tetrahedron

(b) Silicon-Oxygen Tetrahedron expanded

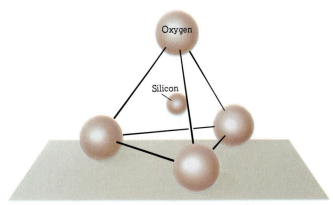

(c) Side view

(d) Top view

2.15 The silicon-oxygen tetrahedron is the most important complex ion in geology because it is the central building unit of nearly 90 percent of the minerals in the Earth's crust. Four O^{2-} ions are arranged around a single, and much smaller, Si^{4+} ion. To visualize this arrangement imagine a four-sided geometrical form, a tetrahedron, as shown in (a). One side is a triangular base. The other three sides are also triangles and rise from the edges of the base to a point forming a three-sided pyramid. In (b) an oxygen ion is located at each corner of the tetrahedron and a single silicon ion floats in the center of the tetrahedron. The positive charge of the silicon ion and the negative charges of the oxygen ions pull the ions closely together, with the silicon snuggly fitting into the space formed by the packing of the oxygen ions. A side view of the tetrahedron is shown in (c) and a top view in (d). In these views the silicon ion is obscured by the closely packed oxygen ions.

TABLE 2.2
Silicate Classification

Arrangement of SiO_4 tetrahedra	Si/O ratio	Rock-forming minerals
Isolated	1:4	Olivine
Rings	1:3	Beryl (beryllium aluminosilicate)
Chains (single)	1:3	Augite (pyroxene family)
Chains (double)	4:11	Hornblende (amphibole family)
Sheets	2:5	Biotite (black mica)
		Muscovite (white mica)
Frameworks	3:8	Othoclase (potassium feldspar)
	3:8	Plagioclase (calcium-sodium feldspar)
	1:2	Quartz

Olivine is composed of isolated silicon-oxygen tetrahedra held together by positive ions of magnesium and iron. Because there are no planes of weakness and its elements are so firmly held together by the ionic bonds, olivine exhibits no cleavage, but rather fractures when struck a blow. It is a relatively hard mineral, 6.5 to 7.

Olivine is an example of a mineral that undergoes compositional changes. Its formula is $(Mg, Fe)_2SiO_4$. The proportions of magnesium and iron vary, so their symbols are in parentheses. The proportion of silicon and oxygen remains constant. Iron and magnesium substitute for one another quite freely in olivine's crystalline structure as they each have two electrons in the outer shell and their ionic radii are very similar. The end members of the olivine series are forsterite and fayalite: When the positive ions are all magnesium, the mineral is forsterite, Mg_2SiO_4; when they are all iron, the mineral is fayalite, Fe_2SiO_4.

Olivine is geologically important as it makes up several percent of the mafic crustal rocks at the surface of the Earth and is believed to predominate in the heavier and deeper-seated rocks of the Earth's mantle. Its specific gravity is 3.27 to 4.37, increasing with iron content (Figure 2.16). Its color is usually green to yellowish green, but it can be olive to grayish green, sometimes brown. This mineral generally occurs in grains or granular masses.

Pyroxene is a mineral group that has a crystalline structure based on single chains of tetrahedra,[1] as shown in Figures 2.17 and 2.18, joined by ions of iron, magnesium, calcium, sodium, and aluminum. It is dark green to black, with a colorless streak. Its hardness is 5 to 6, and its specific gravity ranges from 3.2 to 3.4. It has rather poor cleavage along two planes almost at **right** angles to each other. This cleavage angle is important in distinguishing augite from hornblende. The cleavage results from the contrast between the stronger covalent bonding *within* the chains of silica tetrahedra and the weaker ionic bonding *between* adjacent chains and the intervening cations.

Augite is the commonest species of this pyroxene mineral group (see Table A.8 in Appendix A) and is characterized by a fundamental crystalline structure built with single parallel chains of tetrahedra. Pyroxenes crystallize and form a series whose members are closely analogous chemically to members of the amphibole group (see below).

Amphibole is a mineral group that has a crystalline structure based on double chains of tetrahedra, as shown in Figure 2.19, joined by the iron and magnesium ions common to all ferromagnesians and by ions of calcium, sodium, and aluminum. The amphiboles closely parallel the pyroxenes in composition but contain hydroxyl, OH. Amphi-

bole's color is dark green to black, like that of pyroxene; its streak, colorless; its hardness, 5 to 6; and its specific gravity, 3.2. Two directions of good cleavage meet at angles of approximately 56° (degrees) and 124°, which helps distinguish amphiboles from pyroxenes.

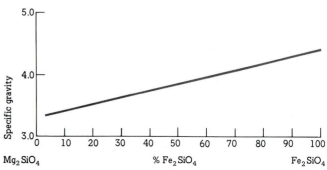

2.16 Specific gravity of olivine increases with increasing iron content. Thus at 20 on the horizontal scale olivine is 20 percent Fe_2SiO_4 and 80 percent Mg_2SiO_4 and has a specific gravity of about 3.5. The specific gravity increases to about 4 where olivine is 70 percent Fe_2SiO_4 and 30 percent Mg_2SiO_4.

2.17 Pyroxene single chain of tetrahedra viewed from above. Each silicon ion has two of the four oxygen ions of its tetrahedron bonded exclusively to itself, and it shares the other two with neighboring tetrahedra fore and aft. The resulting individual chains are in turn bonded to one another by positive metallic ions such as Mg^{++} and/or Fe^{++}. Because these bonds are weaker than the silicon-oxygen bonds that form each chain, cleavage develops parallel to the chains.

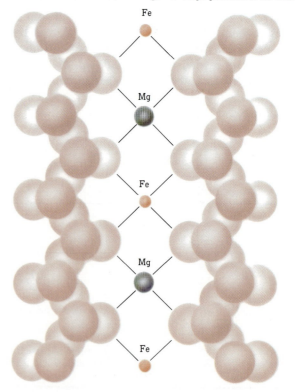

[1] In the remainder of this chapter "tetrahedra" is equivalent to "silicon-oxygen tetrahedra."

2.18 Pyroxene crystals, Stirling Hills, New Jersey. Maximum dimension is 15 cm. [Smithsonian.]

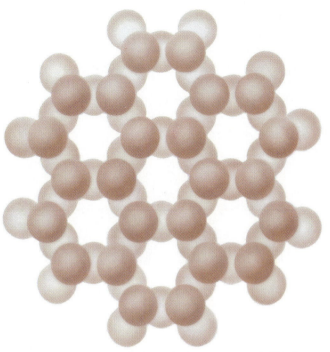

2.20 Tetrahedral sheets. Each tetrahedron is surrounded by three others, and each silicon ion has one of the four oxygen ions to itself, sharing the other three with its neighbors. (Each sheet would be loosely bonded to the next sheet by cations such as Fe^{++}, Mg^{++}, Al^{+++} and/or hydroxyl OH^- ions.)

2.19 Double chain of tetrahedra viewed from above. The doubling of the chain of Figure 2.17 is accomplished by the sharing of oxygen atoms by adjacent chains. (These chains would be bonded to similar chains by Fe^{++} and/or Mg^{++} or similar cations.)

Hornblende is an important and widely distributed rock-forming mineral. It is the commonest of the amphiboles. The minerals classified as hornblende are in reality a complex solid-solution series from anthophyllite to hornblende (see Table A.8 in Appendix A).

Biotite, black mica, is a potassium-magnesium-iron-aluminum silicate, $K(Mg,Fe)_3[AlSi_3O_{10}(OH)_2]$. It is constructed of tetrahedra in sheets, as shown in Figure 2.20. Each silicon ion shares three oxygen ions with adjacent silicon ions to form a pattern like wire netting. The fourth, unshared oxygen ion of each tetrahedron stands above the plane of all the others. The basic structural unit of mica consists of two of these sheets of tetrahedra with their flat surfaces facing outward and their inner surfaces held together by cations. In biotite the ions are magnesium and iron. These basic double sheets of mica, in turn, are loosely joined together by positive ions of potassium.

Layers of biotite or of any of the other micas can be peeled off easily (see Figure 2.21) because there is a distinct contrast between the stronger covalent bonding *within* a sheet, and the weaker ionic bonding *between* adjacent sheets and with the intervening cations. In thick blocks biotite is usually dark green or brown to black. Its hardness is 2.5 to 3, and its specific gravity is 2.8 to 3.2.

Nonferromagnesians (Felsic, Sialic Minerals) The other common rock-forming silicates, the **nonferromagne-**

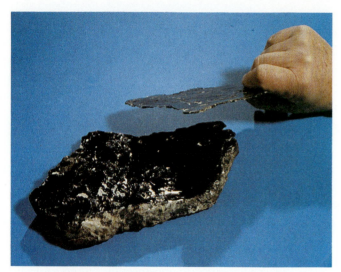

2.21 Mica cleavage. The block (or "book") is bounded on the sides by crystal faces. Cleavage fragments separate parallel to the basal plane of the crystal. [John Simpson.]

Diagnostic ion	Name	Formula[a]
K^+	Orthoclase or K-spar (potassium feldspar)	$K(AlSi_3O_8)$
Na^+	Albite (sodium plagioclase)	$Na(AlSi_3O_8)$
Ca^{2+}	Anorthite (calcium plagioclase)	$Ca(Al_2Si_2O_8)$

[a] In these formulas the symbols inside the parentheses indicate the tetrahedra. The symbols outside the parentheses indicate the diagnostic ions that are worked in among the tetrahedra.

sian minerals (such as muscovite, feldspars, and quartz), do *not* contain magnesium or iron. They are all marked by their light colors and relatively low specific gravities, ranging from 2.6 to 2.9. They are also called **felsic** or **sialic** minerals.

Muscovite, so named because it was once used as a substitute for glass in old Russia (Muscovy), is white mica. It has the same basic crystalline structure as biotite, but in muscovite each pair of tetrahedra sheets is tightly cemented together by ions of aluminum. As in biotite, however, the double sheets are held together loosely by potassium ions, along which cleavage readily takes place. Muscovite has a formula of, essentially, $KAl_2[AlSi_3O_{10}(OH)_2]$. In thick blocks its color is light yellow, brown, or green. Its hardness ranges from 2 to 2.5, and its specific gravity from 2.8 to 3.1.

Feldspars are the most abundant rock-forming silicates. Their name comes from the German *feld,* "field," and "spath," a term used by miners for various nonmetallic minerals. The name reflects the abundance of these minerals: "field minerals," or minerals found in any field. Feldspars make up nearly 54 percent of the minerals in the Earth's crust.

The feldspars are silicates of aluminum with potassium, sodium, or calcium. Crystals of feldspars of different systems resemble each other closely in angles and crystal habit. They all show good cleavages in two directions, which make an angle of 90° or close to 90° with each other. Their hardness is about 6, and their specific gravity ranges from 2.55 to 2.76.

In feldspars all the oxygen ions in the tetrahedra are shared by adjoining silicon ions in a three-dimensional network. However, in the centers of one-quarter to one-half of the tetrahedra, aluminum ions with a radius of 0.51 Å and an electric charge of 3+ have replaced silicon ions with their radius of 0.42 Å and electric charge of 4+. The nega-

tive electric charge resulting from this substitution is corrected by the entry of K^+, Na^+, or Ca^{2+} into the crystalline structures. If Ca^{2+} is added, it does the job of two cations having only one positive charge each.

Two feldspars are **potassium feldspar** (**K feldspar** or simply **K-spar**) and **plagioclase** (with sodium or calcium) (see Table 2.3). K-spar is polymorphous, having three polymorphs whose presence in rocks reflects their rate of cooling. Aluminum replaces silicon in one-fourth of the tetrahedra, and positive ions of potassium correct the electrical unbalance. The introduction of aluminum into the tetrahedra should not be regarded as solid solution. Aluminum is a necessary constituent of K-spar. K-spar is white, gray, or pinkish; its streak is white; and its specific gravity is 2.57.

Plagioclase ("oblique-breaking") feldspars have cleavage planes that intersect at about 86°. Plagioclase feldspars may be colorless, white, blue-gray, or black, and some samples show a very striking play of colors called **labradorescence.**

The plagioclase feldspars albite and anorthite are end members of a solid-solution series. The relative amounts of calcium and sodium vary in proportion to the amount of aluminum in order to maintain electrical neutrality; the more aluminum, the greater the amount of calcium. Here the variation in amount of aluminum may be properly regarded as ionic substitution. For each Ca^{2+} ion replacing a Na^+ ion an additional Si^{4+} is replaced by an Al^{3+} in the tetrahedral framework. Sodium can substitute quite easily for calcium in a mineral's crystalline structure because the ionic radii of these elements are almost identical, 0.97 and 0.99 Å.

In albite aluminum replaces silicon in every fourth tetrahedron, and positive ions of sodium correct the electrical unbalance. The specific gravity of albite is 2.62. In anorthite aluminum replaces silicon in every second tetrahedron, and positive ions of calcium correct the electrical unbalance. The specific gravity of anorthite is 2.76.

Although the plagioclase feldspars have different compositions, they have the same crystal form and are isomorphous. At the same time, they can be thought of as solutions of albite in anorthite in varying proportions, or the reverse. For that reason, they are examples of solid solution. Intermediate minerals have compositions that are conveniently described as mixtures of the pure end members of the series.

Quartz is the most common rock-forming silicate mineral that is composed "exclusively" of tetrahedra. Every oxygen ion is shared by adjacent silicon ions, which means that there are two ions of oxygen for every ion of silicon. This relationship is represented by the formula SiO_2. Because of its crystalline structure, quartz is a relatively hard mineral, 7. Since there are no planes of weakness in its crystalline structure, when struck it does not exhibit cleavage but fractures along curved surfaces known as conchoidal fractures. Being composed of fairly light elements, its specific gravity is 2.65.

Of all the rock-forming minerals, quartz is most nearly a pure chemical compound. However, spectrographic analyses show that even in its most perfect crystals there are traces of other elements. Quartz usually appears smoky to clear in color, but less common varieties include purple or violet amethyst, massive rose-red or pink-rose quartz, smoky yellow to brown smoky quartz, and milky quartz. These color differences are caused by the other elements present as minor impurities, gas bubbles, or liquid inclusions, or even by voids in the molecular lattice.

2.6

ORGANIZATION AND ASSOCIATION OF MINERALS

We know that minerals are special combinations of elements or compounds (see summarizing Table 2.4) and now we can complete our definition of a mineral: (1) It is a naturally occurring element or compound in the solid state;[2] (2) it has a composition that is fixed or varies within narrow limits; and (3) it has a characteristic crystalline structure.

ASSOCIATIONS OF MINERALS

Certain groups of minerals are most commonly found associated in specific geological settings. For example, some minerals formed at high temperatures commonly occur in igneous rocks that crystallize deep within the Earth. Other mineral assemblages reflect their origin under the higher temperatures *and* increased pressures associated with metamorphic events. Finally, there are some mineral groupings that usually develop under conditions approaching the atmospheric temperatures and pressures found at the Earth's surface. Such mineral associations are useful indicators of past conditions at the time of either their origin or their recombination to form new minerals, where temperature, pressure, or chemical environment has changed.

2.7

RESHUFFLING THE ELEMENTS AND MINERALS

As we noted in Chapter 1, rocks continuously undergo changes during their passage through time, and the minerals making up rocks are thus changed. Likewise, the elements constituting these minerals are rearranged in combinations with new elements in varying proportions.

---

[2] Some mineralogists do not restrict the definition to the solid state but include such substances as water and mercury.

TABLE 2.4
Summary of the Organization of Some Common Minerals[a]

Elemental minerals	Compound Minerals					
	Silicates [elements + $(SiO_4)^{4-}$]	Oxides (elements + O)	Sulfides (elements + S)	Carbonates [elements + $(CO_3)^{2-}$]	Sulfates [elements + $(SO_4)^{2-}$]	Halides (elements + Cl⁻, I⁻, Br⁻, or F⁻)
Copper	Ferromagnesian	Cassiterite	Chalcocite	Calcite	Anhydrite	Halite
Diamond	Augite	Corundum	Galena	Dolomite	Gypsum	Sylvite
Gold	Biotite	Hematite	Pyrite	Magnesite		
Graphite	Hornblende	Ice	Sphalerite			
Platinum	Olivine	Magnetite				
Silver	Nonferromagnesian					
Sulfur	Feldspars					
	Orthoclase					
	or K-spar					
	Plagioclase					
	Albite					
	Anorthite					
	Muscovite					
	Quartz					

[a] Minerals may be either elements or compounds, but not all elements or compounds are minerals.

One location where much of this reorganization occurs is along plate boundaries within the Earth's lithosphere (see Section 1.4). For example, at the trailing edge of a plate, new material is welling up from the deeper portion of the Earth's mantle, solidifying and being added as new plate material.

On the other hand, at the leading edge of a lithospheric plate other possibilities exist. The plate may be moving down into the mantle along a subduction zone. Here the increased temperature and pressure cause the minerals in the rocks to undergo changes into other minerals and even to be melted if carried deep enough.

Another possibility exists at the leading edge of a lithospheric plate. The plate may carry a continent upon its back, and this continental mass may encounter another similar mass carried by a plate approaching from the opposite direction. The resulting continent-to-continent collision produces both uplift of mountains and downwarp of "roots" under those mountains (see Chapter 11). Minerals and rocks found in the mountains will undergo changes during uplift, including weathering and erosion. Similarly, those materials found in the roots of the mountains will undergo changes resulting from increased temperature and pressure: melting or remelting of the constituents.

In this way Earth materials are constantly undergoing a reshuffling in terms of their combinations, relative proportions, and associations. New minerals are formed from old; new concentrations of elements and minerals develop where they did not exist previously; and old minerals are broken down into their constituent parts (elements or combinations of elements) and recombined in differing arrangements of elements, minerals, and rocks.

Summary

Minerals are naturally occurring elements or compounds with an orderly, repetitive internal atomic arrangement.

Matter, the substance of the physical universe, is described in terms of atoms with mass and electric charge.
    Electric charge is of two kinds, arbitrarily called **positive** and **negative.**
    Atoms are composed of protons, neutrons, and electrons.
        Atomic structure is expressed by a nucleus of protons and neutrons surrounded by a cloud of electrons.
        For the most common elements, **atomic size,** determined by the protective shield of electrons, is about 2 Å in diameter around a nucleus of 10^{-5} to 10^{-4} Å.
        Atomic mass is 99.95 percent in the nucleus.
    Elements are particular combinations of protons, neutrons, and electrons.
        Isotopes are forms of elements with the same number of protons in the nucleus but different mass because of their different number of neutrons.
        Ions are electrically unbalanced forms of atoms or groups of atoms.

Mineral structure is the internal orderly arrangement of atoms, which is unique for each mineral.
    Compounds are combinations of atoms of different elements.
        A **molecule** is the smallest unit of a compound that displays all the properties of the compound.
    Ionic compounds occur when electrons are lost or added to atoms.
    Covalent bonding results from sharing electrons rather than from gaining or losing them.
    Metallic bonds are responsible for metals being such good conductors of heat and electricity.
    Van der Waals' bonds are weak attractive forces operating between neutral atoms and molecules; this bonding is caused by the momentary distribution of electrons in their orbits.

Distribution of elements in nature depends on the location and characteristics of their origin and dispersal.
    Element abundance on Earth differs from that of the universe as a whole in having a much larger proportion of heavy elements and a much lower proportion of lighter elements, especially hydrogen and helium.

Mineral identification is accomplished by examination of chemical and physical properties, including crystal habit, cleavage, twinning, striations, hardness, specific gravity, color, streak, and luster.
    Crystal habit is the external shape produced by a mineral's internal crystalline structure.
    Cleavage is the tendency of a mineral to break in certain preferred directions along smooth plane surfaces.
    Twinning is the intergrowth of two or more single crystals of the same mineral with different geometric orientations.
    Striations are parallel, threadlike lines or narrow bands running across crystal faces or cleavage surfaces.
    Hardness is governed by the internal atomic arrangement of the elements of a mineral.
    Specific gravity is a number that compares the weight of a given volume of a mineral with the weight of the same volume of water.
    Color is not a reliable property for identifying most minerals, but minerals containing iron are usually dark-colored and those containing aluminum are light-colored.
    Streak is the color of a mineral in finely powdered form.
    Luster is the appearance of a mineral in reflected light. Other mineral properties include magnetism, electrical

properties, fluorescence, fusibility, solubility, fracture, and tenacity.

The most common minerals among the 2,000 or so found in nature comprise only a handful of minerals, with 90 percent belonging to the **silicates.**

Native elements occur without combining with any other elements or compounds.

Halides form from combinations of positive ions with chlorine, iodine, bromine, and fluorine.

Oxide minerals are formed by the direct union of an element with oxygen.

Sulfide minerals are formed by the direct union of an element with sulfur.

Carbonate and **sulfate** minerals are built around the complex ions $(CO_3)^{2-}$ and $(SO_4)^{2-}$, respectively.

Silicate minerals are based on a complex ion called the **silicon-oxygen tetrahedron.**

Ferromagnesian silicates include olivine, pyroxenes, amphiboles, and biotite—all containing iron and magnesium.

Nonferromagnesian silicates do not contain iron; they include muscovite, the feldspars, and quartz.

Organization of minerals lies in naturally occurring combinations of elements or compounds in the solid state, each with diagnostic composition and unique crystalline structure as well as certain physical properties.

Associations of minerals commonly occur in specific geologic settings and reflect the conditions of temperature, pressure, or chemical environment at the time of their origin or when recombined into new minerals.

New combinations of reshuffled elements and compounds produce new minerals, especially along edges of lithospheric plates and in the weathering zones.

Questions

1. How do we know the probable arrangements of protons, neutrons, and electrons in atoms? Draw a diagram to illustrate these arrangements.

2. Why do we assume that matter consists of particles that behave essentially as electrical particles?

3. How do isotopes of one particular element behave similarly to and how do they behave differently from one another?

4. What is the principal property of minerals that controls most of their physical properties? Explain your answer.

5. How are atoms combined with one another?

6. What is the basic building block of the silicates, the most common of all the rock-forming minerals?

7. Explain how solid solution works. Give examples from a common mineral family.

8. When Al^{+3} substitutes for some Si^{+4} ions in silicate minerals, how is electrical neutrality maintained? Give a specific example.

9. After examining the Periodic Chart in Appendix B, explain why certain elements with similar physical properties occur at systematic intervals within that chart.

10. How do associations of minerals reflect the conditions of temperature at the time of their origin?

11. Why are edges of lithospheric plates likely locations for the occurrence of newly combined elements and compounds?

Supplementary Readings

BERRY, L. G., B. H. MASON, AND R. V. DIETRICH *Mineralogy: Concepts, Descriptions, Determinations,* 2nd ed., W. H. Freeman, New York, 1983.
Introduction to crystallography, physical and chemical properties, and genesis of minerals.

BLACKBURN, W. H., AND W. H. DENNEN *Principles of Mineralogy,* W. C. Brown, Dubuque, Iowa, 1988.
Integrates basic science information into treatment of mineralogy as a tool for other geological studies.

HURLBUT, CORNELIUS S., JR. *Minerals and Man,* Random House, New York, 1969.
For the layperson a lively, balanced presentation of expert information on the nature, origin, and properties of 150 of the world's principal minerals, together with stories of how they have been used throughout the ages; with 217 illustrations, including 160 in color.

KLEIN, C., AND C. S. HURLBUT, JR. *Manual of Mineralogy* after J. D. Dana, 20th ed. John Wiley & Sons, New York, 1985.
The classical treatment of minerals from which most other studies have evolved.

NASSAU, KURT "The Causes of Color," *Sci. Am.,* vol. 243, no. 4, 1980.
Discusses the electronic responses to different wavelengths of light and the ultimate colors produced by differential absorption of parts of the energy spectrum.

SHIGLEY, J. E., AND A. R. KAMPF "Gem-bearing Pegmatites: A Review," *Gems & Gemology,* vol. 20, no. 2, 1984.
Discusses our current understanding of pegmatites, the source of many important gem minerals, including aquamarine, tourmaline, and topaz.

In this satellite image a granitic intrusion in Chile's Atacama desert near Chañaral shows up as a bright area against the darker metamorphic rocks that it has intruded. Width of view about 100 km. [NASA.]

CHAPTER ■ 3

ORIGIN AND OCCURRENCE OF INTRUSIVE IGNEOUS ROCKS

"However silent, these large, simple shapes [of granite] have a wordless language. They say that the granite has grown in secure and secluded depths, guided wholly by its own laws; and that no disturbing outside influence has intervened in the slow tranquil growth of the crystals."

Hans Cloos, Conversation with the Earth, *p. 105, translated from the German by E. B. Garside, Alfred A. Knopf, New York, 1953*

3.1

THE EARTH'S HEAT

The principal requirement for igneous activity is heat, and evidence has accumulated that there is a great deal of heat in the Earth's interior. That the Earth's temperature increases with depth was first noted as early as the seventeenth century, when German miners reported a steady rise in temperature with increasing depth in their mine shafts. Further proof came from the temperature measurements made in other mines and from holes drilled deep into the Earth.

The increase in temperature with depth is called the **geothermal gradient.** Reported rates of increase vary from place to place, even in areas far removed from obvious igneous activity. The geothermal gradient can range from less than 10°C/km to as much as 50°C/km or even greater under regions such as Yellowstone Park (Figure 3.1). A geothermal gradient of 30°C/km seems to be a reasonable average. These figures get us "into hot water," so to speak, very quickly. Within 50 km of depth, a geothermal gradient of 30°C/km would result in a temperature of 1500°C, believed to be high enough to melt rocks (at pressure equal to that at the Earth's surface). We do know, however, that the melting point of rocks increases as the pressure increases. Therefore at great depths within the Earth the rocks can be quite hot and still not be melted because of the great pressures found at those depths. The average geothermal gradient (but *not the temperature*) probably diminishes somewhat after depths of several tens of kilometers.

Heat travels extremely slowly through soil and rock. Measurements and calculations have shown that the Earth's **heat flow,** which is the product of the temperature gradient and the thermal conductivity of Earth materials, is

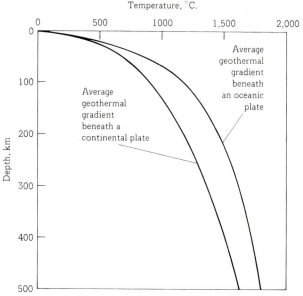

3.1 Geothermal gradients under different geologic settings. [Various sources.]

reasonably constant. Its average value is 1.2 microcalories per square centimeter per second (μcal/cm^2-s) or 4.9 watts of heat per square centimeter. Present indications are that there is some difference in the heat flow from a number of different land and sea areas. The average heat flow represents about 40 cal/cm^2-year or 167.5 watts per square centimeter, which does not seem like much. But in its steady, quiet way it totals more energy in a year than all that released by seismic and volcanic activity. Some parts of the Earth's surface, such as over midocean ridge areas, have much higher heat flow than do others. This variation seems to depend on local tectonic and volcanic activity associated, in part, with lithospheric plate margins.

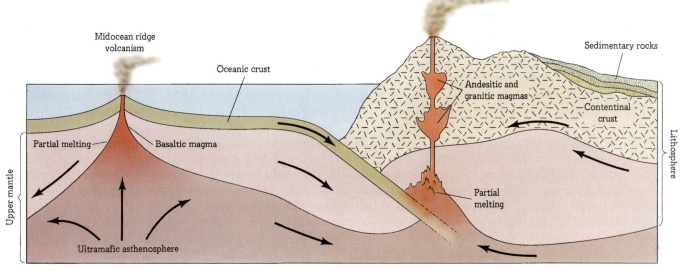

3.2 Converging plate margin showing origin of various magmas by partial melting at different settings: Basaltic magma may form by partial melting of ultramafic material from the upper mantle beneath oceanic crust; granitic magma may form by partial melting of material beneath continental crusts; andesitic or intermediate-composition magma may form by partial melting of sediments and oceanic crust undergoing subduction.

3.2

FORMATION OF IGNEOUS ROCKS

Igneous rocks at the surface today, as well as those found at depth, have been formed from **magma.** Molten rock in the ground is called magma; when it is extruded on the surface, it is termed **lava;** and when solidified pieces of magma are blown out, they are known as **pyroclastic** (from the Greek *pyros,* "fire," and clasts or fragments) **debris.**

Magmas result from the partial melting of portions of the mantle or lower crust (Figure 3.2). Some may be mixtures from both the mantle and the crust. The melting processes that form magmas are complex, involving high temperatures and pressures, plus gases (principally steam and carbon dioxide). These magmas rise toward the surface because they are less dense than the surrounding cooler, more viscous or more solid rocks.

A magma reservoir, or **chamber,** is *not* a big holding tank full of just liquid but rather a mass that includes solids as well as interstitial melt.

All igneous rocks have been formed from the solidification of magma. Magma, extruded as lava at the surface, cools and solidifies to form volcanic rocks. The offshoots of magma that work their way into surrounding rock below the surface cool more slowly and also solidify to form plutonic rocks.

CRYSTALLIZATION OF MAGMA

Magma solidifies through the process of **crystallization.** Magma is simply a **melt,** a liquid solution of elements at high temperature. With a decrease in temperature, the melt starts to solidify. Slowly mineral grains begin to grow. In the latter stages of this process gases are released. Though these gases quickly move away from the magma, for the time being we no longer have a complete liquid but rather a liquid mixed with solid and gaseous materials.[1] As the temperature continues to fall, the mixture solidifies until solid igneous rock is formed.

Igneous rocks usually consist of interlocking grains of a mixture of several minerals but in a few cases may consist of just a single mineral. As indicated in Chapter 2, the most common rock-forming minerals are the silicates — olivine, pyroxene, amphibole, biotite, muscovite, potassium feldspar, plagioclase, and quartz.

BOWEN'S REACTION PRINCIPLE

Magma is a solution of elements, but it does not crystallize the way ordinary solutions do. When dealing with most common chemical reactions in the laboratory, we have noted that solutions of a certain composition always crystallize into a solid of that same composition. This happens regardless of conditions during solidification. If crystallization of a magma were similar, it would always yield rock of the same composition as that of the original magma. However, magma of a certain composition is able to crystallize into a number of different kinds of rock. This phenomenon is the basis for the diversity of igneous rocks that can be formed from the crystallization of, say, a basaltic magma.

---

[1] To reflect this changing picture we define a magma as any naturally occurring silicate melt, whether or not it contains suspended crystals or dissolved gases.

When magma cools rapidly, there is no time for minerals to settle or to react with the remaining liquid. Rapid cooling occurs when a partially crystallized magma is extruded onto the Earth's surface or injected along fractures into cool rocks near the Earth's surface. But when a large body of magma cools slowly deep within the crust, a high degree of chemical reaction may take place. In 1922 N. L. Bowen of the Geophysical Laboratory of the Carnegie Institution of Washington proposed that the differences in end products depend on the rate at which the magma cools and on whether early-formed minerals remain in or settle out of the remaining liquid during its crystallization. He suggested that as a magma cools, the first-formed minerals undergo modification with the liquid remaining after they crystallized. He called this process **reaction.** This reaction principle is the key to magmatic crystallization.

Experiments showed that some minerals crystallizing from a silicate melt can react with the remaining melt. The reactions are of two different types: **discontinuous reaction series,** in which the early-formed minerals react to form new minerals; and **continuous reaction series,** in which a mineral continuously changes composition but maintains its crystal structure.

Bowen was able to show that both reactions occur simultaneously as a given magma cools. Taking a basaltic magma (rich in iron and magnesium), he showed that the first minerals to crystallize are olivine and calcium-rich plagioclase. With further cooling of the magma, the olivine reacts with the melt to form a pyroxene mineral, which will, upon still further cooling, react with the melt to form an amphibole mineral. If carried to completion, this stepwise discontinuous reaction series will result first in the complete elimination of olivine, then in a similar elimination of pyroxene, and so forth (Figure 3.3). Recall from Chapter 2 that olivine has the simplest arrangement of silica tetrahedra (isolated), pyroxenes have the next simplest (single chains of tetrahedra), amphiboles the next (double chains of tetrahedra), and biotite has the most complex arrangement of all the ferromagnesian silicates (sheets of tetrahedra).

During the same crystallization interval, as the magma cools, the calcium-rich plagioclase is independently crystallizing out. With continual cooling, the melt becomes more sodic so the early-formed plagioclase crystals are out of equilibrium with the liquid. The Ca-rich plagioclase reacts with the liquid by exchanging Ca^{++} and Al^{+++} atoms in the crystals for Na^+ and Si^{++++} atoms in the melt. The plagioclase thus becomes more and more sodium-rich in this continuous reaction series. However, the crystal forms in this plagioclase series do not undergo the abrupt changes typical of the discontinuous reaction series of the ferromagnesian silicate minerals.

Early-formed minerals may separate from the remaining melt in processes known as **fractional crystallization.** (The term **magmatic differentiation** is sometimes used for these processes.) The two most important are **crystal settling** (or **crystal floating**) and **filter pressing.** Early-formed minerals that differ in density from the rest of the magma may separate either by sinking or by floating. These processes result in rocks known as **cumulates,** whose composition is different from the parent magma and also from the residual magma, which may undergo further differentiation. In filter pressing, tectonic or other forces squeeze the remaining liquid out of a crystallizing magma. This leaves a residual mesh of early-formed minerals and a liquid that may crystallize immediately or may undergo further differentiation. In either case this fractional crystallization prevents the early-formed minerals from equilibrating with the liquid from which they grew, resulting in a series of residual liquids of more extreme compositions than would have resulted from **equilibrium crystallization.** Thus the resulting rocks will be different in composition from the parent magma.

We see, therefore, that a parent magma could result in a large number of diverse kinds of igneous rocks, depending on the history of cooling, reaction, and crystal separation (see Figure 3.4). Starting with a magma of any bulk composition, we would find that, upon cooling, reactions would occur as described by Bowen. His reaction principle explains how a magma rich in iron and magnesium may

3.3 Bowen's reaction series, showing variations in discontinuous ferromagnesian minerals and continuous plagioclase feldspar minerals during crystallization and differentiation of basaltic magma as the temperature of the magma decreases.

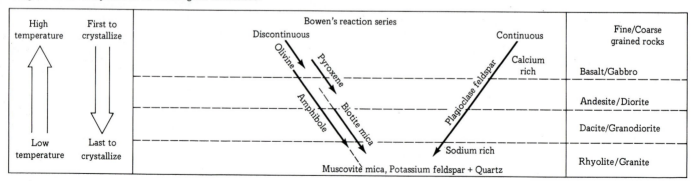

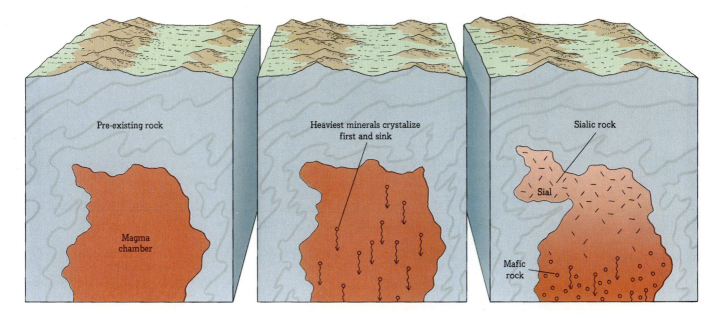

3.4 Fractionation of a magma after being injected into cool country rock. (*a*) Newly injected liquid magma. (*b*) First minerals crystallizing as their freezing temperature is reached (probably olivine, pyroxene, and calcic plagioclase feldspar crystals); as these minerals are separated from the magma, the melt becomes depleted in iron, magnesium, and calcium, and enriched in sodium, potassium, and pure silica. (*c*) If part of this fractionated melt is separated into an adjacent pocket, it may develop into a magma and resulting igneous rock closer in composition to a granite.

solidify as one rock type or may produce several rock types. It might crystallize to a gabbro that is composed approximately half of ferromagnesian minerals and half of plagioclase if the entire magma crystallizes without further fractionation (under equilibrium crystallization). Similarly, it might produce a silica-rich residual melt by magmatic differentiation from which a granite could result. Another possibility is a rock of intermediate composition.

The rock formed from the crystallization of a magma depends on the extent to which early-formed minerals were removed from further reaction with the melt and on the rate of cooling.

Although Bowen's reaction principle has largely withstood seven decades of scrutiny, some objections have been raised. According to his reaction principle, only about 5 or 10 percent of an original basaltic magma could solidify into granite or granodiorite (a composition between that of granite and of diorite). This extremely small percentage cannot account for the large masses of such rocks found on many continents. Where are the earlier-formed, more basic rocks that should be associated with these large masses? Although there is overwhelming experimental evidence that the reaction series is basically correct, we do know that not all granites need be derived from mafic magmas. We shall examine this problem further in Section 7.5.

3.3
TEXTURE OF IGNEOUS ROCKS

Texture, a term derived from the Latin *texere,* "to weave," is a physical characteristic of all rocks. The term refers to the general appearance of rocks. In referring to the texture of igneous rocks, we mean specifically the size, shape, and arrangement of their interlocking mineral grains.

GRANULAR (PHANERITIC) TEXTURE

If magma has cooled at a relatively slow rate, it will have had time to develop grains that the unaided eye can see in hand specimens. Rocks composed of such large mineral grains are called **granular** or **phaneritic** (Greek *phaneros,* "visible"). (See Figure 3.5.)

However, the rate of cooling, though important, is not the only factor that affects the texture of an igneous rock. For example, if a magma is of low viscosity — that is, if it is thin and watery and flows readily — large, coarse grains may form even though the cooling is relatively rapid, for in a magma of this sort the ions can move easily and quickly into their rock-forming mineral combinations.

3.5 Rock can be ground down to a translucent thinness. Here is a photograph taken through a microscope of such a *thin section*. This rock is composed of interlocking crystals of plagioclase (the minerals marked by parallel stripes) and olivine (the more or less equidimensional grains of varying colors). The colors are due to polarized light passing through the minerals and affected by their crystal structure and orientation. Field of view is about 0.5 cm across.

APHANITIC TEXTURE

The rate at which a magma cools depends on the size and shape of the magma body, as well as on its depth below the surface. For example, a small body of magma with a large surface area—that is, a body that is much longer and broader than it is thick—surrounded by cool, solid rock loses its heat more rapidly than would the same volume of magma in a spherical reservoir. And because rapid cooling usually prevents large grains from forming, the igneous rocks that result have **fine-grained** or **aphanitic** textures. Individual minerals are present but are too small to be identified without a microscope.

GLASSY TEXTURE

If magma is suddenly ejected from a volcano or a fissure at the Earth's surface, it may cool so rapidly that there is no time for minerals to form. The result is a **glass,** which by a rigid application of our definition is not really a rock, though it is generally treated as one. Glass is a special type of solid in which the ions are not arranged in an orderly manner. Instead, they are disorganized, like the ions in a liquid. And yet they are frozen in place by the quick change of temperature. Some glasses have many vesicles (or gas holes) resulting from the high gas content of the lava (or magma prior to its extrusion). Others are massive glasses, such as obsidian (a glass formed from rapid cooling of a granitic magma).

PORPHYRITIC TEXTURE

Occasionally a magma cools at variable rates—slowly at first, then more rapidly. It may start to cool under conditions that permit large mineral grains to form in the early stages, and then it may move into a new environment where more rapid cooling freezes the large grains in a **groundmass** of finer-grained texture (see Figure 3.6). The large minerals are called **phenocrysts.** The resulting texture is said to be **porphyritic. Porphyry,** from the Greek word for "purple," was originally applied to rocks containing phenocrysts in a dark red or purple groundmass. Depending on its mineral composition, a rock with over 25 percent phenocrysts is called a porphyry (e.g., rhyolite porphyry or andesite porphyry).

In rare cases magma may be suddenly expelled at the surface after large mineral grains have already formed. Then the final cooling is so rapid that the phenocrysts become embedded in a glassy groundmass.

PEGMATITIC TEXTURE

The solutions that develop late in the cooling of a magma are called **hydrothermal.** From these solutions there crystallize exceptionally granular igneous rock with **pegmatitic** texture (from the Greek *pēgmat-,* "fastened together"). **Pegmatites** embody the chief minerals to form from the hydrothermal solutions: potassium feldspar and quartz. They are distinguished solely on the basis of extraordinary

3.6 Basalt porphyry. This rock is composed of crystals of plagioclase (lighter lath shapes) and olivine and pyroxene (darker minerals). Specimen is 10 cm across. [John Simpson.]

grain size. A few pegmatites consist of intimately intergrown grains of potassium feldspar and quartz, which form essentially a single unit. The quartz is darker than the feldspar, and the overall pattern suggests the wedge-shaped characters in the writing of ancient Assyria, Babylonia, and Persia. As a result, this has become known as **graphic structure** (from the Greek *graphein,* "to write") (see Figure 3.7).

Pegmatites are found at the margins of large intrusives. They range in length from a few centimeters to a few hundred meters and contain crystals of very large size. In fact, some of the largest crystals known have been found in pegmatite. Crystals of spondumene (a lithium mineral) 12 m in length have been found in the Black Hills of South Dakota; crystals of beryl (a silicate of beryllium and aluminum) that measure 1 by 5 m have been discovered in Albany, Maine. Great masses of potassium feldspar, weighing over 2,000 t, yet showing the characteristics of a single crystal, have been mined from pegmatite in the Karelo-Finnish Soviet Socialist Republic.

Nearly 90 percent of all pegmatite is **simple pegmatite,** composed of quartz, orthoclase, and unimportant percentages of micas. It is more generally called **granite pegmatite** because the composition is that of granite and the texture that of pegmatite. The remaining 10 percent are **complex pegmatites.** The major components of complex pegmatites are the same sialic minerals that we find in simple pegmatites, but in addition they contain a variety of rare minerals: lepidolite, tourmaline (best known as a semiprecious gem), topaz (also a gem), tantalite, and uraninite (pitchblende).

3.7 Graphic granite. (Specimen approximately 7 cm by 11 cm.) [Walter R. Fleischer.]

Simple pegmatites are common in some regions and complex pegmatites in others. In southwestern New Zealand, for example, pegmatites are uniformly simple, but throughout the Appalachian regions of North America complex pegmatites are more abundant.

3.4

CLASSIFICATION OF IGNEOUS ROCKS

Several systems have been proposed for the classification of igneous rocks. All are artificial in one detail or another, and all rely on certain characteristics that cannot be determined in the field or by examination of hand specimens. For our present purposes we shall emphasize texture and composition. Such a classification is entirely adequate for an introductory study of physical geology and even for many advanced phases of geology.

This classification appears in tabular form in Figure 3.8, together with a graph that shows the proportions of silicates in each type of igneous rock. The graph demonstrates the *continuous progression from rock types in which dark-colored minerals predominate* (on the right side of the chart) *to rock types in which light-colored minerals predominate* (on the left). (Here "dark-colored minerals" refers to those minerals of ferromagnesian composition and calcium-rich plagioclase, which generally are black, dark gray, or dark blue-black in color. "Light-colored minerals" refers to quartz, potassium feldspar, and sodium-rich plagioclase, which generally are white, pink, or light gray in color. However, some of these "light-colored minerals" may be dark gray or, in rare instances, almost black in color.)

The names of rocks in Figure 3.8 have been arbitrarily assigned on the basis of average mineral composition and texture. Some intermediate types are indicated by such names as "granodiorite." There are many more igneous rocks than are shown in this figure.

Though igneous rocks vary considerably in composition, they seem to have similarities within certain geologic settings. Some types seem to occur primarily on continents, others in ocean regions, while still others appear to be confined to mountain-building settings. So we find a general correlation between igneous rock composition and the geologic setting.

DARK-COLORED IGNEOUS ROCKS

The most abundant igneous rock is **basalt** (an extrusive form of **gabbro**). It has occurred throughout geologic history on continents and in ocean basins. There are two different types of gabbro based upon chemical composition: **tholeiite gabbro** and **alkali olivine gabbro.** Tholeiite gabbro contains two pyroxenes, plagioclase feldspar, and perhaps olivine. Alkali olivine gabbro has only one pyroxene (au-

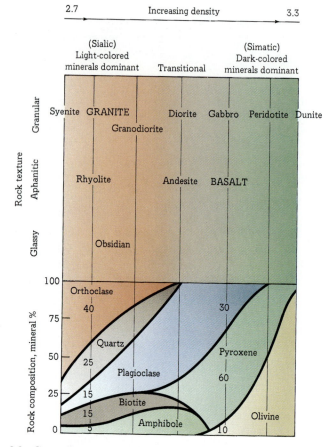

3.8 General composition of igneous rocks is indicated by a line from the name to the composition chart. Granite and rhyolite consist of about 40 percent orthoclase, 25 percent quartz, and 15 percent plagioclase feldspars and 20 percent ferromagnesian minerals. Granite is the most important granular rock, and basalt the most important aphanitic rock. Basalt (and its coarse-grained equivalent gabbro) consists of 30 percent plagioclase (Ca-rich) and 70 percent ferromagnesians (60 percent pyroxene and 10 percent olivine). [Composition chart modified from L. Pirsson and A. Knopf, *Rocks and Rock Minerals*, p. 144, John Wiley & Sons, New York, 1926.]

gite), plus plagioclase feldspar and olivine. Other differences in trace elements commonly occur between the two types of gabbros.

Tholeiitic gabbros are primarily found in the deep ocean and along oceanic ridges, but they also occur as flood gabbros (really flood basalts) in some continental regions (including the Columbia Plateau of northwest United States and the Deccan Plateau in India). Alkali olivine gabbros commonly occur on oceanic islands. See Table 3.1 for chemical compositions of the common igneous rocks.

A popular synonym for gabbro (especially basalt) is **trap rock,** from a Swedish word meaning "step." This name refers to the tendency of certain basalts to form columns that look like stairways in some outcrops (see Figure 3.9). The "steps" are actually joints formed as a tensional product of the cooling process that have become more evident after weathering.

Olivine-rich rocks (**dunites** and **peridotites**) commonly occur in mountain fold belts and along modern

TABLE 3.1
Composition of Typical Igneous Rocks

"Average" rocks	Component minerals			
	Typical volume (%)			
	Ferromag-nesians	Plagioclase feldspar	Potassium feldspar	Quartz
Gabbro/basalt	70	30	0	0
Diorite/andesite	40	60	0	0
Granite/rhyolite	20	15	40	25

mineral side of the classification chart. **Diorite** is the name given to granular (phaneritic) igneous rocks that are intermediate in composition between granite and gabbro. These rocks were first identified in their fine-grained (aphanitic) form in the Andes Mountains of South America, and in this form they are known as **andesites.** Diorites and andesites are mostly found in areas associated with present or former subduction zones (where one plate descends under another plate). Thus we find many andesites and diorites in such areas as the region surrounding the Pacific Ocean.

areas of potential mountain-building sites, such as island arcs. Dunites are named for the type found in the locality of Dun Mountain, New Zealand. Peridotites are coarse-grained igneous rocks composed largely of ferromagnesian minerals.

INTERMEDIATE TYPES

Igneous-rock compositions blend continuously from one to another as we go from the dark- to the light-colored

LIGHT-COLORED IGNEOUS ROCKS

The second most common igneous rock is **granite,** a granular (phaneritic) rock; its fine-grained equivalent is called **rhyolite.** The origin and history of some granites are still under debate; here we use the term to indicate composition and texture, not origin. Most of the igneous rocks that have solidified from magmas within continental regions are granitic in composition. They consist primarily of quartz and potassium-feldspar, with some sodium-rich plagioclase and minor amounts of micas and ferromagnesian minerals.

3.9   Columnar jointing, a pattern often developed in basalt as it cools, outlines this series of prisms in a volcanic neck in Devil's Tower, Wyoming.

3.10 Obsidian usually has a composition similar to granite (also pumice). Specimen approximately 10 cm across [John Simpson.]

The glassy equivalent of granite is **obsidian** (see Figure 3.10). Although this rock is listed near the light side of the chart, it is usually pitch black or brown in appearance because of the finely disseminated iron-oxide minerals dispersed throughout the glass. However, pieces of obsidian thin enough to be translucent are smoky white to almost clear against a light background.

By examining the representative igneous rocks, we can come to some reasonable estimates about the composition of the magmas and lavas from which they were crystallized (see Section 3.5). The original composition of magma may be altered, however, by loss of gases during its rise and by contamination from the country rock as parts of the wall and roof are incorporated into the rising magma. The gases will be discussed further in Chapter 4, but we should note here that they commonly consist mainly of water (steam), with carbon and sulfur gases (including CO_2, CO, HCl, HF, H_2S, SO_2, and H_2) making up lesser proportions.

3.5
ORIGIN OF MAGMAS

Primary magmas result from the partial melting of portions of the lower crust and upper mantle. **Secondary magmas** result from fractional crystallization or other means of producing changes in a primary magma.

Igneous petrologists suggest there are three different primary magmas: **basaltic, granitic,** and probably **andesitic** (there is not unanimous agreement on the last as a distinct primary magma). We will discuss the relationship of magmas to their tectonic settings in this section.

How do we get different magmas from parent material of fairly uniform composition? Studies by many geochemists and petrologists suggest that several factors interact, including: (1) the depth at which a magma undergoes mineral formation; (2) the degree of partial melting that has occurred; (3) the extent to which a magma undergoes later fractional crystallization (see discussion of Bowen's reaction series in Section 3.2); and (4) the amount of interaction of the rising magma with the wall rocks through which it passes.

RELATIONSHIP OF MAGMAS TO TECTONIC SETTINGS

The rigid plates making up the Earth's lithosphere interact with one another and move over the underlying asthenosphere. Some igneous rocks seem to prefer one tectonic setting to another.

Tholeiitic basalts, found in oceanic ridges, may have resulted from partial melting at shallow depths in the asthenosphere. Alkaline olivine basalts commonly occur in oceanic islands removed from lithospheric plate boundaries. These result from magmas generated at greater depths in the asthenosphere than are the tholeiite basalt magmas.

Where one plate is overridden by another plate and descends into the mantle along a subduction zone, magmas of varying composition occur. Basaltic magmas can form from the asthenosphere next to a descending plate. Higher silica andesites may result from the partial melting of both oceanic crust and sediments attached to the descending oceanic plate. Ultramafic rocks may have been emplaced by tectonic activity following partial melting of the asthenosphere.

Some igneous rocks on the continents are related to tectonic activity and others are not—some igneous rocks are located far from active tectonic zones. Basalts and alkalic rocks are found along continental rift zones, which may occur at diverging plate boundaries as a continent begins to break apart. The large dome-shaped granitic bodies along the Pacific Ocean may have been formed by partial melting of continental crust as magmas rose from a subduction zone dipping under this region.

3.6
MASSES OF IGNEOUS ROCKS

When magma within the crust solidifies, it forms igneous rock masses of varying shapes and sizes. Today such rocks can be seen at the surface on continents where previously overlying rocks have been worn away by erosion. For example, the first internal part of a volcano to be exposed by erosion is the plug that formed when magma solidified in the vent. Revealed next are dikes filling the channels through which the magma moved to the surface. Finally, in some regions of ancient activity, the crust has been so elevated (see Chapter 9) and eroded that the reservoir that once stored the magma can now be seen as a solid rock mass at the surface. Solidified offshoots of magma from the reservoir, having intruded themselves into other rocks within the crust, are included in these masses, which were not necessarily associated with the eruption of a volcano.

PLUTONS

All igneous rock masses that were formed when magma solidified within the Earth's crust are the results of **plutonic igneous** activity. When rocks intruded by such **plutons** have a definite layering, we may speak of the magma that invades them as **concordant** if its boundaries are parallel to the layering, or as **discordant** if its boundaries cut across the layering.

Plutons are classified according to their size, shape, and relationship to surrounding rocks. They include sills, dikes, lopoliths, laccoliths, and batholiths (Figure 3.11).

Tabular Plutons   A pluton with a thickness that is small relative to its other dimensions is called a **tabular** pluton.

Sills.   A tabular concordant pluton is called a **sill.** It may be horizontal, inclined, or vertical, depending on the attitude of the rock structure with which it is concordant.

Sills range in size from sheets less than 1 cm in thickness to tabular masses hundreds of meters thick. Because a sill is an intrusive form—that is, has found its way into already existing rocks—it is always younger than the rocks into which it was injected. A sill must not be confused with an ordinary lava flow that was later buried by other rocks. There are fairly reliable ways of distinguishing between the two types: A buried lava flow usually has a rolling or wavy top pocked by the scars of vanished gas bubbles and showing evidence of erosion, whereas a sill has a more even and unweathered surface. Also, a sill may contain fragments of rock that were broken off when the magma forced its way into the surrounding structures.

The Palisades along the west bank of the Hudson River near New York are the remnants of a sill that was several hundred meters thick. Here the magma was originally intruded into flat-lying sedimentary rocks. These are now inclined at a low angle toward the west.

3.11   Plutons and landforms associated with igneous activity.

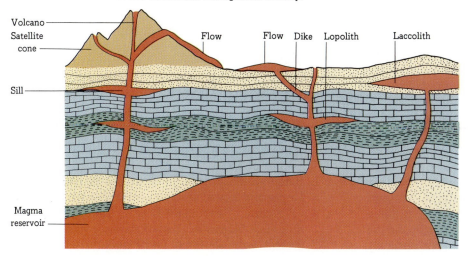

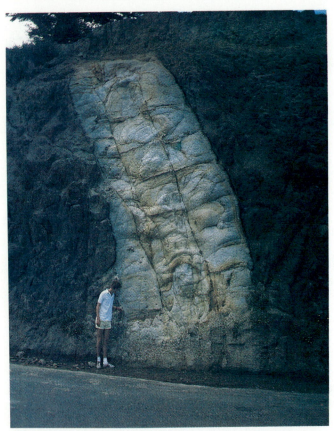

3.12 Dike of rhyolitic composition, light-colored, cutting older darker basaltic material near Dunedin, South Island, New Zealand.

Dikes. A tabular discordant pluton is called a **dike** (see Figure 3.12). Dikes originated when magma forced its way through the fractures of adjacent rocks.

The width of individual dikes ranges from a few centimeters to many meters. The Medford dike near Boston, Massachusetts, is 150 m wide in places. Just how far we can trace the course of a dike across the countryside depends in part on how much of it has been exposed by erosion. In Iceland dikes 15 km long are common, and many can be traced for 50 km; at least one is known to be 100 km long.

As magma forces its way upward, it sometimes pushes out a cylindrical section of the crust. Today, as a result, we find exposed at the surface some roughly circular or elliptical masses of rock that outline the cylindrical sections of the crust. These solidified bodies of magma are called **ring dikes.** Large ring dikes may be many kilometers around and hundreds or thousands of meters deep. Ring dikes have been mapped with widths of 500 to 1,200 m and diameters ranging from 2 to 25 km.

Some dikes occur in concentric sets. These originated in fractures that outline an inverted cone, with the apex pointing down into the former magma source; the fractures were caused by the upward pressure of the intruding magma. These dikes are called **cone sheets.** In Scotland the dip of certain cone sheets suggests an apex approximately

5 km below the present surface of the Earth. Dikes are also found in approximately parallel groups called **dike swarms.**

Lopoliths. **Lopoliths** are tabular concordant plutons shaped like a spoon, with both the roof and the floor sagging downward.

A well-known example is the Duluth lopolith, which crops out on both sides of Lake Superior's western end and appears to continue beneath the lake. It has been computed to be 250 km across and 15 km deep, with a volume of 200,000 km³.

Most lopoliths are composed of rock that has been differentiated into alternating layers of dark and light minerals, presenting the appearance of thinly bedded sedimentary rock.

Massive Plutons Any pluton that is not tabular in shape is classified as a **massive** pluton.

Laccoliths. A massive concordant pluton that was created when magma pushed up the overlying rock structures into a dome is called a **laccolith.**

A classic development of laccoliths is found in the Henry, La Sal, and Abajo Mountains of southeastern Utah, where their features are exposed on the Colorado Plateau, a famous geological showplace.

Batholiths. A large discordant pluton that increases in size as it extends downward is called a **batholith.** The term *large* in this connection is generally taken to mean a surface exposure of more than 100 km². A pluton that has a smaller surface exposure but exhibits the other features of a batholith is called a **stock.**

Batholiths are exposed thousands of meters above sea level, where they have been lifted by forces operating in the Earth's crust. Thousands of meters of rock that covered the batholiths have been stripped away by the erosion of millions of years. We can observe these roots of mountain ranges in the White Mountains of New Hampshire and in the Sierra Nevada.

Although batholiths provide us with some valuable data, they also raise a host of unsolved problems, all of which bear directly on our understanding of igneous processes and the complex events that accompany the folding, rupture, and eventual elevation of sediments to form mountains. (We shall discuss these mountain-forming processes in Chapters 9 and 11.)

We can summarize what we know about batholiths as follows:

1. Batholiths are located in mountain ranges — for example, the Sierra Nevadas — or in regions that were once parts of mountain ranges. Although in some mountain ranges no batholiths are exposed at the present time, we rarely find batholiths that are not

associated with mountain ranges. In any given mountain range the number and size of the batholiths seem to be directly related to the intensity of the folding and crumpling that has taken place. This does not mean, however, that the batholiths caused the folding and crumpling. Actually, there is convincing evidence to the contrary, as we shall see in some of the following features.

2. The long axes of batholiths (which usually are elliptical) generally run parallel to the axes of mountain ranges.

3. Batholiths have been intruded across the folds, indicating that they were formed after the folding of the mountains—although the folding may have continued after the batholiths were formed.

4. Batholiths have irregular dome-shaped roofs. This characteristic shape is related to **stoping,** one of the mechanisms by which magma moves upward into the crust. As the magma moves upward, blocks of rock are broken off from the structures into which it is intruding. At low levels, when the magma is still very hot, the stoped blocks may be melted and **assimilated** by the magma reservoir. Higher in the crust, as the magma approaches stability and its heat diminishes, the stoped blocks are frozen in the intrusion as **xenoliths,** that is, "strange rocks" (see Figure 3.13).

5. Batholiths are composed primarily of granite or granodiorite (a plutonic rock consisting of quartz, sodic plagioclase, orthoclase, and some mafic constituents such as pyroxene, hornblende, and/or biotite).

6. Batholiths give the impression of having replaced the rocks into which they have intruded instead of having pushed them aside or upward. But if that is what really took place, what happened to the great volumes of rock that the batholiths appear to have replaced? Here we come up against the problem of the origins of batholiths—in fact, against the whole mystery of igneous activity. Some observers have even questioned whether granitic batholiths were formed from true magmas at all. The suggestion has been made that the batholiths may have been formed through a process called **granitization,** in which solutions from magmas and from the rocks being changed move into solid rocks, exchange ions with them, and convert them into rocks that have the characteristics of granite but have never actually existed as magma. We shall return to this highly controversial proposal in Chapter 7.

7. Batholiths contain a great volume of rock. The Sierra Nevada complex of California is 650 by 60 to 100 km and consists of a dozen or so plutons with compositional and age differences. A partially exposed complex in southern California and Baja California is probably 1,600 by 100 km.

8. Gravity measurements have been interpreted as indicating that the downward extent, or "thickness," of many batholiths is some 10 to 15 km.

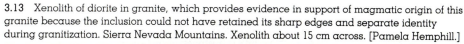

3.13 Xenolith of diorite in granite, which provides evidence in support of magmatic origin of this granite because the inclusion could not have retained its sharp edges and separate identity during granitization. Sierra Nevada Mountains. Xenolith about 15 cm across. [Pamela Hemphill.]

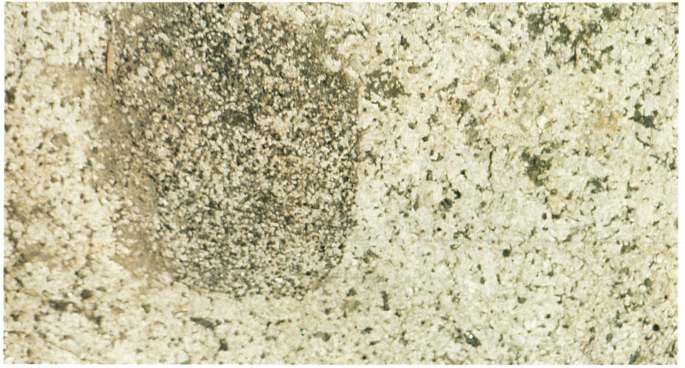

Summary

Igneous rocks are formed from the solidification of molten matter.

Igneous activity consists of movements of molten rock inside and outside the Earth and the variety of effects associated with these movements.
  Cause of igneous activity seems to be the same internal force that elevates mountains, causes earthquakes, and causes metamorphism.

The Earth's heat is indicated by the **geothermal gradient** to be more than adequate for igneous activity. **Speculations on the origin of the Earth's heat** have included original heat and radioactivity.

Igneous rocks at the surface today were formed from magma.
  Magma solidifies through the process of crystallization.
  Bowen's reacton series are incorporated in a hypothesis accounting for all igneous rocks coming from an olivine basaltic magma.
    Limitations of Bowen's hypothesis include the failure to account for large undifferentiated masses of granite.
    The rate of crystallization is an important control over the rocks that form.

Texture of igneous rocks is determined by the size, shape, and arrangement of their interlocking mineral grains.
  Granular texture includes large mineral grains from slow-cooling or low-viscostiy magma.
  Aphanitic texture from rapid cooling consists of individual minerals so small that they cannot be identified without the aid of a microscope.
  Glassy texture results from ions disorganized as in a liquid but frozen in place by quick cooling.
  Porphyritic texture is a mixture of large mineral grains in an aphanitic or glassy groundmass.
  Pegmatitic texture is an exceptionally large granular mass of crystals formed by hydrothermal solutions late in the cooling of a magma.

Types of igneous rock are arbitrarily defined in terms of texture and composition.
  Dark-colored igneous rocks (intrusive **gabbros**, extrusive **basalts**) constitute 98 percent of rock formed from magma that has poured out onto the Earth's surface.
  Intermediate types of compostition are given arbitrary names, such as **andesite** and **diorite,** because igneous rock compositions blend continuously from one to another from the dark to the light side of the classification chart.
  Light-colored igneous rocks, sometimes called **sialic,** are dominated by **granites** and **granodiorites.**

Origin of magmas varies from partial melting of portions of the lower crust and upper mantle that produces primary magmas, to fractional crystallization or other changes that result in secondary magmas.
  Three **primary magmas** are basaltic, granitic, and andesitic.

Relationship of magmas to tectonic settings is demonstrated by different magmas occurring at different plate boundaries.

Masses of igneous rocks are called **plutons,** which are classified according to size, shape, and relationships to surrounding rocks.
  Sills are concordant tabular plutons.
  Dikes are discordant tabular plutons.
  Lopoliths are tabular concordant plutons shaped like a spoon.
  Laccoliths are massive concordant plutons with domed tops.
  Batholiths are massive discordant plutons 10 to 15 km thick.

Questions

1. What is the basic cause of igneous activity?
2. Is the temperature of the Earth's crust essentially the same around the globe or does it differ? Why or why not? How do we know? Define geothermal gradient.
3. Explain in step-by-step fashion what occurs during the fractional crystallization of an olivine basaltic magma according to Bowen's hypothesis.
4. How does the texture of an igneous rock permit us to interpret much about its history of development?

Supplementary Readings

BARKER, D. S. *Igneous Rocks,* Prentice Hall, Englewood Cliffs, N.J., 1983.
A basic treatment of igneous rocks for undergraduate geology majors with some chemistry.

BEST, M. G. *Igneous and Metamorphic Petrology,* Prentice Hall, Englewood Cliffs, N.J., 1982.
An intermediate treatment of igneous and metamorphic products and processes.

DIETRICH, R. V., AND B. J. SKINNER *Rocks and Rock Minerals,* John Wiley & Sons, New York, 1979.
Textbook and reference for studying igneous, metamorphic, and sedimentary rocks in hand-specimens and thin sections.

HALL, A. *Igneous Petrology,* John Wiley & Sons, 1987.
Discusses occurrence, emplacement, and detailed composition of igneous rocks. Origins of main magma types are related to their occurrence in present-day volcanic settings.

McBIRNEY, A. R. *Igneous Petrology,* Freeman, Cooper & Co., San Francisco, Cal., 1984.
Written for students with an elementary background in petrology.

Puu O'o, Kilauea, erupting July 1984. [Katharine Cashman.]

CHAPTER 4

Volcanism and the Occurrence of Extrusive Igneous Rocks

4.1 VOLCANOES

4.2 BASALT PLATEAUS

4.3 IGNEOUS ACTIVITY AND EARTHQUAKES

"Ashes began to fall on his ships, thicker and hotter as they approached land. Cinders and pumice, and also black fragments of rock cracked by heat, fell around them. . . . But meanwhile there began to break out from Vesuvius, in many spots, high and wide-shooting flames, whose brilliancy was heightened by the darkness of approaching night."

Pliny the Younger, Letters, *vol. VI, p. 16 (a description of the eruption of Vesuvius in* A.D. 79)

OVERVIEW

We saw in Chapters 1 and 3 that much of the Earth's igneous activity is concentrated along the boundaries of the lithospheric plates. In this chapter we look at the surface expressions of this subsurface activity, the *volcanoes*, and their associated *extrusive igneous rocks*.

Most volcanoes can be classed either as *shield volcanoes* or as *composite volcanoes*, and the *lavas* are *basaltic*, *andesitic*, or *rhyolitic*. Explosive volcanic activity produces fragments of rock of different types and this *pyroclastic debris* forms deposits that in many places are associated with lavas.

The *history of volcanoes* is instructive for understanding volcanic eruptions, so we examine some case histories of individual volcanoes. One particular type of eruption, a *fissure eruption*, has happened but rarely in historical time. In the geologic past, however, such activity was responsible for the formation of vast areas of *flood basalts*, as, for instance, the Columbia Plateau in northwestern United States.

Earthquakes associated with volcanic eruptions most commonly begin to wane before the onset of the main eruptive event.

4.1

VOLCANOES

Volcanoes are surface expressions of subsurface igneous activity. A volcano may grow in size until it becomes a mountain (Figure 4.1). Commonly cone-shaped, it has a pit at the summit, which may be either a crater or a caldera. A **crater** is a steep-walled depression out of which volcanic materials are ejected. Its floor is seldom over 2 km in diameter; its depth may be as much as 200 m. A crater may be at the top of a volcano or on its flank. The much larger **caldera** is a basin-shaped depression, more or less circular, with a diameter many times greater than that of the included volcanic vent or vents. Most calderas, in fact, are more than 1,500 m in diameter; some are several kilometers across and several hundred meters deep.

The largest volcano on Earth in Mauna Loa, part of the island of Hawaii. It is 600 km around the base, and its summit towers nearly 10 km above the surrounding ocean bottom. This and the rest of the island represent accumulations from eruptions that have gone on for more than 1 million years. The largest volcano yet found anywhere is Olympus Mons ("Mount Olympus") on Mars. Photographed originally from the *Mariner 9* spacecraft, this volcano may be over 23 km high, with a caldera 65 km across (See Figure 20.9).

Between eruptions a volcano's vent may become choked with rock congealed from a past eruption. Sometimes small jets of gas come out through cracks in this rock plug, or dikes may shoot out from this center. The eroded remnant of an old volcanic plug and radiating dikes may be seen at Shiprock, New Mexico (Figure 4.2). A volcano is built by, and remains active because of, materials coming from a large deep-seated reservoir of molten rock. While in the ground, this molten rock is called **magma;** when extruded onto the surface, it is called **lava.**

4.1 Nighttime eruption of Stromboli, a volcanic island north of Sicily often called the "Lighthouse of the Mediterranean" because of its continuous activity. [Andrea Borgia.]

In its reservoir far below the Earth's surface magma is composed of elements in solution, phenocrysts of early-formed minerals, and volatiles. These volatile components play an extremely important role in igneous activity and are the primary agents in producing a **volcanic eruption.** As the magma nears the surface, volatiles tend to separate from the other components and migrate through them to the top of the moving mass. The gases accumulate if the volcanic vent is blocked. Pressure builds up until it can no longer be confined. Then it pushes out. With temperatures of 1000°C or higher, the gases expand several thousandfold as they escape, shattering rock blocking the vent and throwing the rock fragments and magma into the air. After the explosion, the magma still in the ground is left poorer in volatile components, although it may be fluid enough to pour out.

TABLE 4.1
Volcanic Characteristics

Type of activity	Form	Magma composition	Mineral composition	Rock	Location
Quiet	Fissure flows and flood basalts; shield volcanoes; some cinder cones; gentle slopes	Mafic; low in SiO_2; low viscosity	Plagioclase; Pyroxene; (olivine)	Basalt; few cinders and ash	Diverging oceanic plates; intraoceanic plate
Explosive	Composite; strato-cones; steep slopes	Silicic; high in SiO_2; high viscosity	Orthoclase; Quartz; Na-plagioclase	Rhyolite or andesite, ash, pumice, cinders and welded ash	Converging boundaries involving continental crust

Volcanic eruptions are of different types, from relatively quiet outpourings of lava to violent explosions accompanied by showers of volcanic debris. The type of eruption depends to a great extent on the magma's composition and its volatile content. Lavas rich in ferromagnesians but low in silica (SiO_2) tend to flow relatively quietly, whereas those high in silica are viscous and explosive.

CLASSIFICATION OF VOLCANOES

Volcanoes are classified according to the materials that have accumulated around their vents. Thus we have shield volcanoes, composite volcanoes, and cinder cones (see Table 4.1).

When the extruded material consists almost exclusively of lava poured out in quiet eruptions from a central vent or from closely related fissures, a dome builds up that is much broader than it is high, with slopes seldom steeper than 10° at the summit and 2° at the base. Such a dome is called a **shield volcano.** The five volcanoes of the island of Hawaii are shields.

Sometimes a cone is built up of a combination of pyroclastic material and lava flows around the vent. This form is called a **composite volcano** and is characterized by slopes of close to 30° at the summit, tapering off to 5° near the base. Mayon, on Luzon, is one of the finest examples of a composite cone.

A single volcano may develop as a shield volcano during part of its history and as a composite volcano later. Mount Etna is an example of such a volcano.

Finally, small cones consisting mostly of pyroclastic debris, particularly cinders, are called **cinder cones.** They achieve slopes of 30° to 40° and seldom exceed 500 m in

4.2 Shiprock, New Mexico, showing the eroded central volcanic neck, with several dikes emanating from the core. [Sue Cox.]

4.3   Ropy lava or pahoehoe, from the Valley of Fire, New Mexico. [Sue Cox.]

height. Many cinder cones have flows of basalt issuing from their base. Parícutin, in Mexico, is an example of a cinder cone that has developed in modern times, as we shall describe later in this section.

COMPOSITION OF LAVAS

Lavas are of three principal types (similar to the three magma types discussed in Chapter 3), classified primarily by their proportions of silica: basic (or basaltic or mafic), intermediate (or andesitic), and silicic (or rhyolitic). The basaltic lavas generally have less than 50 percent silica; intermediate lavas, 50 to 70 percent; and silicic lavas, more than 70 percent.

Although all lavas have many similar components, no two volcanoes erupt lavas of exactly the same composition. In fact, the composition of lava may vary from one eruption to another in the same volcano. Volcanoes erupting primary basaltic lavas are common in ocean basins and in rift zones of continents. Intermediate-type lavas erupt from volcanoes ringing the Pacific, where subduction zones occur. These lava have been affected by coming up through materials of the continental margin and having their original magmatic compositions changed in the process. Silicic (or rhyolitic) lavas most commonly occur within continental regions, far from plate margins. In the Mediterranean Sea a number of Italian volcanoes erupt lavas believed to

have been contaminated by the ingestion of large quantities of limestone, producing carbonatite lavas; hence their compositions differ from those of most other volcanoes. The world's only active carbonatitic volcano, Ol Doinyo Lengai, occurs in the East African rift valley in Tanzania. It continues in a mild eruptive phase that started in 1983. The lava is essentially liquid sodium carbonate that erupts black but soon weathers to a white color.

The composition of volcanic eruptions will therefore vary, depending on the source of the magma and the nature of the rocks through which it passes on its upward journey to the Earth's surface. Magma generated entirely within material of uniform composition will be composed of that single rock type (such as basaltic types along midoceanic ridges or rhyolitic lavas within continental regions or andesitic rocks above subduction zones).

The surfaces of basaltic lava flows commonly have one of two contrasting shapes, for which the Hawaiian names of **pahoehoe** and **aa** are used. In the pahoehoe type, sometimes known as "corded" or "ropy," the surface is smooth and billowy and frequently molded into forms that resemble huge coils of rope (Figure 4.3). In the aa type the surface of the lava is covered with a random mass of angular, jagged blocks (Figure 4.4). The surface that develops depends on the viscosity of the lava, with viscosity increasing as lavas become more silicic. Pahoehoe is the shape produced on the surface of a basaltic lava, and aa is the shape often found with flank eruptions where the magma generally is cooler at the time of eruption.

TEMPERATURE OF LAVAS

Measurements of lava temperature have been made at some volcanic vents. These show that basic lavas have an average temperature of around 1100°C. The hottest lavas known are the Hawaiian lavas measured at Kilauea. Measurements were made in the lava lake Halema'uma'u by the use of Seger cones (ceramic material constructed to melt at various temperatures). The cones were placed in pipes, and then the pipes were lowered into the lake of lava. At a depth of 13 m the lava had a temperature of 1175°C. This decreased to 860°C at a depth of 1 m and increased again to 1000°C at the lava surface. Temperatures as high as 1350°C were recorded above the lava surface, where volcanic gases were reacting with atmospheric oxygen.

In 1910 Frank A. Perret recorded temperatures between 900 and 1000°C from the lava at Mount Etna in Sicily, and in 1916 to 1918 observed lavas in Mount Vesuvius from 1015 to 1040°C. Minimum temperatures observed on lavas are around 750°C.

VOLCANIC GASES

As one might expect, taking an accurate sample of volcanic gases is not an easy job. We can, however, make a few generalizations from measurements taken at Kilauea. Close to 70 percent of the volume of gases collected directly from a molten lake of lava at Kilauea was steam. Next in abundance were carbon dioxide, nitrogen, and sulfur gases, with smaller amounts of carbon monoxide and hydrochloric acid. In large quantities any of these can be dangerous, as was the cloud of carbon dioxide that flowed from the crater of Lake Nios, Cameroon, in 1986, killing an estimated 1,700 people. During its June 6, 1912, eruption Novarupta poured out nearly 10 times as much gas, ash, and pumice as the 1980 eruption of Mount St. Helens. The upper Ukak River valley in the Katmai region of southwestern Alaska, an area greater than 65 km², formed vents, domes, and fissures in the hot volcanic deposits. Large volumes of hydrochloric and hydrofluoric acid produced a scene called the Valley of Ten Thousand Smokes by early explorers.

When any igneous rock is heated, it yields some quantity of gases. Water vapor predominates, and measurements indicate that it constitutes about 1 percent of fresh —that is, unweathered—igneous rocks. Estimates of the average water content of actual magma range from about 1 to 8 percent, with the weight of opinion centering around 2 percent. A silicate melt will not hold more than about 11 percent of volatiles under any circumstances.

PYROCLASTIC DEBRIS

Fragments blown out by explosive eruptions and subsequently deposited on the ground are called **pyroclastic debris.** The term **tephra** covers all the pyroclastic debris that

4.4 Aa, or clinkery lava, on Mt. Etna. Cinder cones in middle distance. [Pamela Hemphill.]

accumulates through vertical airfall. The finest of these constitute **dust,** which is made up of pieces of the order of 10^{-4} cm in diameter. When volcanic dust is blown into the upper atmosphere, it can remain there for months, traveling great distances. The following tephra fragments settle around or near the volcanic crater:

- Ash. Fragments consisting of sharply angular glass particles. Smaller than cinders.
- Cinders. Small, slaglike, solidified pieces of magma 0.5 to 2.5 cm across. Cinder **cones** may develop from the accumulation of these fragments.
- Lapilli. Pieces about the size of walnuts.
- Blocks. Coarse, angular pieces of the cone or masses broken away from rock that blocks the vent.
- Bombs. Rounded masses that congeal from magma as it travels through the air.
- Pumice. Pieces of frothy magma up to several centimeters across that trap the bubbles of steam or other magmatic gases as they are thrown out (see Figure 4.5). After these solidify, they are honeycombed with gas-bubble holes, which give them sufficient buoyancy to float on water.

Fiery Clouds (Pyroclastic Flows) During some volcanic eruptions a great avalanche of incandescent ash mixed with steam and other gases is extruded. Heavier than air, this highly heated mixture rolls down the mountain slope. Masses of such material are called **fiery clouds (pyroclastic flows),** sometimes referred to by the French equivalent, *nuées ardentes.* The volume of material extruded during the Katmai Valley eruption was so great that it covered a valley of 140 km² to a depth of 30 m. For the next 10 years the steam and gases kept erupting from this extruded material through a great number of holes, called **fumaroles.**

As the pyroclastic material accumulates on the volcanic slopes, it commonly becomes unstable. Mudflows or debris flows, called **lahars,** result from the mixing of the pyroclastic material with water derived from melting snow

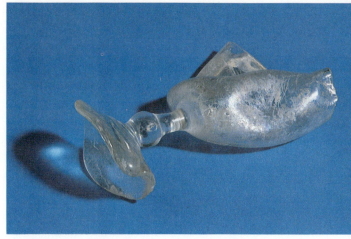

4.6   Partially melted goblet, 15 cm tall, as affected by the fiery cloud event on Martinique in May 1902. [John Simpson.]

and ice, heavy rains, rivers and lakes, or simply by being shaken loose by tremors.

Fiery clouds have characterized eruptions at Mount Pelée on Martinique Island in the West Indies to such an extent that they have come to be known as Peléan types of eruption. Other characteristics are magma expelled as pumice and ash. In the final stage of eruption a mass of viscous lava is accumulated in a domelike form when the gas content is so reduced that it no longer shatters the magma on reaching the surface.

At a few minutes before 8:00 A.M. on May 8, 1902, a gigantic explosion occurred through one side of Pelée. A fiery cloud at temperatures around 800°C swept down the mountainside and engulfed the city of Saint Pierre, wiping out its 25,000 inhabitants and many refugees from other parts of the island who had gathered there during the preceding days, when the eruption was building up with minor explosions and earthquakes. Estimates of the death toll ran as high as 40,000 (see Figure 4.6). By the middle of October a dome of lava too stiff to flow had formed in the crater, and from it a spine extruded like a great blunted needle, with a diameter of 100 to 200 m and at its maximum a height of 300 m above the crater floor.

4.5   Beds of light, porous pumice in Valles Caldera, Northern New Mexico.

Worldwide Effects   In 1783 Asama in Japan and Laki in Iceland had explosive volcanic eruptions and large quantities of dust were blown into the upper atmosphere. The Sun's effectiveness in heating the Earth's surface was so reduced that the winter of 1783 and 1784 was one of the severest on record. Benjamin Franklin was the first person to connect the unusual weather with the volcanic eruptions, publishing his ideas in May 1784.

During 1814 and 1815 the Earth's temperature was reduced following volcanic eruptions of Mayon and Mount Tambora on Sumbawa Island, east of Java. The eruption of Tambora threw so much dust into the air that for 3 days there was absolute darkness over a distance of 500 km. With this dust and the dust erupted from Mayon in the atmosphere, the amount of the Sun's heat reaching the Earth's surface was significantly reduced. The year 1815 became known as the "year without summer," marked throughout the world by long twilights and spectacular sunsets caused by the dust in the stratosphere.

A more recent example of the effect of volcanism on climate is detailed in Box 4.1.

BOX 4.1   Sulfur-Rich El Chichon Volcano Modifies Global Climate

During the spring of 1982 there was an eruption in the Mexican volcano El Chichon. What was so remarkable about this eruption was not the amount of volcanic material thrown out, for it was a relatively small eruption, but the enormous amount of fine mist of sulfuric acid droplets injected into the stratosphere — more than after the great eruption of Krakatoa in 1883. Within weeks after the El Chichon eruption westward-moving winds had produced a narrow zone of fine material completely encircling the globe (Figure B4.1.1). In less than a year the cloud of sulfur-rich material had covered the entire Northern Hemisphere and a major part of the Southern Hemisphere.

Though the origin of the sulfur is still a mystery, intensive geologic investigations of the area surrounding El Chichon are promising to solve it. It seems the sulfur may have originated either from sulfur-rich sediments underlying the volcano or from sulfide deposits associated with a subducting tectonic plate that is undergoing partial melting (see Figures B4.1.2–3).

Whatever the origin of the sulfur, the sulfuric acid mist was formed by photochemical reactions of sulfur gases with water vapor in the stratosphere. This mist was dense enough to effect brief climatic changes, including a decrease in the mean global temperature, by absorbing and reflecting solar radiation. The mist began to disperse, but long after the eruption it could still be detected by satellite instruments. It lasted in the stratosphere for several years.

The study of the record of ice layers in Greenland and Antarctica suggests that the period from about 1350 to 1700 (known as the Little Ice Age because of advances in most glaciers and relatively cooler climates) produced unusually high acidity in those ice layers. Thus there may be a very important correlation between sulfur-rich volcanic eruptions and long-term global climatic changes. Knowledge of this correlation may be the legacy left to us by the eruption of El Chichon.

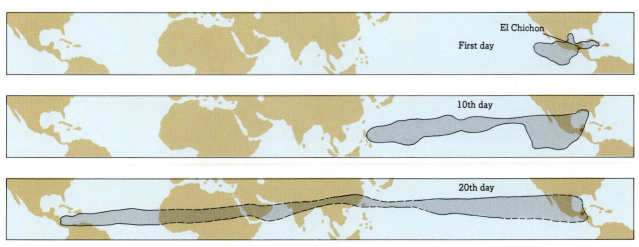

B4.1.1 World maps showing distribution of sulfur-rich volcanic cloud following eruption of El Chichon, April 4, 1982. Information obtained from satellite imagery. [After M.R. Rampino and S. Self. "The Atmospheric Effects of El Chichon," *Scientific American*, 1984, v. 205, pp. 48–57.]

(continued)

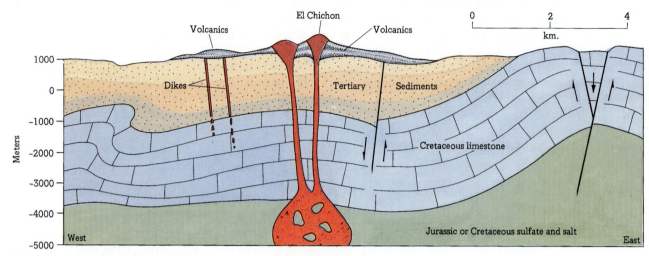

B4.1.2   Geologic cross section of El Chichon region, showing possible source of sulfur-rich clouds from partial melting of sulfate deposits at depth. [After W.A. Duffield, R.I. Tilling, and R. Canul. "Geology of El Chichon Volcano, Chiapas, Mexico," *J. Volcanology and Geothermal Research*, 1983, v. 20, No. ½, p. 117–132.]

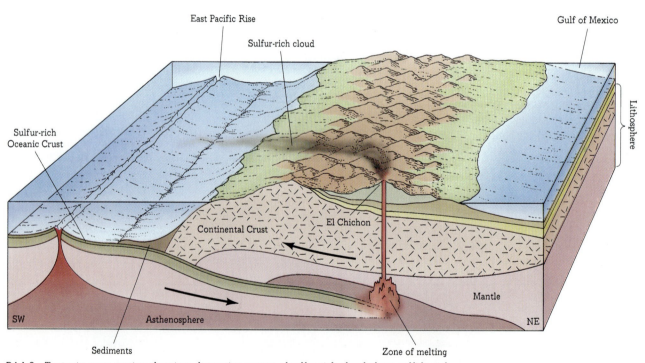

B4.1.3   Tectonic cross section showing alternative source of sulfur-rich clouds from sulfide-rich layer at top of downgoing slab of lithosphere. [Various sources.]

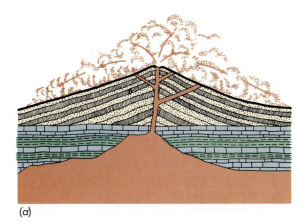

(a)

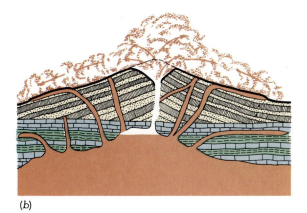

(b)

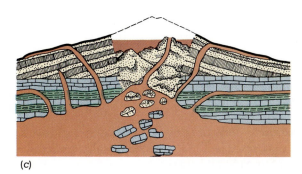

(c)

FORMATION OF CALDERAS

Calderas are formed by collapse (Figure 4.7) in combination with some explosive activity. It is often difficult to determine just which mechanism is primarily responsible.

The caldera on Kilauea was probably formed by the collapse of the summit. As great quantities of magma escaped from the reservoir beneath the volcano, support for the summit was withdrawn and large blocks of it fell in, forming the caldera.

Crater Lake, in southern Oregon, lies in a basin that is an almost perfect example of the caldera shape. This is circular, with a diameter of a little more than 10 km and a maximum depth of 1,200 m, and it is surrounded by a cliff that rises 750 to 1,200 m. Crater Lake itself is about 600 m deep. The caldera was formed when the top of a volcanic cone, Mount Mazama, vanished during an eruption. Geologists have studied the deposits on the slopes and have tried to piece together the history of the basin (See Figure 4.8.)

First, a cone consisting of alternating layers of pyroclastic material and lava flows was slowly built to a height of around 3,600 m. Then glaciers formed, moving down from the crest and grooving the slopes as they traveled. An explosive eruption occurred 6,600 years ago, and the caldera was formed. Later activity built up a small cone inside the caldera, which now protrudes above the surface of Crater Lake as Wizard Island.

4.7 Sequence of events proposed by one hypothesis for the formation of a caldera. (a) An eruption begins with fiery clouds and dust clouds distributing materials on slopes and surrounding country. (b) Eruption continues; part of cone is blown away; and lava flows join in draining the magma reservoir. (c) Most of cone collapses into the reservoir; later activity forms cinder cone in caldera. [After H. Williams, "Calderas and Their Origin," *Bull. Univ. Calif. Dept. Geol. Sci.,* vol, 25, pp. 239–346, 1941.]

4.8 Crater Lake, Oregon, partially fills the caldera of an inactive volcano in the Cascade Range. The lake is 10 km across and is nearly 600 m deep. Wizard Island is a volcanic cone formed after the creation of the caldera. See also Figure 4.7. [John Patterson.]

Not all observers agree on the origin of the Crater Lake caldera itself, however. The question is whether all or nearly all the missing material from the cone was actually blown out during an eruption or whether the caldera was created when the summit collapsed. The answer to the mystery should be provided by an analysis of the unconsolidated material found in the vicinity. Does this material consist of pyroclastics formed during an eruption or of the broken remnants of Mazama's collapsed summit? One investigator has concluded that, of the 70 km³ of Mazama that disappeared, only 8 km³ are represented in the materials now lying on the immediate slopes and that the rest of it was dropped into the volcano when the roof of an underlying chamber collapsed. The chamber may have been partially emptied by the ejection of large volumes of material during an eruption. The investigator found evidence that ash spread over a radius of nearly 50 km. Others have reported Mount Mazama ash as far east and northeast as Alberta and eastern Montana, and some of the magma may have worked its way beneath the surface into adjoining areas. The explanation in favor of collapse seems to have been verified by the studies of K. R. Cranson.

HISTORY OF SOME VOLCANOES

A volcano is considered **active** if there is some historical record of its having erupted. If a volcano has not erupted in historical time and yet shows only minor erosional alteration, it is considered **dormant,** or merely "sleeping," and capable of renewed activity. If a volcano not only has not erupted within historic time but also shows wearing away by erosion and no signs of activity (such as escaping steam or local earthquakes), it is considered **extinct.**

Vesuvius   Vesuvius, on the shore of the Bay of Naples, has supplied us with a classic example of the reawakening of a dormant volcano (Figure 4.9). At the time of Christ Vesuvius was a vine-clad mountain, Mount Somma, a vacation spot in southwest Italy favored by wealthy Romans. For centuries it had given no sign of its true nature. Then, in A.D. 63, a series of strong earthquakes shook the area; and around noon on August 24, A.D. 79, Somma started to erupt. The catastrophe of that August day lay hidden for nearly 17 centuries, until the remains of Herculaneum and Pompeii — cities that had been buried by the eruption — were uncovered (Figure 4.10). Then a story to grip the world's imagination was revealed: Roman sentries had been buried at their posts; family groups, in the supposed safety of subterranean vaults, had been cast in molds of volcanic mud cemented to a rocklike hardness, along with their jewels, candelabra, and the food they had hoped would sustain them through the emergency.

Somma is believed to have erupted first about 10,000 years ago as a submarine volcano in the Bay of Naples (Figure 4.11). It then emerged as an island and finally filled in so much of the bay around it that it became a part of the mainland. It is the youngest volcano found in that vicintiy.

There must have been an exceedingly long interval of quiet before the eruption of the year 79 because no earlier historical records of volcanic activity exist. But in 79 part of the old Somma cone was destroyed, and the cone now called Vesuvius started. During this eruption Pompeii was buried by pyroclastic debris; people were asphyxiated by gases from the ash and suffocated from the dust. Herculaneum was overwhelmed by mudflows of water-soaked ash deposited to a considerable depth.

Eruptions of pyroclastic debris occurred at intervals after 79. The longest period of quiescence lasted 494 years; it was followed by the eruption of 1631, which poured out the first lava in historic time. This rejuvenation is attributed to chemical changes accompanying the magmatic ingestion of great quantities of limestone in a reservoir with a roof about 5 km below sea level.

Krakatoa   One of the world's greatest explosive eruptions took place in 1883 at Krakatoa (also spelled Krakatau), in Sunda Strait between Java and Sumatra. Krakatoa

4.9   Vesuvius with the jagged remnant of Mount Somma, seen from the East. [Pamela Hemphill.]

Modern Herculaneum

Volcanic

Roman Herculaneum

Roman road

had once been a single island, a volcanic mountain built up from the sea bottom. At a remote period in the past it split apart during an eruption. By 1883, after a long period of rebuilding, three cones had risen above sea level and had merged. These cones, named Rakata, Danan, and Perboewatan, and various unnamed shoals made up the outline of Krakatoa.

On August 26, 1883, a series of explosions began. The next day, at 10:20 A.M., a gigantic explosion blew the two cones Danan and Perboewatan to bits. A part of the island that had formerly stood 800 m high was left covered by 300 m of water. The noise of the eruption was heard on Rodrigues Island, 5,000 km across the Indian Ocean, and a wave of pressure in the air was recorded by barographs around the world. A great flood of water created by the activity drowned 36,500 persons in the low coastal villages of western Java and southern Sumatra. Columns of ash and pumice soared kilometers into the air, and fine dust rose to such heights that it was distributed around the globe and took more than 2 years to fall. During that time sunsets were abnormally colored all over the world. A reddish-brown circle, known as "bishop's ring," which was seen

4.10 Roman Herculaneum, like Pompeii, was buried as a result of the eruption of Vesuvius in 79 A.D. This view shows an excavated portion of the old town beneath volcanic debris on which the new town has been built.

4.11 Map showing geographic relationships in the vicinity of the Somma-Vesuvius volcano.

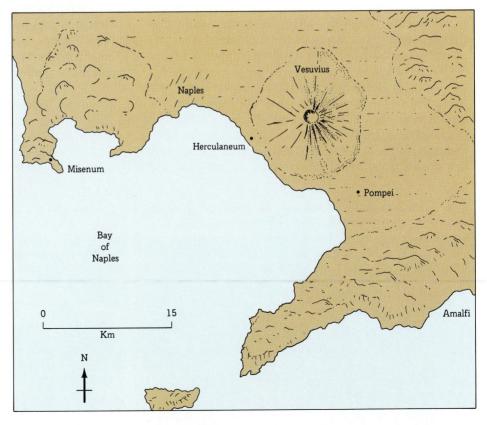

around the Sun under favorable conditions, gave evidence not only of the continued presence of dust in the upper air but also of the approximate size of the pieces—just under 0.002 mm. Since 1883 Krakatoa has revealed from time to time that it is in the process of actively rebuilding.

Parícutin   About 320 km west of Mexico City (19.50°N, 102.05°W), Parícutin sprang into being on February 20, 1943. Nine years later it had become quite inactive, but during its life it was studied more closely than any other newborn vent in history, until the eruption of Mount St. Helens in 1980.

Many stories of Parícutin's first hours have been told. According to the version now generally regarded as the most reliable, the volcano began about noon as a thin wisp of smoke rising from a cornfield that was being plowed by Dionisio Pulido. By 4 P.M. explosions were occurring every few seconds, dense clouds of ash were rising, and a cone had

begun to build up. Within 5 days the cone was 100 m high, and after 1 year it had risen to 425 m. Two days after the eruption began, the first lava flowed from a fissure in the field about 300 m north of the center of the cone. At the end of 7 weeks this flow had advanced about 1.5 km. Some 15 weeks after the first explosion lava had also begun to flow from the flanks of the cone itself.

After 9 years of activity Parícutin abruptly stopped its eruptions and became just another of the many small "dead" cones in the neighborhood. The histories of these other cones, parasites of Toncítaro or of neighboring major volcanoes, undoubtedly parallel the story of Parícutin.

Mount St. Helens   A sleeping giant awoke on May 18, 1980, with an enormous blast that blew off the top of Mount St. Helens in the State of Washington (see Figures 4.12 and 4.13). Warnings of potentially severe volcanic activity had occurred in the form of earthquakes and vent-

4.12   (a) Mount St. Helens as most southwest Washington residents remember it prior to the explosive eruption on May 18, 1980. (b) After the eruption. [USGS.]

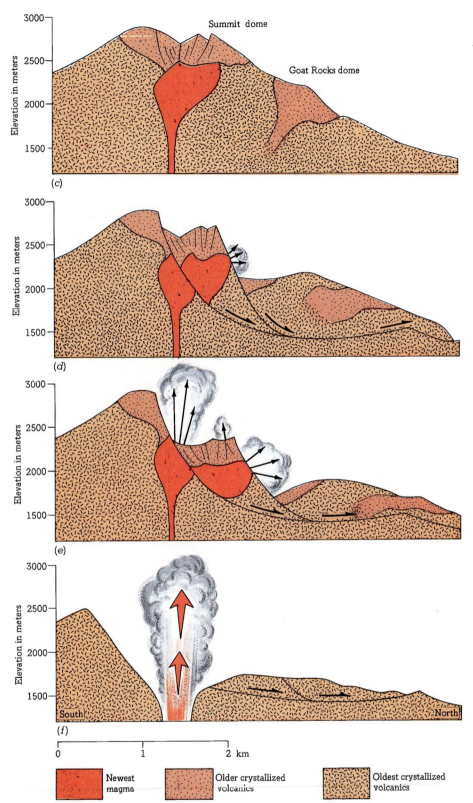

4.13 Changes in the profile of Mount St. Helens during the morning of May 18, 1980. (a) Before eruption: Magma (dark red color) had intruded beneath older domes (light red color). (b) The eruption started at 8:32 A.M. when an earthquake caused the bulge on the volcano's north side to give way. About 20 seconds after the landslide started, explosions began where the new magma was exposed. (c) The landslide released steam and lava inside the volcano. About 30 seconds after the landslide, massive explosions from side and top of the fractured block created the lateral blast that went north from where the bulge had been. (d) After landsliding exposed the main volcanic conduit, a vertical eruption column of steam and ash rose more than 25 kilometers. (From J. G. Moore and W. C. Albee, *The 1980 Eruptions of Mount St. Helens, Washington,* U.S. Geological Survey, 1981.)

ing of steam for nearly 2 months prior to the eruption. The last previous eruption of this formerly dormant volcano began in 1831 and was followed by 15 to 20 years of lava and mudflows, which built the surrounding topography.

Formerly standing at a height of 2,950 m above sea level, the top of the majestic cone was beveled off at 2,560 m as rock, ash, and clouds of gases were spewn over a tremendously large area, killing trees, livestock, and at least 70 people.

Mount St. Helens is located along a chain of volcanoes extending from northern California through Oregon and Washington into southern British Columbia. This chain marks the boundary between two colliding plates, the American plate and the very small Juan de Fuca plate, lying just offshore. (See Section 11.6) Volcanism has been occurring along the Cascade Mountains for over 50 million years, but most of the volcanic rocks are less than 50 million years old. Mount St. Helens is relatively young, with a history going back only about 40,000 years, and most of the present cone is less than 10,000 years old. During the past 4,500 years Mount St. Helens has been the most explosive volcano in the conterminous United States. Its eruption should not have been such a surprise!

Hawaii   The Hawaiian Islands are peaks of volcanoes projecting above the ocean and strung out along a line extending roughly 2,400 km in a northwesterly direction. The Marquesas, Society, Tuamotu, Tubuai, Samoan, and other volcanic groups of the South Pacific form lines roughly parallel to the Hawaiian Islands.

At the northwestern end of the Hawaiian chain are the low Kure and Midway islands. At the southeastern end is Hawaii, the largest of the group (140 km long and 122 km wide) and the tallest deep-sea island in the world. It is composed of five volcanoes—Kohala, Hualalai, Mauna Kea, Mauna Loa, and Kilauea (see Figure 4.14). Each has developed independently, and each has its own geologic history. Kohala has been extinct for many years, but Mauna Kea shows evidence of having been active in the recent geological past, although not within recorded history.

Investigation of the Hawaiian Island chain suggests that it is the result of the relative northwesterly motion of the Pacific plate over a zone of melting in the Earth's mantle (a **plume**), which causes crustal extension and produces a "hot spot" in the Earth's crust (according to J. Tuzo Wilson and W. Jason Morgan). Lavas of basalt composition pour out at these **hot spots,** forming islands, which in turn become inactive as motion of the plate continues (Figure 4.15). Radiometric ages of the Hawaiian Island basalts support this hypothesis, with the ages becoming younger toward the hot spot (the southeastern end of the chain) and older away from it. In fact, recent volcanic activity is occurring on the ocean floor southeast of Hawaii. The Emperor seamount chain is a continuation of the Hawaiian Islands, and indicates even older ages and a shift in the direction of plate migration over this hot spot (see Figure 4.16) and Section 16.4, subsection on aseismic ridges).

Measurements made on tilt meters show that both Mauna Loa and Kilauea swell up during the period when magma is rising from below. The uplift reaches a maximum just before an eruption, but after the eruption the mountains shrink back again. Such measurements of Kilauea

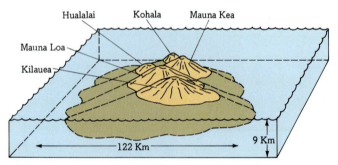

4.14   Schematic drawing of the five volcanoes that have been built up from the sea floor to merge and form the island of Hawaii (viewed here from the southeast.)

4.15   Schematic section and plan view of a plume similar to that postulated for producing the volcanoes of the Hawaiian Islands. As the lithospheric plate moves slowly over a cylindrical upwelling of magmatic material, a new volcano develops from time to time, older volcanoes becoming inactive as the plate moves away from this hot spot.

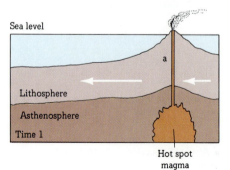
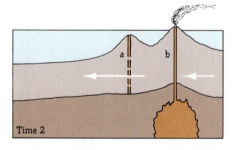
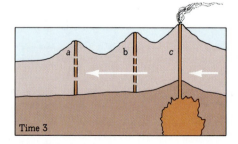

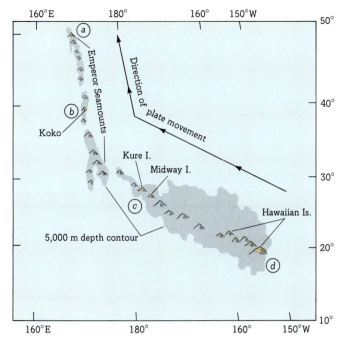

4.16 Bend in seamount line (Emperor and Hawaiian chains) marks change in movement of Pacific plate over hot spot in mantle. Volcanoes of progressively younger age developed as the plate moved first in a northwesterly direction and later in a more westerly direction. An age of 70 million years has been obtained from rocks near the northwestern extremity of the Emperor chain (a), 46 million years for Koko seamount (b), 20 million years for Midway Island (c), and less than 1 million years for the volcano Kilauea on Hawaii (d).

show that slow swelling over a period of months (about 16 on average) produces a maximum uplift of 40 to 135 cm. Flank eruptions of the volcano allow sudden collapse of the summit over a period of days, amounting to 20 to 50 cm of subsidence per eruption. The swelling extends over an area about 10 km in diameter centered on, or just south of, Kilauea caldera. From the area and the amount of swelling involved, it was computed that in the 1960 eruption each centimeter of uplift or subsidence represented deformation of 1 million m^3 of Earth materials. Total volumes of 20 to 30 million m^3 have been computed for larger eruptions. Records of tilt can also be used to forecast an impending eruption.

When Kilauea is active, magma rises up within the mountain and floods out as lava into a pit in the floor of the caldera. Occasionally the lava flows out over the rim of the pit onto the floor and gradually raises its level. Usually, however, the lava is confined to the pit, forming what is termed a **lava lake.** This lava lake may last for years and then disappear completely for equally long periods. The level of the lake falls when lava flows from the flanks of the volcano, both above and below sea level. From time to time the system is drained, and the caldera floor collapses. Then the magma rises again, lava floods into the caldera, and the process is repeated.

Disappearing Islands   Submarine volcanoes, like Krakatoa, that build themselves up above sea level, erupt violently, erode away by wave action, and then rebuild produce the so-called disappearing islands of the Pacific Ocean. In 1913, for example, Falcon island (20.4°S, 175.6°W), in the South Pacific, suddenly disappeared after an explosive eruption. On October 4, 1927, accompanied by a series of violent explosions, it just as suddenly reappeared. The island of Bogosloff (about 56°N, 168°W), in the Aleutians, was first reported in 1826 and has been playing hide-and-seek with mapmakers ever since.

On December 11, 1967, a volcanic eruption started at Metis Shoal of the Tonga Archipelago. After 20 days a new island about 700 m long and 100 m wide had poked up 20 m above the sea. From the start of the activity the island stayed above the surface for a total of 58 days. By February 1 it was a few "jagged rocks and water washing across." On February 19 there were "very high breakers on subsurface rocks" at the site; and by April 1 the shoal was completely underwater, and there were no observed breakers.

Submarine Volcanism   Enormous volumes of basalt are extruded on ocean floors. As these submarine lavas extrude, they take on characteristic bulbous or rounded "pillow" shapes. Molten lava is rapidly cooled by the cold seawater and a thin hard crust develops. As pressure from rising material within these pillows builds to the breaking point, the crust fractures, molten material bursts forth, and a new "pillow" rolls down the slope. When it comes to rest on the sea floor, it forms a hardened crust with rounded top and irregularly shaped bottom where the pillow conforms to the round-shaped tops of previously produced pillows. This material has been called **midocean ridge basalt,** or simply **MORB.**

DISTRIBUTION OF ACTIVE VOLCANOES

We find evidence of volcanic eruptions in rocks of all ages. Apparently igneous activity has been going on throughout geologic time and has occurred in the highest mountain ranges, on the bottom of the ocean, and on open plains. There are more than 500 active volcanoes in the world today (some are shown in Figure 4.17).

As we discussed in Chapter 3, magmas vary in composition and viscosity. Different volcanic types with different morphologies result from such differences in magmas. Most of the world's composite volcanoes are located at convergent-plate boundaries (Fuji, Mayon, Vesuvius, the Cascade volcanoes, Central American volcanoes, and South American Andean volcanoes, for example). The big shield volcanoes are found within the oceanic plates (the Hawaiian volcanoes, for example).

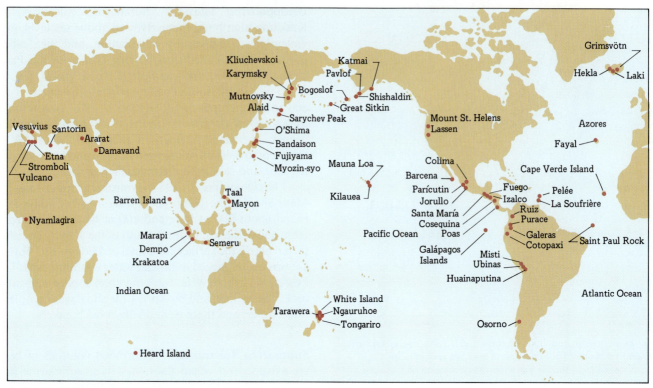

4.17   Location of some active volcanoes. [Various sources.]

4.2

BASALT PLATEAUS

On June 11, 1783, after a series of violent earthquakes on Iceland near Mount Skapta, an immense outpouring of lava began along a 16-km line, the Laki Fissure. Lava poured into the Skapta River, drying up the water and overflowing the stream's canyon, which was 150 to 200 m deep and 60 m wide in places. Soon the Skapta's tributaries were dammed, and many villages in adjoining areas were flooded. The lava flow was followed by another a week later and a third on August 3. So great was the volume of the lava flows that they filled a former lake and an abyss at the foot of a waterfall. They spread out in great tongues 20 to 25 km wide and 30 m deep. As the lava flows diminished and the Laki Fissure began to choke up, 22 small cones formed along its length, relieving the waning pressures and serving as outlets for the final extrusion of debris.

This is one of the few authenticated instances within historical time of the mechanism known as **fissure eruption,** or **lava flood.** There is strong evidence, however, that floods of this sort occurred on a gigantic scale in the geological past. The rocks produced by lava floods are known as **flood basalts,** or **plateau basalts,** because of the tendency to form great plateaus. The low viscosity required for lava to flood freely over such great areas is characteristic only of lavas that have basaltic composition (Figure 4.18).

Iceland itself is actually a remnant of extensive lava floods that have been going on for over 50 million years and that have blanketed 0.5 million km². The congealed lava is believed to be at least 2,700 m thick in this area. The Antrim Plateau of northeastern Ireland, the Inner Hebrides, the Faeroes, and southern Greenland are also remnants of this great North Atlantic, or Britoarctic, Plateau.

Of equal magnitude is the Columbia Plateau in Washington, Oregon, Idaho, and northeastern California. In some sections more than 1,500 m of rock have been built up by a series of fissure eruptions. Individual eruptions deposited layers ranging from 3 to 30 m thick, with an occasional greater thickness. In the canyon of the Snake River, Idaho, granite hills from 600 to 750 m high are covered by 300 to 450 m of basalt from these flows. The Columbia Plateau has been built up during the past 17 million years. The prinicipal activity took place between 16.5 and 13.5 million years ago in northeastern California and Idaho, but some flows in the Craters of the Moon National Monument in southern Idaho, probably the most recent of United States fissure eruptions, are believed to have occurred within the last 250 to 1,000 years.

Other extensive areas built up by fissure eruptions include north-central Siberia, the Deccan Plateau of India, Ethiopia, around Victoria Falls on the Zambezi River in Africa, and parts of Australia.

Some of the most extensive flood basalts found anywhere occur as mare basalts on the Moon and on both Mars and Mercury (Chapter 20).

4.3
IGNEOUS ACTIVITY AND EARTHQUAKES

Most volcanic eruptions are associated with earthquakes. In recent years records of this seismic activity have given warnings of impending eruptions. The relationship between local earthquakes and eruptions is important enough to be described in detail in the following case.

Raoul Island, a volcanic island, is 8 km across and is the largest of the Kermadec group, located approximately 1,000 km northeast of New Zealand. It lies in a very active zone of earthquake activity extending from Tonga Island to New Zealand.

Unusual seismic activity began on November 10, 1964, when the first of a swarm of local earthquakes was recorded. In the 10 weeks before this there had been only 1 earthquake. Then, within 4 h (hours), over 80 earthquakes per hour were being recorded. On November 11 earthquakes became less frequent, but the ground began to shake continuously, a phenomenon called **volcanic tremor.** Within 1 day the volcanic tremors were large enough to mask records of small individual earthquakes. By November 13 the tremors dropped off to about one-half the maximum intensity, and earthquakes were again being recorded at about 30 to 40 per hour. At 21 h 57 min (minutes) Greenwich time the largest earthquake of the series occurred. Raoul's seismograph was temporarily out of order, but this earthquake was large enough to be recorded on other seismographs, some as far away as North America. The United States Coast and Geodetic Survey determined it to be a moderate-sized earthquake and located its center at 20 km west of the island at a depth of 77 km. Volcanic tremor and frequency of earthquakes decreased then, until on November 20 only 10 to 15 per hour were being recorded.

4.18   Plateau basalts exposed in the gorge of the Rio Grande River, northern New Mexico.

Suddenly, on November 20, just before 18 h Greenwich time, the amplitude of volcanic tremor increased, and there was a "sound of a big landslide" accompanied by a cloud of steam, followed by "another roaring noise," which heralded the appearance of a great column of black mud that shot up to 1,000 m in the center of the cloud with rocks flying out of the column and falling back into the crater. During the eruption a new crater 100 m in diameter was blown out within the main crater. Within 1 h the level of volcanic tremor settled down, with about 10 earthquakes per hour.

The Raoul eruption followed the not unusual pattern of occurring when the associated earthquake activity was on the wane after an extremely sudden rise. This kind of earthquake commonly occurs as magma moves upward into the upper crust from the lower crust or the mantle.

Seismographs are being watched carefully in the hope that they will provide forewarning of any future outbursts of volcanic activity.

DEPTH OF MAGMA SOURCE

Magma supplying eruptive materials of active volcanoes comes from deep within the Earth. The depths of earthquakes and earth tremors associated with eruptions are giving us some clues to the depths of the magma.

Swarms of earthquakes and "harmonic tremor" with foci 60 km below the summit of Kilauea preceded the 1959 eruptions, which suggests that 60 km may be the depth at which magma enters the "plumbing system" of Kilauea.

Summary

Volcanoes are surface piles of material that have accumulated around vents during successive eruptions.
   A **volcanic eruption** is started and maintained by gases, of which steam is the most important.
   Classification of volcanoes according to the materials that have accumulated around their vents designates them as **shield, composite,** and **cinder cones.**
   Temperatures of lavas have been measured from 750 to 1175°C.
   Volcanic gases are two-thirds steam but include carbon monoxide and dioxide, nitrogen, sulfur, and hydrochloric acid.
   Pyroclastic debris consists of fragments blown out by explosive eruptions and includes ash, cinders, lapilli, blocks, bombs, and pumice.
   Fiery clouds of ash mixed with gases are heavier than air and flow down a volcano's side.
   Worldwide effects of volcanoes include volcanic dust in the stratosphere and reduction of solar energy reaching the Earth's surface.

Formation of calderas is, by collapse, combined with some explosive activity.

History of some volcanoes shows they may be dormant for long periods and again become active.
   Vesuvius was dormant for centuries and then in A.D. 79 started to erupt again.

Krakatoa erupted in 1883 with one of the world's greatest explosions.
Parícutin sprang into being on February 10, 1943, and was active for 9 years.
Mount St. Helens abruptly came alive May 18, 1980, and became the most widely studied volcano in history.
Hawaiian Islands are peaks of volcanoes projecting above sea level and strung out along a line running 2,400 km to the northwest.
Disappearing islands are submarine volcanoes that build up above sea level, blow their heads off, become inactive, and erode to below sea level.
Submarine lavas occur along ocean ridges, forming pillow structures, in what we call midocean ridge basalt, or **MORB.**
Distribution of active volcanoes locates many in the circum-Pacific, the Alpine-Himalayan, and the midoceanic belts and others in the Pacific Ocean, the Atlantic Ocean, the Indian Ocean, Africa, and the Antarctic, for a total of more than 500.

Basalt plateaus are built by fissure eruptions.

Igneous activity and earthquakes seem to be closely related.
   Depth of magma source suggested by associated earthquakes may be 60 km at Kilauea.

Questions

1. What are some of the kinds of activity that commonly occur before the eruption of a volcano?

2. What mechanisms are responsible for the kinds of activity described in Question 1? What is actually happening within the Earth beneath the site of the eventual volcano?

3. Why do the volcanic eruptions differ from one location to another on the same continent? Why do continental and oceanic eruptions differ?

4. What is the difference between an active, a dormant, and an extinct volcano?

5. What are the principal reasons for human deaths associated with volcanic eruptions such as the eruption of Mount St. Helens?

6. Why do most volcanoes occur in certain specific belts around the Earth? Why are a few found at other locations?

Supplementary Readings

CHESTER, D. K., ET AL. *Mount Etna: The Anatomy of a Volcano,* Stanford University Press, Stanford, Cal., 1985.
Summarizes the history, geological setting, volcanic processes, internal plumbing, and volcanic hazards on Etna and human reactions to them.

MACDONALD, G. A., A. T. ABBOTT, AND F. L. PELMAN *Volcanoes in the Sea,* 2nd ed., University of Hawaii Press, Honolulu, 1983
A comprehensive treatment of the geology and volcanic history of Hawaii.

SHANE, S. *Discovering Mount St. Helens: A Guide to the National Volcanic Monument.* University of Washington Press, Seattle, Wash., 1985.

Guide for visitors to observe the history of its eruption, recovery, and promise.

SIMKIN, T., AND R. S. FISKE *Krakatau, 1883: The Volcanic Eruption and Its Effects,* Smithsonian Institution Press, Washington, D.C., 1983.
Clear historic account of this most widely known catastrophic volcanic event.

SIMKIN, T., ET AL. *Volcanoes of the World,* Smithsonian Institution, Hutchinson Ross Publication Co., Stroudsburg, Pa., 1981.
A regional directory, gazetteer, and chronology of volcanism during the last 10,000 years.

Differential weathering has produced Rainbow Bridge, Lake Powell, Utah. [Sue Cox.]

CHAPTER ■5

WEATHERING AND SOILS

"The hills are shadows, and they flow
From form to form and nothing stands;
They melt like mists the solid lands,
Like clouds they form themselves and go"

Alfred, Lord Tennyson, from "In Memoriam"

Weathering is such an everyday phenomenon that we can easily overlook how important it is in the Earth's scheme of things. Here we examine the process of weathering and its products, including soils.

Energy from the Sun drives the weathering process, and *energy from the Earth,* as expressed in plate tectonics, raises rock into the realm of weathering.

Weathering is both *mechanical,* the change in shape and size of rocks and minerals, and *chemical,* the change of preexisting minerals into entirely new ones. In this chapter we examine the factors controlling both of these processes. The chemical weathering of the common igneous rock-forming minerals provides examples of the creation of most of the sedimentary rock-forming minerals.

The *rate of weathering* varies with the mineral type. We will also find that the rate and the *depth of weathering* vary with temperature and moisture.

Erosion, the removal of weathered material, is the subject of several later chapters, but here we provide a first glimpse of this process by examining some *rates of erosion* as well as the subject of *differential erosion.*

We close the chapter with a section on soils, including *classification, soil horizons,* and *soil types.*

5.1

ENERGY AND WEATHERING

Weathering is the process of change that takes place in surface or near-surface material in response to air, water, and living matter. The energy that drives the weathering process comes both from within the Earth and from outside. From time to time, motions originating inside the Earth elevate some portions of its surface above other portions. These are the same motions that express themselves in earthquakes (Chapter 10), plate tectonics and mountain building (Chapter 11). Whatever their cause, they arrange the Earth's surface materials so that gravity can be effective in the breakdown of rock material. Thus rock raised in mountain building has potential energy because of its high elevation. This may be transformed to kinetic energy, the energy of motion, if gravity is strong enough to pull rock downward to a lower level. The shattering of this rock as it falls is truly a form of weathering.

We have already seen that there is a measurable amount of heat that flows from the Earth's interior to its surface and is there dissipated. Far more heat is received at the Earth's surface from the Sun. It is the distribution of this heat that causes the differential heating of the atmosphere and of the oceans. This differential heating brings about circulation of atmosphere and ocean, causes our weather and climate, and determines the pattern of organic activity. All of these in turn help to modify the Earth's surface materials—in short, to determine the process of weathering.

How much solar energy is available at the Earth's surface? The Sun radiates about 100,000 cal/cm²-min from its surface. A plane at the Earth's outer surface perpendicular to the Sun's rays would receive approximately 2 cal/cm²-min, which is 2×10^{-5} the amount of heat generated by an equivalent area on the Sun's surface. This figure is the **solar constant.**

If our hypothetical plane perpendicular to the Sun's rays at the edge of the outer atmosphere has the same diameter as the Earth, geometry tells us that its area must be one-quarter that of the Earth. Thus we can express the solar constant for the Earth's surface as 0.5 cal/cm²-min. This is an average for the entire globe. But of this energy, approximately one-third is unavailable to heat either the atmosphere or the Earth because it is reflected back into space from clouds or from the Earth, or because it is scattered into space from particles within the atmosphere. Therefore approximately 0.3 cal/cm²-min is what is available to heat the atmosphere and the Earth.

Obviously this energy is not evenly distributed over the Earth's surface. Less is available toward either pole than toward the equator. We shall have occasion to return to this distribution later in the chapter when we consider the process of chemical weathering.

5.2

TYPES OF WEATHERING

There are two general types of weathering: **mechanical** and **chemical.** It is difficult to separate these two types in nature because they often go hand in hand, although in some environments one or the other predominates. Still, for our purposes here, it is more convenient to discuss them separately.

MECHANICAL WEATHERING

Mechanical weathering, also called physical weathering or disaggregation or **disintegration,** is the process by which rock is broken down into smaller and smaller fragments as the result of energy developed by physical forces. For example, when water freezes in a fractured rock, sufficient energy may develop from the pressure caused by expansion of the frozen water to split off pieces of the rock. Or a boulder moved by gravity down a rocky slope may be shattered into smaller fragments.

Expansion and Contraction   Changes in temperature, if they are rapid enough and great enough, may bring about the mechanical weathering of rock. In areas where bare rock is exposed at the surface and is unprotected by a cloak of soil, forest or brush fires can generate heat adequate to break up the rock. The rapid and violent heating of the exterior zone of the rock causes it to expand; and if the expansion is sufficiently great, flakes and larger fragments of the rock are split off. Lightning often starts such fires and, in rare instances, may even shatter exposed rock by means of a direct strike.

The debate continues concerning whether variations in temperature from day to night or from summer to winter are enough to cause mechanical weathering. Theoretically, such variations would cause disintegration. For instance, we know that the different minerals forming a granite expand and contract at different rates as they react to rising and falling temperatures. We suspect that even minor expansion and contraction of adjacent minerals would, over long periods of time, weaken the bonds between mineral grains and that it would thus be possible for disintegration to occur along these boundaries. In deserts we may find fragments of a single stone lying close beside one another. Obviously the stone has split. But how? Many think the cause lies in expansion and contraction caused by heating and cooling.

But laboratory evidence to support these speculations is inconclusive. In one laboratory experiment coarse-grained granite was subjected to temperatures ranging from 14.5° to 135.5°C every 15 min. This alternate heating and cooling eventually simulated 244 years of daily heating and cooling; yet the granite showed no signs of disintegration. Perhaps experiments extended over longer periods of time would produce observable effects. In any event, we are still uncertain of the mechanical effect of daily or seasonal temperature changes; if these fluctuations bring about the disintegration of rock, they must do so very slowly.

On the lunar surface we find many examples of rock debris resulting from mechanical weathering due to meteor impact. Moreover, it is believed that heating and cooling add significantly to the weathering of such lunar particles.

Frost Action   Frost is much more effective than heat in producing mechanical weathering. When water trickles down into the cracks, crevices, and pores of a rock mass and then freezes, its volume increases about 9 percent. This expansion of water as it passes from the liquid to the solid state sets up pressures that are directed outward from the inside of the rock and **frost wedging** results. These pressures can dislodge fragments from the rock's surface.

The dislodged fragments of mechanically weathered rock are angular, and their size depends largely on the nature of the bedrock from which they have been displaced. Usually the fragments are only a few centimeters in maximum dimension, but in some places—along the cliffs bordering Devil's Lake, Wisconsin, for instance—they reach sizes up to 3 m.

Certain conditions must exist before frost action can take place: There must be an adequate supply of moisture; the moisture must be able to enter the rock or soil; and temperatures must move back and forth across the freezing line. As we might expect, frost action is more pronounced in high mountains and in moist regions where temperatures fluctuate across the freezing line either daily or seasonally.

Exfoliation   Exfoliation is a mechanical weathering process in which curved plates of rock are stripped from a larger rock mass by the action of physical forces (Figure 5.1). This process produces features that are fairly common in the landscape: large domelike hills, called **exfoliation domes.**

Let us look at the manner in which exfoliation domes develop. Fractures, or parting planes, called **joints,** occur in many massive rocks. These joints are broadly curved and run more or less parallel to the rock surface. The distance

5.1 Exfoliation of the Black Hill norite, a Cambro-Ordovician intrusion in southern Australia. [William C. Bradley.]

5.2 Dome-shaped landform, a gray-granite erosional remnant in the Sierra Nevada Mountains. The mountain stands about 300 m above the adjacent valley. Exfoliation plays a major role in forming most dome-shaped mountains. [Pamela Hemphill.]

between joints is only a few centimeters near the surface, but it increases to several meters as we move deeper into the rock (see Chapter 9). Under certain conditions, one after another of the curved slabs between the joints is spalled, or sloughed, off the rock mass. Finally, a broadly curved hill of bedrock develops, as shown in Figure 5.2.

Just how these slabs of rock come into being in the first place is still a matter of dispute. Most observers believe that as erosion strips away the surface cover, the downward pressure on the underlying rock is reduced. Then, as the rock mass begins to expand upward, lines of fracture develop, marking off the slabs that later fall away. Precise measurements made on granite blocks in New England quarries provide some support for this theory. Selected blocks were accurately measured and then removed from the quarry face, away from the confining pressures of the enclosing rock mass. When the free-standing blocks were measured again, it was found that they had increased in size by a small but measurable amount. Massive rock does expand, then, as confining pressures are reduced, and this slight degree of expansion may be enough to start the exfoliation process.

Well-known examples of exfoliation domes are Stone Mountain, Georgia, the domes of Yosemite Park, California, and Sugar Loaf in the harbor of Rio de Janeiro, Brazil.

OTHER TYPES OF MECHANICAL WEATHERING

Plants also play some role in mechanical weathering, although their principal effect is through profoundly different microchemical environments around their roots, the development of humic acids, and similar changes, all of which are classified as chemical weathering. However, the roots of trees and shrubs growing in rock crevices sometimes exert sufficient pressure to dislodge previously loosened fragments of rock, much as tree roots heave and crack sidewalk pavements (Figure 5.3). More important, though, is the mechanical mixing of the soil by ants, worms, rodents, and other small animals. Constant activity of this sort makes the soil particles more susceptible to chemical weathering and may even assist in the mechanical breakdown of the particles.

Abrasion is also a major factor in mechanical weathering. Particles moved by water, ice, and air are remarkably effective in wearing away rock. And in the process of movement these particles themselves become abraded and broken down into smaller and smaller pieces (see discussion of these surficial processes in Chapters 13, 15, 16 and 17).

CHEMICAL WEATHERING

Chemical weathering, sometimes called **decomposition,** is a more complex process than mechanical weathering. As we have seen, mechanical weathering merely breaks rock material down into smaller and smaller particles, without changing its composition. Chemical weathering, however, actually transforms the original material into something different. The chemical weathering of the mineral feldspar, for example, produces the clay minerals that have a different composition and different physical characteristics from those of the original feldspar. Sometimes the products of chemical weathering have no mineral form at all, as in the salty solution that results when halite is dissolved.

Dissolution   Some weatherng reactions involve **dissolution** of solid material in natural fluids. As an example, gypsum undergoes the reaction

$$CaSo_4 \cdot 2H_2O \rightarrow Ca^{2+} + SO_4^{2-} + 2H_2O$$

5.3   A tree growing in a crevice pries a large block apart in the Beartooth Mountains on the Montana-Wyoming border.

as water flows through an evaporite deposit. This leaves no residue behind. All weathering products end up in soluton.

Carbonatization

Some weathering reactions involve carbon dioxide as an agent. These can be called **carbonatization** reactions. For example, limestone weathering involves a first step in which atmospheric CO_2 is dissolved in water to produce bicarbonate

$$H_2O + CO_2 \rightarrow H_2CO_3 \rightarrow H^+ + HCO_3^-$$

followed by a second step in which bicarbonate reacts with calcite

$$H^+ + HCO_3^- + CaCO_3 \rightarrow Ca^{2+} + 2HCO_3^-$$

Hydration

There are some weathering reactions that involve the structural addition of water to a solid to form a hydrated solid residual product. A good example of these **hydration** reactions is the action of water on potassium feldspar (K-spar) to form a clay plus silica:

$$2KAlSi_3O_8 + H_2O + 2H^+ \rightarrow 2K^+ + Al_2Si_2O_5(OH)_4 + 4SiO_2$$

Notice that in this reaction and the previous one, one of the reactants is H^+, which may be derived from the carbon dioxide–water equilibrium, or may be the product of organic decay or other inorganic reactions (like the weathering of pyrite, for example). (H^+ is a synonym for acid when used in this context.)

Oxidation

Some reactions involve oxygen; for example, in the **oxidation** of iron-bearing minerals

$$6H_2O + 2Fe_2SiO_4 + O_2 \rightarrow 4Fe(OH)_3 + 2SiO_2$$

Notice that all these chemical weathering processes involve common agents of weathering: water, carbon dioxide, oxygen, and "acid" or H^+.

Particle Size

The size of the individual particles of rock is an extremely important factor in chemical weathering because substances can react chemically only where they come into contact with one another. The greater the surface area of a particle, the more vulnerable it is to chemical attack. If we were to take a pebble, for example, and grind it up into a fine powder, the total surface area exposed would be greatly increased. As a result, the materials that make up the pebble would undergo more rapid chemical weathering.

Figure 5.4 shows how the surface area of a 1-cm (or any other unit) cube increases as we cut it into smaller and smaller cubes. The initial cube has a surface area of 6 cm^2 and a volume of 1 cm^3. If we divide the cube into smaller cubes, each 0.5 cm on a side, the total surface area increases to 12 cm^2, although, of course, the total volume remains the same. Further subdivision into 0.25-cm cubes increases the surface to 24 cm^2. And if we divide the original cube into units 0.125 cm on a side, the surface area increases to 48 cm^2. As we have seen, this same process is performed by mechanical weathering. It reduces the size of the individual particles of rock, increases the surface area exposed, and thus promotes more rapid chemical weathering.

Other Factors

The rate of chemical weathering is affected by other factors as well—the composition of the original mineral, for example. As we shall see later in this section, a mineral such as quartz, SiO_2, responds much more slowly to chemical weathering than does a mineral such as olivine, $(Fe, Mg)_2SiO_4$.

Climate also plays a key role in chemical weathering. Moisture, particularly when it is accompanied by warmth, speeds up the rate of chemical weathering; conversely, dryness slows it down. Finally, plants and animals contribute directly or indirectly to chemical weathering because their life processes produce oxygen, carbon dioxide, and certain acids that enter into chemical reactions with Earth materials.

5.4 Relationships of volume, particle size, and surface area. In this illustration a cube 1 cm (or any other unit) on a side is divided into smaller and smaller units. The volume remains unchanged, but as the particle size decreases, the surface area increases. Because chemical weathering is confined to surfaces, the more finely a given volume of material is divided, the greater is the surface area exposed to chemical activity and the more rapid is the process of chemical weathering.

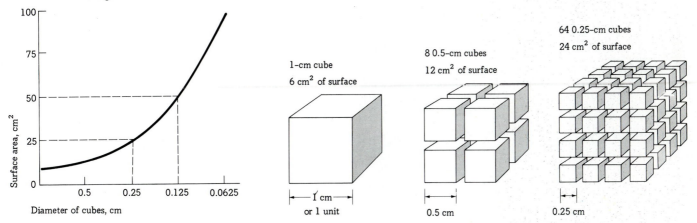

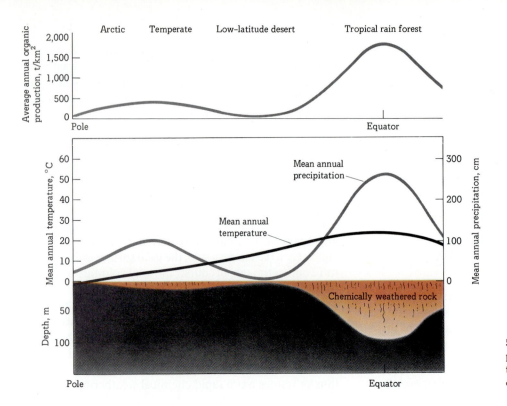

5.5 Variation in temperature, precipitation, and organic matter from the poles to the tropics is related to the depth of chemical weathering.

The interrelation of some of these factors is shown in Figure 5.5. This is a generalized section from the pole to the equator and shows the fluctuation of precipitation, temperature, and amount of vegetation. At the same time, the figure shows the relative depth of weathering as these three factors vary. Thus weathering is most pronounced in the equatorial zone, where the factors of precipitaton, temperature, and vegetation reach a maximum. Weathering is least in the desert and semidesert areas of the subtropics and in the far north. A secondary zone of maximum weathering exists in the zone of temperate climates. Here both the precipitation and the vegetation reach secondary maxima.

SPHEROIDAL WEATHERING

Now let us look at some examples of spheroidally weathered boulders. These boulders have been rounded by the spalling off of a series of concentric shells of rock (Figures 5.6 and 5.7). But here the shells develop from pressures set up within the rock by chemical weathering rather than from the lessening of pressure from outside by erosion. As we have already seen, when certain minerals are chemically weathered, the resulting products occupy a greater volume than does the original material, and it is this increase in volume that creates the pressures responsible for spheroidal

5.6 Spheroidally weathered granitic boulders are almost completely separated from the bedrock on the eastern side of the Sierra Nevada.

5.7 This cross section through a spheroidally weathered boulder suggests the stresses set up within the rock. The stress is thought to develop as a result of the change in volume as feldspar is converted to clay. (See also Figure 5.6.)

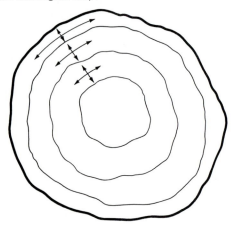

weathering. The surface area of a rock is important in this process of spheroidal weathering. The surface area gradually diminishes as spheroidal weathering progresses, yielding the minimum surface area for a particular volume. Because most chemical weathering takes place in the portions of the rock most exposed to air and moisture, it is there that we find the most expansion and hence the greatest number of shells. Spheroidally weathered boulders are sometimes produced by the crumbling off of concentric shells. If the cohesive strength of the rock is low, individual grains are partially weathered and dissociated, and the rock simply crumbles away. The underlying process is the same in both cases, however.

Certain types of rocks are more vulnerable to spheroidal weathering than are others. Igneous rocks such as granite, diorite, and gabbro are particularly susceptible because they contain large amounts of the mineral feldspar, which, when chemically weathered, produces new minerals of greater volume.

CHEMICAL WEATHERING OF COMMON ROCKS AND MINERALS

In Chapter 2 we found that the most common minerals in rocks are silicates and that the most important silicates are quartz, the feldspars, and certain ferromagnesian minerals. Let us see how chemical weathering acts on each of these three types.

Weathering of Quartz   Chemical weathering affects quartz very slowly, and for this reason we speak of quartz as a relatively stable mineral. When a rock such as granite, which contains a high percentage of quartz, decomposes, a great deal of unaltered quartz is left behind. The quartz grains found in the weathered debris of granite are the same as those that appeared in the unweathered granite.

When these quartz grains are first set free from the mother rock, they are sharp and angular; but because even quartz responds (slowly) to chemical weathering, the grains become more or less rounded as time passes. After many years of weathering, they look as though they had been abraded and worn by the action along a stream bed or a beach. And yet the change may have come about solely through chemical action. Indeed, the presence of silica in natural waters of lakes and rivers reminds us that the silicate minerals are soluble and that some of this silica may come from the chemical weathering of quartz.

Weathering of Feldspars   In the Bowen reaction series (Section 3.2) we saw that when a magma cools to form an igneous rock such as granite, feldspars crystallize before quartz. When granite is exposed to weathering at the Earth's surface, the feldspars are the first minerals to be broken down. Mineralogists and soil scientists still do not understand the precise process by which feldspars weather, and some of the end products of this action—the clay minerals—offer many puzzles. But the general direction and results of the process seem fairly clear.

Let us review the decomposition of potassium feldspar, a good example of the chemical weathering of the feldspar group of silicates. In this instance, a source of hydrogen ions is necessary to the weathering process. Several naturally occurring substances have been increased by human activity in the last century—notably carbon dioxide (CO_2) and sulfur dioxide (SO_2). Each occurs in the atmosphere and, when combined with moisture, produces acids. (See Box 5.1.)

The atmosphere contains small amounts of carbon dioxide, and the soil contains much greater amounts. Because carbon dioxide is extremely soluble in water, it unites with rainwater and water in the soil to form the weak acid H_2CO_3, carbonic acid. This ionizes to form hydrogen and bicarbonate ions as follows:

$$\underset{\text{Water}}{H_2O} \; plus \; \underset{\text{carbon dioxide}}{CO_2} \; yields \; \underset{\text{carbonic acid}}{H_2CO_3} \; yields \; \underset{\text{hydrogen ion}}{H^+} \; plus \; \underset{\text{bicarbonate ion}}{(HCO_3)^-}$$

And when K-spar comes into contact with hydrogen ions, the following reaction takes place (as shown previously):

$$\underset{\text{Potassium feldspar}}{2K(AlSi_3O_8)} \; plus \; \underset{\text{hydrogen ions}}{2H^+} \; plus \; \text{water} \; yields \longrightarrow$$

$$\underset{\text{clay}}{Al_2Si_2O_5(OH)_4} \; plus \; \underset{\text{potassium ions}}{2K^+} \; plus \; \underset{\text{silica in solution}}{4SiO_2}$$

BOX 5.1   Acid Rain

Many coals contain sulfur in small amounts. When coal is burned, it enters the air as sulfur dioxide gas, SO_2, which can be harmful to life at certain concentrations. In addition, SO_2 combines readily with moisture in the atmosphere to form sulfuric acid, H_2SO_4. Precipitation washes the H_2SO_4 out of the atmosphere, causing rain of higher than normal acidity. Acid rain is a significant problem for some regions downwind from coal- and oil-burning plants. It has made some lakes too acid for much fish life to survive and flourish. It has also had degrading effects on some forests, especially those in the northeastern United States, southeastern Canada, some parts of the Rocky Mountains, Scandinavia, and Germany.

In this reaction the hydrogen ions from the water force the potassium out of the feldspar, disrupting its crystal structure. The hydrogen ion combines with the aluminum silicate radical of the feldspar to form the new clay minerals. (The process by which water combines chemically with other molecules we called **hydration.**) The disruption of the orthoclase crystal yields a second product, potassium ions. These may join with the bicarbonate ions formed by the ionizaton of the carbonic acid to form potassium bicarbonate. The third product, silica, is formed by the silicon and oxygen that are left over from the original potassium feldspar after the production of the clay.

The action of living plants may also bring about the chemical breakdown of potassium feldspar. A plant root in the soil is negatively charged and is surrounded by a swarm of hydrogen ions, H^+. If there happens to be a fragment of potassium feldspar lying nearby, these positive ions may change places with the potassium of the feldspar and disrupt its crystal structure (see Figure 5.8). Once again, a clay mineral is formed, as in the equation above.

Now let us look more closely at each of the three products of the decomposition of potassium feldspar: first, the clay minerals. At the start, these minerals develop from a suspension of matter, sometimes of **colloidal** size, a size variously estimated as between 0.2 and 1 μm (micrometer) or 0.001 mm. Immediately after it is formed, the aluminum silicate may be amorphous; that is, its atoms are not arranged in any orderly pattern. It seems more probable, however, that even at this stage the atoms are arranged according to the definite pattern of a true crystal. In any event, as time passes, the small individual particles join together to form larger crystals, which, when analyzed by such means as X-rays, exhibit the crystalline pattern of true minerals.

5.8   The conversion of orthoclase to a clay mineral by plant roots. In this diagram a swarm of hydrogen ions (positive) is shown surrounding a negatively charged plant rootlet. The suggestion has been made that a hydrogen ion from the rootlet may replace a potassium ion in a nearby orthoclase fragment and there bond with the oxygen within the original mineral, to begin the conversion of the orthoclase to clay. The potassium ion thus ejected replaces the hydrogen ion along the negatively charged rootlet and is eventually utilized in plant growth. [After W. D. Keller and A. F. Frederickson, "Role of Plants and Colloidal Acids in the Mechanism of Weathering," *Am. J. Sci.,* vol. 250, p. 603, 1952.]

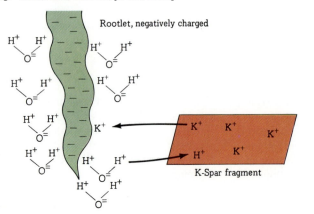

Aluminum silicate, derived from the chemical breakdown of original feldspars, combines with water to form hydrous aluminum silicate, which is the basis for another group of silicate minerals, the clays.

Clay minerals are sheet silicates like the micas, and may have come from a **metamorphic parent,** a **sedimentary parent,** or an **igneous parent material.** There are many different clay minerals and each has its own chemical behavior, physical structure, and evolution. Most of the clay minerals fall into three major groups: **kaolinite, montmorillonite** (or smectite), and **illite.** Kaolinite is derived from the Chinese *Kao-ling,* "High Hill," the name of the mountain from which the first kaolinite was shipped to Europe for ceramic uses; the mineral montmorillonite was first described from samples collected near Montmorillon, a town in west-central France; and the name "illite" was selected by geologists of the Illinois Geological Survey in honor of their state. Like the micas shown in Figure 2.20, the clay minerals are built up of silicon-oxygen tetrahedra linked together in sheets. These sheets combine in different ways with sheets composed of aluminum atoms and hydroxyl ions (OH^-). (For this reason we refer to the clay minerals as **hydrous aluminum silicates.**) In addition, montmorillonite may contain magnesium and some sodium and calcium, and illite contains potassium, occasionally with some magnesium and iron. Some clay minerals alter to other clay minerals. For example, under deep burial the important clay mineral illite is produced from other clays.

We still do not understand exactly what factors determine which clay minerals will form when a feldspar is weathered. Climate is important, for we know that kaolinite tends to form as a result of the intense chemical weathering in warm, humid climates and that illite and montmorillonite seem to develop more commonly in cooler, drier climates. The history of the rock also seems to be influential. For example, when a soil forms from a sedimentary rock in which a clay has been incorporated, we often find that the soil contains the same type of clay as does the parent rock. The analysis of a number of sedimentary rocks and of the soils developed on these rocks has shown that when illite is present in the original rock, it is usually the dominant clay in the soil, regardless of climate. Clearly, then, **both environment and inheritance** influence the type of clay that will develop from the chemical weathering of a feldspar.

Let us look back for a minute to the equation for the decomposition of potassium feldspar. Notice that the second product is potassium ions. We might expect that these would be carried off by water percolating through the ground and that all the potassium would eventually find its way to the rivers and finally to the sea. Yet analyses show that not nearly so much potassium is present in river and ocean water as we should expect. What happens to the rest of it? Some of it is used by growing plants before it can be carried away in solution, and some of it is absorbed by clay minerals or even taken into their crystal structure.

The third product resulting from the decomposition

TABLE 5.1
Chemical-Weathering Products of Common Rock-Forming Silicate Minerals

Mineral	Composition	Important decomposition products	
		Minerals	Others
Quartz	SiO_2	Quartz grains	Minor silica in solution
Feldspars:			
Orthoclase (or K-spar)	$K(AlSi_3O_8)$	Clay Silica	Potassium in solution Some silica in solution
Albite (sodium plagioclase)	$Na(AlSi_3O_8)$	Clay	Some silica and sodium in solution
Anorthite (calcium plagioclase)	$Ca(Al_2Si_2O_8)$	Silica Calcite	Aluminum and calcium in solution
Ferromagnesians:			
Biotite Augite Hornblende	Fe, Mg, Ca silicates of Al	Clay Hematite Limonite Silica Calcite	Calcium and magnesium in solution Some silica in solution
Olivine	$(Fe, Mg)_2SiO_4$	Hematite Limonite Silica	Iron and magnesium in solution Some silica in solution

of potassium feldspar is silica, which appears either in solution (for even silica is slightly soluble in water) or in the size range of the colloids. In the colloidal state silica is amorphous and may exhibit some of the properties of silica in solution.

So far we have been talking about the weathering of only potassium feldspar. But the products of the chemical weathering of plagioclase feldspars are very much the same. Instead of potassium carbonate, however, either sodium or calcium carbonate is produced, depending on whether the feldspar is the sodium albite or the calcium anorthite (see Table 5.1). As we found in Section 2.5 (see the subsection on silicate minerals), the plagioclase feldspars almost invariably contain both sodium and calcium. The carbonates of sodium and calcium are soluble in water and may eventually reach the sea. We should note here, however, that calcium carbonate also forms the mineral calcite (see subsection on carbonate and sulfate minerals in Section 2.5). Calcite, in turn, forms the greater part of limestone (a sedimentary rock) and marble (a metamorphic rock). Both limestone and marble are discussed in subsequent chapters.

Weathering of Ferromagnesians Now let us turn to the chemical weathering of the third group of common minerals in igneous rocks: the ferromagnesian silicates. The chemical weathering of these minerals produces the same products as the weathering of the feldspars: clay, soluble salts, and finely divided silica. But the presence of iron and magnesium in the ferromagnesian minerals makes possible certain other products as well.

The iron may be incorporated into one of the clay minerals or into an iron carbonate mineral. Usually, however, it unites with oxygen to form hematite, Fe_2O_3, one of the most common of the iron oxides. Hematite commonly has a deep red color, and in powdered form it is always red; this characteristic gives it its name, from the Greek *haimatitēs,* "bloodlike." Sometimes the iron unites with oxygen and a hydroxyl ion to form goethite, $FeO(OH)$, which is generally brownish in color. (Goethite was named after the German poet Goethe, because of his lively scientific interests.) Chemical weathering of the ferromagnesian minerals often produces a substance called **limonite,** which is yellowish to brownish in color and is referred to in everyday language as just plain "rust." Limonite is not a true mineral because its composition is not fixed within narrow limits, but the term is universally applied to the iron oxides of uncertain composition that contain a variable amount of water. Limonite and some of the other iron oxides are responsible for the characteristic colors of most soils.

What happens to the magnesium produced by the weathering of the ferromagnesian minerals? Some of it may be removed in solution as a carbonate, but most of it tends to stay behind in newly formed minerals, particularly in the illite and montmorillonite clays.

SUMMARY OF WEATHERING PRODUCTS

If we know the mineral composition of an igneous rock, we can determine in a general way the products that the chemical weathering of that rock will yield. Conversely, when we examine the weathering products, we can gain a good idea of the kind of rock from which those products have come.

The chemical-weathering products of the common rock-forming minerals are listed in Table 5.1. These products include the minerals that make up most of our sedimentary rocks, and we shall discuss them again in Chapter 6.

We can generalize as follows:

1. Any silicate mineral that contains aluminum will yield clay.
2. Any iron-bearing mineral will oxidize to iron oxides or take water into its chemical makeup.
3. Chemical weathering commonly yields a collection of dissolved constituents of potassium, sodium, silicon, magnesium, and calcium, which go off in solution, as well as a collection of residual solids. The solids go to make clastic (or fragmental) sedimentary rocks; the dissolved constituents go to make chemical precipitates.
4. All the reactions listed in Table 5.1 assume continuation until completion. Because these reactions take much time, erosion commonly removes debris before the reactions have gone to completion. In fact, the debris may even be incorporated into sedimentary rocks before weathering processes have gone to completion.

5.3
RATES OF WEATHERING

Some rocks weather very rapidly and others only slowly. The rate of weathering is governed by the type of rock and a variety of other factors, from minerals and moisture, temperature and topography, to plant and animal activity.

RATE OF MINERAL WEATHERING

On the basis of field observations and laboratory experiments, the minerals commonly found in igneous rocks can be arranged according to the order in which they are chemically decomposed at the surface. We are not sure of all the details, and different investigators report different conclusions, but we can make the following general observations:

1. Quartz is highly resistant to chemical weathering.
2. The plagioclase feldspars weather more rapidly than orthoclase feldspar does.
3. Calcium plagioclase (anorthite) tends to weather more rapidly than sodium plagioclase (albite).
4. Olivine is less resistant than augite, and in many instances augite seems to weather more rapidly than hornblende.
5. Biotite mica weathers more slowly than do the other dark minerals, and muscovite mica is more resistant to weathering than biotite.

Notice that these points suggest a pattern (Figure 5.9) similar to that of Bowen's reaction series for crystallization from magma (discussed in Section 3.2 and illustrated in Figure 3.3). But there is one important difference: In weathering, the successive minerals formed do **not** react with one another as they do in a continuous reaction series. The relative resistance of these minerals to decomposition may reflect the difference between the surface conditions under which they weather and the conditions that existed when they were formed. Olivine, for example, forms at high temperatures and pressures, early in the crystallization of a melt. Consequently, as we might expect, it is extremely

5.9 Relative rapidity of chemical weathering of the common igneous rock-forming minerals. The rate of weathering is most rapid at the bottom and decreases toward the top. Note that this table is in the same order as Bowen's reaction series (left). The discrepancy in the rate of chemical weathering between, for instance, olivine and quartz is explained by the fact that in the zone of weathering olivine is farther from its environment of formation than is quartz. It therefore reacts more rapidly than quartz does to its new environment and thus weathers more rapidly.

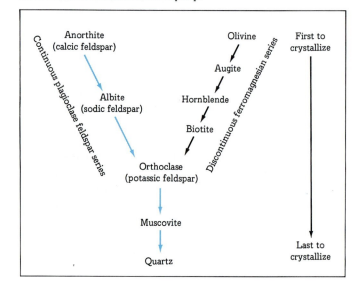

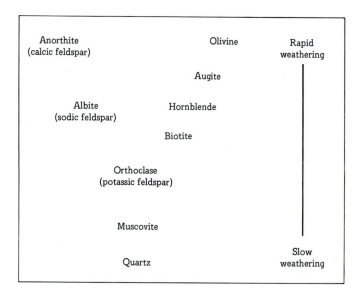

unstable under the low temperatures and pressures that prevail at the surface, so it weathers quite rapidly. On the other hand, quartz forms late in the reacton series, under considerably lower temperatures and pressures. Because these conditions are more similar to those at the surface, quartz is relatively stable and is very resistant to weathering.

Now we can qualify slightly the definition of weathering given at the beginning of this chapter. We have found that weathering disrupts the equlibrium that existed while the minerals were still buried in the Earth's crust and that this disruption converts them into new minerals. We may revise our definition as follows: **Weathering is the response of materials that were once in equilibrium within the Earth's crust to new conditions at or near contact with air, water, and living matter.** Only at the temperatures and pressures of formation are minerals truly stable. At any other temperature or pressure they will respond, change, or alter to reestablish an equilibrium with the new conditions.

DEPTH AND RAPIDITY OF WEATHERING

Most weathering takes place in the upper few meters or tens of meters of the Earth's crust, where rock is in closest contact with air, moisture, and organic matter. But some factors operating well below the surface permit weathering to penetrate to great depths. For instance, when erosion strips away great quantities of material from the surface, the underlying rocks are free to expand. As a result, parting planes, or fractures—the joints that we spoke of earlier in the chapter—develop hundreds of meters below the surface.

Then, too, great quantities of water move through the soil and deep undergound, transforming some of the materials there long before they are ever exposed at the surface. Rock salt that is located far below the surface in the form of a sedimentary rock often undergoes exactly this transformation. If a large quantity of underground water is circulating through, the salt is dissolved and carried off long before erosion can expose it.

Weathering is sometimes so rapid that it can actually be recorded. The Krakatoa eruption of August 1883, described in Chapter 4, threw great quantities of volcanic ash into the air and deposited it to a depth exceeding 30 m on the nearby island of Long-Eiland. By 1928, 45 years later, a soil nearly 35 cm deep had developed from the upper part of the deposit, and laboratory analyses showed that a significant change had taken place in the original materials. Chemical weathering had removed a measurable amount of the original potassium and sodium. Furthermore, mechanical weathering or chemical weathering or both had broken down the original particles so that they were generally smaller than the particles in the unweathered ash beneath.

In a more recent study scientists from Ohio State University have demonstrated the nature and rate of soil development in unweathered material that has been exposed to the atmosphere with the retreat of Muir glacier in Glacier

5.10 Weathering of a marble headstone, Madison Cemetery, Connecticut. The monument illustrates the instability and rapid weathering of calcite (the predominant mineral in marble) in a humid climate (1980).

Bay National Park in southeastern Alaska. Geological and human records establish the successive positions of the retreating ice front since about 1700. Between this date and 1965, the front of the glacier retreated approximately 65 km. The age of soils at various points along this line of retreat can thus be demonstrated. Over a period of around 250 years a soil about 35 cm thick has developed. Of this, the upper half is represented chiefly by the accumulation of organic material. The lower half, however, shows changes in the materials left by the retreating ice. After 250 years virtually all the calcite and dolomite had been removed to a depth of at least 15 cm, and the soil acidity had increased markedly. The amount of iron oxide in the form of hematite had increased measurably, particularly in the lowest 10 cm.

Graveyards provide many fine examples of weathering within historic time. Calcite in the headstone pictured in Figure 5.10 has weathered so rapidly that the inscription is only partially legible after more than a century and a half. Undoubtedly the rate of weathering has increased with time, for two reasons. First, continued weathering roughens the marble surface, exposing more and more of it to chemical attack and quickening the rate of decomposition. Second, as the number of factories and dwellings in the area has increased, the amount of carbon dioxide in the atmosphere has also increased. Consequently, rainwater in the twentieth century carries more carbonic acid than it did in the nineteenth, and it attacks calcite more rapidly.

In contrast to the strongly weathered marble is the headstone of slate shown in Figure 5.11. Many years later the inscription is still plainly visible. Slate is usually a metamorphosed shale, which, in turn, is composed largely of clay minerals formed by the weathering of feldspars. The clay minerals in the headstone were originally formed in the zone of weathering. Slight metamorphism has since changed many of the clay minerals to muscovite, a white mica, and the shale to slate (see Section 7.3). The muscovite, however, is relatively stable in the zone of weathering, as indicated in Figure 5.9.

These examples show, then, that weathering often occurs rapidly enough to be measured during a lifetime. Let us now turn to the process that moves these products of weathering: erosion.

5.11   Weathering of a slate headstone in the burying ground at Rheems, Pennsylvania. The legibility of the inscription testifies to the durability of the slate (1974).

5.4
RATES OF EROSION

We found that chemical and mechanical weathering produces certain material from the rocks of the Earth. It is these materials that the agents of erosion move from one place to another until they finally reach the settling basins of the world's oceans. Several agents are involved in this movement, including gravity, water, ice, and wind. These agents are individually treated in Chapters 12 to 17, but here it will be instructive to pause and consider how much material is being removed from the continents and at what rate.

Sometimes ancient ruins provide an index to the rates of erosion. Thus Figure 5.12 is a photograph of the remains of a cistern built 60 km north of Rome, Italy, in the second century. The footings exposed at the base of the finished wall indicate the amount of erosion that has occurred here since the structure was built. The rate of erosion from then to the present averages 30 cm/1,000 years.

This, however, is only a spot measurement at a specific place. What method can we use to measure the rates over large areas? One is to measure the amount of material carried by a stream each year from its drainage basin (all the area drained by the stream and all its tributaries); the amount averaged over the area of the basin gives an average figure of erosion. Now, obviously, the rate of erosion is not the same at every place in the river basin — in some places the material will be removed more rapidly than at others. Furthermore, there will be places at which deposition takes place, and the material is temporarily halted on its way out of the basin. Nevertheless, this method gives us an average figure for a unit area within the drainage basin.

A stream carries material in solution as well as solid material in the form of sediments. Most of the solid material is buoyed up by the flow of water and is said to be carried in **suspension.** A relatively smaller amount is pushed and bounced along the stream bottom in what is known as **bed load** or **traction load.** Suspended and dissolved matter can be measured without much difficulty. It is very difficult to measure traction load. It is generally small and usually considered to be about 10 percent of the suspended load in the average stream, though it can vary considerably depending on climatic variations. The relative importance of bed load, suspended load, and dissolved load varies stongly with climatic factors. Bed load can even exceed supended load under special circumstances of climatic extremes, such as some glacial settings.

With these facts in mind, we may ask how much material is carried annually by streams out of the drainage basin in which the ancient Roman cistern in Figure 5.12 stands. The major stream here is the Tiber River. If we measure the load of the Tiber at Rome, we find that it is, on average, removing sediments of about 7.5 million t of solid material each year from the area upstream. This converts to an erosion rate of 17 cm/1,000 years over the entire basin. If

5.12 Ruins of a cistern built in the second century on a Roman farm located 60 km north of Rome. The exposed footings of the building measure 1.3 m, as indicated by the staff. The volcanic ash at the bottom of the building measures 30 cm. Therefore, the amount of erosion since the cistern was built is 1.6 m.

we had data on bed load and dissolved load for the Tiber, the figures would be higher but not by much.

Turning to the United States, we can examine Table 5.2, which describes the erosion going on in the major regions of the country. Not included here is the drainage to Hudson Bay and the Saint Lawrence River area. Also, the table does not include the area in the western states known as the Great Basin, where topography and rainfall are such that streams do not reach the sea. We see that the rates vary from region to region but average approximately 6 cm/1,000 years.

Data from the Amazon River, the world's largest, indicate that it is removing material from its basin at the rate of 4.7 cm/1,000 years. Another large tropical river, the Congo, is carrying enough material out of its basin each year to reduce its drainage basin by approximately 2 cm/1,000 years.

A single large drainage basin integrates many factors affecting the rapidity of erosion. One we should consider carefully is the human influence. There is enough information from disciplines as diverse as archaeology and nuclear physics to indicate that when people occupy an area intensively and turn it to cropland, they increase the erosion rate 10 to 100 times over that of a naturally forested or grassed area. Destroying the natural cover exposes soil to much more rapid erosion and removal by running water, wind, and animals.

How much material is transported each year to the oceans? We have already cited some figures for portions of the Earth. These figures, however, cover only 10 percent of the land surface. Therefore any estimate of the total amount of erosion of the Earth's surface per unit of time can be no better than that, a mere estimate. The information now at hand suggests, however, that something ap-

TABLE 5.2
Rates of Regional Erosion in the United States[a]

Drainage region	Drainage area,[b] thousands of km²	Runoff, thousands of m³/s	Load, t/km²-yr			Erosion, cm/1,000 yr	Area sampled, %
			Dissolved	Solid	Total		
Colorado	629	0.6	23	417	440	17	56
Pacific Slopes, California	303	2.3	36	209	245	9	44
Western Gulf	829	1.6	41	101	142	5	9
Mississippi	3,238	17.5	39	94	133	5	99
South Atlantic and Eastern Gulf	736	9.2	61	48	109	4	19
North Atlantic	383	5.9	57	69	126	5	10
Columbia	679	9.8	57	44	101	4	39
Total (overall)	6,797	46.9	43	119	162	6	

[a] After Sheldon Judson and D. F. Ritter, "Rates of Regional Denudation in the United States," *J. Geophys. Res.*, vol. 69, p. 3399, 1964.
[b] Great Basin, Saint Lawrence, and Hudson Bay drainage not considered.

proaching 10 billion t/year was being delivered to the oceans before human beings became effective geological agents. This does not include windblown material, although that is thought to be negligible. Glacier ice is also not considered, but it probably would not appreciably affect our figure. The rate of 10 billion t/year corresponds to a continental reduction of approximately 2.5 cm/1,000 years. Taking into consideration the human effect on erosion, it is estimated that at present approximately 24 billion t of material are moved annually by the rivers to the oceans. This is 2.5 times the amount moved before our intervention.

DIFFERENTIAL EROSION

Differential erosion is the process by which different rock masses or different sections of the same rock mass erode at different rates. Almost all rock masses of any size weather in this manner. The results vary from the boldly sculptured forms of the Grand Canyon to the slightly uneven surface of a marble tombstone. Unequal rates of erosion are chiefly caused by variations in the composition of the rock. The more resistant zones stand out as ridges, ribs, or pinnacles above the more rapidly weathered rock on either side (Figures 5.13, 5.14, and 5.15). A resistant caprock may protect much less resistant rocks from being removed. This may produce features called "voodoos" (Figure 5.16).

A second cause of differential erosion is simply that the intensity of weathering varies from one section to another in the same rock. This may result from differences in the amount of sunlight hitting the rock, causing variation in the effect of moisture and frost action on parts of the rock exposure.

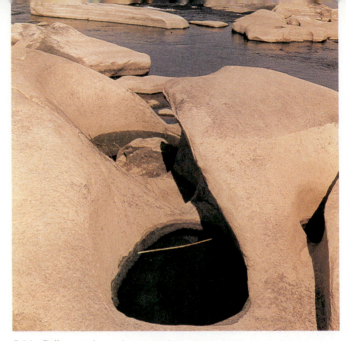

5.14 Differential weathering to form "potholes" in basaltic dike crossing the Susquehanna River 20 km south of Harrisburg, Pennsylvania.

5.15 Differential weathering of sandstone and shale sequence in Monument Valley. [Sue Cox.]

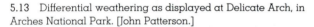

5.13 Differential weathering as displayed at Delicate Arch, in Arches National Park. [John Patterson.]

5.16 "Voodoos" produced by differential weathering in the San Juan Formation west of Regina, New Mexico. Resistant sandstone forms the cap and less resistant claystone forms the pedestal beneath.

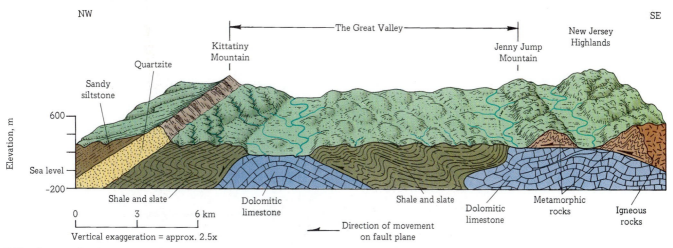

NW

SE

The Great Valley

New Jersey
Highlands

Kittatiny
Mountain

Jenny Jump
Mountain

Quartzite

Sandy
siltstone

600

Elevation, m

Sea level

-200

Shale and slate

Dolomitic
limestone

Shale and slate

Dolomitic
limestone

Metamorphic
rocks

Igneous
rocks

0   3   6 km

Vertical exaggeration = approx. 2.5x

Direction of movement
on fault plane

5.17 A generalized cross section in northeastern Pennsylvania and northwestern New Jersey (in the vicinity of the Delaware Water Gap) to show the relation of topography to the different rock types below the surface. See the text for discussion.

On a larger scale, differential erosion is shown in the cross section in Figure 5.17. The section is drawn across a portion of northeastern Pennsylvania near the Delaware Water Gap. Note that the highest ridge is underlain by a tough quartzite, which erodes very slowly. The major portion of the Great Valley is underlain by a shaley rock, which erodes easily. In this same valley are still lower spots underlain by a dolomitic limestone, which is susceptible to rapid chemical weathering in this humid climate. To the southeast lie the New Jersey Highlands, held up by resistant units of both igneous and metamorphic rocks.

5.5
SOILS

Solid rock, or bedrock, underlies all parts of the land surface. In some places it is exposed at the surface, but in other areas it is covered by **soil,** the surface accumulation of sand, clay, and decayed plant material (called **humus**) that sustains plant life. **Residual** soils develop in place on the bedrock from which they are derived. Other soils may be **transported** from elsewhere, including glacially and river-derived (fluviatile) soils.

So far we have been discussing the ways in which weathering acts to break down existing rocks and to provide the material for new rocks. But weathering also plays a crucial role in the creation of soils. In fairly recent years the study of soils has developed into the science of **pedology** (from the Greek *pedon,* "soil," with *logos,* "reason," hence "soil science").[1]

---

[1] In the United States the term **pedology** is sometimes confused with words based on *ped-* and *pedi-,* combining forms meaning "foot," or with words based on *ped-* and *pedo-,* combining forms meaning "boy," "child," as in **pediatrics,** the medical science that treats of the hygiene and disease of children. Consequently there is a tendency in this country to use **soil science** instead of **pedology.**

SOIL CLASSIFICATION

In the early years of soil study in the United States, researchers thought that the parent material almost wholly determined the type of soil that would result from it. Thus, they reasoned, granite would weather to one type of soil and limestone to another.

It is true that a soil reflects to some degree the material from which it developed, and in some instances one can even map the distribution of rocks on the basis of the types of soil that lie above them. But as more and more information became available, it became apparent that the bedrock is not the only factor determining soil type. Russian soil scientists, following the pioneering work of V. V. Dokuchaev (1846 – 1903), have demonstrated that different soils develop over identical bedrock material in different areas when the climate varies from one area to another. The idea that climate exerts a major control over soil formation was introduced into this country in the 1920s by C. F. Marbut (1863 – 1935), for many years chief of the United States Soil Survey in the Department of Agriculture. Since that time soil scientists have discovered that still other factors exercise important influences on soil development. For instance, the relief of the land surface plays a significant role. The soil on the crest of a hill is somewhat different from the soil on the slope, which in turn differs from the soil on the level ground at the foot of the hill; yet all three soils rest on identical bedrock. The passage of time is another factor: A soil that has only begun to form differs from one that has been developing for thousands of years, although the climate, bedrock, and topography are the same in each instance. Finally, the vegetation in an area influences the type of soil that develops there: One type of soil will form beneath a pine forest, another beneath a forest of deciduous trees, and yet another on a grass-covered prairie.

Exactly what is a soil? We have said it is a natural surficial material that supports plant life. But since each soil exhibits certain properties that are determined by climate and living organisms operating over periods of time on Earth materials and on landscape of varying relief, and because all these factors are combined in various ways all over the land areas of the globe, the number of possible soil types is almost unlimited.

Soil scientists have developed a detailed classification system for soils of various types (discussed later). Geologically, we are interested primarily in what materials are removed or added during the soil-forming process.

Certain valid generalizations can be made about soils. We know, for example, that the composition of a soil varies with depth. A natural or artificial exposure of a soil reveals a series of zones, each recognizably different from the one above. Each of these zones is called a **soil horizon,** or, more simply a **horizon.** The three major zones or horizons in a typical soil, shown in Figure 5.18, may be described as follows from the bottom upward.

C Horizon The *C* horizon is a zone of partially disintegrated and decomposed rock material. Some of the original bedrock minerals are still present, but others have been transformed into new materials. The *C* horizon grades downward into the unweathered rock material. It should be noted that soils can form from the weathering of unconsolidated sediments as well as from solid bedrock.

B Horizon The *B* horizon lies directly above the *C* horizon. Weathering here has proceeded further than in the underlying zone, and only those minerals of the parent rock that are most resistant to the decomposition (quartz, for example) are still recognizable. The others have been con-

5.19 Soil profile on limestone bedrock showing 12 cm of dark-grayish crown (*A* horizon) and 30 cm of red clay (*B* horizon) on *C* horizon of weathered limey-clay grading into limestone bedrock. [U.S. Department of Agriculture.]

verted into new minerals or into soluble salts. In moist climates the *B* horizon contains an accumulation of clayey material and iron oxides delivered by water percolating downward from the surface. In dry climates we generally find, in addition to the clay and iron oxides, deposits of more soluble minerals, such as calcite. This mineral too may have been brought down from above, but some is brought into the *B* horizon from below, as soil water is drawn upward by high evaporation rates. Because material is deposited in the *B* horizon, it is known as the "zone of accumulation" (see Figure 5.19).

In many soils there develops a dense subsurface layer, called a **fragipan,** whose resistance and near-impermeability to fluids are chiefly due to its extreme compactness (as in a **claypan**) or to cementation (as in a **hardpan**). When cemented by carbonate, it is called **caliche** (a Spanish word derived from the Latin *calix,* "lime"), a whitish accumulation made up largely of calcium carbonate. When cemented by iron oxides, it is called an **ironpan.** In some instances it may be cemented by silica or other materials. In many cases the pan is so hard that it has the character of and behaves like firm rock.

A Horizon The *A* horizon is the uppermost zone—the one into which we sink a spade when we dig a garden. This is

5.18 The three major horizons of a soil. In many places it is possible to subdivide the zones themselves. Here the soil is shown as having developed from limestone.

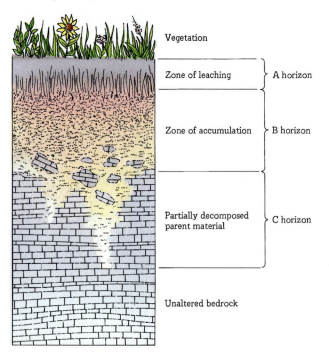

Vegetation

Zone of leaching } A horizon

Zone of accumulation } B horizon

Partially decomposed parent material } C horizon

Unaltered bedrock

the zone from which the iron oxides have been carried to the B horizon, and in dry climates it is the source of some soluble material that may be deposited in the B horizon. The process by which these materials have been moved downward by soil water is called **leaching**, and the A horizon is sometimes called the "zone of leaching" (Figure 5.19). Varying amounts of organic material tend to give the A horizon a gray-to-black color.

The three soil horizons have all developed from the underlying parent material. When this material is first exposed at the surface, the upper portion is subjected to intense weathering and decomposition proceeds rapidly. As the decomposed material builds up, downward-percolating water begins to leach out some of the minerals and to deposit them lower down. Gradually the A horizon and the B horizon build up, and weathering continues (at a slower rate now) on the underlying parent material, giving rise to the C horizon. With the passage of time, the C horizon reaches deeper and deeper into the unweathered material below, the B horizon keeps moving downward, and the A horizon in turn encroaches on the upper portion of the B horizon. Finally a **mature soil** is built up—that is, a soil with fully developed soil horizons, as contrasted with an **immature soil**, one that lacks well-developed horizons and resembles the parent material.

The thickness of the soil formed depends on many factors. In the northern United States and southern Canada the material that was first exposed to weathering after the retreat of the last ice sheet some 10,000 years ago is now topped by a soil 60 to 80 cm thick. Farther south, where the surface was uncovered by the ice at an earlier time, the soils are thicker. In some places, especially where glaciation has not been a factor, the processes of weathering have left their mark 5 to 10 m below the present surface.

SOME SOIL TYPES

We can understand the farmer's interest in soil, but why is it important for the geologist to understand soils and the processes by which they are formed? There are several reasons. First, soils provide clues to the environment in which they were originally formed. By analyzing an ancient soil buried in the rock record, we may be able to determine the climate and physical conditions that prevailed when it was formed. Second, some soils are sources of valuable mineral deposits (see Chapter 19), and the weathering process often enriches otherwise low-grade mineral deposits, making them profitable to mine. An understanding of soils and soil-forming processes, therefore, can serve as a guide in the search for ores. Third, because a soil reflects to some degree the nature of the rock material from which it has developed, we can sometimes determine the nature of the underlying rock by performing an analysis of the soil.

But most important of all, soils are the source of many of the sediments that are eventually converted into sedimentary rocks. And these in turn may be transformed into metamorphic rocks or, following another path in the rock cycle, may be converted into new soils. If we understand the processes and results of soil formation, we are in a better position to interpret the origin and evolution of many rock types.

Earlier soil classifications dealt with three major types of soil. Two of them, the **pedalfers** and the **pedocals**, are typical of the middle latitudes. The third group, referred to as **laterites**, is found in tropical climates.

A **pedalfer** is a soil in which iron oxides, clays, or both have accumulated in the B horizon. The name is derived from *pedon*, Greek for "the ground," and the symbols Al and Fe for aluminum and iron. In general, soluble materials such as calcium carbonate or magnesium carbonate do not occur in the pedalfers but rather stay in solution in groundwater or surface waters. Pedalfers are commonly found in temperate, humid climates, usually beneath a forest vegetation.

Notice in Figure 5.20 how the original minerals of the granite are transformed as weathering progresses. The left side of the figure shows increases in quartz, plagioclase, and orthoclase, which were released directly from the granite. Then, as weathering progresses, we find (on the right side of the figure) that the amount of kaolinite increases at the

5.20 A pedalfer soil that has developed on a granite. Note the transition from unaltered granite, upward through partially decomposed granite of the C horizon, into the B horizon, where no trace of the original granite structure remains, and finally into the A horizon, just below the surface.

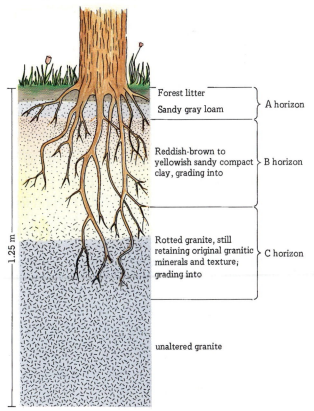

Forest litter
Sandy gray loam } A horizon

Reddish-brown to yellowish sandy compact clay, grading into } B horizon

Rotted granite, still retaining original granitic minerals and texture; grading into } C horizon

unaltered granite

1.25 m

5.21   Profile of a podsol soil, Cape Cod, Massachusetts.

expense of the original minerals. The initial rise in the amount of quartz and orthoclase indicates simply that these minerals tend to accumulate in the soil because of their greater resistance to decomposition. Iron oxides also increase with weathering as the iron-bearing silicates decompose.

There are several varieties of soils in the pedalfer group, including the red and yellow soils of the southeastern states as well as **podsol** (from the Russian for "ashy gray soil") and the gray-brown podsolic soils of the northeastern quarter of the United States and of southern and eastern Canada (Figure 5.21). Prairie soils are transitional varieties between the pedalfers of the East and the pedocals of the West.

Pedocals are soils that contain an accumulation of calcium carbonate. Their name is derived from a combination of *pedon* and calcium. The soils of this major group are found in the temperate zones where the temperature is relatively high, the rainfall is low, and the vegetation is mostly grass or brush.

In the formation of pedocals, calcium carbonate and, to a lesser extent, magnesium carbonate are deposited in the soil profile, particularly in the *B* horizon. This process occurs in areas where the temperature is high, the rainfall scant, and the upper level of the soil hot and dry most of the time. Water evaporates before it can remove carbonates from the soil. Consequently these compounds are precipitated as caliche. The occasional rain may carry the soluble material down from the *A* horizon into the *B* horizon, where it is later precipitated as the water evaporates. Soluble material may also move up into the soil from below. In this case, water beneath the soil or in its lower portion rises toward the surface through small capillary openings. Then, as the water in the upper portions evaporates, the dissolved materials will be precipitated out.

The term **laterite** is applied to many tropical soils that are rich in hydrated aluminum and iron oxides. The name itself, from the Latin for "brick," suggests the characteristic color produced by the iron in these soils. The formation of laterites is not well understood.

TABLE 5.3
Soil Orders of the United States Soil Classification System[a]

Soil order	Soil characteristics
Alfisols	Subhumid climates, precipitation 15–125 cm, mostly under forest cover. Clay accumulation in *B* horizon. Leached upper gray-brown layer; red-brown lower layer, with aluminum and iron silicate concentrations.
Aridisols	Dry climates with lime or gypsum accumulations, salt layers; accumulation of Ca, Mg, and K ions.
Entisols	Recent soils with no recognizable horizons developed; many recent river floodplains, volcanic ash deposits, and recent sands.
Histosols	Rich in organic content, including peats and mucks; accumulated plant remains in bogs and swamps.
Inceptisols	Weakly developed horizons showing just the "inception" of soil development.
Mollisols	Grassland soils; black, organic-rich near surface; high concentration of lime.
Oxisols	Excessively weathered soils, often over 3 m deep; of low fertility; lots of iron and aluminum oxide clays.
Spodosols	Sandy, leached soils of cool forests, mostly conifers; acid soil; lots of organic matter and/or Fe and Al clays.
Ultisols	Strongly acid, well-weathered soils of tropical and subtropical climates; accumulation of residual silicates.
Vertisols	High content of shrinking and expanding clays; very old weathered soils; deep wide vertical cracks when dry.

[a] After R. L. Donahue, R. W. Miller, and J. C. Shickluna, *Soils: An Introduction to Soils and Plant Growth;* 4th ed., Prentice Hall, Englewood Cliffs, N.J., 1977.

In the development of the laterites iron and aluminum accumulate in what is presumed to be the *B* horizon. The aluminum is in the form of $Al_2O_3 \cdot nH_2O$, which is generally called **bauxite,** an ore of aluminum. This ore appears to be developed when intense and prolonged weathering removes the silica from the clay minerals and leaves a residuum of hydrous aluminum oxide (that is, bauxite). In some laterites the concentration of iron oxides in the presumed *B* horizon is so great that it is profitable to mine them for iron.

UNITED STATES SOIL CLASSIFICATION SYSTEM

During the last two decades soil scientists have developed a soil classification system based on variations in precipitation, temperature, vegetative cover, and drainage. Certain characteristics are common to each soil order, although "pedalfers" (soils rich in iron and aluminum) may be as diverse as the spodosols of the Canadian forest and the oxisols of Florida, which do not look or behave alike.

The soil orders are summarized in Table 5.3.

PALEOSOLS

Not all soils occur at the surface. Some are buried beneath younger material and are known as **paleosols** (see Figure 5.22). Such buried soil zones are useful indicators of past periods of erosion and weathering.

5.22 Buried soil zones in the Hondo Formation, Magdalena Valley, Colombia. Note human figures for scale along the dark soils developed on successive units of stream deposits. [Franklyn B. Van Houten.]

Summary

Weathering is the response of surface or near-surface material to contact with water, air, and living matter.

The energy that drives the weathering process comes from motions within the Earth that raise portions of its surface and from outside the Earth as solar energy.

Types of weathering are mechanical and chemical.
    Mechanical weathering (or disintegration or physical weathering) involves a reduction in the size of rock and mineral particles but no change in composition.
    Chemical weathering (decomposition) involves a change in the composition of the material weathered. The rate of chemical weathering increases with the decrease in the size of particles and with an increase in moisture and temperature (except for calcite).

Rates of weathering vary with the type of material weathered and the environment. For example, olivine chemically weathers more rapidly than does quartz, and limestone weathers very rapidly in a moist climate but very slowly in a dry climate.

Rates of erosion indicate that before human beings began to use the landscape intensively, rivers annually transported about 10 billion t of material to the seas. Today it is two to three times that, chiefly because of human activities.

Differential erosion is the process by which different rock masses or different sections of the same rock mass erode at different rates. For example, limestone is more resistant to erosion in a dry climate than is mudstone.

Soil is a naturally occurring surface material that supports life and generally is the product of weathering.
    Soil zones are the *A, B,* and *C* **horizons** from the surface downward.
    Fragipans are dense, nearly impermeable subsurface layers that are extremely compact **(claypan)** or tightly cemented **(hardpan); caliche** is cemented by calcium carbonate, **ironpan** by iron oxides.

Mature soils may develop, given sufficient time; otherwise **immature soils** result.
    Soil types include pedalfers in moist temperate climates, pedocals in dry temperate climates, and laterites in moist tropical climates.

The **United States Soil Classification System** is based on variations in precipitation, temperature, vegetative cover, and drainage.
    Paleosols are buried soil zones indicative of past periods of erosion and weathering.

Questions

1. Where do we find bedrock and why does it appear at the surface in some instances and not in others?
2. What would be the end products of the chemical weathering of a typical granite?
3. What property of H_2O makes it such an important agent in weathering? How is this different from almost every other chemical substance?
4. Explain the process by which bedrock that has not been moved at all can come to be rounded, looking like so many boulders that have been rolled around to produce their rounded edges?
5. How is climate important in the chemical weathering of minerals and rocks?
6. How does the rate of chemical weathering of the minerals of an igneous rock compare with the pattern of crystallization from its original magma? What are the reasons for this relationship?
7. How can we determine *average* rates of erosion for whole continents?
8. What besides the bedrock itself will influence the kind of soil that develops in a given area?
9. Draw and label a typical "soil profile."
10. Of what use are buried soil zones or paleosols in interpreting the geologic history of an area?
11. With physical and chemical weathering taking place constantly, why do the weathering products not build up to staggering amounts on all continents? What happens to them?

Supplementary Readings

BIRKELAND, P. W. *Soils and Geomorphology,* Oxford University Press, New York, 1984.
Study of soils in their natural setting—the field—with emphasis on the use of soils to date deposits on the basis of soil development. Attempts are made to reconstruct the environment that occurred during soil formation.

EL-SWAIFY, S. A., ED. *Soil Erosion and Conservation,* Soil Conservation Society of America, Ankeny, Iowa, 1985.
Based on international conference on soil erosion and conservation held in Honolulu, Hawaii.

JUDSON, SHELDON "Erosion of the Land—or What's Happening to Our Continents?" *Am. Sci.,* vol. 56, pp. 356–379, 1968.
An overview of worldwide erosion.

KRAUSKOPF, K. B. *Introduction to Geochemistry,* 2nd ed., McGraw-Hill, New York, 1979.
A basic look at chemical activity in geologic settings. Includes an analysis of the chemical activity of soils.

McRAE, S. G. *Practical Pedology: Studying Soils in the Field,* John Wiley & Sons, 1988.
Describing, naming, and classifying soils are approached from a field perspective, with practical methods for soil mapping given.

RITTER, D. F. *Process Geomorphology,* 2nd ed., William C. Brown Co., Dubuque, Iowa, 1986.
A modern examination of the Earth's surficial processes. Excellent coverage of physical and chemical weathering in the production of soils.

Jurassic/Triassic Chinle Formation, Ghost Ranch, New Mexico.

CHAPTER 6

ORIGIN AND OCCURRENCE OF SEDIMENTS AND SEDIMENTARY ROCKS

"Clay and silt, sand and gravel, testimonials to the wreckage of the continents."

Anonymous

OVERVIEW

Most of us have dug our toes into a sandy beach, or picked our way over the gravels of a rushing stream, or perhaps slogged through the mud of a swamp. All of these—sand, gravel, and mud—are products of weathering, the subject of the last chapter, and none of them immediately suggests hard, solid rock. Yet deposits of this sort, or very similar ones, are the stuff from which 75 percent of the rocks exposed at the Earth's surface are made.

Sedimentary rocks form both from deposits of rock fragments and minerals created by weathering—*detrital sedimentary rocks*—and from chemical deposition—*chemical sedimentary rocks*. The process by which these materials are laid down is called *sedimentation*.

In this chapter we examine the *mineral composition* of the sedimentary deposits as well as their *texture*. These deposits will undergo changes in a process called *diagenesis*, and one of these changes is *lithification*, the process of converting an unconsolidated deposit to a firm rock.

As we seek a method of *classification*, we will find it useful to use *detrital* and *chemical* groups, *composition*, and *texture*. Once we have acquired a basis for recognizing the various sedimentary rocks, we will look at some of their special features, including *bedding*, the single most characteristic feature of the family.

We conclude the chapter with a discussion of *geosynclines*, and the relation of *plate boundaries to sedimentation*.

6.1

FORMATION OF SEDIMENTARY ROCKS

We found in Chapter 3 that igneous rocks harden from molten material that originates beneath the surface, under the high temperatures and pressures that prevail there. In contrast, sedimentary rocks form at the much lower temperatures and pressures that prevail at or near the Earth's surface (Figure 6.1).

ORIGIN OF MATERIAL

Sedimentary rocks or metamorphic rocks derived from them make up about 75 percent of the rocks exposed at the surface of the earth (see Box 6.1).

The material from which sedimentary rocks are fashioned originates in two ways. First, the deposits may be accumulations of minerals and rock fragments derived either from the erosion of existing rock or from the weathering products of these rocks. Deposits of this type are called

6.1 Grand Canyon from south rim showing flat-lying sedimentary rocks into which the Colorado River has cut its valley.

BOX 6.1   Relative Abundance of Crystalline and Sedimentary Rocks

The graphs show relative abundance of sedimentary rocks and crystalline (igneous and metamorphic) rocks. (*a*) The great bulk (95 percent) of the outer 10 km of the Earth is made up of crystalline rocks. Only a small proportion (5 percent) is sedimentary. (*b*) In contrast, the areal extent of sedimentary rocks at the surface of the Earth's continents is three times that of crystalline rocks.

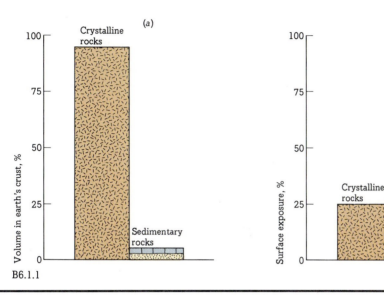

B6.1.1

detrital (from the Latin for "worn down"), and sedimentary rocks formed from them are called **detrital sedimentary rocks.** Second, the deposits may be produced by chemical processes. We refer to these deposits as **nondetrital** or **chemical deposits** and to the rocks formed from them as **chemical sedimentary rocks.**

Gravel, sand, silt, and clay derived from the weathering and erosion of a land area are examples of detrital sediments. Let us take a specific example. The quartz grains freed by the weathering of a granite may be winnowed out by the flowing water of a stream and swept into the ocean. There they settle out as beds of sand, a detrital deposit. Later, when this deposit is cemented to form a hard rock, we have a sandstone, a detrital rock.

Chemically formed deposits usually develop from the chemical precipitation of material dissolved in water. This process may take place either directly, through inorganic processes, or indirectly, through the intervention of plants and animals. The salt left behind after a salty body of water has evaporated is an example of a deposit laid down by inorganic chemical precipitation. On the other hand, certain organisms, such as the corals, extract calcium carbonate from the seawater and use it to build up skeletons of calcite. When the animals die, their skeletons collect as a biochemical (from the Greek for "life") deposit, and the rock that subsequently forms is called a **biochemical rock** —in this case, limestone.

Although we distinguish between the two general groups of sedimentary rocks—detrital and chemical—many sedimentary rocks are mixtures of the two. We commonly find that a chemically formed rock contains a certain amount of detrital material. In similar fashion, predominantly detrital rocks include some material that has been chemically deposited.

Each year the streams of the world deliver staggering amounts of material to the oceans. An estimate of this volume is given in Box 6.2.

Geologists use various terms to describe the environment in which a sediment originally accumulated. For example, if a limestone contains fossils of an animal that is known to have lived only in the sea, the rock is known as a **marine** limestone. **Fluvial,** from the Latin for "river," is applied to rocks formed by deposits laid down by a river. (**Deltaic** deposits occur near the mouth of major streams, whose overall shape appears like the Greek letter Δ.) **Eolian,** derived from Aeolus, the Greek wind god, describes rock made up of wind-deposited material. Rocks formed from lake deposits are termed **lacustrine,** from the Latin word for "lake." **Glacial** deposits include both those deposited directly by the ice and those deposited by meltwater from the glaciers (**glaciofluvial**).

Detrital and chemical, however, are the main divisions of sedimentary rocks based on the origin of material, and, as we shall see later, they form the two major divisions in the classification of sedimentary rocks.

BOX 6.2   Mass of Material Delivered to Oceans

We have some estimates of the total mass of sedimentary rocks on the Earth. Probably the best available estimate was made some time ago by Arie Poldervaart[1] of Columbia University, who calculated their weight as $1{,}702 \times 10^{15}$ t. Of this total, he estimated that 480×10^{15} t are presently on the continents and that the rest are in the oceans.

We have seen that sedimentary rocks are formed from the materials weathered from preexisting rocks. Eventually this material reaches the deep ocean basins. What estimate can we make about the amount of material delivered

[1] A. Poldervaart (ed.), "Crust of the Earth," *Geol. Soc. Am. Spec. Paper,* no. 62, 1955.

Order-of-Magnitude Estimates by Source of Materials Delivered Annually to the Oceans[a]

Source	t/yr
Rivers	10 billion (10^{10})
Glaciers	100 million to 1 billion ($10^8 - 10^9$)
Wind	100 million (10^8)
Extraterrestrial	0.03 to 0.3 (3×10^{-2} to 3×10^{-1})

[a] Estimated from various sources.

each year to the oceans? The figures presented in this box are order-of-magnitude estimates. Therefore a statement of the approximate amount of material delivered annually to the oceans— where most sedimentary rocks form— is approximately 10^{10} t. As these figures suggest, and as we implied in earlier chapters, the great bulk of this material is carried by rivers. That contributed by wind, by ice, or by extraterrestrial sources does not appreciably change the total amount of material deposited in the oceans.

SEDIMENTATION

The general process by which rock-forming material is laid down is called **sedimentation,** or deposition. The factors controlling sedimentation are easy to visualize: To have any deposition at all, there must obviously be something to deposit—which is another way of saying that a source of sediments must exist. We also need some process to transport this sediment. Finally, there must be some place and some process for the deposition of the sedimentary material.

Methods of Transportation   Water—in streams and below glaciers, underground, and in ocean currents—is the principal means of transporting material from one place to another. In a stream the coarsest material is carried along the bed of the stream by rolling and sliding; the medium-sized material is carried partially within the flowing water, at times falling to the bottom only to bounce back up into the current; and the finer material is carried suspended within the flowing water. The material dissolved from the weathering of minerals is carried by the stream in solution and ultimately adds to the salinity of the ocean into which it finally empties.

Landslides and other movements induced by gravity also play a role, as do wind and ice. We shall look more closely at these processes in Chapters 12 through 17.

Processes of Sedimentation   Detrital material is **deposited** when its agent of transportation no longer has sufficient energy to move it farther. For example, a stream flowing along at a certain velocity possesses energy to move particles up to a certain maximum size. If the stream loses velocity, it also loses energy, and it is no longer able to transport all the material that it has been carrying at the higher velocity. The solid particles, beginning with the heaviest, start to settle to the bottom. The effect is much the same when a wind that has been driving sand across a desert suddenly dies—a loss of energy accompanies the loss in velocity.

Material that has been carried in solution is deposited in a different way, that is, by **precipitation,** a chemical process by which dissolved material is converted into a solid and separated from the liquid solvent. As already noted, precipitation may be caused by chemical interaction and evaporation concentration.

Although at first glance the whole process of sedimentation seems quite simple, it is actually as complex as nature itself. Many factors are involved, and they can interact in a variety of ways. Consequently the manner in which sedimentation takes place and the sediments that result from it differ greatly from one situation to another (see Figures 6.2 and 6.3). Think, for instance, of the different ways in which materials settle out of water. A swift, narrow mountain stream may deposit coarse to medium particles along its bed, but farther downstream, as the valley widens, the same stream may overflow its banks and spread fine particles, including mud, over the surrounding country. A lake provides a different environment, varying from the delta of the inflowing stream to the deep lake bottom and the shallow, sandy shore zones. In the oceans, too, environment and sedimentation vary, from the brackish tidal lagoon to the zone of plunging surf and out to the broad, submerged shelves of the continents and to the ocean depths beyond.

6.2 The environment of this quiet pond, Watt Pond, Putney, Vermont, favors the deposition of fine-grained sediments, largely silt and mud. [Lee Krohn.]

6.3 Exposed to the direct attack of ocean surf, the environment of this cliffed New Zealand coastline favors the deposition of coarse sand and gravel.

MINERAL COMPOSITION OF SEDIMENTARY ROCKS

As we discovered in the chapter on weathering, an igneous rock made up of (1) quartz, (2) feldspar, and (3) ferromagnesian silicates will break down to those specific minerals that are stable at the Earth's surface—namely, quartz, some feldspars, and clay minerals. Detrital sedimentary rocks are accumulations of these minerals combined with precipitated mineral matter that serves to cement the grains together. The grains of many sandstones are predominantly quartz, with some feldspar and other minor accessory minerals. The cementing material may be calcite, dolomite, silica, or iron oxide. Nondetrital sedimentary rocks are dominated by limestones and dolomites. Most sedimentary rocks are mixtures of more than one mineral, although one may dominate. Limestone, for example, is composed mostly of calcite, but even the purest limestone contains small amounts of other minerals, such as clay or quartz.

Clay In an earlier chapter we described how clay minerals develop from the weathering of the silicates, particularly the feldspars. These clays may be subsequently incorporated into sedimentary rocks; they may, for example, form an important constituent of mudstone and shale. Examination of recent and ancient marine deposits shows that the kaolinite and illite clays (illites are increasingly abundant in older rocks) are the most common clays in sedimentary rocks and that the smectite clays are relatively rare.

Quartz An important component of sedimentary rocks is silica, including the very common mineral quartz. The mechanical and chemical weathering of an igneous rock such as granite sets free individual grains of quartz that eventually may be incorporated into sediments. These quartz grains produce the detrital forms of silica and account for most of the volume of the sedimentary rock sand-stone. But silica in solution is also produced by the weathering of an igneous rock. This silica may be precipitated or deposited in the form of quartz, particularly as a cementing agent in coarse-grained sedimentary rocks.

Calcite The chief constituent of the sedimentary rock limestone, calcite, $CaCO_3$, is also the most common cementing material in the coarse-grained sedimentary rocks. The calcium is derived from igneous rocks that contain calcium-bearing minerals, such as calcium plagioclase and some of the ferromagnesian minerals. It also comes from weathering of carbonate sedimentary rocks. Calcium is carried from the zone of weathering as calcium bicarbonate, $Ca(HCO_3)_2$, and is eventually precipitated as $CaCO_3$ through the intervention of plants, animals, or inorganic processes. The carbonate is ultimately derived from water and carbon dioxide.

Other Materials in Sedimentary Rocks Accumulations of clay, quartz, and calcite, either alone or in combination, account for all but a very small percentage of the sedimentary rocks, but certain other materials occur in quantities large enough to form distinct strata. The mineral dolomite, $CaMg(CO_3)_2$, for example, is usually intimately associated with calcite, although it is far less abundant. (Dolomite is named after an eighteenth-century French geologist, Déodat de Dolomieu.) When the mineral is present in large amounts in a rock, the rock itself is also known as dolomite or dolostone. The mineral dolomite is easily confused with calcite, and because they often occur together, distinguishing them is important. Calcite effervesces freely in dilute hydrochloric acid; dolomite effervesces very slowly or not at all unless it is finely ground or powdered. The more rapid chemical activity results from the increase in surface area, an example of the general principle discussed in Section 5.2.

The feldspars and micas are abundant in some sedimentary rocks. We have found that chemical weathering converts these minerals into new minerals at a relatively rapid rate. Therefore when we find mica and feldspar in a sedimentary rock, chances are that it was predominantly mechanical, rather than chemical, weathering that originally made them available for incorporation in the rock.

Iron produced by chemical weathering of the ferromagnesian minerals in igneous rocks may be caught up again in new minerals and be incorporated into sedimentary deposits. The iron-bearing minerals that occur most frequently in sedimentary rocks are hematite, goethite, and limonite. In some deposits these minerals predominate, but more commonly they act simply as coloring matter or as a cementing material.

Halite (NaCl) and gypsum ($CaSO_4 \cdot 2H_2O$) are minerals precipitated from solution by evaporation of the water in which they were dissolved. The salinity of the water—that is, the proportion of the dissolved material to the water—determines the type of mineral that will precipitate out. The gypsum begins to separate from seawater when the salinity (at 30°C) reaches a little over three times its normal value. Then, when the salinity of the seawater has increased to about ten times its normal value, halite begins to precipitate.

Pyroclastic rocks, mentioned in Chapter 4, are sedimentary rocks composed mostly of fragments blown from volcanoes. The fragments may be large pieces that have fallen close to the volcano or extremely fine ash that has been carried by the wind and deposited up to many hundreds of kilometers from the volcanic eruption.

Finally, organic matter may be present in sedimentary rocks. In the sedimentary rock known as coal, plant materials are almost the only components. More commonly, however, organic matter is very sparsely disseminated throughout sedimentary deposits and the resulting rocks.

TEXTURE

Texture refers to the size, shape, and arrangement of the particles that make up a rock. There are two major types of texture in sedimentary rocks: clastic (or detrital) and nonclastic (crystalline).

Clastic (or Detrital) Texture   The term **clastic** is derived from the Greek for "broken" or "fragmental," and rocks that have been formed from detritus of mineral and rock fragments are said to have clastic or detrital texture. The size and shape of the original particles have a direct influence on the nature of the resulting texture. A rock formed from a bed of gravel and sand has a coarse, rubble-like texture that is very different from the sugary texture of a rock developed from a deposit of rounded, uniform sand grains. Furthermore, the process by which a sediment is

deposited also affects the texture of the sedimentary rock that develops from it. Thus the debris dumped by a glacier is composed of a jumbled assortment of rock material ranging from particles of clay size to large boulders. A rock that develops from such a deposit has a very different texture from one that develops from a deposit of windblown sand, for instance, in which all the particles are approximately 0.15 to 0.30 mm in diameter.

Chemical sedimentary rocks may also show a clastic texture. A rock made up predominantly of shell fragments from a biochemical deposit has a clastic texture that is just as recognizable as the texture of a rock formed from sand deposits (see Figure 6.4).

One of the most useful factors in classifying sedimentary rocks is the size of the individual particles. In practice, we usually express the size of a particle in terms of its diameter rather than in terms of its volume, weight, or surface area. When we speak of "diameter," we may seem to imply that the particle is a sphere; but it is very unlikely that any fragment in a sedimentary rock is a true sphere. In geological measurements the term simply means the diameter that an irregularly shaped particle *would* have if it were a sphere of equivalent volume. Obviously, it would be a time-consuming, if not impossible, task to determine the volume of each sand grain or pebble in a rock and then to convert these measurements into appropriate diameters. So the diameters we use for particles are commonly determined by such rapid techniques as **sieving.** We can think of an irregularly shaped grain as having three mutually perpendicular axes or diameters; often these will be a long, an intermediate, and a short diameter. When we examine the "size" of large quantities of sand grains, we commonly pass the samples through nested sets of sieves, with the largest openings

6.4   Large number of oyster, clam, and snail shells make up this Miocene conglomerate of shell beds near Plum Point, Maryland.

in the sieve at the top of the stack, progressing downward to sieves with smaller and smaller openings at the bottom of the stack. The sand grains are bounced back and forth and up and down until they fit through the sieve with openings just a bit larger than the intermediate diameter and come to rest on the next sieve with openings just a bit smaller than that size grain. (Note that such sieve openings measure the intermediate diameter—not the long and not the short diameters!) By weighing the contents of each sieve we can demonstrate the distribution of the sand in various size classes and in this way compare and contrast one sample with others.

Several scales have been proposed to describe particles ranging in size from large boulders to minerals of microscopic dimensions. The Wentworth scale, presented in Table 6.1, is used widely, though not universally, by American and Canadian geologists. Notice that although the term **clay** is used in the table to designate all particles below $\frac{1}{256}$ (0.004) mm in diameter, the same term is also used to describe certain minerals. To avoid confusion, therefore, we must always refer specifically to either "clay size" or "clay mineral" unless the context makes the meaning clear.

Because determining the size of particles calls for the use of special equipment, the procedure is normally carried out only in the laboratory. In the examination of specimens in the field, therefore, comparisons are made to samples of known size or to scales prepared for that purpose.

Nonclastic Texture   Some—but not all—sedimentary rocks formed by chemical processes have **nonclastic** texture, in which the grains are interlocked. These rocks have somewhat the same appearance as igneous rocks with crystalline texture. Actually, most of the sedimentary rocks with nonclastic texture have crystalline structure, although a few of them, such as opal, do not exhibit this structure.

The mineral crystals that precipitate from an aqueous solution are usually small. Because the fluid in which they form has a very low density, they usually settle out rapidly and accumulate on the bottom as fine sediment. Eventu-ally, under the weight of additional sediments, this material is compacted more and more. Now the size of the individual crystals may begin to increase. Their growth may be induced by added pressure, which causes the favorably oriented grains to grow at the expense of less favorably oriented neighboring grains. Or crystals may grow as more and more mineral matter is added to them from the saturated solutions trapped in the original mud. In any event, the resulting rock is made up of interlocking crystals and has a texture similar to that of crystalline igneous rocks. Depending on the size of the crystals, we refer to these nonclastic textures as **fine-grained** (or finely crystalline), **medium-grained,** or **coarse-grained.** A coarse-grained texture has grains larger than 5 mm in diameter, and a fine-grained texture has grains less than 1 mm in diameter.

Sorting   Clastic or detrital rocks commonly have a variety of different-sized materials mixed together. We refer to the range of different sizes that are present as the **sorting** of the sediment or sedimentary rock. A rock with a narrow range of sizes is called "well-sorted"; one with very wide range of sizes is considered to be poorly sorted. The degree of sorting can have significant ramifications in the search for fluids such as water, gas, and oil. A well-sorted sandstone has little or no fine silt- or clay-sized material filling the pores found among the sand grains, and therefore fluids may fill those pores. A poorly sorted sandstone, on the other hand, has fine-sized particles in those potential voids, resulting in a low porosity and less chance for water, gas, or oil to find pore spaces to occupy.

The degree of sorting varies with the depositional environment. Where currents are strong, for example, finer particles are swept away, leaving only the coarsest clasts, and a well-sorted sediment. In quiet environments only fine particles may be carried into the basin. Again, the sorting could be good. Where different transporting media provide different sizes of clasts, the sorting is poorer. Glacial environments produce poorly sorted sediments, for example. Wind-blown sediments tend to be very well sorted.

Shape of Particles   Another component of texture is the shape of individual particles making up the sedimentary deposit. We refer to the roundness and sphericity of particles when describing their shape. **Roundness** is the degree to which the edges and corners have been ground off. **Sphericity** is the degree to which the shape of a fragment approaches the form of a sphere. Notice that a particle can have a high degree of roundness but a very low sphericity if it is bladelike or pencil-shaped, as long as the corners and edges have been rounded. It is somewhat less likely for a pebble or any other particle to be shaped almost like a sphere without having a concurrent high degree of roundness. (Exceptions include equidimensional minerals like garnet, magnetite, and ilmenite, which are very angular but would have a high sphericity.) Particles are shaped by the method and extent of transport to which they have been

TABLE 6.1
Wentworth Scale of Particle Sizes for Clastic Sediments[a]

Wentworth scale		For next larger size multiply by
Size, mm	Fragment	
	Boulder	
256	Cobble	4
64	Pebble, Gravel	16
4	Granule	2
2	Sand	32
0.06	Silt	16
0.004	Clay	

[a] Modified after C. K. Wentworth, "A Scale of Grade and Class Terms for Clastic Sediments," *J. Geol.*, vol. 30, p. 381, 1922.

subjected. Stream-transported particles of sand size and larger tend to get better rounded with higher sphericity the longer and farther they are transported. Fine windblown silt tends to remain fairly angular even with long-distance transport because of the lack of grain contact, unlike bedload or saltation carpet. Clay-sized particles, because they commonly are clay minerals of sheetlike shapes, do not tend to become either rounded or spherical.

DIAGENESIS

After deposition a sediment or sedimentary rock may undergo a series of physical, chemical, and biologic changes that we call **diagenesis.** These changes may be in the recombination or rearrangement of a mineral, resulting in the formation of a new mineral. Such changes are specifically limited to those that occur after the sediment has been deposited, during and after its lithification, but *before* any significant increases in temperature and pressure associated with metamorphism. This process is also limited to changes occurring before the sediment or sedimentary rock has been lifted up to come into contact with the atmosphere. Changes that occur under these conditions have already been discussed in the chapter on weathering. Some examples of diagenesis include dewatering and decrease in porosity associated with compaction of sands, silts, and claymuds; bacterial decomposition of organic matter; the distortion and destruction of bedding laminations by burrowing organisms; the removal of soluble materials such as calcium carbonate by dissolution in deep-sea sediments; the formation of concretions; the formation of a variety of **authigenic minerals** (those **born** or **formed in place**), such as orthoclase feldspar, illite, and other clay materials, and even quartz. During diagenesis many sedimentary rocks may undergo (1) **dissolution,** producing voids within the rock; (2) **cementation,** with the growth of new crystals into void space; or (3) **replacement,** involving the nearly simultaneous dissolution of existing minerals and the precipitation of new minerals.

LITHIFICATION

The process of **lithification** converts unconsolidated rock-forming materials into consolidated, coherent rock. The term is derived from the Greek *lithos,* meaning "rock," and the Latin *facere,* "to make." In the following subsections we shall discuss the various ways in which sedimentary deposits are lithified.

Compaction and Desiccation   As sediments accumulate in any basin of deposition (such as the ocean, a lake, a pond, along the floodplain of a river, or on an open plain), they undergo compaction. In **compaction** the pore space between adjacent grains is gradually reduced by the pressure of overlying sediments or by pressures resulting from

Earth movement. Coarse deposits of sand and gravel undergo some compaction, but fine-grained deposits of silt and clay respond much more readily because they contain a large volume of pore fluids. As the individual particles are pressed closer and closer together, the thickness of the deposit is reduced by expulsion of the fluids and its coherence is increased. It has been estimated that deposits of clay-sized particles buried to depths of 1 km have been compacted to about 60 percent of their original volume. This fluid expulsion forces out much of the water, gas, or oil that may have been present.

In **desiccation** the fluid that originally filled the pore spaces of water-laid clay and silt deposits is forced out (or drawn out by evaporation). Sometimes this is the direct result of compaction, but desiccation also takes place when a deposit is simply exposed to the air and the water evaporates.

Cementation   Pressure alone does not produce an indurated rock. Fluids in pores of sediment contain varying amounts of dissolved matter in solutions and some of this material is dissolved by pressure solution at contacts between adjacent minerals. In **cementation** this dissolved material is precipitated in the spaces between individual particles binding them together. Of the many minerals that serve as cementing agents, the most common are calcite, dolomite, and quartz. Others include iron oxide, opal, chalcedony, anhydrite, pyrite, and especially the clay minerals. Apparently the cementing material is carried in solution by water that percolates through the open spaces among the particles of the deposit. Then some factor in the new environment causes the mineral to be deposited, and the former unconsolidated deposit is cemented into a sedimentary rock.

In coarse-grained deposits there are relatively large interconnecting spaces among the particles. As we should expect, these deposits are very susceptible to cementation because the percolating water can move through them with great ease. Deposits of sand and gravel are transformed by cementation into the sedimentary rocks sandstone and conglomerate.

Crystallization   The crystallization of certain chemical deposits is in itself a form of lithification. Crystallization also serves to harden deposits that have been laid down by mechanical processes of sedimentation. For example, new minerals may crystallize within a deposit, or the crystals of existing minerals may increase in size. New minerals are sometimes produced by chemical reactions among amorphous, colloidal materials in fine-grained muds. Exactly how and when these reactions occur is not yet generally understood, but that new crystals have formed after the deposit was initially laid down becomes increasingly apparent as we make more and more detailed studies of sedimentary rocks. Furthermore, it seems clear that this crystallization promotes the process of lithification, particularly in the finer sediments.

6.2

TYPES OF SEDIMENTARY ROCK

CLASSIFICATION

Having examined some of the factors involved in the formation of sedimentary rocks, we are in a position to consider a classification for this rock family. The classification presented in Table 6.2 represents only one of several possible schemes, but it will serve our purposes very adequately. Notice there are two main groups—detrital and chemical—based on the origin of the sediment, and that the chemical category is further split into inorganic and biochemical. All the detrital rocks have clastic texture, whereas the chemical rocks have either clastic or nonclastic texture. We use particle size to subdivide the detrital rocks, and composition to subdivide the chemical rocks.

6.6 When the fragments in a conglomerate are more angular than rounded, the rock is called a *breccia*. When the rock is made up of volcanic materials, as is this one in Iceland, it is called a *volcanic breccia*. [Roland Hellmann.]

Conglomerate   A **conglomerate**, referred to by the common name *puddingstone*, is a detrital rock made up of more or less rounded fragments (Figure 6.5), an appreciable percentage of which are of gravel size (greater than 2 mm). If the fragments are more angular than rounded, the rock is called a **breccia** (Figure 6.6). An example of a conglomerate is **tillite**, a rock formed by the lithification of deposits laid down directly by glacier ice (see Chapter 15). The large particles in a conglomerate are usually rock fragments, and the finer particles are usually minerals derived from the weathering and erosion of preexisting rocks.

6.5 Semirounded fragments in consolidated "conglomerate"; Precambrian from Great Slave Lake district of Canada.

Sandstone   A sandstone is formed by the consolidation of grains of sand size between $\frac{1}{16}$ (0.06) and 2 mm in diame-

TABLE 6.2
Classification of Sedimentary Rocks (Excluding Pyroclastic Rocks)

Origin	Texture	Particle size or composition	Rock name
Detrital	Clastic	Granule or larger	Conglomerate (round grains) breccia (angular grains)
		Sand	Sandstone
		Silt	Siltstone
		Clay	Mudstone and shale
Chemical:			
Inorganic	Clastic or nonclastic	Calcite, $CaCO_3$	Limestone
		Dolomite, $CaMg(CO_3)_2$	Dolostone
		Halite, $NaCl$	Salt
		Gypsum, $CaSO_4 \cdot 2H_2O$	Gypsum
Biochemical	Clastic or nonclastic	$CaCO_3$ shells	Limestone, chalk, coquina
		SiO_2 diatoms	Diatomite
		Plant remains	Coal

ter. Sandstone is thus intermediate between coarse-grained conglomerate and fine-grained mudrocks. The size of the grains varies sufficiently from one sandstone to another to allow us to speak of coarse-grained, medium-grained, and fine-grained sandstone.

Commonly the grains of a sandstone are almost all quartz. When this is the case, the rock is called a **quartz arenite** (from the Latin for "sand"). The name **quartzose sandstone** is also used. If the minerals are predominantly quartz and feldspar, the sandstone is called an **arkose,** a French word for the rock formed by the consolidation of debris derived from the mechanical weathering of a granite. Another variety of sandstone, called **lithic sandstone** or **graywacke,** is characterized by dark color and by angular grains of quartz, feldspar, and small fragments of rock set in a matrix of clay-sized particles. Arkosic and lithic sandstones result from more rapid sedimentation and burial than quartz-rich sandstones. These "immature" sediments have not normally been transported any great distance and hence have not been subjected to long periods of weathering, which would have decomposed the feldspars and lithic fragments.

Siltstone   Particles smaller than 1/16 (.06) mm but larger than 1/256 (.004) mm are termed **silt;** when lithified, they form **siltstones.** The grains may be of any composition, but commonly consist of quartz and feldspar.

Mudrocks   Fine-grained detrital rocks composed of clay-sized particles (less than 0.004 mm in diameter) are termed **mudrocks,** or **pelites.** Mudstones are fine-grained rocks with a massive or blocky aspect; shales are fine-grained rocks that split into small pieces more or less parallel to the bedding; the term **shale** is sometimes used for any very fine grained detrital sedimentary rocks. The particles in these rocks are so small that it is difficult to examine them visually; to determine the precise mineral composition, we

need to use X-ray diffraction and related techniques. We do know that they contain not only clay minerals but also clay-sized particles of quartz, feldspar, calcite, and dolomite, to mention but a few.

CHEMICAL SEDIMENTARY ROCKS OF INORGANIC ORIGIN

Limestone   Limestone is a sedimentary rock that is made up chiefly of the mineral calcite that has been deposited by either inorganic or organic chemical processes. Many limestones have a clastic texture, but nonclastic, particularly crystalline, textures are common.

Inorganically formed limestone is made up of calcite that has been precipitated from solution by inorganic processes. Some calcite is precipitated from the fresh water of streams, springs, and caves, although the total amount of rock formed in this way is negligible. When calcium-bearing rocks undergo chemical weathering, calcium bicarbonate, $Ca(HCO_3)_2$, is produced in solution. If enough of the water evaporates, or if the temperature rises, or if the CO_2 pressure falls, or if the water is agitated, calcite is precipitated from this solution. For example, most **dripstone,** or **travertine,** is formed in caves by the evaporation of water that is carrying calcium bicarbonate in solution. And **tufa** (from the Italian for "soft rock") is a spongy, porous limestone formed by the precipitation of calcite from the water of streams and springs (see Figure 6.7).

Although geologists understand the inorganic processes by which limestone is formed by precipitation from fresh water, they are not quite sure how important these processes are in precipitation from seawater. Some observers have questioned whether they operate at all. On the floors of modern oceans and in rocks formed in ancient oceans, however, we find small spheroidal grains called

6.7 Hot springs deposits, known as "tufa," Thermopolis, Wyoming.

oölites, the size of sand and often composed of calcite; these grains are thought to be formed by the inorganic precipitation of calcium carbonate from seawater. (The term comes from the Greek for "egg" because an accumulation of oölites resembles a cluster of fish roe.) Cross sections show that many oölites, though not all, have grown around a mineral grain or around a small fragment of shell that acts as a nucleus. Some limestones are made up largely of oölites. One, widely used for building, is the so-called Indiana or Spergen Limestone.

A much larger percentage of limestone is of biochemical origin and will be discussed under that heading below.

Dolostone   In discussing the mineral dolomite, $CaMg(CO_3)_2$, we mentioned that when it occurs in large concentrations, it forms a rock that is called **dolostone,** or **dolomite.** Extensive deposits of dolostone, the origin of which is not fully understood, appear to have been formed by replacement of preexisting deposits of calcite. There is now increasing agreement that most dolostone is in some way related to the local increase of the amount of magnesium in solution. Previously deposited calcite is modified by the movement through it of these magnesium-rich solutions. Field and laboratory observations show that in shallow-water intertidal zones evaporation of seawater may cause precipitation of calcium-bearing deposits. The waters may be increased by an order of magnitude in their content of magnesium relative to calcite. Such high-magnesium-content waters may then circulate through underlying calcite deposits, replacing some of the calcium with magnesium and thus converting limestone to dolostone. In supratidal zones, just above high tide, thin crusts of primary dolomite crystals form in tropical and subtropical regions today. Such dolostones have been found in the Bahamas and in the Florida Keys.

Evaporites   An evaporite is a sedimentary rock composed of minerals that were precipitated from solution with the evaporation of the liquid in which they were dissolved. Rock salt (composed of the mineral halite) and gypsum are the most abundant evaporites. Anhydrite (from the Greek *anhydros,* "waterless") is an evaporite composed of the mineral of the same name, which is simply gypsum without its crystallization water, $CaSO_4$. Most evaporite deposits seem to have been precipitated from seawater according to a definite sequence. The less highly soluble minerals are the first to drop out of solution. Thus gypsum and anhydrite, both less soluble than halite, are deposited first. Then, as evaporation progresses, the more soluble halite is precipitated.

In the United States the most extensive deposits of evaporites are found in Texas, Louisiana, and New Mexico. Here gypsum, anhydrite, and rock salt make up over 90 percent of the Castile Formation, which has a maximum thickness of nearly 1,200 m. In central New York State there are thick deposits of rock salt; and in central Michigan there are layers of rock salt and gypsum. Some evaporite deposits are mined for their mineral content, and in certain areas, particularly in the Gulf Coast states, deposits of rock salt have pushed upward toward the surface to form salt domes producing commercially important reservoirs of petroleum (see Chapter 18).

Silica   Silica occurs in sedimentary rocks in a variety of forms other than the typical clastic grains of quartz. It is a chemical constituent of many rocks, precipitated as opal, a hydrous silica, $SiO_2 \cdot nH_2O$. Opal is slightly softer than common quartz and has no specific crystal structure.

Silica also occurs in sedimentary rocks in a form called **cryptocrystalline.** This term (from the Greek *kryptos,* "hidden," and "crystalline") indicates crystalline structure so fine that it cannot be seen under most ordinary microscopes. The microscope does reveal, however, that some cryptocrystalline silica has a granular pattern and that some has a fibrous pattern. To the naked eye the surface of the granular form is somewhat duller than that of the fibrous form. Among the dull-surfaced, or granular, varieties is flint, usually dark in color. Flint is commonly found in certain limestone beds—the chalk beds of southern England, for example. **Chert** is the general term for these rocks. Jasper is a red variety of granular cryptocrystalline silica. The general term *chalcedony* is often applied to the fibrous types of cryptocrystalline silica, which have a higher, more waxy luster than the granular varieties. Sometimes the term is also used to describe a specific variety of brown translucent cryptocrystalline silica. Agate is a variegated form of silica, its bands of chalcedony alternating with bands of either opal or some variety of granular cryptocrystalline silica, such as jasper (Figure 6.8).

6.8 Agate showing banded character of rock. Maximum diameter is 15 cm. [John Simpson.]

BIOCHEMICAL ROCKS

A significant quantity of sedimentary rocks is the direct result of the accumulation of shells, plant fragments, and other remains of organisms. Together these are called **biochemical** rocks. For example, diatoms are microscopic, single-celled plants that grow in marine or fresh water and secrete siliceous shells. When these shells accumulate in great numbers on the bottom of a basin, they may ultimately form a sedimentary rock called **diatomite.**

Limestone   Biochemically formed limestones are created by the action of plants and animals that extract calcium carbonate from the water in which they live. The calcium carbonate may be either incorporated into the skeleton of the organism or precipitated directly. In any event, when the organism dies, it leaves behind a quantity of calcium carbonate, and over a long period of time thick deposits of this material may be built up. Reefs, ancient and modern, are well-known examples of such accumulations. The most important builders of modern reefs are algae, mollusks, corals, and one-celled animals, the same animals whose ancestors built up the reefs of ancient seas—the reefs, now old and deeply buried, that are often valuable reservoirs of petroleum.

There are virtual limestone "factories" in tropical regions like the Bahamas, the Florida Keys and Bay, Shark Bay in western Australia, parts of the East and West Indies—all shallow-water continental shelf regions dominated by lime muds and lacking in terrigenous (land-derived) detritus. All these shelves have in common a warm climate and warm-surface ocean water that appears to be supersaturated with $CaCO_3$. In these regions organisms live and thrive, especially those that are particularly efficient in fixing large amounts of $CaCO_3$ in shells or other skeletal parts.

Carbonate-utilizing algae grow in abundance on shallow shelves, contributing enormous quantities of sand- and silt-sized carbonate sediments when they die and their stalks and branching arms disaggregate. It has been estimated that several centimeters of carbonate sediments can build up over a few hundred years. No wonder such regions have been called **carbonate factories.**

Deep-sea oozes commonly contain microscopic calcareous shells of pelagic organisms (free-swimmers and floaters). These calcareous oozes occur over wide regions of the oceans, except in the North Pacific, Arctic, and Antarctic. They accumulate slowly in water with maximum depths of 3 to 5 km. In deeper, colder waters these carbonate shells dissolve and are not found as deposits on the sea floor. (The depth at which carbonates disappear from bottom sediments is called the **carbonate compensation depth.**)

Freshwater carbonates occur in lakes of temperate regions. Here again, organisms play a role in the accumulation, concentration, and deposition of calcium carbonate. Freshwater limestones commonly have fairly large admixtures of fine clastic detritus, causing them to be calcareous muds (marls) or muddy limestones.

Chalk is made up in part of biochemically derived calcite in the form of the skeletal fragments of microscopic oceanic plants and animals. These organic remains are found mixed with very fine-grained calcite deposits of either biochemical or inorganic chemical origin. A much coarser type of limestone composed of organic remains is known as **coquina** (from the Spanish for "shellfish" or "cockle") and is characterized by the accumulation of many large fragments of shells.

Coal   Coal is a rock composed of combustible matter derived from the partial decomposition of plants. We shall consider coal as a biochemically formed sedimentary rock, although some geologists prefer to think of it as a metamorphic rock because it passes through various stages.

The process of coal formation begins with an accumulation of plant remains in a swamp. This accumulation is known as **peat,** a soft, spongy, brownish deposit in which plant structures are easily recognizable. Time, coupled with the pressure produced by deep burial and sometimes by Earth movement, gradually transforms the organic matter into coal. During this process the percentage of carbon increases as the volatile hydrocarbons and water are forced out of the deposit. Coals are ranked according to the percentage of carbon they contain. Peat, with the least amount of carbon, is the lowest ranking; then come lignite, or brown coal; bituminous, or soft coal; and finally anthracite, or hard coal, which has the highest percentage of carbon of all the coals (see Section 18.2).

RELATIVE ABUNDANCE OF SEDIMENTARY ROCKS

Sandstone, mudstone and shale, and limestone constitute about 99 percent of all sedimentary rocks. Of these, mudstone and shale are the most abundant. On the basis of extrapolation from measurements made in the field, the estimates of the percentages of mudstone and shale approximate 50 percent of all sedimentary rocks. Similar calculations for limestone and sandstone suggest that the limestone forms about 22 percent of these rock types and that sandstone accounts for the remaining 28 percent. These percentages, however, do not agree with theoretical determinations of relative abundances. They are based on the determination of the products to be expected from the weathering of an average igneous rock. If these weathering products are assigned to the three major sedimentary rock types, then we find that shale should be considerably more important volumetrically than it appears to be on the basis of field measurements. On such theoretical grounds, mudstone and shale constitute approximately 75 percent of the three major sedimentary rock types. Sandstone and limestone are approximately of equivalent volume and together constitute the other 25 percent. The discrepancy between the two estimates has not yet been resolved, though it is partly due to the loss of mud to the deep sea, where it is only rarely observed again in the sedimentary rock record.

TABLE 6.3
Average Composition of All Sediments (as Oxides)[a]

Oxide	Wt. %
SiO_2	44.5
TiO_2	0.6
Al_2O_3	10.9
Fe_2O_3	4.0
FeO	0.9
MnO	0.3
MgO	2.6
CaO	19.7
Na_2O	1.1
K_2O	1.9
P_2O_5	0.1
CO_2	13.4
Total	100.0

[a] From Arie Poldervaart, "Chemistry of the Earth's Crust," *Geol. Soc. Am. Spec. Paper*, no. 62, p. 132, 1955.

Estimates have been made of the average chemical composition of the world's sediments. One of these estimates is presented in Table 6.3.

6.3
FEATURES OF SEDIMENTARY ROCKS

We have mentioned that the stratification, or bedding, of sedimentary rocks is their single most characteristic feature. Now we shall look more closely at this feature, along with certain other characteristics of sedimentary rocks, including mud cracks and ripple marks, nodules, concretions, geodes, fossils, and color.

BEDDING

The beds, or layers, of sedimentary rocks are separated by **bedding planes,** along which the rocks tend to separate or break (see Figure 6.9). The varying thickness of the layers in a given sedimentary rock reflects the changing conditions that prevailed when each deposit was laid down. In general, each bedding plane marks the termination of one deposit and the beginning of another. Within each bed the texture and composition tend to be approximately uniform. Across bedding planes, however, something changes — the texture decreases from sand size to mud size or the composition goes from quartzose sandstone to limestone, for example.

As an illustration, let us imagine a bay of an ocean into which rivers normally carry fine silt from the nearby land. This silt settles out from the seawater to form a layer or bed of mud. Now heavy rains or melting snows may cause the river suddenly to flood and thereby pick up coarser material, such as sand, from the river bed. This material will be carried along and dumped into the bay. There it settles to the bottom and blankets the silt that was deposited earlier. The plane of contact between the silt and the sand represents a bedding plane. If, later on, the silt and the sand are lithified into siltstone and sandstone, the bedding plane persists in the sedimentary rock. In fact, it marks a plane of weakness along which the rock tends to break.

Bedding planes are usually horizontal, and, when parallel, the bedding is called **parallel bedding.** Closely spaced parallel bedding planes form **laminated bedding.** But some beds are laid down at an angle to the horizontal, and such bedding is variously called **cross bedding,** or **false bedding** (see Figure 6.10). Such nonhorizontal bedding can occur in several situations. For example, the bedding in sand dunes may have high angles on the leeward side of the dune (see Chapter 17). The deposits laid down at the growing edge of a delta may be inclined from 5° to 30°, and such beds are usually given a special name, **foreset beds** (see Chapter 13).

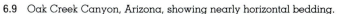

6.9 Oak Creek Canyon, Arizona, showing nearly horizontal bedding.

(a)

(b)

6.10 (a) Cross bedding in the Navajo Sandstone, Zion National Park, Utah. Individual beds are up to 50 cm in thickness but the packages of similarly inclined beds measure up to 5 m from top to bottom. These wind-deposited beds represent ancient sand dunes. [Robert Key.] (b) Detailed look at the cross-bedding. [Robert Key.]

In flowing water scouring of the stream floor by turbulent water may create small depressions in the channel deposits, which later will be refilled by inclined beds. Alternatively, cross bedding can occur as flowing water passes to somewhat quieter conditions and drops its load in a way similar to that in which a stream builds a foreset bed in a delta.

When the particles in a sedimentary bed vary from coarse at the bottom to fine at the top, the bedding is said to be **graded.** Such bedding is characteristic of rapid deposition from a water mass turbid with sedimentary particles of differing sizes. The largest particles settle most rapidly, and the finest most slowly. Some graded beds form **turbidites,** a

term from the Latin *turbidus* for "disturbed," referring to the stirring up of sediments and water (see also Chapter 16).

Bedding may be distorted before the material becomes consolidated into firm rock (see Figure 6.11). Thus unconsolidated sediments may sometimes slump or flow. In addition, animals may burrow or tunnel through unconsolidated layers to produce **disturbed,** or **mottled, bedding,** or, in extreme cases, to destroy all vestiges of the original bedding.

6.11 Slump structure in laminated marly limestone beds cored 916 m below the ocean floor on Deep Sea Drilling Project, Leg 40, Site 364, in the Angola Basin. Core width is 6 cm. [Courtesy Deep Sea Drilling Project.]

RIPPLE MARKS, MUD CRACKS, AND SOLE MARKS

Ripple marks are the little waves of sand that commonly develop on the surface of a sand dune, along a beach, or on the bottom of a stream. **Mud cracks** are familiar on the dried surface of mud left exposed by the subsiding waters of a river or pond. **Sole marks** are the casts, or filling, of a primary sedimentary structure such as a groove or track found on the underside or lower surface of a siltstone or sandstone that had overlain such a structure in soft underlying sediment.

Ripple marks preserved in sedimentary rocks often furnish clues to the conditions that prevailed when a sediment was originally deposited. For instance, if the ripple marks are symmetric, with sharp or slightly rounded ridges separated by more gently rounded troughs, we are fairly safe in assuming that they were formed by the back-and-forth movement of water such as we find along a seacoast outside the surf zone. These marks are called **ripple marks of oscillation.** If, on the other hand, the ripple marks are asymmetric, we can assume that they were formed by air or water moving more or less continuously in one direction. These marks are called **current ripple marks** (Figure 6.12).

6.12 (a) Modern current ripple marks from Barnegat Bay, New Jersey. (b) Ancient current ripples preserved in a Permian sandstone. The color is due to oxidation of iron-bearing minerals. [Franklyn B. Van Houten.]

(a)

(b)

(a)

(b)

6.13 (a) Polygonal pattern of modern mud cracks developed when fine-grained sediments of a temporary lake dried, shrank, and cracked. (b) Mud cracks preserved in limy shale at Rheems, Pennsylvania, in the Ordovician Beekmantown Formation.

Mud cracks form when a deposit of silt or clay dries out and shrinks (see Figure 6.13). The cracks outline roughly polygonal areas, making the surface of the deposit look like a section cut through a large honeycomb. Eventually another deposit may come along to bury the first. If the deposits are later lithified, the outlines of the cracks may be accurately preserved for millions of years. Then, when the rock is split along the bedding plane between the two deposits, the cracks will be found much as they appeared when they were first formed, providing evidence that the original deposit underwent alternate flooding and drying.

Sole marks develop when a groove or trail is formed on the surface of soft sediment by a stick or other object carried along in the current. Under ideal conditions this depression will be preserved and filled by silt or sand. When lithified, this sequence of sedimentary beds can be used to decipher current directions, tops and bottoms of strata, and conditions of the environment.

NODULES, CONCRETIONS, AND GEODES

Many sedimentary rocks contain structures that were formed only *after* the original sediment was deposited. Among these are nodules, concretions, and geodes.

A **nodule** is an irregular, knobby-surfaced body of mineral matter that differs in composition from the sedimentary rock in which it has formed. It usually lies parallel to the bedding planes of the enclosing rock. Some adjoining nodules coalesce to form a nearly continuous bed. Nodules average about 30 cm in maximum dimension. Most nodules are thought to have formed when some mineral matter replaced some of the materials of the original deposit; some, however, may consist of silica that was deposited at the same time the main beds were laid down.

A **concretion** is a depositional concentration, sometimes with concentric layering. Concretions range in size from a fraction of a centimeter to a meter or more in maximum dimension (see Figure 6.14). Most are shaped like simple spheres or disks, although some have fantastic and complex forms. When the cementing material entered the unconsolidated sediment, it tended to concentrate around a common center point or along a common center line. The particles of the resulting concretion are cemented together more firmly than the particles of the host rock that surrounds it. The cementing material usually consists of calcite, dolomite, iron oxide, or silica—in other words, the same cementing materials that we find in the sedimentary rocks themselves.

Geodes, more eye-catching than either concretions or nodules, are roughly spherical hollow structures up to 30 cm or more in diameter. In some siliceous geodes an outer layer of chalcedony is lined with crystals that project inward toward the hollow center. The crystals, often perfectly formed, are usually quartz, although crystals of calcite and dolomite have been found and, more rarely, crystals of other minerals. Geodes occur most commonly in limestone but they also occur in shale.

How does a geode form? First, a water-filled pocket develops in a sedimentary deposit, probably as a result of the decay of some plant or animal that was buried in the sediments. As the deposit begins to consolidate into a sedimentary rock, a wall of silica with a jellylike consistency forms around the water, isolating it from the surrounding material. As time passes, fresh water may enter the sediments. The water inside the pocket has a higher salt concentration than does the water outside. To equalize the concentrations, there is a slow mixing of the two liquids through the silica wall or membrane that separates them. This process of mixing is called **osmosis.** As long as the osmotic action continues, pressure is exerted outward toward the surrounding rock. The original pocket expands bit by bit until the salt concentrations of the liquids inside and outside are equalized. At this point osmosis stops, the outward pressure ceases, and the pocket stops growing. Now the silica wall dries, crystallizes to form chalcedony, contracts, and cracks. If, at some later time, mineral-bearing water finds its way into the deposit, it may seep in through the cracks in the wall of chalcedony. There the minerals are precipitated, and crystals begin to grow inward, toward the center, from the interior walls. Finally, we have a crystal-lined geode imbedded in the surrounding rock.

Other geodes are known to have occurred in a very different way. Replacement of evaporite (anhydrite) nodules by silica produces silicified geodes. The evaporite nodules can occur under conditions of extreme imbalance between precipitation and evaporation. After they have formed, silica-rich fluids may come into contact with them and result in the silicification as molecule-by-molecule replacement.

FOSSILS

The word **fossil** (derived from the Latin *fodere,* "to dig up") originally referred to anything that was dug from the ground, particularly a mineral or some inexplicable form. It is still used in that sense occasionally, as in the term **fossil fuel** (see Chapter 18). But today fossil generally means any direct evidence of past life—for example, the bones of a dinosaur, the shell of an ancient clam, the footprints of a long-extinct animal, or the delicate impression of a leaf (see Figures 6.15 and 6.16).

Fossils are usually found in sedimentary rocks, although they sometimes turn up in igneous and metamorphic rocks. They are most abundant in mudstone, shale, and limestone, but are also found in sandstone, dolomite, and conglomerate. Fossils account for almost the entire volume of certain rocks, such as the coquina and limestones that have been formed from ancient reefs.

The remains of plants and animals are completely destroyed if they are left exposed on the Earth's surface; but if they are somehow protected from destructive forces, they may become incorporated in a sedimentary deposit, where they will be preserved for millions of years. In the quiet water of the ocean, for example, the remains of starfish, snails, and fish may be buried by sediments as they settle slowly to the bottom. If these sediments are subsequently lithified, the remains are preserved as fossils that tell us about the sort of life that existed at the time when the sediments were laid down.

Fossils are also preserved in deposits that have settled out of fresh water. Countless remains of land animals, large and small, have been dug from the beds of such once-watery environments as extinct lakes, floodplains, and swamps.

(a)

(b)

6.14 Giant concretions from Eocene beds along east coast of New Zealand, 20 km north of Dunedin; locally known as *Moeraki boulders.* Largest concretion approximately 2 m in diameter. Note yellow-brown calcite filling expansion cracks near nucleus of concretion.

In order to have a fossil, the original shell or plant fragment does not have to be preserved, nor even the mold or cast left by that animal or plant. We can discern much about these forms of life merely from tracks, trails, or burrows left behind as the organism walked, crawled, foraged, or browsed through the sediment. *All* these evidences of past life are called *trace fossils,* the study of which forms a subdiscipline within paleontology. The detailed story of the development of life as recorded by fossils is properly a part of historical geology, and although we do not have the time to trace it here, in Chapter 8 we note that fossils are very useful in subdividing geologic time and constructing the geologic column.

Fossils are also useful in telling us about the environment of deposition. The remains of certain kinds of animals and plants indicate specific conditions and can provide clues to the environment in which the sediments enclosing those fossils were deposited. In addition, the orientation and method of preservation may tell us much about how the organisms lived.

6.15a Fossil clam, *Pecten*-like form. [Smithsonian.]

6.15c Fossil oyster, *Trigonia.* [Smithsonian.]

6.15b Fossil oyster, *Exogyra*. [Smithsonian.]

6.15d Cambrian olenellid trilobite. [Smithsonian.]

6.15 Some different types of fossils. [Smithsonian.] (*a*) Fossil bivalve (*Pecten*-like form), 2.5 cm in diameter. (*b*) Fossil bivalve (*Exogyra*), maximum dimension 11 cm. (*c*) Fossil bivalve *(Trigonia),* long dimension 6.5 cm. (*d*) Trilobite *(Paradoxides hicksi)*, 530 million years old, from rocks of Middle Cambrian age, Newfoundland; fossil is 12 cm long. (*e*) Slab of rock with trilobites *(Phacops rana)*, 365 million years old, from rocks of Middle Devonian age, Ontario; trilobites average 4 cm in length. (*f*) Fossil bird-reptile link (*Archaeopteryx*); fossil is 30 cm long. (*g*) Fernlike plants from rocks of Pennsylvanian age; slab of rock is 25 cm long. (*h*) Fossil leaf-ferns from Mazon Creek locality, Illinois; specimen is 15 cm long.

6.15e Slab of trilobites; Middle Devonian *Phacops rana.* [Smithsonian.]

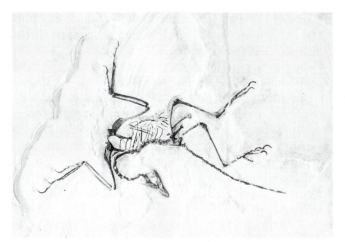

6.15f Fossil of reptile-bird link, *Archaeopteryx*. [Smithsonian.]

6.15g Pennsylvanian fern-like plants, central Pennsylvania. [Smithsonian.]

6.15h Fossil leaf-ferns in concretion from Mazon Creek, Illinois. [Smithsonian.]

6.16 Animal footprints in the Coconino Sandstone of Aubrey Cliffs, Coconino County, Arizona. Ruler for scale. [Edwin D. McKee.]

COLOR OF SEDIMENTARY ROCKS

Throughout the western and southwestern areas of the United States bare cliffs and steep-walled canyons provide a brilliant display of the great variety of colors exhibited by sedimentary rocks. The Grand Canyon of the Colorado River in Arizona cuts through rocks that vary in color from gray, through purple and red, to brown, buff, and green. Bryce Canyon in southern Utah is fashioned of rocks tinted a delicate pink, and the Painted Desert, farther south in Arizona, exhibits a wide range of colors, including red, gray, purple, and pink.

We commonly use colors to make first assumptions regarding the physical environment of sedimentary rocks. For example, most rocks that are red had their origin as continental beds—river, floodplain, or delta environments. Black sediments commonly indicate reducing environments, such as lagoons or deep basins, low in oxygenated waters. Such generalizations may prove false in some cases, but they are good first approximations.

The most important sources of color in sedimentary rocks are the iron oxides. Hematite, Fe_2O_3, for example, gives rocks a red or pink color, and limonite or goethite produces tones of yellow and brown. Some of the green, purple, and black colors may be caused by iron, though in exactly what form is not completely understood. Only a very small amount of iron oxide is needed to color a rock. In fact, few sedimentary rocks contain more than 6 percent of iron, and most contain very much less.

Organic matter, when present, may also contribute to the coloring of sedimentary rocks, usually making them gray to black. Generally, but not always, the higher the organic content, the darker the sedimentary rock.

The size of the individual particles in a rock also influences the color, or at least the intensity of the color. For example, fine-grained clastic rocks are usually somewhat darker than coarse-grained rocks of the same mineral composition.

SEDIMENTARY FACIES

If we examine the environments of deposition that exist at any one time over a wide area, we find that they differ from place to place. Thus the freshwater environment of a river changes to a brackish-water environment as the river nears the ocean. In the ocean itself marine conditions prevail. But even in the ocean the marine environment changes—from shallow water to deep water, for example. And as the environment changes, the nature of the sediments that are laid down also changes. The deposits in one environment show characteristics that are different from the characteristics of deposits laid down at the same time in another environment. This change in the "look" of the sediments is called a change in **sedimentary facies,** the latter word derived from the Latin for the notion of "aspect" or "form."

We may define sedimentary facies as an accumulation of deposits that exhibits specific characteristics, reflecting a particular depositional environment, and grades laterally into other sedimentary accumulations formed at the same time but exhibiting different characteristics. The concept of facies is widely used in studying sedimentary rocks and the conditions that gave rise to metamorphic rocks (see Chapter 7). The concept is generally not used in referring to igneous rocks.

Let us consider a specific example of facies. Figure 6.17 shows a coastline where rivers from the land empty into a lagoon. The lagoon is separated from the open ocean by a sandbar. The fine silts and clays dumped into the quiet waters of the lagoon settle to the bottom as a layer of mud. At the same time, waves are eroding coarse sand from a nearby headland outside the lagoon. This sand is transported by currents and waves and deposited as a sandy layer seaward of the sandbar. Different environments exist inside and outside the lagoon; therefore different deposits are being laid down simultaneously. Notice that the mud and the sand grade into each other along the sandbar. Now imagine that these deposits have been consolidated into rock and then exposed to view at the Earth's surface. We should find a shale layer grading into sandstone—that is, one sedimentary facies grading into another.

But the picture is not always so simple. Recent sediments range from sand, through a sandy mud, to mud, and in a few areas there are limy deposits. Where the sea floor is rocky, little or no recent sedimentation has taken place.

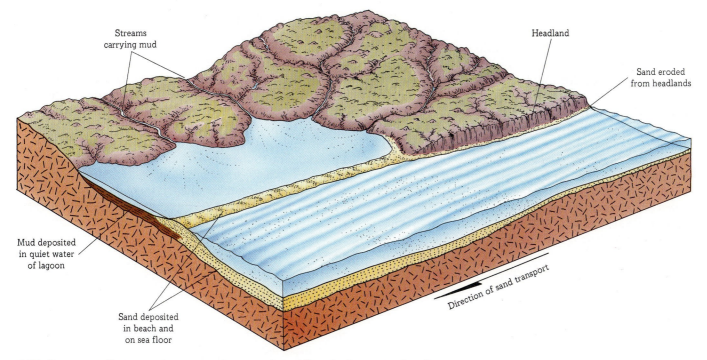

Streams
carrying mud

Headland

Sand eroded
from headlands

Mud deposited
in quiet water
of lagoon

Sand deposited
in beach and
on sea floor

Direction of sand transport

6.17 Diagram to illustrate a change in sedimentary facies. Here the fine-grained muds are deposited in a lagoon close to shore. A sandbar separates them from sand deposits farther away from shore. The sand in this instance has been derived from a sea cliff and transported by waves and currents.

Should the soft sediments become lithified, a sedimentary rock ranging from sandstone, through sandy shale, to shale and limestone would result. Ancient sedimentary deposits show exactly this kind of variation in facies.

6.4

PLATE TECTONICS AND SEDIMENTATION

As noted in Chapter 1, we are just beginning to understand the crustal motions of the colossal portions of the lithosphere known as **plates**. Areas exist on the Earth's crust where sediments accumulate to great thicknesses. Many of these regions are ultimately pushed up into mountains, developing most commonly along the margins of the plates.

Thick accumulations of sediments may occur either where there is a deep basin into which the sediments are dumped or where the crust subsides as more and more sediments are deposited in relatively shallow water. We have evidence of both in the geologic record, but the overwhelming bulk of sedimentary rocks found in mountain ranges apparently formed originally as shallow-water sediments, as proved by such indicators as certain kinds of ripple marks, mud cracks, and shallow-water organisms.

As continents are raised, the character of the resulting sediments changes markedly. The size of clasts accumulating at a given location increases and the lithologies change as newly exposed rock units supply sediments to the basin of deposition (Figure 6.18).

GEOSYNCLINES

Following the work of James Hall, a nineteenth-century pioneer in American geology, Marshall Kay, J. D. Dana, and other twentieth-century proponents of the idea, the concept of geosynclines developed into a dominant theme

6.18 St. Peter Sandstone; beach sand to sandstone. In the early part of Middle Ordovician time, about 475 million years ago, eastern North America was flooded gradually by a shallow sea, which moved toward the northwest (shorelines 1 to 5). Aided by currents that brought sand from the northeast (arrows), the sea spread a thin sheet of quartz sand across the region. This sheet became the St. Peter Sandstone of today—a pure and practically limitless deposit of silica sand. [Robert L. Bates and Julia A. Jackson, *Our Modern Stone Age*, William Kaufmann, Inc., p. 40, 1982.]

in the interpretation of many mountain belts. The concept held that folded rocks in mountain ranges of the world apparently developed from thick deposits of sediments that accumulated in these geosynclines — literally, "Earth syncline," or large-scale downwarp in the sedimentary rocks found in the Earth's crust. A geosyncline is thus a large sediment-filled elongate basin. (We examine the problem of deformation of these sedimentary rocks and the actual formation of mountains in Chapters 9 and 11.)

Parallel adjacent belts characterize geosynclines. These have been called **miogeosyncline** (or simply **miogeocline** by Robert Dietz and others) where sediments are deposited on continental basement and where volcanic rocks and deeper-water sediments are lacking; and **eugeosyncline** (or **eugeocline**) where sediments are deposited on noncontinental basement and where there is an abundance of interstratified volcanic rocks and deeper-water sediments.

PLATE BOUNDARIES AND SEDIMENTATION

As discussed in Chapter 1, there are three basic types of plate boundaries: **convergent,** where old lithosphere is carried downward along subduction zones and magma rises to form volcanic regions, as the west coast of South America or the Aleutians; **divergent,** where old lithosphere pulls apart and new lithosphere forms along midoceanic spreading centers, as the Midatlantic Ridge; and **transform,** or **parallel,** where two plates slide past one another with no new lithosphere forming and no old lithosphere being destroyed, as along the San Andreas fault.

In relating sedimentation to plate tectonic theory, those holding to the geosynclinal concept might interpret the miogeosynclines as representing thick sediment wedges forming on divergent continental margins and the eugeosynclines as representing sediments deposited farther offshore or in areas marked by volcanic island arcs, perhaps at a convergent plate boundary. Transform margins are marked by numerous basins that are later filled with thick sequences of sediments similar to those found along the San Andreas fault zone.

The relative motion of the plates on or between which the sediments have accumulated will determine the ultimate structure of the mountain systems that later form from these sediments and sedimentary rocks (see Chapter 11).

Summary

Sedimentary rocks cover about 75 percent of the Earth's surface and make up about 5 percent by volume of the outer 10 km of the solid Earth.

Formation of sedimentary rocks takes place at or near the Earth's surface.
   Detrital material worn from the landmasses and **chemical** deposits precipitated from solution are the two chief types of sediments.
   Sedimentation is the process by which rock-forming materials are laid down; the resulting deposits vary with the source of material, the methods of transportation, the processes of deposition, and the environment of deposition.
   Clay, quartz, calcite and some **feldspars** are the most common minerals in sedimentary rocks. Other minerals include dolomite, goethite, hematite, limonite, mica, halite, and gypsum.
   Texture depends on the size, shape, and arrangement of the particles of a rock. Texture may be clastic or nonclastic.
   Diagenesis includes all the physical, chemical, and biological changes that occur after burial but prior to or immediately following lithification.
   Lithification converts unconsolidated sediments to firm rock by cementation, compaction, desiccation, crystallization.

Types of sedimentary rocks include detrital, chemical, and biochemical forms.
   Detrital rocks include conglomerate, sandstone, siltstone, and mudrock.
   Chemical rocks include limestone, dolostone, and evaporites.
   Biochemical rocks include chalk, coquina, diatomite, coal, and some limestones.
   Most abundant are shale and mudstone, sandstone, and limestone, in that order. They form 99 percent of the sedimentary rock family.

Features of sedimentary rocks include bedding, ripple marks, mud cracks, nodules, concretions, geodes, and fossils.
   Color of sedimentary rocks is due largely to small amounts of the iron oxide minerals and, less importantly, to organic matter.
   Sedimentary facies refer to an accumulation of deposits that exhibits specific characteristics and that grades laterally into other accumulations formed at the same time but showing different characteristics.

Sedimentation is closely related to **plate boundaries,** with thick accumulations along some margins.

Questions

1. Stratification is the single most characteristic feature of sedimentary rocks. How does it form?
2. How do detrital sedimentary rocks differ from chemical rocks?
3. What are the most common mineral components of the majority of all sedimentary rocks?
4. How does iron affect the color of sedimentary rocks?
5. Why do some very ancient sediments never undergo lithification, whereas some very recent sediments are lithified almost overnight?
6. Where are limestones forming today?
7. Where are many evaporites forming today?
8. Why do we consider coal to be a sedimentary rock?

What is the principal component of that rock?
9. When we examine the sedimentary rocks at the Earth's surface, we find three principal kinds. What are these three rocks and why do you suppose they are so abundant?
10. Under what circumstances will sediments accumulate in a condition other than horizontal?
11. What is meant by the term *sedimentary facies?* Give one example, other than those given in the text, where we might get several sedimentary facies adjacent to one another.
12. How is it possible to get great thicknesses of sediments, all of which bear evidence of being accumulated in very shallow water?

Supplementary Readings

ALLEN, J. R. L. *Principles of Physical Sedimentology,* Allen & Unwin, London, 1985.
   Written for geology majors, it gives an overview of sedimentary processes with field and laboratory experiments.

BLATT, H., G. MIDDLETON, AND R. MURRAY *Origin of Sedimentary Rocks,* 3rd ed., Prentice Hall, Englewood Cliffs, N.J., 1986.
   A modern text that emphasizes physical and chemical mechanisms and processes of sedimentation; good treatment of facies and environments of deposition.

FRASER, G. S. *Stratigraphic Evolution of Clastic Depositional Sequences,* Prentice Hall, Englewood Cliffs, N.J., 1989.
   Intermediate-level study showing how geomorphic and sedimentologic processes act together during the formation of sedimentary rock sequences.

GREENSMITH, J. T. *Petrology of the Sedimentary Rocks.* Allen & Unwin, London, 1980.
   A modern treatment of sedimentary rocks in hand samples, in
thin-sections, and as representatives of specific origins and environments of deposition.

LAPORTE, LEO F. *Ancient Environments,* Prentice Hall, Englewood Cliffs, N.J., 1968.
   An excellent treatment of the way sedimentary materials can be used to determine environments.

LINDHOLM, R. *A Practical Approach to Sedimentology,* Allen & Unwin, London, 1987.
   This field-oriented book bridges the gap between the principles and processes of sedimentology and the practical methods used to collect and evaluate sedimentological data.

REINECK, H. E., AND I. B. SINGH *Depositional Sedimentary Environments,* 2nd ed., Springer-Verlag, New York, 1980.
   Profusely illustrated textbook on modern environmental studies, with reference to examples of ancient environments as found in the stratigraphic record.

Gneiss along track to Franz Josef Glacier, west coast of South Island, New Zealand.

METAMORPHISM AND METAMORPHIC ROCKS

"Observe always that everything is the result of a change. . . ."

Marcus Aurelius, Meditations, *vol. IV, p. 36*

In earlier chapters we found that minerals tend to be in equilibrium with their environments. We saw (Chapter 3) that magmas cooled and different minerals formed at temperatures that varied with their specific freezing points—the temperatures at which these minerals were in equilibrium with their environments. When uplifted into the realm of weathering, they were no longer in equilibrium and began to change (Chapter 5). During sedimentation minerals deep in the accumulating pile of sediments started to undergo diagenetic changes (Chapter 6). *Metamorphism*, the subject of the present chapter, is a logical extension of this pattern. It is a process whereby rocks undergo physical or chemical changes or both in order to achieve equilibrium with conditions other than those under which the rocks were originally formed.

The primary agents of metamorphism are *heat, pressure,* and *fluids.* Metamorphism takes place locally as *contact metamorphism* where magma comes into contact with preexisting rock. *Regional metamorphism* is usually associated with mountain building and extends over thousands of square kilometers.

Metamorphic rocks are classified on the basis of *texture* as either *foliated* or *unfoliated.* Mineral composition is useful (and used), but of secondary importance. Present-day metamorphism occurs along plate boundaries, and, not unexpectedly, different types of metamorphism and metamorphic products develop along the different types of plate boundaries. Finally, we look at the process called *granitization* and at another called *metasomatism,* both of which are extreme forms of the processes that produce metamorphism.

7.1

METAMORPHIC PROCESSES AND METAMORPHIC CHANGES

Metamorphism is defined as a set of processes, involving the application of heat and (usually) pressure and commonly chemical fluids, in which rocks undergo a change in mineralogy, texture, or both (Figure 7.1).

Metamorphism occurs in response to pronounced changes in temperature, pressure, and chemical environment within the Earth's crust, below the zone of weathering and cementation but under conditions insufficient to produce melting. In this environment rocks undergo chemical and structural changes to adjust to conditions different from those under which they were originally formed. Because metamorphism normally takes place within the crust, it cannot be directly studied, as can weathering, sedimentation, and volcanic activity. Nevertheless, laboratory simulations of temperature and pressure conditions permit us to examine changes that occur during the processes of metamorphism. An early experiment involving high temperature and pressure is described in Box 7.1.

HEAT

Heat may be the most essential agent of metamorphism. Some geologists question whether pressure alone could produce changes in rocks without a simultaneous increase in

7.1 Banded gneiss (metamorphic rock) showing alignment of previously unoriented minerals. The light-colored minerals are mainly quartz and potassium feldspar; the dark streaks are biotite and other ferromagnesian minerals. The bulk composition is that of granite, but in contrast to the random mixing in granite, the minerals here are distributed in relatively systematic patterns.

BOX 7.1   Marble from a Gun

The first laboratory experiments to simulate the Earth's subsurface conditions were conducted by Sir James Hall (1761–1832), a Scottish chemist and geologist who was a younger colleague of James Hutton.

Hutton's theory of the Earth held that heat had acted on all rocks at some past time and turned the original materials into new forms—that is, metamorphosed them. This troubled Hall, who reported that he had "almost daily warfare with Dr. Hutton on the subject of his theory." Hall thought, for example, that "the fate of any geological theory must depend upon its successful application to the various conditions" of "Carbonate of Lime" because its "history is . . . interwoven

in such a manner with that of the mineral kingdom at large." Specifically, it was general knowledge that heat turned calcium carbonate ($CaCO_3$), the predominant mineral in limestone and marble, to a powdery product we know as lime (CaO), which bore little resemblance to firm rock.

But Hutton had visualized heat operating at depth and under the confining pressure of overlying rock masses. Once Hall realized that, he viewed Hutton's "fundamental principles with less and less repugnance."

In 1801, four years after Hutton's death, Hall began a series of experiments that were effective, although, by

today's standards, certainly crude. He introduced powdered carbonate into the breech of a gun, and then sealed both the barrel and the touch-hole of the breech. He then stuck the breech into a furnace and heated it. Thus the calcium carbonate was heated under pressure. In several tries the breech of the gun yielded to the stress. In others, however, the powdered limestone turned to a solid. In one, a small amount of the powder "was found to be completely crystallized, having acquired the rhomboidal fracture of calcareous spar."

Hall had created marble from powdered limestone under the two conditions necessary for metamorphism —heat and pressure.

---

temperature. In fact, they maintain that metamorphism is invariably controlled by temperature.

Heat aids in recrystallization and dehydration. Thermal energy breaks chemical bonds and increases the rates of reactions in general. Heat, by itself or in conjunction with the fluids released during heating, increases mobility in rocks. Thus new minerals, and commonly those with simplified mineralogy, are formed, as well as rocks with larger crystals and coarser textures.

PRESSURE

Under the influence of **pressure,** changes take place to reduce the space occupied by the mineral components of a rock mass or to reorient the minerals. Pressure may produce a closer atomic packing of the elements in a mineral, recrystallization of the mineral, or formation of new minerals.

When rocks are buried to depths of several kilometers, they gradually become plastic and responsive to the heat and deforming forces that are active in the Earth's crust and mantle. When plastic, they deform by intergranular motion, by the formation of minute shear planes within the rock, by changes in texture, by reorientation of grains, and by crystal growth. The type of the original rock has an important effect on the results achieved by burial and deformation.

We should differentiate two kinds of pressure that are

effective on minerals and rocks. If pressure is equidimensional (for example, where hydrostatic forces act on a body under water), the most likely changes will be reduction in volume so that minerals of greater density will be more at equilibrium. We call this kind of pressure **lithostatic,** or **uniform pressure.** When differential pressures act on buried minerals and rocks (with greater pressures in certain directions than in others), we tend to get **foliated** or layered metamorphic rocks. This is known as **directed pressure.**

CHEMICAL FLUIDS

Water plays a very important role in metamorphism. Often charged with dissolved ions and gases (commonly CO_2), water may join the metamorphic process from any of three common sources: (1) formation waters, (2) water of hydration, and (3) water from magmatic bodies. These sources vary in significance, depending on the setting of metamorphism. Hydrothermal solutions released in the solidification of magma often percolate beyond the margins of the magmatic body and react on surrounding rocks. Sometimes these solutions remove ions and substitute others; at other times they add ions to the rock minerals to produce new minerals. Chemical reactions within the rock or the introduction of ions from an external source may cause one mineral to grow or change into another of different composition.

Fine-grained rocks are more readily changed than

others because they expose greater areas of grain surface to chemically active fluids. Some of the chemically active fluid of metamorphism is the liquid already trapped in the pores of a buried sediment.

Another possible metamorphic change involves the progressive dehydration of hydrous sedimentary minerals. Certain minerals have water of hydration as part of their chemical makeup; examples are gypsum, a hydrous calcium sulfate; and kaolinite, a hydrous aluminum silicate. During metamorphism of such substances the water tends to be driven off, producing anhydrous forms of these minerals.

7.2
TYPES OF METAMORPHISM

Several types of metamorphism occur, but we shall concern ourselves here with the two basic ones: **contact metamorphism** and **regional metamorphism**. A third type, **cataclastic metamorphism,** results from the grinding related to faulting and extreme folding, where minerals and rocks are broken and even pulverized, producing coarse-grained **fault breccias** or finer-grained varieties called **mylonites.** These rocks show the intensity of the pressures and temperatures associated with faulting and folding. A fourth type of metamorphism, **impact metamorphism,** results from the intense and rapid increase in pressure, with some increase in heat, associated with events such as meteor impacts.

CONTACT METAMORPHISM

The alteration of rocks at or near their contact with a body of magma intruding into the Earth's crust is called **contact metamorphism.** Minerals formed by this process are called **contact metamorphic minerals.** Most intrusions occur at shallow depths in the crust; therefore the pressure plays only a minor role. Baking produces more anhydrous minerals of larger size. No reorientation of minerals normally occurs. The type of reaction depends on the temperature, the composition of the intruding mass, and the properties of the intruded rock. At the actual surface of contact all the elements of a rock may be changed or replaced by other elements introduced by the hydrothermal solutions escaping from the magma. Farther away, the replacement may be only partial.

Contact metamorphism occurs in zones called **aureoles** ("halos"), which seldom measure more than a few hundred meters in width and may be only a fraction of a centimeter wide (Figure 7.2). Aureoles are found bordering plutons: sills, dikes, laccoliths, stocks, lopoliths, and batholiths. During contact metamorphism temperatures may range from 300° to 800°C, and load pressures may range from 100 to 3,000 atm (atmospheres) or bars.[1]

Two kinds of contact metamorphic mineral are recognized: those produced by heating up the intruded rock and those produced by the hydrothermal solutions reacting with the intruded rock. Hydrothermal metamorphism develops at relatively shallow depths. It is only late in the cooling of a body of magma, when it nears the surface, that large quantities of hydrothermal solutions are released.

Contact Metamorphic Minerals   Many new minerals are formed during contact metamorphism. For example, when an impure limestone is subjected to thermal contact metamorphism, its dolomite, clay, or quartz may be changed to new minerals. Calcite and quartz may combine to form wollastonite plus carbon dioxide:

calcite *plus* quartz *yields* wollastonite *plus* carbon dioxide

$$CaCO_3 + SiO_2 \rightarrow CaSiO_3 + CO_2$$

Dolomite may react with the quartz to form diopside plus carbon dioxide:

dolomite *plus* quartz *yields* diopside *plus* carbon dioxide

$$CaMg(CO_3)_2 + 2SiO_2 \rightarrow CaMg(Si_2O_6) + 2CO_2$$

---

[1] One atmosphere (or approximately one bar) is the "unit of pressure equal to the pressure exerted per unit area at sea level by the column of air from sea level to the top of the Earth's atmosphere."

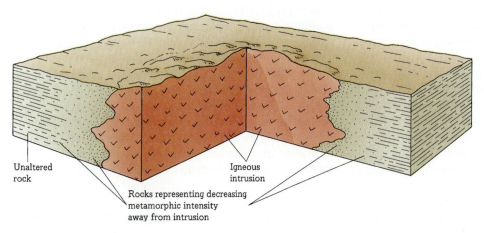

Unaltered rock

Igneous intrusion

Rocks representing decreasing metamorphic intensity away from intrusion

7.2 Intrusive igneous rock can alter the older rock that it intrudes. The intensity of alteration decreases away from the intrusion and forms zones of different metamorphic rocks arranged concentrically around the igneous rocks. Subsequent erosion, as suggested in this block diagram, exposes the concentric zones — aureoles — of metamorphic rocks. At some distance from the intrusion the effect of metamorphism dies out and metamorphic rock gives way to unaltered rock.

Aluminum and silica in clay will react to form andalusite, spinel, or garnet. If carbonaceous materials are present, they may be converted to graphite. Many other minerals are produced by hydrothermal contact metamorphism. Solutions given off by the magma react to produce new minerals containing elements not present in the limestone. In this type of metamorphism oxide and sulfide minerals are frequently formed and may constitute ore deposits of economic importance (see Chapter 19).

REGIONAL METAMORPHISM

Regional metamorphism results from intense compression, deep burial, and large-scale areal effects of large intrusions—all of which are commonly associated with mountain-building events. Regional metamorphic belts usually extend over thousands of square kilometers. Intense heat, high pressure, and chemically reactive fluids can yield a broad spectrum of mineral suites. Because of directed pressures, new mineral orientations are likely with attendant foliations. Regionally metamorphosed rocks are found in the root regions of old mountains and in the ancient continental shields. Thousands of meters of overlying rock must commonly be eroded away in order to expose these rocks.

Regional Metamorphic Facies   As we have seen, temperature and pressure are agents of metamorphism. A rock undergoing metamorphism experiences chemical reactions that produce a set of minerals we can use as a "fingerprint" of the temperature and pressure. Each rock (each bulk composition) will have its own set of fingerprint minerals for a given temperature and pressure. When we take all such fingerprint assemblages, though, they define what is known as a **metamorphic facies** (just as there are sedimentary facies).

In Chapter 6 we defined facies as it applies to sedimentary rocks. Now we see that it has broader application: **A facies is an assemblage of mineral or rock (or fossil) features reflecting the environment in which the rock was formed.**

Metamorphic facies are collections of assemblages of minerals that reached equilibrium during metamorphism under a specific set of environmental conditions. Each facies is named after a common metamorphic rock that belongs to it, and every metamorphic rock is assigned to a facies according to the conditions that attended its formation, not according to its composition. Although it is not always possible to assign a rock to a particular metamorphic facies on the basis of a single hand specimen, assignment can usually be made after examining various rocks in the region. Note also that different names are sometimes applied by different geologists to a given assemblage because many different systems of facies names have been suggested.

Widely recognized regional metamorphic facies are shown in Table 7.1. They are the zeolite, greenschist, amphibolite, granulite, blueschist (glaucophane schist), and eclogite facies. The table shows typical mineral assemblages for metamorphism of a shale and of a basalt.

Pressure-temperature fields for all metamorphic facies are shown in Figure 7.3. Also shown in this figure are those temperature and pressure conditions below which metamorphic changes do not commonly occur, known as the **realm of diagenesis.** Note also the lower-pressure region in which moderate-to-high-temperature contact metamorphism produces hornfels.

Table 7.2 shows the general relationships among zones and resulting metamorphic rocks in the progressive metamorphism of several original rock types.

Regional Metamorphic Minerals   During regional metamorphism new minerals are developed as rocks re-

TABLE 7.1
Regional Metamorphic Facies and Minerals Typically Produced
in the Metamorphism of a Shale and a Basalt[a]

Metamorphic facies	Mineral assemblages developed from:	
	Shale	Basalt
Zeolite	Illite, chlorite, quartz	Zeolites
Greenschist	Muscovite, chlorite, quartz	Albite, epidote, chlorite, actinolite
Amphibolite	Muscovite, biotite, garnet, quartz, plagioclase	Amphibole, plagioclase, garnet
Granulite	Garnet, sillimanite, plagioclase, quartz, biotite, pyroxene	Calcic pyroxene, plagioclase
Blueschist (glaucophane schist)	Muscovite, chlorite, quartz manganese-aluminum garnet	Glaucophane, quartz, lawsonite (high-pressure alteration of plagioclase)
Eclogite	Not known	Sodic pyroxene, magnesian garnet

[a] Modified from various sources.

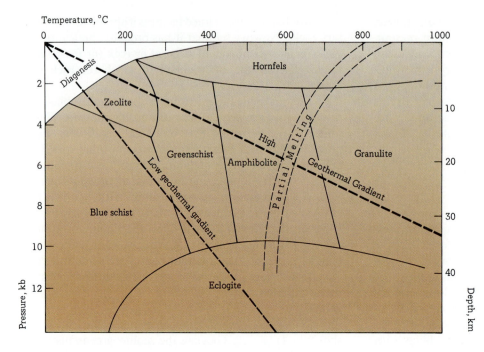

Temperature, °C

7.3 Generalized pressure-temperature fields for metamorphic facies. Boundaries are not sharp and distinct lines but extend over considerable width. [Modified from various sources.]

spond to increases in temperature and pressure. These include some silicate minerals not *normally* found in igneous and sedimentary rocks, such as sillimanite, kyanite, andalusite, staurolite, almandite, brown biotite, epidote, and chlorite. (Chlorite *is* a fairly common sedimentary mineral.)

The first three of these new minerals are aluminosilicates and have the formula Al_2SiO_5. Their independent SiO_4 tetrahedra are bound together by positive ions of aluminum. Sillimanite develops in long, slender crystals that are white, green, or brown. Kyanite forms bladelike blue crystals. Andalusite forms coarse, nearly square prisms.

Staurolite is a silicate composed of independent tetrahedra bound together by positive ions of iron and aluminum. It has a unique crystal habit, which is striking and easy to recognize when twinned, with six-sided prisms that intersect either at 90°, forming a cross, or at 60°, forming an X (see Figure 7.4).

Garnets are a group of metamorphic silicate minerals. All have the same atomic structure of independent SiO_4 tetrahedra, but a wide variety of chemical compositions is produced by the many positive ions that bind the tetrahedra together. These ions may be iron, magnesium, aluminum, calcium, manganese, or chromium. But whatever the

TABLE 7.2
Products of Regional Metamorphism[a]

	Metamorphic rocks resulting from increasing regional metamorphism				
Original rock	Chlorite zone	Biotite zone	Almandite (garnet) zone	Staurolite zone	Sillimanite zone
Shale	Slate	Biotite phyllite	Biotite-garnet phyllite	Biotite-garnet-staurolite schist	Sillimanite schist or gneiss
Clayey sandstone	Micaceous sandstone	Quartz-mica schist	Quartz-mica-garnet schist or gneiss	Quartz-mica-garnet schist or gneiss	Quartz-mica-garnet schist or gneiss
Quartz sandstone	Quartzite	Quartzite	Quartzite	Quartzite	Quartzite
Limestone and dolomite	Limestone and dolomite	Marble	Marble	Marble	Marble
Basalt	Chlorite-epidote-albite schist (greenschists)	Chlorite-epidote-albite schist (greenschists)	Albite-epidote amphibolite	Amphibolite	Amphibolite
Granite	Granite	Granite gneiss	Granite gneiss	Granite gneiss	Granite gneiss
Rhyolite	Rhyolite	Biotite schist or gneiss	Biotite schist or gneiss	Biotite schist or gneiss	Biotite schist or gneiss

[a] After Marland P. Billings, *The Geology of New Hampshire. Part II: Bedrock Geology*, p. 139, Granite State Press, Manchester, 1956.

7.4 Twinned crystals of staurolite, 2 cm across. [John Simpson.]

chemical composition, garnets appear as distinctive 12-sided or 24-sided fully developed crystals. Actually, it is difficult to distinguish one kind of garnet from another without resorting to chemical analysis. A common deep-red garnet containing iron and aluminum is called almandite.

Epidote is a complex silicate of calcium, aluminum, and iron in which the tetrahedra are in pairs independent of each other. This mineral is pistachio green or yellowish to greenish black.

Chlorite is a sheet silicate of calcium, magnesium, aluminum, and iron. The characteristic green color of chlorite was the basis for its name, from the Greek *chlōros,* "green." Chlorite exhibits a cleavage similar to that of mica, but the small scales produced by the cleavage are not elastic, like those of mica. Chlorite occurs either as aggregates of minute scales or as individual scales that will be scattered throughout a rock.

Regional Metamorphic Zones   Some minerals, like staurolite and chlorite, are more useful as indicators of particular facies than other minerals, such as quartz and calcite; therefore they are called **index minerals.** Metamorphic zones are identified by these index minerals.

Let us imagine a mapping exercise in which we record the presence or absence of index minerals as they appear on the ground. We can draw lines between each set of index minerals that define boundaries, or zones, of equal grade of metamorphism. These boundary lines, or **isograds,** can be interpreted as connecting similar temperature-pressure values. In fact, on geologic maps isograds distinguish one zone of index materials from another.

The temperature and pressure conditions that de-

velop under regional metamorphism may result in rocks characteristic of different zones: high-grade, middle-grade, and low-grade. **High-grade** metamorphism occurs in rocks that have been subjected to the most intense pressure and temperature. **Low-grade** metamorphic rocks have undergone the least change and blend along their outer margins into unchanged rocks. **Medium-grade** metamorphism falls between these two extremes.

Zones of regional metamorphism reflect the varied mineralogical response of chemically similar rocks to different physical conditions. Each index mineral gives an indication of the conditions at the time of its formation. For minerals composed of Al_2SiO_5, for example, the temperature and pressure conditions will determine whether the resulting mineral will be kyanite (low temperature with a wide range of pressures), andalusite (moderate temperature with low pressures), or sillimanite (high temperature with a wide range of pressures), as shown in Figure 7.5.

The first occurrence of chlorite tells us that we are at the beginning of a low-grade metamorphic zone. The first occurrence of almandite garnet is evidence of the beginning of a middle-grade metamorphic zone. And the first occurrence of sillimanite marks a high-grade zone. Other minerals sometimes occur in association with each of these index minerals (see Table 7.3), but they are usually of little help in determining the degree of metamorphism of a given zone.

By noting the occurrence of the minerals that are characteristic of each metamorphic zone, it is possible to draw a map of the regional metamorphism of an entire area. Of course, the rocks must have the proper chemical composition to allow these minerals to form.

7.5   Temperatures and pressures at which Al_2SiO_5 forms different minerals. [Modified from various sources.]

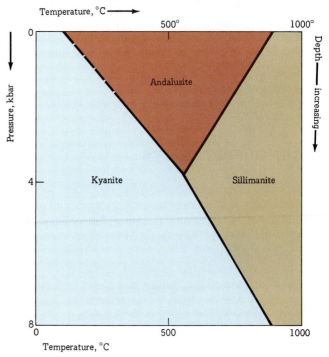

TABLE 7.3
Regional Metamorphic Minerals

	Zone	Grade of metamorphism	Minerals
Increasing	Chlorite	Low	Chlorite, muscovite, quartz
	Biotite	Low	Biotite, muscovite, chlorite, quartz
	Garnet	Middle	Garnet (almandite), muscovite, biotite, quartz
	Staurolite	Middle	Staurolite, garnet, biotite, muscovite, quartz
	Kyanite	Middle	Kyanite, garnet, biotite, muscovite, quartz
metamorphism	Sillimanite	High	Sillimanite, quartz, garnet, muscovite, biotite, oligoclase, orthoclase

LOW-GRADE REGIONAL METAMORPHISM IN PROGRESS

Studies on the cuttings and cores of several wells in sedimentary deposits of the Imperial Valley of California have indicated that low-grade metamorphism is in progress on a regional scale and at relatively shallow depth. These wells, in a geothermal field at the southeastern end of the Salton Sea, are part of a group of ten wells that have been drilled since 1960 to extract the elements contained in solution in the hot brines of the region and to tap the heat energy.

In the Salton Trough region the sedimentary deposits are geologically young. The oldest deposits in the Trough were deposited in the early Pliocene Epoch, and those in the geothermal field during the Pleistocene Epoch. They consist of poorly sorted sandstones and siltstones of the Colorado River delta. The original dominant minerals were quartz, calcite, subordinate dolomite, plagioclase, potassium feldspar, montmorillonite, illite, and kaolinite.

Samples from these drilled wells show increasing hardness and a regular sequence of mineral changes with depth as the temperature increases. Some detrital minerals drop out, and other minerals are formed. The rocks look like ordinary sedimentary rocks, but all have undergone mineralogical changes at depth since deposition. This change is brought about by progressive metamorphism.

7.3

METAMORPHIC ROCKS

Contact and regional metamorphic rocks are found in mountain ranges, at roots of mountain ranges, and on continental shields. The regional metamorphic rocks are by far the most widespread. They vary greatly in appearance, texture, and composition.

Even when different regional metamorphic rocks have all been formed by changes of a single uniform rock type, such as a shale, they are sometimes so drastically changed that they are thought to be unrelated.

These and other characteristics of regionally metamorphosed shale are well illustrated in New Hampshire.

Along the Connecticut River, west of the White Mountains, rocks that were originally shales are found in a low-grade metamorphic zone as slate, with chlorite. Southeast of these rocks the original shale is now found as phyllite, grading into schist. New metamorphic minerals appear one after the other toward the southeast: almandite, staurolite, and then sillimanite. These metamorphosed shales occur in belts surrounding certain granitic bodies. The closer the area is to the intrusion, the higher is its grade of regional metamorphism. This correlation is indicated by the presence of the index minerals found in the metamorphic rock.

CLASSIFICATION OF METAMORPHIC ROCKS

The compositional aspect to classification is secondary for metamorphic rocks. In most rocks that have been subjected to heat and deforming pressures during regional metamorphism, the minerals tend to be arranged in parallel layers of flat or elongated grains (see Figure 7.6). This arrangement gives the rocks a property called **foliation** (from the Latin *foliatus,* "leaved" or "leafy," hence consisting of thin sheets). Recrystallization results from directed pressure being applied to the rocks. The consequently produced linear and planar features, in general, are at an angle to bedding in sedimentary rocks or flow banding in the parent igneous rocks. Foliation gives geologists a way of interpreting the mechanical environment of metamorphism.

The textures most commonly used to classify metamorphic rocks are **nonfoliated** (either aphanitic or granular) and **foliated.** Let us look first at the nonfoliated textures. In rocks with aphanitic texture the individual grains cannot be distinguished by the unaided eye, and these rocks do not exhibit rock cleavage. Previously we used the term **cleavage** to describe the relative ease with which a mineral breaks along parallel planes. But notice here that we are using the modifier "rock" to distinguish rock cleavage from mineral cleavage. In rocks with granular texture the individual grains are clearly visible, but again, no rock cleavage is evident.

Foliated Rocks   If we start with a sedimentary rock of diverse composition and a very fine grain size (a mudstone

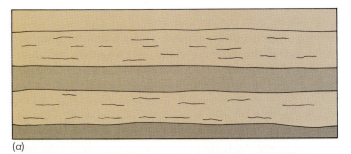

(a)

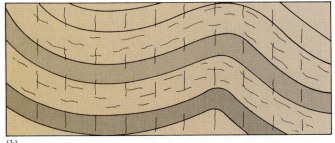

(b)

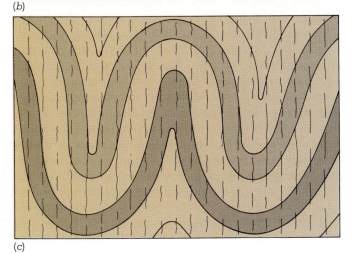

(c)

7.6 Development of schistosity in argillaceous sedimentary rocks: (a) Nearly horizontal bedding soon after deposition. (b) Moderately folded rocks with beginning of schistosity produced by realignment and growth of clay minerals perpendicular to regional stresses (coming from the sides). Note angle between bedding and schistosity at various places around folds. (c) Strongly folded rocks with highly developed schistosity nearly parallel to axial planes of folds and nearly parallel to most of the beds.

or a shale), we can demonstrate the changes that take place with increasing degrees of regional metamorphism. In this case, there are four foliated metamorphic rocks:

1. **Slate** (from the old French *esclat,* "fragment," "splinter"), in which the foliation occurs along planes separated by distances of microscopic dimensions.
2. **Phyllite** (from the Greek *phyllon,* "leaf"), in which the foliation produces flakes barely visible to the unaided eye. Phyllite produces fragments thicker than those of slate.
3. **Schist** (from the Greek *schistos,* "divided" or "divisible"), in which the foliation produces flakes that are clearly visible. Here surfaces are rougher than in slate or phyllite.

7.7 Slate quarry, showing how rock cleavage controls breaking. Pen Argyl, Pennsylvania.

4. **Gneiss** (from the Greek *gneis,* "spark," for the luster of certain of the components), in which the minerals occur in bands or layers.

Slate is a metamorphic rock that has been produced from the low-grade metamorphism of shale or pyroclastic igneous rock (see Figure 7.7). It is very fine grained with a slaty foliation caused by the alignment of platy minerals by the stress field. Some of the clay minerals in the original shale have been transformed by heat into chlorite and mica. In fact, slate is composed predominantly of small colorless mica flakes and some chlorite. It occurs in a wide variety of colors. Dark-colored slate owes its color to the presence of carbonaceous material or iron sulfides.

Phyllite is a metamorphic rock with much the same composition as slate, but its minerals exist in larger units. Phyllite is actually slate that has undergone further metamorphism. When slate is subjected to heat greater than 250° to 300°C, the chlorite and mica minerals of which it is composed develop large flakes, giving the resulting rock its characteristic phyllitic cleavage and a silky sheen on freshly broken surfaces. The predominant minerals in phyllite are chlorite and muscovite. This rock usually contains the same impurities as slate, but sometimes a new metamorphic mineral, such as tourmaline or magnesium garnet, makes its appearance.

Schist is the most abundant of the metamorphic rocks formed by regional metamorphism. There are many varieties of schist, for it can be derived from many igneous, sedimentary, or low-grade metamorphic rocks. However, all schists are dominated by clearly visible flakes of some platy material, such as mica, talc, chlorite, or hematite. Fibrous minerals are commonly present as well. Schist tends to break between the platy or fibrous minerals, giving the rock its characteristic schistose cleavage.

Schists often contain large quantities of quartz and feldspar as well as lesser amounts of minerals such as augite, hornblende, garnet, epidote, and magnetite. A green schistose rock produced by low-grade metamorphism, sometimes called a **greenschist,** owes its color to the presence of the minerals chlorite and epidote.

TABLE 7.4
Common Schists

Variety	Rock from which derived
Chlorite schist	Shale
Mica schist	Shale
Hornblende schist	Basalt or gabbro
Biotite schist	Basalt or gabbro
Quartz schist	Impure sandstone
Calc-schist	Impure limestone

Table 7.4 lists some of the more common varieties of schist, together with the names of the rocks from which they are derived.

Gneiss, a granular metamorphic rock, has a banded appearance that makes it easy to recognize (see Figure 7.8). In gneiss derived from igneous rocks such as granite, gabbro, or diorite, the component minerals are arranged in parallel layers: The quartz and the feldspars alternate with the ferromagnesians. In gneiss formed from the metamorphism of clayey sedimentary rocks such as graywackes, bands of quartz or feldspar usually alternate with layers of platy or fibrous minerals.

Other foliated metamorphic rocks do occur, but with much less frequency than the four just described. One fairly common foliated rock is **amphibolite.** Composed mainly of hornblende and plagioclase, it has some foliation due to alignment of hornblende grains, but the foliation is less conspicuous than in schists. Amphibolites may be green, gray, or black and sometimes contain such minerals as

epidote, green augite, biotite, and almandite. They are products of the medium-grade to high-grade metamorphism of ferromagnesian igneous rocks and of some impure calcareous sediments.

Nonfoliated Rocks   Some other metamorphic rocks do not show this foliated character and are not formed in the same manner. Increased grain size is commonly the only indication that some granular rocks have been metamorphosed at all. Regional metamorphism of mono-mineralic rocks can produce such granular textures, as can the contact metamorphism of almost any rocks. A few examples of nonfoliated metamorphic rocks follow.

Marble, a familiar metamorphic rock, is composed predominantly of calcite or dolomite, is granular, and was derived during the contact or regional metamorphism of limestone or dolomite. It does not exhibit rock cleavage. Marble differs from the original rock in having larger mineral grains. In most marble the crystallographic direction of its calcite is nearly parallel; this is in response to the metamorphic pressures to which it was subjected. The rock shows no foliation, however, because the grains have the same color, and lineation does not show up.

Although the purest variety of marble is snow white, many marbles contain small percentages of other minerals that were formed during metamorphism from impurities in the original sedimentary rock. These impurities account for the wide variety of color in marble. Black marbles are colored by bituminous matter; green marbles by diopside, hornblende, serpentine, or talc; red marbles by the iron oxide hematite; and brown marbles by the iron oxide limonite. Garnets have often been found in marble, as have rubies on rare occasions.

Marble occurs most commonly in areas of regional metamorphism, where it is often found in layers between mica schists or phyllites.

Quartzite is formed by the metamorphism of quartz-rich sandstone. The quartz in the original sandstone has become firmly bonded by the entry of silica into the pore spaces. Although siliceous sandstones are cemented by percolating water under the temperatures and pressures of ordinary sedimentary processes working near the surface of the Earth, quartzites are true metamorphic rocks and are formed by metamorphism of any grade.

Quartzite is generally nonfoliated and is distinguishable from sandstone in two ways: (1) there are fewer pore spaces in the quartzite; and (2) the rock breaks right through the sand grains that make it up, rather than around them. The structure of quartzite cannot be recognized without a microscope; but when we cut it into thin sections, we can identify the original rounded sand grains, the silica that has filled the old pore spaces, and the interlocking texture along sutured boundaries (see Figure 7.9). Most quartzites show evidence of metamorphic strain.

Pure quartzite is white, but iron or other impurities

7.8   Under the microscope this thin section of biotite gneiss shows the alignment of different minerals. The bands of "salt and pepper" texture contain quartz and potassium feldspar. The bands of larger pinkish-colored crystals are made chiefly of biotites and other ferromagnesian minerals. The rock has a bulk composition similar to that of granite but, in contrast to the random mixing in granite, the minerals here are segregated into systematic patterns that produce a foliation so characteristic of metamorphic rocks. Field of view about 0.5 cm.

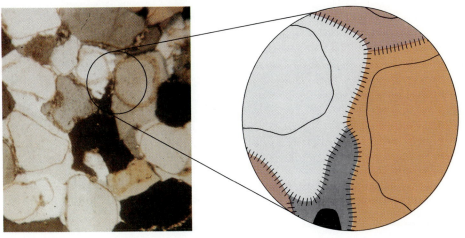

7.9 This thin section of a quartzite shows individual sand grains averaging about 1 mm in diameter. The color of each grain depends upon the orientation of its crystal structure to the polarized light that passes through the thin section. If you look carefully, you will see that individual, rounded grains are outlined by a faint line and that more mineral matter continues beyond this until it meets material surrounding an adjacent rounded grain. An enlarged drawing of a portion of the thin section shows this in more detail. The sequence of events recorded here begins with the deposition of quartz sand grains. These originally loose grains became cemented by more quartz deposited from fluids circulating through the pore spaces between the grains. The quartz cement was deposited first on the surface of the grains and took on the same crystal orientation as the host grain. The cement continued to grow until it met cement growing on adjacent grains and the pore space was closed.

sometimes give the rock a reddish or dark color. Among the minor minerals that often occur in quartzite are feldspar, muscovite, chlorite, zircon, tourmaline, garnet, biotite, epidote, hornblende, and sillimanite.

Hornfels is any fine-grained rock that has undergone metamorphism. Its composition varies depending upon the composition of the original rock. Commonly the density of the rock increases during metamorphism.

7.4
METAMORPHISM AND PLATE BOUNDARIES

Different types of metamorphism and metamorphic products can be expected to develop along the margins of lithospheric plates, depending on the kind of boundary and its geologic setting.

CONVERGENT-BOUNDARY METAMORPHISM

Crustal plates are commonly subducted at trenches under convergent island-arc regions (such as the island arcs of the western Pacific) and at convergent ocean-continent boundaries (such as the Andean coast of South America). In these regions subducting slabs of lithosphere interact with sediments that have accumulated on the sea floor. These sediments tend to consist of a **mélange** (heterogeneous mixture) of fine-grained deep-sea facies combined with exotic blocks of varying ages up to several kilometers in size. A good example of such an assemblage is the Franciscan Formation (Jurassic to Eocene in age) of California, which consists

of a mélange of deep-sea sediments, oceanic-crust fragments, and various metamorphic assemblages (blue schist facies).

A second convergent-boundary setting involves the downwarping of thick piles of sediments that have accumulated between an island arc and an adjacent continent or on the ocean edge of a continental block. Low-grade regional metamorphism commonly occurs at the bottom of such a wedge of sediments. Plutonic activity also develops, so that contact metamorphism affects the rocks adjacent to these intrusions.

Much of the regional metamorphism discussed earlier in this chapter is related to the third principal convergent boundary, continent-continent plate convergence. The Himalayas, for example, have broad areas of metamorphism associated with the crustal thickening between the Australo-Indian and the Asian plates.

In all these settings descending slabs of lithosphere undergo metamorphism caused by increasing temperature and pressure. Such effects can be calculated for slabs of a given composition, for varying thicknesses and sizes of slabs, and for different rates of movement. These measurements become very complicated but are possible within certain limiting conditions for these parameters.

The lithospheric slabs apparently retain their mechanical competence and are relatively cool in contrast to their surroundings until thermal equilibrium is reached (Figure 7.10). One analysis estimated that a slab would reach thermal equilibrium with the surrounding mantle at a depth of about 650 km for a lithospheric plate being subducted at a rate of 8 cm/year. When subduction has extended to a given depth, melting can occur, producing magmas that rise to supply surface volcanoes (Figure 7.11).

Different pressure-temperature conditions prevail at

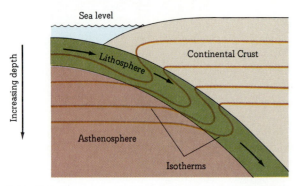

7.10 Relative temperature increases at the convergent boundary. The lithospheric plate undergoing subduction retains its cooler temperature, deflecting the isotherms downward. [Various sources.]

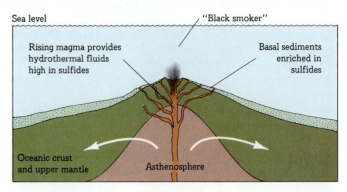

7.12 Cross section of divergent boundary at spreading center, showing hydrothermal fluids and their contact-metamorphosed sediments.

different parts of this descending plate: (1) A zone of low heat flow has been observed over what is considered to be the upper portion of this descending slab. Here low-temperature metamorphic minerals are created. (2) An intermediate zone of relatively higher temperature and pressure occurs along the slab at intermediate depths. (3) A zone of high heat flow occurs over the deepest part of the downward-moving lithosphere, where melting occurs and heat is transferred upward by rising magmas. High-temperature–high-pressure metamorphic minerals are created here.

DIVERGENT-BOUNDARY METAMORPHISM

Metamorphism (and attendant mineralization) occurs along divergent plate boundaries, notably at midocean spreading centers such as the Midatlantic Ridge. Outflowings of basaltic lavas heat rocks to produce contact metamorphic minerals.

Seawater over midocean ridges forms hydrothermal solutions, penetrating fissures, dissolving minerals from rocks under the ridge crests, and precipitating those minerals in highly concentrated deposits often of economic importance (see Chapter 19 and Figure 7.12).

Metallic sulfides have been found in rocks dredged from midoceanic ridges at a number of places, including the Midatlantic and Indian ridges.

The island of Cyprus is famous for a variety of rich mineral deposits, first mined by the Phoenicians. The origin of this island is closely related to sea-floor spreading from a midoceanic ridge that existed in the earliest development of the eastern Mediterranean.

The Red Sea is perhaps the very early product of a divergent plate boundary currently forming between the African and Arabian plates. Very salty hot brines have been found associated with sediments containing significant quantities of sulfide minerals, including iron, zinc, and copper sulfides, plus some silver and gold. The brines are thought to be hydrothermal solutions from which these minerals precipitate or into which minerals are introduced from volcanic sources beneath the Red Sea or from sediments with high mineral content found adjacent to the basins.

7.11 Thermal and metamorphic zones associated with descending lithospheric plate under a converging island-arc region. The outer zone near the trench is a zone of relatively low heat flow. The zone associated with the deepest portion of the descending plate is a zone of high heat flow, with melting and the production of magma, contact metamorphism, and overlying volcanic island chain. Types of metamorphism are indicated for various parts of the system. [Modified from various sources.]

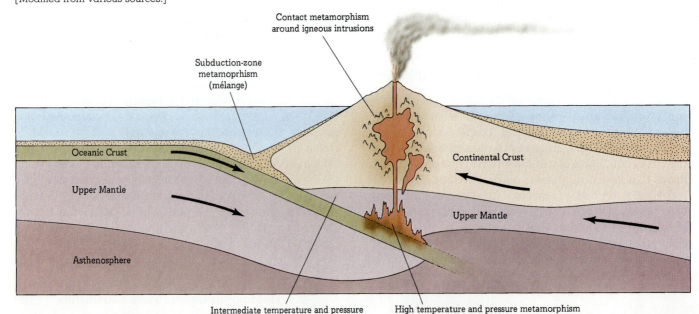

Fractures occur along transform boundaries as plates slide past one another. These fractures and faults provide zones of weakness along which magma can rise. Volcanoes occur along such zones in oceanic regions. Traces of plates moving past one another can also be observed on land, for example, along the San Andreas fault system. Relatively low temperatures and high pressures develop as rock masses moving past one another produce metamorphic minerals. This is commonly called **dynamic metamorphism.**

7.5

GRANITIZATION VS. METASOMATISM

The eighteenth-century geologist James Hutton once stated that granite was produced by the crystallization of minerals from a molten mass, and ever since, most geologists have accepted the magmatic origin of granite. However, several investigators have questioned this conclusion, suggesting instead that granite is a metamorphic rock produced from preexisting rocks by a process called **granitization.**

In discussing batholiths we mentioned that one of the reasons for questioning the magmatic origin of granite was the mystery of what happened to the great mass of rock that must have been displaced by the intrusion of the granite batholiths. This so-called space problem has led some geologists to conclude that batholiths actually represent preexisting rocks transformed into granite by metasomatic processes (see discussion below). Certain rock formations support this theory. These sedimentary rocks were originally formed in a continuous layer but now grade into schists and then into **migmatites** ("mixed rocks"), apparently formed when magma squeezed in between the layers of schist.

The concept of migmatite as a rock group was introduced by J. J. Sederholm in 1907. It is a generic term that covers a large number of petrological combinations and is to be applied to rock formed from two principal ingredients: the host of early-formed material and the material introduced by permeation, metasomatism, or injection of liquids. The average composition of a migmatite is granitic. Its diagnostic characteristics are structure, texture, and broad regional relationships.

The migmatites, in turn, grade into rocks that contain the large, abundant feldspars characteristic of granite but that also seem to show shadowy remnants of schistose structure. Finally, these rocks grade into pure granite. The proponents of the granitization theory say that the granite is the result of extreme metasomatism and that the schists, migmatites, and granitelike rocks with schistose structure are way stations in transforming sedimentary rocks into granite.

What mechanism could have brought about granitization? Perhaps ions migrated through the original solid rock, building up the elements characteristic of granite, such as sodium and potassium, and removing superfluous elements, such as calcium, iron, and magnesium. The limit to which the migrating ions are supposed to have deposited the sodium and potassium is called the **granitic front.** The limit to which the migrating ions are supposed to have carried the calcium, iron, and magnesium is called the **simatic front.** As we saw in Section 7.1, chemical reactions within the rock or the introduction of ions from an external source may cause one mineral to grow or change into another of different composition. The process may eventually extend beyond metamorphism to the process called **metasomatism,** producing massive changes in a body of rocks that result in an igneous mass.

In the middle of the twentieth century geologists were carrying on an enthusiastic debate over the origin of granite, but they had reached agreement on one fundamental point: Various rocks with the composition and structure of granite may have different histories. In other words, some may be igneous and others metasomatic. So the debate between "magmatists" and "granitizationists" has been reduced to the question of what percentage of the world's granite is metasomatic and what percentage is magmatic. Those who favor magmatic origin admit that perhaps as much as a quarter of the granite exposed at the Earth's surface is metasomatic. But the granitizationists reverse the percentages and insist that about three-quarters or more of all granite is metasomatic and only one-quarter or less of magmatic origin.

Thus we see that partial melting at the base of a thick sedimentary pile approaches Bowen's reaction series "from the bottom up." This is "ultrametamorphism." (We are not talking about forming a magma, however, which would be beyond the field of metamorphism.) Field relationships in some cases suggest that great granite bodies have formed as second-generation igneous rocks in the cores of mountain ranges. (This could resolve the space problem in large granitic batholiths.)

7.6

GEOTHERMOMETERS

Geologists have long sought out ways to determine what conditions have existed at various times in the past— conditions of pressure, climate, wind direction, stream flow direction, and especially **temperature.** Any mechanism that can be used to determine the Earth's temperature at a specific time and place is known as a **geothermometer.** These geothermometers may be animal, plant, or mineral. The changes that rocks and minerals undergo during metamorphism are caused in part by increasing temperatures. These same temperature increases can cause changes in the enclosed plants and animals, changes that in some cases can be instructive in telling us just what temperatures were encountered during metamorphism (see Table 7.5).

TABLE 7.5
Geothermometers, Showing Changes That Occur
as Temperature Increases

Temperature range (°C)[a]	Metamorphic grade	Pelitic rocks	Carbonate rocks	Coals	Conodonts	Hydrocarbons
20° —						
	D I A G E N E S I S	Shales contain clay minerals	Calcium and magnesium carbonates	Change from peat to brown coal to lignite	First color change from pale yellow to very pale brown	Beginning stages of oil generation
50° —		Submicroscopic changes in clay minerals	Unchanged	Sub-bituminous to bituminous	Very pale brown to light brown	Peak stages of oil generation; early generation of gas
100° —		Microscopic changes in clay minerals	Unchanged	Bituminous to semianthracite	Brown to dark brown	Phase-out of oil generation; peak of gas generation
200° —	Very low-grade metamorphism	Change to slates	Recrystallized	Anthracite	Very dark grayish brown to dark reddish brown	Hydrocarbons are metamorphosed and destroyed
300° —		Change to phyllites and chlorite schists	Coarsely crystalline marbles	Meta-anthracite	Black	
400° —	Low-grade metamorphism	Biotite to garnet schist	Marbles; impure carbonates contain talc and tremolite	Graphite	Very black	
500° —	Medium-grade metamorphism	Andalusite to kyanite schist	Marbles; impure carbonates contain tremolite and actinolite	Graphite	Colors fade	
600° —	High-grade metamorphism	Sillimanite schist	Replaced by calcium silicate (e.g., wollastonite)	Graphite to diamonds (if sufficient pressure)	Clear	

[a] Arbitrary boundaries; summarized from various sources, especially A. G. Harris, U.S. Geological Survey, personal communication.

Summary

Metamorphism produces metamorphic rocks by changing igneous and sedimentary (and other metamorphic) rocks while they are in the solid state.

Agents of metamorphism are heat, pressure, and chemically active fluids.

Heat may be the essential agent.

Pressure may be great enough to induce plastic deformation.

Chemically active fluids, particularly those released late in the solidification of magma, react on surrounding rocks.

Types of metamorphism are contact and regional.

Contact metamorphism occurs at or near an intrusive body of magma.

Contact metamorphic minerals include wollastonite, diopside, and some oxides and sulfides constituting ore minerals.

Regional metamorphism is developed over extensive areas and is related to the formation of some mountain ranges.

Regional metamorphic facies is an assemblage of minerals that reached equilibrium during metamorphism under a specific set of conditions.

Regional metamorphic minerals include sillimanite, kyanite, andalusite, staurolite, almandite garnet, brown biotite, epidote, and chlorite.

Regional metamorphic zones are identified by diagnostic index minerals.

Metamorphic rocks are found in mountain ranges, at mountain roots, and on continental shields.

Textures of metamorphic rocks are nonfoliated and foliated.

Nonfoliated rocks do not exhibit rock cleavage.

Foliated rocks exhibit rock cleavage as slaty, phyllitic, schistose, or gneissic.

Metamorphism and plate boundaries are closely related.

Convergent boundaries show increased temperature and pressure effects on subducting slabs of lithosphere with sediments intruded by batholiths, downwarped thick sedimentary sequences, and plates that have collided.

Divergent boundaries have high heat flow at spreading centers with mineralization including sulfides and evaporites.

Transform or parallel boundaries have less igneous activity but include some metamorphism and some economic mineralization.

Granitization vs. metasomatism refers to the debate over what percentage of the world's granite is produced by each of these processes.

Geothermometers are animal, plant, or mineral indicators of changes in rock temperature.

Questions

1. What limits are placed on conditions under which the process of metamorphism is considered to occur?
2. Why is heat considered to be the most essential agent of metamorphism?
3. How does pressure produce changes during metamorphism?
4. How do contact aureoles develop around igneous intrusions?
5. Regional metamorphism produces zones characterized by which index materials?
6. What is a regional metamorphic facies? How is it similar to a sedimentary facies? How does it differ?
7. Why is diagenesis considered by some to be the "final stage" before metamorphism?

8. Where and under what conditions do we find low-grade regional metamorphism taking place today?
9. What is the typical progression in the increasing regional metamorphism of a shaly rock? Rearrange these metamorphic rocks in their proper order: phyllite, schist, shale, slate.
10. Quartzite is nonfoliated and can be distinguished from a sandstone in what two ways?
11. Why do temperature isograds make sharp bends along subducted plates as shown in Figure 7.10?
12. How is it possible to produce a granite by extension of the processes that have produced metamorphism?
13. What is meant by a geothermometer? Give an example of a plant, a mineral, and an animal geothermometer.

Supplementary Readings

BEST, M. G. *Igneous and Metamorphic Petrology,* W. H. Freeman, San Francisco, 1982.
An integration of igneous and metamorphic processes and products.

ERNST, W. G. *Metamorphism and Crustal Evolution of the Western United States,* Prentice Hall, Englewood Cliffs, N.J., 1988.
Detailed description of the geological development of the western North American continent and the movement of plates.

MASON, ROGER *Petrology of the Metamorphic Rocks,* 2nd ed., Unwin Hyman, Winchester, Mass., 1989.
An intermediate treatment of metamorphic products and processes.

PHILPOTTS, A. *Petrography of Igneous and Metamorphic Rocks,* Prentice Hall, Englewood Cliffs, N.J., 1989.
Introductory-level description of igneous and metamorphic rocks, including classification.

OXBURGH, E. R., B. W. D. YARDLEY, AND P. C. ENGLAND *Tectonic Settings of Regional Metamorphism,* Cambridge University Press, Cambridge, 1988.

Shows how our understanding of global plate tectonics can cast light on the evolution of the large mountain belts and the metamorphism associated with mountains.

READ, H. H. *The Granite Controversy,* Thomas Murby and Co., London, 1957.
Eight addresses, delivered between 1939 and 1954, concerned with the origin of granitic and associated rocks.

TURNER, FRANCIS J. *Metamorphic Petrology,* 2nd ed. McGraw-Hill Book Co., New York, 1981.
A general text for advanced students. Embodies a survey of mineralogical aspects of metamorphism.

WINKLER, H. G. F. *Petrogenesis of Metamorphic Rocks,* 4th ed., Springer-Verlag, New York, 1976.
Stresses the mineralogical-chemical and physiochemical aspects of metamorphism, especially as related to specific rock groups of common composition.

Sunset, Grand Canyon National Park, Arizona [Harald Sund, The Image Bank.]

CHAPTER ■ 8

GEOLOGIC TIME

"If I were asked as a geologist what is the single greatest contribution of the science of geology to modern civilized thought, the answer would be the realization of the immense length of time. So vast is the span of time recorded in the history of the earth that it is generally distinguished from the more modest kinds of time by being called 'geologic time.'"

Adolph Knopf, Time and Its Mysteries, *ser. 3, p. 33, New York University Press, New York,* 1949

OVERVIEW	In Chapter 1 we stressed the vastness of geologic time and emphasized its importance to an understanding of the processes of geology. In this chapter we look at some of the ways of measuring geologic time and at how the geologic-time scale has developed.

We will differentiate between *relative time* and *absolute time*. Relative time, the "relative" position of one event in time to another without regard to either event's age in actual years, is established by the *law of superposition* and the *law of cross-cutting relationships*. In a process called *correlation* these relative ages are extended from one area to another.

Also useful in establishing relative time are features known as *unconformities*, which mark gaps in the time recorded by the rock record.

We demonstrate that absolute time can be determined in several ways but that the most important method is that of *radioactivity*.

The *geologic-time scale* was first established on the basis of relative time. Subsequently it was found that absolute dates could be inserted into the scale.

Finally, we consider a type of dating of rocks known as *magnetostratigraphy*, which has a number of uses but is particularly valuable in establishing the reality of sea-floor spreading and the development and age of the ocean basins.

8.1

RELATIVE TIME

Before geologists knew how to determine absolute time, they had discovered events in Earth history that convinced them of the great length of geologic time. In putting these events in chronological order, they found themselves subdividing geologic time on a relative basis and using certain labels to indicate relative time. You have probably picked up a newspaper or magazine and read of the discovery of a dinosaur that lived 100 million years ago during the Cretaceous Period or of the development of a coal deposit in strata formed 290 million years ago in the Pennsylvanian Period. The names of the periods are terms used by geologists to designate certain units of **relative geologic time.** In this section we look at how such units have been set up and how absolute dates have been suggested for them.

Relative geologic time has been determined largely by the relative positions of sedimentary rocks. Remember that a given layer of sedimentary rock represents a certain amount of time—the time it took for the original deposit to accumulate. By arranging various sedimentary rocks in their proper chronological sequence, we are in effect arranging units of time in *their* proper order. Our first task in constructing a relative-time scale, then, is to arrange the sedimentary rocks in their proper order.

THE LAW OF SUPERPOSITION

The principle used to determine whether one sedimentary rock is older than another is very simple, and is known as the **law of superposition.** Here is an example: A deposit of mud laid down this year in, say, the Gulf of Mexico will rest

on top of a layer that was deposited last year. Last year's deposit, in turn, rests on successively older deposits that extend backward into time for as long as deposition has been going on in the gulf. If we could slice through these deposits, we would expose a chronological record, with the oldest deposit on the bottom and the youngest on top. This sequence would illustrate the law of superposition: **If a series of sedimentary rocks has not been overturned, the topmost layer is always the youngest and the lowermost layer is always the oldest.**

The law of superposition was first derived and used by Nicolaus Steno (1638–1687), a physician and naturalist who later turned theologian. His geologic studies were carried on in Tuscany, in northern Italy, and were published in 1669. On first glance, the principle worked out by Steno is absurdly simple. For instance, in a cliff of sedimentary rocks, with one layer lying on top of another, it is perfectly obvious that the oldest is on the bottom and the youngest on top. We can quickly determine the relative age of any one layer in the cliff in relation to any other layer. The difficulty, however, lies in the fact that unknown tens of thousands of meters of sedimentary rock have been deposited during geologic time and that there is no one cliff, no one area, where *all* these rocks are exposed to view. The rocks in one place may be older or younger or of the same age as those in some other place. The task is to find out how the rocks around the world fit into some kind of relative time frame.

THE LAW OF CROSS-CUTTING RELATIONSHIPS

A geologic feature that cuts across another geologic feature (e.g., a rock unit, a fault) must be younger than that feature. The basic rule here is called the **law of cross-cutting rela-**

tionships, which states that **a rock is younger than any rock that it cuts across** (or a fault is younger than the rocks it cuts). Although very simple in concept, this rule is one of the building stones for determining relative geologic time.

CORRELATION OF SEDIMENTARY ROCKS

Because we cannot find sedimentary rocks representing all of Earth time neatly arranged in one convenient area, we must piece together the rock sequence from locality to locality. This process of tying one rock sequence in one place to another in some other place is known as **correlation,** from the Latin for "together" plus "relate."

Correlation by Physical Features When sedimentary rocks show rather constant and distinctive features over a wide geographic area, we can sometimes connect sequences of rock layers from different localities. Figure 8.1 illustrates how this is done. Here is a series of sedimentary rocks exposed in a sea cliff. The topmost, and hence the youngest, is a sandstone. Beneath the sandstone we first find a shale, then a seam of coal, and then more shale extending down to the level of the modern beach. We can trace these layers for some distance along the cliff face, but how are they related to other rocks farther inland?

Along the rim of a canyon that lies inland from the cliff, we find that limestone rocks have been exposed. Are they older or younger than the sandstone in the cliff face? Scrambling down the canyon walls, we come to a ledge of sandstone that looks very much like the sandstone in the cliff. If it *is* the same, then the limestone must be younger

because it lies above it. The only trouble is that we cannot be certain that the two sandstone beds *are* the same; so we continue down to the bottom of the canyon and there find some shale beds very similar to the shale beds exposed in the sea cliff beneath the sandstone. We feel fairly confident that the sandstone and the shale in the canyon are the same beds as the sandstone and upper shale in the sea cliff, but we must admit the possibility that we are dealing with different rocks.

In searching for further data, we find a well drilled still farther inland, its drill bit cutting through the same limestone that we saw in the canyon walls. As the bit cuts deeper and deeper, it encounters sandstone, shale, a coal seam, more shale, and then a bed of conglomerate before the drilling finally stops in another shale bed. This sequence duplicates the one we observed in the sea cliff and reveals a limestone and an underlying conglomerate and shale that we have not seen before. We may now feel justified in correlating the sandstone in the sea cliff with that in the canyon and that in the well hole. The limestone is the youngest rock in the area; the conglomerate and lowest shale are the oldest rocks. This correlation is shown in Figure 8.1(*b*).

Many sedimentary formations are correlated in just this way, especially when physical features are our only keys to rock correlation. But as we extend the range of our correlation over a wider and wider area, physical features become more and more difficult to use. Clearly it is impossible to use physical characteristics to determine the relative age of two sequences of layered rocks in, say, England and the eastern United States. Fortunately we have another method of correlation, a method that involves the use of fossils.

8.1 (*a*) Diagram to illustrate the data that might be used to correlate sedimentary rocks (right) in a sea cliff with those (center) in a stream valley and with those (left) encountered in a well-drilling operation. (*b*) Similar lithologies and sequences of beds in the three different localities of (*a*) suggest the correlation of rock layers shown in this diagram.

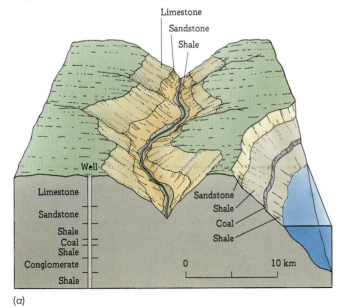

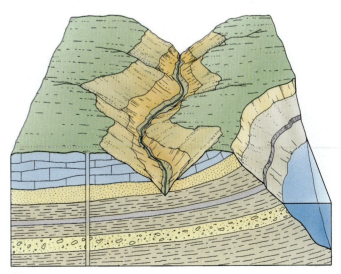

(*a*)

(*b*)

Correlation by Fossils   Around the turn of the nine-teenth century an English surveyor and civil engineer named William Smith (1769–1839) became impressed with the relationship of rock strata to the success of various engineering projects, particularly the building of canals. As he investigated rock strata from place to place, he noticed that many of them contained fossils. Furthermore, he observed that (no matter where found) some rock layers contained identical fossils, whereas the fossils in rock layers above or below were different. Eventually Smith became so skillful that when confronted with a fossil, he could name the rock from which it had come.

At about the same time, two French geologists, Georges Cuvier (1769–1832) and Alexandre Brongniart (1770–1847), were studying and mapping the fossil-bearing strata that surround Paris. They too found that certain fossils were restricted to certain rock layers, and they had also used the law of superposition to arrange the rocks of the Paris area in chronological order, just as Smith had done in England. Then Cuvier and Brongniart arranged their collection of fossils in the same order as the rocks from which the fossils had been dug. They discovered that the fossil assemblages varied in a systematic way with the chronological positions of the rocks. In comparing the fossil forms with modern forms of life, Cuvier and Brongniart discovered that the fossils from the higher rock layers bore a closer resemblance to modern forms than did the fossils from rocks lower down.

From all of these observations it became evident that the **relative age** of a layer of sedimentary rock could be determined by the nature of the fossils that it contained. This fact has been verified time and again by other workers throughout the world. It has become an axiom in geology that **fossils are a key to correlating rocks** and that **rocks containing the same fossil assemblages are similar in age.**

Through countless observations of the occurrence of fossils in different rock units throughout the world, we find that: **Groups of fossil animals** (and plants) **have succeeded one another in a definite and discernible order and each geologic time period can be recognized by the fossils found in rocks of that age.** This is known as the **principle of faunal succession.**

Superposition and faunal succession go hand in hand to work out the geologic history of any region. In fact, superposition is basic to faunal succession and proves its validity. When we examine the assemblages of fossils, we find that older rocks commonly have one association, younger rocks another. Faunal succession shows that life has changed with the passage of time. By examining fossil assemblages from rocks in one area and comparing them with fossil assemblages from rocks in another area, we are able to demonstrate the correlation of rock formations with their included fossil assemblages from region to region and even from continent to continent.

UNCONFORMITIES

There is no known place on Earth where sedimentation has been continuous throughout geologic time. Periods of quiet stability of the Earth's crust were interrupted by tectonic uplift and downwarping or subsidence. During the formation of a continent, large sections of the crust are raised out of the shallow seas in which its rocks were formed, subjected to erosion, and then lowered again to levels where deposition of sediments is renewed. Activity of this sort produces a buried erosion surface with younger rocks overlying older rocks. Some surfaces of this kind can be seen today because

8.2 (a) A striking angular unconformity between steeply dipping and slightly metamorphosed sedimentary rocks and nearly flat overlying sedimentary beds on the southern coast of England. The underlying rocks, originally flat-lying, have been tilted, metamorphosed, and then truncated by erosion. Deposition of the presently horizontal beds took place on this erosional surface, and created the angular unconformity. (b) A sketch of the major features shown in (a).

(a)

(b)

of another cycle of uplift. A buried erosion surface separating two rock masses of which the older was exposed to erosion for a long interval of time before deposition of the younger is called an **unconformity.**

The time represented by an unconformity is important geologic evidence of the history of a region, marking an interval when the surface was above the sea and sediments were not being deposited. Some unconformities represent gaps of a few thousand years; others, as many as billions of years.

There are three principal types of unconformity: angular unconformity, disconformity, and nonconformity.

Angular Unconformity   An unconformity in which the older strata dip at an angle different from that of the younger strata is called an **angular unconformity.** Along the southern coast of England can be seen some layers of sedimentary rocks dipping steeply. Above them are some nearly horizontal sedimentary rocks (Box 8.1). The older rocks were deposited under the waters of the ocean and were then folded and uplifted above the water. While exposed at the surface, the tilted beds were beveled by erosion. Then, as time passed, these tilted and eroded rocks sank again and were covered by new layers of sediments. Both were later elevated and exposed to view (see Figure 8.2). On the basis of fossil evidence we know that the second sedimentation began during the Pleistocene Epoch, indicating that many millions of years elapsed between the two periods of sedimentation. The older rocks meet the younger rocks at an angle, so this unconformity is an angular unconformity.

Disconformity   An unconformity with parallel strata above and below is called a **disconformity.** It is formed when layered rocks are elevated, exposed to erosion, and then lowered to undergo further deposition without being folded. Careful study and long experience are required to recognize a disconformity because the younger beds are parallel to the older ones. Geologists rely heavily on fossils to determine ages of beds above and below a disconformity and establish whether a significant interval of time is unrepresented. Fossils are a useful tool because some groups lived during a specific geologic time range.

Nonconformity   An unconformity between profoundly different rocks is called a **nonconformity.** Nonconformities are formed where intrusive igneous rocks or metamorphic rocks are exposed to erosion and then are downwarped to be covered by sedimentary rocks. A structure of this sort is illustrated in Figure 8.3, which shows a sandstone deposit lying on top of an eroded surface of metamorphic rocks.

Unconformities in the Grand Canyon   The Grand Canyon of the Colorado exposes several unconformities (see Figure 8.4). This region has undergone three major sequences of uplift and erosion, subsidence and deposition, and crustal deformation and crustal stability. These have produced three major rock units: Vishnu Schist, Grand Canyon Series, and Paleozoic Series.

The Vishnu Schist is the base rock of the canyon. It is composed of highly metamorphosed sedimentary rock and some volcanics. It has yielded no fossils and may be 2 billion years old.

8.3   A nonconformity between dark-colored Precambrian schists and gneisses (Vishnu Schist) in the gorge of the Grand Canyon and the overlying horizontal sedimentary beds of the Cambrian Tapeats Sandstone, forming the flat Tonto Plateau at the top of the gorge.

BOX 8.1   Siccar Point—Hutton's Classic Unconformity

Nearly 200 years ago James Hutton, John Playfair and Sir James Hall discovered gently dipping strata of the Devonian Old Red Sandstone resting upon nearly vertical graywackes of Silurian age at Siccar Point, Scotland (Figure B8.1.1). In commenting on the significance of this discovery Playfair wrote, "The mind seemed to grow giddy by looking so far into the abyss of time. . . ."

Siccar Point is on the east coast of Scotland about 4 km east of Cockburnspath. From an historical viewpoint the unconformity at Siccar Point is of great significance. Here was solid evidence of the vast amount of time that many were just beginning to recognize as characteristic of Earth history. The exposure showed that a dark sandstone, a graywacke, (Hutton called it a schistus) had been deposited, then lithified and then tipped on end. Erosion followed and the vertical beds then submerged in a depositional basin in which sand was deposited, then lithified and finally uplifted to be exposed once again to erosion. How many millions of years were involved no one then knew. But the 18th century geologists grasped the fact that an enormous amount of time was most certainly involved. Here are portions of Playfair's account of the discovery:

B8.1.1   The unconformity at Siccar Point, Scotland, shows gently dipping beds of sandstone overlying the eroded ends of nearly vertical beds of a much older sandstone. [Edward A. Hay.]

". . . .We made for a high rocky point or head-land, the Siccar, near which, from our observations on shore, we knew that the object we were in search of was likely to be discovered. On landing at this point, we found that we actually trode on the primeval rock, which forms alternately the base and the summit of the present land. It is here a micaceous schistus, in beds nearly vertical, highly indurated, and stretching from southeast to northwest. The surface of this rock runs with a moderate ascent from the level of low-water, where the schistus has a thin covering of red horizontal sandstone laid over it; and this sandstone, at the distance of a few yards

farther back, rises into a very high perpendicular cliff. Here, therefore, the immediate contact of the two rocks is not only visible, but is curiously dissected and laid open by the action of the waves. The rugged tops of the schistus are seen penetrating into the horizontal beds of sandstone, and the lowest of these last form a breccia containing fragments of schistus, some round and others angular, united by an arenaceous cement.

Dr. Hutton was highly pleased with appearances that set in so clear a light the different formations of the parts which compose the exterior crust of the earth, and where all the circumstances were combined that could render the observation sat-

isfactory and precise. On us who saw these phenomena for the first time, the impression made will not easily be forgotten. The palpable evidence presented to us, of one of the most extraordinary and important facts in the natural history of the earth, gave a reality and substance to those theoretical speculations, which, however probable, had never till now been directly authenticated by the testimony of the senses. We often said to ourselves, "What clearer evidence could we have had of the different formation of these rocks, and of the long interval which separated their formation, had we actually seen them emerging from the bosom of the deep?" We felt ourselves necessarily carried back to the time when the schistus on which we stood was yet at the bottom of the sea, and when the sandstone before us was only beginning to be deposited, in the shape of sand or mud, from the waters of a superincumbent ocean. An epoch still more remote presented itself, when even the most ancient of these rocks instead of standing upright in vertical beds, lay in horizontal planes at the bottom of the sea, and was not yet disturbed by that immeasurable force which has burst asunder the solid pavement of the globe. Revolutions still more remote appeared in the distance of this extraordinary perspective. The mind seemed to grow giddy by looking so far into the abyss of time; and while we listened with earnestness and admiration to the philosopher who was now unfolding to us the order and series of these wonderful events, we became sensible how much farther reason may sometimes go than imagination can venture to follow. . . ." (*Biographical Account of the Late Dr. James Hutton*, Transactions, Royal Society of Edinburgh, vol. 5, p. 39–99, 1805).

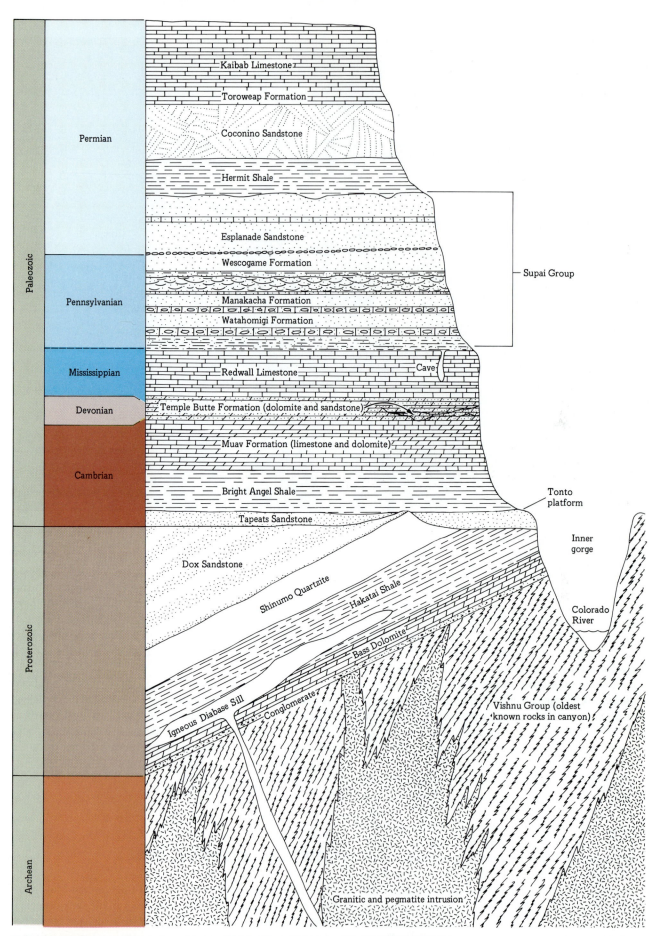

8.4 Diagrammatic representation of one wall of the Grand Canyon along the Kaibab Trail, showing unconformities. [After W. J. Breed and E. Roat (eds.), *Geology of the Grand Canyon*, Museum of Northern Arizona — Grand Canyon Natural History Association, Flagstaff and Grand Canyon, Ariz., 1976.]

8.5   Paleozoic section of the Grand Canyon from the Permian Kaibab Limestone at the top to the Cambrian Tapeats Sandstone at the base.

Resting unconformably on the Vishnu Schist in some places are tilted beds of the **Precambrian Grand Canyon Series**. The beds consist of 75 m of Bass Limestone, 180 to 240 m of sandy shale, and 300 m of relatively resistant beds of Shinumo Sandstone, topped off by 600 m of shaly sandstone. It has been estimated that an additional 3,000 m of Grand Canyon sediments were deposited on the beds before the region was elevated, deformed, and then eroded to the surface on which Paleozoic sediments were deposited. An uplift of at least 3,600 m had to take place to bring the lowest Grand Canyon layers above sea level. Erosion then removed all but the lower sections of some tilted blocks. It is not possible at the present time to say how long it took to complete the downwarping of the Vishnu Schist, deposition, sedimentation, deformation, elevation, and erosion.

After the Grand Canyon Series had been eroded to a surface of low relief and then submerged, Paleozoic sediments were deposited. These sediments are found today in horizontal layers along the upper part of the canyon walls, indicating uplifting of the region without deformation. The Paleozoic Series contain at least four discontinuities with the longest time gap between the Muav Limestone and the Temple Butte Formation. The Muav Limestone was deposited in Cambrian time. The Temple Butte Formation immediately overlying these and essentially parallel to them contains fossils of primitive armored fish that occur only in Devonian rocks. Apparently there was no deposition during Ordovician or Silurian time, a lapse of some 80 million years (Figure 8.5).

In areas where all the Grand Canyon Series had been eroded away before deposition of Paleozoic sediments, we find the "great unconformity." Its time gap is estimated as several hundred million years. Because sedimentary rocks are then found resting on metamorphic and igneous rocks, this is an example of a nonconformity.

8.2
THE GEOLOGIC COLUMN

Using the law of superposition and the concept that fossils are an index to time, geologists have made chronological arrangements of sedimentary rocks from all over the world. They have pictured the rocks as forming a great column,

TABLE 8.1
The Geologic Column

Era	System	Series	Aspects of the life record		
			Some major events	Dominant life form[a]	
Cenozoic	Quaternary[b]	Holocene (Recent) Pleistocene			Man
	Tertiary	Pliocene Miocene	Grasses become abundant	Mammals	Flowering plants
		Oligocene Eocene Paleocene	Horses first appear		
Mesozoic	Cretaceous Jurassic Triassic	—[c]	Extinction of dinosaurs Birds first appear Dinosaurs first appear	Reptiles	Conifer and cycad plants
Paleozoic	Permian Pennsylvanian Mississippian	—[c]	Coal-forming swamps	Spore-bearing land plants	
	Devonian Silurian		Vertebrates first appear (fish)		Fish · Marine invertebrates
	Ordovician Cambrian		First abundant fossil record (marine invertebrates)		Marine plants
Proterozoic Archean	—[d]			Primitive marine plants and invertebrates One-celled organisms	

[a] This column does not give the complete time range of the forms listed. For example, fish are known from pre-Silurian rocks and obviously exist today; but when the Silurian and Devonian rocks were being formed, fish represented the most advanced form of animal life.
[b] Some geologists prefer to use the term Cenozoic for the Quaternary and the Tertiary.
[c] Many Mesozoic and Paleozoic series are distinguished but are not necessary here.
[d] Precambrian rocks are abundant, but worldwide subdivisions are not generally agreed upon.

with the oldest at the bottom and the youngest at the top. The pioneer work in developing this pattern was carried out in the British Isles and in Western Europe, where modern geology had its birth. There geologists recognized that the change in the fossil record between one layer and the next was not gradual but sudden. It seemed as if whole segments recording the slow change of plants and animals had been left out of the rock sequence. These breaks, or gaps, in the fossil record served as boundaries between adjoining strata. The names assigned to the various groups of sedimentary rocks are given in the **geologic column** of Table 8.1. (See Box 8.2 and also front endpaper.)

Notice that the oldest rocks are called **Precambrian,** a general term applied to all the rocks that lie beneath the Cambrian rocks. Although the Precambrian rocks represent the great bulk of geologic time, we have yet to work out satisfactory subdivisions based on fossils because Precambrian rocks do not contain the variety of hard-bodied life forms that can be used to correlate from one region to another. Correlation has been accomplished primarily by physical features and by dates obtained from radioactive minerals. Physical features have been useful for establishing local sequences of Precambrian rocks, but these sequences cannot be extended to worldwide subdivisions. On the other hand, radiometric dates are not yet numerous enough to permit subdivisions of the Precambrian rocks, although such dates will become increasingly important as more are assembled.

All of geologic time since the Precambrian is known as the Phanerozoic (from the Greek *phaneros,* "visible," plus *zoic,* "life"); that is, when fossil forms are more readily found in the rock record.

We have said that the geologic column was originally separated into different groups of rocks on the basis of apparent gaps in the fossil record. But as geologic research progressed and as the area of investigation spread from Europe to other continents, new discoveries narrowed the gaps in the fossil record. It is now apparent that the change in fossil forms has been continuous and that the original gaps can be filled in with data from other localities. This increase in information has made it more and more difficult to draw clear boundaries between groups of rocks. Yet, despite the increasing number of "boundary" problems, the broad framework of the geologic column is still valid.

BOX 8.2   The Geologic Calendar

Elsewhere we have said that evidence points to about 4.6 billion years for the age of the solid Earth. During the last 200 years our knowledge about our planet has increased enormously, and as the information grew, geologists were able to divide geologic time into smaller and more precisely defined units, particularly for the last half billion years or so. The resulting geologic calendar is, in its broader aspects, understood by geologists the world over. However, the student who comes upon the geologic-time scale for the first time finds the various units of geologic time an almost meaningless mishmash of terms, and their proper chronological sequence daunting to master.

The newcomer's difficulties derive in part from the fact that the geologic calendar represents an accretion of names and terms over two centuries. New terms have replaced old ones, old terms have acquired new meanings, and the very basis for the choice of terms has varied. A helpful analogy of how the geologic calendar evolved is the evolution of terms in our present-day calendar. Thus March comes from Mars, the Roman god of war; August is named for the Roman emperor Augustus; and the name of our twelfth month, December, comes from the Latin *decem,* for "ten," because the new year once began in March and December was then the tenth month.

The names in the geologic calendar were initially based on the belief that rock types were an index to their age. Thus the igneous rocks, such as granite and basalt, were thought to be the oldest rocks and so were called **Primary** in age. The layered sedimentary rocks,

such as limestone, sandstone, and shale, were considered younger than the igneous rocks and were therefore called **Secondary.** Finally, the loosely consolidated sediments were termed **Tertiary** and the loose, unconsolidated sediments were called **Quaternary.** Eventually, as the significance of fossils and the laws of superposition and cross-cutting relationships became known and appreciated, rock type was abandoned as a basis for age determination. Instead, the presence of past life (fossils) in many of the rock units was used to denote relative age. **Paleozoic** (ancient life) replaced Primary; **Mesozoic** (middle life) replaced Secondary; and Tertiary and Quaternary became **Cenozoic** (recent life). These terms are still in use today.

We now subdivide these eras of geologic time into smaller and smaller units. As will be seen in the examples below, at times the names for these units have been chosen on the basis of geography, at other times because of rock type, and in some instances to honor ancient tribes.

Let us begin with the Paleozoic. Its oldest time division is the **Cambrian,** derived from the Roman name for Wales, where rocks of this particular age were first described. The Cambrian is followed by the **Ordovician** and **Silurian,** from the names of early tribes in northern Wales and the eastern Welch borderlands, respectively. Next comes the **Devonian,** after Devon in southern England, where these rocks were first described. Then come the **Mississippian,** after the upper Mississippi valley of the midwestern states, and the **Pennsylvanian,** after western Pennsylvania. In England rocks of

these two ages are called **Carboniferous** because of the coal beds they contain. The **Permian** ends the Paleozoic Era and is named for a region in the Ural Mountains of the Soviet Union.

The Mesozoic era, the age of dinosaurs, begins with the **Triassic** Period, so named because of the characteristic threefold division of rocks of this age in Germany, where sandstone is often succeeded by limestone and that by shale. **Jurassic** time, which follows, is based on rocks in the Jura Mountains of eastern France and northwestern Switzerland. **Cretaceous** time saw the end of both the Mesozoic Era and the dinosaurs; it takes its name from the Latin *creta,* for the chalk found in the White Cliffs of Dover and similar rocks across the English Channel in France.

The old terms *Tertiary* and *Quaternary* are still used today, but as subdivisions of the Cenozoic.

But what about all the Earth time that we know existed before the Paleozoic began 570 million years ago? It was clear that there were rocks older than the Cambrian, the initial period of the Paleozoic. What time terms should they be given? The first answer was a simple one and the obvious name of **Precambrian** was used for the time between the formation of the Earth and the Cambrian, the beginning of the Paleozoic. The subdivision of Precambrian time has proved to be difficult, even though this period lasted nearly 4 billion years. For our purposes here, we can recognize two major subdivisions: The younger is called the **Proterozoic** (for "earlier life"), which began 2.5 billion years ago; and the older is called the **Archean.**

8.3
ABSOLUTE TIME

We Earthlings use two basic units of time: (1) the day, the interval required for our planet (in the present epoch) to complete one rotation on its axis; and (2) the year, the interval required for the Earth to complete one revolution around the Sun. In geology the problem is to determine how many of these units of time elapsed in the dim past when nobody was around to count and record them. Our most valuable clues in solving this problem are provided by the decay rates of radioactive elements.

RADIOACTIVITY

The nuclei of certain elements spontaneously emit particles that change in form and produce new elements. This process is known as **radioactivity.** Shortly after the turn of the twentieth century researchers suggested that minerals containing radioactive isotopes (see Chapter 2) of certain elements could be used to determine the age of other minerals in terms of absolute time (see Chapter 1).

We will take a single radioactive element as an example of how this method works. Regardless of what element we use, we must know what products result from its radioactive decay. Let us choose uranium 238, which we will designate in shorthand fashion as $^{238}_{92}U$ to symbolically illustrate that this uranium isotope has an atomic number of 92 and an atomic mass of 238. $^{238}_{92}U$ is known to yield helium and lead, $^{206}_{82}Pb$, as end products. We know, too, the rate at which uranium 238 decays. As far as we can determine, this rate is constant and is unaffected by any known chemical or physical agency. The rate at which a radioactive element decays is expressed in terms of what we call its **half-life** — the time required for half of the nuclei in a sample of that element to decay. The half-life of uranium 238 is 4.51×10^9 years, which means that if we start with 1 g of uranium 238, there will be only 0.5 g left after 4,510 million years (Figure 8.6).

The history of 1 g of uranium 238 may be recorded as shown in Table 8.2. At any instant during this process there is a unique ratio of lead 206 to uranium 238. (Remember that lead 206 is formed only by uranium 238 decay and therefore *accumulates* with time.) This ratio depends on the length of time decay has been going on. Theoretically, then, we may find the age of a uranium mineral by determining how much lead 206 is present and how much uranium 238 is present. The ratio of lead to uranium then serves as an index to the age of the mineral. (Though Figure 8.6 shows this change as a smooth curve going from one parent element to a second daughter element, it is important to note that radioactive decay is a stepwise process. One radioactive isotope decays to a given product, which in turn may decay to another and another before the final stable end product is reached.)

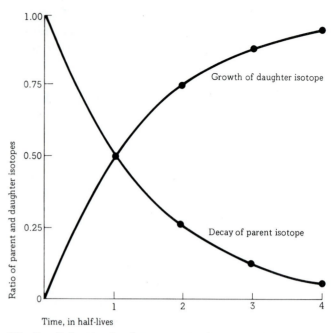

8.6 Graph showing the change in ratio of parent and daughter isotopes plotted against half-lives.

A radioactive element decays in one of several ways. The nucleus may lose an alpha particle — the nucleus of the helium atom, which, as we saw in Chapter 2, consists of two protons and two neutrons. This process is called **alpha decay,** and the mass of the element decreases by 4 (two protons plus two neutrons), and the atomic number decreases by 2 (two protons). An element undergoing **beta decay** loses an electron (a beta particle) from one of the neutrons of the nucleus. The neutron thereby becomes a proton, and the atomic number is increased by 1. The mass — the sum of all protons and neutrons — remains the same. In a third type of decay, **electron-capture decay,** the nucleus changes by picking up an electron from its orbital electrons. This electron is added to a proton within the nucleus, converting the proton to a neutron and decreasing the atomic number by 1.

Many elements, in decaying from a radioactive to a nonradioactive state, go through a series of transformations until one or more stable end products are reached. Thus uranium 238 begins to decay by alpha emissions before lead 206 is produced.

TABLE 8.2
History of 1 g of Uranium 238[a]

Age, millions of yr	$^{206}_{82}Pb$ formed, g	$^{238}_{92}U$ remaining, g
100	0.013	0.985
1,000	0.116	0.825
2,000	0.219	0.747
3,000	0.306	0.646
4,500 (1 half-life)	0.433	0.500
9,000 (2 half-lives)	0.650	0.250
13,500 (3 half-lives)	0.758	0.125

[a] From various sources.

Some elements may follow two or more different paths in their decay. Potassium 40 is an example. It may decay either by beta emission to form calcium 40 or by electron capture to form argon 40.

Methods of determining age by radioactivity have produced tens of thousands of dates for events in Earth history, and new ones are constantly being reported. The most reliable dates utilize more than one isotopic pair and yield consistent results.

Rocks from southwestern Minnesota and Greenland have been dated as approximately 3.8 billion years old. Field relations show that other, still older rocks exist. A granite about 3.2 billion years old from Pretoria, South Africa, contains inclusions of a quartzite, positive indication that older sedimentary rocks existed before the intrusion of the granite. An age of 4.2 billion years was obtained from samples of zircon in Australian rocks. It is interesting to note that these zircons are not *now* in the rocks in which they crystallized. They are, in effect, 4.2-billion-year-old "fossil grains" recovered from metamorphosed sediments that accumulated half a billion years later. These are exciting as well as tantalizing results because they tell us something about an early continental crust that is no longer around.

It should be noted that we are able to check radioactive age dates against each other. **Because decay rates are unique to each stable isotope, the probability that two or more methods will agree randomly is very low.** The fact that several clocks regularly agree indicates that radioactive dating is self-consistent and reassures us that we are measuring real ages.

Today's discoveries, then, have fully vindicated the assumptions made over a century and a half ago that geologic time *is* vast and that within Earth history there *is* abundant time for slow processes to accomplish prodigious feats. The exact age of the Earth itself is still undetermined, but several lines of evidence converge to suggest an age of about 4.6 billion years.

Radiocarbon Dating   There are scores of radioactive isotopes, but only a few are useful in geologic dating (see Table 8.3). For older rocks, potassium 40, rubidium 87, uranium 235, and uranium 238 have proved most important; a relatively new method of age determination compares argon 40 to argon 39. Potassium 40 has also proved

useful in dating some of the more recent events of Earth history, events younger than 2 million years. For very recent events, however, the radioactive isotope of carbon, carbon 14, $^{14}_{6}C$, has proved most versatile. It is usable *only* on organic material that is around 50,000 years old or younger. We will show in detail how this method works, even though we recognize that **many other methods of geologic age dating are more useful.**

The carbon 14 method, first developed at the University of Chicago by Willard F. Libby, works as follows. When neutrons from outer space, sometimes called **cosmic rays,** bombard nitrogen in the outer atmosphere, they knock a proton out of the nitrogen nucleus, thereby forming **carbon 14:**

Neutron

$^{14}_{7}N$

Proton $^{14}_{6}C$

The carbon 14 combines with oxygen to form a special carbon dioxide, $^{14}CO_2$, which circulates in the atmosphere and eventually reaches the Earth's surface, where it is absorbed by living matter. It has been found that the distribution of carbon 14 around the world is almost constant. Its abundance is independent of longitude, latitude, altitude, and the type of habitat of living matter.

The bulk of carbon in living material is the stable isotope carbon 12. Nevertheless, there is a certain small amount of carbon 14 in all living matter. And when the organism—whether it is a plant or an animal—dies, its supply of carbon 14 is, of course, no longer replenished by life processes. Instead, the carbon 14, with a half-life of about 5,730 years, begins spontaneously to change back to $^{14}_{7}N$ by beta decay. The longer the time that has elapsed since the death of the organism, the less the amount of carbon 14 that remains. So when we find carbon 14 in a buried piece of wood or in a charred bone, by comparing the amount present with the universal modern $^{14}C/^{12}C$ ratio, we can calculate the amount of time that has elapsed since the material ceased to take in carbon 14, that is, since the organism died.

We have seen that there are assumptions on which we base our evaluation of the validity of radioactive dating. These assumptions are reasonable and have some support

TABLE 8.3
Some of the More Useful Elements for Radioactive Dating

Parent element	Half-life, yr	Daughter element	Type of decay
Carbon 14	5,730	Nitrogen 14	Beta
Potassium 40	1,300 million	Argon 40	Electron capture
Rubidium 87	47,000 million	Strontium 87	Beta
Uranium 235	713 million	Lead 207	Seven alpha and four beta
Uranium 238	4,510 million	Lead 206	Eight alpha and six beta

from observation. Nevertheless, the more we use radiometric methods, the more sophisticated our understanding of them—which is another way of saying that we become aware of the problems involved. Let us take carbon 14 as an example of how increased knowledge reveals increased complexity.

When Libby did his initial work on carbon 14, its half-life was determined to be 5,570 years. More refined measurements now indicate it to be about 5,730 years. By common agreement, however, 5,570 is still used as the "accepted value," and therefore all published dates should be adjusted upward 3 percent. Beyond this, Hans Suess, while working with the United States Geological Survey, demonstrated that since about 1850, the amount of nonradioactive, or "dead," carbon 12 poured into the atmosphere by the combustion of coal and petroleum has diluted the amount of carbon 14 to 98 percent of its original concentration before human beings began tampering with the atmosphere. Working in the opposite direction, we have created a great amount of new carbon 14 since 1945, when we began atmospheric nuclear explosions. In fact, this source of contamination has doubled the carbon 14 activity of the atmosphere.

Of more historical interest, perhaps, is the discovery that radiocarbon dates obtained on materials from about the time of Christ and going back for another 5,000 years, at least, are younger than we should expect when we compare them with carbon 14 dates of samples of known age dated independently of carbon 14. The latter have included historically dated material from the Mediterranean area, particularly from Egypt, and tree rings of the bristlecone pine from the western United States. This tree, which includes the world's oldest known living material, has provided us with a set of tree rings (like the modern tree rings in Figure 8.7) that goes back over 7,000 years and is securely tied to our modern solar calendar. When we subject tree rings older than 2,000 years to carbon 14 dating, the radiocarbon date is younger than the actual date (Box 8.3). The farther back we go, the greater is the divergence. For example, by the time the tree ring count reaches 3200 B.C., the radiocarbon age is only 2700 B.C. (see Figure 8.8). We do not yet know the cause of this divergence or how far back into time

8.7 Tree rings exposed in a cross section of a lodgepole pine stump. The variations in ring width suggest that conditions in the swamp alternated between dry and wet several times during the life of this tree.

it may continue. But it suggests that we are well advised to use the term **radiocarbon years** to indicate that radiocarbon dates are not absolutely synonymous with dates in our solar calendar.

Despite these difficulties, the thousands of radiocarbon dates now available have revolutionized many aspects of archaeology as well as the study of the last 50,000 years of Earth history.

Fission-Track Dating   A slightly different method of absolute age dating is made possible by the spontaneous fission of certain heavy elements, particularly the relatively abundant isotope uranium 238. As this isotope undergoes fission, the resulting parts of its nucleus leave characteristic tracks in the crystals of rocks where they occur. The number

8.8 Radiocarbon analyses of tree rings of known age show that data postulate ages that are too young for the period from the birth of Christ back at least into the sixth millennium B.C. The discrepancy increases from zero to about 750 years in 5000 B.C. [Modified from E. K. Ralph, H. N. Michael, and M. C. Han, *Masca Newsletter*, vol. 9, Fig. 8, p. 5, 1973.]

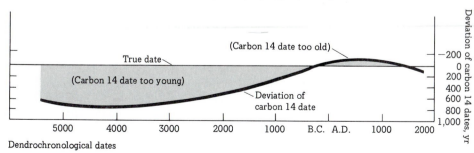

BOX 8.3   Dendrochronology—Reading the Rings of Trees

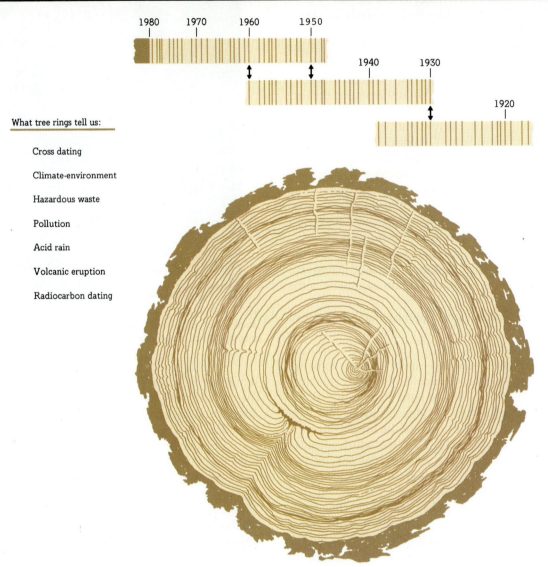

B8.3.1   Some types of information that can be obtained from tree rings.

By reading tree rings, scientists have been able to trace back a continuous record of at least 8,000 years of the Earth's history—its climate, volcanic eruptions, fires, droughts, earthquakes, solar cycles, periods of glaciation, pollution, and other changes in the atmosphere. Tree rings allow for more precise radiocarbon age dating. (See Figure B8.3.1.)

Dendrochronology (from the Greek words *dendron,* meaning "tree," *chronis,* meaning "time," and *logos,* meaning "study") involves matching the rings of one tree to corresponding rings of other trees, dead or alive, through a process known as **cross dating.** By using overlapping sections of living trees and even of trees that died long ago, dendrochronologists have been able to piece together this unbroken record of more than 8,000 years.

That record tells us much about the climate and environment in which the tree grew. Each ring records details about the atmosphere for each given year—whether wet or dry, hot or cold, polluted or clean. Many trees are sensitive to stresses and record them as thicker or thinner rings, burn scars from fires, irregularities due to insect infestations, and chemical scars from natural or manufactured products.

Recent detailed studies around a hazardous waste dump site have demonstrated the ability of scientists to trace the movement of pollution through an area. Tom Yanosky, a U.S. Geological Survey hydrologist-botanist, in such a study at the Aberdeen Proving Ground in Maryland, compared the distribution of elements such as arsenic, chlorine and iron, from ring to ring, and tree to tree, and succeeded of such tracks gives us a clue to the absolute ages of those rocks, ages that can be measured from a few tens of years ago to the beginning of the solar system. This tremendous range makes the fission-track dating method extremely useful.

RADIOACTIVITY AND SEDIMENTARY ROCKS

Until recently radioactive minerals suitable for dating geologic events were sought chiefly in the igneous rocks. These rocks were usually the uranium- and thorium-bearing minerals. Even today the great bulk of the radioactively dated rocks are igneous in origin. The development of new techniques, however, particularly the use of radioactive potassium, has extended radioactive dating to some of the sedimentary rocks.

Some sandstones and, more rarely, shales contain glauconite, a silicate mineral similar to biotite. Glauconite, formed in certain marine environments when the sedimentary layers are deposited, contains radioactive potassium, another geologic hourglass that reveals the age of the mineral and hence the age of the rock. The end products of the decay of radioactive potassium are argon and calcium. Age determinations are based on the ratio of potassium to argon.

Some pyroclastic rocks are composed mostly or completely of volcanic ash. Biotite in these rocks includes radioactive potassium and so offers a way of dating the biotite and sometimes the rock itself.

SEDIMENTATION AND ABSOLUTE TIME

Another way of establishing absolute dates for sedimentary strata is to determine the rate of their deposition.

Certain sedimentary rocks show a succession of thinly laminated beds. Various lines of evidence suggest that, in some instances at least, each one of these beds represents a single year of deposition. Therefore, by counting the beds, we can determine the total time it took for the rock to be deposited.

Unfortunately, we have been able to link this kind of information to our modern calendar in only a very few places, such as the Scandinavian countries. Here the Swedish geologist Baron de Geer counted the annual deposits, or laminations, that formed in extinct glacial lakes. These laminations, called **varves** (see Figure 8.9 and Chapter 15), enable us to piece together some of the geologic events of the last 20,000 years or so in the countries ringing the Baltic Sea.

Much longer sequences of laminated sediments have been found in other places, but they tell us only the **total length of time** during which sedimentation took place, **not how long ago** it happened in absolute time. One excellent example of absolute-time sequence is recorded in the Green

8.9 Varves from Puget Lowland, Washington. Strips of paper are used to record thickness and succession of the various layers. Each pair of light (summer) and dark (winter) layers is thought to represent a single year of sedimentation. (Note pocket knife holding uppermost paper.) [J. Hoover Mackin.]

River shales of Wyoming. Here each bed, interpreted as an annual deposit, is less than 0.02 cm thick, and the total thickness of the layers is about 980 m. These shales, then, represent approximately 5 million years of time.

8.4

THE GEOLOGIC-TIME SCALE

The names in the geologic column refer to rock units that have been arranged in a chronological sequence from oldest to youngest. Because each of the units was formed during a definite interval of time, they provide a basis for setting up time divisions in geologic history.

In effect, the terms we apply to time units are the terms that were originally used to distinguish rock units. Thus we speak either of **Cambrian time** or of **Cambrian rocks.** When we speak of **time units,** we are referring to the geologic-time scale. When we speak of **rock units,** we are referring to the **geologic column.**

The geologic-time scale is given in Table 8.4 as well as in the front endpaper. Notice the terms **Era, Period,** and **Epoch** across the top of the table. These are general time terms. Thus we can speak of the Paleozoic Era, or the Permian Period, or the Pleistocene Epoch. In Table 8.1 the terms **System** and **Series,** used as general terms for rock units, correspond to the time units of "period" and "epoch," respectively. The term **Erathem** is becoming the generally accepted rock term equivalent to the "era" of the geologic-time scale.

ABSOLUTE DATES IN THE GEOLOGIC-TIME SCALE

We have found that the geologic-time scale is made up of units of relative time and that these units can be arranged in proper order without the use of any designations of absolute time. This relative-time scale has been constructed on the

TABLE 8.4
The Geologic-time Scale[a]

Era	Period	Epoch	Millions of yr	
			Duration	Before present
CENOZOIC	QUATERNARY	Holocene	0.01	
		Pleistocene	2	
MESOZOIC	TERTIARY	Pliocene	64	
		Miocene		
		Oligocene		
PALEOZOIC		Eocene		
		Paleocene		66
	Cretaceous		78	
	Jurassic		64	
	Triassic		37	245
	Permian		41	
PRECAMBRIAN	Pennsylvanian		34	
	Mississippian		40	
	Devonian		48	
	Silurian		30	
	Ordovician		67	
	Cambrian		65	570
	Proterozoic		1930	
	Archean[b]		2000	2500

[a] From A. R. Palmer, "The Decade of North American Geology, 1983 Geologic Time Scale," *Geology,* vol. 11, pp. 503–504, 1983.
[b] The oldest dated rock material is about 4.2 billion years old; the Earth itself is about 4.6 billion years old.

basis of sedimentary rocks. As noted earlier, geologists have only recently been able to date some sedimentary rocks by radioactive methods, and most dates have come from the igneous rocks. How can the dates obtained from igneous rocks be fitted in with the relative-time units from sedimentary rocks?

We use the absolute age of igneous rocks to bracket the probable age of sedimentary units. To do this, we must know the relative-time relationships between the sedimentary and igneous rocks. Let us recall once again the **law of cross-cutting relationships,** which states that **a rock is younger than any rock that it cuts across.**

Consider the example given in Figure 8.10, which shows a hypothetical section of the Earth's crust with both igneous and sedimentary rocks exposed. The sedimentary rocks are arranged in three assemblages, numbered 1, 3, and 5, from oldest to youngest. The igneous rocks are 2 and 4, also from older to younger.

The sedimentary rocks labeled 1 are the oldest rocks in the diagram. First they were folded by Earth forces; then a dike of igneous rock was injected into them. Because the sedimentary rocks had to be present before the dike could cut across them, they must be older than the dike. After the first igneous intrusion, erosion beveled both the sedimentary rocks and the dike, and across this surface were deposited the sedimentary rocks labeled 3. At some later time the batholith, labeled 4, cut across all the older rocks. In time, this batholith and the sedimentary rocks labeled 3 were also beveled by erosion, and the sedimentary rocks labeled 5 were laid across this surface. We now have established the relative ages of the rocks, from oldest to youngest, as 1, 2, 3, 4, and 5.

Now, if we can date the igneous rocks by means of radioactive minerals, we can fit these dates into the rela-

TABLE 8.5
Relative, Absolute, and Approximate Ages of Rocks[a]

Event	Relative	Age Absolute, millions of yr	Age Approximate, millions of yr
Sedimentary rocks	5[b]		<230
Erosion			<230
Batholith	4	230	
Sedimentary rocks	3		>230, <310
Erosion			>230, <310
Dike	2	310	
Folding			>310
Sedimentary rocks	1		>310

[a] See Figure 8.10 and the text.
[b] Youngest.

tive-time sequence. If we establish that the batholith is 230 million years old and that the dike is 310 million years old, the ages of the sedimentary rocks may be expressed in relation to the known dates; the final arrangement will be as shown in Table 8.5.

By this general method, approximate dates have been assigned to the relative-time units of the geologic-time scale, as shown in Table 8.4. These dates may be revised and refined as new techniques for dating develop. As suggested earlier, one of the most exciting developments in this field lies in the direct application of dating methods to the radioactive minerals of sedimentary rocks.

CHANGING ESTIMATES OF THE AGE OF THE EARTH

Estimates of the age of the Earth as made by Western scholars have increased dramatically since the middle of the seventeenth century. The most widely quoted of the early estimates is that of Archbishop James Ussher, who in the 1760s used the Mosaic account in the Bible to calculate the origin of earth as 4004 B.C., approximately 6,000 years ago. Other eighteenth-century scholars, however, turned to an examination of the record of the Earth itself and were impressed by the demands it made for long periods of time, well in excess of that suggested by the biblical account. Few attempted a precise judgment, although the Comte de Buffon, Georges Louis Leclerc (1707–1788), an able and remarkable French naturalist, publicly suggested in 1778 an age of 75,000 years for the Earth, a most extravagant claim for the time. Interestingly, Buffon made the even more extravagant claim of 3 million years in his unpublished studies. Well into the nineteenth century, however, most students of the Earth were content to see, as James Hutton had, "no vestige of a beginning, no prospect of an end," although they were prepared to accept Earth time as spanning millions, perhaps hundreds of millions, of years.

8.10 This diagram illustrates the law of cross-cutting relationships, which states that a rock is younger than any rock that it cuts across. The rock units are arranged in order of decreasing age from 1 to 5. The manner in which radioactive ages of the igneous rock (2 and 4 of the diagram) are used to give approximate ages in terms of years for the sedimentary rocks (1, 3, and 5 of the diagram) is discussed in the text and shown in Table 8.5.

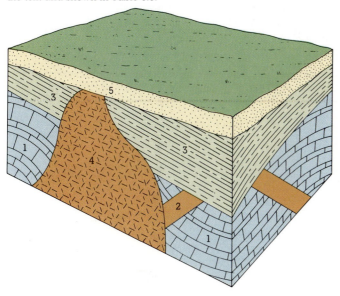

In the second half of the nineteenth century some physicists and geologists attempted more exact estimates for the age of the Sun, the Earth, and of segments of Earth time. Thus the Baron Hermann von Helmholtz suggested that the Sun owed its luminosity to gravitational attraction as particles fell in toward the center of the huge solar mass, and on this basis estimated the Sun and the Earth to be between 20 and 40 million years old. Shortly thereafter William Thomson (later Lord Kelvin) stated that the Sun, whatever its origin, was a cooling body whose energy would be dissipated in a finite length of time.

Charles Darwin, in the first edition of *Origin of Species* in 1859, suggested that 300 million years had passed since the end of what we now call the Mesozoic Era, a figure nearly five times greater than that now accepted. This prompted Kelvin to take issue with those who were advancing what seemed to him exorbitant estimates of Earth time. In 1864, on the basis of his calculations of a presumably originally molten Earth, he warned his geological colleagues that time was not infinite and that the Earth was at most 400 million years old and very probably much less. In 1868 Kelvin lowered his 400-million-year maximum to 100 million years. A number of geologists, assuming a constant rate for the deposition of the sedimentary rocks of the Earth, arrived at various estimates centered on 100 million years. The 100-million-year figure was repeated in 1899 by John Joly, an Irish geologist, who estimated the age of the oceans according to their total salt content and the annual rate at which salt was contributed. Of course we know today the salts of the ocean are derived by a complex of methods; dissolved salts are contributed not only by erosion of the continents but also through recycling of waters in plate tectonic processes (see Chapter 11).

Even as these end-of-the-century estimates appeared, the basis for an entirely new approach to dating Earth history was taking shape, an approach that was to expand our concept of the age of the Earth 1,000- to 10,000-fold. The French chemist Henri Becquerel discovered that uranium-bearing substances were the source of a previously unknown form of energy. Intrigued by the discovery, Marie Curie began analyzing pitchblende, which we know as an ore of uranium. In 1898, in conjunction with her husband, Pierre, she isolated and identified two new elements, radium and polonium, and applied the term **radioactivity** to describe them. The next link was provided by Ernest Rutherford (later Lord Rutherford), a physicist on the faculty of McGill University, to which he had come, via Cambridge University, from his native Australia. Rutherford held that the atoms of radioactive elements were unstable, that they disintegrated at a constant rate by the emission of alpha and beta particles, thus producing energy and wholly new and lighter elements. In 1904 Rutherford summarized our knowledge of the subject to that time in a volume called *Radio-Activity* and pointed out the potential usefulness of radioactive elements and their daughter products in dating Earth materials.

A British chemist, J. W. Strutt, obtained in 1905 an age of 2 billion years for the Earth, based on the helium content of a radium-bearing rock. About the same time Bertrand B. Boltwood, a Yale University chemist, after analyzing the ratio of uranium to lead in a number of samples, obtained an Earth age between 400 million years and 2.2 billion years. By present-day analytical standards these were crude dates, but they showed how measuring radioactivity could greatly enlarge our understanding of Earth time. As the twentieth century advanced, so did our understanding of nuclear physics and refinement of laboratory techniques. In 1931 the British geologist Arthur Holmes summarized the consensus of knowledge to that time and concluded that the Earth was older than 1 billion years — probably about 1.6 billion years and presumably less than 3.3 billion years.

Since World War II rocks older than 3 billion years have been dated from such widely separated locales as Minnesota, Greenland, and South Africa. Obviously the Earth must be considerably older than these rocks. Dating of meteorites gives us ages between 4.5 and 4.8 billion years for the origin of the solar system, and thus of our planet.

This brief history might suggest that future scientific developments will extend even further our estimates for the age of the Earth (see Figure 8.11). Such an inference could be very misleading and should not be the necessary outcome of such speculation!

8.11 Changing estimates of the age of the Earth with changing methods of age determination. See text for discussion.

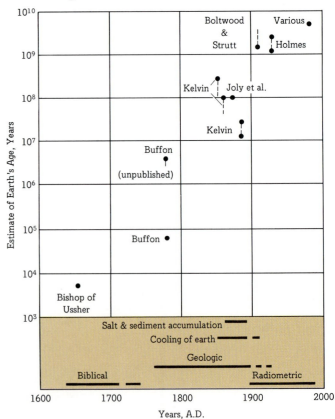

8.5

MAGNETOSTRATIGRAPHY

The direction of the Earth's magnetic field in past geologic ages can be determined by examining lava flows on land, deep-sea cores of sediments, and mafic igneous rocks from many parts of the Earth's surface (see Chapter 11). These rocks contain iron minerals (magnetite, ilmenite, hematite) that become magnetized at the time of their formation, thus creating a "magnetic fingerprint" in the rock for a certain instant of geologic time. These rock records indicate that the Earth's magnetic field has changed directions many times, with the present field representing **normal** directions for north and south. During intervals of **reversed polarity** the poles apparently switch from south to north and vice versa. These changes have taken place fairly rapidly (geologically speaking) and have affected newly forming magnetic minerals. The process of reversal takes about 2,000 years according to studies conducted at the University of Pittsburgh.

Reversals have apparently occurred during at least the last billion years. Studies of magnetism in rocks from various parts of the world can be used with the geologic-time scale to work out a magnetic time chart. A pattern emerges for the polarity reversals through geologic time, with **polarity epochs**[1] lasting roughly 1 million years. Within these epochs there occur several **polarity events**[2] in which the Earth's polarity switches from one direction to another and back to the first, with durations on the order of 100,000 years for each polarity. Such a pattern has existed for some time (Figure 8.12): Polarity episodes of rapid change alternating with long intervals of little change occurred as far back as a billion years ago, according to present understanding. As we go farther back in geologic time, of course, the difficulties of age dating and measuring magnetic directions increase enormously. Nevertheless a distinct pattern is emerging for the use of magnetized rocks in determining a **magnetic stratigraphy (magnetostratigraphy)** for the Earth.

---

[1] Called polarity **zones** by the International Subcommittee on Stratigraphic Classification (ISSC), *Geology*, vol. 7, December 1979, pp. 578–583.
[2] Called polarity **subzones** by ISSC; ibid.

8.12 Chart of magnetostratigraphy, showing magnetic-anomaly numbering system (based on marine magnetic-anomaly analysis), polarity epochs (based on potassium-argon dating and paleomagnetic studies of outcropping Pliocene and Pleistocene igneous rocks), and the polarity-epoch number system (based on micropaleontological and paleomagnetic analysis of piston cores.) See also Figure 11.11.
[Modified from N. D. Watkins, *Geotimes*, April 1976.]

Time, millions of yr	Magnetostratigraphic units		Geologic-time units			
	Magnetic-anomaly number	Paleomagnetic polarity (polarity epoch)	Epoch		Period	Era
	1	Brunhes 1	Pleistocene	Late	Quaternary	
	2	Matuyama 2		Early		
	3	Gauss 3	Pliocene	Late		
		Gilbert 4		Early		
5		5		Late		
		7				
	4	6 8				
	5	9				
10		11 10				
		12		Middle		
	5a	13	Miocene			
15						

Reversed polarity

☐ Normal polarity

Summary

Geologic time can be expressed in either relative or absolute terms.

Relative time has been determined largely by the relative position of sedimentary rocks to each other.
   The rocks are arranged in proper chronological position according to the **law of superposition.**
   A rock is younger than any rock it cuts across according to the **law of cross-cutting relationships.**
   Correlation of sedimentary rocks from one area to another allows the extension of a chronology of relative time from region to region and continent to continent. Correlation is based on **physical features,** but most importantly on **fossils.**

An **unconformity** is a buried erosion surface separating two rock masses where the older was exposed to erosion for a long interval of time before deposition of the younger.
   Angular unconformity is one in which the older strata dip at an angle different from that of the younger strata.
   Disconformity is an unconformity with parallel strata above and below.
   Nonconformity is an unconformity between profoundly different rocks.

The geologic column is a chronological sequence of rocks from oldest to youngest. The units are rock units.

Absolute time is expressed in terms of years and is measured by the rate of decay of radioactive elements. Among the most useful such elements are **carbon 14, potassium 40, rubidium 87, uranium 235,** and **uranium 238.**

The geologic-time scale is the chronological sequence of units of Earth time represented by the rock units of the world geologic column. Absolute dates in the geologic-time scale are based on the relation of radioactively dated igneous, sedimentary, and metamorphic rocks of the geologic column. Position in the time scale of events dated from radioactive rocks is determined by the relative position of those rocks and the fossils they contain.

Magnetic reversals have occurred at intervals during the geologic past and can be used to subdivide Earth history into **epochs of magnetostratigraphy.**

Questions

1. Reword and simplify the law of superposition. What does it mean in simple terms?
2. How are rocks correlated from one place to another by physical features? By fossils?
3. How does a disconformity differ from a nonconformity? How might one merge into the other?
4. Explain briefly the carbon 14 dating method.
5. How are radioactive dating techniques possible on sedimentary rocks?
6. What proportion of the total age of the Earth is represented by Precambrian time? By Phanerozoic (that is, post-Precambrian) time?
7. Explain how dating of igneous intrusions aids in the establishment of a geologic-time scale.
8. What is meant by normally polarized rocks? By reversed polarity?
9. What causes reversals in the Earth's polarity?

Supplementary Readings

BEDINI, S. A. *Thomas Jefferson and American Vertebrate Paleontology,* Virginia Division of Mineral Resources, 1985.
   Details how Jefferson had a greater impact than any other American of his time on the collecting and study of fossil remains of extinct animals.

HARLAND, W. B., A. V. COX, P. G. LLEWELLYN, C. A. G. PICTON, H. G. SMITH, AND R. WALTERS *A Geologic Time Scale,* Cambridge University Press, Cambridge, 1982.
   A systematic presentation of geologic time and stratigraphic principles.

Libby, W. F. "Radiocarbon Dating," *Am. Sci.,* vol. 44, no. 1, 1956, pp. 98–112.
A summary of the method written by the man who discovered it.

Ojakangas, R. W., and D. G. Darby *The Earth, Past and Present,* McGraw-Hill Book Co., New York, 1976.
A condensed review of the Earth through time, including biologic and physical developments.

Ralph, E. K., and H. N. Michael "Twenty-five Years of Radiocarbon Dating," *Am. Sci.,* vol. 62, no. 5, 1974, pp. 553–560.
A review of errors inherent in radiocarbon dating and the use of long-lived bristlecone pines to correct radiocarbon dates.

Stanley, S. M. *Earth and Life through Time,* W. H. Freeman & Co., New York, 1986.
Deals with the interplay between the physical history of the Earth and the history of life, from the beginning to the present. Integrates plate tectonics, mountain building, climatic change, regional geology, and the history of life.

Watkins, N. D. "Polarity Subcommission Sets Up Some Guidelines," *Geotimes,* vol. 21, no. 4, 1976, pp. 18–20.
A short history of its development and presentation of data and chart for magnetostratigraphy.

Atacama Fault, paralleling coast of Chile, with both normal and strike-slip movements. [NASA.]

CHAPTER ▪9

DEFORMATION

"[I]t is . . . the geologist who must furnish the essential data which define the nature of the physical processes that have produced the present structure of the Earth's crust."

Walter H. Bucher, American Geophysical Union Transactions, *p. 163, 1940*

In Chapter 5 we discovered that rocks of the Earth's crust were being weathered and thus changed into sediments by surface and near-surface processes. The crust is also being changed in other ways by forces acting within it. These forces lift rocks to positions where weathering and erosion can operate in the global recycling program—the rock cycle.

In our examination of crustal deformation we begin with a discussion of *stress* and *strain* and find that rocks may behave as *elastic, plastic,* or *brittle* substances. When deformation takes place in rocks, it is expressed in a number of different features. These include *folds* which are of several kinds, and fractures, which include *joints* and *faults,* the latter being of several different kinds.

The final section of this chapter discusses the very *abrupt movements* of the crust, both vertical and horizontal; the very *slow movement* of the crust; and those movements that create *broad surface features.*

9.1

STRESS, STRAIN, AND STRENGTH OF ROCKS

We can speak of deformation of the Earth's crust in terms of changes of volume, changes of shape, and a combination of these two, as well as in terms of displacements of rock units on either side of a fracture in those rocks. We can illustrate changes in shape without changes in volume by shearing a deck of cards (Figure 9.1). If the bottom card is held fixed and the others are slid forward by amounts proportional to their distances from the bottom, the shape of the space occupied by the deck is altered, but its volume (the total amount of cardboard) remains the same. Moreover, if shape remains unchanged but rock volume is decreased or increased, we have what is sometimes called **volume deformation.** Some geologists do not regard this as true deformation because the shape still looks the same. Nevertheless, as we saw in our discussion of densities in Chapter 2, reduction in volume produces a denser rock, often with mineralogic changes. This chapter will concentrate on shape deformation (folding) and on displaced rock masses (faulting).

9.1 Deformation may produce change in volume without change in shape, or change in shape without change in volume, or a combination of the two.

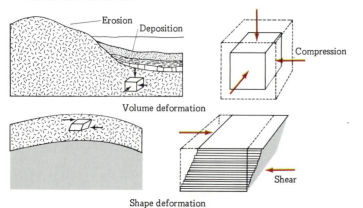

9.2 Graph of stress versus strain for any hypothetical material. Three stages of deformation are illustrated: elastic, where stress and strain are proportional, giving a straight-line relationship; plastic range, where strain increases rapidly with only minor increases in stress; and rupture or fracture, where failure occurs. The yield point marks the boundary between elastic and plastic ranges. [From various sources.]

STRESS AND STRAIN

A force applied to material that tends to change the material's dimensions is called **stress.** This is most often described in terms of **unit stress,** which is defined as the total force divided by the area over which that force is applied.

The *effect* of the deforming forces on earth materials, produced by the application of stress, is known as **strain. Unit strain** is the fraction change in length, width, or height. Graphs of stress versus strain are usually set up with unit stress and unit strain as the axes (Figures 9.2 and 9.3). In most materials, as stress increases, for a time the **strain is proportional to the stress.** If the stress is removed, the strain goes back to zero. Deformation in this interval is an example of **elastic deformation.** If the stress is increased, a **yield point** is eventually reached, beyond which, even with the complete removal of stress, the material will no longer return to its original size or shape. We then have **plastic deformation.** It is a fact that rocks, hard and brittle as they appear at the Earth's surface, can indeed be bent and squeezed

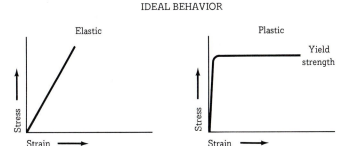

IDEAL BEHAVIOR

Elastic

Plastic

Yield strength

Stress

Strain

Stress

Strain

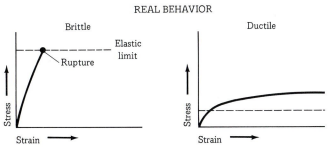

REAL BEHAVIOR

Brittle

Elastic limit

Rupture

Ductile

Stress

Strain

Stress

Strain

9.3 Responses to applied stress. The dashed line represents the approximate yield strength for plastic behavior. [Modified from Myron G. Best.]

without breaking, provided that deformation occurs under high confining pressure, such as at depths within the Earth. If so much stress is applied that the material can no longer resist breaking, **rupture** occurs.

Elastic Deformation   Elastic deformation of rock is recoverable. A solid is deformed elastically if it can recover its size and shape after the deforming force has been removed. An elastic solid resists deformation up to its yield point. A material's resistance to elastic shear is termed its **rigidity.**

   Retardation of recovery.   Rocks are said to be deformed elastically if the deformation disappears when the stress is removed. However, they do not ordinarily regain their former shape the instant the stress is removed. There is a time lag over which recovery takes place.

Plastic Deformation   Plastic deformation is permanent. It involves a property of rock called **viscosity.** A material that is deforming plastically does so by the propagation and movement of dislocations (or small structural defects in the material), by intragranular shear, or by recrystallization (especially in response to pressure solution). If the rate of the flow is proportional to the stress causing it, the material is said to be **viscous,** and such a substance is known as Newtonian. If we plot the rate of flow against the stress, the slope of the graph is the viscosity. Viscosity is an important property in some geological processes. It governs the ability of magma to flow during igneous activity — it is the *resistance* to flow.

   Plastic deformation is equivalent to flowage (like toothpaste squeezed from the tube), but a plastic solid differs from a fluid in having a yield strength. (A fluid de-

forms at even the slightest application of stress, whereas a plastic solid does not deform until some significant stress has been applied and the threshold of the yield point has been reached.)

Brittle Deformation   As we have seen, if stresses are too great, a rock will break. If there is differential movement on either side of such a break, it is called a **fault.** Brittle deformation results in fracturing and faulting.

Types of Stress   There are three types of stress: **tensional, compressional,** and **shear.** Tension is a stretching stress and can increase the volume of a material. Compression, on the other hand, tends to decrease the volume. A shear stress produces changes in shape.

Time Factor   Within the Earth materials show a duality of response. They behave in an elastic fashion in response to some stresses but move plastically during folding. They may behave in a brittle fashion when stresses are applied very rapidly. Here the important factor, which cannot easily be reproduced in a laboratory, is time. Material may be elastic when subjected to short-term stresses but respond plastically to stresses maintained over a longer time.

   In the short term a tuning fork formed from pitch can ring as though it were the purest steel. Yet, given sufficient time, the fork will flow into a formless blob under its own weight. A steel bar supported at its ends and loaded in the middle appears to have strength entirely adequate to support the load but will in time slowly bend. Likewise, many metamorphic rocks testify to deformation under prolonged periods of stress, increased temperature, and fluids promoting recrystallization. On a geologic-time scale, the rocks of the Earth's crust have shown themselves to be weak.

STRENGTH OF ROCKS

Rocks possess strength. **Strength** is the limiting stress that a solid can withstand without failing by rupture or continuous plastic flow.

   Rocks possess several types of strength, as they respond differently to different stresses. Laboratory test results for rocks generally give **compression strength, shear strength,** and **tension strength** (corresponding to the three types of stress discussed above). These tests have shown that a rock's strength in tension is smaller than its strength in compression. When the difference between these two is large, we say the material is **brittle.** When it is small, we say the material is **ductile.** As the confining pressure increases, a brittle rock tends to become ductile (Figure 9.4). (At sufficient depths within the crust the load of overlying rocks produces **lithostatic pressure,** an all-sided confining pressure uniform in all directions, similar to **hydrostatic pressure** at significant depths in the ocean.) Brittle rock does not rupture in the ordinary sense of disintegration but flows

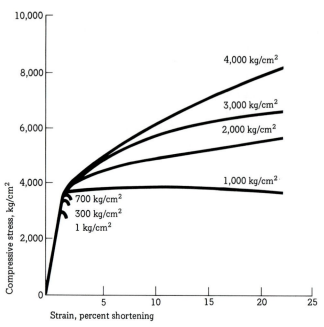

9.4 Effect of confining pressure on behavior of limestone under compression. [After E. C. Robertson, "Experimental Study of the Strength of Rocks," *Bull. Geol. Soc. Am.*, vol. 66, pp. 1294–1314, 1955.]

along an indefinite number of shear planes. Rupture in tension may be important near the Earth's surface, but it will not occur at great depth. The strength in compression for brittle rocks is about 30 times larger than the strength in tension.

How a rock mass responds to stress depends on temperature and depth of burial. At depth rocks typically respond to stress by flowing before they rupture, whereas at the Earth's surface they usually break. The stresses at the Earth's surface are quite different from those at depth, where temperatures and pressures increase considerably. Whether a rock breaks or flows under a given stress depends on the prevailing pressures and temperatures. Rocks are generally weaker at high temperatures than they are at low ones. If a rock is surrounded by equal pressure on all sides, it tends to increase in strength. Such pressure strengthens a rock, and increased temperature weakens it.

Different rocks have different compressive strengths. Hard, brittle rocks such as sandstone have a different strength than clay or shale. The Earth's crust responds to an addition or subtraction of mass on a regional scale. This process is called **isostasy** (see Box 9.1).

BOX 9.1   Isostasy and Equilibrium

A mass standing high above its surroundings will exert a gravitational attraction that can be computed and measured. One device for measuring this attraction makes use of a plumb bob suspended on a plumb line. Like every other object on the globe, the suspended bob is pulled by gravity. On the surface of a perfect sphere with uniform density, the bob would be pulled straight down, and the plumb line would point directly toward the center of the sphere. But if there is any variation from these ideal conditions—that is, if there are surface irregularities on the sphere or variations in rock density—the plumb line will be deflected as the bob is attracted by their concentrations of mass.

In 1749 Pierre Bouguer found that plumb bobs were deflected by the Chimborazo Mountain in the Andes, but by amounts much less than calculated. In 1849, in the Pyrenees, British surveyors actually found that the plumb bob appeared to be deflected away from the mountains. Similar

discrepancies between calculated and measured values of gravitational attraction of mountains were also observed in the middle of the nineteenth century by British surveyors in India. They were using the plumb line to sight stars, in an attempt to fix the latitude of Kaliana, near Delhi in northern India, and of Kalianpur, about 600 km due south. They observed that the difference in latitude between these two stations was 5°23′37.06″. Then they checked the difference directly by standard surveying methods. The difference computed from these measurements was 5°23′42.29″. Thus there was a discrepancy of 5.23″, or about 150 m, between measurements. That may not seem very much over a distance of 600 km, but it was too large to be explained by errors of observation. Scientists then concluded that the plumb line at Kaliana had been deflected more by the attraction of the Himalaya Range than it had been at Kalianpur farther south. Actually, however, the discrepancy should have been three times as large on the

assumption that the mountains were of the same average density as the surrounding terrain and that they were resting as a dead load on the Earth's crust.

Two quite different explanations were proposed to explain the discrepancies between computed and observed values of gravity. The first was made by G. B. Airy in 1855. He regarded the Earth's crust as having the same density everywhere and suggested that differences in elevation resulted from differences in the thickness of the outer layer. Lithostatic equilibrium was achieved by lighter material's floating in a denser substratum. The depth of compensation is variable. Continents, mountains, and other topographic features are in equilibrium because they have roots like icebergs and are floating in a denser material.

J. H. Pratt, on the other hand, proposed that all portions of the Earth's crust have the same total mass above a certain uniform level, called the **level of compensation.** Any section with an

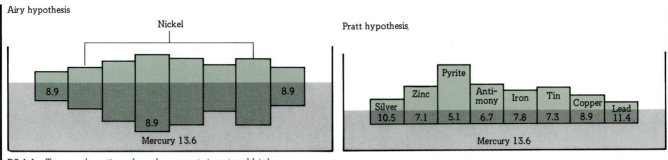

Airy hypothesis

Nickel

Pratt hypothesis

| Silver 10.5 | Zinc 7.1 | Pyrite 5.1 | Anti-mony 6.7 | Iron 7.8 | Tin 7.3 | Copper 8.9 | Lead 11.4 |

Mercury 13.6

Mercury 13.6

B9.1.1 Two explanations for why mountains stand high.

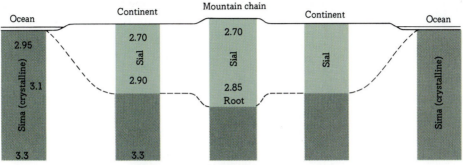

B9.1.2 Hypothesis explaining isostasy, with variations in density of crustal material. [After R. C. Daly, *Strength and Structure of the Earth*, Prentice Hall, Englewood Cliffs, N.J., 1940, p. 61.]

elevation higher than its surroundings would have a proportionately lower density. These ideas of Airy and Pratt are illustrated in Figure B9.1.1.

In 1889 C. E. Dutton suggested that different portions of the Earth's crust should balance out, depending on differences in their volume and specific gravity (see Figure B9.1.2). He called this **isostasy**. In his words:

> If the Earth were composed of homogeneous matter its normal figure of equilibrium without strain would be a true spheroid of revolution; but if heteroge-

neous, if some parts were denser or lighter than others, its normal figure would no longer be spheroidal. Where the lighter matter was accumulated there would be a tendency to bulge, and where the denser matter existed there would be a tendency to flatten or depress the surface. For this condition of equilibrium of figure, to which gravitation tends to reduce a planetary body, irrespective of whether it be homogeneous or not, I propose the name *isostasy* (from the Greek *isostasios,* meaning "in

equipoise with"; compare *isos,* equal, and *statikos,* stable). I would have preferred the word *isobary,* but it is preoccupied. We may also use the corresponding adjective *isostatic.* An isostatic Earth, composed of homogeneous matter and without rotation, would be truly spherical. . . .[1]

[1] "On Some of the Greater Problems of Physical Geology," *Bull. Phil. Soc. Wash.,* vol. 11, p. 51, 1889. Reprinted in *J. Wash. Acad. Sci.,* vol. 15, p. 359, 1925; also in *Bull. Natl. Res. Council (U.S.),* vol. 78, p. 203, 1931.

9.2

FEATURES OF PLASTIC DEFORMATION

Structural geology deals primarily with deformed masses of rock, their shapes, and stresses that caused the deformation. Sedimentary rocks and some igneous rocks form in approximately horizontal layers. When these are found tilted, folded, or broken, they provide evidence of deformation of the Earth's crust. Not uncommonly, erosion has stripped

away as much as several kilometers of uplifted portions of the crust, revealing structures that were once buried deeply. Deformation features include folds, joints, and faults.

In dealing with structural features geologists commonly describe the orientation of planar features in terms of **dip** and **strike** (Figure 9.5). These can more easily be described with reference to layered rocks. If a rock layer is not horizontal, the angle that the layer makes with the horizontal is the **magnitude** of its dip (see Figure 9.6). Dip direction is the direction of the maximum amount of **inclination**

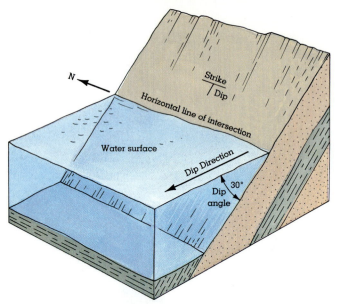

(at right angles to the strike). The strike of the beds is the course, or bearing, or any horizontal line on the inclined bed or structure. There can be an infinite number of such lines, but they will all be parallel to one another and therefore will all have the same bearing or strike. If the rock layer is tilted so that it disappears below the surface but protrudes somewhat because of its resistance to weathering and erosion, as in Figure 9.7, the strike is the direction in which the resulting ridge trends. A bed that dips either east or west has a north-south strike (usually designated by convention simply as north). A bed that dips either to the north or to the south has an east-west strike (Figure 9.8). Flat-lying strata, like some in the area of the Grand Canyon, have neither strike nor dip; they are simply described as being horizontal.

9.5   Block diagram illustrating the meaning of dip and strike of a plane. These planes, representing bedding planes, strike north and dip 30° to the west. Their attitudes can also be described as dipping 30° in the direction 270°.

9.6   Vertically dipping beds. This wall of vertically dipping sandstone is near Radussa, Sicily. [Pamela Hemphill.]

9.7   Dipping beds. Originally deposited horizontally, these sandstones and shales were later tilted so they now dip at an angle to the horizontal.

9.8   Aerial view of McDonnell Range, Australia. Folded beds are late Precambrian to early Paleozoic, with east-west trending axes. Width of view about 75 km. [NASA.]

FOLDS

Folds are a common feature of rock deformation. They are produced by compressive stress, and range in size from microscopic wrinkles in a piece of metamorphic rock to huge folds involving thousands of meters of thickness for distances of hundreds of kilometers. Folds are seldom isolated structures but rather generally occur in closely related groups. Sometimes they are very broad, and sometimes they are tight and narrow. Sometimes they are tilted to one side, and sometimes they are tilted on end.

In some areas sedimentary layers are only slightly bent, whereas in others, usually associated with mountain structures, they are intensely deformed. The difference in the kind of folding mainly depends on the amount of deforming stress in relation to the strength of the rocks and on the temperature and confining pressure during the deformation.

Mechanism of Folding   Folding can be classified as **concentric** or **flow. Concentric folding** is basically an elastic bending of an originally horizontal sheet. It occurs in surface layers of the crust. The size of a concentric fold is determined by the thickness of the beds involved and by their elastic properties. Beds are buckled during concentric folding, but the thickness of the beds and the volume remain the same. Large-amplitude folds form from thick strata and small-amplitude folds from thin strata. The distance between crests is also controlled by the thickness of the beds.

Flow is a type of deformation that occurs in rocks when they are in the plastic state. It involves internal movement that may take any direction. It occurs in weak rocks near the surface and in both strong and weak rocks at depth. Flow is due to differential stress operating at high pressures and high temperatures. It is the only true *plastic deformation* of rocks.

Monoclines   Monoclines are relatively simple examples of deformation, involving an elastic bending of sedimentary layers. A **monocline** is a double flexure connecting strata at one level with the same strata at another level. Extensive horizontal layers are bent down and pass beneath younger horizontal strata. The flexed layers did not accumulate with the bend in them. As bending progressed, erosion worked to strip away the higher portion.

There are many monoclines in the Colorado Plateau, some involving displacements as great as 4,000 m and lengths of 250 km. The Waterpocket fold in southern Utah is one of the better known. It involves displacements of close to 2,000 m. Estimates of the rate of its deformation range from a few centimeters to possibly a few meters per century.

Anticlines and Synclines   Arching of layered rocks produces a structure called an **anticline** (Figures 9.9 to 9.11). The two sides of the fold, called its **limbs,** normally dip away from each other. An imaginary plane bisecting the fold into as nearly equal halves as possible is called the **axial plane.** The **axis** of an anticline is the direction of an imaginary line drawn on the surface of a single layer parallel to its intersection with the axial plane. The axes of most geological folds are inclined. The angle of dip of its axis is the **plunge** of the fold. Parts of a fold are illustrated in Figure 9.10.

9.9 Broad anticlinal fold in Devonian rocks along the southwestern coast of England (opening approximately 1 m high and 4 m wide).

Anticlines can be spectacular when their anatomy has been exposed by erosion (see Figure 9.11). After erosional truncation the oldest rocks are exposed at the core and the youngest on the outer flanks.

A **syncline** is a downward fold, the opposite of an anticline. The limbs dip toward the central axis. Synclines are best seen in mountainous areas (such as the Ridge and Valley Province of the Appalachians [Figure 9.12] or the Jura Mountains in Western Europe), where there has been uplift along with the folding. The youngest rocks are in the center of the fold, with successively older rocks farther out (Figure 9.13).

9.10 The parts of a fold, shown on a plunging anticline but applicable to a syncline also. The axis is the line joining the places of sharpest folding. The axial plane (Ap) includes the axis and divides the fold as symmetrically as possible. If the axis is not horizontal, the fold is plunging, and the angle between the axis and the horizontal is the angle of plunge. The sides of a fold are the limbs. Figure walking from younger beds toward older beds, with oldest bed in center of anticline at the axial plane.

9.11 Folds in basement rocks, Flinders Range, Australia, showing resistant beds of eroded plunging synclines and anticlines. Width of area in photograph about 50 km. [NASA.]

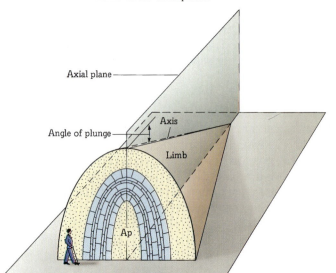

9.12 Photograph of a syncline along Interstate Highway 81 in northern Pennsylvania.

9.13 View of plunging syncline from top of cliff overlooking tightly folded Devonian rocks along the southwestern coast of England. See figures on beach for scale.

Anticlines and synclines commonly occur together in elongate groups like belts of wrinkles in the Earth's outer skin. Anticlines and synclines are said to be **symmetrical** if their opposite sides have approximately equal dips, and **asymmetrical** if one limb is steeper than the other. Types of folds are listed in Table 9.1 and illustrated in Figures 9.14 and 9.15.

Structural Domes   Structural domes are the result of pressure acting upward from below to produce an uplifted portion of the crust with beds dipping outward on all sides. Where erosion has removed the top of a dome, concentric ridges are commonly formed by one or more relatively resistant beds. The troughs are formed by erosion of weaker rocks. It is possible to project the dip and strike of the beds upward over the dome and reconstruct its whole shape before erosion removed any parts.

Some structural domes are formed by the upwelling of plastic material such as salt or magma. It is impossible to know whether slight folding started the upwelling or whether hydrostatic adjustment of loads on the crust was the primary cause. Where there is a local thickening or thinning of the crust, the plastic material starts to rise and

TABLE 9.1
Types of Folds

Name	Description
Anticline	Upfold, or arch
Syncline	Downfold, or trough
Monocline	Local steepening of an otherwise uniform dip
Dome	Anticline roughly as wide as it is long, with dips outward in all directions from a high point
Basin	Syncline roughly as wide as it is long, with dips inward in all directions toward a low point
Asymmetrical	Strata of one "limb of the fold" dip more steeply than those of the other
Overturned	One limb tilted beyond the vertical; both dip in the same direction, although perhaps not the same amount
Recumbent anticline	Beds on the lower limb upside down, axial plane nearly horizontal
Recumbent syncline	Upper limb inverted, axial plane nearly horizontal
Isoclinal	Beds on both limbs nearly parallel, whether fold upright, overturned, or recumbent.

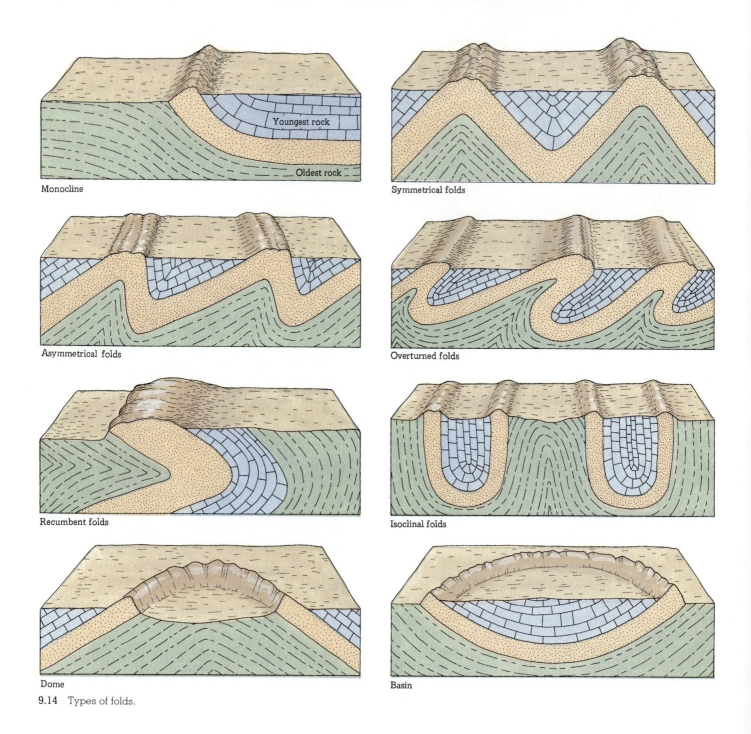

Monocline

Youngest rock

Oldest rock

Symmetrical folds

Asymmetrical folds

Overturned folds

Recumbent folds

Isoclinal folds

Dome

Basin

9.14 Types of folds.

continues as a result of the unequal static load. Most domes show marked upward bending of the surrounding beds, often accompanied by faulting. The beds above are domed and stretched by the push from below and often exhibit an intricate pattern of normal tension faults.

Structural Basins Downward fold beds dipping inward in all directions form structural basins, synclinal structures equivalent to, but opposite of, structural domes.

9.3
FEATURES OF BRITTLE DEFORMATION

JOINTS

Masses of rock may be broken, with or without movement along the cracks. If there has been no slippage along the fracture surface, the crack is called a **joint.** If there has been displacement, the structural features is called a **fault.**

(a)

(b)

9.15 (a) A recumbent fold in carbonate rocks in southeastern Pennsylvania. The fold has been tipped over on its side until the axial plane is nearly horizontal. Maximum dimension of the fold is 1 m. (b) Larger-scale recumbent fold in Beekmantown limestones of the Rheems Quarry, Pennsylvania.

Joints occur in all kinds of rock and, together with bedding, are the most common features of rocks exposed at the surface. They usually occur in sets, the spacing between them ranging from just a few centimeters to a few meters. As a rule, the joints in any given set are almost parallel to one another, but the whole set may run in any direction — vertically, horizontally, or at some angle commonly controlling streams and waterfalls (see Figure 9.16). Most rock masses are traversed by more than one set of joints, often with two sets intersecting at approximately right angles. Such a combination of intersecting joint sets is called a **joint system.** A regional pattern of joint systems may occur over areas of hundreds of square kilometers in a given type of rock exposed at the surface. As a rule, the kinds of rock have a marked influence on what joint systems develop. Massive sandstone and sandy shale, for example, each show characteristic jointing directions and spacing. See Figure 9.17 for illustration of sheeting, joints that occur fairly close together near the surface; and Figure 9.18 for illustrations of columnar jointing, a special pattern of jointing produced by the mechanism of cooling.

9.17   Sheet joints formed in granite rocks of the Sierra Nevada batholith, California. [William Bradley]

9.18   (a) Columnar jointing in Devils Tower. This volcanic remnant stands 390 m above the valley of the Belle Fourche River, northeastern Wyoming. (b) Columnar jointing in Devil's Post Pile, California. Note curving columns. [Pamela Hemphill.]

(a)

9.16   Bridal Veil Falls along Franz Josef glacial valley, South Island, New Zealand, with joint-controlled paths for waterfalls.

(b)

It is rarely possible to determine the age or origin of a joint set. It is generally accepted, however, that different sets in a joint system were probably made at different times and under different conditions. In sedimentary rocks one set could develop while the rock was being compressed by the weight of overlying rocks during burial and consolidation. Another set could be produced when pressure is released during unloading by erosion. In other cases, sets may be developed during deformation by tension. Some joint sets are known to be formed during the cooling of a mass of igneous rock, and others during movements of magma.

FAULTS

Faults reflect brittle deformation in which the sections on each side of the break move relative to each other. (Types of faults are listed in Table 9.2 and illustrated in Figure 9.19.) A useful basis for the classification of faults is the nature of the relative movement of the rock masses on opposite sides of the fault. If displacement is in the direction of dip, the fault is a **dip-slip fault.** The sections separated by the fault were named by miners who explored mines by following the faults underground. The block overhead is called the **hanging wall,** and the one beneath is the **footwall.** Dip-slip faults are classified according to the relative movement of these blocks. If the hanging wall seems to have moved downward in relation to the footwall, the fault is called **normal.** Normal faults occur when the vertical stress is greater than the horizontal stress.

Crustal movements that cause the surface to stretch may produce a series of related normal faults. A **graben** is an elongated, trenchlike structural form bounded by parallel striking normal faults created when the block that forms the

TABLE 9.2
Types of Faults

Type	Predominant relative movement
Dip-slip	Parallel to the dip
Normal	Hanging wall down
Thrust	Hanging wall up
Strike-slip (transcurrent)	Parallel to the strike
Right-lateral	Offset to the right
Left-lateral	Offset to the left
Oblique-slip	Components along both strike and dip
Hinge	Displacement dies out perceptibly along strike and ends at definite point

trench floor has moved downward relative to the blocks that form the sides. The term comes from the German for "trough" or "ditch" and is the same in both singular and plural. A **horst** is an elongated block bounded by parallel striking normal faults in such a way that it stands above the blocks on both sides. The term comes from the German and is used figuratively for a crag or height. The type of movement involved in the formation of a graben and a horst is illustrated in Figure 9.20.

A dip-slip fault in which the hanging wall appears to have moved upward in relation to the foot wall is a **thrust fault** (or **reverse fault** if high-angle, generally greater than 45°). Thrust and reverse faults are the result of largely horizontal compressive stresses.

A fault along which the movement has slipped parallel to the strike of the fault is called a **strike-slip fault.** Such faults have also been called **transcurrent faults.** Strike-slip faults are further designated as **right-lateral** or **left-lateral,** depending on how the ground opposite you appears to have moved when you stand looking across the fault.

9.19 Types of faults.

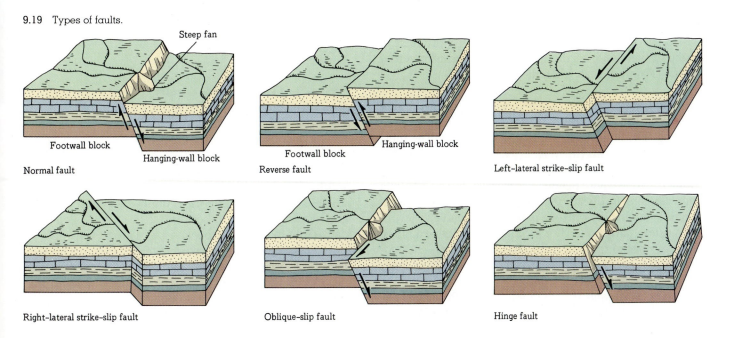

Normal fault

Steep fan

Footwall block

Hanging-wall block

Reverse fault

Footwall block

Hanging-wall block

Left-lateral strike–slip fault

Right-lateral strike–slip fault

Oblique–slip fault

Hinge fault

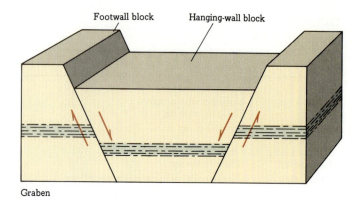

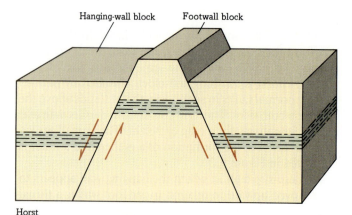

9.20 Horst and graben faulting.

The directions of movement involved in faulting are entirely relative. It is convenient to indicate one block as having moved upward or downward or as moving left or right. However, the absolute direction of movement usually cannot be determined, and the best that can be done is to indicate relative movements. During the upward warping of a large region, for example, all blocks could have moved upward, but some may have moved less than others. These lagging blocks could be said to have dropped relative to their neighbors, even though all parts at the finish were at higher elevations than when they started.

Normal Faults   In the Earth's crust there are large zones that have repeatedly been disturbed by large-scale normal faulting. The faults form horsts and rift valleys. The most famous, the African rift zone, extends over 6,000 km in a north-south direction. In some areas volcanoes are found along the zone where magma has worked its way up along fault surfaces. In Europe there are several zones of normal faults, including the rift zone of the Upper Rhine valley. In the western part of the United States we find an extensive zone of normal faulting in the Basin and Range Physiographic Province. Innumerable smaller normal fault zones can be found all over the world.

The Basin and Range Physiographic Province consists of tilted fault blocks. Its eastern limit is the Wasatch

fault. The main faults in the vicinity of Wasatch Mountain dip 50° to 55°. Net slip on major fault planes in the Wasatch Range is from 2,200 to 2,600 m. Farther west are the faulted Oquirrh Range, the Stansbury Range, and the Cedar Range. Each range is a tilted fault block some 30 km in breadth. The western limit is the great normal fault that bounds the tilted Sierra Nevada block on the east.

Thrust Faults and Reverse Faults   Thrust faults are usually closely associated with the folding process. Big, steep thrust faults are characteristic of the margins of mountain chains and also occur in their uplifted central blocks of crystalline rocks. Rocks above the fault plane seem to have been pushed up and over those below. Along the fault plane rocks are considerably crushed and sheared. Thrust faulting is most clearly seen in eroded regions where older rocks that were once deeply buried now overlie younger rocks that were nearer the surface. Older rocks are generally thrust over younger ones.

The total movement along a thrust fault can only be estimated in most cases, but some faults involve total vertical displacements as much as 1,000 m, combined with horizontal movements of tens of kilometers. The dips of thrust faults vary from less than 10° up to 45°. **Reverse faults** dip more steeply than thrust faults—generally 45° to 60° or even more. **Overthrusts** are far-traveled, low-angle thrusts.

Some well-known regions of thrust faulting are in the Alps, the southern Appalachians, the central and northern Rocky Mountains, and southern Nevada. Besides thrust faults of considerable displacement, every deformed region contains numerous small thrust faults (Figure 9.21). These may have displacements as small as 2 m but do not usually appear on geologic maps unless they occur in a mining area.

Thrust faults are often associated with asymmetric anticlines where rupture occurs in the highest part of the fold. A thrust fault of this type is a superficial structure because it extends down only to the basal plane and occurs in brittle rocks near the Earth's surface. Some are found in the Ridge and Valley Physiographic Province of the Appalachians, the best known being the Cumberland overthrust in Virginia.

Strike-Slip (Transcurrent) Faults   Strike-slip faults are found all over the world. A well-known example of left-lateral strike-slip faulting is the Great Glen fault, which intersects Scotland from coast to coast. There is a string of lakes, including Loch Ness, along its eroded trace, which is marked by a belt of crushed, sheared rock up to 1.5 km in width. Horizontal displacements along the fault of as much as 100 km have been suggested on the basis of correlating geological structures on both sides. The Great Glen fault dates from the Late Paleozoic.

Several left-lateral strike-slip faults rip across folds of the Jura Mountains, not far from Lake Geneva, and Lake Neuchâtel. They clearly originated during the folding process.

The San Andreas fault in California (Figure 9.22) is a

9.21   Thrust fault. Note movement was up on the side toward which the fault dips.

9.22   The scar of the San Andreas fault in the Carrizo Plains, 200 km northwest of Los Angeles. [Robert E. Wallace, U.S.G.S.]

right-lateral strike-slip fault that has been traced for nearly 1,000 km. Terrace deposits cut by the fault show offsets of as much as 10 km, and stream channels show shifts of 25 km within relatively short geological intervals. The total strike slip has been debated. Evidence has been interpreted as showing a displacement of 16 km since the Pleistocene, of 370 km since late Eocene time, and 580 km since pre-Cretaceous time. An offshoot from the San Andreas, the Garlock fault separates two clusters of dike swarms that seem pretty clearly to have been intruded as a single event. They are now separated by 80 km in a sense of movement that makes the Garlock a left-lateral strike-slip fault.

Transform Faults   Deformation of the Earth's ocean crust produces offsets of midocean ridges along strike-slip faults with large horizontal displacements. Ocean ridge segments are separated by what are called transform faults (because they "transform" from active faults between ridge axes to inactive fracture zones past the ridge axes). Figure 9.23 suggests the relationships that exist along a transform fault connecting two segments of a fragmented oceanic ridge and compares them with the relationships along a transversely faulted ridge. Analysis of earthquakes along the faults connecting segments of ridges indicates the motion to be as shown in Figure 9.23(a). This is exactly oppo-

site to that which would result from a simple offsetting of the ridge by transcurrent faults, as shown in Figure 9.23(b). The original offset of the ridges along transform faults may well have been due to transcurrent faulting. But with the establishment of transform motion, the ridge crests do not change position with time. In the case of continued transcurrent faulting, the offset increases with time. Many of the faults across the ocean ridges have been interpreted as transform faults: for example, those of the equatorial Mid-atlantic Ridge shown in Figure 9.24.

9.4
DEFORMATION OF THE EARTH'S CRUST

Evidence derived from rock outcroppings at the Earth's surface permit us to evaluate deforming forces at work deep within the Earth. These forces elevate and depress large landmasses. As we shall detail in the next chapter, such forces may be driving the huge plates of the outer part of the Earth. The resulting structures may produce vertical and/or horizontal displacement during rapid, abrupt movements or during long, imperceptibly slow movements.

9.23 Comparison between transform and transcurrent faults. The perspective sketch and plan in (a) show a ridge-to-ridge transform fault. New crustal material is postulated as being continuously added at the crest and transferred laterally away from the crest. There is no net displacement of the forms of the ridge segments with time. Transcurrent faults are shown in perspective and plan in (b). The ridge fragments are increasingly offset with time.

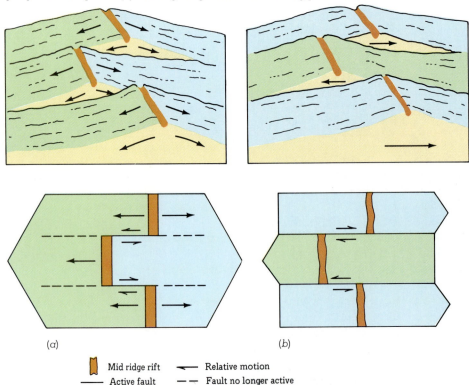

(a)   (b)

- ▮ Mid ridge rift   ← Relative motion
- —— Active fault   -- Fault no longer active

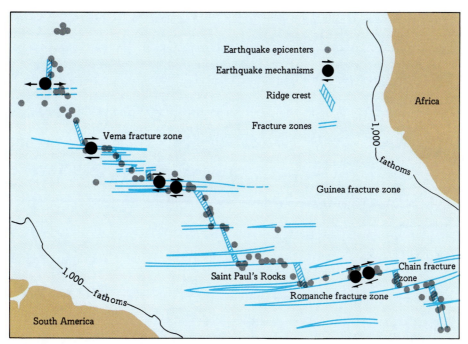

9.24 Equatorial section of the Midatlantic Ridge showing ridge crest and fracture zones. [After L. R. Sykes, "Seismological Evidence for Transform Faults, Sea Floor Spreading and Continental Drift," in R. A. Phinney, ed., *The History of the Earth's Crust*, Princeton University Press, Princeton, 1968. © 1968 by Princeton University Press. Reprinted by permission.]

ABRUPT MOVEMENTS

Crustal deformation has produced some large earthquakes, which have resulted in measurable displacements in historic times; surveys have revealed that large portions of the Earth's crust have been warped, tilted, or moved horizontally. Earthquakes have also altered the Earth's surface by triggering landslides.

Three major earthquakes in 1811 and 1812 were centered near New Madrid, Missouri, on December 16, January 23, and February 7. These were accompanied by changes in topography over an area of 8,000 to 13,000 km². There were also hundreds of small earthquakes, and at one point the ground shook almost continuously for weeks. Compaction of sand and mud produced sinking of 1.5 to 6 m in places, forming swamps and lakes and drowning forests. The largest lake to form was Reelfoot Lake in Tennessee.

The Alaskan Prince William Sound earthquake of 1964 was accompanied by considerable vertical and horizontal displacements. These occurred over more than 120,000 km². The Coast and Geodetic Survey found that there was a general south-to-north rotation of a portion of the Earth's crust centered at the northern coast of Prince William Sound. A maximum uplift of 13 m occurred off the southwestern tip of Montague Island. The largest subsidence of 2 m occurred 43 km north of Glennallen. Maximum horizontal movements were found between Montague and Latouche Islands, which, as we shall see, were 5 to

7 m closer together after the earthquake than before. Surveys showed that the southwest side of Montague Island moved northwest with respect to Latouche Island. One of the largest changes took place at Valdez, when the earthquake triggered landslides with deepening of the harbor by as much as 100 m in one place.

Vertical Displacements Vertical displacements during any one earthquake cannot account for the amount of elevation of the Earth's crust associated with mountains or for the large depressions of sedimentary layers found in deep coal mines. However, if displacements occur often enough throughout geologic time, the total can become significant. For instance, on the shore of Sagami Bay, Japan, not far from Yokohama, a cliff reared up during an earthquake on September 1, 1923. The amount of movement was measured by using some marine bivalves called *Lithophaga* ("rock eaters") as a reference. These little animals scoop out small caves for their 5-cm shells at mean sea level and spend their lives waiting for the sea to bring food at each rise of the tide. After the 1923 earthquake rows of *Lithophaga* were found starved to death 5 m above the waters that used to feed them. Other rows of *Lithophaga* holes in this same cliff were found and correlated with quakes in A.D. 1703, 818, and 33 (see Figure 9.25). The total elevation over that 2,000-year interval was 15 m. At this rate, the elevation over a geologic time interval of 2 million years would be 15,000 m.

9.25 Drawing showing elevated cliff on Sagami Bay, Japan. Inset, *Lithophaga* shells.

Horizontal Displacements The San Andreas fault is a large scar in the Earth's crust approximately 1,000 km long extending in a line from Cape Mendocino southeastward to the Gulf of California. Displacements have occurred along portions of the fault at the times of some earthquakes, but not over its entire length during any one quake (see Figure 9.26). At the time of the earthquake of April 18, 1906, there were horizontal displacements along 400 km of the northerly end from Point Arena to San Juan Bautista. The maximum slippage was 7 m on the shore near Tomales Bay. This displacement died out in both directions.

SLOW MOVEMENTS

Not all crustal movements are accompanied by earthquakes. Slow movements of the crust are going on today. These are being measured along faults in California by government geologists and university investigators in a broad, continuing program. They use cross-fault strain meters, which show that deformation occurs in zones up to 10 m across. Creep is in progress along the Hayward fault from Richmond to south of Fremont, along the Calaveras fault in the Hollister area, and along the San Andreas fault from San Juan Bautista to Cholame. It is interesting to note that at San Juan Bautista and at the Cienega Winery, the creep

occurs as distinct events, whereas the pattern of movement is different at Stone Canyon Observatory. Creep does not occur simultaneously at San Juan Bautista and Cienega Winery but may be several days apart. The difference in time at which creep starts suggests that it propagates from San Juan Bautista to the Cienega Winery at a rate of approximately 10 km/day. The average movement along a 100-km section of the San Andreas in the vicinity of Parkfield is 5 cm/year, with the Pacific side of the fault moving northwestward.

Another example is that of Mount Suribachi on Iwo Jima — the scene of the famous flag raising by United States Marines in 1945. According to geologist Takeyo Kosaka, pressure from underlying magma caused the surface to rise 7 m in 23 years from that date.

ROCK DEFORMATION AND PLATE MOTION

Plate boundaries have associated types of deformation. For example, overthrusts, tight-fold mountains, and other compressional features are associated with convergent plate margins. Normal faults, including horsts and grabens, rift valleys, and other extensional features, are associated with divergent plate boundaries. Strike-slip faults, like the San Andreas, are associated with transform boundaries. (For details see Chapter 11.)

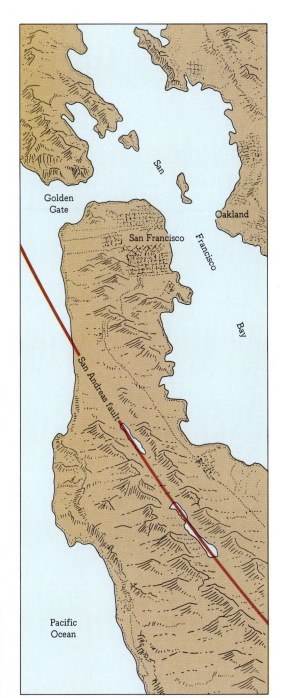

(a)

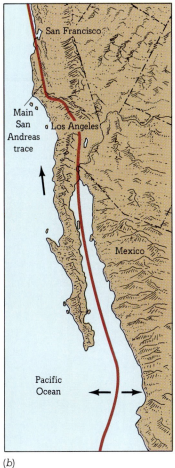

(b)

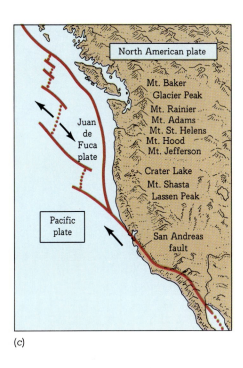

(c)

9.26 (a) "The San Andreas fault separates the North American and Pacific plates. (b) Movement along the fault has been responsible for many earthquakes, including the San Francisco quake of 1906 and the San Fernando quake of 1971 (see Figure 10.16). (c) Magma is generated along the subduction zone east of the Juan de Fuca crest. Such magma produces the material for the volcanoes of the Washington and Oregon coastal ranges." [Modified from Raymo, Chet, 1983, *The Crust of Our Earth: An Armchair Traveler's Guide to the New Geology*; Prentice Hall, Englewood Cliffs, N.J.]

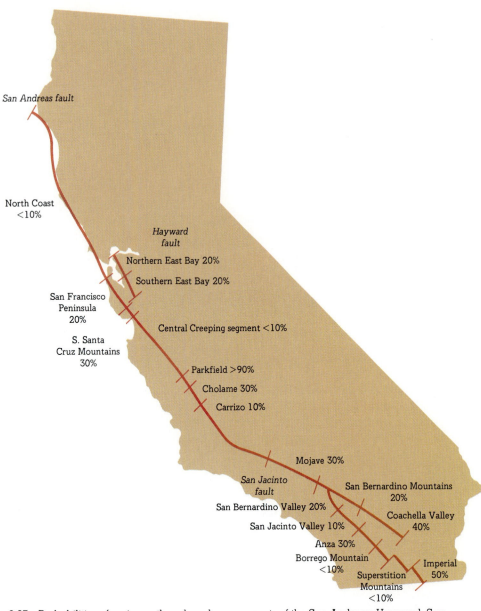

9.27 Probabilities of major earthquakes along segments of the San Andreas, Hayward, San Jacinto, and Imperial faults for the 30 years 1988–2018. (After Wayne Thatcher, U.S. Geological Survey Yearbook 1988).

Summary

Kinds of deformation are described in terms of change of volume, change of shape, or a combination of both.

Stress-strain relationships are used to define a material's response to stress.

Elastic deformation is recoverable.

Plastic deformation is permanent and involves a property of rock called **viscosity.**

Types of stress are **tensional, compressional,** and **shear.**

Time factor sometimes results in a material's being elastic for a short time but plastic over a longer time.

Strength of rocks is the stress at which they begin to be permanently deformed.

Structural features are the shapes of deformed masses of rock, including folds, joints, and faults.

Folds are a common feature of rock deformation and are produced by compressional stresses.

Mechanism of folding is elastic bending or plastic flow.

Monoclines are double flexures.

Anticlines and **synclines** occur together in elongate groups like belts of wrinkles in the Earth's outer skin.

Structural domes are circular-shaped anticlines.

Structural basins are circular-shaped synclines.

Joints usually occur in sets.

Sheeting is a pattern of essentially horizontal joints.

Columnar jointing is a set of cracks produced by the mechanism of cooling certain tabular plutons, flaws, and welded tuffs to produce columns.

Faults are deformation by rupture in which the sections on each side move relative to each other.

Normal faults have the hanging wall apparently dropped relative to the footwall.

Thrust faults have the hanging wall apparently moving higher than the footwall.

Strike-slip faults have movement predominantly horizontal, parallel to the strike of the fault.

Transform faults are strike-slip faults that terminate in a midocean ridge or deep-sea trench.

Evidence of deformation of the Earth's crust is best preserved in sedimentary and metamorphic rocks.

Abrupt movements of the Earth's crust have occurred at the times of some larger earthquakes.

Slow movements are being measured instrumentally in several places today, and on the San Andreas fault near Parkfield they average 5 cm/year.

Certain types of deformation are associated with the different types of plate boundaries.

Questions

1. Distinguish between stress and strain; between elastic and plastic deformation.
2. Define and illustrate strength as related to rocks.
3. How do we know that most folding of rocks takes place far below the Earth's surface rather than near the surface?
4. Differentiate between transcurrent and transform faults.
5. What are the surface patterns of outcrops for plunging anticlines and synclines?
6. How are normal and reverse faults different? How does each form?
7. Which displacements are greatest during earthquakes —vertical or horizontal? Explain your answer.
8. What differences are noted at the Earth's surface between crustal movements that are slow as compared to those that are rapid?
9. Why do we still have mountains when weathering and erosion have been at work on them for eons?

Supplementary Readings

DAVIS, G. H. *Structural Geology of Rocks and Regions,* John Wiley & Sons, New York, 1984.
Presents a practical understanding of geologic structures from hand specimen to outcrop to regional synthesis.

DE SITTER, L. U. *Structural Geology,* 2nd ed., McGraw-Hill Book Co., New York, 1964.
A textbook presupposing a certain familiarity with the elements of structural geology and its terminology.

KLIGFIELD, R., AND P. A. GEISER *Geology and Geometry of Thrust Belts,* Unwin Hyman, Winchester, Mass., 1989.
A summary of those regions of the world where major thrusting has occurred, with detailed descriptions of faulting.

RAGAN, D. M. *Structural Geology: An Introduction to Geometrical Techniques,* 3rd ed., John Wiley & Sons, New York, 1984.
A manual demonstrating structural methods and procedures.

SUPPE, JOHN *Structural Geology,* Prentice Hall, Englewood Cliffs, N.J., 1985.
An excellent reference with major attention to principles of deformation, classes of structures, and regional structural geology, the last being illustrated by the structural history of the Appalachian and Cordilleran mountain belts of the United States and Canada.

UMBGROVE, J. H. F. *The Pulse of the Earth,* 2nd ed., Martinus Nijhoff, The Hague, 1947.
A treatise that must be rated a classic for its impact on the broad philosophical aspects of structural geology.

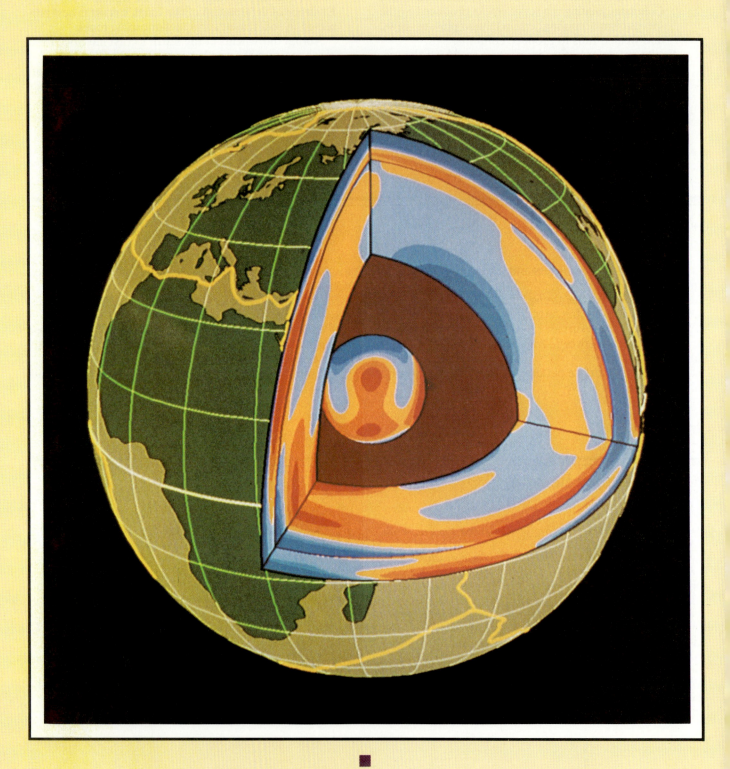

A composite schematic representation of the inner core. The color scale is such that red colors indicate lower than average wave velocities at a given depth and blue indicates higher than average velocities. [Andrea Morelli et al. 1986, Geophys. Res. Lett., *13*, 1545–1548].

CHAPTER ■ 10

EARTHQUAKES AND THE EARTH'S INTERIOR

" . . . oft the teeming earth
Is with a kind of colic pinch'd and vex'd
By the imprisoning of unruly wind
Within her womb; which for enlargement striving,
Shakes the old beldam earth and topples down
Steeples and moss-grown towers."

William Shakespeare, King Henry IV, Part I, *Act III, Scene I*

OVERVIEW

We now know that earthquakes occur not because of an "unruly wind within [Earth's] womb," as young Hotspur suggested, but because of the movement of the Earth's plates.

We begin our discussion of earthquakes with a look at *early instrumental observations* and follow this with a description of *modern instruments*. We find that the different types of *earth waves* travel different paths and at different speeds and behave differently in different layers of the Earth. These differences allow us to construct *time-distance charts* that help us to locate and interpret the Earth's structures as well as to establish the *location of earthquakes*.

We always want to know how big an earthquake is, and we can express that by its *intensity*, but more often by its *magnitude* as measured on the Richter scale.

This chapter reviews some early thoughts on the *cause of earthquakes*, which leads to some case studies of *earthquakes caused by plate movements*.

The *effects of earthquakes* can be catastrophic and we examine some, including *fire, damage to structures, seismic sea waves, landslides, land movement*, and *liquefaction*.

Can we *predict* and *control* earthquakes? We find that some progress has been made in this area.

The chapter closes by showing how we have used earthquakes to sketch the *structure of the Earth* from outer crust to inner core.

10.1

SEISMOLOGY

The scientific study of earthquakes is called **seismology,** from the Greek *seismos,* "earthquake," and *logos,* "reason" or "speech." At the turn of the twentieth century there were approximately a dozen scientists in the world who could have been classified as professional seismologists. The subject matured, however, until by midcentury there were close to 400 seismograph stations that had seismologists recording and studying earthquakes and other ground vibrations. Data from seismology have become an integral part of physical geology in its growth from a descriptive natural history to a science that includes geophysics, a category that cuts across the groupings of knowledge labeled "physics" and "geology."

10.2

EARLY INSTRUMENTAL OBSERVATIONS

Much information about the Earth's interior has come from the study of the records of earth waves. Later in this chapter we learn that some large earthquakes involve violent displacements of the Earth's crust. These movements use up some of the energy released at the time of the earthquake. The rest is carried to great distances by earth waves,

which may travel through the whole Earth and around its entire surface if the earthquake is large enough. At great distances sensitive instruments are needed to detect the passage of these waves. By studying the manner in which earth waves travel out from earthquakes, supplemented by the study of waves generated by dynamite and nuclear explosions, geologists have assembled a wealth of information about the structure of the globe from surface to center. But before we look at the results of these studies, let us review some of the methods used to obtain information.

The earliest instrument used to detect an earthquake was built around A.D. 136. Its design is credited to a Chinese philosopher, Chang Hêng. His instrument was said to resemble a wine jar about 2 m in diameter. Evenly spaced around this were eight dragon heads under each of which was a toad with head back and mouth open. In each dragon's mouth was a ball (see Figure 10.1). When there was an earthquake, one of the balls was supposed to fall into a toad's mouth. There is no known record of what was inside the jar. Speculations over the centuries have assumed a pendulum of some sort, which would swing as the ground moved, knocking the ball out of a dragon's mouth in line with the swing. Chang Hêng had the notion that if a frog on the south side caught a ball, the earthquake had happened to the north of the instrument. His instrument was intended to record the occurrence of an earthquake and show the direction of its origin from the observer. It was a seismoscope, not a seismograph, as it had no provision for a written record of the motion.

10.1 Chang Hêng's seismoscope, as visualized by Wang Chen-To.

In 1703 J. de la Haute Feuille had the same general idea. He proposed a seismoscope design that he felt would respond to the tilting of the Earth's surface at the time of an earthquake. His instrument consisted of a bowl of mercury so placed that when the Earth's surface tilted, mercury would spill over the rim and fall into one of eight channels. Each channel represented a principal direction of the compass; the spilled mercury was to show the direction of the earthquake, and the amount of spillage the earthquake's intensity. However, there is no record that this seismoscope was ever built.

The first European to use a mechanical device for studying earthquakes was Nicholas Cirillo. He used his device to observe the motion of the ground from a series of earthquakes in Naples in 1731. His instrument was a simple pendulum whose swing indicated the amplitude of ground motion. He made observations at different distances from earthquakes to see how the motion died out with distance. He found that the amplitude decreased with the inverse square of the distance.

In 1783 there occurred a series of earthquakes in Calabria, Italy, that resulted in 50,000 deaths. These stimulated the development of other mechanical devices for studying such events. They also led to the appointment of the first earthquake commission. Shortly after the first large quake a clockmaker and mechanic named D. Domenico Salsano had an instrument operating in Naples approximately 320 km from Calabria. It consisted of a long common pendulum with a brush attached that was supposed to trace out the motion with slow-drying ink on an ivory slab, the first attempt to preserve a written record, or **seismogram.** His instrument was also equipped with a bell that was to ring when motions were large enough. It has been reported that the bell rang on several occasions. A similar instrument was reported in use in Cincinnati around the time of the New Madrid, Missouri, earthquakes (1811 and 1812).

A series of small earthquakes near Comrie, Scotland, led to the development of an inverted-pendulum seismometer by James Forbes in 1844. His instrument consisted of a vertical metal rod with a movable mass on it. This was supported by a steel wire. The stiffness of the wire and the position of the mass could be adjusted to alter the free period of the pendulum. Forbes was trying to give his pendulum a long period so that it would remain stationary when the ground moved during an earthquake. At one end of the metal rod he attached a pencil that came into contact with a stationary paper-lined dome above it, to preserve a written record of the movement (see Figure 10.2). By moving the pencil some distance from the mass, he was able to magnify the motion of the pendulum two or three times. This instrument, a crude **seismograph,** was not successful because friction between the writing pencil and the recording surface reduced the sensitivity of the instrument, ruining recording of felt local quakes.

The first record of a distant earthquake was obtained accidentally by Ernst von Rebeur-Paschwitz at Potsdam, Germany, on April 17, 1889. The earthquake was felt in Tokyo about 1 h before it was recorded in Germany. The motion was too small and was on a time scale that was too compressed to give much information. However, it showed that the energy from an earthquake could travel halfway around the world. And because the record gave the time of day, it indicated how long the waves took to travel that far, how long the motion lasted, and that the motion of the ground had a horizontal component.

10.2 Forbes's seismometer. Screws, *E*, acting on a support, *D*, were to help set the pendulum in an upright position.

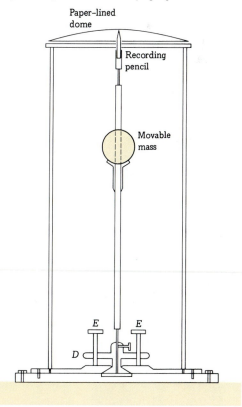

Paper-lined dome

Recording pencil

Movable mass

E   *E*

D

10.3

MODERN SEISMOLOGY

With the discovery that earth waves could travel to such great distances from an earthquake, modern seismology was born. The first records of earthquake motion registered on a tilt instrument were too fuzzy to be analyzed accurately. They were too small in amplitude, so the events of 1 h had to be compressed into the space of 6 mm. To obtain meaningful data about earthquakes, it was necessary to develop seismographs with increased recording speed and sensitivity so that the fuzzy lines would be spread out and enlarged.

A group of British professors teaching in Japan in the late nineteenth century initiated the activities that devel-

oped into all aspects of seismology as we find it today. Chief among these in his influence on the entire field was John Milne. He was the one mainly responsible for developing a seismograph capable of picking up and recording intelligible signals from distant earthquakes. Milne was also responsible for promoting the first worldwide network of seismograph stations for systematic, continuous registration and interpretation of their records. By 1900 Milne's seismographs were operating on all the inhabited continents (see Figure 10.3). Sixteen stations were regularly recording and sending their data to Milne.

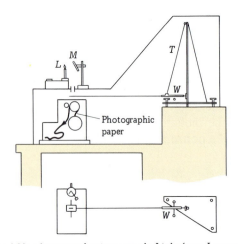

10.3 The Milne horizontal seismograph. Light from *L* was reflected by *M* onto photographic paper through the intersection of two crossed slits. The lower illustration is a top view of the instrument with its outer case removed. *T* is a flexible wire holding up the boom. The weight, *W*, was pivoted on the boom.

MODERN INSTRUMENTS

A modern seismograph consists of three basic parts: an inertia member, a transducer, and a recorder.

The inertia member is a weight suspended by a wire, or spring, so that it acts like a pendulum but is so constructed that it can move in only one direction. The inertia member tends to remain at rest as earth waves pass by. It has to be damped so that the mass will not swing freely.

The transducer is a device that picks up the relative motion between the mass and the ground. It converts this into a recordable form. It may be a mechanical lever or an electrodynamic system. In an electrodynamic system a coil of wire moves back and forth in a magnetic field. This movement creates an electric current that passes through a galvanometer to be recorded on a sheet of paper.

To record motion in all directions, it is necessary to have three seismographs. One records vertical motion, and two record horizontal motion in directions at right angles to each other. A well-equipped seismograph station will have a set of three components: vertical, north-south, and east-west.

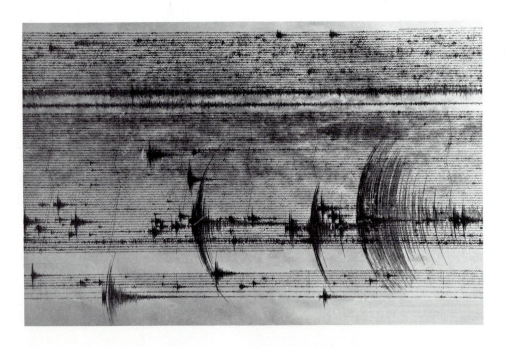

10.4 Seismogram of shallow-level earthquake activity in the southwest rift zone of Kilauea Volcano, Hawaii. [U.S. National Park Service.]

A mass on a spring will oscillate if displaced and then released. The time required to complete one oscillation is called the **period.** If the ground under this system oscillates with a shorter period, the mass hangs still in space, or nearly so. It then serves as a point of reference from which to measure the Earth's motion. The inertia members are built to stand still during the passage of waves of a certain selected period range. Generally two sets are used, one for short-period waves of 5 s or less and one for long-period waves of 5 to 60 s. Because the motion of the ground at distant stations is microscopic, the motion must be magnified to make a visible record. Therefore the transducer includes a system that will serve to magnify the ground motion.

A standard recorder is a cylindrical drum on which is wrapped a sheet of recording paper. It rotates at a constant speed and by a helical drive moves sidewise as it turns. This produces a continuous record, a series of parallel lines. These lines appear straight when there is no ground motion; but if there is motion, the lines move in response. A clock-controlled device jogs the recording line briefly once per minute and for a longer time each hour. These jogs on the record identify the time of day. Each sheet of paper has the capacity for 24 h of continuous recording (see Figure 10.4).

10.4
EARTH WAVES

When Earth materials rupture and cause an earthquake, some of the energy released travels away by means of earth waves. The manner in which earth waves transmit energy can be illustrated by the behavior of waves on the surface of water. A pebble dropped into a quiet pool creates ripples that travel outward over the water's surface in concentric circles. These ripples carry away part of the energy that the pebble possessed as it struck the water. A listening device at some distant point beneath the surface can detect the noise of impact. The noise is transmitted through the body of the water by sound waves, which are far different from surface waves and are not visible by ordinary means.

Just as with water-borne waves, there are two general classes of earth waves: **body waves,** which travel through the interior of the mass in which they are generated; and **surface waves,** which travel only along the surface.

BODY WAVES

Body waves are of two general types: **push-pull,** or compressional; and **shear,** or side-to-side. Each is defined by its manner of moving particles as it travels along.

Push-pull waves can travel through any material—solid, liquid, or gas. They move the particles forward and backward; consequently the materials in the path of these

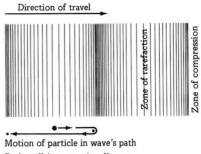

Motion of particle in wave's path
Push-pull (compressional) wave

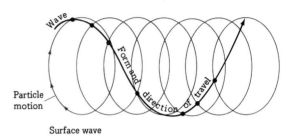

Surface wave

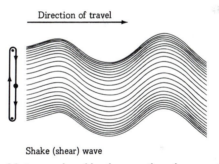

Shake (shear) wave

10.5   Motion produced by three earthquake wave types.

waves are alternately compressed and rarefied (see Figure 10.5). For example, when we strike a tuning fork sharply, the prongs vibrate back and forth, first pushing and then pulling the molecules of air with which they come in contact. Each molecule bumps the next one, and a wave of pressure is set in motion through the air. If the molecules next to your eardrum are compressed at the rate of 440 times/s, you hear a tone that is called "middle A."

Shear waves can travel only through materials that resist a change in shape. These waves move the particles in their path at right angles to the direction of their advance. Imagine that you are holding one end of a rope fastened to a wall. If you move your hand up and down regularly, a series of waves will travel along the rope to the wall. As each wave moves along, the particles in the rope move up and down, just as the particles in your hand did. In other words, the particles move at **right angles** to the direction of the wave's advance. The same is true when you move your hand from side to side instead of "vibrating" it up and down (Figure 10.5).

SURFACE WAVES

Surface waves can travel along the surface of any material. Let us look again at the manner in which waves transmit energy along the surface of water. If you stand on the shore and throw a pebble into a quiet pool, setting up surface waves, some of the water seems to be moving toward you. Actually, though, what is coming toward you is **energy** in the form of waves. The particles of water move in a definite pattern as each wave advances: up, forward, down, and back, in a small circle. We can observe this pattern by dropping a small cork on the surface into the path of the waves (see Figure 10.5 and Section 16.6).

When surface waves are generated in rock, one common type of particle motion is just the reverse of the water-particle motion—that is: forward, up, back, and down.

10.5
RECORDS OF EARTHQUAKE WAVES

As we have seen, the first records of earthquake motion were difficult to analyze accurately. But when the drum speed was increased to spread the events of 1 h over a space of 1 m, a sharp pattern emerged.

This pattern consisted of three sets of earth waves. The first waves to arrive at the recording station were named **primary;** the second to arrive were named **second-**ary; and the last to arrive were named **large waves.** The symbols P, S, and L are commonly used for these three types. Closer study revealed that the P waves are push-pull waves and travel with speeds that vary inversely with the density of the material. It also revealed that the S waves are shear waves and likewise travel at speeds that vary inversely with density.

The P and S waves travel from the origin of an earthquake through the interior of the Earth to the recording station. The L waves are surface waves that travel along the surface to the recording station from the area directly above the focus.

The P waves arrive at a station before the S waves because, although they travel along the same general paths, they go at different speeds. The push-pull mechanism by which P waves travel generates more rapid speed than does the shear mechanism of the S waves. The S waves travel at about two-fifths the speed of P waves in any given Earth material. The L waves are last to arrive because they travel at slower speeds and over longer routes (see Figure 10.6).

TIME-DISTANCE GRAPHS

As data accumulated, Milne began plotting the arrival time of the first P, S, and L waves from felt earthquakes. When he set up time-distance graphs for each wave, he found that all three waves took longer and longer to travel to greater and greater distances. He also found that the time between

10.6   A record of an earthquake on a seismogram at Harvard, Massachusetts. All waves started in Rumania at the same instant. They arrived as indicated below because of different speeds and paths. The distance was 7,445 km.

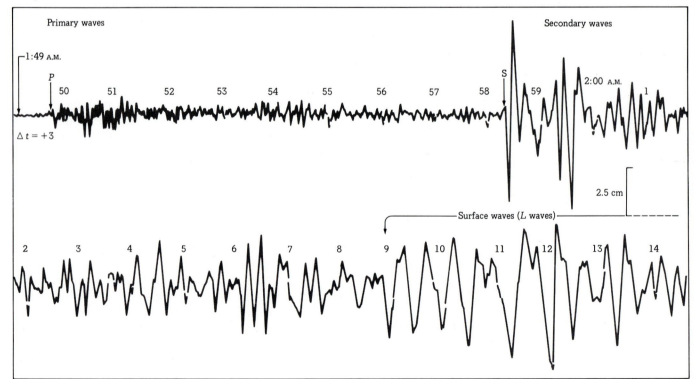

the arrival of each wave type increased as the distance from the source increased.

Travel Times   In Chapter 9 we defined elasticity as a property of materials that determines the extent to which they resist small deformations, from which they recover completely when the deforming force is removed. If earth waves travel through a material of uniform elasticity, their paths will be straight lines as they go at constant speed to greater and greater distances. However, if there is a progressive increase in elasticity with depth, they will gradually increase their speed, and their paths will become smooth curves. When there is a sharp increase in elasticity, the wave directions are changed, and this is reflected by sharp changes of direction in the line on a time-distance graph.

Distance is sometimes expressed as the number of kilometers for the length of a great-circle arc between two surface points or as the number of degrees for the angle at the Earth's center subtended by that arc. For example, one-quarter of the way around the Earth is 10,000 km, or 90°.

Less Than 11,000 km   From thousands of measurements the world over, it has been learned that P, S, and L waves have regular travel schedules for distances up to slightly more than 11,000 km. From an earthquake in San Francisco, for example, we can predict that P will reach El Paso, 1,600 km away, in 3 min 22 s and S in 6 min 3 s; P will reach Indianapolis, 3,220 km away, in 5 min 56 s and S in 10 min 48 s; P will reach Costa Rica, 4,800 km away, in 8 min 1 s, and S in 14 min 28 s.

The travel schedules move along systematically out to a distance of 11,000 km, as shown in Table 10.1.

Beyond 11,000 km   Beyond 11,000 km, however, something happens to the schedule, and the P waves are delayed. By 16,000 km they are 3 min late. When we consider that up to 11,000 km we could predict their arrival by seconds, a 3-min delay is significant.

The fate of the S waves is even more spectacular: They disappear altogether, never to be heard from again.

When the strange case of the late P waves and the missing S waves was first recognized, seismologists became excited, for they realized that they were not just recording earthquakes but were developing a picture of the interior of the Earth. (See Section 10.11). Experiments show that P waves slow down when passing through liquids, and shear waves, with their side-to-side motion, do not pass through liquids at all. The seismic data from earthquakes beyond 11,000 km suggests that the Earth has a liquid portion to its core (but more about that later).

EARTHQUAKE FOCUS AND EPICENTER

In seismology the term **focus** designates the source of a given set of earthquake waves. Just what is this source? As we know, the waves that constitute an earthquake are generated by the rupture of Earth materials. When these waves are recorded by instruments at distant points, their pattern indicates that they originated within a limited region. Most sources have dimensions probably closer to 50 km in length and breadth than 5 or 500 km. A few of the largest earthquakes may involve up to 1,000 km. But trying to fix these dimensions more accurately offers a real problem, one that has not yet been solved.

In any event, the focus of an earthquake is usually at some depth below the surface of the Earth. An area on the surface vertically above the focus is called the **epicentral area,** or **epicenter,** from the Greek epi, "above," and "center" (Figure 10.7).

Foci have been located at all depths down to 700 km, a little more than one-tenth of the Earth's radius (see Table 10.2). On some continental margins foci have clustered along a plane dipping toward the interior of the continent, as in Figure 10.8. The deepest have been limited to the Tonga-Fiji area and the Andes. (These are at the bottom of active subduction zones; see Section 11.5.)

10.7   Diagram showing the positions of the focus and epicenter of an earthquake.

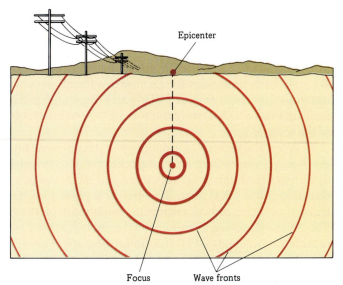

TABLE 10.1
Sample Timetables for P and S

Distance from source, km	Travel time P		Travel time S		Interval between P and S (S–P)	
	min	s	min	s	min	s
2,000	4	06	7	25	3	19
4,000	6	58	12	36	5	38
6,000	9	21	16	56	7	35
8,000	11	23	20	45	9	22
10,000	12	57	23	56	10	59
11,000	13	39	25	18	11	39

TABLE 10.2
Focal Depths for a 7-Year Period[a]

Depth, km	Foci
<75	27,650
100 ± 25	3,533
150 ± 25	1,942
200 ± 25	974
250 ± 25	447
300 ± 25	223
350 ± 25	196
400 ± 25	200
450 ± 25	205
500 ± 25	292
550 ± 25	488
600 ± 25	448
650 ± 25	154
>650 ± 25	8
Total	36,760

[a] Summarized from United States Coast and Geodetic Survey's preliminary determination of epicenter data (personal communication from Leonard M. Murphy, Chief, Seismology Division).

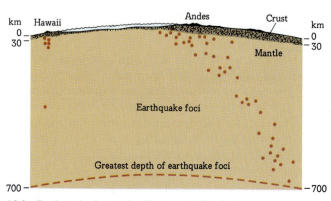

10.8 Earthquake foci under Hawaii and South America.

LOCATING EARTHQUAKES

We now have timetables for earth waves for all possible distances from an earthquake. These are represented in the graph of Figure 10.9. Data of this sort are the essential tools of the seismologist.

When the records of a station give clear evidence of the P and S waves from an earthquake, the observer first determines the interval between the P and the S waves, as shown in Table 10.3. Next, this interval is plotted on the travel-time curve, always keeping the lower mark on the P line. At one place — and only one — the distance representing this time interval (S-P) will occur on the graph. This will happen at the place that corresponds to the quake's distance, in this case 7,440 km, or 67°. The graph also shows

that P travels 7,440 km in 8 min 52 s. The seismologist subtracts this travel time from the arrival time of P and has the time of the earthquake at its source:

P arrived	15 h 09 min 06 s
P had traveled	8 min 52 s
P started at	15 h 0 min 14 s

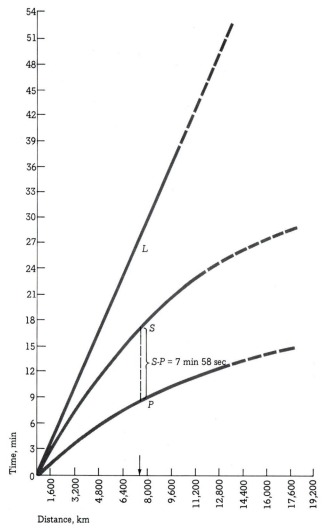

10.9 Travel-time curves for earth waves. An example is shown for an earthquake whose epicenter is approximately 7,500 km from the seismograph station. The delay between arrival of the primary wave and the secondary wave is demonstrated by the dashed line (S-P).

TABLE 10.3
Interval Between Arriving Waves

Time, beginning with P			Interval, time after P	
h	min	s	min	s
15	09	06		
15	17	04	7	58

This process is carried out for several seismograph stations that have recorded the quake. Then an arc is drawn on a globe to represent the computed distance from each station. The point where all the arcs intersect indicates the center of the disturbance.

The example in Table 10.3 and Figure 10.9 was based on a record taken at the station in Alert, Canada. Palisades, New York, recorded it at 3,580 km, and Pasadena, California, at 2,675 km. The United States Coast and Geodetic Survey located it in south-central Mexico, where they reported extensive property damage, 3 deaths, and 16 injured.

Although this whole procedure is essentially very simple, some have found its accuracy hard to believe. On December 16, 1920, seismologists all over the world found the record of an exceptionally severe earthquake on their seismographs. Each computed the distance of the quake and sent the information along to the central bureaus, where the various reports were assembled. The next day the location of the earthquake was announced to the press, unlike many lesser shocks that fail to make the news. The announcement stated that a very severe earthquake had occurred at 5 min 43 s after 12:00, Greenwich time, December 16, 1920, in the vicinity of 35.6°N, 105.7°E. That placed it in the Chinese province of Kansu, about 1,600 km inland from Shanghai, on the border of Tibet. This area is densely populated but quite isolated. No reports of damage came in, however, and the matter was soon forgotten by the general public. But it was not forgotten by the members of the press, who were sure that they had been misinformed. Then, 3 months later, a survivor staggered into the range of modern communications with a story of a catastrophe in Kansu on the day and at the time announced, a catastrophe that had killed more than 100,000 persons and had created untold havoc by causing great landslides.

10.6
INTERPRETING EARTHQUAKES

By examining the location and strength of earthquakes, we can learn much about how they do their damage.

WORLDWIDE STANDARD SEISMOGRAPH NETWORK

In 1961 the United States Coast and Geodetic Survey began establishing a worldwide network of 125 standard seismograph stations. Since then, seismological knowledge has been expanding rapidly because of the quantity and quality of earthquake measurements. Also, computers have helped to speed the processing of data for determining earthquake locations, depths of foci, and size. The United States Coast and Geodetic Survey and the United States Geological Sur-

vey now locate approximately 6,000 quakes per year, whereas before the network was set up, the number was one-tenth that. This increase does not mean that the seismicity of the Earth has increased, but rather that the improved equipment in the increased number of recording stations makes it possible to gather more and better data.

From this worldwide seismic network and from other seismograph stations come measurements that give us a new perspective on the sequences of earthquake activity associated with large shocks and information on the size of the area involved.

EARTHQUAKE INTENSITY

How to specify the size of an earthquake has always posed a problem. Descriptions were crude and subjective until in 1883 M. S. de Rossi of Rome and F. A. Forel of Lausanne, Switzerland, combined their efforts and developed an intensity scale. For half a century the Rossi-Forel intensity scale was widely used throughout the world. According to this scale, earthquake effects were classified in terms of 10 degrees of intensity. It had definite limitations, however, from a scientific standpoint. For example, the definition of the sixth degree of intensity included "general awakening of those asleep; general ringing of bells; oscillation of chandeliers; stopping of clocks; some startled persons leave their dwellings." But an earthquake that produced those effects in Italy or Switzerland might not even wake a baby in Japan, though it might cause a general stampede in Boston. The Rossi-Forel scale also relied on the presence of certain structures, like chimneys, steeples, and other tall, unstable constructions, which obviously do not exist all over the Earth.

Objections to the Rossi-Forel intensity scale led L. Mercalli of Italy to set up a new scale in 1902. This, as modified in 1931 by H. O. Wood and Frank Neumann, is the scale currently in use by the United States Geological Survey to evaluate earthquake intensity. It has 12 degrees of intensity and takes into account varying types of construction (see Box 10.1).

By means of postcards, letters, and interviews, investigators for the United States Geological Survey study the effects caused by each earthquake. They then determine what places have been shaken equally. These places are plotted on a map, and the points are contoured using **isoseismic lines,** or lines of equal shaking.

EARTHQUAKE MAGNITUDE AND ENERGY

Having to rely on people's impressions is a very unsatisfactory way of compiling accurate information on the actual size of an earthquake. Therefore a scale was devised in 1935, based on instrumental records. By computation based on the amount of motion in certain waves, it ascribed

BOX 10.1   Comparing Magnitude and Intensity

Magnitude and intensity are quite different. Intensity is linked with the particular ground and structural conditions of a given area, depth of focus, and distance from the epicenter. Magnitude depends only on the energy released at the focus of the earthquake.

THE RICHTER SCALE

To determine the magnitude of an earthquake we connect on the chart:

(a)   the maximum amplitude recorded by a standard seismometer; and

(b)   the distance of that seismometer from the epicenter of the earthquake (or the difference in times of arrival of the *P* and *S* waves) by a straight line* that crosses the center scale at the magnitude (5 on this example).

---

* In this case, 210 km in distance.

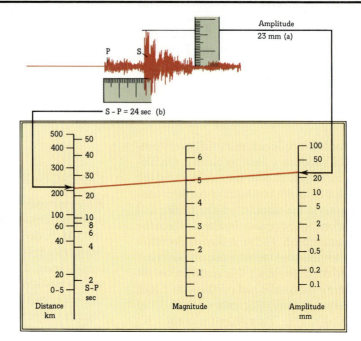

M	Intensity near epicenter	Damage near epicenter
2	I–II	Usually detected only by instruments
3	III	Felt indoors
4	IV–V	Felt by most people, slight damage
5	VI–VII	Felt by all, many frightened and run outdoors, damage minor to moderate
6	VII–VIII	Everybody runs outdoors, damage moderate to major
7	IX–X	Major damage
8+	X–XII	Total and major damage

B10.1   Definition of a magnitude 3 earthquake. (Modified from *California Geology*, February 1979.)

to each earthquake a number called the earthquake's **magnitude,** an index of the quake's energy at its source. This was refined in 1956 by B. Gutenberg and C. F. Richter, and again in 1967 by an International Committee on Magnitude. According to this scale, shallow earthquakes near inhabited areas can have the following effects: One of magnitude 2.5 is just large enough to be felt nearby; one of magnitude 4.5 or over is capable of causing some very local damage; one of 6 or over is potentially destructive. A magnitude 7 or over represents a major earthquake.

It should be noted that the magnitude calculation for any given earthquake is corrected for distance from the epicenter, so that seismic stations all over the world calculate the same magnitude for any particular event.

The **Richter** magnitude varies logarithmically with the wave amplitude of the quake as recorded on the seismo-

graph. Each higher whole number on the scale represents an increase of 10 times in the measured wave amplitude of an earthquake. Therefore the amplitude of a magnitude 8.0 earthquake is not twice as large as one of magnitude 4.0 but 10,000 times as large.

We can also estimate the **amount of energy** released during a quake by use of the Richter scale magnitudes. An increase of one unit on the magnitude scale has been calculated to represent a 31-fold increase in energy released. Therefore a quake of magnitude 6 releases 31 times as much energy as one of magnitude 5, and over 900 times as much as one of magnitude 4. In the example from the previous paragraph, a magnitude 8.0 earthquake releases almost 1 million times more energy than one of magnitude 4.0 on the Richter scale.

An earthquake of magnitude 5 releases approximately the same amount of energy as did the first atomic bomb when it was tested on the New Mexico desert on July 16, 1945. A megaton nuclear device releases the same amount of energy as an earthquake of magnitude 6. Of course, the energy is applied in quite different ways—highly concentrated in the atomic device, widely dispersed in the earthquake—so the results are correspondingly different.

Historically, some earthquakes that have proved major catastrophes in a human and property sense have not been of the greatest magnitude. Near the coast of northern Peru, at 9.2°S, 78.8°W, on May 31, 1970, a great quake killed an estimated 50,000 persons and left 800,000 homeless while causing $230 million in damage. Hualas Canyon towns were flooded by burst dams and were buried under landslides and mudslides. The quake was felt along 1,000 km of Peru; yet its magnitude was 7.8, less than that of all the quakes listed in Table 10.4.

The largest magnitude assigned to an earthquake to date has been 8.6. As Table 10.4 shows, four of that size have occurred during the modern observational period: in Alaska, September 10, 1899; in Colombia, January 31, 1906; in the Pakistan-Tibet-Burma region, August 15, 1950; and in Alaska, March 27, 1964. The only earthquake in history that might have been larger—to judge from the reported effects—was the Lisbon quake of 1755. Possibly

TABLE 10.4
Great Earthquakes

Date	Location	Magnitude
September 10, 1899	Alaska, 60°N, 140°E	8.6
January 31, 1906	Colombia, 1°N, 82°W	8.6
August 17, 1906	Chile, 33°S, 72°W	8.4
January 3, 1911	Soviet Union–China, 44°N, 78°E	8.4
December 16, 1920	China, 36°N, 106°E	8.5
March 2, 1933	Japan, 39°N, 145°E	8.5
August 15, 1950	Pakistan-Tibet-Burma, 29°N, 97°E	8.6
May 22, 1960	Chile, 38°S, 73.5°W	8.4
March 27, 1964	Alaska, 61.1°N, 147.7°W	8.6

its magnitude was between 8.7 and 9.0. An earthquake with a magnitude of over 10 should theoretically be perceptible in scattered areas over the entire Earth, but such an occurrence has never been reported.

It should be clear that the location of earthquake epicenters, with respect to population density, can account for lower-magnitude events doing more damage than some larger events. In addition, local rock conditions and style of construction are major determinants in loss of life and property damage. (See Box 10.2.)

BOX 10.2 List of the Twentieth Century's Most Deadly Natural Disasters

This list of the twentieth century's natural disasters that have caused more than 10,000 deaths shows that a very large proportion have been associated with earthquakes.

Estimated Deaths

- 3.7 million—flood of Yellow River in China, 1931
- 655,000—earthquake in Tangshan, China, 1976
- 500,000—cyclone, tidal wave in Bangladesh, 1970
- 200,000—floods in China, 1939
- 180,000—earthquake, landslides in Gansu, China, 1920
- 160,000—earthquake in Messina, Sicily, 1908
- 143,000—earthquake, fire in Tokyo-Yokohama, Japan, 1923
- 100,000—floods in North Vietnam, 1971
- 100,000—earthquake in Gansu, China, 1927

- 100,000—flood in Canton, China, 1915
- 70,000—earthquake in Gansu, China, 1932
- 66,794—earthquake in Yungay, Peru, 1970
- 57,000—flood of Yangtze River in China, 1949
- 56,000—earthquake in Quetta, India, 1935
- 55,000—earthquake in Soviet Armenia, 1988
- 50,000—tidal wave in Italy, 1908
- 50,000—earthquake in Chile, 1939
- 50,000—earthquake in Turkey, 1939
- 50,000—flood of Yellow River in China, 1933
- 40,000—volcano eruptions in Martinique, 1902
- 40,000—cyclone in Bengal, India, 1942
- 40,000—flood of Yangtze River in China, 1954

- 30,000—earthquake in Avezzano, Italy, 1915
- 30,000—flood of Yellow and Yangtze rivers in China, 1935
- 25,000—earthquakes in northeast Iran, 1978
- 24,047—cyclones, tidal waves in East Pakistan, 1965
- 23,000—mudflows from volcanic eruption, Nevada del Ruiż, Colombia, 1985
- 22,778—earthquake in Guatemala, 1976
- 22,000—storm in Chittagong, East Pakistan, 1963
- 19,000—earthquake, flood in Kangra, India, 1905
- 15,000—cold in Inner Mongolia and Suiyuan, China, 1930
- 14,000—earthquakes in Central Asia, 1907
- 12,200—earthquake in Guatemala, 1902
- 12,000—earthquake in Iran, 1968

Modified from *Los Angeles Times*, May 28, 1985.

TABLE 10.5
Average Annual Number and Magnitude of Earthquakes[a]

	Magnitude	Av. number	Release of energy, approx. explosive equiv.
Actually observed:			
Great	7.7–8.6	2	50,000 1-megaton (hydrogen) bombs
Major	7.0–7.6	12	
Potentially destructive	6.0–6.9	108	
Estimates based on sampling special regions			1-megaton bomb
	5.0–5.9	800	
			1 small atom bomb (20,000 t TNT)
	4.0–4.9	6,200	
	3.0–3.9	49,000	
	2.5–2.9	100,000	
			0.5 kg TNT
	<2.5	700,000	

[a] Modified from B. Gutenberg and C. F. Richter, *Seismicity of the Earth,* Princeton University Press, Princeton, N.J., 1949.

From year to year there are wide variations in the total energy released by earthquakes, as well as in the number of individual shocks. For example, 1906 showed 6 times the average energy released for the first half of the twentieth century and 40 times the minimum; 1950 was another very active year, second only to 1906; and 1976, one of the most active of all years for earthquakes, was dubbed by some the "Year of the Killer Quakes."

Most of the energy released is concentrated in a relatively small number of very large earthquakes. A single earthquake of magnitude 8.4 releases just about as much energy as was released, on the average, each year during the first half of the twentieth century. It is not unusual for the energy of one great earthquake to exceed that of all the others in a given year or of all the earthquakes of several years put together. The nine great earthquakes listed in Table 10.4 represent nearly a quarter of the total energy released from 1899 through 1964.

The maximum energy released by earthquakes becomes progressively less as depth of focus increases. The nine earthquakes in the table were relatively shallow and had the magnitudes shown; the five largest intermediate-depth shocks over the same interval had magnitudes of 8.1, 8.2, 8.1, 7.9, and 8.0; and the three largest deep shocks had magnitudes of only 8.0, 7.75, and 7.75.

The average annual number of earthquakes from 1904 through 1946 was estimated by Gutenberg and Richter (Table 10.5), who placed the number of earth-

10.10   Locations of earthquakes during a 9-year period, with focal depths of 0 to 100 km. [Data from National Earthquake Information Center.]

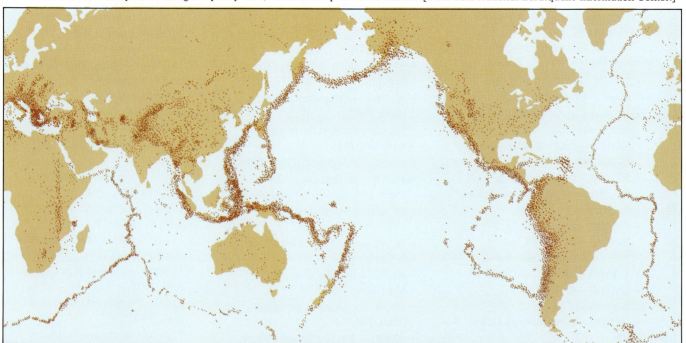

10.11   Locations of earthquakes during a 9-year period, with focal depths of 100 to 700 km. [Data from National Earthquake Information Center.]

quakes large enough to be felt by someone nearby at more than 150,000 per year. Gutenberg and Richter estimated that the total number of earthquakes "may well be of the order of a million each year." Seismograph stations throughout the world are now supplying data leading to the location of close to 6,000 earthquakes per year. Figures 10.10 and 10.11 show the distribution of quakes for a 9-year period, standardized networks and computer-programming techniques, including locations that have been noted from seismographic records since 1899. These active zones have not changed significantly through human history.

Earthquakes tend to occur in belts, or zones, many of which are associated with belts of active volcanoes. The earthquakes that occur in the most active zone, around the borders of the Pacific Ocean, account for a little over 80 percent of the total energy released throughout the world. The greatest activity is near Japan, western Mexico, Mela-

nesia, and the Philippines. The loop of islands bordering the Pacific has a high proportion of great shocks at all focal depths. (See also the discussion of earthquakes resulting from plate movements, Section 10.8.) Another 15 percent of the total energy released by all earthquakes is in a zone that extends from Burma through the Himalaya Range, into Baluchistan, across Iran, and westerly through the Alpine structures of Mediterranean Europe. This is sometimes called the "Mediterranean and trans-Asiatic" zone. Earthquakes in this zone have foci aligned along mountain chains. The remaining 5 percent of the energy is released throughout the rest of the world. Narrow belts of activity are found to follow the oceanic ridge systems (see Figure 10.10).

Maps showing zones of earthquake expectancy in the United States have been issued by the United States government (Figures 10.12 and 10.13). The zones in Figure 10.12

10.12   Seismic risk map for the United States, issued January 1969, shows earthquake damage zones of reasonable expectancy in the next 100 years. [Developed by U.S. Coast and Geodetic Survey, *ESSA Rel. ES-1*, January 14, 1969.]

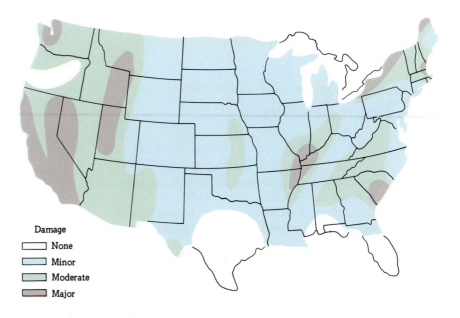

Damage

☐ None
☐ Minor
☐ Moderate
☐ Major

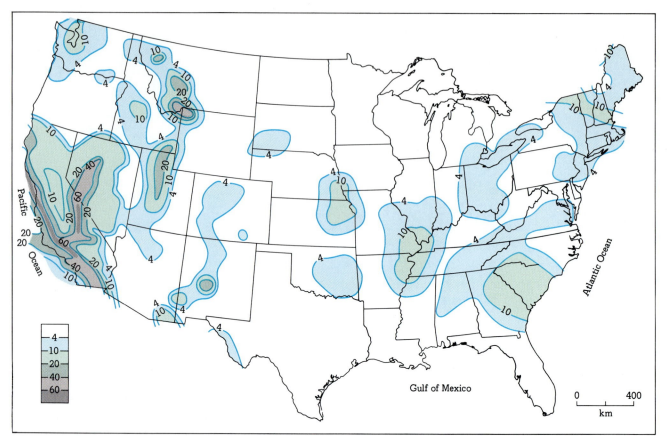

10.13   Map of expectable levels of earthquake shaking hazards. Levels of ground shaking for different regions are shown by contour lines, which express in percentages of the force of gravity the maximum amount of shaking likely to occur at least once in a 50-year period. [Modified from S. T. Algermissen, and D. M. Perkins, "A Probabilistic Estimate of Maximum Acceleration in the Contiguous United States," *U.S. Geol. Surv. Open-File Rep. 76-416*, July 1976.]

were outlined after a 2-year study of 28,000 earthquakes and show where earthquakes may be expected in the next 100 years. Figure 10.13 gives a quick method for evaluating relative earthquake hazards: Levels of ground shaking are shown by contour lines that indicate the percentages of the force of gravity likely to occur at least once in a 50-year period. For example, a contour of 40 percent of gravity means one can be relatively certain that the region will not experience ground acceleration more than 40 percent of the force of gravity. (All percentages are given with 90 percent relative certainty, or at the 90 percent probability level.)

10.7

CAUSE OF EARTHQUAKES

Centuries ago (and in some places today) people believed that the mysterious shakings of the Earth were caused by the restlessness of a monster that was supposed to be supporting the globe. In Japan it was first a great spider, and then a giant catfish; in some parts of South America it was a whale; and some of the North American Indians decided that the Earth rested on the back of a giant tortoise. The lamas of

Mongolia had another idea. They assured their devout followers that after God had made the Earth, he placed it on the back of an immense frog; every time the frog shook his head or stretched one of his feet, an earthquake occurred immediately above the moving part (see Figure 10.14).

10.14   Cause of earthquakes, according to the lamas of Mongolia. It was believed that when the frog lifted a foot, there was an earthquake immediately above the part that had moved.

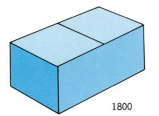 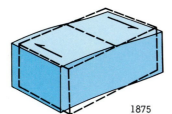 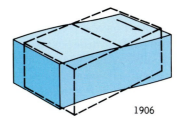 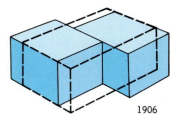

1800   1875   1906   1906

10.15   Reid's proposed representation of the movements on the San Andreas fault associated with the 1906 earthquake, illustrating the elastic-rebound hypothesis. The colored blocks represent the continued distortion of a segment of the Earth's crust. The block in dashed line represents the segment as it was in 1800.

The Greek philosopher Aristotle (384–322 B.C.) held that all earthquakes were caused by air or gases struggling to escape from the subterranean cavities. Because the wind must first have been forced into the cavities, he explained that just before an earthquake the atmosphere became close and stifling. As time passed, people began to refer to "earthquake weather," and to this day some insist that the air turns humid and stuffy when an earthquake is about to occur.

These early ideas have long been abandoned as we have learned more about the behavior of gases, the structure of the Earth's crust, and finally the depths at which earthquakes occur. What, then, causes earthquakes? Most are caused by deforming forces in the Earth; and the immediate cause is the **sudden rupture** of Earth materials distorted beyond the limit of their strength.

ELASTIC-REBOUND HYPOTHESIS

After the California earthquake of 1906 it was generally accepted that the immediate cause of earthquakes was faulting due to rupture of rocks of the Earth's crust. The displacements along the San Andreas fault provided an unusual opportunity to study the mechanics of faulting and led to the formulation of Reid's **elastic-rebound hypothesis.** Surveys of part of the fault that broke in 1906 had been made during the years preceding the earthquake. H. F. Reid, of Johns Hopkins University, analyzed these measurements in three groups: 1851 to 1865, 1874 to 1892, and 1906 to 1907. The first two groups showed that the ground was twisting in the area of the fault. The third showed displacements that occurred at the time of the earthquake. From these Reid reconstructed a history of the movement.

Although there was no direct evidence, Reid assumed that the rocks in the Earth's crust in the vicinity of the fault had been storing elastic energy at a uniform rate over the entire interval and that the region had started from an unstrained condition approximately a century before the earthquake. As the years passed, a line that in 1800 had cut straight across the fault at right angles was assumed to have become progressively more and more warped. When the relative movement on either side of the fault became as great as 6 m in places, the strength of the rock was exceeded, and rupture occurred. The blocks snapped back toward an unstrained position, driven by the stored elastic energy (see Figure 10.15).

The San Andreas fault runs roughly from northwest to southeast. Land on the western, or Pacific Ocean, side of the fault has moved northwest relative to land on the eastern side. In 1906 a section of the fault 450 km long broke. The strains and adjustments were greatest within a zone extending 10 km on each side of the fault. Imagine a straight line 20 km long crossing this zone at right angles to the fault, which was in the center of the zone. After the earthquake this line was broken at the fault and was shifted into two curves. The broken ends were separated by 6 m at the break, but the other ends were unmoved. Fences, roads, and rows of vegetation provided short sections of lines by which the fault displacement could be gauged.

Movement continues along branches of the San Andreas, resulting in numerous earthquakes in southern California (Figure 10.16).

10.16   Disturbed ground and broken sidewalk, Olive View Hospital, San Fernando earthquake of February 9, 1971. [R. Kachadoorian, USGS.]

Elastic rebound is the result of faulting and associated loss of confinement. This idea was based on observed surface movements. It was also based on our knowledge of how rocks behave when they have been subjected to deforming forces. In the laboratory rocks have been subjected to pressures that are equivalent to the pressures in the Earth's crust. Under these pressures the rocks gradually change shape; but they resist more and more as the pressure builds up, until finally they reach the breaking point. Then they tear apart and snap back into unstrained positions. This snapping back is called **elastic rebound** and was thought by Reid to be the mechanism associated with the generation of earthquakes.

SEISMIC AND VOLCANIC ACTIVITY

One of the results of increased data collection on the number and characteristics of earthquakes is that the locations of volcanoes relative to major earthquake belts have come into focus.

At the scene of Chile's 1960 earthquake series there is a chain of 16 active volcanoes paralleling the coast. One of these, Puyehue, erupted the day after the largest earthquake of the series, and 10 were variously active during the following year. Figure 10.17 shows the affected region and the volcanoes.

It has been estimated that approximately 6 percent of the world's large earthquakes occur in the Aleutian Islands and continental Alaska. The Aleutian Islands arc is more than 3,200 km in length and contains 76 volcanoes, 36 active since 1760. Two large earthquakes occurred off the adjacent coast after the 1912 eruption of Katmai.

It has been fashionable for decades to think of "volcanic earthquakes" as a special feature of explosive eruptions, small in energy and seldom damaging. But there is emerging a realization that volcanoes and earthquakes may have a common ultimate cause in deep movements of mantle materials, and the coincidence of belts of major earthquake activity with belts that include active volcanoes supports the idea. The most obvious common cause of seismic and volcanic activity relates to plate interactions (see Chapter 1 and 11) where fracture zones afford access for volcanic materials welling up from the lower crust and mantle. These boundaries are also areas where earthquakes would naturally occur by interaction of the plates in zones of convergence, or divergence, or where two plates slide past one another along parallel boundaries. (See Section 10.8.) On the other hand, there are **aseismic ridges** (see Section 16.4), such as the Hawaiian Island–Emperor Seamount Ridge (Figure 4.16), along which there may be little or no seismic activity. (Minor earthquakes there are related to volcanism and the movement of magma toward volcanic centers.)

10.17 Location of volcanoes of Chile in the region affected by the earthquake of 1960.

10.8
EARTHQUAKES RESULTING FROM PLATE MOVEMENTS

With the widespread acceptance of the theory of plate tectonics (Chapters 1 and 11), it has become evident that most earthquakes occur along the margins of plates, where one plate comes into contact with another and stresses develop. The release of energy associated with these stress conditions is in the form of earthquakes. There are, however, examples of significant earthquakes apparently not associated with plate boundaries. These include the great quake near New Madrid, Missouri, in 1811–1812; the Charleston, S.C. quake of 1886; quakes near Boston Massachusetts, and Lancaster, Pennsylvania; and others associated with fault movements far removed from present-day plate boundaries.

As the American and Pacific plates slide past each other, enormous amounts of stress build up within the plates. This stress is commonly released in sudden slipping action, producing the numerous earthquakes along this fault zone (the San Andreas being the most famous of such faults).

There are, however, sections of this zone that have become locked across the fault. This produces a **seismic gap** or region of the fault zone along which little earthquake activity has occurred for many years. Such gaps occur at a number of localities along the contact between the American and Pacific plates. One is near the Parkfield region and has led to concern that stresses have built to the point where a major earthquake may be in the offing. (See Box 10.3.)

Another gap occurs in southern Alaska near Cape Yakataga. There have been no major earthquakes in this area during this century but it is feared that an earthquake of magnitude 8 or more could occur in the next few decades. An earthquake of magnitude 7.7 did occur in February 1979 at St. Elias, about 100 km to the east of this gap; and the great 1964 earthquake of magnitude 8.6 occurred as close as 150 km to the west. These earthquakes relieved the stresses that had built up in adjacent areas, but it is still believed the Cape Yakataga gap may experience an earthquake of magnitude 8 or so in the near future.

Before instruments were developed, reports of earthquakes were dependent on population distribution. Alaska was settled in 1783 by Russians on Kodiak Island, who first reported an earthquake occurring south of the Alaskan peninsula on July 27, 1788. From that time, more than 1,500 earthquakes have been felt by persons living in the area. Information on early earthquakes has been obtained from historical records, newspapers, and ships' logs. From 1899 on, there have been more than 300 instrumentally recorded earthquakes of which the magnitude has been 6 or greater (see Figure 10.18).

BOX 10.3 The Parkfield Earthquake-Prediction Experiment*

Moderate-sized earthquakes occurred along the 15-mile-long segment of the San Andreas fault at Parkfield, California, in 1881, 1901, 1922, 1934, and 1966 — an average 21-to-22-year recurrence interval. If the next Parkfield earthquake conforms to this pattern, it should occur in the late 1980s to early 1990s.

Besides their predictable intervals, Parkfield earthquakes have other similar features; namely, magnitude, length, epicenter, and location along the San Andreas fault zone. Because Parkfield quakes have been so similar, they seemed ideal for an unusual experiment to observe the stages leading up to the next earthquake, so in the mid-1980s an earthquake-prediction experiment was tailored to their characteristic features.

In the 6 months before June 28, 1966, all earthquakes near Parkfield occurred on the San Andreas fault northwest of the zone of rupture. The foreshocks (smaller quakes occurring near the zone shortly before the main shock) occurred within a mile or so of the main shock epicenter, which was at the northwest end of the rupture zone, an area named the "preparation zone."

At the time of the 1934 earthquake a similar pattern of foreshock and main shock occurred in this preparation zone. Quakes within the critical preparation zone are much less frequent than other small quakes on the San Andreas fault northwest of the rupture zone. The U.S. Geological Survey set up a closely spaced network of seismograph stations around Parkfield to measure seismicity near the preparation zone so that any increase in seismic activity could be observed very closely.

It is believed by some geologists that earthquakes are preceded by fault creep (small movements without detectable quakes). A pipeline crossing the fault near Parkfield had broken without detectable quakes about 9 hours before the 1966 earthquake. If significant and unusual fault creep were to precede a quake in the Parkfield area, the sensitive measuring devices should record it. Surface fault movement was continuously measured by creepmeters, 30-foot-long devices set across the fault trace. Also, 2- to 5-mile-long lines of sight that cross the fault are measured daily on laser distance-measuring geodimeters. These devices can resolve changes of less than 1 cm over a distance of 10 km. In addition, strainmeters, balloonlike instruments installed in boreholes, measure very small changes in strain. Together, these instruments should confirm changes in seismicity and fault creep, deformation processes leading up to the next earthquake.

Remember, this experiment was set up 2 or 3 years prior to the expected quake, as an important part of the U.S. Geological Survey's earthquake-prediction program. It was hoped that important information could be gathered to assist in the successful design of earthquake-prediction techniques for other earthquake-prone regions.

(If the Parkfield quake has *not yet* occurred when you read this, be sure to keep alert for its occurrence. Geologists are quite certain that it will occur, although they do not know just when. If the Parkfield quake *has already* occurred, try to determine what was learned from it, what changes are to be made in our earthquake-prediction techniques, and what areas are likely candidates for future study by similar methods.)

---

* U.S. Geological Survey, *Report of Research — 1985.*

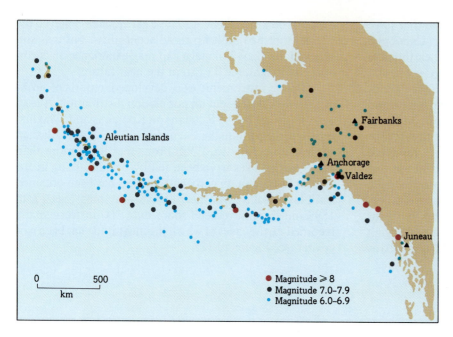

10.18 Earthquakes in Alaska with a magnitude of 6 or greater during a 65-year period.

Legend on map:
● Magnitude ≥ 8
● Magnitude 7.0–7.9
• Magnitude 6.0–6.9

Map labels: Aleutian Islands, Fairbanks, Anchorage, Valdez, Juneau

Scale: 0 — 500 km

The Prince William Sound Earthquake of 1964   On March 27, 1964, at 5:36 P.M. Alaskan standard time (or 3 h 36 min 13 s Greenwich mean time, March 28, 1964), an earthquake of magnitude 8.6 rocked the Prince William Sound area in south-central Alaska. It released twice as much energy as the great San Francisco earthquake of 1906 and was the most destructive earthquake ever recorded on the North American continent — one of our greatest natural disasters (Figure 10.19).

Its focus was at a depth of 33 km. The earthquake was felt over 1.3 million km², and major damage to buildings, ports, and transportation occurred over an area of 130,000 km². Although only 10 percent of the land area of Alaska was involved in the effects of the earthquake, this tenth was occupied by 50 percent of Alaska's total population and included all the major seaports, highways, and railroads. Property damage was estimated at $86 million. In spite of the substantial damage, only nine lives were lost in Anchorage and six elsewhere from building collapse and landslides. If the earthquake had been earlier in the day, the casualties could have been much higher. A seismic sea wave generated by the earthquake accounted for another 98 deaths in Alaska; 11 in Crescent City, California; and 1 in Seaside, Oregon.

The seismic sea wave was the first major one associated with an earthquake whose epicenter was on land. However, it was started when submarine landslides and vertical displacements disturbed a large area of the sea bottom. The wave was highly destructive around the Gulf of Alaska, causing extensive damage to many of Alaska's principal harbors: Anchorage, Cordova, Kodiak, Seward, Valdez, and Whittier. It also caused damage to the west coast of Canada and Crescent City, California. The wave accounted for the greatest loss of life from the 1964 earthquake.

In Anchorage, 120 km from the center of the quake, a substantial portion of earthquake damage was attributable to landslides. The largest and most damaging slide occurred in the Turnagain Heights section, a residential area built on a bluff overlooking an arm of Cook Inlet.

Valdez, 80 km from the epicenter, suffered extensive damage. The town was built on an alluvial fan that had grown out into the head of a deep, steep-sided fiord. The earthquake caused a major landslide, which totally destroyed the waterfront. There were extensive shoreline changes, and nearly 6 months later the land was found to be still shifting. The town was therefore relocated approximately 6.5 km northwest on more stable ground.

Aftershocks   More than 7,500 shocks were instrumentally recorded in the months that followed. These occurred in a belt 300 km wide, stretching 800 km from the main shock southwestward to a region off the southwest coast of Kodiak Island. Their focal depths ranged from 20 to 60 km.

Mexico City Earthquake of 1985   On September 19, 1985, a great earthquake occurred near Mexico City. At least 4,000 deaths resulted from this 8.1-magnitude quake, with most of the damage occurring in the center of Mexico City, nearly 400 km from the epicenter. The ground shook even more violently in the suburbs, which are underlain by soft ancient lake beds. The physical characteristics of the sediments, unfortunately, were such that the vibrations of the strongest seismic waves were greatly amplified as they passed through, shaking the ground back and forth about 40 cm every 2 seconds for nearly 2 minutes. The poorly designed buildings, especially those between 5 and 15 stories, which swayed in rhythm with the quake, were destroyed (Figure 10.20). This quake occurred along the western boundary of the American plate.

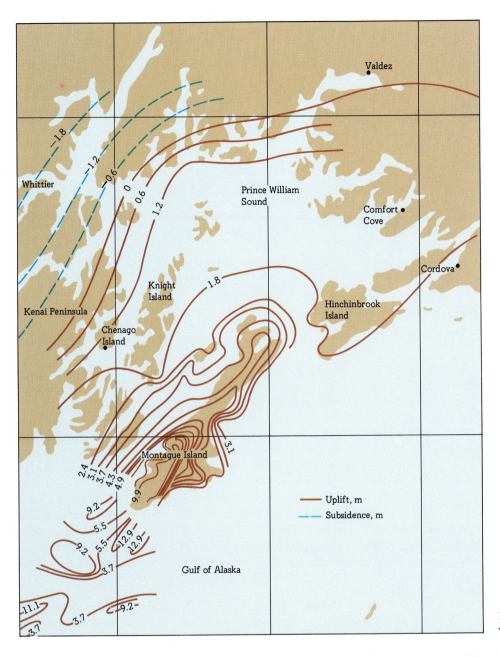

10.19 Uplift and subsidence associated with Alaskan quake of March 27, 1964.

10.20 Mexico city earthquake, September 19, 1985. Nuevo Leon, a 15-story reinforced concrete structure, collapsed because of severe shaking of unconsolidated subsurface material. [M. Celibi, USGS.]

10.21   The remains of an ancient cathedral, that had stood for centuries, lies amid the rubble in Armenia's second largest city. [Mark Porubcansky, AP/Wide World Photos.]

Armenian Earthquake of 1988   On December 7, 1988, a quake registering 6.9 on the Richter scale virtually demolished the town of Spitak and destroyed more than half the buildings in the cities of Kirovakan and Leninakan in Soviet Armenia. More than 55,000 people were killed and more than half a million were left homeless. This quake occurred along the Soviet-Turkish border on the boundary between the Arabian and Eurasian plates. This was the largest and most devastating of a dozen or so quakes in the region during this century.

The damage was very heavy for a 6.9 earthquake, in part because buildings in this region are made of stone, with clay and mud filler, and many were built on unconsolidated material. Most of the casualties in this and other quakes result from the collapse of buildings (Figure 10.21).

CLUSTERS OF SEVERE EARTHQUAKES

There appear to be times when the worldwide occurrence of severe earthquakes is more frequent than usual. The year 1976 was a time of several rather severe earthquakes in widely separated parts of the Earth, all occurring in active earthquake-prone regions along plate boundaries.

Guatemala   A severe quake, measuring 7.5 on the Richter scale, jolted much of Central America on February 4, 1976. More than 22,000 persons died. This earthquake was apparently caused by movement of plates in the Caribbean region (Figure 10.22 and Box 10.2). The American plate moves westward relative to the Caribbean plate; the northern portion of the American plate carries Canada, the United States, and Mexico, while the southern portion carries most of South America in this westerly direction. A small plate off the southwest coast of Central America, the

Cocos plate, moves northeasterly and slides beneath Central America, forming a deep trench along a subduction zone just off the west coast.

Movement of these plates has averaged about 4 to 6 cm/year, building up enormous stress over the years. There had been no major movement along these plates in Guatemala for at least 200 years. Suddenly the stresses could no longer be constrained, and rupture occurred. The American plate slipped westward as much as a meter along a zone extending more than 150 km. This break occurred along the Motagua fault zone, where a number of faults had been mapped, but where the exact location of the boundary between the American plate and the Caribbean plate had never before been confirmed.

China   On July 27, 1976, a series of very severe earthquakes hit parts of northeast China in the province of Hopei, approximately 125 to 150 km east-southeast of Beijing. This series of shocks devastated Tangshan, an industrial city of a million and a half, and caused casualties and damage in cities as far away as Beijing. The number of deaths was estimated at more than half a million. The initial shock registered 8.2 on the Richter scale. This was followed by many aftershocks, some registering as high as 7.9. Although apparently no precautionary measures had been taken prior to this particular series of quakes, accurate prediction of earlier quakes that had hit other provinces in China in February 1975 and May 1976 had permitted authorities to take measures to minimize casualties. China has been a leader in the study of the prediction of earthquakes (see Section 10.10).

Turkey   A quake of magnitude 7.6 struck near the Turkish-Soviet border on November 24, 1976, destroying over 100 villages and killing almost 4,000 persons. This quake was the strongest in Turkey since tremors in 1939 that

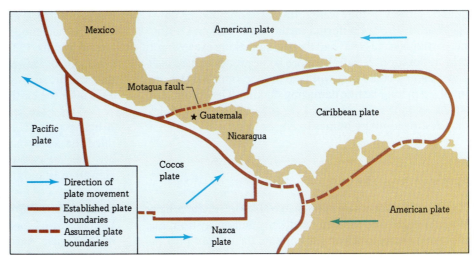

10.22 Guatemala earthquake of February 1976. The quake was the result of movement along the Motagua fault zone at the boundary between the American plate and the Caribbean plate.

registered 7.9 and killed 30,000. The province hit by the 1976 quake lies along the Anatolia fault, stretching inland from Turkey's Aegean coast. It is part of the major Mediterranean boundary zone between the African and Eurasian plates.

10.9
EFFECTS OF EARTHQUAKES

Earthquakes are interesting to most people because of their effects on human beings and structures. Their geological effects may also be profound. However, of all the earthquakes that occur every year, only one or two are likely to produce spectacular geological effects such as landslides or the elevation or depression of large landmasses. A hundred or so may be strong enough near their sources to destroy human life and property. The rest are too small to have serious effects.

FIRE

When an earthquake occurs near a modern city, fire can be a greater hazard than the shaking of the ground. In fact, fire has been responsible for an estimated 95 percent of the total loss caused by some earthquakes. One reason for the rapid spread of fire after an earthquake is that the vibrations often disrupt the gas and water systems in the area. In San Francisco, for example, some 23,000 service pipes were broken by the great earthquake of 1906 (often called the "San Francisco Fire" rather than "Earthquake"). Water pressure throughout the city fell so sharply that when the hoses were attached to fire hydrants, only a small stream of water trickled out. Since that time, a system of valves has been installed to isolate any affected area and keep water pressure high in unbroken pipes in the rest of the city.

DAMAGE TO STRUCTURES

Modern, well-designed buildings of steel-frame construction have withstood the shaking of even some of the most severe earthquakes. Chimneys have often been particularly sensitive to earthquake vibrations, however, because they tend to shake in one direction while the building shakes in another. Consequently chimneys often break off at the roof line. In contrast, tunnels and other underground structures are little affected by the vibrations of even the largest earthquakes because they move as a unit with the surrounding ground.

In some earthquakes the extent to which buildings are affected by vibrations depends in part on the type of ground on which they stand. For example, in the San Francisco earthquake of 1906 buildings set on water-soaked sand, gravel, or clay suffered up to ten times as much damage as similar structures built on solid rock nearby (see subsection on liquefaction below).

It has been found that the duration as well as the intensity of ground motion is a factor in causing damage to buildings. Continued motion may cause damage even if it does not do so at the start. Reinforced concrete buildings start to show hairline shear cracks at the beginning of damaging motion. With continuing motion the cracks become enlarged, and eventually disintegration results if shaking continues long enough. Repeated strong movements will bring destruction even to steel buildings.

In the Alaskan earthquake of March 27, 1964, the duration of damaging motion was approximately 3 min, or three times as long as the damaging shaking in San Francisco in 1906. Many buildings withstood the early vibrations, only to collapse in the later stages. Had the buildings in Alaska been subjected to vibrations for only 1 min, many probably would not have collapsed. Some of the buildings in Alaska that suffered severe damage had been built to conform to the earthquake provisions of the Uniform Building Code recommended by the Structural Engineers Association of California. These provisions were intended to safeguard against major structural failures during earthquakes. However, they were formulated on building damage resulting from California earthquakes, and no California earthquake has ever had damaging shaking exceeding 1 min in duration. It will therefore be necessary to establish new building codes that will take into account the stresses involved in long-lasting damaging shaking. The duration of strong shaking is also blamed for the extensive landslides that destroyed many homes in Anchorage.

SEISMIC SEA WAVES

If you are ever fortunate enough to be basking in the sun on Waikiki Beach and the water suddenly pulls away from the shore and disappears over the horizon, do not start picking up seashells or digging clams—a seismic sea wave is coming, and the withdrawal of the water may be the first warning of its approach.

Some submarine earthquakes abruptly elevate or lower portions of the sea bottom, setting up great sea waves in the water. The same effect may also be produced by submarine landslides at the time of a quake. These giant waves are called **seismic sea waves,** or **tsunami** (a Japanese term that has the same form for both singular and plural). Seismic sea waves may be generated by volcanic eruptions, as in 1883, when Krakatoa erupted. These waves killed 36,500 people in the East Indies. Seismic sea waves have devastated oceanic islands and continental coastlines from time to time throughout history. The Hawaiian Islands have been hit 37 times since being discovered by Captain Cook in 1778.

On April 1, 1946, a severe earthquake occurred at 53.5°N, 163°W, 130 km southeast of Unimak Island, Alaska, in the Aleutian Trench. Here the ocean is 4,000 m deep. Minutes after the earthquake occurred, waves more than 33 m high smashed the lighthouse at Scotch Cap, Unimak Island, killing five persons. About 4.5 h later the first seismic sea wave from this quake reached Oahu, Hawaii, after traveling 3,600 km at 800 km/h. Several more followed.

Starting in the Alaskan waters where the earthquake occurred, these seismic sea waves had spread out much as waves do when a rock is thrown into a pond. But these waves were tremendous in length. Their crests were about 160 km apart. They swept out into the Pacific Ocean, moving at terrific speeds of over 800 km/h. As they passed ships, however, these waves went unnoticed. They were about 1 m high in water 3,000 m deep, with 160 km between crests; so their effect was similar to the ground level rising 1 m as you walk 80 km. But when they reached Oahu and other Pacific shores, the effect was dramatic. The energy that moved thousands of meters of water in the open ocean became concentrated on moving a few meters of water at a shallow shore. There the water curled into giant crests that increased in height until they washed up over shores meters above high tide: 12 m on Oahu and 18 m on Hawaii. These same seismic sea waves swept on down the Pacific. They

reached Valparaiso, Chile, 18 h after their launching, when they still had enough energy to cause 2-m rises of the water after traveling 13,000 km. Some even returned to hit the other side of Hawaii 18 h later. In fact, tide gauges showed that seismic sea waves sloshed around the Pacific basin for days after the earthquake was over.

This 1946 seismic sea wave was one of the most destructive to hit Hawaii. It came without warning and was rated the worst natural disaster in Hawaii's history, but it was the last destructive seismic sea wave to surprise the Hawaiian Islands.

Seismic-Sea-Wave Warning System In 1948 the United States Coast and Geodetic Survey established a seismic-sea-wave warning system (SSWWS) for Hawaii (see Figure 10.23). It operates continuously, recording visible seismograms at its seismological observatory in Honolulu. These are equipped with an automatic alarm system that sounds whenever an unusually strong earthquake is recorded. Tide stations have also been set up to detect waves with characteristics of seismic sea waves. These gauges filter out the normal tides and the wind waves. When the alarm rings, requests are made for immediate readings from other seismograph stations around the Pacific. Within approximately 1 h the earthquake's location is determined. If it is found that the earthquake was in the Pacific Ocean or near its perimeter, tide gauges are then checked to see whether they show the existence of a seismic sea wave. If they do, an estimate of its arrival time is made, an alert is sounded, and people are warned to evacuate coastal areas. Of course, a warning system is only effective if it is heeded.

The Chilean seismic sea wave of May 22, 1960, was the most destructive in recent history, causing deaths and extensive damage in Chile, Hawaii, the Philippines, Okinawa, and Japan. As usual, Honolulu had computed the location and time of the earthquake; so when word was flashed that seismic sea waves were racing out over the Pacific, the Honolulu observatory issued warnings of the danger, urging evacuation of coastal areas and correctly predicting the arrival time of the first waves. These arrived on schedule, 6 h after the warnings were broadcast, 15 h after they started off South America. Through failure of many people to heed the warnings, however, 61 lives were lost in Hilo, Hawaii. Also, 229 dwellings were destroyed or severely damaged. In Japan no general seismic-sea-waves warning was issued, for it was not realized that a seismic sea wave of such distant origin would prove so destructive. About 8 h after hitting Hawaii, more than 22 h after the earthquake that started them, the seismic sea waves roared up the coasts of Honshu and Hokkaido. There, more than 17,000 km from where they started, they brought death to 180 people and caused extensive damage. The waves also did considerable damage in New Zealand. All Chilean coastal towns between the 36th and 44th parallels were destroyed or severely damaged.

Fortunately, only an extremely small fraction of all submarine or coastal earthquakes cause seismic sea waves.

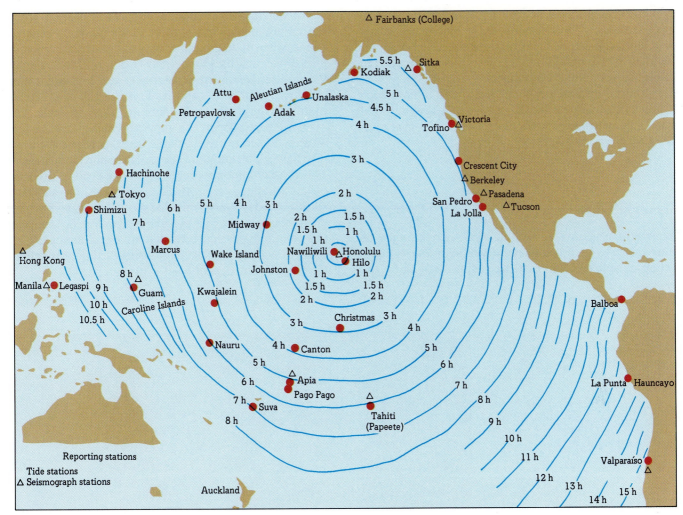

10.23 Seismic-sea-wave travel times to Honolulu.

LANDSLIDES

In regions where there are many hills with steep slopes or areas with special soil conditions sensitive to vibration, earthquakes are often accompanied by landslides. These slides occur within a zone seldom exceeding 40 to 50 km in radius, although the very largest earthquakes have affected areas as far away as 150 km. The Alaskan quake of 1964 did this.

One of the worst earthquake-caused landslides on record occurred on June 7, 1692. More than 20,000 lives were lost and much property destroyed in a large section of the then-bustling town of Port Royal, Jamaica. The whole waterfront was launched into the sea, together with several streets of two- to four-story brick houses. The houses and other buildings had been built on loose sands, gravel, and filled land. Shaken loose by the quake, the underlying gravel and sand gave way and slid into the sea; two-thirds of the town, consisting of the government buildings, wharf, streets, homes, and people, went along.

In the province of Kansu, China, in deposits of loess (wind-deposited silt), an earthquake on December 16, 1920, caused some of the most spectacular landslides on record. The death toll was 100,000. Great masses of surface material moved nearly 2 km, and some of the blocks carried along damaged roads, trees, and houses.

Several landslides occurred in the vicinity of Hebgen Reservoir, Montana, when the area was shaken by an earthquake just before midnight on August 17, 1959. The largest slide dammed the Madison River to form Earthquake Lake, as shown in Figures 10.24 and 10.25.

LAND MOVEMENTS

Some earthquakes are accompanied by significant vertical and horizontal movement of the land surface. The surface sinks in some places, rises in others, and is often tilted, and portions can even be moved laterally great distances. Furthermore, there are stories about great gaping cracks in the ground (see Box 10.4).

Fault creep is slow, periodic movement of the land on opposite sides of a fault (see Chapter 9). Measurable movement along the San Andreas is on the order of 1 or 2 cm/year. During the past 20 million years this has resulted in offsets of several hundred kilometers.

Vertical displacements in excess of 10 m have been recorded in a single earthquake, and areas of thousands of square kilometers may be affected. (Details of some examples were given in Chapter 9.)

LIQUEFACTION

Damage to structures as a result of earthquakes may be caused more by foundation failure and sliding due to **liquefaction** of saturated soil or loose sediments than by the actual shaking of the ground. Liquefaction often occurs several minutes to tens of minutes after a quake strikes a region, when underground sewers, storage tanks, pipes, and piles driven into the ground may be observed to float up to the surface. Nearby structures may settle several meters into the ground. Water and sand may be ejected a meter or more into the air for several minutes after the quake.

All these phenomena apparently result from turning what had been relatively stable deposits near the surface into liquefied material incapable of supporting structures. This most commonly occurs where the deposits are thoroughly saturated, such as along seacoasts, the shores of lakes or ponds, or anywhere that subsurface water has filled the fractures, voids, and pores in the ground. So it appears that soils and loose deposits that are free-draining are less likely to undergo liquefaction during earthquakes and are therefore better foundations for structures.

10.24 Region of Earthquake Lake, Montana: earthquake of August 17, 1959. [After John H. Hodgson, *Earthquakes and Earth Structure*, Prentice Hall, Englewood Cliffs, N.J., 1964.]

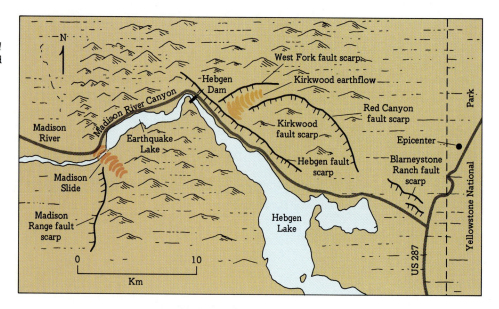

10.25 Earthquake Lake, Montana, as seen from the landslide that blocked the Madison River as the result of an earthquake on August 17, 1959. (See also Figures 10.24, 12.18, and 12.19.) [Noel Potter.]

BOX 10.4   Cracks in the Ground

One of the most persistent fears about earthquakes is that the Earth is likely to open up and swallow everyone and everything in the vicinity. Such fears have been nourished by a good many tall tales, as well as by pictures showing fault scarps after an earthquake.

One account of the Lisbon, Portugal, earthquake of November 1, 1755, claimed that about 40 km from Lisbon, the Earth opened up and swallowed a village's 10,000 inhabitants with all their cattle and belongings and then closed again. The story probably got its start when a landslide buried some village in the area.

In Japan widespread ideas about the Earth's ability to swallow people may have sprung from an allusion in one of the Buddhist scriptures: When Devadatta, one of Sakya Muni's disciples, turned against his master and even made attempts on his life, Heaven punished him by consigning him to Hades, whereupon the ground opened and immediately swallowed him up.

It is true that landslides do bury people and buildings, and under special conditions they may even open small, shallow cracks. In California, in 1906, a cow did fall into such a crack and was buried with only her tail protruding. But there is no authenticated case in which solid rock has yawned open and swallowed anything.

SOUND

When an earthquake occurs, the vibrations in the ground often disturb the air and produce sound waves that are within the range of the human ear. These are known as **earthquake sounds.** They have been variously described, usually as low, booming noises. Very near the source of an earthquake sharp snaps are sometimes audible, suggesting the tearing apart of great blocks of rock. Farther away the sounds have been likened to heavy vehicles passing rapidly over hard ground, the dragging of heavy boxes or furniture over the floor, a loud, distinct clap of thunder, an explosion, the boom of a distant cannon, or the fall of heavy bodies or great loads of stone. The true earthquake sound, of course, is quite distinct from the rumble and roar of shaking buildings, but in some cases the sounds are probably confused.

10.10

EARTHQUAKE PREDICTION AND CONTROL

From time to time observers have tried to relate the occurrence of earthquakes to sunspots, tides, positions of heavenly bodies, and other phenomena; but they have all ignored the facts in one way or another. Occasionally someone steps forward to predict earthquakes on the basis of some completely fancied correlation, usually keeping the supporting data secret. One such prophet stated that on each of certain days during three months of one year there would be "an earthquake in the southwest Pacific." Because he did not specify magnitude, time, or place in this highly active zone, statistically speaking, it was not possible for him to miss.

During the last decades, however, great strides have been made toward scientifically accurate measurement, interpretation, and potentially precise prediction of earthquakes. This had been an elusive goal for seismologists and astrologers alike for centuries. But recent advances in the Earth sciences provide some hope that this dream will someday become a reality.

EARTHQUAKE PREDICTORS

Many different kinds of short-term and long-term effects have been observed to precede earthquakes. These **predictors** include crustal movements, unexpected changes in surface tilting, and changes in fluid pressure and in electrical and magnetic fields, as well as local small quakes.

One predictor of earthquake activity is animal behavior. It has long been alleged that dogs, cattle, and other animals behave strangely immediately prior to the occurrence of an earthquake. It now appears that these animals are not soothsayers but are merely sensing minute changes in the Earth and its activities that tend to escape human senses.

Another predictor involves seismic waves. Observations for a number of earthquakes in Russia, Japan, and the Adirondack region of the United States have shown that prior to each earthquake there is a marked change in the velocity of seismic waves passing through the rocks near the quake area.

Laboratory experiments long ago demonstrated that rocks undergo **dilatancy,** or increase in volume, before failure. This is apparently produced by the formation and gradual outward movement of cracks within the rock. This dilatancy begins to occur at stresses as low as half the breaking strength of the rock. Water is released as the cracks expand. The rocks become more fragile and ultimately are too weak to withstand the stress; rupture results. If these changes could be observed in sufficient time, it should be possible to predict the earthquake.

A bowing up of the Earth's crust commonly occurs in the region of the earthquake epicenter. This changes the tilt of the surface, which can often be measured with sensitive instruments, thus furnishing us with another possible earthquake predictor.

Finally, time of occurrence of earthquakes can be plotted from historic and stratigraphic data, and analysis of such information allows us to estimate the probable recurrence intervals for many major quake zones. Some zones of major earthquake activity have a few segments where little or no seismicity has occurred for a significant period of time. Such seismic gaps may be considered "ripe" for future quakes.

So far, only about 10 earthquakes have been successfully predicted prior to their actual occurrence. With continued close observation of earthquakes and their associated events, however, earthquake prediction could soon become more reliable. It seems that no single predictor is sufficient. Earthquakes may be preceded by quite different physical events and changes in different geological regions, and earthquake-prediction methods will thus vary according to different sets of criteria for different regions.

CONTROL OF EARTHQUAKES

A fortuitous discovery made some years ago has led to the distinct possibility of controlling or modifying at least some earthquakes.

As waste fluids were injected into a deep well near Denver, Colorado, over a period of several years, a series of minor earthquakes developed, which were subsequently demonstrated to coincide with periods of fluid injection. During lulls in the injections, fewer or no quakes occurred. Since then, laboratory and field investigations have shown that movement along a fault zone can be initiated by reducing frictional resistance across the fault. This can be accomplished by injecting fluids along the fault. Conversely, faults can be locked when fluids are withdrawn. In the Rangely oil field in Colorado some United States Geological Survey workers demonstrated this ability to turn the seismic activity in an area on and off.

Although it is not likely that a major active fault zone will be treated in this manner in the near future, this method of control over fault movements may be possible for future generations.

10.11

STRUCTURE OF THE EARTH

Studies of the travel habits of body waves through the Earth and of surface waves around the Earth have given us information about its structure from surface to center (see Figure 10.26). These studies have been made possible by our

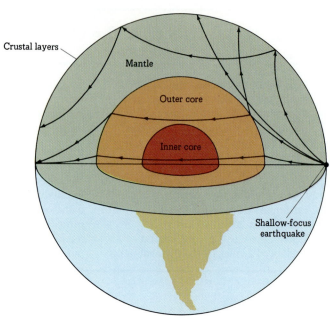

10.26 The paths of some of the more common earthquake waves.

knowledge of the speed of these earth waves and of their behavior in different materials (for example, body waves travel at slower speeds through denser materials than through less dense materials).

When body waves move from one kind of material to another, they are refracted (bent) and reflected. These waves have revealed several places within the Earth where there are changes in physical properties. Such changes could be due to changes in composition, atomic structure, or atomic state. The boundary where change takes place is called a **discontinuity.**

For body waves to reach greater and greater distances on the surface, they must penetrate deeper and deeper into the Earth's interior. Thus in traveling from an earthquake in San Francisco to a station in Dallas, a surface distance of 2,400 km, the body waves penetrate to 480 km below the surface. This holds true for any other 2,400-km surface distance. To reach a station 11,200 km away, the body waves dip into the interior to a maximum depth of 2,900 km and bring us information from that depth. (See Section 10.5).

On the basis of data assembled from studies of the travel habits of earth waves, the Earth has been divided into three major zones: **crust, mantle,** and **core** (see Figure 10.27).

CRUST

Information on the Earth's **crust** comes primarily through observation of the velocities of P and S waves from local earthquakes (within 1,000 km) and from dynamite and nuclear blasts. One of the first things revealed is that the Earth's crust is solid rock. Early in the history of crustal

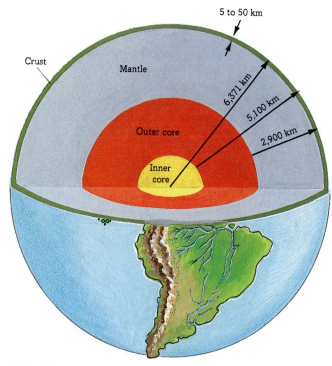

10.27 Structure of the crust, mantle, and core of the Earth.

M discontinuity). The Moho marks the base of the Earth's crust separating it from the mantle.

Crust of the Continents   The depth to the Moho varies in different parts of the continents. However, it is generally greater under mountain regions than it is under lowlands. The crust is also variable in composition.

In the United States the continental crust has an average thickness of 33 km but varies between 20 and 60 km (see Figure 10.28). Under the intermountain plateaus and the Great Basin Ranges, it averages less than 30 km; under the Great Valley of California, 20 km; under the Sierra Nevada, more than 50 km. It is thickest under the eastern front of the Rocky Mountain ranges. The crust under New England is about 36 km thick.

The average velocity of waves traveling through the upper crustal layer is similar to the velocity expected for granite, granodiorite, or gneiss. Because these rocks are rich in silica and aluminum, we say that the material of the upper crust is **sialic** in composition; **sial** ("Si" for silicon and "Al" for aluminum) is generally used in speaking of this layer, which is found only on the continental areas of the Earth. The second layer is mafic in composition and is believed to be basalt. These darker, heavier rocks are sometimes designated collectively as **sima,** a name coined from "Si" for silicon and "ma" for magnesium. A simatic layer encircles the Earth, presumably underlying the sial of the continents, and is the outermost rock layer under deep ocean basins.

Crust Under Oceans   Our knowledge of the structure of the crust beneath the oceans is based on observations of rocks exposed on volcanic islands, on deep sea islands and on studies of the velocities of L waves from earthquakes, supplemented by dynamite-wave profiles and magnetic anomalies (Chapters 1 and 11).

On the basis of seismic-wave velocities, it appears that

studies a seismologist in Yugoslavia, A. Mohorovičic, made a study of records of the earth waves from an earthquake on October 8, 1909, in the Kulpa Valley, Croatia. He observed two P waves and two S waves and concluded that the first P and S waves had encountered something that caused some of their energy to be reflected to the surface. He also concluded that velocities of P and S waves increased abruptly below a depth of several tens of kilometers. This abrupt change in the speed of P and S waves indicated a change in material and became known as the **Mohorovičic discontinuity** (for convenience, now referred to as the **Moho,** or the

10.28 Crustal thickness of United States, determined from seismic data. [After L. C. Pakiser and I. Zietz.]

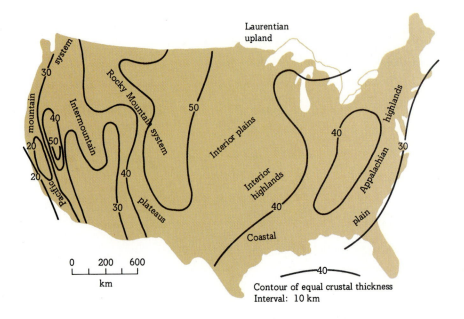

Contour of equal crustal thickness
Interval: 10 km

the crust under the Pacific basin is not layered and is appreciably thinner than the crust of the continents. Its thickness also varies but averages about 5 km. It is thinner under the Gulf of California and in the northeast Pacific. It is composed of simatic rocks.

MANTLE

Below the Earth's crust is a second major zone, the **mantle,** which extends to a depth of approximately 2,900 km into the interior of the Earth. Our knowledge of the mantle is based in part on evidence supplied by the behavior of P and S waves recorded between 1,100 and 11,000 km (Figure 10.27).

At the Moho, the speeds of P and S waves increase sharply, an indication that the composition of the material suddenly changes. We have no direct evidence of the new material's nature, but the change in speed suggests that it may contain more ferromagnesian minerals than does the crust. Scientists have accepted the idea that the mantle is solid because it is capable of transmitting S waves. To explain mountain-building processes, observers have emphasized that the mantle material undergoes slow flow as it adapts to changing conditions on the surface.

There are also variations in the composition of the upper mantle. We know this from variations in the velocity of P_n, the wave that travels through the upper portion of the mantle. Speeds of P_n vary from 7.5 to 8.5 km/s. This difference may be due in part to variation in elastic properties of the mantle material.

LITHOSPHERE AND ASTHENOSPHERE

As already defined in Chapter 1, the **lithosphere** is the rigid outer layer of the Earth, including the crust and the upper part of the mantle; the **asthenosphere** is a zone within the Earth's mantle that is particularly plastic. It lies below the lithosphere, extending from about 50–100 km below the Earth's surface downward to 500 km. There is a seismic discontinuity in the mantle. It occurs at the so-called low-velocity zone, which varies from about 50 to 80 km under oceans to about 100 km under continents (Figure 10.29). This apparently marks the boundary between the upper, colder portion of the mantle and the lower, hotter mantle. It also coincides, therefore, with the boundary between the lithosphere and the asthenosphere.

When we speak of plates, we are essentially referring to this lithosphere, which moves over the asthenosphere. It consists of crust *and* upper mantle, and is the part of the Earth that moves slowly but persistently. In this model part of the mantle is in motion, carrying the lithospheric plates on its top in piggyback fashion. We may also consider the cold lithosphere as likely to sink—the leading edge, in a sense, pulling the rest of the plate downward and away from the spreading center.

CORE

We come now to the **core,** a zone that extends from the 2,900-km inner limit of the mantle to the center of the Earth at a depth of 6,370 km. An analysis of seismographic

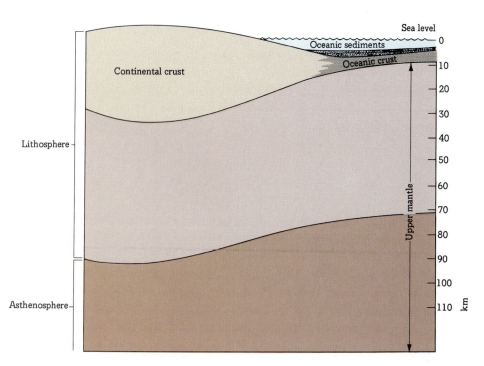

10.29 Crust, mantle, lithosphere, and asthenosphere.

records from earthquakes 11,200 km or more distant reveals that the core has two parts, an outer zone 2,200 km thick and an inner zone with a radius of 1,270 km.

In traveling between two points on the surface 11,200 km apart, *P* and *S* waves penetrate 2,900 km into the interior. But after they go deeper than that, they enter a material that delays *P* and eliminates *S* altogether (Figure 10.30). There have been various suggestions to explain these observations. The most widely favored theory postulates that the outer core is liquid. *S* waves are capable of traveling only through substances that possess rigidity; and since rigidity is a property of the solid state, it is assumed that the outer core, not transmitting *S* waves, is behaving as a liquid.

P waves speed up again at a depth of about 5,100 km, indicating that the inner core is solid.

By its orbiting behavior and influence on other planets the Earth is known to have a specific gravity of 5.5, but geologists have found that the average specific gravity of rocks exposed at the surface is less than 3.0. And even if rocks with this same specific gravity were squeezed under 2,900 km of similar rocks, their specific gravity would increase to only 5.7. Geophysicists have computed that the specific gravity of the core must be about 15.0 to give the whole globe an average of 5.5. To meet this requirement, it has been suggested that the core may be composed primarily of iron, possibly mixed with about 8 percent of nickel and some cobalt, in the same proportions found in metallic meteorites, and some oxygen, sulfur, silica, and carbon.

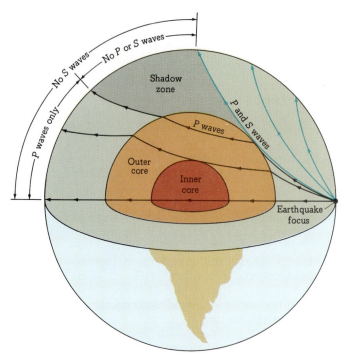

10.30   Interior of Earth as deduced from interpretation of seismic waves. Shadow zone occurs 103° to 143° away from each earthquake. No *P* or *S* waves penetrate this region. The last ray to miss the outer core arrives just below 103°. The next ray catches the outer core: Its *S* wave is not transmitted through the liquid outer core; its *P* wave is refracted to a position just beyond 143° from the quake. Only *P* waves penetrate the outer core to arrive at the surface on the opposite side of the Earth.

BOX 10.5   The Loma Prieta, California Quake

At 5:04 PM, Pacific Daylight Time, on October 17, 1989, an earthquake of magnitude 6.9, with its epicenter between Santa Cruz and San Jose, shook the San Francisco Bay area. About 100 people died, hundreds more were injured, and damage was estimated at more than 7 billion dollars.

The quake was caused, as was that of 1906, by a sudden movement along the San Andreas fault as the Pacific plate shifted northwest in relation to the American plate to the east. Although the 1989 earthquake was a repeat of the San Francisco quake 83 years earlier, it was on a smaller scale. The 1906 quake had an estimated magnitude between

7.9 and 8.3 and thus released well over 30 times the energy of the 1989 event. Furthermore, in 1906 the movement along the fault occurred over a distance of about 300 km in contrast to approximately 40 km in 1989.

Damage was greatest to older structures which had not been built to the stricter earthquake codes of recent years. Many buildings in the Marina district were either lost or severely damaged, in part, because they were old and in part, because they were built on unstable fill land. In Santa Cruz many of the older buildings in the downtown business district were lost. The greatest loss of life occurred along the double-decked

Nimitz Freeway in Oakland built between 1949 and 1954. Here along a ten-block section the upper deck collapsed onto cars and their occupants moving along the lower deck.

The 1989 quake released some of the stress building up along the San Andreas fault in the Bay area. Some seismologists feel, however, that there is still enough additional stress accumulated along the fault since 1906 to suggest that a quake bigger than the 1989 quake will occur sooner or later. Seismologists also point to the Hayward fault across the bay from San Francisco as potentially capable of causing a disastrous quake in the future.

Summary

Earthquakes are vibrations caused by movement of the Earth's plates.

Seismology is the scientific study of earthquakes.

Early instrumental observations were mostly by seismoscopes; the first record of a distant earthquake was made in 1889.

Modern seismology had its beginnings with a group of British professors teaching in Japan in the late nineteenth century.
   Modern instruments consist of an inertia member, a transducer, and a recorder.

Earth waves carry away some of the energy released when Earth materials rupture.
   Body waves travel through the interior of the Earth.
   Surface waves travel along the surface.

Records of earthquake waves are characterized by three sets of earth waves: P, S, and L waves.
   Time-distance graphs are bases for analyzing the history of earth waves.
      Travel times are governed by the materials through which waves have traveled.
   Earthquake focus is a term used to designate the source of a given set of earthquake waves; the **epicenter** is the area on the surface vertically above the focus.
   Locating earthquakes requires coordinating data from several stations.

Earthquakes are interpreted and evaluated from different facts and scales.
   Worldwide standard seismograph network is supervised by the Coast and Geodetic Survey.
   Earthquake-intensity scales are used to estimate the amount of shaking at different places.
   Earthquake magnitude evaluates the amount of energy at the earthquake's source.

Cause of earthquakes is the sudden rupture of Earth materials distorted beyond the limit of their strength.
   Elastic-rebound hypothesis proposed snapping back of distorted and ruptured rocks as the cause of earthquakes.

Plate movements generate a large proportion of all earthquakes. **Rates of movement** average 4 to 6 cm/year.

Effects of earthquakes include fire, damage to structures, seismic sea waves, landslides, cracks in the ground, changes in land level, and sound.
   Fire has caused an estimated 95 percent of the total loss from some earthquakes.
   Damage to structures depends on intensity and duration of the shaking.
   Seismic sea waves are now being detected, and a **seismic-sea-wave warning system** operated by the United States Coast and Geodetic Survey alerts threatened communities.
   Landslides often accompany the largest earthquakes.
   Land movements result in slow creep or rapid offsets both horizontally and vertically.
   Liquefaction of soil and loose sediments often follows earthquakes and commonly causes more damage than the actual shaking by the quake.
   Sounds associated with earthquakes vary from inaudible to booming and thunderous.

Earthquake prediction may now be possible, primarily by observing and interpreting **predictors**, including **crustal movements, tilt of Earth's surface**, and **changes in fluid pressure** and **in electrical and magnetic fields.**
   The **dilatancy model** is based on volumetric increases in rocks immediately prior to their rupture.
   Control of earthquakes may be achieved by injection and removal of fluids from potentially active fault zones.

Interior of the Earth is known to us from studying the records of earth waves.

Structure of the Earth is made of crust, mantle, and core.
   The Earth's **crust** is separated from the mantle at a **discontinuity** called the **Mohorovičic**, or the **Moho**, or simply the **M discontinuity.**
   The **crust of the continental United States** varies between 20 and 60 km in thickness.
   The **crust under the oceans** averages about 5 km in thickness.
   The **mantle** extends to 2,900 km in depth.
   The **lithosphere** contains the crust and the upper, colder part of the mantle.
   The **asthenosphere** extends downward from its border with the lithosphere to about 500 km.
   The **core** is thought to have a liquid outer zone and a solid center; the outer zone behaves as a liquid in not allowing S waves to penetrate.

Questions

1. What are the principal earthquake waves? How are they similar? How do they differ from one another?

2. What difference is there between an earthquake's focus and its epicenter? Make a sketch to show the relationship.

3. Explain briefly how the epicenter of an earthquake can be located accurately from three or more recording seismograph stations.

4. How do we distinguish between an earthquake's intensity and its magnitude? Which is more objective? Why?

5. How does the distribution of earthquakes compare to the distribution of volcanoes? What is the reason for this relationship?

6. What is the elastic-rebound hypothesis for the cause of earthquakes?

7. Why are there "seismic gaps" along some major fault zones, such as the San Andreas?

8. What is the greatest hazard associated with earthquakes? Why is this so?

9. How does the type of ground affect the probability of a building's undergoing severe damage during an earthquake?

10. Why are oceanfront structures thousands of kilometers from an earthquake likely to be in danger?

11. What produces the impetus for large landslides associated with earthquakes?

12. What are some of the principal kinds of earthquake predictors? What may be the reason or reasons for their occurrence?

13. How close are we to being able to predict earthquakes? Explain.

14. How close are we to being able to control earthquakes? Explain.

15. What have earthquake studies taught us about the structure of the Earth's interior? Make a simplified sketch to show the various parts of the Earth.

Supplementary Readings

ALLEN, C. R. (CHM.) *Predicting Earthquakes: A Scientific and Technical Evaluation—With Implications for Society,* Panel on Earthquake Prediction of the Committee on Seismology, National Academy of Sciences, Washington, D.C., 1976.
Evaluates the status of earthquake prediction and previews future trends. Presents urgent scientific and technical recommendations for further study and implementation of earthquake-prediction capabilities.

BOLT, B. A. *Earthquakes,* W. H. Freeman & Co., 1987.
Explores the latest theories of earthquake causes, prediction, and environmental effects, providing a basis for understanding attempts to forecast the time and strength of earthquakes.

CLARK, S. P. *Structure of the Earth,* Prentice Hall, Englewood Cliffs, N.J., 1970.
An excellent treatment of the subject.

DEWEY, JAMES, AND PERRY BYERLY "The Early History of Seismometry (to 1900)," *Bull. Seis. Soc. Am.,* vol. 59, 1969, pp. 183–227.
Describes the first instruments tried for recording ground motion from earthquakes, the first successful record of a distant earthquake, and early research that was the forerunner of all aspects of modern seismology.

HODGSON, JOHN H. *Earthquakes and Earth Structure,* Prentice Hall, Englewood Cliffs, N.J., 1964.
A good account written for nonspecialists.

OLSON, RICHARD STUART (WITH BRUNO PODESTA AND JOANNE M. NIGG) *The Politics of Earthquake Prediction,* Princeton University Press, Princeton, N.J., 1989.
The story—really the drama—of an ill-fated 1976-1981 earthquake prediction for Lima, Peru. It is a detailed account of how governments, their agencies, the scientific establishment and plain individuals became enmeshed in a scientific controversy. This detailed case history, written by a social scientist, is reading that is often depressing, always fascinating and throughout scarily instructive for the non-scientist and scientist alike.

PRESS, FRANK "Earthquake Prediction," *Sci. Am.,* vol. 232, no. 5, 1975, pp. 14–23.
A summary of recent technical advances, which have brought the United States and several other countries into a position of nearly achieving reliable long-term and short-term forecasts within a decade.

WOOD, FERGUS J. (ED.) *The Prince William Sound, Alaska, Earthquake of 1974 and Aftershocks,* vols. I–III, Publication 10-3, U.S. Coast and Geodetic Survey, Washington, D.C., 1966, 1967, 1969.
Definitive reports on the greatest earthquake ever recorded on the North American continent, making it also the most thoroughly studied.

WOOD, R. M. *Earthquake Geology,* Unwin Hyman, Winchester, Mass., 1989.
An introduction to the ways in which earthquakes may be used to interpret the geology of selected regions.

Sinai Peninsula, showing the Red Sea (bottom right), the Gulf of Aqaba (smaller water body in center right), and the Gulf of Suez (on the left), reminding us that continents can be rifted apart. [NASA.]

CHAPTER ▪ 11

PLATE TECTONICS AND MOUNTAINS

"The first notion of the displacement of continents came to me in 1910 when, on studying the map of the world, I was impressed with the congruency of both of the Atlantic coasts, but I disregarded it at the time because I did not consider it probable."

Alfred Wegener, The Origin of Continents and Oceans, *p. 5, trans. from the 3rd German edition by J. G. A. Skerl, Methuen & Company, London, 1924*

The apparent fit between the coasts of Africa and South America has long been noted. Indeed, one distinguished geologist, while admitting to skepticism about continental drift, found this fit so credible that he opined, "If the fit between South America and Africa is not genetic, surely it is the device of Satan for our frustration."[1] Half a century after Alfred Wegener put forth his theory on continental drift (he called it "continental displacement"), the concept began to achieve wide acceptance, albeit in altered form from Wegener's original model.

In this chapter we first review some of the *early arguments for continental drift*. Then we turn to *magnetism and paleomagnetism*. We consider the origin and causes of these phenomena and how they help demonstrate the movement of continents.

Sea-floor spreading is an important component of the twentieth-century model of continental drift, and we examine this process before turning to *plate tectonics*, which pictures the Earth's crust as a jigsaw puzzle of rigid plates jostling one another, growing in some places and decaying in others. We examine *plate boundaries* and the *movement of plates* as well as *microplates*, slivers of the Earth's crust that have become welded to larger plates.

We close this chapter with a consideration of mountain building, and particularly the role that plate tectonics plays in the process.

11.1

ORIGIN AND PERMANENCE OF THE CONTINENTS AND OCEAN BASINS

How and when did the continents and ocean basins form? Were they always distinct and separate entities, or have they only recently developed their characteristic identities? Have the continents always been approximately the same size and in the same positions relative to one another? These questions are of primary importance to geology. In this chapter we summarize some of the evidence, some of the clues to these mysteries that have plagued students of the Earth for many years.

We are awed by the immense amount of time and energy required to build mountains, erode continents, and shift shorelines. Yet we have come to accept these changes because they are documented by evidence we can see. But change has occurred on an even larger scale, challenging not only our sense of direction but our concept of geographic permanency as well. In Section 11.3 we show evidence that the North and South Poles have shifted position through time. We also show that our landmasses have been torn from their moorings and have wandered over the Earth's surface, a phenomenon known as **continental drift.**

One of the favored mechanisms for continental drift is the opening up of ocean basins through the process we call **sea-floor spreading.** Actually, the sea floor does not really spread. Rather, new sea floor is formed at midocean ridge areas and older sea floor dives back down into the mantle, all the while carrying continents in a piggyback fashion.

The concept of **plate tectonics** pictures the Earth's crust as a jigsaw puzzle of rigid plates jostling one another, growing in some places and decaying in others. In Section 11.5 we show how attempts have been made to determine the movement rates and to detect the signs of the most recent movements, and we discuss the implications of such global tectonics.

The present ocean basins are geologically quite young. The oldest features found in the sedimentary cover on the ocean floor have been dated as Jurassic in age. This is not to say that there were no earlier ocean basins with sediments, but rather that any older sediments that once existed were lost as these former ocean floors were consumed along subduction zones.

11.2

CONTINENTAL DRIFT

We now have good evidence that the continents have not always been in their present positions but have moved about throughout most, if not all, of geologic time. Acceptance of the concept of continental drift by the Earth science community was slow in coming. In fact, some still find it difficult to accept the enormity of the concept that ocean basins and continents have not always been where we find them today.

[1] Chester Longwell, "My Estimate of the Continental Drift Concept," in *Continental Drift: A Symposium,* p. 10, Tasmania University Press, Hobart, Australia, 1958.

ARGUMENTS FOR CONTINENTAL DRIFT

Wegener's Theory   The first coherent theory that our continents have moved was presented early in this century by Alfred Wegener (1880–1930), a German meteorologist and a student of the Earth. Wegener, as many before and after him, was intrigued by the apparent relationship in form between the opposing coasts of South America and Africa. Was it possible, he wondered, that these two landmasses were once part of the same general continent but have since drifted apart?

Beginning about 1911 and culminating in his 1915 book *The Origin of Continents and Oceans,* Wegener presented his theory of continental drift. Besides comparing the shapes of Africa and South America, he noted similarities in the geology, fossils, and glaciation on continents now widely separated. He pictured the dry land of the Earth as included in a single vast continent that he named **Pangaea** (from the Greek for "all" and "Earth"). This primeval continent, he argued, began to split asunder near the beginning of the Mesozoic Era, and the continental fragments began a slow drift across the Earth's face until by the Pleistocene they had taken up their positions as our modern continents.

Wegener's emphatic assertions started a spirited discussion that has continued to the present. The early evidence for his theory was drawn almost entirely from the terrestrial geologic record. Wegener visualized the continents as plowing through the oceanic crust, pushing up mountains along their margins in the process. He postulated that tidal forces from the Sun and moon acting upon the Earth might be the driving mechanism. Soon physicists began calculating the necessary forces to account for such enormous disruptions in the Earth's crust. In 1925 Sir Harold Jeffreys, an English scientist, presented calculations showing the great strength of the Earth and the forces that would be required to allow continents to move. He "proved it was physically impossible" for continents to drift, or so the scientific community thought at that time. In fact, the theories of tidal forces and centripetal forces that had been advanced as possible mechanisms for continental drift were so soundly rebuked by Jeffrey's calculations that it was not until the 1950s that serious discussion regarding continental drift was revived.

In the mid-twentieth century the development of geophysical techniques brought new data to bear on the discussion of continental drift. This began the various lines of evidence to allow for the development of the "new global tectonics." But first, let's look at a more or less historic reconstruction of some of the evidence for continental drift.

Shape of Continents   As we observed at the outset of this chapter, anyone observing a map of Africa and South America is struck by the jigsaw-puzzle match of the two continents. If we fit the eastern nose of South America into the large western bight of Africa, the two continents show a near-perfect match. And if we examine the outlines of the other continents, we can perhaps be persuaded that these landmasses too can be reassembled into a single large landmass, as suggested by Wegener. If the continents are fitted at the margin of the continental slope (a level of about 900 m below sea level), there is a very impressive match indeed, as shown in Figure 11.1. (Wegener himself first suggested using the edge of the continental shelf, *not* the edge of the continent at present sea level.)

Evidence from Ancient Climates   Much geologic evidence cited in support of continental drift and polar wandering is based on the reconstruction of climates of the past. Their use is justified by the fact that modern climatic belts are arranged in zones roughly parallel to lines of latitude and symmetrical about the equator (Figure 11.2). Although climates at various times in Earth history have been both colder and warmer than they are at present, we assume that basic climatic controls have remained the same and, therefore, that the climatic belts of the past have always paralleled the equator and been concentric outward from the poles.

Evidence from the geologic record allows us to map the distribution of ancient climates in a very general way. The pattern of some ancient climatic zones suggests that they were related to poles and an equator with locations different from those of today and that, therefore, the landmasses and the poles have varied relative to each other since those climates existed.

On the Indian peninsula lies a sequence of rocks known as the Gondwana System, reaching in age from the late Paleozoic to the early Cretaceous. Beds of similar nature and age have been recorded in South Africa, Malagasy, South America, the Falkland Islands, Australia, and Antarctica. In these other localities they are known by other names, but we can still correlate them to the Gondwana System.

Geologists who have worked on the Gondwana formations have discovered many similarities among the rocks of the various continents despite their wide geographic separation. Some of these similarities are so striking that many accept only one interpretation: that the various southern lands must once have been part of a single landmass, a great southern continent, early called **Gondwanaland** by these geologists.

The distribution of ancient glacial deposits was one of Wegener's most convincing lines of evidence for continental drift in these southern lands. During late Paleozoic time continental ice sheets covered sections of what are now South America, Africa, the Falkland Islands, India, Antarctica, and Australia. In southwestern Africa deposits related to these ancient glaciers are as much as 600 m thick. In many places the now-lithified deposits (tillites) rest on older rocks striated and polished by these vanished glaciers.

If we plot the distribution of these deposits and the direction of ice flow on a map (see Figure 11.3[a]), we can make two immediate observations. First, these traces of

11.1 | A statistically determined fit of North America, South America, Africa, and Europe at the 900-m depth in the ocean. Zones of overlap are shown in the dark tone; zones of gap are shown in the lighter tone. Present-day lines of latitude and longitude are distorted when continents are reassembled to their pre-drift positions. [After Edward Bullard et al., "The Fit of the Continents around the Atlantic," *Phil. Trans Roy. Soc. London*, vol. 1088, pp. 41–51, 1965.]

Legend:
- Continental margins to 900-m depth
- Zones of gap
- Zones of overlap
- Line of 900-m depth

11.2 Present-day climatic boundaries are arranged concentrically around the poles and thus are approximately parallel to lines of latitude.

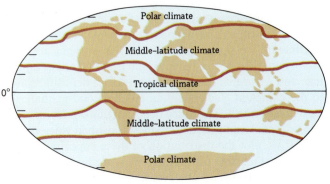

Polar climate

Middle-latitude climate

Tropical climate

Middle-latitude climate

Polar climate

Paleozoic ice sheets occur in areas where no ice sheets (except for Antarctica) exist now or did exist during the glacial epochs of the Pleistocene Ice Age. Only an occasional towering peak may bear the scars of modern or Pleistocene ice. Continental ice sheets cannot exist in these latitudes today. Second, the direction of glacier flow is such that we can imagine the ice of Africa, South America, India, and Antarctica to have been part of the same ice sheet when the continents were one (Figure 11.3[b]). If we use the present geographic continental positions, the source of the ice would have to be in the southwestern part of the Indian Ocean. But if we reconstruct Gondwanaland, this source site would be a landmass near the South Pole (during that time).

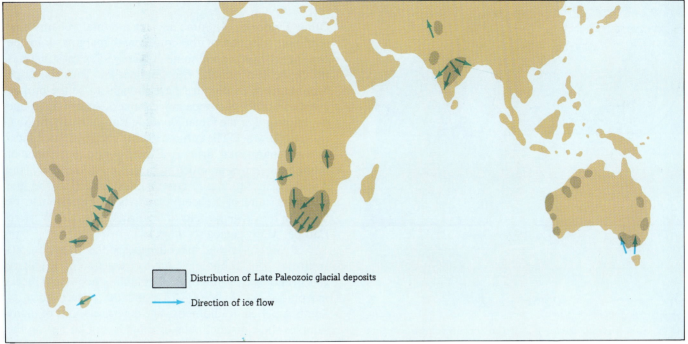

(a)

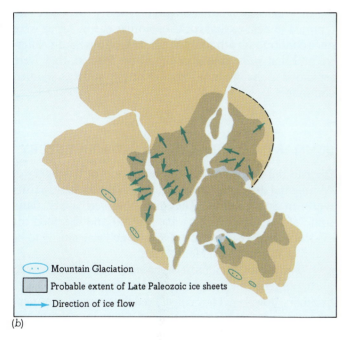

(b)

11.3 (a) Direction of movement of late Paleozoic glaciers and distribution of some known late Paleozoic tillites. (b) If we fit the Gondwana continents back together, the directions of the striations suggest a generally outward direction of flow of ice on the late Paleozoic landmass. If the Gondwana continents are restored to their Carboniferous positions, the glacial striations fit an understandable pattern, with movement out from a common center.

Distribution of Late Paleozoic glacial deposits

Direction of ice flow

Mountain Glaciation
Probable extent of Late Paleozoic ice sheets
Direction of ice flow

Evaporites, sedimentary rocks composed of minerals that have been precipitated from solutions concentrated by the evaporation of water, are generally accepted as evidence of an arid climate. The ancient evaporite deposits represent the great arid belts of the past. The present hot-arid belts are located in the zones of subtropical high pressure, centered at about 30° north and south of the equator. In the Northern Hemisphere an evaporite belt has shifted through time from a near-polar location in the Ordovician and Silurian Periods to its present position in the modern desert belts. This again suggests a relative motion of pole and landmasses of the past as compared with present-day conditions.

Turning to another line of evidence, we find that

corals mark a climate shifting geographically through time. Today true coral reefs are restricted to warm, clear marine waters between 30° north and 30° south of the equator. If we assume that ancient reef-forming corals had similar restrictions, then plotting their distribution in the past should show the distribution of tropical waters of the past and the location of the past equator. When we do this, we find that reef-forming corals did not approximate their present distribution until halfway through the Mesozoic Era. Prior to that time, they lay askew to the present equator.

Plants, Reptiles, and Continental Drift   Shortly after the disappearance of the Southern Hemisphere's late Paleozoic ice sheets, an assemblage of primitive land plants became widespread. This group of plants, known as *Glossopteris* flora, named for the tonguelike leaves of the seed fern *Glossopteris* (Figure 11.4), has been found in South America, South Africa, Australia, India, and within 480 km of the South Pole in Antarctica. The *Glossopteris* flora is very uniform in its composition and differs markedly from the more varied contemporary flora of the Southern Hemisphere. Wegener and subsequent geologists have argued that the uniformity of the *Glossopteris* flora could not have been achieved across the wide expanse of deep open ocean water now separating the different collecting localities. In other words, in one way or another, there

11.4 These leaves of the fossil plant *Glossopteris* come from strata of Permian age in Australia. The *Glossopteris* flora is found also in South America, Africa, India, and Antarctica. The widespread occurrence of this very uniform flora has been used as evidence by both opponents and proponents of continental drift. Diameter of detail is 2.5 cm. [Specimen from Princeton University Paleobotanical Collections, Willard Starks.]

must have been continuous or near-continuous land connections between now-separate continents. To many this suggests that a single continent with a single uniform flora has been split apart into smaller continents that have since migrated to their present position. As we shall see, this conclusion has not gone unchallenged.

Fossil discoveries since Wegener's time have tended to substantiate his early theory. Among the vertebrate fossils of the Paleozoic and earliest Mesozoic we find a great number of different reptilian types. Two of them provide us with arguments for continental drift. *Mesosaurus,* a toothed early reptile of the late Permian, lived in the water and thus far is known only from Brazil, South Africa, and Antarctica. Although it was aquatic, most paleontologists do not believe that it could have made the trip across the South Atlantic. If this is true, then *Mesosaurus* may offer evidence for a closer proximity of South America and Africa in the past, and thus for continental drift. A second reptile, *Lystrosaurus,* dates from the Triassic Period, *Lystrosaurus,* which is about 1 m in length, was adapted to an aquatic but nonmarine environment, and its remains are reported from South Africa and from Antarctica's Alesandra Range. It is argued that *Lystrosaurus* could not have made the journey across the Antarctic Ocean separating South Africa from Antarctica, and that this implies the former joining of the two continents.

We know also that not only did the Mesozoic rocks of the various Southern Hemisphere continents develop differently after the theorized Mesozoic breakup, but also that the flora and fauna evolved independently from this time on.

Other Evidence Some ancient mountain chains now terminate abruptly at the continental margins. Join the continents together and some of these geological structures match up between the two landmasses. Thus the Cape Mountains of South Africa are thought to be the broken extension of the Sierra de la Ventana of Argentina in one direction and of the Great Dividing Range in eastern Australia in the other. The entire stretch is cited by some adherents of continental drift as a once-continuous chain of mountains now segmented and separated. Similarly, the Appalachian Mountain system ends in the sea on the northern shore of Newfoundland. Is its extension to be found in the orogenic belts of the British Isles and Western Europe? Many geologists think so.

P. M. Hurley and his colleagues at the Massachusetts Institute of Technology have carried on extensive geochronologic studies of Precambrian igneous and metamorphic rocks in West Africa and the eastern bulge of South America. Radiometric-age determinations of many samples show distinct belts of roughly similar age on the two continents. If the continents are shifted back together, the several provinces of Precambrian rocks of differing ages on the rejoined landmasses match fairly well. We are tempted to believe that rocks of similar age in Africa were once continuous with rocks of the same age in South America and that sometime after they were formed, they were separated so that we now see them on two widely separated continents.

SUMMARY OF THE PROBLEM AT MIDCENTURY

Prior to the 1940s most of the supporters of the theory of continental drift lived in the Southern Hemisphere, particularly in South Africa. They had observed at first hand some of the similarities we have noted and accepted the evidence that Wegener postulated, which was mostly to be found in the Southern Hemisphere.

The reliability of most of the early evidence advanced to support the concept of continental drift was questioned by others, however. They considered the shape of continental margins to be merely fortuitous. Some thought the paleoclimatic evidence was inconclusive, others attacked the fossil evidence as fragmentary and open to too many interpretations to be considered diagnostic.

Starting in the 1950s, the techniques of geophysics, particularly those of paleomagnetic measurement, brought new data into the discussion. This new evidence reopened the entire question of continental drift for American and European geologists. A series of projects to extensively examine the Earth's ocean floor began in the 1960s. Magnetic measurements at sea have since revealed a characteristic pattern of anomalous magnetic intensities. These magnetic anomalies are most often associated with ocean ridges and are arranged in stripes of alternating intensity parallel to the ridge. An example is provided by the magnetics of the Reykjanes Ridge, a portion of the Midatlantic Ridge southwest of Iceland. Figure 11.5 shows the stripes parallel with the ridge and shows also a bilateral symmetry, the axis for which is the rift valley along the crest of the ridge.

In the 1950s Keith Runcorn, of Newcastle upon Tyne, and his associates developed the concept of paleomagnetism (see Section 11.3). By 1962, Harry Hess of Princeton had developed the ideas leading to the concept of sea-floor spreading. He postulated the concept that the ocean floor had an age of no greater than Mesozoic.

In 1963 two British geophysicists, Fred Vine and D. H. Matthews of the University of Cambridge, suggested that the alternating stripes of high- and low-intensity magnetism are in reality zones of normal and reverse polarity in the rock of the sea floor. From this suggestion, now widely accepted, have flowed some intriguing concepts about the sea floor, the drift of continents, and the building of mountains.

The Deep Sea Drilling Project (DSDP) gave us our first extensive information about deep-sea sediments and the upper portion of the oceanic basaltic crust. This was followed by more data in the 1980s from the Ocean Drilling Program (ODP), operated by Texas A&M University through the Joint Oceanographic Institutions Inc. (J.O.I.), a nonprofit consortium of 10 major oceanographic institutions. Joint Oceanographic Institutions for Deep Earth Sampling (JOIDES) consists of an international group of scientists who give planning and program advice.

Additional evidence for the development of continents through time has been gathered since the 1970s by such joint ventures as Cornell's Consortium for Continental Reflection Profiling (COCORP) and the more recent Continental Drilling Project (CDP). Each of these has as its principal mission to gain new insights about the origin and development of the continents. COCORP gets its information by indirect geophysical examination of seismic waves generated by "thumper" machines. Returning waves are recorded by a series of listening devices ("geophones") and the subsurface structure is interpreted. A mid-1980s COCORP project in North America, using the "vibroseis" technique, has demonstrated that much of the southern Appalachian mountain belt is detached from basement rocks. The COCORP project confirmed much of the inferred cycle of opening and closing of the Atlantic Ocean basin. CDP is gaining first-hand information by drilling into the continental crust and examining actual rock samples returned to the surface.

The pattern of the Earth's magnetism, as recorded in the rock record, together with the age of the oldest sediments found in various parts of the ocean basins, gave us new and compelling evidence regarding the development of the ocean basins. The result was that nearly all students of the Earth finally became convinced of Wegener's concept of continental drift. There is, however, one major difference between the earlier concept and today's theory: Whereas Wegener postulated that the continents push their way through the oceanic crust, today we believe the continents are passive riders on the backs of the moving plates. These plates include **both** continental and oceanic components.

11.5 Magnetic anomalies over the Reykjanes Ridge southwest of Iceland. The patterned areas display normal polarity and are separated by areas of reversed polarity. Note that the belts of normal and reversed polarity are generally symmetric about the axis of the Midatlantic Ridge. Ages of the various polarity belts are based on a time scale derived from a study of magnetic anomalies on land and on the sea floor. [After F. J. Vine, "Magnetic Anomalies Associated with Ocean Ridges," in R. A. Phinney (ed.), *The History of the Earth's Crust*, Princeton University Press, Princeton, 1968. © 1968 by Princeton University Press. Reprinted by permission.]

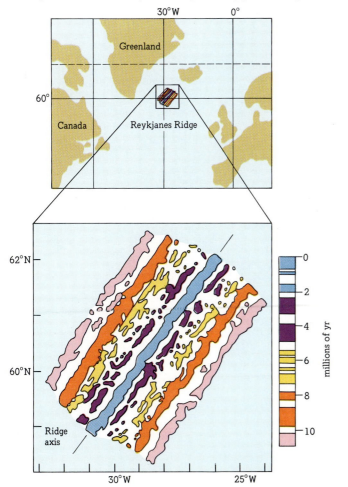

11.3

MAGNETISM AND PALEOMAGNETISM

It is general knowledge that the Earth acts as a large magnet, which is the cause of the behavior (and usefulness) of the common compass. In this section we take an extensive look at Earth magnetism and particularly at magnetism of the geologic past—**paleomagnetism.** The data derived from paleomagnetic studies not only gave the subject of continental drift a new push in the early 1950s but also provided proponents of continental drift with some of their strongest arguments. Furthermore, these studies have helped demonstrate the youthfulness of the ocean basins as compared with the continents and have been instrumental in the development of the idea of plate tectonics (Figure 11.6).

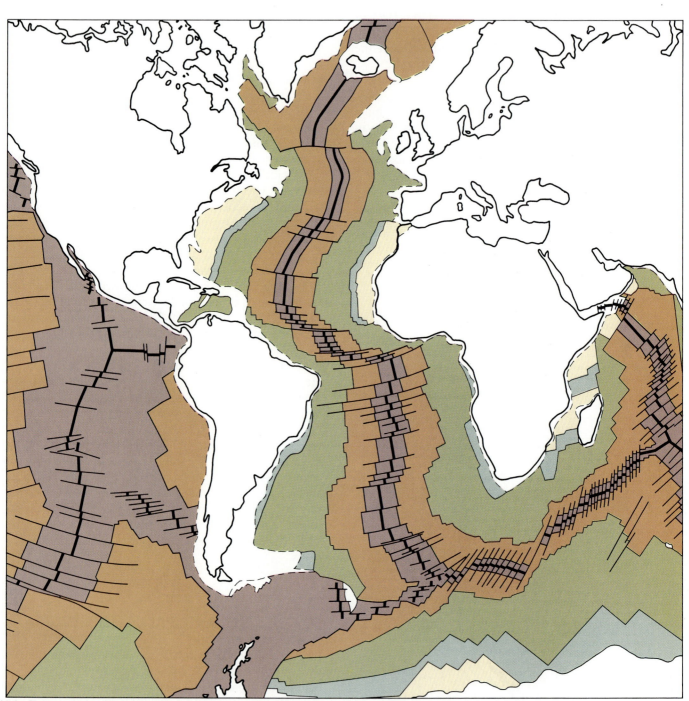

11.6 The age of the ocean floor is shown as bands of different colors, with the youngest ocean floor near midoceanic spreading centers and the oldest farthest away. The edge of the deep ocean floor is marked by the dashed line. [Modified from R. L. Larson and W. C. Pitman III, *The Bedrock Geology of the World*, W. H. Freeman and Co., New York, 1988.]

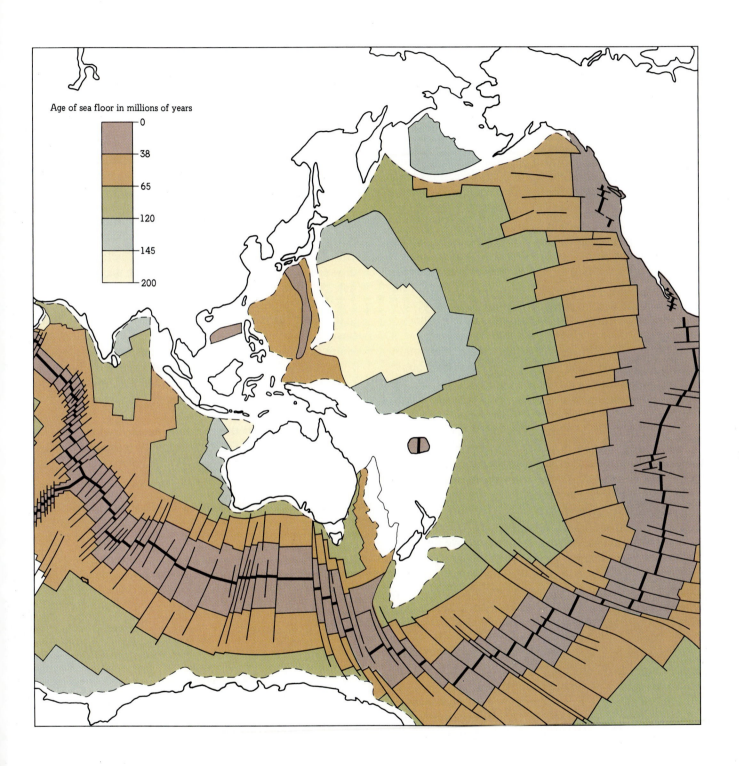

Age of sea floor in millions of years

0
38
65
120
145
200

CAUSE OF THE EARTH'S MAGNETISM

The cause of the Earth's magnetism has remained one of the most vexing problems of Earth study. A completely satisfactory answer to this question is still to be advanced.

We have long known that the Earth's magnetic field is composed of both internal and external components. The external portion of the field is due largely to the activity of the Sun. This activity affects the ionosphere and appears to explain magnetic storms. The changes and effects of the external field may be rapid and dramatic, but they have little effect on the internal field of the Earth, which is of greatest concern to us.

William Gilbert, a British physicist of the sixteenth century, first showed that the Earth behaves as if it were a magnet. He suggested that the Earth's magnetic field results from a large mass of permanently magnetized material beneath the surface. This notion might at first seem attractive, not only because large quantities of magnetic minerals have been found in the Earth's crust, but also because geologists think that the Earth's core is made largely of iron. But since close examination reveals that the average intensity of the Earth's magnetization is greater than that of the observable crustal rocks, we must look deeper for the source of magnetism.

The first difficulty we face is that materials normally magnetic at the Earth's surface lose their magnetism above a certain temperature. This temperature is called the **Curie temperature** and varies with each material. The Curie temperature for pure iron is about 760°C; for hematite, 680°C; for magnetite, 580°C; and for nickel, 350°C. The temperature gradient for the Earth's crust is estimated to average about 30°C/km, and at this rate of increase the temperature should approximate the Curie temperature of iron about 25 km below the surface and exceed the Curie temperatures for most normally magnetic materials. Therefore we should not expect Earth materials to be magnetic below 25 km, and permanent magnetism can exist only above this level.

On the other hand, if all the Earth's magnetism were concentrated in the crustal rocks, then the intensity of magnetism of these rocks would have to be some 80 times that of the Earth as a whole. And yet, we know that the magnetic intensity of the surface rocks is less than that of the Earth's average intensity.

From these observations we must conclude that the Earth's magnetic field is not due to permanently magnetized masses either at depth or near the surface.

The theory of Earth magnetism most widely entertained at present is that the Earth's core acts as a self-exciting dynamo. One model of the Earth's core pictures its outer portion as a fluid composed largely of iron. This core therefore not only is an excellent conductor of electrical currents but also exists in a physical state in which motions can easily occur. Electromagnetic currents are pictured as being generated and then amplified by motions within the current-conducting liquid. The energy to drive the fluid is thought to come from convection, which in turn results from temperature differences. The dynamo theory further requires that the random convective motions and their accompanying electromagnetic fields be ordered to produce a single, united magnetic field. It is thought that the rotation of the Earth can impose such an order. The dynamo theory of Earth magnetism still remains only a theory, but so far it has proved the most satisfactory explanation of the Earth's magnetism.

THE EARTH'S MAGNETISM

We have learned that the Earth behaves as if it were a magnet and that consequently the compass needle seeks the north magnetic pole. We can picture the Earth's magnetic field as a series of lines of force. A magnet free to move in space will align itself parallel to one of these lines. At a point north of the Prince of Wales Island, at about 75°N and 100°W, the north-seeking end of the magnetic needle will dip vertically downward. This is the **north magnetic** or **dip, pole.** Near the coast of Antarctica, at about 67°S and 143°E, the same end of our needle points directly skyward at the **south magnetic,** or **dip, pole.** Between these dip poles the magnetic needle assumes positions of intermediate tilt. Halfway between the dip poles the magnetic needle is horizontal and lies on the **magnetic equator.** Here the intensity of the Earth's field is least; intensity increases toward the dip poles, where the field is approximately twice as strong as it is at the magnetic equator (Figure 11.7).

The angle that the magnetic needle makes with the surface of the Earth is called the **magnetic inclination,** or **dip.** The north and south dip poles do not correspond with the

11.7 The Earth's magnetic field can be pictured as a series of lines of force. The arrows indicate positions that would be taken by a magnetic needle free to move in space and located at various places in the Earth's field. The magnetic, or dip, poles do not coincide with the geographic poles, nor are the north and south magnetic poles directly opposite each other.

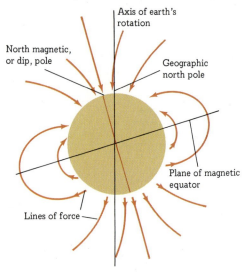

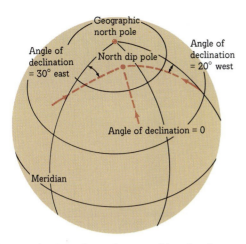

11.8 Because the dip poles and geographic poles do not coincide, the compass needle does not point to true north. The angle of divergence of the compass from the geographic pole is the declination and is measured east and west of true north.

true north and south geographic poles as defined by the Earth's rotation. Because of this, the direction of the magnetized needle will in most instances diverge from the true geographic poles. The angle of this divergence between a geographic meridian and the magnetic meridian is called the **magnetic declination,** and it is measured in degrees east and west of geographic north (see Figure 11.8).

PALEOMAGNETISM

Some rocks, such as iron ores of hematite or magnetite, are strongly magnetic. Most rocks, however, are only weakly so. Actually the magnetism of a rock is located in its individual minerals, and we should be more correct to speak of the magnetism of minerals rather than of the rock. By convention, however, we refer to rock magnetism. This magnetism is called the rock's **natural remanent magnetism (NRM).**

Let us examine some of the ways an igneous rock acquires its magnetism. As a melt cools, minerals begin to crystallize. Those that are magnetically susceptible, such as rocks containing iron or nickel, acquire a permanent magnetism as they cool below their Curie temperatures. This magnetism has the orientation of the Earth's field at the time of crystallization. It is called **thermoremanent magnetism (TRM).** This remanent magnetism remains with the minerals—and hence with the rock—unless the rock is reheated past the Curie temperatures of the minerals involved. This new heating destroys the original magnetism; and when the temperature again drops below the Curie temperatures of the magnetic minerals, the rock acquires a new TRM.

This remanent magnetism may or may not agree with the present orientation of the Earth's field and may have been acquired by the rock in many different ways. Identify-

ing, measuring, and interpreting the different components of a rock's NRM form the basis of **paleomagnetism,** the study of the Earth's magnetic field in the geological past.

Sedimentary rocks acquire remanent magnetism in a different way than do igneous rocks. Magnetic particles such as hematite tend to orient themselves in the Earth's magnetic field as they are deposited. Because the sedimentary particles are laid down on a nearly flat surface (such as the sea floor), they are limited in the position they can take. Therefore such particles normally reflect only the Earth's magnetic declination and *not* its inclination. This orientation is retained as the soft sediments are lithified. This magnetism, known as **depositional remanent magnetism (DRM),** records the Earth's field at the time the rock particles were deposited.

Geomagnetic Intensity, Variations, and Reversals

As long ago as the mid-seventeenth century it was known that the magnetic declination changed with time. Because such changes are detectable only with long historical records, they are called *secular changes,* from the Latin *saeculum,* meaning "age" or "generation," implying a long period of time. Such secular variation in the position of the geomagnetic pole over the last several hundred years (Figure 11.9) has been accompanied by variation in intensity of the Earth's magnetic field. Thus by 1965 the intensity had decreased nearly 6 percent from the time it was first successfully analyzed in 1835 by K. F. Gauss, a German mathematician.[2]

We have been able to extend our knowledge of the variation in the field's intensity by techniques first successfully used by the French physicist Emile Thellier. Samples that can be precisely dated by historical or radiocarbon

11.9 Magnetic inclination and declination vary with time. At Paris we have a continuous record of these changes since 1540. The data for inclination and declination for Gallo-Roman times (colored squares) are based on magnetic measurements of archaeological materials. [After Emile Thellier, "Recherches sur le champ magnétique terrestre," L'Astronomie, vol. 71, p. 182, 1957.]

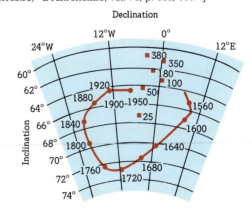

Declination

Inclination

---

[2] The present magnetic field at the Earth's surface is about 0.5 G (gauss), where "gauss" (or "oersted") is the unit of measure for magnetic-field intensity.

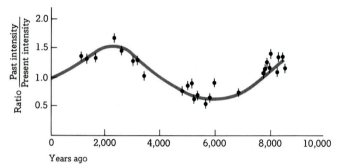

11.10 Intensity variation of the Earth's magnetic field, from Czech archaeological samples. Each circle represents a sample or samples of the same age. The vertical bars represent standard deviation from the mean ratio. [After V. Bucha, *Archaeometry*, vol. 10, p. 20, 1967.]

techniques show that the intensity of the field has been decreasing since about 2,000 years ago, when it was about 1.6 times its present intensity. Previously it had risen from a low value of about half the present intensity around 3500 B.C., this having been preceded by a steady decline from an intensity of about 1.5 times that of the present around 6500 B.C. (Figure 11.10).

If the decay of magnetic intensity were to continue at its present rate, it has been estimated by Allan Cox, a Stanford University geophysicist, the intensity would reach zero about 2,000 years from now. Thereafter the direction of the magnetic field might reverse itself. There is no assurance that this will happen; the intensity may just as well fluctuate as it has for the last several thousand years. But if we turn further back into Earth history, we find that the poles have been reversed—not once, but many times. So we assume that such a reversal is possible in the future.

We now have a great deal of information about past reversals of the magnetic field. That is to say, if we were able to take our ordinary magnetic compass back into time, we should find some periods in which the north-seeking end of the needle would point toward the North Pole as it does today. But there would be many other times during which the needle would point to the South Pole instead. The present polarity is called **normal.** A polarity at 180° to it is called **reversed.** A succession of normal and reversed fields can be pieced together for much of the recent geologic past. We call the longer intervals of dominance by a particular polarity a **polarity epoch,** and we have named these intervals after distinguished students of Earth magnetism. During each polarity epoch are shorter periods in which the polarity is in the opposite direction, and these time intervals are referred to as **polarity events** (see Figure 11.11).

In summary we have seen that iron-bearing minerals acquire magnetism as they pass through the Curie point, that this magnetism can be measured, that the Earth's magnetic poles can thus be determined, and that the Earth's magnetic field has reversed itself many times during geologic history.

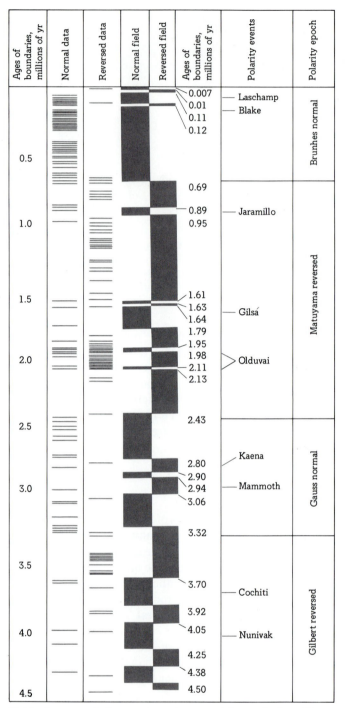

11.11 A time scale for geomagnetic reversal during the last 4.5 million years is based on extrusive igneous rocks. Each horizontal line in the two left-hand columns represents a rock sample, whether its polarity is normal or reversed. In the normal-field column normal-polarity intervals are shown in black; in the reversed-field column reversed-polarity intervals are shown in black. (See also Figure 8.11.) [After Allan Cox, "Geomagnetic Reversals," *Science*, vol. 168, pp. 237–245, 1969.]

RESULTS OF PALEOMAGNETIC STUDIES

If we can measure the paleomagnetism of a rock and relate it to the Earth's present field, we can determine to what extent the orientation of the Earth's magnetic field has varied at that spot through time.

Studies of ancient pole positions assume that the Earth's field has been dipolar and, further, that this dipole has approximated the Earth's axis of rotation. A consequence of these assumptions is that the Earth's geographic poles must have coincided in the past with the Earth's geomagnetic poles. Clearly this is not so at present. Why, then, should we think that it was true in the past? A partial answer to the question lies in the dynamo theory of Earth magnetism. If the dynamo theory is correct, then theoretical considerations suggest that the rotation of the Earth should orient the axis of the magnetic field parallel to the axis of rotation.

Observational data do indeed support the theoretical considerations. When we plot the geomagnetic poles of changing fields recorded over long periods at magnetic observatories around the world, we find a tendency for these pole positions to group around the geographic poles. More convincing are the paleomagnetic poles calculated on the basis of magnetic measurements of rocks of Pleistocene and Recent age. These materials, including lava flows and varves, reveal pole positions clustered around the present geographic pole rather than the present geomagnetic pole (see Figure 11.12).

As a result of theoretical and observational data, therefore, most authorities feel that the apparent present-day discrepancy between magnetic and rotational poles would disappear if measurements were averaged out over a

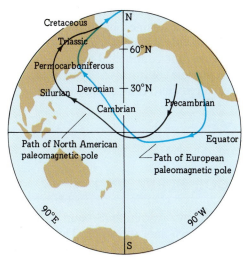

11.13 "Wander curves" of rocks from North America and Europe show the paths followed by the magnetic poles of these two continents from Precambrian times to the present. [After Allan Cox and R. R. Doell, "Review of Paleomagnetism," *Bull. Geol. Soc. Am.*, vol. 71, p. 758, 1960.]

span of approximately 2,000 years. The same principle would apply for any 2,000-year period throughout geologic time. Thus when we speak of a paleomagnetic pole, we have some confidence that it had essentially the same location as the true geographic pole of the time.

Paleomagnetic data derived from rocks of Tertiary age indicate fairly conclusively that there has been no significant shift of the geomagnetic poles from the Oligocene to the present. The farther backward we go in time beyond the Oligocene, however, the more convincing becomes the case for a changing position of many continents with respect to the magnetic (and therefore geographic) poles. The magnitude of subsequent change is suggested in Figure 11.13, which shows that Europe and North America were in different positions and have rotated since Precambrian time. Movement since then has seen the migration of these continents to their present position on the globe.

The apparent migration of the magnetic poles (and, by extension, the geographic poles), combined with the observation that the paths of polar migration of different continents fail to coincide, adds further support to the concept that the continents have drifted.

Evidence for Continental Drift   It has been the increasing body of information on rock magnetism and paleomagnetism that has kindled new interest in the old theory of continental drift. Consider the position of the North Pole as determined from rocks for Europe and for North America. We saw in Figure 11.13 how the paleomagnetic poles have shifted during the last 500 million years. For these two continents there is a divergence between the ancient poles as we go back in time. If we fit North America and Europe back together, however, the separate polar wandering paths superpose nicely.

11.12   The virtual geomagnetic poles determined by magnetic measurements of Earth materials of Pleistocene and Recent age cluster around the modern geographic pole rather than around the present geomagnetic pole. [After Allen Cox and R. R. Doell, "Review of Paleomagnetism," *Bull. Geol. Soc. Am.*, vol. 71, p. 758, 1960.]

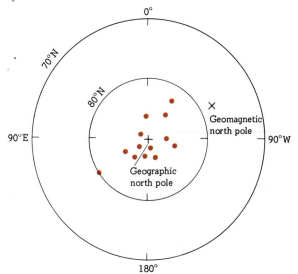

The divergence of the paths of polar migration among continents can be explained by the shifting of the land-masses in relation to each other. The course of polar migration suggests that the continental movement consists of a general drift of these continents away from each other, and in some instances there is an additional continental rotation.

11.4

SEA-FLOOR SPREADING

The Vine-Matthews hypothesis (Section 11.2) leads us to a consideration of sea-floor spreading. This is a mechanism that involves the active spreading of the sea floor outward, away from the crests of the main ocean ridges. As material moves from the ridge, new material replaces it along the ridge crest by welling upward from the mantle. As this mantle material cools below the Curie temperature, it takes on the magnetization of the Earth's field at that time. Room for the new material is made along the ridge by the continued pulling apart of the crust and its movement laterally away from the ridge. At times of change in the Earth's polarity, then, newly added material along the crest will record this change. Spreading laterally, it carries this record with it, to be followed at a later time by the record of the next polarity change. We can visualize this as a gigantic magnetic tape preserving a history of the Earth's changing magnetic polarity. A diagram to suggest the mechanism is shown in Figure 11.14.

Data from the JOIDES project (discussed in Section 11.2) confirm that the oldest sediments on the sea floor occur along the continental margins. Progressively younger sediments are found as one moves toward the ocean ridges. This sedimentary layer is thin or nonexistent adjacent to the spreading centers and thickens progressively away from the ridge as a result of the outward movement of the crust away from the spreading center. The oldest crust receives sediments for periods of time longer than does the younger crust.

The rates at which new sea floor forms at spreading centers is seen to vary. In the North Atlantic it is about 1 cm/year, and in the South Atlantic about 2.3 cm/year. Along the Juan de Fuca Ridge off California the rate is 3 cm/year, and on the east Pacific rise it is calculated to be 4.6 cm/year, reaching a rate of 9 cm/year in some sections. (These are all "half-spreading rates": Each side of the ocean floor is moving at these rates, so the continents on either side of the ocean basins would be moving apart at **twice** these rates.)

Figure 11.15 shows the effects of magnetic reversals on deep-sea sediments and associated oceanic lithosphere. If there were no sea-floor spreading, the layers of sediments would simply show alternating polarity as the Earth's magnetic field changed. With sea-floor spreading, each sedimentary layer begins at the ridge with the magnetization of that particular time in the Earth's history. As the litho-

11.14 A diagram to suggest the development of magnetic anomalies related to an oceanic spreading ridge. New oceanic crust forms from basaltic magma at the axis of the ridge. As it cools below the Curie temperature, the new crust acquires the magnetic characteristics of the Earth's field at that time. As the crust moves laterally away from the ridge crest, changes in the Earth's magnetic field are recorded in stripes of the oceanic crust. Each change in the direction of the field produces a pair of similar stripes on either side of the ridge. [After F. J. Vine, "Sea-floor Spreading—New Evidence," *J. Geol. Educ.*, vol. 17, pp. 6–16, 1969.]

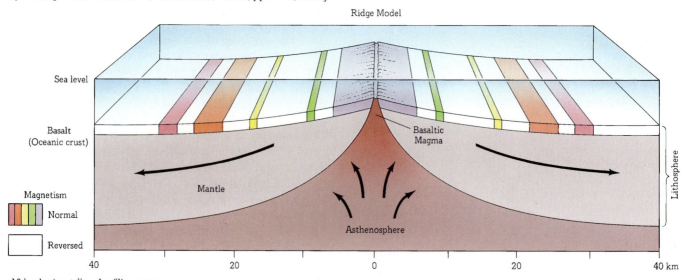

Ridge Model

Sea level

Basalt (Oceanic crust)

Basaltic Magma

Lithosphere

Mantle

Magnetism

Normal

Reversed

Asthenosphere

40   20   0   20   40 km

10 km horizontally = 1 million years
Crust moves away from ridge 1 cm/yr

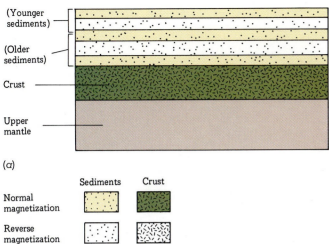

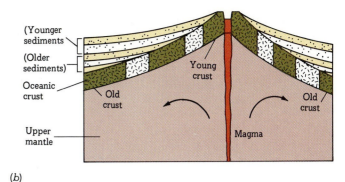

(a)

(b)

Sediments   Crust

Normal
magnetization

Reverse
magnetization

11.15   Magnetization of marine sediments: (a) If there were no sea-floor spreading, the layers would show alternating polarity in continuous sheets; (b) with sea-floor spreading, the alternate normal and reversed stripes in the crust have correspondingly magnetized sediments, beginning adjacent to that part of the crust formed at the ridge crest of the spreading center. [After P. J. Wyllie, *The Way the Earth Works*, John Wiley & Sons, New York, 1976.]

sphere migrates away from this spreading center, the sediments are carried farther and farther away.

Notice that while Wegener believed the ocean was passive — the continental lands actively moved through the sea floor — modern theory holds the opposite — the continents are merely passive riders on the backs of moving plates. The concept of sea-floor spreading indicates that the ocean floor is a dynamic system, and in terms of continental drift it suggests that expansion of ocean basins may also affect the location of continents. Furthermore, it raises many questions. For example: What drives the ocean floor, and where do the moving ocean floors go? The proposed answers to these questions are considered in the next section.

11.5
PLATE TECTONICS

Given the nonrandom distribution of earthquakes, major sites of weakness or flaws in the Earth's crust can be recognized. Geologists have discovered in recent years that the Earth's outer surface can be divided along these flaws into large units, or plates. We suggested at the beginning of this chapter and elsewhere that the origin, movement, and interrelation of these plates can be coordinated with sea-floor spreading, earthquake belts, crustal movement, and continental drift.

THE WORLD SYSTEM OF PLATES

At least six major plates and a host of minor ones are recognized, as shown in Figure 11.16. The major plates include

the Indian, Pacific, Antarctic, American, African, and Eurasian plates. The boundaries of these plates are loci of present-day earthquakes and volcanic activity. Consider for example the Pacific plate. On its southern and southeastern side it is bounded by an oceanic rise that undergoes active spreading. On its northern and northwestern borders it is marked by island arcs, deep-sea trenches, and tectonic and volcanic activity. Its northeastern side along western United States is defined by the San Andreas fault, where the Pacific plate moves generally northwestward and jostles the American plate in the process. Thus we have movements of plates relative to one another, movements that produce divergent, convergent, and strike-slip (or transform) plate boundaries.

The relationship of plates to the distribution of earthquakes is striking when one compares Figure 11.16 with Figures 10.10 and 10.11. In general, the shallow quakes mark the ocean ridges; deep earthquakes are missing in these zones. Both shallow and deep quakes are found along the arcs and trenches and along collision zones, as the Himalayas (see Table 11.1).

The plates may include both continental and oceanic areas, as, for example, the American plate, which embraces North and South America and Greenland as well as the western Atlantic Ocean. On the other hand, the Pacific plate is restricted to oceanic area.

MAKEUP OF PLATES

The lithosphere is the rigid outer layer of the Earth, including the crust and upper mantle. It comprises the plates that ride over the warmer, more plastic asthenosphere. The latter zone begins 50 to 100 km below the Earth's surface and extends perhaps to 400 to 500 km.

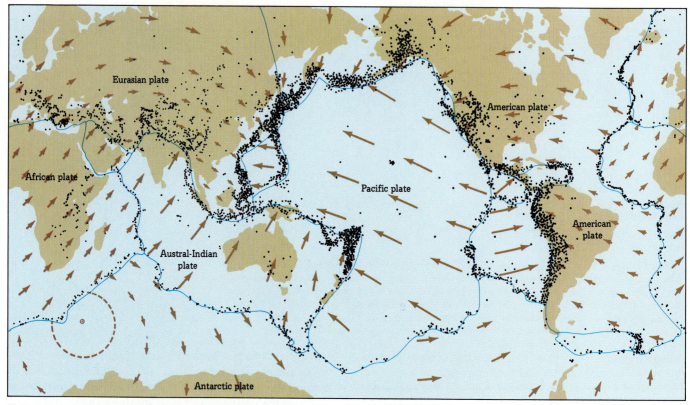

11.16 Crustal plates are outlined by zones of intensive earthquake activity. The directions of plate motion are shown by arrows. Each plate moves the distance equivalent to the length of the arrow in 20 million years. [After Sheldon Judson, Kenneth S. Deffeyes, and Robert B. Hargraves, *Physical Geology*, Prentice Hall, Englewood Cliffs, N.J., 1976.]

The rigid lithospheric plates consist of several distinct layers. The lowest part comprises the cooler, solid upper part of the mantle. The Moho discontinuity marks the upper contact of this layer with the crust. Above this lowermost layer is the principal part of the crust. As we have explained, under the oceans this crustal layer may be only a few kilometers thick, consisting of coarse gabbrolike simatic rock (rock rich in silica plus iron and magnesium). It is capped by lava flows—solidified magma extruded along the midocean ridges, rises, and other volcanic centers. Under continents this layer may be 30 km or more thick.

The lower crust may include a continuation of the oceanic gabbroic layer. The upper crust is sialic (rich in silica and alumina) and includes the crystalline granitic masses of the roots of mountains covered by a relatively thin layer of sedimentary and metamorphic rocks at the continental surfaces.

Topping this layer-cake structure in the oceanic regions are the sediments of the sea floor, consisting of fine inorganic debris plus the remains of organisms that lived in the upper portion of the ocean, died, and settled to the bottom, where they accumulated.

TABLE 11.1
Plate Boundaries and Their Interactions with Other Plates under
Tension, Compression, or Neither

Boundary	Divergent (tensional)	Convergent (compressional)	Transform (neither tensional nor compressional)
Ocean-ocean	Ridge crest; S[a] earthquakes in narrow belt; submarine lavas	Ocean trench and volcanic island arc; S, I, and D quakes over wide belt; volcanoes common	Ridge and valley fracture zone; S quakes in narrow belt only between offset ridges; volcanoes absent
Continent-continent	Rift valley; S quakes over wide zone; volcanoes present	Young mountain ranges; S and I quakes (or just S) over wide zone; no volcanoes	Fault zone; S quakes over broad zone; no volcanoes
Ocean-continent		Ocean trench and young mountain range; S and I quakes over wide belt; D quakes occasionally; volcanoes common	

[a] S, shallow-focus earthquakes; I, intermediate focus; D, deep focus.

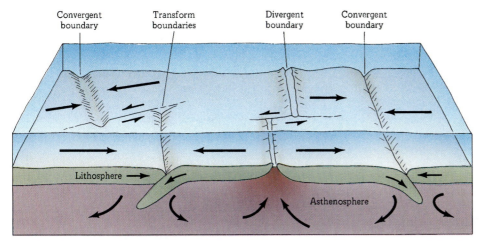

11.17 Generalized features of sea-floor spreading with the three principal types of plate boundary: convergent, divergent, and transform (or parallel). [After B. Isacks, J. Oliver, and L. R. Sykes, "Seismology and the New Global Tectonics," *J. Geophys. Res.*, vol. 73, pp. 5855–5899, 1968.]

CLASSIFICATION OF PLATE BOUNDARIES

Lithospheric plates originate at oceanic ridges and spread laterally toward subduction zones. The motion of these plates with respect to other adjacent plates results in three principal kinds of boundaries (Figures 11.17 and 11.18). Where plates move away from one another, **divergent** boundaries result; where they move toward each other, **convergent** boundaries result; and where one plate moves past another in a nearly parallel fashion, **transform** boundaries result.

The plates may be entirely dense oceanic lithosphere or they may include the lighter rafts of continental lithosphere. Movement of these plates may involve ocean-ocean, ocean-continent, or continent-continent interactions.

Because divergent boundaries represent conditions of **tension,** like the Midatlantic Ridge or the Red Sea area, shallow-focus earthquakes and volcanic emanations occur along the ridge crests of ocean-ocean regions and along the rift valleys of continent-continent zones (Table 11.1).

Compression characterizes convergent boundaries like the western margin of South America, and results in trenches and island-arc volcanoes, with shallow, intermediate, and deep-focus quakes along ocean-ocean contacts. Trenches and young mountain ranges occur with volcanoes, and shallow and intermediate-focus quakes (with or without deep-focus quakes) occur at ocean-continent boundaries.

Where continental plates meet other continental plates, like the Indian and Eurasian plates, young mountain ranges result from the piling up, or doubling, of the lighter continental masses. It appears that continental crust cannot be subducted, for it is too light to be carried down into the much denser material of the mantle. Volcanoes are rare or nonexistent under these conditions, but shallow-focus quakes are common. Intermediate-focus quakes are rare, and deep-focus quakes entirely absent.

Neither tension nor compression dominates along transform boundaries, where the plates appear to be moving past one another. In fact, in some locations at least, one plate may be partially subducted under the other. Continent-continent interactions are characterized by fault zones and shallow-focus earthquakes over a rather broad zone, such as along the San Andreas fault in California. Fracture zones with shallow-focus earthquakes in narrow belts mark ocean-ocean transform boundaries. Little volcanic activity is likely along any transform boundaries.

Plate boundaries are the locales for most of the Earth's volcanic and tectonic activity—for example, magmatism, batholith formation, folding, and uplift.

MOVEMENT OF PLATES

We saw in Figure 11.16 that some of the zones between adjacent plates are zones of active sea-floor spreading, a process described in Section 11.4. Along such zones, therefore, crustal material is considered as continuously forming, and to make room for it the sea floor is pictured as moving laterally away from the ridge crest. Transform faults, described in Chapter 9 and illustrated in Figure 9.23, cut across the ridge and produce its offset. An analysis of the first movement of the crust during earthquakes on the ridge shows two types of movement. Along the rifts that mark the axes of the oceanic ridges, motion during earthquakes indicates tensional stress, normal faulting, and dropping down of the blocks associated with the grabens along the ridges. In contrast is the movement along the fractures offsetting the ridges. Along these structures analysis of earthquake data shows that the movement is of a strike-slip nature on a plane at right angles to the ridge; and movement is away from the ridge crests, as in Figure 9.24 of the Midatlantic Ridge near the equator.

If plates are actually moving away from each other through a process of sea-floor spreading, two questions arise: What drives the ocean floor? Where does the moving ocean floor go?

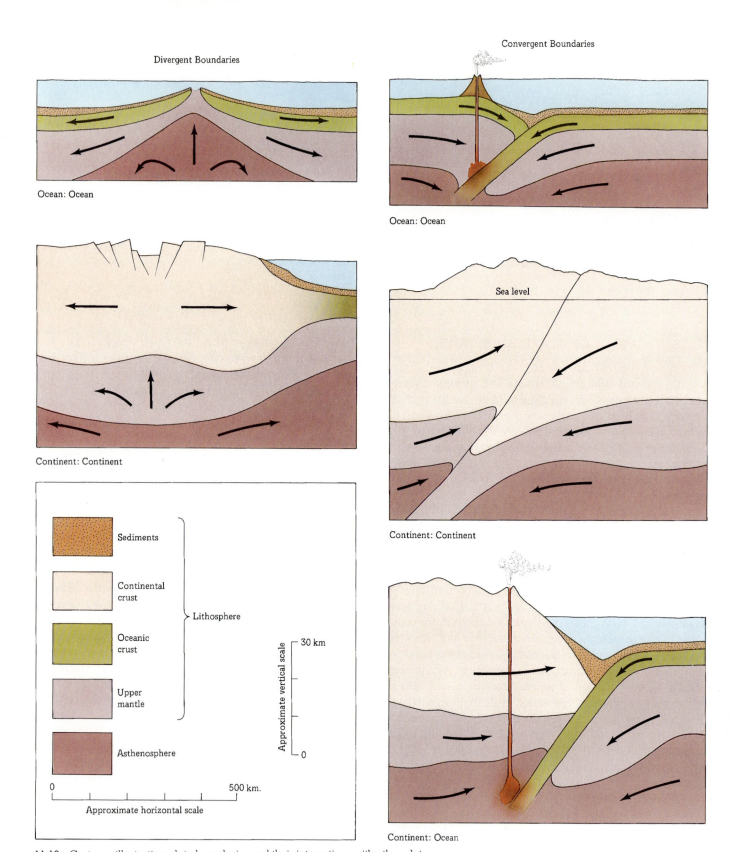

Divergent Boundaries

Ocean: Ocean

Continent: Continent

Convergent Boundaries

Ocean: Ocean

Sea level

Continent: Continent

Sediments

Continental crust

Oceanic crust

Upper mantle

Lithosphere

Asthenosphere

Approximate vertical scale

30 km

0

0   500 km.

Approximate horizontal scale

Continent: Ocean

11.18   Cartoons illustrating plate boundaries and their interactions with other plates.

Convection Cells   Some students of the problem consider it likely that convection currents, moving in cells, provide a mechanism for moving the ocean floor. The ocean ridges are seen to lie over a rising convection current. Seismic and volcanic activity are generally high on the ridges and heat flow from the interior is higher at the ridge crest than it is at most other places. The convection current is thought to lie in the mantle, well below the lithosphere, and to represent ascending limbs of adjacent convection cells. At some point below the lithosphere the rising current is pictured as splitting into two currents flowing laterally away from the spreading ridge center.

Plume Theory   A more likely driving mechanism for the movement of plates has been suggested by Jason Morgan (of Princeton University), who earlier had pieced together much of the evidence in support of plate tectonics.

He infers the existence of hot spots, or plumes, of hot rising solid mantle in zones a few hundred kilometers in diameter. This is an extension of an earlier hypothesis by J. Tuzo Wilson of Toronto, who suggested similar structures to account for volcanic island chains unrelated to plate boundaries (see Figure 4.14). Over 100 hot spots have been active during the past 10 million years according to one study. By this hypothesis, hot mantle rock rises and spreads out laterally in the asthenosphere in about 20 major thermal plumes around the world (see Figure 11.19), creating hot spots of volcanic activity. The rest of the mantle undergoes very slow downward movement to balance the upward flow in the plumes. The movement of any plate is caused by the combination of forces exerted by the upward flow in the plume, the outward lateral flow away from the plume, and the return downward flow of the mantle. (See Box 11.1.)

11.19   Some of the major mantle plumes. Thin black lines represent volcanic trails, or linear volcanic chains, tracing the hot spots. [After W. J. Morgan, "Plate Motions and Deep Mantle Convection," in R. Shagam (ed.), *Studies in Earth and Space Sciences*, pp. 7–22, Geological Society of America Memoir 132, Boulder, Colo., 1972; and K. C. Condie, *Plate Tectonics and Crustal Evolution*, Pergamon Press, Elmsford, N.Y., 1976.]

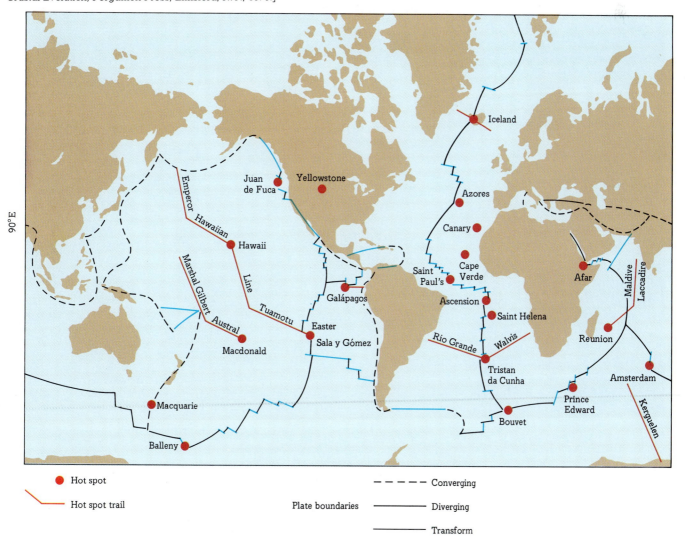

BOX 11.1   Earth Cat Scan

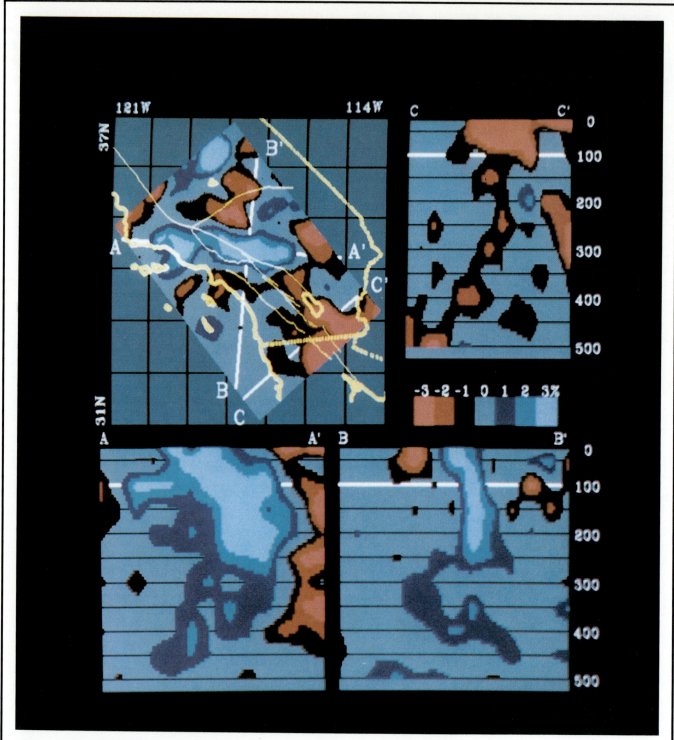

B11.1.1 Earth CAT scan images of Southwestern United States. Map view in upper left shows images of the Earth's mantle at a depth of about 100 km, with outline of U.S. state boundaries and part of San Andreas fault system superimposed on map. Blue indicates cold dense rocks through which seismic waves travel quickly; red shows hot less dense rocks through which waves travel more slowly. The Cal Tech geophysicists who developed this technique call it "seismic tomography." Cross-sections A-A', B-B', and C-C' show third dimension from surface to depths of about 500 km. Notice slablike mass in blue extending down 250 km below the surface in A-A' and B-B'. Also note shallow hot region in red on C-C'. Combination of red and blue regions leads geophysicists to suggest these features represent convection cells. [R. Clayton.]

The Earth is getting a CAT scan by the California Institute of Technology in Pasadena, much as doctors use sound waves to create internal images of humans. **Seismic tomography** is what Caltech geophysicists Robert Clayton, Thomas Hearn, and Eugene Humphreys call the powerful new imaging technique that uses seismic waves to give subsurface three-dimensional geological models.

Data from thousands of earthquakes from 200 Caltech–U.S. Geological Survey seismic stations in southern California were analyzed to produce the first two crustal maps from differing seismic velocities under southern California. One map is of images 100 km deep and measures discontinuities about the middle area of the crust. (See Figure B11.1.1.) The other is of images 30 km down, at the boundary between the crust and the mantle.

Other data show on a 10-km seismic tomography map the sharp boundaries of the San Andreas fault, where each side is of different origin: The eastern side of the fault is Mojave Desert rock characteristic of the American plate,

while on the western side the rock is that of the Pacific plate. The 10-km map indicates that the Transverse Ranges, including the San Gabriel and San Bernardino Mountains, consist of rock that has migrated from elsewhere. The mountain ranges do not extend deep into the Earth's crust.

According to the 30-km seismic tomography map, the thickness of the crust varies widely throughout southern California, with the thinnest area, about 22 km deep, around the Colorado River. Slower-velocity seismic waves are characteristic of sediment such as the 10-km-thick layer that lies within the Los Angeles and San Bernadino basins, while the dense volcanic rock underlying the sediment in the Salton Sea region causes high-velocity waves.

The Caltech scientists think the San Andreas fault is not as sharp at 30 km because it acts more broadly at such a depth and deviates from the surface fault line.

The weight of the mountains causes deformation of the Moho boundary, producing roots that, according to the seismic images, appear in the Trans-

verse Range beneath the San Bernardino Mountains, but not under the San Gabriel Mountains. Also, there appears to be no major root underneath the Peninsular Ranges, which may indicate that less dense materials underlie and compensate for the weight of the mountains.

While studying similar images drawn for the Earth's mantle 800 km down, Clayton's attention was drawn to two features. One is a seismically fast or thermally cold slablike feature extending down some 250 km into the Earth (Figure B11.1.1A and B). The other feature he noticed was a shallow, hot region (Figure B11.1.1C). According to Clayton, these two features combined are the first seismic evidence of a convection cell.

The Caltech geophysicists believe the seismic CAT scans will provide images that may help us achieve a greater understanding of the forces and events that shaped the Earth.

These early studies have led to similar studies of the entire globe by such methods. The chapter-opening photo is such a summary by John Woodhouse of Harvard University.

Other Possibilities Other mechanisms to drive the plates across the Earth's surface have been suggested. Most of them depend on the thermal differences that exist within the Earth and between the colder plates and hotter mantle. One theory holds that the subducting ocean plate is colder and denser than the surrounding mantle, so that it sinks into the mantle and actively pulls the remainder of the slab along with it. This "slab-pull" mechanism may have a corollary "ridge-push" operating in conjunction with it or independently. The midocean ridges stand higher than the surrounding ocean basins and might tend to "push" the plates away from these spreading centers.

Another possibility depends on the Earth's expanding from a smaller spheroid to its present size since the beginning of the Mesozoic. When this theory was postulated by S. Warren Carey of the University of Tasmania more than two decades ago, it did not catch the interest of many other geologists. Carey and Hugh Owen of the British Museum suggest that an Earth with only 80 percent of its present diameter and 50 percent of its present volume would allow an even better fit of the continents. In support of their theory they point out that measurements of the number of days in a year done for corals and other fossils from many

geologic ages have determined that there were about 400 days (rotations of the Earth) in each year nearly 400 million years ago. An expanding Earth would account for the slowing down in rotation that has produced a fewer number of days in the present Earth's year. (This phenomenon of slowing down could also be explained by gravitational tidal Earth-Moon friction and by changing Earth-Moon distances.)

Many more mechanisms will probably be suggested before we finally know the principal mechanism that has produced the phenomenon of continental drift. But the phenomenon itself has been accepted, to the ultimate vindication of Alfred Wegener.

RECYCLING THE SEA FLOOR

What happens to the old sea-floor crust when it is displaced from the ridge crest by younger and younger material? Those who explain the motion by convection see the crust and a part of the upper mantle carried laterally by convection currents toward the margins of the ocean basins. This is

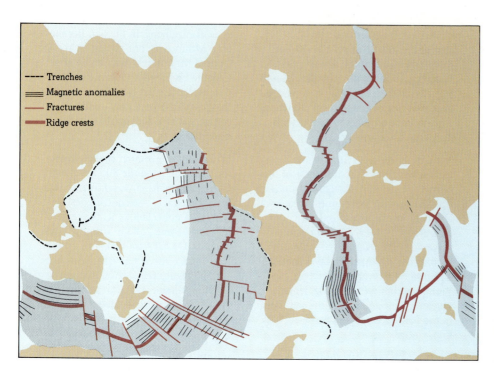

11.20 The shaded portion of the basins represents the area of oceanic crust formed within the last 65 million years. [After F. J. Vine, "Sea-Floor Spreading— New Evidence," *J. Geol. Educ.*, vol. 17, pp. 6–16, 1969.]

Legend:
- - - Trenches
≡ Magnetic anomalies
— Fractures
━ Ridge crests

supported not only by the paleomagnetic measurements already discussed but also by studies of oceanic sediments. These studies indicate not only that the sedimentary deposits are thinner in the zones of the midoceanic ridge but also that, as they become thicker away from the ridge, the basal layers of the sedimentary accumulations become older. All this suggests that the oldest sea floor lies farthest from the oceanic ridges.

If we follow this line of reasoning, we should expect to find that belts of varying widths along the active ridges mark the most recently formed oceanic crust. If we accept the reconstruction shown in Figure 11.20, then an area approaching 50 percent of the ocean floor was formed during the Cenozoic—that is, in approximately the last 65 million years. In this regard, it is interesting to note that no rocks older than 150 million years have as yet been taken from the deep ocean basins. Moreover, present-day sedimentation rates in the oceans can account for the sedimen-

tary accumulations there within the last 100 to 200 million years. All this leads to the suggestion that the ocean basins, as we know them, are young. It then follows that although ocean basins have existed through most of geologic time, they have been continuously reconstituted and recycled. Furthermore, their shapes and their geographic locations have been shifting constantly (Figure 11.21).

OBDUCTION

In the process of subduction some of the subsiding plate may be broken off and thrust up onto the overriding plate during plate collision. The subducted plate may have mafic and ultramafic igneous rocks and deep-sea sediments attached to it. In this way, formerly deep-ocean rocks can be thrust up in mountain masses and be exposed at the Earth's surface by this process called **obduction.**

11.21 In this diagram convection cells in the asthenosphere of the upper mantle are shown as providing the energy to drive the lithospheric plates. An ocean floor opens up by spreading from an ocean ridge and is in turn consumed beneath a trench in a subduction zone. The leading edge of a drifting continent becomes an active margin, and the trailing edge, an inactive margin.

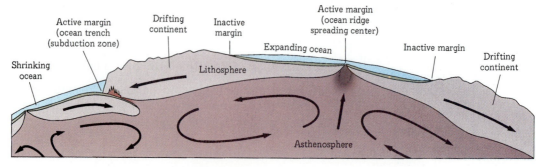

DATING THE MOVEMENTS

Interpretation of earthquake data tells us lithospheric plates have been carried as distinct entities downward along subduction zones to depths of 700 km or more. Although we can only tentatively decipher the present position of plate borders and their most recent directions and rates of movement, we know that these movements have occurred at least long enough for the generation of 700 km of new lithosphere at the spreading centers (otherwise a gap would exist in the ocean where plates have moved away from the midoceanic-ridge spreading centers).

Rates of sea-floor spreading have been determined from analyzing magnetic anomalies (Section 11.3). By relating these rates and distances of movement, we can determine the ages for various portions of the sea floor. The initial opening of several parts of the ocean basins has thus been determined for the Northern Hemisphere landmass (Laurasia) and for the Southern Hemisphere Gondwanaland. Opening of the ocean basins coincides with, and results in, the drifting of the associated continents.

Physical and biological events continuing on a given landmass might be rather abruptly terminated by the rifting apart of that landmass. Dating of these events would permit the determination of a minimum age for the initial breaking apart of such a continent. By a combination of such methods and by the dating of the ocean-floor materials, it is possible to establish the principal sequence of events in the breakup of the supercontinents, as listed in Table 11.2.

TABLE 11.2

Major Events in Continental Drift and Sea-Floor Spreading (Mesozoic and Cenozoic Episodes)[a]

Era or period	Age, millions of yr	Southern Hemisphere (Gondwanaland)	Northern Hemisphere (Laurasia)
	—		
	—		Gulf of California opens
CENOZOIC	—	Gulf of Aden opens	Spreading direction in eastern Pacific changes
	—	Red Sea opens	Iceland begins to form
	50 —	India collides with Eurasia	Arctic Basin opens
	—		———— 65 m yr
	—		Greenland begins to separate from Norway
	—		
CRETACEOUS	—	Australia begins to separate from Antarctica	
	100 —		
	—		
	—	Madagascar begins to separate from Africa, and South America from Africa	North America separates from Eurasia
	—		———— 136 m yr
	—		
	150 —		
JURASSIC	—		
	—		
	—	Africa begins to separate from India and Australia-Antarctica	North America begins to separate from Africa
	—		———— 190 m yr
TRIASSIC	200 —		

[a] From various sources.

Note: There is growing evidence to suggest the Atlantic region opened into a proto-Atlantic Ocean (called **Iapetus Ocean**) in early Paleozoic and closed in late Paleozoic.

The idea of plate tectonics can be used to integrate some of the major features of the ocean basins and landmasses. Thus the oceanic ridges and rises, with their central rift valleys and a sedimentary cover that is young and thin (or lacking), reflect the upwelling of mantle material along the ascending currents of adjacent convection cells. The large fracture systems that offset the midoceanic ridges are transform faults. These faults come into being as the rising convection currents finally diverge laterally, carrying with them the overlying crustal plates and upper mantle. Different portions of these plates are moved at different rates so that tear faults develop at their lateral margins. As the underlying currents cool and turn downward, they carry with them the oceanic crust with its overlying sediments, as well as the underlying upper mantle. To many students of the problem, the deep arcuate trenches of the oceans are the surface manifestations of the downturning of the convection currents along subduction zones. Great crustal plates, then, are seen as moving within a worldwide framework. The leading edges of the plates are consumed in the trench-island arc systems of the Pacific basin and along the Mediterranean-Himalayan belt of Eurasia. Along the trailing edges, new crust is generated in the active ridges. The drifting of continents, then, becomes a consequence of the movement of crustal plates and the appearance and disappearance of oceanic crust (see Box 11.2).

If the mantle wells up beneath a continent, we can expect that the continent will be rifted and that the two fragments will drift apart as a new ocean basin forms. This is what we believe happened during the early Mesozoic to form the present Atlantic Ocean. As the ocean basins enlarge and the continents ("tied" to the underlying but moving upper mantle) drift away from each other, their leading edges may be fronted by trenches.

A modern example presents itself in the western margin of South America. Here Pacific oceanic crust appears to spread eastward toward the continent. South America's margin is marked by a deep ocean trench, a narrow continental shelf, and the towering Andes. Earthquakes occur, becoming deeper in origin inland, the deepest being about 700 km below the surface. The earthquake foci lie along the Benioff zone that dips beneath the South American continent.[3]

The elevation of the Andean chain is interpreted in part as the result of underflow of oceanic crust. Seismic and igneous activity are seen as related to the motion between the Pacific and American plates along the Benioff zone. Farther north, the Mesozoic history of the western United States has been interpreted in terms of the underflow of 2,000 km of Pacific material beneath the American plate.

And the early Paleozoic mountain-building episodes in Newfoundland and Scotland have been ascribed by some workers to the interplay between two plates, one underthrusting the other.

Thus, according to plate-tectonic theory, mountain ranges, ocean basins, and most major features of the Earth's surface owe their origin to the movement and interaction of lithospheric plates.

In the following section we look more specifically at mountain building and in Section 16.4 we consider the topography of the ocean floors.

11.6

MOUNTAIN BUILDING

Overwhelmingly, mountains form along convergent plate boundaries and are the result of the movement of plates with respect to each other. It is here that the stresses necessary to fold and break rocks are generated, as are the temperatures and pressures that cause igneous activity and metamorphism. Present-day mountain building is concentrated along the convergent junctions between plates, as in the Himalaya Mountains and the Andes. Plate interaction also produced most of the old and inactive mountains, chains such as the Appalachian Mountains of eastern United States and Canada.

As we will see, the processes along convergent boundaries thicken the Earth's crust. We will also discover that this thicker-than-average crust accounts for the elevation of the great mountain chains. We can rightly say that the primary conditions needed to create the truly great major mountain systems of the world are convergent plate boundaries and a thickened crust.

With but few exceptions mountain building does not occur along transform (side-by-side) plate boundaries, although we will consider one such exception in later paragraphs. Diverging plates produce highland topography, but this bears little resemblance to the world's great mountain chains. (See Section 16.4 and Figure 16.11.)

CRATONS AND OROGENY

Geologists believe that mountain building, due to plate interaction similar to that of the present, has been going on for at least the last 2.5 billion years. Nevertheless continents have areas that have escaped this activity for a billion years or more. Such an area is called a **craton** (from the Greek *kratos,* meaning "power" or "strength"). It is the stable part of a continent.

We recognize on all continents two subdivisions of a craton: the **continental shield** and the **continental platform** (see Figure 11.22). The continental shield is the area of the craton in which the old, Precambrian igneous and meta-

---

[3] A seismic zone dipping beneath a continental margin or an island arc, and having a deep-sea trench as its surface expression, is called a **Benioff zone,** after Hugo Benioff, the American seismologist who first described the feature. It is the location of our subduction zone.

BOX 11.2　Microplates—"Exotic Terranes"

In the 1980s much interest centered on the discovery that parts of Alaska, California, Asia, and perhaps many of the Earth's major plates contain small **microplates.** The origin of these microplates is varied, but generally they appear to be far-traveled from their original settings, having become incorporated into other major plates. They could provide evidence on how at least some of the larger plates grew by accreting many of these "exotic terranes" to a more ancient central mass (craton or shield).

One model of this new "microplate tectonics" suggests that ancient Pacific plateaus were broken apart and transported on lithospheric plates toward subduction zones around the northern margin of the Pacific Ocean. (Though we often talk about six or seven continents, there are really several dozen continental masses, most very small and almost entirely submarine.) These smaller pieces of continental lithosphere are known as **microcontinents.** The microcontinents include New Zealand and the Lord Howe Rise to the north, Madagascar, Japan, Corsica, Sardinia, some part of the Philippines, and several lesser-known fragments. Most of them are the result of rifting from larger continental masses and are very important to the tectonic development of many mountain belts. Commonly the microcontinents move around until they encounter compressive plate boundaries, are accreted to larger continents, and become integral parts of those landmasses.

In the case of the Pacific microplates referred to above, the fragments were accreted to the western margin of North America and to Japan and other parts of eastern Asia. Detailed examination of Alaska, for example, shows that the rocks of southern Alaska are in a mixed state of disorder and of mixed age and origin. It is generally held that most continents grew by adding progressively consolidated belts of ever-younger rocks along continental margins as sedimentary material was deposited by erosion from the continental uplands.

Study of rocks in southern Alaska from the Jurassic and Cretaceous Periods, 190 to 65 million years old, suggests that some of them were moved into their present positions from more southerly latitudes. This has been determined from the magnetism preserved in them from their original site and orientation.

Magnetic data from most of the rocks sampled indicate a consistent northward movement that began about 170 million years ago. These data show that many pieces of the crust were not in their present position until about 20 million years ago or even later. Some of the separate pieces must have been joined while they were still considerably farther south than present-day Alaska. Other pieces collided with these pieces of exotic terranes after they joined mainland Alaska.

This reconstruction of the geologic history of southern Alaska raises questions about what actually constitutes the original or ancestral Alaska and what is the origin of Arctic Alaska. The only portion of the state that can be considered part of ancestral North America is that near the Yukon River and the Canadian border—a geologic unit from 225 to 570 million years old or older—and even that is open to question.

Current explanations for the origin of Arctic Alaska are that it rotated away from the Canadian Arctic, or that it somehow got up there from the south. In any event it was not there before about 135 million years ago, when what is now southern Alaska was much farther south. Thus we have no framework for reconstructing the last 200 million years of geologic history of Alaska.

Near the Golden Gate Bridge in San Francisco are found at least 10 different terranes, linking fragments of continents that once were thousands of miles apart. One terrane includes the Marin Headlands, consisting of rocks containing microscopic marine fossils that suggest a deep-sea origin for this terrane. Another terrane includes the island of Alcatraz, also formed in oceanic conditions but with a nearby continental source of its quartz-rich sandstones and some shales. A third major terrane, called Yollabolly, includes Angel Island to the north of Alcatraz. Rocks from this terrane are highly metamorphosed, folded and deformed in a tectonically active setting. Other bits and pieces of exotic terranes give evidence of widely varied original settings of these fragments that have been united into a single geographic locale.

Similar exotic terranes have been detected in the Canadian/peninsular Alaskan region along the British Columbia coast and in various other parts of the world. Much of the Canadian Shield may consist of small exotic terranes, or microcontinents. These discoveries suggest that microplates may be the rule rather than the exception in the geological development of the major world plates.

---

morphic rocks are actually exposed at the surface. On the continental platform similar types of rocks are covered by a veneer of generally flat-lying sedimentary rocks. This cover ranges in thickness from a feather edge at its contact with the shield to 3 km.

Mountain building has been concentrated around the margins of these stable cratons since Precambrian time.

The process is called **orogeny** (from the Greek *oros* for "mountain" + *gen* for "production of"). It involves crustal thickening, intense folding, faulting, igneous activity, and metamorphism. We find it concentrated in long, narrow belts along the unstable zones marking the junctions between plates, and most importantly between convergent plates. These zones are called **orogenic belts** or **orogens.**

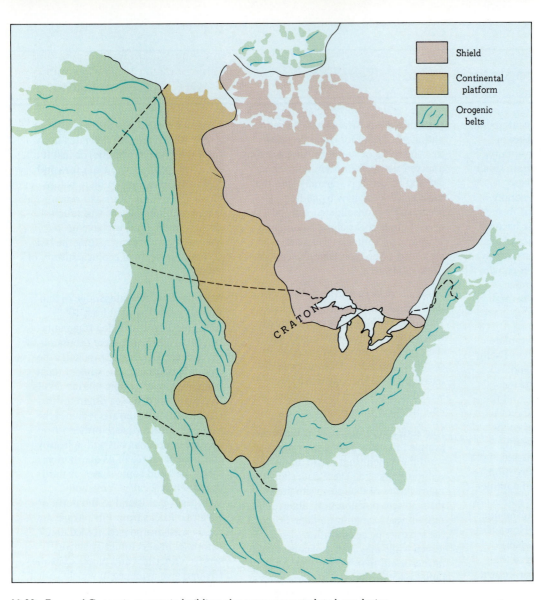

Shield

Continental
platform

Orogenic
belts

C R A T O N

11.22 Orogenic belts, both past and
present, have formed around the
North American craton, the stable [core]
of the continent.

11.23 Zones of Cenozoic mountain building along convergent plate boundaries.

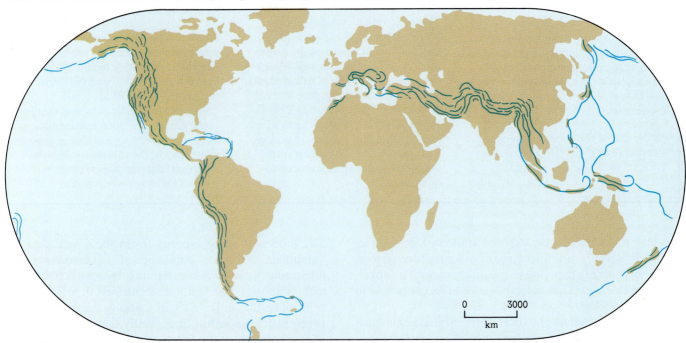

0   3000
km

OROGENY ALONG CONVERGENT PLATE BOUNDARIES

The patterns of convergent plate boundaries along which most orogeny occurs are basically linear; some are straight, others are arcuate. This helps explain why most major mountain ranges are themselves longer than they are wide and have linear to gracefully curved traces on maps of world or continental scale (see Figure 11.23).

Convergent (compressive) plate boundaries are responsible for the major mountain ranges of both the present and the past. There are, however, differences among plate boundaries. Another look at Figure 11.18 will remind us of the differences between **ocean-continent, continent-continent,** and **ocean-ocean** plate convergence.

In Section 11.5 we cited the Andes Mountains as an example of modern ocean-continent convergence. The Cascade Mountains of northwestern United States and extreme southwestern British Columbia provide another.

Figure 11.24 is a map of the Cascade Mountains and of the area just offshore. A line of volcanoes is strung along

11.24 The distribution of the volcanoes of the Cascade Mountains of British Columbia, Washington, Oregon, and northern California is shown in relation to the line of convergence between the Juan de Fuca and the American plates.

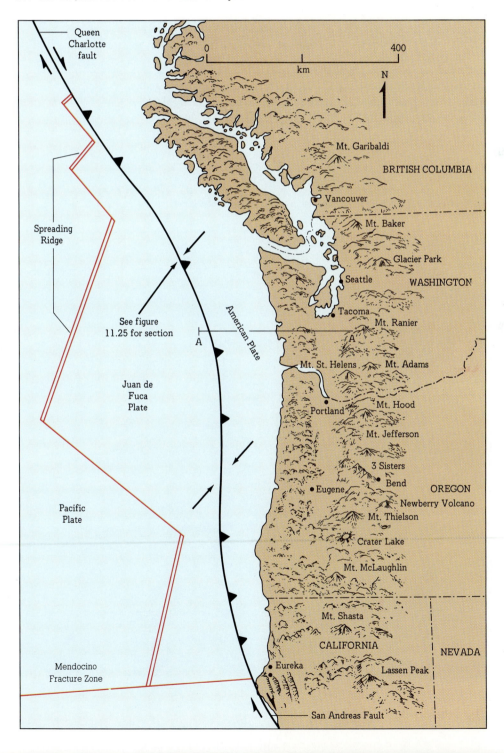

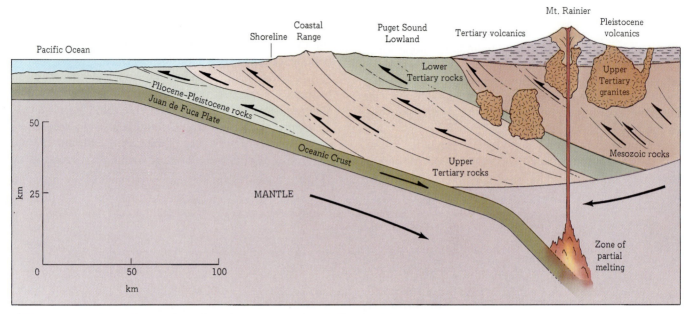

11.25 An interpretation of the geology along the converging Juan de Fuca and American plates in the latitude of Mt. Rainer, Washington. [After Darrel Cowan et al., *Centennial Continent/ Ocean Transect No. 9, B-3, Juan de Fuca Spreading Ridge to Montana Thrust Belt*, Geological Society of America, 1986.]

11.26 These Himalayan peaks along the Bhutan skyline rise over 7 km above sea level. [Lincoln Hollister.]

the axis of the Cascades. It more or less parallels the junction, west of the shoreline, of the small Juan de Fuca plate with the American plate.

The Cascades illustrate an important principle in mountain building: the thickening of the crust along the orogen. With the help of Figure 11.25 we can visualize how this happens. First, when the subducting oceanic plate reaches a great enough depth and high enough temperature, partial melting begins. The magma created rises and is added to the continental plate, both as plutonic bodies and volcanic deposits. Because the oceanic plate continues to subduct beneath the continental plate, its melting provides a continuous supply of new material to the crust. Second, the compressive stresses along the boundary between the two plates shorten and thicken the crust by folding and thrusting the rocks caught in the orogen. This stacks up sheet after sheet in an imbricate pile of rock slices. The material includes both sedimentary deposits and large pieces of the subducting oceanic crust. Finally, and probably least important to the process of mountain building, sediments are continually brought to the convergent zone by the oceanic plate. We return to this subject of crustal thickening in later paragraphs when we consider the roots of mountains.

The **Himalaya Mountains** (from the Sanskrit *hima* for "snow" and *alaya* for "abode") provide a spectacular example of a continent-continent convergence. This mountain system includes the great 2,700-km arc that forms the boundary between India and Tibet, as well as the vast Tibetan Plateau. Elevations of the peaks of the Himalayas are in many places 6, 7, and 8 km (see Figure 11.26) and Mt. Everest approaches 9 km. The Tibetan Plateau itself is generally over 5 km high, an elevation that exceeds that of any of the mountains of North America except that of Alaska's Mt. McKinley.

The Himalayas lie along the compressional boundary between the great Eurasian plate and the subcontinent of India, a part of the larger Austro-Indian plate. They had their beginnings in the breakup of the ancient continent of Gondwanaland. One of the fragments of this breakup was a plate that included both oceanic crust and the continental crust that is now India. It was moving northward by 80 million years ago (see Figure 11.27).

The ocean segment of the northward-drifting Indian plate subducted beneath the continental Asiatic plate in a fashion similar to that of the oceanic crust we saw subducting beneath the Cascades. With the arrival of the continental segment of the plate, the nature of convergence changed. The Indian continental mass, being of a density similar to that of the Asiatic continent, was unable to subduct beneath it, as had the oceanic section of the plate. But some of the Indian continental crust has jammed beneath Asia, as suggested in Figure 11.28. The net effect of this was to double the continental crust beneath what is now the Tibetan Plateau. To the south extensive folding and thrusting thickened the crust beneath what we now know as the Himalaya Mountains.

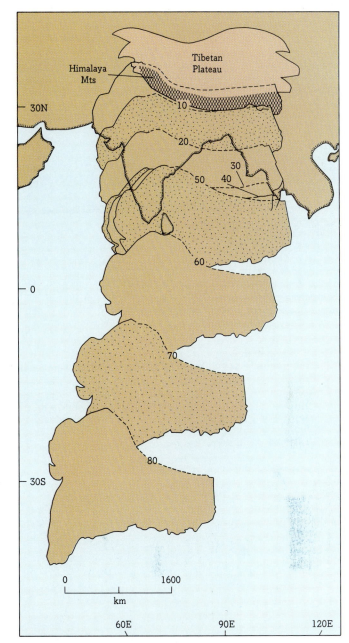

11.27 India was carried northward during the Cenozoic as part of a plate broken from the larger continent of Gondwanaland. It moved northward until it ran into southern Asia about 45 million years ago. As the two plates continued to push against each other, the Himalayas formed. [After C. McA. Powell and B. D. Johnson, "Constraints on the Cenozoic Position of Sundaland," *Tectonophysics*, vol. 63, Fig. 4, p. 97, 1980.]

The result of this thrusting and folding, combined with the underwedging of the Asiatic plate, produced a continental crust over 70 km thick. This is in contrast to the 30 to 35-km average thickness of the greater part of the Indian crust to the south. As a result, not only are the Himalaya Mountains the highest on Earth but they also have the deepest root.

Whereas the Cascades, Andes, and Himalayas are all examples of modern mountain building along convergent

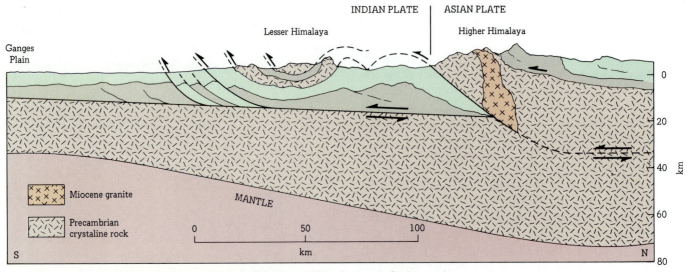

INDIAN PLATE | ASIAN PLATE

Lesser Himalaya   Higher Himalaya

Ganges
Plain

⌗ Miocene granite

⌗ Precambrian
crystaline rock

MANTLE

0

20

40

60

80

km

0   50   100

km

S   N

11.28 An interpretation of the geology beneath the Himalayas. [After Leonardo Seeber et al.,
"Seismicity and Continental Subduction in the Himalayan Arc," in H. K. Gupta and F. M. Delany
(eds.), *Zagros-Hindu Kush-Himalaya Geodynamic Evolution*, Geodynamics Series Vol. 3, Fig. 11,
p. 227. Copyright American Geophysical Union, Washington, D.C., 1981.]

11.29 Major geographic divisions of the southern and central Appalachian Mountains.

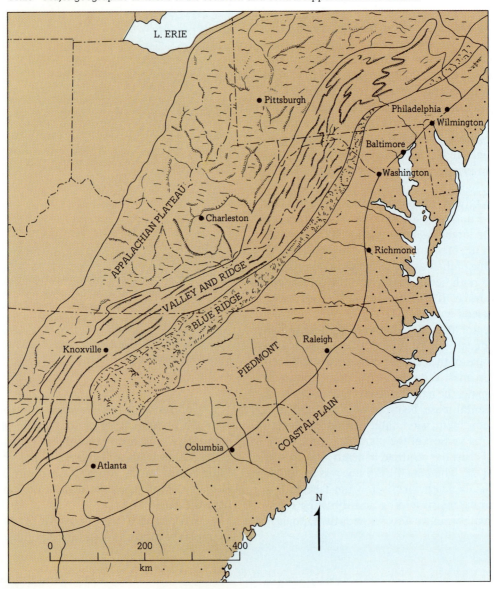

L. ERIE

• Pittsburgh

• Philadelphia
• Wilmington

Baltimore •

Washington •

APPALACHIAN PLATEAU

• Charleston

VALLEY AND RIDGE

Richmond •

BLUE RIDGE

• Raleigh

Knoxville •

PIEDMONT

• Atlanta

Columbia •

COASTAL PLAIN

N

0   200   400

km

plate boundaries, the Appalachian Mountains of eastern North America are examples of ancient mountain building along a junction between convergent plates.

The Appalachians extend for 3,200 km from Newfoundland to Alabama and have extensions beneath the lower Mississippi valley to the Ouachita Mountains of Arkansas and Oklahoma. Furthermore, fragments of the original range drifted off during the breakup of the supercontinent of Pangaea and are now found in Greenland, the northern British Isles, and northwestern Scandinavia.

When orogeny created the Appalachians the mountains had zones of folding, thrusting, and igneous and metamorphic activity, just as do the mountains forming today. Orogenic activity, however, came to an end 225 million years ago. Since then, weathering and erosion have removed thousands of meters of rock and reduced the mountains' elevation drastically. At the same time these processes have etched out the geologically different zones along the old mountain range and left us with some distinctive geographic regions. Figure 11.29 shows these regions in the area from northern New Jersey and eastern Pennsylvania to Alabama. The satellite view in Figure 11.30 and the companion geologic section in Figure 11.31 help clarify the picture.

The Appalachian Plateau lies along the northwestern side of the old orogen. Locally, it has such names (from north to south) as the Catskill and Pocono Mountains, the Allegheny Plateau, and the Cumberland Mountains. The sequence of Paleozoic sedimentary rocks that forms the Appalachian Plateau escaped most of the compressive pressures that characterized Appalachian orogenesis.

To the southeast of the Appalachian Plateau, however, Paleozoic sedimentary rocks were caught up in the orogenic movement directed northwestward. This zone is now the Valley and Ridge Province. The deformational style is one of thrusting and folding. Thrust faults are best displayed in the southern Appalachians. To the north, in West Virginia and eastern Pennsylvania, the structures seen at the surface are chiefly folds. The topography differs with the structure. In Pennsylvania and West Virginia differential erosion has fashioned the anticlines and synclines into long, elegantly sweeping ridges and intervening valleys (see Figure 11.32). To the south erosion has also etched the geologic structures into valleys and ridges, but here the features are straight and more closely crowded together (see Figure 11.30).

The Blue Ridge Province to the southeast is underlain chiefly by Precambrian crystalline rocks torn from the margin of the North American craton and thrust northwestward over younger sedimentary rocks. The province is

11.30 Satellite image of the southern Appalachians. Annotations indicate locations of the Appalachian Plateau, the Valley and Ridge Province, the Blue Ridge Province, and a portion of the Piedmont Province. [NASA.]

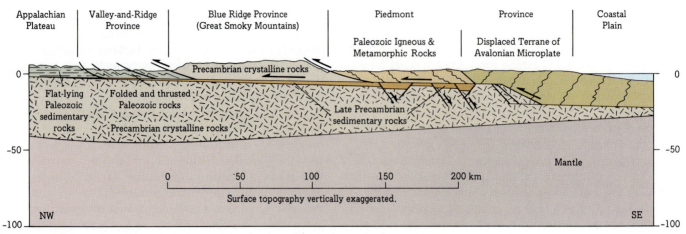

Appalachian
Plateau

Valley-and-Ridge
Province

Blue Ridge Province
(Great Smoky Mountains)

Piedmont

Province

Coastal
Plain

Paleozoic Igneous &
Metamorphic Rocks

Displaced Terrane of
Avalonian Microplate

Precambrian crystalline rocks

Flat-lying
Paleozoic
sedimentary
rocks

Folded and thrusted
Paleozoic rocks

Precambrian crystalline rocks

Late Precambrian
sedimentary rocks

Mantle

0

50 100 150 200 km

Surface topography vertically exaggerated.

NW

SE

11.31   An interpretation of the geology beneath the Southern Appalachians. Surface topography is greatly exaggerated. [After John Suppe, *Principles of Structural Geology*, Fig. 13.20, p. 482, Prentice Hall, Englewood Cliffs, N.J., 1985.]

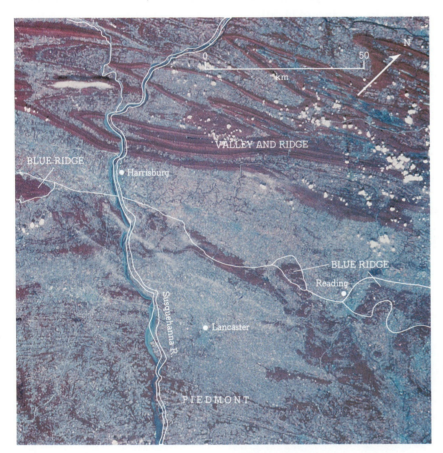

VALLEY AND RIDGE

BLUE RIDGE

Harrisburg

BLUE RIDGE

Reading

Susquehanna R.

Lancaster

PIEDMONT

11.32   A satellite image of the Appalachian Mountains in southeastern Pennsylvania. The heavy vegetation of the ridges and higher areas shows as red in this infrared image. White spots are clouds. Note that the Blue Ridge Province is separated here so that the Piedmont lies directly next to the Valley and Ridge. [NASA].

highest in the Great Smoky Mountains of the South. There 2037-m Mt. Mitchell in North Carolina is the highest point in the Appalachians.

The Piedmont Province, 200 km wide in the south, narrows northward. Its relief is moderate throughout and its landforms are gentle. Here are the igneous and meta-morphic rocks, attributable to an ancient zone of igneous and metamorphic activity. Here also are fragments of old ocean floor. Furthermore, the southeastern portion of the province is thought to be the remains of a microcontinent welded to growing North America during the orogeny. All of this complex was thrust northwestward, and since then, erosion has reduced it to its present unimposing state. Southeastward the Piedmont disappears beneath the Coastal Plain, a wedge of sediments laid down on top of the subsiding margin of the postorogenic Appalachians.

ROOTS OF MOUNTAINS

In discussing mountains formed along convergent boundaries we emphasized that one result of orogeny was to thicken the Earth's crust. One of the reasons that mountains stand high is because of this accumulation of material. We now know not only that thickening of the Earth's crust leads to greater elevation but also that part of the thickened crust pushes downward into the underlying mantle. In other words, mountains have roots. This is the process of isostasy discussed in Box 9.1. It is the same as the familiar response, although on a different scale, of an iceberg floating in seawater. The ice, with a density that may vary, depending on the amount of trapped air bubbles, from 0.9 to 0.7 g/cc^3, floats in seawater with a density of little over 1 g/cc^3. Most of the iceberg is below water; only a small proportion is above.

If mountains are really the tops of a mass of thickened crust floating in the mantle beneath, then the mantle must adjust itself to the increased load. It does so by flowing to make room for the additional crust. The mantle need not be a liquid in the ordinary sense. It could be a rock in a state not unlike silicone putty, which, when shaped into a ball, bounces, but which under a very slight load—even its own weight—gradually loses its original shape. Hence this rock would be rigid enough when subjected to stresses of short duration but would have very little strength when subjected to similar stresses over long periods of time.

We can carry the silicone putty analogy farther. A small piece of erasing gum nine-tenths as dense as silicone putty would, in time, sink into the putty with only one-tenth of its volume above the surface. Likewise, a mountain range with an average density of 2.8 can sink into plastically behaving mantle with a density of 3.2. In sinking into the mantle it displaces a volume of mantle equal to a volume of crustal material having the same mass as the displaced mantle.

If we add material along an orogenic belt, the crust thickens and the mountain will rise isostatically. Because isostasy applies, however, the increase in elevation will not be as great as that represented by the material added. Some of the newly thickened crust will become part of the root as the mountain mass adjusts by floating deeper in the mantle. The thicker we make the crust, the higher the mountain, but also the deeper its root. For example, the Himalaya Mountains have a root that reaches about 70 km into the mantle, which compares with their 5-km average elevation. If we were able to add another kilometer of crust to the

Himalayas, their elevation would increase by only about 125 m and the root would be deeper by 875 m.

This process works the other way as well. Material eroded from a mountain range will reduce the elevation of the mountain, but not by as much as that represented by the material eroded. Some of the lost mountain would be compensated for by a rise of the mountain mass and a shallower root. Again, take the Himalayas as an example. Let us assume that orogeny has stopped, the mountains are no longer growing, and erosion is proceeding at an extremely rapid rate of 1 m/1000 y. To erode 1 km from the mountain range would take a million years. But because of isostasy the mountain would not be a kilometer lower at the end of a million years, but only 125 m lower. To lower the elevation of the Himalayas by a kilometer would take much longer, about 8 million years. In the process some 8 km of rock would be removed and the root would be shallower by about 7 km. This explains why mountains are such persistent features on the Earth's surface, and why, for example, the Appalachians still exist, even in lowered form, a quarter of a billion years after their formation.

DEFORMATION ALONG TRANSFORM AND DIVERGENT BOUNDARIES

Mountain building is not generally associated with transform boundaries. The major direction of movement is neither compressional nor extensional but shearing, side-by-side translation. Occasionally, however, the geometry of the fault is such that compressive pressures build up and vertical movements take place. The **Transverse Ranges** of southern California provide examples.

The Transverse Ranges are so named because they are oriented east-west, transverse to the northwest-southeast orientation of most of the major geologic and topographic features of California. Just north of the Los Angeles basin the San Andreas fault bends to the east and then turns southward again. This large bend in the San Andreas, along with its right lateral slip motion, places this area in compression (see Figure 11.33). The result is the appearance of the Transverse Ranges. One of these, the **San Gabriel Mountains,** rises abruptly above the Los Angeles basin. Its steep southern front is a thrust fault that has brought up granitic rock, which forms the mountain core. The northern boundary is the San Andreas fault.

Figure 11.33 also shows that if the sense of motion along the fault were in the opposite direction (left-lateral), extensional stress would occur. Pull-apart basins would then develop along normal faults. Death Valley is an example.

Along divergent boundaries uplift of the crust is due primarily to high heat flow from the mantle. This gives buoyancy to the crust and raises it in a broad, linear welt.

TABLE 11.3
Characteristics of Orogenic Activity Arranged by Type of Plate Boundary[a]

Characteristics of Orogenic Activity	Convergent — Continent–Ocean	Convergent — Continent–Continent	Convergent — Ocean–Ocean	Transform — Strike–Slip	Transform — Compressive	Transform — Tension	Divergent — Continent–Continent	Divergent — Ocean–Ocean
Examples	Andes, Cascades	Himalayas, Zagros (Iran), Appalachian (late Paleozoic fold and thrust belt)	Japan, Aleutian Islands, Marianas	San Andreas fault zone	Transverse Ranges (California)	Death Valley (California)	East African Rifts, Red Sea, Basin and Range Province (U.S.A.–Mexico)	Midatlantic Ridge, Juan de Fuca Ridge, East Pacific Rise
Dominant type of stress	Compressional			Shear	Local compression	Local extension	Extensional	
Igneous activity — Volcanic	Andesitic, often explosive	Basaltic to andesitic		None	Generally None	Basalt to rhyolite	Basalt to rhyolite cones, flows, and calderas	Basaltic
Plutonic	Granitic		Gabbroic, some diorite along volcanic arc axis			Generally mafic	Gabbroic to granitic	Gabbroic
Seismicity	Foci range from 70–100 km deep. Can be destructive.			Foci generally less than 15 km deep. Can be destructive.			Foci generally shallow	Most foci less than 10 km deep
Dominant structures	Thrust sheets and folds			Strike–slip faults	Strike–slip and thrust faults	Strike–slip and normal faults	Normal faults	
Metamorphism	Zeolite facies in shallow crust. Greenschist, low-rank amphibolite deeper. Blue schist along subduction zone. High-rank amphibolite and granulites near base of volcanic arc. Contact	Contact	Same as preceding but high-rank and granulites absent.	None to mild			Contact	Hydrothermal alteration due to seawater along fractures. Zeolite facies in shallow-crust and Greenschist facies deeper.

[a] After Gregory R. Wessel, *The Geology of Plate Margins*. Geological Society of America, Map and Chart Series MC-59, 1986.

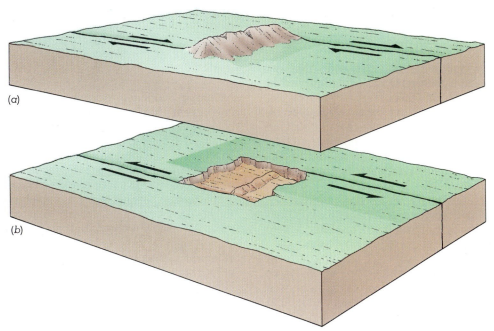

11.33 Usually along a transform fault there is neither compressional nor extensional stress across the fault. These diagrams show a bend in the fault and, depending on the direction of movement along the fault, local zones of compression or extension can develop. (a) Diagram models the situation of the San Andreas fault in the Los Angeles area. The transform is a right lateral fault and compression builds along the bend. The Transverse Ranges are the result of such a condition. (b) Diagram shows a left lateral movement along a similar fault line. In this case the stress is pull-apart in the region of the bend.

Examples are the Midatlantic Ridge and the highlands along either side of the Red Sea. If the heat source disappears, the excess surface elevation is lost as well.

Recapitulation   Table 11.3 summarizes several aspects of orogeny found along different plate boundaries.

Summary

Oldest continental crust has been dated at 3.8 billion years from highly metamorphosed rocks from the Baltic Shield, Greenland, and Minnesota.

Wegener, as early as 1911, presented ideas on **continental drift,** including shape of continents, ancient climatic patterns, disrupted geologic structures, glacial evidence, and similar stratigraphy in Gondwanaland continents.

The Earth behaves as if it were a huge **magnet;** its changes as recorded in rocks are closely tied to the case in favor of continental drift.

    Cause of magnetism is both internal and external in relation to the Earth. A small part is external, caused by the Sun. The largest part of the Earth's field originates within the Earth and is best explained as caused by circulation within the Earth's liquid core, which acts as if it were a giant dynamo. The **direction** and **intensity** of the magnetic field vary with time. If the intensity decreases far enough, it may eventually pass a zero point and the Earth's field may become polarized in a sense the reverse of that of the present.

    Paleomagnetism is the study of the Earth's field in past time. **Thermoremanent magnetism** (TRM) occurs in igneous rock-forming minerals as they cool below the Curie temperature.

    Depositional remanent magnetism (DRM) forms as magnetic particles align parallel to the Earth's field at time of deposition. DRM (and TRM) reflects magnetic field existing at time of rock formation.

    Secular variation of the magnetic field has been known through long-term changes in magnetic declination since the mid-seventeenth century.

    The geomagnetic pole at the Earth's surface does not coincide with the geographic pole.

Sea-floor spreading has been demonstrated through parallel magnetic stripes coincident about ocean-ridge spreading centers and deep-sea sediment cores.

Plate tectonics is a hypothesis that postulates that the Earth's crust is broken into large blocks being renewed in some zones and destroyed in others.

Lithospheric plate boundaries may be **convergent, divergent** or **transform,** depending on whether the driving forces are compressional, tensional, or parallel.

Movement of plates may be by means of convection cells or by interactions of mantle material over **hot spots, or plumes.**

Obduction may bring deep-sea sediments and mafic and ultramafic igneous rocks up to the Earth's surface, where they may be exposed in cores of eroded mountains.

Dating the movement of plates can be done by analyzing magnetic anomalies and rates of sea-floor spreading, as well as by dating structural trends and stratigraphic units that developed on continents prior to their rifting.

Plate-tectonic theory can integrate notions of continental drift, sea-floor spreading, seismic zones, continents, and ocean basins.

Microplates or **exotic terranes** are fragments of larger continents that have broken loose and are rafted along by convection currents or plumes, or are pulled by sinking crustal slabs, until they intersect and are welded to other continents.

Mountain building or **orogeny** takes place primarily along convergent plate boundaries. Thickened crust, needed for a major mountain chain, is created there. Orogenic belts are arranged around **cratons,** the stable Precambrian cores of continents. The Andes and the Cascades are forming along ocean-continent convergent boundaries and the Himalayas along a continent-continent convergent boundary. The Appalachians formed along a convergent plate boundary during long-continued activity that ceased 225 million years ago. Erosion has worn them to their present levels.

Roots of mountains are made of crust thickened during compressive orogeny. Mountains stand high isostatically because they are buoyed up by their underlying roots embedded in the mantle.

Mountain building along transform boundaries is rare. On **diverging boundaries high heat flow raises the crust** in long, broad welts.

Questions

1. What lines of evidence did Wegener first use to support the idea of continental drift? Why did so few Northern Hemisphere geologists believe his story?
2. What is the approximate age of the present ocean basins? How do we know? Were there earlier oceans? What is the evidence?
3. What causes the Earth's magnetism? Is it constant? If not, how does it vary?
4. Why is the Curie temperature of a mineral extremely critical in the use of paleomagnetic determinations?
5. What is meant by paleomagnetism? How is it determined?
6. What are the principal forms of remanent magnetism? How are they similar? How do they differ?
7. Have the Earth's magnetic poles always been where they are in relation to present-day latitude and longitude? If not, how have they changed and how do we know that?
8. How is sea-floor spreading established as fact? How does it occur?
9. What do ancient plants and animals contribute to the discussion of continental drift?
10. What are the principal kinds of plate boundaries? Make simple sketches to show them. What stress conditions occur at each type of boundary?
11. What causes the lithospheric plates to move?
12. How fast are these plates moving? How do we measure these rates?

13. What are some of the features associated with orogeny along an ocean-continent plate boundary?
14. How does orogeny along an ocean-continent plate convergence differ from orogeny along a continent-continent convergence?
15. Why do mountains have roots?
16. If we add 2 km of rock to a growing mountain range, how much higher will the mountain be?

(Our answer was 250 m. We used a density of 2.8 for the mountain mass and 3.2 for the mantle. That meant we had to substitute 2 km of low-density crustal material for an equal volume of mantle material with a density of 3.2. That in turn meant that the volume of the mantle material displaced would be less than the mountain material substituted. We used the following formula):

$$H \times D_m = A (D_m - D_{mr})$$

where H = increased height of mountain

D_m = density of mantle

A = material added

D_{mr} = density of mountain range

17. Continuing question 16, what would be the increase in the depth of the root?
(We got 1750 m.)

18. Work problems 16 and 17 in reverse. Assume that the mountain is no longer orogenically active. How much will we lower the height of the mountain by eroding away 3 km of rock? What will be the difference in the depth of the mountain root?

(We got a decrease in elevation of 375 m and a decrease in the depth of the mountain root of 2625 m.)

19. Oceanic crust 5 km thick is being pushed under the Andes Mountains at a rate of 50 mm/y. Ten percent of this crust is being melted to make new igneous rocks. These are added to a north-south magmatic belt parallel to the coast and about 100 km wide. What is the vertical thickness of new igneous rock added each year to the Andes?

(We got 0.25 mm/y. We used a section 1 m wide, east to west through the Andes. So the volume of new rock in that 1 m wide section would be):

$$km \times 50 \text{ mm/y} \times 1m \times 10\%$$

which converts to:

$$m \times 0.05 \text{ m} \times 1 \text{ m} \times 0.1 = 25 \text{ m}^3/\text{y}$$

Spread this volume over an area 100 km × 1 m wide and the thickness comes out to be:

$$\frac{m^3/y}{km \times 1000 \text{ m/km} \times 1 \text{ m}} = 25 \times 10^{-5} \text{ m/y} = 0.25 \text{ mm/y}$$

20. Now go back to question 4. What would be the annual increase in elevation of the Andes?

(We got a little over 0.03 mm/y.)

Supplementary Readings

CONDIE, K. C. *Plate Tectonics and Crustal Evolution*, Pergamon Press, Elmsford, N.Y., 1976.
A good basic summation of the development of oceanic and continental crust in light of plate-tectonic theory; well illustrated.

LE GRAND, H. E. *The Modern Revolution in Geology and Scientific Change*, Cambridge University Press, Cambridge, 1988.
Describes how the modern revolution in geology saw the triumph of the global theory of plate tectonics.

PRICE, LARRY W. *Mountains and Man: A Study of Process and Environment*, University of California Press, Berkeley, 1981.
Don't let a sexist title put you off. This natural history of mountains is a book for the general reader on the origin, climate, vegetation, wild life, and agriculture of mountains, as well as a discussion of human attitudes toward mountains and human impact on them.

VAN ANDEL, T. H. *New Views on an Old Planet: Continental Drift and the History of the Earth*, Cambridge University Press, Cambridge, 1985.
A general and popularized history of the Earth and the evolution of our ideas concerning it.

WEGENER, ALFRED *The Origin of the Continents and Oceans*, trans. from the 3rd German edition by J. G. A. Skerl, Methuen & Co., 1924.
Alfred Wegener is very properly more closely associated than anyone else with the concept of continental drift.

WINDLEY, BRIAN F. *The Evolving Continents*, 2d ed., John Wiley & Sons, New York, 1984.
An excellent summary, albeit advanced, of the evolution of continents from Precambrian times to the present, with considerable attention given to plate boundaries and mountains.

WOOD, R. M. *The Dark Side of the Earth: The Battle for the Earth Sciences, 1800–1980*, Allen & Unwin, Winchester, Mass., 1985.
Describes the scientific revolution resulting in our acceptance and understanding of global tectonics.

WYLLIE, PETER J. *The Way the Earth Works*, John Wiley & Sons, New York, 1976.
A highly readable and well-illustrated paperback on tectonic developments, using as a central theme the new global geology, with good coverage of plate tectonics and continental drift.

Hillslopes, such as these in the Snowdonia National Park of Wales, provide the pathways down which water and Earth materials find their way to the rivers and hence to the world ocean.

CHAPTER ■ 12

HILLSLOPES: PROCESS AND FORM

". . . some of these times, these earthquake tremors that are coming so often are going to hit at about the right time when the mountain is wooziest, and down she'll come."

"Uncle Billy" Bierer, quoted in Barry Voight, "Lower Gros Ventre Slide, Wyoming USA," in Barry Voight, (ed.), Rockslides and Avalanches, *Vol. 1, p. 116, Elsevier Publishing Co., Amsterdam, 1978*

The Earth's surface is a collection of slopes, some gentle, some steep; some long, some short; some cloaked with vegetation, some bare of plants; some veneered with soil and some made of naked rock. They lead from hill and mountain crests down to streams and rivers and to basins, some lake-filled, some dry. They continue downward past the ocean's edge and across the shelving sea floor into the great deeps of the world's ocean. To us they are important because together they make up the forms of the Earth's landmasses and the sea floor.

Slopes are also important because down them move rainwater and the water of melting snow and ice to collect in rivulets, brooks, and rivers that flow to the oceans. Down them also travels material, some of which, via the rivers, finally gets to the sea. The bulk of material moves down slopes under the pull of gravity. We call this *mass movement.* Generally it goes on unnoticed. But at times this mass movement can be catastrophic, as for instance in landslides and avalanches, which can cause great loss of life and property. Even the slower movements can express themselves over tens of years in damage to buildings, roads, and underground installations.

First we look briefly at some of the *factors* that help prepare slope material for its movement by gravity, and then at the *behavior of material* in motion. As we seek to understand the various types of movement, we will find that they fall naturally into three groups, which we can define by their rates of movement. *Slow* movements may be as low as 1 mm/year. Some landslides and avalanches are very *rapid* and may reach over 100 km/hour. Transitional between these extremes is a family of movements of *moderate* velocity whose rates range from centimeters per second to centimeters per day.

In the latter part of this chapter we look at the shapes of slopes and the relation of these shapes to such factors as rock types, geologic structure, climate, surface processes, and time.

12.1

MOVEMENT OF MATERIAL ON HILLSLOPES

Hillslopes provide the paths for the movement of material, weathered and unweathered, to lower elevations and eventually to the stream channels through which materials are carried to that great settling basin, the world ocean. Of course, the topography of the ocean floor is itself defined by the slopes of the submarine hills, mountains, and valleys, but we leave its consideration to Section 16.4.

On the continents the movement of material down hillslopes is one of two types. Water flowing in thin sheets or shallow rills may carry solid material such as individual sand grains or may move material in dissolved form. But more important is the movement of Earth materials downhill in solid or fairly viscous masses. The latter type of movement is referred to as **mass movement,** and it is to this that we now turn.

12.2

FACTORS AFFECTING MASS MOVEMENT

Gravity provides the energy for the downslope movement of surface debris and bedrock, but several other factors, particularly water, augment gravity and ease its work.

Immediately after a heavy rainstorm you may have witnessed a landslide on a steep hillside or on the bank of a river. In many unconsolidated deposits the pore spaces between individual grains are filled partly with moisture and partly with air. So long as this condition persists, the surface tension of the moisture gives a certain cohesion to the soil. When, however, a heavy rain forces all the air out of the pore spaces, this surface tension is completely destroyed. The cohesion of the soil is reduced, and the whole mass becomes more susceptible to downslope movement.

The presence of water also adds weight to the soil on a slope, although in most cases this added weight is probably not a very important factor in promoting mass movement. Water that soaks into the ground and completely fills the pore spaces in the slope material contributes to instability in another way: The water in the pores is under pressure, which tends to push apart individual grains or even whole rock units and to decrease the internal friction or resistance of the material to movement. Here again, water assists in mass movement.

Gravity can move material only when it is able to overcome the material's internal resistance to being set into motion. Clearly, then, any factor that reduces this resistance to the point where gravity can take over contributes to mass movement. The erosive action of a stream, an ocean, or a glacier may so steepen a slope that the Earth material can no longer resist the pull of gravity and is forced to

Sherman Slide

(a) before   (b) after

12.1   (a) On August 26, 1963, 7 months before the Good Friday earthquake in southern Alaska, the Sherman glacier appeared like this in the camera of Austin Post of the U.S. Geological Survey. (b) The earthquake of March 24, 1964, set off a number of major rock slides, one of which occurred across the Sherman glacier. Post's photograph of August 24, 1964, shows the slide as a 4-km-long tongue of dark debris against the bright background of the glacier.

undergo mass movement. In regions of cold climate alternate freezing and thawing of Earth materials may suffice to set them in motion. The impetus needed to trigger movement may also be furnished by earthquakes (Figure 12.1), by excavating or blasting operations, by sonic booms of aircraft, by the gentle activities of burrowing animals, and even by growing plants.

Most of the extremely rapid and spectacular mass movements occur in areas of youthful landscape. Here landmasses raised by tectonism are rapidly carved by erosional processes into regions characterized by steep, unstable slopes. As tectonism fades and ceases, erosion is able to reduce the land to lower and lower levels. Extremely large and rapid mass movements give way to the smaller, less dramatic movements.

BEHAVIOR OF MATERIAL

The rate at which material moves downslope varies dramatically over several orders of magnitude from extremely rapid to extremely slow, from over 100 km/hour to millimeters per year. Relative velocity is one way to think about mass movements, and indeed we use this approach later in this chapter. Before doing that, however, let us suggest some other ways of thinking about the process.

Material moving downslope may behave in the same way that solid rock deforms (Section 9.1): as an **elastic solid,** as a **plastic substance,** or as a **liquid.** In fact, all three varieties of behavior may occur in some types of movement.

A slightly different way of viewing this behavior is to describe the movements as the processes of *slide, fall, flow,* and *heave,* as we do in Figure 12.2 and Table 12.1 and in the discussion below.

During the process of *sliding,* material maintains continuing contact with the surface on which it moves, even as a child does on a playground slide. The movement may range from slow to fast. The material sliding may preserve its original form or it may be extensively deformed.

Flow, like sliding, can be slow or fast. It involves continuous movement. Material behaves in a plastic to semiliquid to liquid manner. Internally individual particles are rearranged during the process.

Material at or near the surface that is pushed upward by the stress of, for example, ground ice or swelling clay usually moves at right angles to the surface. When the pressure is released, the material is pulled vertically downward by gravity. Upward and downward motions are different so in any one cycle there is a net downslope motion.

Fall refers to the free fall of material. A rock pried from a cliff face, for instance, travels downward with no

TABLE 12.1
Types of Mass Movement by Rate of Movement[a]

	Slow	Moderate	Rapid
Names	Creep Frost heave Solifluction Rock glaciers	Slump Earthflows Debris slides	Rock slides (avalanches) Debris flows and mudflows Rock fall (talus)
Rate	1 mm/yr to mm/day	cm/day to cm/s	Over 100 km/hr
Volume	Small to large	Moderate, 10^0 to 10^4 m³	Moderate to very large, 10^6 to 10^{12} m³
Material involved	Mainly soil, bedrock minor	Soil and bedrock (various proportions)	Chiefly bedrock, unconsolidated or weathered
Some causes	Gravity, freeze-thaw, saturation by water	Gravity, slope oversteepening, water saturation of pores, earthquakes, shear along tilted planes of weakness	Gravity, earthquakes, water saturation of pores, slope oversteepening, shear along tilted planes of weakness
Dominant type of movement	Flow, heave	Flow, slide	Slide, flow, fall

[a] Modified from an unpublished table by John A. Klasik

12.2 Material may move down slopes by (a) sliding, (b) flowing, (c) heaving, or (d) falling.

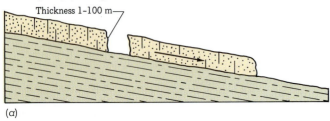

(a)

(b)

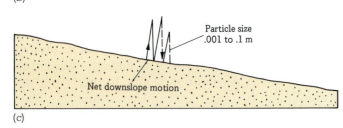

(c)

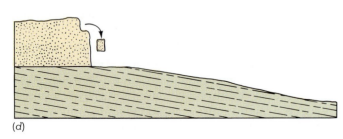

(d)

contact with the surface. It may bounce from one spot to another but in between bounces it is considered to be in free fall.

Most movements are not pure examples of any one of the processes—slide/fall, flow, or heave—but rather a combination of two or more. We can illustrate this by the diagram in Figure 12.3, in which the processes occupy corners of a triangle. There is, for instance, a gradation from

12.3 This triangular diagram shows how most mass movements involve a combination of slide/fall, flow, and heave. The diagram also relates these movements to their relative velocities (base and right arm of triangle) and to the moisture content of the material involved (left arm of triangle). [Adapted from M. A. Carson, and M. J. Kirkby *Hillslope Form and Process*, Cambridge University Press, New York, 1972, Fig. 5.2.]

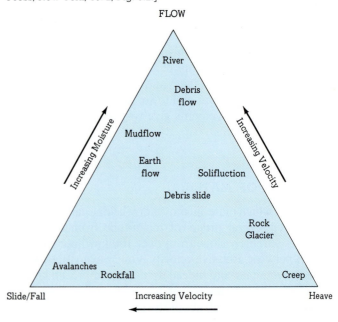

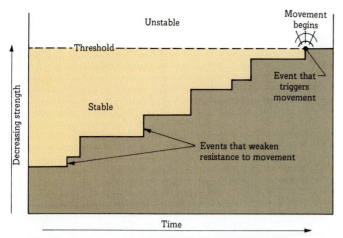

12.4 A series of events may weaken a slope to the point where its resistance to motion crosses a threshold between stability and instability and movement takes place.

almost pure sliding to pure flow as we progress from landsliding and avalanching to river flow. The same diagram suggests the increase in velocities of movement from the slow process of heaving to the more rapid ones of flow and slide/fall.

Still another way of thinking about mass movements is to consider that they begin only after a history during which the slope becomes progressively weaker. A series of events may reduce the resistance of a slope to downward movement. During this time the slope remains stable but continues to approach a **threshold** between stability and instability. Finally, there comes a moment when the strength of the slope is so reduced that some specific event that weakens it further succeeds in pushing it across the threshold. Material on the slope can no longer resist the pull of gravity and movement begins. For instance, a catastrophic landslide is often triggered by some discrete event such as an earthquake, an exceptionally wet season, or the removal of support from the base of the slope. Whatever the initiating event, it is only the last of many events that have contributed, bit by bit, to the weakening of the slope. It is the last event that pushes the slope across the threshold from stability to instability (see Figure 12.4).

We can express this concept of a threshold in a somewhat more precise fashion by first remembering that mass movement begins when the resistance of the slope material to movement is exceeded by the driving force or stress that works to start slope motion. We can state this as a ratio that represents how close the slope is to the threshold of failure:

$$\text{Approach to failure threshold} = \frac{\text{resisting force (resistance)}}{\text{driving force (stress)}}$$

When the ratio is smaller than 1, the slope is stable. The closer it approaches 1, the less stable the slope, and at unity (1) the slope will be at the threshold of movement and very little additional stress will be needed to start motion. Another way of putting it is that the larger the ratio, the smaller the safety factor of the slope.

Some ways in which the safety factor of a slope can be reduced are given in Table 12.2.

TABLE 12.2
Some Factors That Reduce the Safety Factor of a Slope

Some Factors That Increase the Driving Force or Stress on Slope Materials

Removal of lateral support by steepening slope
    Erosion by streams, glaciers, waves
    Road construction, quarrying, terracing
Removal of underlying support
    Solution of rock beneath surface
    Mining
Addition of mass
    Buildings, artificial fills, dumps
    Rain, snow and ice, vegetation
Lateral pressure
    Freezing of water in slope materials
    Swelling of clay minerals

Factors That Decrease the Strength or Resistance of Slope Materials

Weathering
    Mechanical disintegration of granular rocks
    Removal of cementing material in granular rocks
    Drying of clays
Pore water
    Increased pressure of fluid in pore spaces of slope materials
Fracturing of rock
    Rock expansion due to relief of overlying load
    Tectonic movements
Organic activity
    Burrowing by animals
    Prying by plant roots
    Decay of root system
Vibrations
    Earthquakes
    Sonic booms, blasting, traffic

Modified from D.J. Varnes, "Slope Movement: Types and Processes," in R.L. Schuster and R.J. Krizek (eds.), *Landslide Analysis and Control*, National Academy of Sciences Special Report 17611-33, Washington, D.C., 1978.

In the pages that follow, we examine some of the various types of mass movement, beginning with those we can classify as slow movements.

12.3
SLOW MOVEMENTS

Slow movements are difficult to recognize precisely because they are slow, from about 1 mm/year to 1 mm/day. They are extremely important, however, in the sculpturing of the land surface. Geologists believe that slow processes, operating more or less continuously over long periods of time, are probably responsible for the transportation of much more material than are the violent and rapid movements of rock and soil.

Most slow movements occur in unconsolidated materials and thus are near-surface phenomena averaging perhaps a meter or so in depth. Velocity of movement decreases with depth.

Before the end of the nineteenth century William Morris Davis aptly described the nature of slow movements.

The movement of land waste is generally so slow that it is not noticed. But when one has learned that many land forms result from the removal of more or less rock waste, the reality and the importance of the movement are better understood. It is then possible to picture in the imagination a slow washing and creeping of the waste down the land slopes; not bodily or hastily, but grain by grain, inch by inch; yet so patiently that in the course of ages even mountains may be laid low.[1]

CREEP

In temperate and tropical climates a slow downward movement of surface material known as **creep** operates even on gentle slopes with a protective cover of grass and trees. It is hard to realize that this movement is actually taking place because the observer sees no break in the vegetative mat, no large scars or hummocks, and has no reason to suspect that the soil is in motion beneath his feet.

Yet this movement can be demonstrated by exposures in soil profile (Figure 12.5) and by the behavior of tree roots, of large blocks of resistant rock, and of artificial objects such as fences and telephone poles (Figure 12.6). Fig-

[1] William Morris Davis, *Physical Geography*, p. 261, Ginn & Co. Lexington, Mass., 1898.

12.6 Telephone poles set in a talus slope south of Yellowstone National Park record movement of the slope material. Originally the poles stood vertically, but over a period of about 10 years the talus has moved and tilted the two nearest poles 5° and the most distant pole 8°. [Photo, 1968.]

ure 12.7 shows a section through a hillside underlain by flatlying beds of shale, limestone, clay, sandstone, and coal. The slope is covered with rock debris and soil. But notice that the beds near the base of the soil bend downslope and thin out rapidly. These beds are being pulled downslope by gravity and are strung out in ever-thinning bands that may

12.5 Alternating beds of sandstone and shale have been tilted until they stand vertically. A little over 1 m below the surface, they begin to bend downslope under the influence of gravity. The amount of downslope movement increases toward the surface. Haymond Formation in the Marathon region of Texas. [William C. Bradley.]

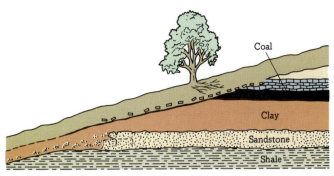

12.7 The partially weathered edges of horizontal sedimentary rocks are dragged downslope by soil creep. The tree is also moving slowly downslope, as is evidenced by the root system spread out behind the more rapidly moving trunk. [Redrawn from C. F. S. Sharpe and E. F. Dosch, "Relation of Soil-Creep to Earthflow in the Appalachian Plateaus," *J. Geomorphol.*, vol. 5, p. 316, 1942. By permission of Columbia University Press.]

extend for hundreds of meters. Eventually they approach the surface and lose their identity in the zone of active chemical weathering.

Figure 12.7 shows other evidence that the soil is moving. Although when viewed from the surface the tree appears to be growing in a normal way, it is actually creeping slowly down the slope. Because the surface of the soil is moving more rapidly than is the soil beneath it, the roots of the tree are unable to keep up with the trunk.

Many other factors cooperate with gravity to produce creep. Probably the most important is moisture in the soil, which works to weaken the soil's resistance to movement. In fact, any process that causes a dislocation in the soil brings about an adjustment of the soil downslope under the pull of gravity. Thus the burrows of animals tend to fill downslope, and the same is true of cavities left by the decay of organic material, such as the root system of a dead tree. The prying action of swaying trees and the tread of animals and even of human beings may also aid in the motion. The result of all these processes is to reduce the resistance of the slope material to the pull of gravity, thereby contributing to the slow and inevitable downslope creep of the surface cover of debris and soil.

We saw in Section 5.2 that alternate freezing and thawing is a powerful process in mechanical weathering. It is also effective in the movement of material downslope. **Frost heaving** usually occurs in fine-grained, unconsolidated deposits rather than in solid rock. Much of the water that falls as rain or snow soaks into the ground, where it freezes during the winter months. If conditions are right, more and more ice accumulates in the zone of freezing as water is added from the atmosphere above and is drawn upward from the unfrozen ground below, much as a blotter soaks up moisture. In time, lens-shaped masses of ice are built up, and the soil above them is heaved upward. Frost heaving of this sort is common on poorly constructed roads, and lawns and gardens are often soft and spongy in the springtime as a result of the soil's heaving during the winter.

During the process of frost heaving, material moves up with each interval of freezing (expansion) and down with each internal of thawing (contraction). Therefore material moves downslope with each freeze-thaw cycle (see Figure 12.2[c]).

SOLIFLUCTION

The term **solifluction** (from the Latin *solum*, "soil," and *fluere*, "to flow") refers to the downslope movement of debris under saturated conditions. Solifluction is most pronounced in high latitudes and high elevations even in low latitudes. Here the soil is strongly affected by alternate freezing and thawing and the ground freezes to great depths; but even moderately deep seasonal freezing promotes solifluction.

PERMANENTLY FROZEN GROUND

Solifluction occurs most extensively in areas of **permanently frozen ground,** known as **permafrost.** In areas where the mean annual temperature is 0°C or less the ground is usually frozen throughout the year. If there is moisture in the soil or rock, ice will persist from year to year. Such regions are commonly called **periglacial** for their near-glacial climate.

Estimates of the total land area underlain by permanently frozen ground vary, but 20 percent is probably not too far off. Permanently frozen ground characterizes some 80 percent of Alaska and 50 percent of Canada. In the high mountains of the 48 contiguous states an estimated 100,000 km² is underlain by permanently frozen ground (Figure 12.8). The thickest reported permafrost is between 1,450 and 1,500 m along the upper reaches of the Markha River in Siberia. On Ellsworth Island, Northwest Territories, Canada, permanently frozen ground is about 1,000 m thick, the thickest known in North America.

The temperature distribution in zones of permanently frozen ground (Figure 12.9) helps to explain the cause of mass movements in such regions. The zone above the permafrost table (that depth at which the maximum annual temperature is 0°C) thaws each summer, while the ground below remains frozen. The permafrost acts as a barrier to the infiltration of meltwater from the seasonally thawed zone. As a result, the heavily water-laden material above the permafrost is very susceptible to downslope movement.

Under these conditions great sheets of debris move slowly down even the gentlest slopes. Moreover, with the passing seasons frost heave is particularly effective. It is small wonder that permanently frozen ground raises formidable obstacles to engineering and thus to extensive settlement of arctic regions.

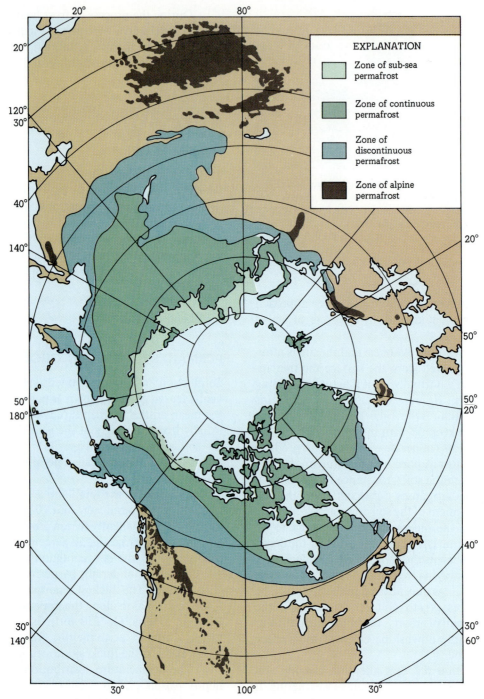

12.8 Distribution of permafrost in the Northern Hemisphere. [Compiled from various sources by Troy L. Péwé. For a list of sources see *Arctic and Alpine Research*, vol. 15, Fig. 1, pp. 146–147, 1983.]

ROCK GLACIERS

Rock glaciers are tongue - and lobate-shaped masses of rock rubble found in many high, cold mountainous areas (Figure 12.10). The rubble is cemented by interstitial ice and sometimes encloses bodies of clear ice. The movement of rock glaciers is due to the deformation of the ice within them. The forward rates of motion of those rock glaciers that have been measured are as slow as 1 mm/year (slower than the rate at which continents drift apart) to as fast as 1.7 m/year. The average is probably between 30 and 40 cm/year.

The surfaces of rock glaciers are characteristically marked by furrows and ridges at right angles to their direc-

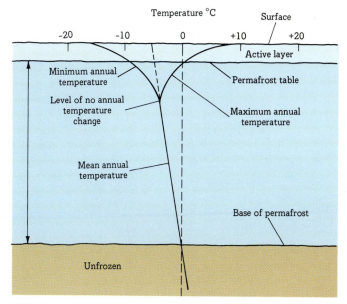

12.9 Temperature distribution in an area of permanently frozen ground.

tion of movement. These are clearly visible from the air and give the impression that the rubble is flowing viscously. The furrow-and-ridge topography is obscured, however, to the ground observer by the huge boulders that litter the surfaces of most rock glaciers. Their snouts are steep-faced (30 to 60 m high), as are their lateral margins.

Active rock glaciers are found above the treeline and the lower limit of the zone of permanently frozen ground. They carry rock rubble fed to them from the toes of steep talus slopes (see Section 12.5) or from the rock-littered margins of stagnant and melting mountain glaciers.

12.10 This rock glacier near the Col de Lauteret in the southern French Alps has a typically steep front and a characteristic location at the foot of an active talus slope. [Roland Hellmann.]

12.4
MOVEMENTS OF MODERATE VELOCITY

A number of types of mass movement are intermediate in velocity between the very slow ones just discussed and the very rapid movements to be described in Section 12.5. Here we turn our attention to these slope processes of moderate velocities—from 1 cm/day to 1 cm/s or more.

SLUMP

Sometimes called **slope failure, slump** is the downward and outward movement of earth traveling as a unit or as a series of units. Slump usually occurs where the original slope has been sharply steepened, either artificially or naturally. The material reacts to the pull of gravity as if it were an elastic solid, and large blocks of the slope move downward and outward along curved planes. The upper surface of each block is tilted backward as it moves.

Figure 12.11 shows the beginning of a slump at Gay Head, Massachusetts. The action of the sea has cut away the unconsolidated material at the base of the slope, steepening it to a point where the earth mass can no longer support itself. Now the large block has begun to move along a single curving plane, as suggested in Figure 12.12.

After a slump has been started, it is often helped along by rainwater collecting in basins between the tilted blocks and the original slope. The water drains down along the spoon-shaped plane on which the block is sliding and promotes further movement.

12.11 The beginning of slump, or slope failure, along the sea cliffs at Gay Head, Massachusetts. Note that the slump block is tilted back slightly, away from the ocean. This slump block eventually moved downward and outward toward the shore along a curving plane, a portion of which is represented by the face of the low scarp in the foreground. [Harvard University, Gardner Collection.]

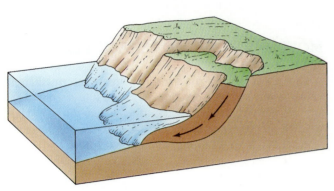

12.12 This diagram shows the type of movement found in a slump similar to that pictured in Figure 12.11. A block of earth material along the steepened cliff has begun to move downward along a plane that curves toward the ocean.

A slump may involve a single block, as shown in Figures 12.11 and 12.12 but very often it consists of a series of slump blocks (Figure 12.13).

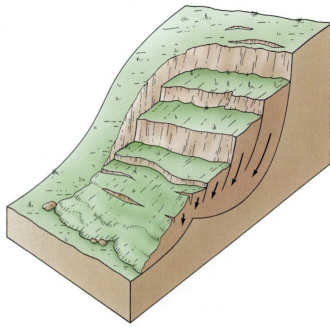

12.13 Many slump movements contain several discrete units, as suggested here.

EARTHFLOWS

Earthflows involve the plastic movement of an unconsolidated mass of soil and weathered rock. Movement is along the bedrock surface beneath. Earthflows move slowly but perceptibly, usually helped along by excessive moisture. The upper limit is usually marked by a pull-away scar. The sides of the earthflow are well defined and downslope it ends in a lobelike form (Figure 12.14). The movement may involve from a few to several million cubic meters of material. With the addition of water an earthflow becomes more fluid, moves more rapidly, and grades into the feature called a **mudflow** discussed in Section 12.5.

DEBRIS SLIDES

Debris slides involve the downslope movement of comparatively dry and unconsolidated material. In many ways they are similar to earthflows, but differ in that the material moved has a larger component of coarse material than does an earthflow and the process does not involve as much water. The material slides and rolls downward and produces a surface of low hummocks with intervening depressions.

Debris slides may be quite common on grassy slopes, particularly after heavy rains, and in unconsolidated material along the steep slopes of stream banks and shorelines. They quite often occur in sparsely vegetated country of steep slopes mantled with loose Earth materials.

With the addition of large amounts of water debris slides are transitional to the very rapidly moving debris flows discussed in Section 12.5.

12.14 An earthflow down a hillslope in California has the typical sharp scarp at its top and the area of soil flow downslope. [Martin Miller.]

12.5

RAPID MOVEMENTS

Rapid movements encompass slope processes in a range from meters per second to over 100 km/h. They are an extension of the processes we have already discussed.

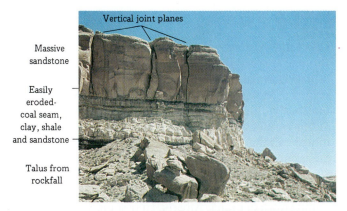

Massive
sandstone

Vertical joint planes

Easily
eroded-
coal seam,
clay, shale
and sandstone

Talus from
rockfall

12.15 Here in Chaco National Monument, New Mexico, a vertically jointed, thick sandstone unit overlies more easily eroded thin beds of sandstone, shale, coal seams, and claystone. Erosion of these less resistant units undermines the overlying sandstone. The sandstone breaks away along the vertical joint planes that define a new cliff face.

12.16 Alternate freezing and thawing has resulted in rapid mechanical weathering that has produced the rubble forming these slopes of talus in the French Alps. The angle of the surface of these slopes is typical, a little over 30°. [Robert Stallard.]

ROCKFALL AND TALUS

A **rockfall** is the sudden fall of one or more large fragments of rock. It is a common occurrence along steep cliffs of massive vertically jointed or layered rock. This massive rock is undermined by erosion of softer material beneath it and breaks off along the joint or bedding or cleavage plane. Figure 12.15 provides an example.

Strictly speaking, a **talus** is a slope built up by an accumulation of rock fragments at the foot of a cliff or a ridge. The rock fragments are sometimes referred to as **rock waste** or **slide rock**. In practice, however, talus is widely used as a synonym for the rock debris itself.

Mechanical weathering produces the rock fragments that make up a talus slope. Most commonly the process of freezing and thawing (Section 5.2) is responsible for breaking up the bedrock. But the transfer of the fragments down a slope to form a talus is in the realm of mass movement.

In the development of a talus rock fragments are loosened from the cliff and clatter downward in a series of free falls, bounces, and slides. As time passes, the rock waste finally builds up a heap or sheet of rock rubble. An individual talus resembles a half-cone with its apex resting against the cliff face in a small gulch or re-entrant (Figure 12.16). A series of these half-cones often forms a girdle around high mountains, completely obscuring their lower portions. Eventually, if the rock waste accumulates more rapidly than it can be destroyed or removed, even the upper cliffs become buried, and the growth of the talus stops. The slope angle of the talus varies with the size and shape of the rock fragments. Although angular material can maintain slopes up to 50°, a talus rarely exceeds angles of 40°

A talus is subject to the normal process of chemical weathering, particularly in a moist climate. The rock waste is decomposed, especially toward its lower limit, or toe, which may grade imperceptibly into a soil.

DEBRIS FLOWS AND MUDFLOWS

A **debris flow**—sometimes called a **mudflow**—is a well-mixed mass of water, earth, sand, and rock that moves down valley slopes. The distinction between them lies largely in the size of particles involved in the movement. A debris slide contains more coarser material than does a mudflow. Debris flows and mudflows are transitional from debris slides and earthflows, respectively. The distinction lies in an increase of water; hence there is greater fluidity in the debris flows and mudflows.

Most debris flows move with a moderate velocity somewhat less than that of a river—that is, less than 1 m/s—but some may be as slow as 1 m/year or as rapid as 160 km/h. In mountainous and arid to semiarid areas debris flows manage to transport tremendous masses of material very rapidly.

A typical debris flow originates on the slopes of a small steep-sided canyon or gulch where the slopes have become covered by loose, unconsolidated and unstable material. A sudden cloudburst in dry country or water from melting snows in high mountains flushes earth and rock from the slopes and carries them to the usually dry stream channel. Here the debris accumulates and may fill the channel until the growing pressure of the water becomes great enough to start the mass moving down the valley. Water and debris mix together with a rolling motion that produces a product with the consistency of wet concrete. The advance may be intermittent, at times slowed by a narrowing of the canyon or an accumulation of old debris; at other times it surges forward, pushing obstacles aside or carrying them along with it.

Eventually the debris flow leaves the canyon mouth for gentler slopes and less confined channels. Here it may splay out, its water may soak into the underground, and the thinning mass may turn into a sheet of inhospitable rubble.

A debris flow can carry boulders weighing 100 tons or more and float houses, cars, and giant trees along as if they were toys. So in inhabited areas debris flows can be a great threat to life and property, as they are on the lower slopes of the San Gabriel Mountains along the northern edge of the Los Angeles metropolitan area.

ROCK SLIDES OR ROCK AVALANCHES

Of all the mass movements the most terrifying and catastrophic are **rock slides,** sometimes called **rock avalanches,** or **debris avalanches** if largely unconsolidated material is involved. Moving at speeds in excess of 100 km/hour, they transport millions of tonnes of earth and broken rock over distances measured in kilometers. This was the type of movement that "Uncle Billy" Bierer was predicting in the quotation at the head of this chapter. (See also Box 12.1.)

One of the first major landslides to be studied in any detail occurred in 1903 and killed 70 people in the coal mining town of Frank, Alberta. Here some 36.8 million m³ of rock crashed down from the crest of Turtle Mountain, which rises over 900 m above the valley. Mining activities may have triggered this movement, but natural causes were basically responsible for it. As Figure 12.17 shows, Turtle Mountain has been sculptured from a series of limestone, sandstone, and shale beds, which have been tilted, folded, and faulted. The diagram shows that the greater part of the

---

BOX 12.1   A Woozy Mountain

In 1925 the Gros Ventre slide in west-central Wyoming (Figure B12.1.1) moved an estimated 40 × 10³ m³ of rock and debris across the Gros Ventre Valley and dammed the Gros Ventre River to form a lake 5 km long and some 70 m deep. Two years later erosion of the dam along its overflow channel caused heavy flooding downstream. A wall of water 5 m high destroyed the village of Kelly 6 km downstream and killed 6 people.

This was the slide that "Uncle Billy" Bierer predicted. His prediction is characterized by accuracy and understanding as well as by an absence of scientific jargon. Here is his story.

Albert Nelson, a local rancher along the Gros Ventre Valley, came upon some springs high up on the valley wall. This was sometime before his neighbor Bierer sold his cabin in 1920, a cabin that lay downslope from the springs and not far from the river. When Nelson spoke with Bierer about the springs, Bierer is reported to have responded:

Yes, I have noticed that and I cannot see where the water can

B12.1.1   The scar left by the Gros Ventre slide is still clearly visible on Sheep Mountain decades later. In the foreground are a tangle of logs left from trees carried by the slide.

be going to unless it is following the formation between the two different stratifications and

coming to the surface at some other water-level point. If not, this mountain side would be a mushy, woozy boil. However it may be, there is a wet line running between these strata and the time will come when the entire mountain will slip down into the canyon below.*

Another member of the Nelson family remembers hearing additional Bierer remarks:

Anywhere on that slope, if I lay my ear to the ground, I can hear water tricklin' and runnin' underneath. It's runnin' between strata and some day, if we have a wet enough spring, that whole mountain is gonna let loose and slide. Give it a wet enough year and all that rocky strata will just slide right down on the gumbo like a beaver's slickery slide.

---

* Barry Voight, "Lower Gros Ventre Slide, Wyoming, USA," in Barry Voight (ed.), *Rockslides and Avalanches,* Vol. 1, p. 116, Elsevier Publishing Co., Amsterdam, 1978.

The causes of the Gros Ventre slide were multiple. They included northward dipping beds of weak clay and weathered sandstone lying parallel to the slope of the south wall of the valley (see Figure B12.1.2); heavy precipitation and snow melt prior to the slide; and one or more small earthquakes. The cabin that "Uncle Billy" sold in 1920 was buried by the slide five years later.

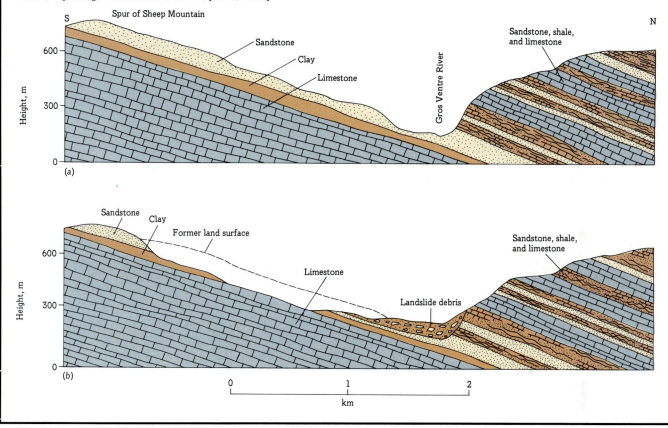

B12.1.2 Diagrams to show the nature of the Gros Ventre slide: (a) Conditions existing before the slide took place; (b) the area of the slide and the location of the debris in the valley bottom. Note that the sedimentary beds dip into the valley from the south. The large section of sandstone slid downward along the clay bed. [Redrawn from William C. Alden, "Landslide and Flood at Gros Ventre, Wyoming," Trans. AIME, vol. 76, p. 348, 1928.]

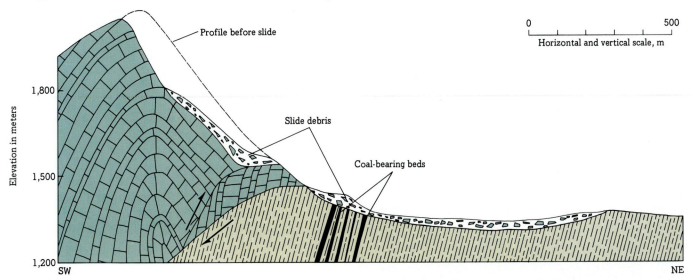

12.17 A cross section to show the conditions at Turtle Mountain that brought about the Frank, Alberta, landslide. [Modified from D. M. Cruden and J. Krahn, "A Reexamination of the Geology of the Frank Slide," Canad. Geotech. J., vol. 10, no. 4, 1973.]

mountain is made of limestone and that the valley below is underlain by less-resistant beds of sandstone, siltstone, shale, and coal. It also shows four factors that contributed to the slide: the steepness of the mountain; the bedding planes in the limestone dipping parallel to the mountain face; the thrust plane that breaks the strata partway down the mountain; and the weak shale, siltstone, and coal beds in the valley.

The steep valley wall enhanced the effectiveness of gravity, and the bedding planes along with the fault planes served as potential planes of movement. The weak shale beds at the base of the mountain probably underwent slow plastic deformation under the weight of the overlying limestone, and as the shale was deformed, the limestone settled lower and lower. The settling action may have been helped along by the coal mining operations in the valley as well as by frost action, rain, melting snows, and earthquake tremors that had shaken the area 2 years before.

In any event, stress finally reached the point where the limestone beds gave way along their bedding planes and pushed outward toward the valley along the uppermost fault; the great mass of rock hurtled down into the valley. The rock material behaved in three different ways: (1) The shales underwent plastic deformation, producing a condition of extreme instability on the mountain slope. (2) When the strata that still held the limestone mass on the slope sheared in the manner of an elastic solid, the slide actually began. (3) Once under way, the rock debris bounced, ricocheted, slid, and rolled down the mountain slope until it was literally "launched" into the air by a ledge of rock. Thereafter it arched outward and downward to the valley below. Once on the valley floor, it moved at high speed up the hills on the far side of the river. In this phase it moved like a viscous fluid with a series of waves spreading out along its front.

12.18   An aerial view west across the Madison Canyon landslide. The scar of the slide is seen on the mountain to the south (left). The slide debris fills the valley bottom and has dammed up a new lake. An outlet channel for the lake has been cut across the landslide dam. The former Madison Canyon campsite lies beneath the slide material on the north (right) side of the valley. [William C. Bradley.]

A more recent slide occurred in southwestern Montana on August 17, 1959. An earthquake whose focus was located just north of West Yellowstone, Montana, triggered a rock slide in the mouth of the Madison River Canyon, about 32 km to the west. A mass of rock estimated to be over 32 million m³ fell from the south wall of the canyon. It climbed over 100 m up the opposite valley wall and dammed a lake 8 km long and over 30 m deep. More than a score of people lost their lives in the Madison River campground area below the slide (Figure 12.18). Survivors reported extremely powerful winds along the margins of the slide. In fact, some persons were blown away, never to be seen again, and a 2-t automobile was observed to have been carried by the wind over 10 m before smashing against a row of trees. It seems clear that when the mass of rock fell to the valley bottom, it trapped and compressed large quantities of air. The high winds represented the extrusion of this mass of air from beneath the moving debris.

The geologic conditions at the site of the Madison Canyon slide are shown in Figure 12.19. The south wall of the canyon is underlain by a dolomite and by gneiss and schist. The dolomite helped to buttress the weaker units of strongly foliated metamorphic rocks. Over the years the thinning of the protective wedge of dolomite by continued canyon cutting brought the slope ever closer to the threshold between stability and instability. The immediate cause of the slide was an earthquake of magnitude 7.1 early on the morning of August 17, 1959, an earthquake about 30 km east of the site and north of the town of West Yellowstone. Under the stress of both direct seismic wave motion and ground subsidence, the barrier of dolomite gave way, the rockslide was triggered, and dolomite and large masses of unsupported schist and gneiss swept down into Madison Canyon, buried the campsite, and formed a dam behind which we now have Earthquake Lake.

There are also vestiges of prehistoric slides of great magnitude. One, at Saidmarreh, Iran, involved the movement of 4,245 million m³ of rock for a distance of over 12 km. Another, the Blackhawk slide of southern California, moved 283 million m³ of material over a distance of 8 km.

The largest recorded movement, modern or prehistoric, took place some 300,000 years ago on the slopes of ancestral Mt. Shasta in what is now northern California. The debris avalanche extended 43 km northwestward from the volcano and involved 17×10^{12} m³ of volcanic material.

Here we pause to examine the movement of catastrophic slides in greater detail. A number of investigators have reported that the shattered and jumbled materials in an avalanche maintain the same general relationships that they had before the bedrock was broken and movement began. In other words, material toward the back of the rock slide does not overtake that in front. The deduction drawn is that the material in the back part of the slide collides with material immediately in front of it, urging it on faster. This process is repeated throughout the slide, so material behind

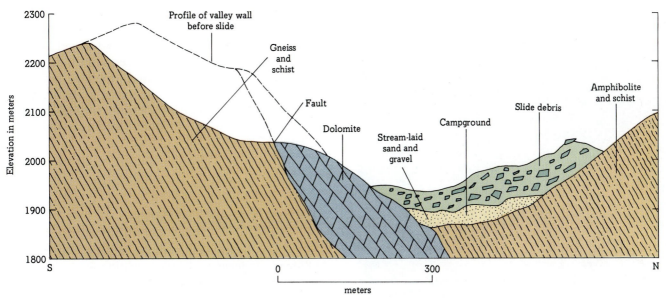

12.19 Geologic conditions at the site of the Madison Canyon slide. See text for discussion. [From Jarvis B. Hadley, *Landslides and Related Phenomena Accompanying the Hebgen Lake Earthquake*, U.S. Geological and Survey Professional Paper 435-K, p. 115, Washington, D.C., 1964.]

never overtakes the debris ahead of it. This transfer of energy through the debris of the slide, called **momentum transfer,** makes possible the progressively more rapid movement of material in the downslope positions within the sliding mass.

Warnings of Rapid Movements   We usually think of a landslide as breaking loose without warning, but it is more accurate to say that people in the area simply ignore or fail to detect the warning.

For example, a disastrous rock slide at Goldau, Switzerland, in 1806, wiped out a whole village, killing 457 people. The few who lived to tell the tale reported that they themselves had no warning of the coming slide but that animals and insects in the region may have been more observant or more sensitive. For several hours before the slide horses and cattle seemed to be extremely nervous, and even the bees abandoned their hives. Some slight preliminary movement probably took place before the rock mass broke loose.

On the other hand, some indications of impending landslides are sometimes noted. For instance, during the spring of 1935 slides took place in clay deposits along a German superhighway that was being built between Munich and Salzburg. The slides came as a complete surprise to the engineers, but for a full week the workmen had been murmuring, *Der Abhang wird lebendig* ("The slope's becoming alive"). Another example is described in Box 12.1.

Landslides like the one on Turtle Mountain are often preceded by slowly widening fissures on the rock near the upward limit of the future movement.

Studies of debris slides in the Santa Monica Mountains and vicinity in southern California indicate that debris slides are coincident with particular patterns of rainfall. When the winter rainfall exceeds 25 cm, the stage is set for

the initiation of debris flows. Then slides are most apt to occur as a result of a brief period of additional rain if the intensity is 0.5 cm/h or greater.

12.6
WATER ON SLOPES

In the preceding sections we considered only the bulk movement (mass movement) of material down slopes. But it is clear that material is also moved from slopes as individual particles or in solution. The processes by which water removes material from slopes include **rainsplash, surface flow,** and **subsurface flow.**

RAINSPLASH

Raindrops falling on unvegetated, or partially vegetated, slopes can be a very effective mechanism for moving particulate material down slopes (Figure 12.20). For instance, in a high-intensity storm raindrops may reach 6 mm in diameter and hit the ground at a velocity of 9 m/s. This is powerful enough to kick into the air particles up to 10 mm in diameter, and to undermine and loosen still larger particles and thus prepare the way for downslope movement.

SURFACE FLOW

Water flowing on a slope will not be concentrated initially in discrete channels but will flow in a sheet of irregular

12.20 Raindrops hitting bare soil throw individual solid particles into the air and down the slope. [Photograph by Wide World.]

thickness. If the vegetative cover is heavy, this flow will be able to move very little particulate material and much of the water will sink into the subsurface. When the rainfall becomes intense or when the vegetative cover is sparse or absent, the water will begin to move particles downslope. Eventually small, shallow channels or rills will form and will shift back and forth across the slope. Advantageously located rills may deepen and become incised into the slope as small gullies (Figure 12.21).

SUBSURFACE FLOW

Some of the surface water on a slope may seep into the subsurface. This will be particularly so if the slope material is permeable and if the vegetative cover is complete enough to slow the surface runoff. As water comes into contact with minerals in the subsurface, it may react chemically with them and move material away in the dissolved form.

12.21 Surface runoff on unvegetated slopes develops a complex system of gullies, as displayed in this view in the petrified National Monument, Arizona. Distance to skyline is about 400 m. [Robert Key.]

Subsurface flow also removes solid material from a slope. This happens if enough water seeps into the slope and the materials are permeable enough to allow flow to be concentrated. Channels may then form beneath the surface. Water is carried to lower elevations within the slope and may break forth as a temporary spring carrying material with it.

12.7
SHAPES OF HILLSLOPES

Thus far we have examined how material moves down hillslopes but have paid little attention to the hillslopes themselves. We will consider this subject in the remaining pages of the chapter.

DEPOSITIONAL VS. EROSIONAL SLOPES

Slopes in a landscape may be either **depositional** or **erosional**. Depositional slopes originate, for example, when a river deposits its sediments to form a floodplain or when basaltic lava flows pile one on top of another to create a fairly smooth, gently sloping plain. Other depositional landforms are much steeper: for instance, volcanic cones built of ash, cinders, and lava such as those of the Cascade Range in Washington, Oregon, and northern California. Other processes also create landforms with depositional slopes, and some of them are listed in Table 12.3.

In general, most depositional slopes are simpler and less intricate than erosional slopes. Moreover, most of the landscapes of the continents are made up of erosional slopes. This should not surprise us for we recognize that as soon as portions of the Earth's surface appear above sea level, they are subjected to the processes of weathering and to the erosive processes that work continuously to wear them back toward sea level. Because the creation of depositional slopes is a process of building the land surface locally upward, we should certainly expect such forms to be short-lived, soon destroyed by the down-wearing forces of erosion. Figure 12.22 is an example of these processes.

TABLE 12.3
Some Landscape Forms Whose Slopes Are
Constructional When First Formed

Volcanic cones
Volcanic flows
Floodplains
Deltas
Alluvial fans
Beaches
Glacial deposits
Sand dunes

Modified from various sources.

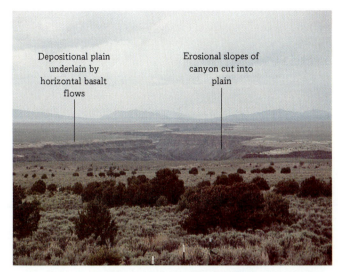

12.22 In the Rio Grande Valley of northern New Mexico a series of horizontally arranged basalt flows has built up a volcanic plain in the valley. The surface of this plain is a smooth depositional slope into which the Rio Grande has cut its canyon. The steep canyon walls are made up of erosional slopes that are being formed at the expense of the depositional slope of the plain.

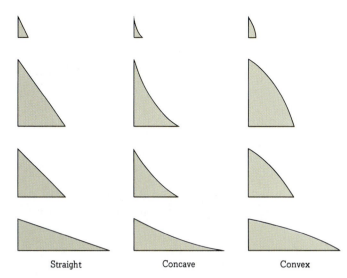

Straight	Concave	Convex

12.23 The basic shapes of hillslopes involve convex, concave, and straight segments. They may have many different lengths and inclinations.

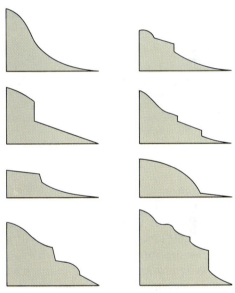

12.24 Different slope segments can be put together in different ways to give a variety of profiles to a hillslope.

BASIC SHAPES

The **shapes** of slopes seem almost endless and the landscape features they form range from extremely simple to extremely complex. Look carefully, however, and you will see that the shape of a complex slope is made up of only three basic elements—**convex, concave,** and **straight** slopes—which are sometimes referred to as *waxing, waning,* and *constant* slopes (Figure 12.23). Put these fundamental shapes together in differing lengths, inclinations, and sequences and you can create a remarkable variety of landscapes, of which Figure 12.24 illustrates but a few, and only in two-dimensional profiles at that. Figure 12.25 combines the slope elements into a three-dimensional framework and shows why landscapes are so varied.

Generally the upper slopes of a hill are convex and dominated by the gravitational process of creep. The lower hillslope is generally concave and dominated by flowing water across the surface or concentrated in rills. The straight slopes may be one of two types. The very steep ones (65° to 90°) are cliff faces from and over which rock material is free to fall to lower elevations. Slopes of lower angle (25° to 35°) are often talus- or bed rock-mantled by thin talus moved by creep.

The basic shapes of slopes vary according to a number of factors. These include **rock type, geologic structure, climate, surficial processes, time,** and **human activity.**

Rock Type Of the several factors that influence the shapes of slopes the relative resistance of the underlying rock to erosion is an obvious one. In general, the more resistant a rock to erosion, the steeper the slope that forms on it. (See Table 12.4.)

TABLE 12.4
General Relationship Between Steepness of Slopes and Underlying Bedrock

Steep slopes	Gentle slopes
Metamorphic rocks such as gneiss, quartzite, and schist	Unconsolidated or weakly consolidated sediments such as clay, silt, and volcanic ash
Most igneous rock such as granite and basalt	Shale, mudstone, and siltstone
Firmly cemented sedimentary rocks such as sandstone siltstone, conglomerate	Limestone and marble in humid climate
Limestone and marble in arid climate	

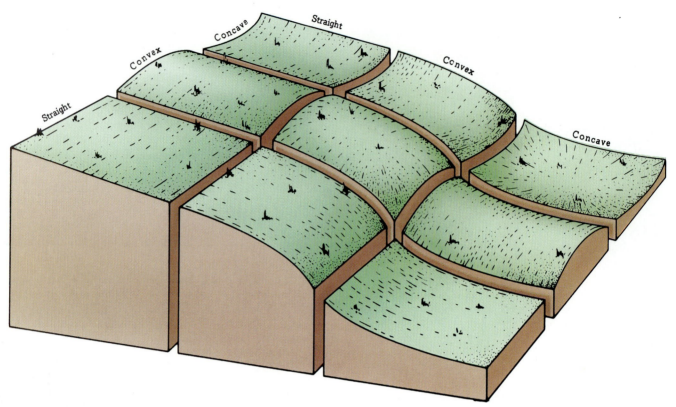

12.25 This hypothetical hillslope is made up of 9 blocks that represent the basic slope elements in three dimensions. In nature the size, inclination, and juxtaposition of these building blocks vary to produce the complexities of hillslopes, but the 9 units exhaust the possible combinations of convex, concave, and straight. [Robert V. Ruhe, *Geomorphology*, copyright © 1975 by Houghton Mifflin Co.; adapted with permission]

Geologic Structure   The way rocks of differing relative resistance to erosion are arranged in relation to one another will help to determine landscape form. Thus erosion of horizontal resistant and nonresistant beds will produce a hill shaped like the small butte shown in Figure 12.26. Folded and faulted rocks will erode into complex and varied forms as slopes outline the twisted and broken arrangement of differing rock types (Figure 12.27).

Climate and Slopes   Climate operates in a number of ways in the development of slopes. In Chapter 5 we found that the higher the temperature and rainfall, the greater the rate of chemical weathering and thus the thicker the soil cover. The presence or absence of an extensive soil cover is one of the major differences between hillslopes in the humid and arid regions, as a cross-country journey from the East Coast to Death Valley would demonstrate. This chapter has reminded us that the rigors of a periglacial environment produce particular types of slopes. Future chapters discuss other examples of the close association of particular landforms with specific climates. These and others are listed in Table 12.5.

Time   Slopes are active, dynamic features that change with time, sometimes rapidly, more often slowly. But

12.26   Horizontally layered sedimentary rocks of differing resistance to erosion produce the different slopes of Fajada Butte in Chaco Canyon National Monument. The cliffs have formed from the massive, vertically jointed sandstone of the Cliff House Formation. The gentler slopes are underlain by the more easily eroded Menefee Formation of shale and thin, discontinuous sandy units, and these are littered with debris from the sandstone cliffs above.

Thick-bedded resistant sandstone

Non-resistant shale and thin-bedded sandstone

12.27 Compressional stress has folded sedimentary rocks of Jurassic and Cretaceous age to form Little Dome, an elliptical structure in the Maverick Springs area of Fremont County, Wyoming. The beds are alternating beds of sandstone and shale. Erosion has eroded the dome downward, etching out the less resistant shales to leave the sandstone beds standing as ridges. [U.S. Geological Survey.]

TABLE 12.5
Some Landscape Characteristics Associated with Particular Climates

Climate	Landscape characteristics
Periglacial (alternate freezing and thawing in presence of water)	Long sweeping slopes produced by creep and solifluction; concave slopes dominant; talus (this chapter)
Glacial	In mountains spectacular cliffs, peaks, valleys, and high lakes of "alpine" topography; in lowlands lakes, swamps, gravel plains, and ridges (Section 15.2.)
Dry	Limestone forms cliffs, rocky slopes, and thin, stony soils, sand dunes (Section 17.2)
Moist	Thick soils (Section 5.5); caves and sinkholes in limestone (Section 15.5); convex slopes dominate

change they do. Precisely how any given set of hillslopes will change is not completely agreed on, but there are some things we can say about their change over time.

In general, it appears that cliffs or very steep hillsides retreat at a constant angle in a variety of climates and at rates that vary two orders of magnitude from between about .03 and 3 mm/year. We know also that the slope angles and elevation of mountainous areas are reduced with time. Thus the geologically young Rocky Mountains of New Mexico, Colorado, Utah, Wyoming, and Montana are marked by spectacular slopes in contrast to the much older Appalachian Mountains of the eastern states. There the mountains are lower and steep slopes, while present, are shorter and fewer.

Here is another way hillslopes may change with time. Earlier in this section we noted how geologic structure can affect slope form. Figure 12.28 depicts an area underlain by a simple arrangement of flat-lying shale and sandstone intruded by a granitic dike. It shows the landscape at one

12.28 These block diagrams illustrate one way in which hillslopes can vary with time, geologic structure, and differing resistance of rocks. See text for discussion.

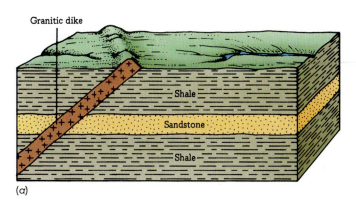

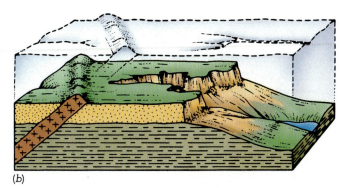

12.29 The plow has rendered this hillslope near Tarquinia, Italy, very vulnerable to erosion.

moment in time (Figure 12.28[a]) and at a time some millions of years later (Figure 12.28[b]). In Figure 12.28(a) erosion has etched out the granitic dike so that it stands as a steep-sided ridge, while the less resistant shale has been eroded into low hills of gentle slope. In Figure 12.28(b) erosion has stripped the rock down to a new surface. The ridge remains but is now displaced to the left. Where the sandstone has been eroded down to the underlying shale steep cliffs have formed. Beyond the sandstone cliffs gentle slopes are formed on the lower shale.

Human Activity The arrival of humans a brief geologic moment ago brought a new geologic agent to the Earth — **human activity.** It has become extremely effective in modifying old slopes and creating entirely new ones. One of the most powerful tools available for slope modification has been the plow. Pulled first by animal power and then by mechanical power, the plow has been in use for an estimated 8,000 years. Breaking the protective vegetative cover renders the slope very susceptible to rapid erosion by water and wind (Figure 12.29) and increases the rate at which material is eroded by two or three orders of magnitude.

The pick and shovel, more recently joined by the bulldozer and power shovel augmented by explosives, have produced new and artificial slopes. They may be the steep slopes of a quarry or an open-pit mine or the less dramatic slopes of a highway cut or embankment. Virtually all building construction produces new or modified slopes. Some are later landscaped with grass and plantings, others are topped by gravel, asphalt, or concrete.

Summary

Hillslopes provide the ramps down which products of weathering and erosion reach the streams and eventually the ocean. Movement is by individual particles and dissolved material carried by **water on slopes** and as bulk transport of solid or viscous material called **mass movement.**

Factors of mass movement that facilitate the work of gravity include water; earthquakes; orientation of bedding, foliation, fractures, faults, or other planes of weakness; vegetative cover; climate; animals; and human activity.
    Behavior of material may be as a **brittle solid,** a **plastic substance,** or a **fluid.** Motion may be described as **slide, fall, flow,** or **heave.** Motion begins as slope material crosses a **threshold** between stability and instability. Instability is reached when the **force (stress)** exerted on the slopes is greater than the **resisting force.**

Slope movements can be divided into **slow movements** (1 mm/year to 1 mm/day; **movements of moderate velocity** (1 cm/day to 1 cm/s or more); and **rapid movements** (meters per second to over 100 km/hour).
    Slow movements involve **creep, frost heaving, solifluction, permanently frozen ground,** and **rock glaciers.**
    Movements of moderate velocity include **slump, earthflows,** and **debris slides.**
    Rapid movements include **debris flows** and **mudflows, rockfalls producing talus,** and **rock slides** or **rock avalanches.**

Water on slopes moves particles and dissolved material by **rainsplash, surface flow,** and **subsurface flow.**

Slopes may be either **depositional** or, much more commonly, **erosional.**

Basic shapes of hillslopes are **convex, concave,** and **straight.** Their precise shapes are determined by **rock type, geologic structure, climate, time,** and **human activity.**

Questions

1. List some of the factors that affect the movement of Earth materials down slopes.

2. Some movements of materials on slopes are very slow compared with, for example, rock slides. List some different types of slow movement and describe one in some detail.

3. Describe at least two types of rapid slope movement, using examples if possible.

4. From your own experience and observation, can you cite examples of the downslope movements of material? If so, describe them and try to determine the causes.

5. A slope 1 km long and 250 m wide ends along the banks of a stream. A soil layer 0.5 m thick moves down the slope at an average rate of 1 mm/year. The stream carries off the slope material as rapidly as it is delivered to it. How long will it take for the material at the top of the slope to be removed by the stream at the base of the slope?

We divided distance traveled by the rate of travel and got time as follows:

$$10^6 \text{ mm} \div 1 \text{ mm/y} = 10^6 \text{ y}$$

What would be the time if the rate of travel were 1 mm/day?

(We followed the same procedure as above, but used a rate of 1 mm/day and got a little over 2,700 years as time of travel.)

In each of the above situations what volume of material would be carried to the stream? What would this represent in metric tons?

(For this last answer you need to know the specific gravity of the material. A specific gravity of 2.2 is a reasonable figure. Our answers were 125,000 m³ and 275,000 t.)

6. What are the three basic elements in the shapes of slopes? Observe a hillslope, draw its profile, and label the slope segments.

7. A hill is underlain from its top to its base by horizontally-layered nonresistant shale, thickly-layered sandstone, and thin-bedded siltstone, in that order. Draw a cross section of a hill to show the distribution of rock types and the slopes associated with them.

8. Distinguish between erosional and depositional slopes and give examples of each.

9. In an arid climate a rock sequence from hilltop to bottom involves shale, limestone, and shale. This sequence has been folded into a syncline whose limbs dip at a maximum angle of 45°. Draw a cross section of the syncline, showing at the top of the profile the topography to be expected.

10. Figure 12.25 gathers hillslope elements into a three-dimensional diagram. Devise a classification that separates the blocks into groups.

Supplementary Readings

CARSON, M. A., AND M. J. KIRKBY *Hillslope Form and Process,* Cambridge University Press, Cambridge, 1972.
One of the first volumes devoted exclusively to the subject and still a good basic text.

CHORLEY, RICHARD J., STANLEY A. SCHUMM, AND DAVID E. SUGDEN *Geomorphology,* Methuen & Co., London, 1984.
A standard textbook on geomorphology in which Chapters 10 and 11 cover in some detail the material of this chapter on hillslopes.

CRANDELL, D. R., ET AL. "Catastrophic Debris Avalanche from Ancestral Mount Shasta Volcano, California," *Geology,* vol. 12, no. 3, 1984, pp. 143–146.
Description of the largest known prehistoric avalanche on Earth.

EMBLETON, CLIFFORD, AND CUCHLAINE A. M. KING *Periglacial Geomorphology,* 2nd ed., John Wiley & Sons, New York, 1975.
A short but authoritative treatment of surface processes found in arctic areas.

GIARDINO, JOHN R., JOHN F. SHRODER, JR., AND JOHN D. VITEK (EDS.) *Rock Glaciers,* Allen & Unwin, Boston, 1987.
A collection of reports on rock glaciers, including a good overview of their features in an informative foreward by Clyde Wahrhaftig.

MCPHEE, JOHN *The Control of Nature,* Farrar, Straus, Giroux, 1989.
Describes the debris flows in the San Gabriel Mountains on the north side of the Los Angeles basin as well as how they are protected against (sometimes), the damage they do, the man-made changes they effect in downslope stream channels, and the demands they place on local government. Good reading.

RITTER, DALE F. *Process Geomorphology,* 2nd ed., Wm. C. Brown Publishers, Dubuque, Iowa, 1986.
A standard textbook in which Chapter 10 is a compact, upper-level discussion of the subject.

SELBY, M. J. *Hillslope Materials and Processes,* Oxford University Press, Oxford, 1982.
An advanced-level treatment that incorporates data from several disciplines, including geology, soil science, and engineering.

SMALL, R. J., AND M. J. CLARK *Slopes and Weathering,* Cambridge University Press, Cambridge, 1982.
A straightforward discussion of mass movement as well as the general characteristics of slopes.

VOIGHT, BARRY (ED.) *Rockslides and Avalanches,* Elsevier Publishing Co., Amsterdam, 1978.
This compilation of studies is in two volumes. Volume 1, Natural Phenomena, *provides an excellent summary of the processes involved in rock slides and avalanches, and gives detailed descriptions of major slides, both historic and prehistoric. Volume 2 focuses on engineering problems.*

WASHBURN, A. LINCOLN *Geocryology,* John Wiley & Sons, New York, 1980.
A very fine review of periglacial processes and environments. Handsomely illustrated.

WILLIAMS, PETER J. *Pipelines and Permafrost: Physical Geography and Development in the Circumpolar North.* Longman, New York, 1979.
An interesting discussion of pipelines in the arctic and subarctic regions that also contains a summary discussion of permanently frozen ground and the processes associated with it.

ZÁRUBA, QUIDO, AND VOJTĚCH MENCL *Landslides and Their Control,* 2nd ed., Elsevier Publishing Co., Amsterdam, 1982.
This volume gives many examples of mass movement, most of them drawn from Central Europe.

The Falls of the Niagara (American Falls in middle distance and Canadian Falls beyond) mark the head of a gorge that the Niagara River continues to push upstream as it flows from Lake Erie to Lake Ontario.

RUNNING WATER

"All the rivers run into the sea; yet the sea is not full; unto the place from whence the rivers come, thither they return again."

Ecclesiastes 1:7

Through the eons of Earth history agents of erosion have been working constantly to reduce the landmasses to the level of the seas. Of these agents *running water* is the most important, for year after year the streams of the Earth move staggering amounts of debris and dissolved material through their valleys (see chapter-opening figure) to the great settling basins, the oceans.

In this chapter we first examine the distribution of the Earth's water supply and how it moves through the *hydrologic cycle,* as foreshadowed by the biblical quotation that opens the chapter.

Most water moves over the land in stream channels and follows certain patterns. We will find that a stream can be thought of as having a certain "economy," a dynamic equilibrium that characterizes all earth processes. For a stream this equilibrium involves, among other things, the very important concept of *base level.*

The work of running water includes *transportation* of material, the *deposition* of this material under the proper conditions, and *channel maintenance.* The end product of much of this activity is recorded in most of the world's landscape. Thus *valley shapes, drainage basins, channel patterns,* and *floodplains,* for instance, reflect to a greater or lesser degree the work of running water.

13.1
WORLD DISTRIBUTION OF WATER

The great bulk of the water of the Earth is contained in the world's ocean, the great body of water made up by the interconnection of the Atlantic, Indian, Pacific, and Arctic Oceans. This world ocean represents 97 percent of our planet's water. Less than 3 percent is located on or beneath the surface of the continents, and most of that is in the form of ice. Atmospheric moisture, the source of all our rain and snow, is less than 0.001 percent at any one time. Even less water is contained in the rivers of the world at any given moment (see Appendix C).

It is clear that water moves. We can watch the rain. We can see the continuous flow of rivers to the ocean, streams that obviously need a continuing supply of water. Furthermore, it is clear that the oceans must somehow lose water to make room for the unending supply of water they receive. This movement of water—this circulation—is part of the **hydrologic cycle.**

13.2
HYDROLOGIC CYCLE

We can conceive of two directions in which the hydrologic cycle might move. Water might move from the oceans through the underground to the headwaters of rivers on the land, whence it could be conveyed back to the oceans. Or water might move from the ocean surface as vapor evaporated into the atmosphere and carried landward to fall as precipitation, then be drained back to the oceans by the rivers. Today we know the second direction is the one that water takes: from ocean to atmosphere to rivers to ocean. But our belief has not always been thus.

SOME HISTORY

The identification of the source of stream water is historically recent. Well into the eighteenth century it was the general belief that streams were replenished by springs that drew their water by some complex system through the underground from the oceans. This belief was nourished by the assumption that rainfall was inadequate to account for the flow observed in rivers, for rivers ran continuously even though rain was intermittent. Furthermore, it was generally held that rainwater could not soak into the ground and replenish springs, a belief that seems to owe its origin to the Roman statesman and philosopher Seneca (3 B.C.–A.D. 65), who based his conclusion on observations made while tending his vineyards.

In the seventeenth century three different types of observation laid the base for the modern concept of the hydrologic cycle, which is diagrammatically represented in Figure 13.1. In 1674 Pierre Perrault (1611–1680), a French lawyer and sometime hydrologist, presented the results of his measurements in the upper drainage basin of the Seine River. Over a 3-year span he collected data on the amount of precipitation in this portion of the basin. At the same time he kept track of the amount of water discharged by the river below the portion of the river basin where he had data on precipitation. The results showed that the flow of the

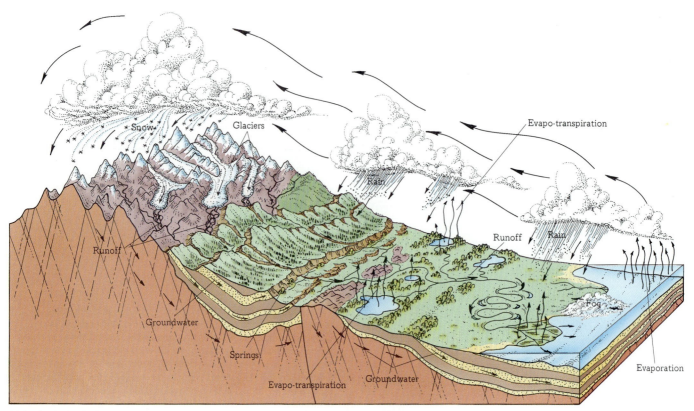

13.1 In the hydrologic cycle water evaporated into the atmosphere reaches the land as rain or snow. Here it may be temporarily stored in glaciers, lakes, or the underground before returning by the rivers to the sea. Or some may be transpired or evaporated directly back into the atmosphere before reaching the sea.

river was surprisingly low compared to the total amount of water available from precipitation. In fact, the annual precipitation was six times the total volume of river flow. There was more than enough rainfall, then, to account for the flow of the Seine in this part of its basin.

At about the same time the French physicist Edmé Mariotte (1620–1684) made more exact studies of discharge in the Seine basin. His publications, which appeared posthumously in 1684, verified Perrault's conclusions. Mariotte further demonstrated, by experimentation at the Paris Observatory, that seepage through the soil was less, but not much less, than the amount of rainfall. He also demonstrated the increase in the flow of springs during rainy weather and the decrease during time of drought. His data proved that the Earth permits penetration of moisture.

In 1693 Edmund Halley (1656–1742), of comet fame, provided data on evaporation in relation to rainfall. He roughly calculated the amount of water discharged annually by rivers into the Mediterranean Sea and added this to the amount that fell directly on the sea's surface. He was then able to compute the approximate amount of water that was being evaporated back into the air each year from the surface of the Mediterranean. Halley found that there was more than enough water being pumped into the atmosphere to feed all the rivers coming into the sea.

HOW MUCH AND HOW FAST?

After water has fallen to the ground as **precipitation** it follows one of several paths in the hydrologic cycle. The greatest amount is taken back into the atmosphere either by **evaporation** directly from open-water bodies or from the soil. An appreciable amount of water is intercepted by plants and returned to the atmosphere by the process of **transpiration** (breathing back). It is difficult to measure the two processes independently, so we generally lump them together as **evapo-transpiration.** A smaller amount, **runoff,** reaches the rivers. Another part of the precipitation may soak into the ground; we call this process **infiltration.** The water that sinks into the ground is called **groundwater** and is the subject of Chapter 14. Almost all groundwater eventually returns to the surface as springs and becomes part of the runoff. We will deal with infiltration by lumping it with runoff. The equation for the hydrologic cycle can then be written as follows:

$$Precipitation = runoff + evapo\text{-}transpiration$$

We have some approximate figures for the United States as a whole (Figure 13.2). Average annual precipitation per unit area is about 75 cm, evapo-transpiration is 52.5 cm, and runoff is calculated as 22.5 cm. If we use these

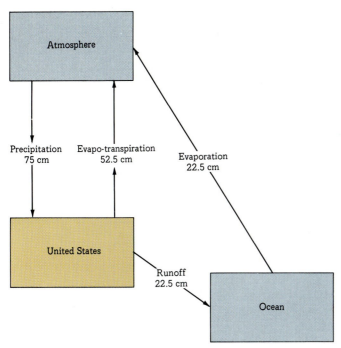

13.2 Water budget for the United States based on an average annual rainfall of 75 cm for the country. Groundwater returned to streams is included in runoff. [Data from the U.S. Geological Survey.]

figures in the equation, it balances. But look again at the equation. Notice that runoff is escaping from the atmosphere-continent system and disappearing into the ocean. Thus there is insufficient moisture being evaporated and transpired from the land back into the atmosphere to support the 75 cm of precipitation. The simple answer: Some of the water evaporated from the oceans is transported back over the land to make up the deficit.

How fast does water move through the system? It would take streams about 40,000 years to replenish the waters of the world ocean. In contrast, the moisture in the air is renewed every 9 or 10 days and stream water about every 2 weeks.

13.3

STREAM FLOW

Consider now the manner in which water in a stream moves. If it moves slowly enough, what we call **laminar flow** occurs. Particles of water move in parallel layers (hence the name *laminar*, from the Latin for "layer"). Each layer moves at a constant velocity, which differs from that of adjacent layers. The layers will separate around obstructions and rejoin on the downstream side, but they will not mix. Seldom will we see laminar flow in streams. If it exists in streams at all, it is in a thin zone near the channel wall.

Very much more important in the behavior of streams is **turbulent flow** (turbulent being derived from the Latin for turmoil), in which, as the velocity of flow in-

creases, the paths of water particles break into the swirls and eddies so familiar to us in ordinary stream flow (Figures 13.3 and 13.4). The change from laminar to turbulent flow is largely a function of increasing velocity. Secondary factors include the decreasing viscosity of the water and the increasing depth and roughness of the channel.

Increasing velocity and roughness

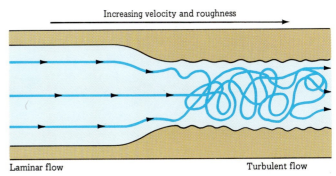

Laminar flow   Turbulent flow

13.3 Diagram showing laminar and turbulent flow of water through a section of pipe. Individual water particles follow paths depicted by the colored lines. In laminar flow the particles follow paths parallel to the containing walls. With increasing velocity and increasing roughness of the confining walls, laminar flow gives way to turbulent flow. The water particles no longer follow straight lines but are deflected into eddies and swirls. Most water flow in streams is turbulent.

13.4 Stream flow is not always as obviously turbulent as it is here in the Isarco River in the Alps near Fortezza, Italy.

VELOCITY, GRADIENT, AND DISCHARGE

The **velocity** of a stream is measured in terms of the distance its water travels in a unit of time. A velocity of 15 cm/s is relatively low, and a velocity of about 625 to 750 cm/s is relatively high.

A stream's velocity is determined by many factors, including the amount of water passing a given point, the roughness of the stream banks, and the **gradient,** or slope, of the stream bed. In general, a stream's gradient decreases from its headwaters toward its mouth; as a result, a stream's longitudinal profile is more or less concave toward the sky. We usually express the gradient of a stream as the vertical distance a stream descends during a fixed distance of horizontal flow. The Mississippi River from Cairo, Illinois, to the mouth of the Red River in Louisiana has a low gradient, for along this stretch the drop varies between 2 and 10 cm/km. On the other hand, the Arkansas River, in its upper reaches through the Rocky Mountains in central Colorado, has a high gradient, for there the drop averages 7.5 m/km. The gradients of other rivers are even higher. The upper 20 km of the Yuba River in California, for example, have an average gradient of 42 m/km; and in the upper 6.5 km of the Uncompahgre River in Colorado, the gradient averages 66 m/km.

The velocity of a stream is checked by the turbulence of its flow, by friction along the banks and bed of its channel, and, to a much smaller extent, by friction with the air above. Therefore, if we were to study a cross section of a stream, we should find that the velocity would vary from point to point. Along a straight stretch of a channel the greatest velocity is achieved toward the center of the stream at, or just below, the surface, as shown in Figure 13.5. Around a curve in the channel, however, the zone of greatest velocity is thrown to the outside of the bend (Figure 13.5).

We then have two opposing forces: the **forward flow** of the water under the influence of gravity and the **friction** developed along the walls and bed of the stream. These are the two forces that create different velocities. Zones of maximum turbulence occur where the different velocities come into closest contact. These zones are very thin because the velocity of the water increases very rapidly as we move into the stream away from its walls and bed; but within these thin zones of great turbulence, a stream shows its highest potential for erosive action (Figures 13.6 and 13.7).

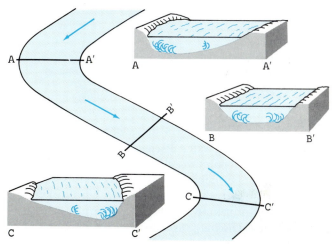

13.6   Zones of maximum turbulence in a stream are shown by the curly symbols in the sections through a river bed. They occur where the change between the two opposing forces—the forward flow and the friction of the stream channel—is most marked. Note that the maximum turbulence along straight stretches of the river is located where the stream banks join the stream floor. On bends the two zones have unequal intensity; the greater turbulence is located on the outside of a curve.

13.7   A small stream meanders across its flood plain near Steamboat Springs, Colorado. Greatest erosion takes place on the outside of the bends, and the result is a steep, cut bank at these places.

13.5   Velocity variations in a stream. Both in plan view and in cross section the velocity is slowest along the stream channel, where the water is slowed by friction. On the surface it is most rapid at the center in straight stretches and toward the outside of a bed where the river curves. Velocity increases upward from the river bottom.

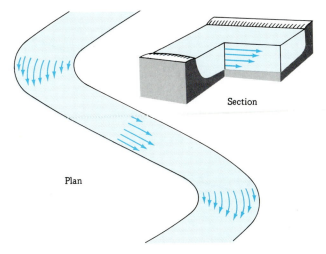

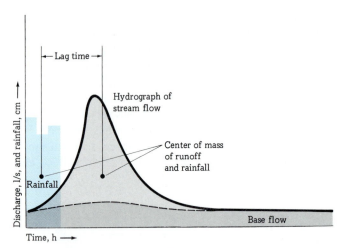

13.8 Hypothetical hydrograph, showing significant characteristics. [After Luna B. Leopold, "Hydrology for Urban Land Planning," *U.S. Geol. Surv. Circ. 554*, p. 3, 1968.]

There is one more term that will be helpful in our discussion of running water: **discharge,** the volume of water that passes a given point in a unit of time. (In this book we shall measure it in cubic meters per second.) Discharge varies not only from one stream to another but also within a single stream from time to time and from place to place along its course. Discharge usually increases downstream as more and more tributaries add their water to the main channel. Spring floods may so greatly increase a stream's discharge that its normally peaceful course becomes a raging torrent.

The Hydrograph   The hydrograph of a river shows the variation of stream discharge with time, as indicated in Figure 13.8. In the example given, we see a period of high flow, peaking in a flood, and a generally low-water stage, which represents that part of stream flow attributable to groundwater, a flow referred to as **base flow.** In addition, the figure indicates a period of rainfall responsible for the high-water, or flood, stage, which occurs after an interval called the **lag time.** The shape of the hydrograph for different streams (and even for different places on the same stream) varies with a number of natural factors, including the infiltration rate, the relief, the geology, and the vegetative cover.

13.4
ECONOMY OF A STREAM

Elsewhere we have seen that Earth processes tend to seek a balance, to establish an equilibrium, and that there is, in the words of James Hutton, "a system of beautiful economy in the works of nature."[1] We found, for example, that weath-

---

[1] James Hutton, *Theory of the Earth*, vol. 2, Edinburgh, 1795.

ering is a response of Earth materials to the new and changing conditions they meet as they are exposed at or near the Earth's surface. On a larger scale we found that the major rock groups—igneous, sedimentary, and metamorphic—reflect certain environments, and that as these environments change, members of one group may be transformed into members of another group. These changes were traced in what we called the *rock cycle* (Figures 1.9 and 1.14). Water running off the land in streams and rivers is no exception to this universal tendency of nature to seek equilibrium.

ADJUSTMENTS OF DISCHARGE, VELOCITY, AND CHANNEL

Just a casual glance tells us that the behavior of a river during its flood stage is very different from its behavior during its low-water stage. For one thing, a river carries more water and moves more swiftly in flood time. Furthermore, the river is generally wider during flood; its level is higher; and we should guess, even without measuring, that it is also deeper. We can relate the discharge of a river to its width, depth, and velocity:

Discharge (m^3/s) = channel width (m) × channel depth (m) × water velocity (m/s)

13.9   As the discharge of a stream increases at a given gauging station, so do its velocity, width, and depth. They increase in an orderly fashion, as shown by these graphs based on data from a gauging station in the Cheyenne River near Eagle Butte, South Dakota. [After Luna B. Leopold and Thomas Maddock, "The Hydraulic Geometry of Stream Channels and Some Physiographic Implications," *U.S. Geol. Surv. Prof. Paper 252*, p. 5, 1953.]

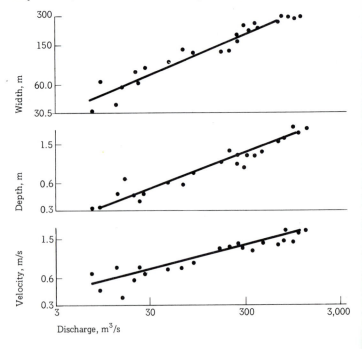

In other words, if the discharge at a given point along a river increases, then the width, depth, velocity, or some combination of these factors must also increase. We now know that variations in width, depth, and velocity are neither random nor unpredictable. In most streams if the discharge increases, then the width, depth, and velocity each increase at a definite rate. The stream maintains a balance between the amount of water it carries, on the one hand, and its depth, width, and velocity, on the other, as shown in Figure 13.9.

Let us turn now from the behavior of a stream at a single locality to the changes that take place along its entire length. From our own observation we know that the discharge of a stream increases downstream as more and more tributaries contribute water to its main channel. We also know that the width and depth increase as we travel downstream. But if we were to go beyond casual observation and gather accurate data on the width, depth, velocity, and discharge of a stream from its headwaters to its mouth for a particular stage of flow — for example, flood or low water — we should find again that the changes follow a definite pattern (Figure 13.10) and that depth and width increase downstream as the discharge increases. And, surprisingly enough, we should also find that the stream's **velocity** increases toward its mouth. This is contrary to our expectations; for we know that the gradients are higher upstream, which suggests that the velocities in the steeper headwater areas should also be higher. But the explanation for this seeming anomaly is simple: To handle the greater discharge downstream, a stream deepens and widens its channel and also accommodates some of the increased discharge by an increase in its velocity.

Floods   When the discharge of a stream exceeds the capacity of its channel, water must rise over the banks of the channel and flood the adjacent low-lying land (Figure 13.11). Floods can be catastrophic in human terms, but even so, we must consider them as part of the natural behavior of streams. A study of long-term records of streams shows that there is some method to the apparent madness of streams. For instance, a discharge that just fills a stream channel, so that it is at the **bankfull stage,** occurs every 1.5 to 2 years. When we look at greater volumes of flow, we find that the frequency of floods, which exceed the bankfull stage, can be statistically anticipated, as discussed in Box 13.1 and shown in Figure B13.1.1.

13.10   Stream velocity and depth and width of a channel increase as the discharge of a stream increases downstream. Measurements in this example were taken at mean annual discharge along a section of the Mississippi–Missouri river system. [After Luna B. Leopold and Thomas Maddock, "The Hydraulic Geometry of Stream Channels and Some Physiographic Implications," *U.S. Geol. Surv. Prof. Paper 252*, p. 13, 1953.]

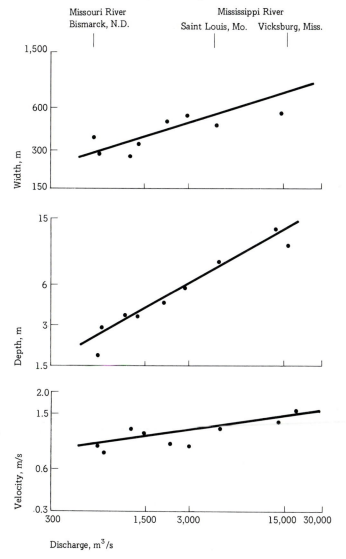

13.11   One of the results of the flooding of the Willamette River on December 23, 1964, north of Salem, Oregon. [A. O. Waananen, U.S. Geological Survey.]

BOX 13.1   Estimating Flood Recurrence

We can estimate how often a flood of a given size might occur by using a simple statistical method. It involves a stream-discharge record of at least 10 years. The first step is to list the maximum flood discharge for each year. The second is to rank each maximum annual flood discharge by size from largest to smallest, designating the largest as 1. Next we compute the flood-recurrence interval by using the following formula:

$$R = \frac{(N+1)}{M}$$

where R equals the flood recurrence in years, N equals the number of years of the record, and M equals the magnitude or rank of each maximum annual flood.

The results from such an analysis for a 93-year record on the Cumberland River at Nashville, Tennessee, are plotted on semilogarithmic paper as shown in Figure B13.1.1. The largest flow, 5,750 m³/s, is entered at 94 years, (93 + 1)/1 in our formula. The size of the second-largest flood, 5,570 m³/s, is plotted at 47 years, (93 + 1)/2, because two floods of this size or larger have occurred in the last 93 years. The third flood is shown as 94/3, or 31 years plus.

On the basis of a record such as that graphed in the figure, we can speak, for instance, of 10-year floods or 50-year floods. Because the graph shows points

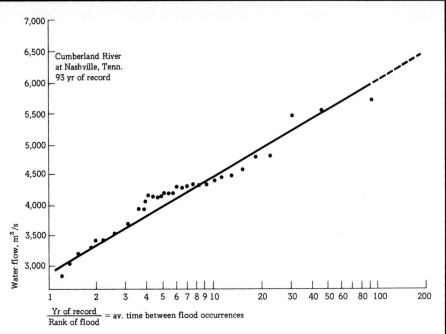

B13.1.1   Average interval between floods of various sizes on the Cumberland River at Nashville, Tennessee, plotted from data in *U.S. Geol. Surv. Water Supply Paper 771.* The vertical scale is an ordinary linear scale, and the horizontal scale is known as a logarithmic scale.

on a roughly straight line, we may extend the line to the upper right to estimate the size of floods to be expected once in 100 years or once in 200 years. We could even estimate the volume of a flood to be expected once in 1,000 years, although this is probably an unwarranted extrapolation of our data.

Flood-recurrence analysis is a useful and defensible technique. But remember, we don't know precisely when a flood of a given size will occur. The historical record merely tells us to expect a flood of a given size sometime within a given span of years.

That floods recur and that the recurrence interval of a flood of a given size is predictable should assure us that floods are a natural, expectable phenomenon. We should expect them every so often. And indeed we view a river's floodplain as part of the river's domain, useful to the river in flood times and certain to be claimed by the river from time to time. Therefore, if we use the floodplain, we must remember that it is only on loan to us. Whatever use we make of it ought to be as compatible as possible with the use to which the river is sure to put it. Experience should teach us not only that streams will top their banks from time to time but also that even the most energetic flood-control plans (including the use of dams, levees, dredging, and modification of channels) will sometimes fail. The problem is partly

an economic one: Against what magnitude flood shall we try to protect the floodplain? The 10-year flood? The 100-year flood? The 500-year flood? Obviously the larger the flood we attempt to defend against, the greater the expense. At what point does the expense of protection outweigh the gain achieved? This is clearly a question of public policy. An alternative to protection by engineering works is to put the floodplain to a use compatible with flooding, such as agriculture or a park system. This practice may in many instances be more appropriate than tampering with the river system.

Plans to modify the flood focus on modifying the hydrograph by lowering the peak of the flood, increasing the lag time between the precipitation and flood crest, and

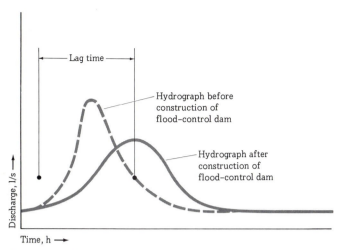

13.12 Hypothetical hydrograph showing the effect on stream flow of a flood-control dam upstream from the hydrograph station.

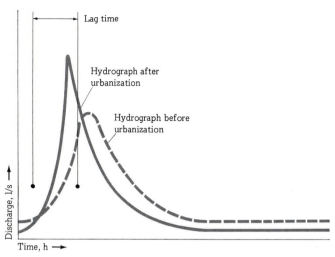

13.13 Hypothetical hydrograph showing the effect of urbanization on stream flow. [After Luna B. Leopold, "Hydrology for Urban Land Planning," *U.S. Geol. Surv. Circ. 554*, p. 3, 1968.]

spreading the flood flow over a longer period of time. A flood-control dam, then, is designed to store water during periods of high runoff and to let the excess water out slowly to downstream areas. The result is to reduce the flood crest downstream and spread the discharge of flood waters over a longer time, as suggested in Figure 13.12.

Urbanization and Suburbanization   Building of cities and their suburbs affects stream flow and changes the hydrograph, although in the opposite direction from that of flood-control dams. Agglomerations of buildings with their associated roads, streets, sidewalks, and paved parking areas achieve the following effects: (1) The amount of water sinking into the underground is curtailed in proportion to the amount of area sealed off by surface veneers of buildings and pavements, and even the areas of ground left open tend to be less permeable because of their compaction by extensive human activity. (2) The rate and amount of surface runoff increase. (3) The groundwater level falls because of the decreased infiltration. The effect on stream flow is shown in Figure 13.13 and can be characterized as follows: (1) The time lag between precipitation and flood peak is shortened because water is not slowed on its way to the stream channel by a vegetative cover nor does any appreciable amount sink into the ground. (2) The peak is higher because the stream must carry more water in a shorter period of time, and also because channels are cleaned and straightened by storm drains. (3) The base flow is generally lower because of a decreased supply of groundwater on which the base flow depends. The net result of these changes is a **flashy stream,** one that has a low base flow and a high, short flood peak.

We have found that the stream channel is adjusted to the stream discharge, and this was expressed as discharge = width × depth × velocity. Observation shows that a stream rises to the limit of its bank on the average of once every 1.5 to 2 years. As noted earlier, this stage of flow

is called the **bankfull stage.** The stream channel is adjusted to handle the volume of flow that occurs every 1.5 to 2 years. When flow exceeds this amount, flooding begins. Studies reported by Luna Leopold of the United States Geological Survey indicate that with an increase of urbanization, the bankfull stage of flow increases. If an area is 50 percent urbanized, experience indicates that the bankfull stage of discharge increases by a factor of 2.7 over that of the same stream in an unurbanized condition. An example is shown in Table 13.1.

In the table the stream has increased the depth and width of its channel as the discharge for bankfull stage has increased. The bankfull stage has increased because of the increase in runoff rates caused by urbanizing one-half of the drainage area. This erosion of the channel produces sediments that are moved downstream, where they may be deposited.

Changes in Vegetative Cover   Changes in vegetative cover can effect changes in the hydrograph. In a general way, a decrease in the vegetative cover acts in the same direction as does an increase in urbanization.

Studies of the effects of selective cutting of forest on stream flow have been carried on by the United States Forest Service in the Fernow Experimental Forest near Par-

TABLE 13.1
Stream Flow and Urbanization

	Before urbanization	After 50% urbanization
Discharge at bankfull stage	1.5 m³/s	4.0 m³/s
Velocity	0.75 m/s	0.75 m/s
Depth of channel	0.5 m	0.9 m
Width of channel	4.0 m	6.0 m
Area of drainage basin	2.5 km²	2.5 km²

sons, West Virginia. The results, not entirely unexpected, are several: The removal of the trees increases the flow of streams during the growing season chiefly because the water ordinarily transpired into the atmosphere by trees makes its way to the streams instead. The increase in stream flow is approximately proportional to the percentage of tree cover removed, the greatest increase—over 100 percent—occurring with complete cutting of the forest cover. Concurrently with the increase of stream flow, the flashiness of streams increases. As the tree cover replaces itself, the flow decreases and the flashiness of streams declines. The return to stream-flow conditions similar to those before cutting is projected to be about 35 years.

BASE LEVEL OF A STREAM

Base level is a key concept in the study of activity of streams. The **base level** is defined as the lowest point to which a stream can erode its channel. Anything that prohibits the stream from lowering its channel serves to create a base level. For example, the velocity of a stream is checked when it enters the standing, quiet waters of a lake. Consequently the stream loses its ability to erode, and it cannot cut below

the level of the lake. The lake's control over the stream is actually effective along the entire course upstream because no part of the stream can erode beneath the level of the lake—at least not until the lake has been destroyed. In a geological sense, every lake is temporary, and therefore after the lake has been destroyed (perhaps by the downcutting of its outlet), it will no longer control the stream's base level, and the stream will be free to continue its downward erosion. Because of its impermanence, the base level formed by a lake is referred to as **temporary.** But even after a stream has been freed from one temporary base level, it will be controlled by others farther downstream. Then, too, its erosive power is always influenced by the ocean, which is the **ultimate** base level. Yet, as we shall see in Chapter 16, the ocean itself is subject to changes in level, so even the so-called ultimate base level is not fixed.

The base level of a stream may be controlled not only by lakes but also by layers of resistant rock at the lip of a waterfall and the level of the main stream into which a tributary drains (see Figure 13.14).

Adjustment to Changing Base Level   We have defined base level as the lowest level to which a stream can erode its channel. If for some reason the base level is either

13.14   Base level for a stream may be determined by natural and artificial lakes, by a resistant rock stratum, by the point at which a tributary stream enters a main stream, and by the ocean. Of these, the ocean is considered the ultimate base level; all the others are temporary base levels.

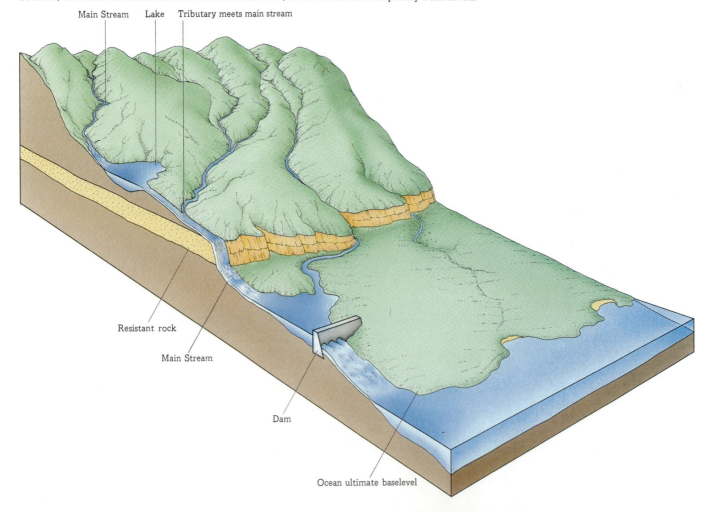

Main Stream   Lake   Tributary meets main stream

Resistant rock

Main Stream

Dam

Ocean ultimate baselevel

raised or lowered, the stream will adjust the level of its channel to adapt to the new situation.

Let us see what happens when we **raise** the base level of a stream by building a dam and creating a lake across its course. The level of the lake serves as a new base level, and the gradient of the stream above the dam is now less steep than it was originally. As a result, the stream's velocity is reduced. Because the stream can no longer carry all the material supplied to it, it begins to deposit sediments at the point where it enters the lake. As time passes, a new river channel is formed with approximately the same slope as the original channel but at a higher level (Figure 13.15).

What happens when we **lower** the base level by removing the dam and hence the lake? The river will now cut down through the sediments it deposited when the lake still existed. In a short time the profile of the channel will be essentially the same as it was before we began to tamper with the stream.

In general, then, we may say that a stream adjusts itself to a rise in base level by building up its channel through sedimentation and that it adjusts to a fall in base level by eroding its channel downward.

13.5

WORK OF RUNNING WATER

The water that flows in streams does several jobs: (1) It **transports** debris, (2) it **erodes** the river channel deeper into the land, and (3) it **deposits** sediments at various points along the valley or delivers them to lakes or oceans. Running water may help to create a chasm like that of the Grand Canyon of the Colorado; or in flood it may spread mud and sand across vast expanses of valley flats; or it may build deltas, as at the mouths of the Nile and Mississippi.

The kind and extent of these activities depend on the kinetic energy of the stream, and this in turn depends on the amount of water in the stream and the gradient of the channel. A stream expends its energy in several ways. By far the greatest part is used up in the friction of the water with the stream channel and in the friction of water with water in the turbulent eddies we discussed above. Relatively little of the stream's energy remains to erode and transport material. Deposition takes place when energy decreases and the stream can no longer move the material it has been carrying.

TRANSPORTATION

The material that a stream picks up directly from its own channel—or that is supplied to it by slope wash, tributaries, or mass movement—is moved downstream toward its eventual goal, the ocean. The amount of material that a stream carries at any one time, which is called its **load,** is usually less than its **capacity**—that is, the total amount it is

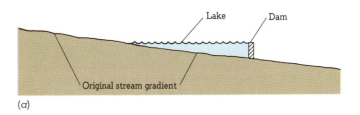

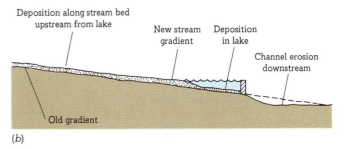

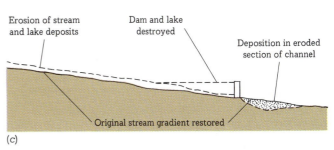

13.15 A stream adjusts its channel to the changing base level. Construction of a dam across a stream raises its base level, imposes a lower velocity on the stream above the dam, and thus causes deposition in this section of the channel. Failure of the dam lowers the base level, increases the velocity, and causes erosion of the previously deposited sediments.

capable of carrying under any given set of conditions. The maximum size of particle that a stream can move measures the **competence** of a stream.

There are three ways in which a stream can transport material: (1) solution, (2) suspension, and (3) bed load.

Solution   In nature no water is completely pure. We have already seen that when water falls and filters down into the ground, it dissolves some of the soil's compounds. Then the water may seep down through openings, pores, and crevices in the bedrock and dissolve additional matter as it moves along. Much of this water eventually finds its way to streams at lower levels. The amount of dissolved matter contained in water varies with climate, season, and geologic setting, and is measured in terms of parts of dissolved matter per million parts of water. Sometimes the amount of dissolved material exceeds 1,000 ppm, but usually it is much less. By far the most common compounds in solution in running water, particularly in arid regions, are calcium and magnesium carbonates. In addition, streams carry small amounts of chlorides, nitrates, sulfates, and silica, with perhaps a trace of potassium. It has been estimated that the total load of dissolved material delivered to the seas every year by the streams of the United States is nearly 300 million t. The rivers of the world average an estimated 115 to 120 ppm of dissolved matter, which means that annually they carry to the sea about 3.9 billion t.

Suspension Particles of solid matter that are swept along in the turbulent current of a stream are said to be in **suspension.** This process of transportation is controlled by two factors: the turbulence of the water and a characteristic known as **terminal velocity of fall** of each individual grain. Terminal velocity is the constant rate of fall that a grain eventually attains when the acceleration caused by gravity is balanced by the resistance of the fluid through which the grain is falling. In this example the fluid is water. If we drop a grain of sand into a quiet pond, it will settle toward the bottom at an ever-increasing rate until the friction of the water on the grain just balances this rate of increase. Thereafter it will settle at a constant rate, its terminal velocity. If we can set up a force that will equal or exceed the terminal velocity of the grain, we can succeed in keeping it in suspension. Turbulent water supplies such a force. The eddies of turbulent water move in a series of orbits, and a grain caught in these eddies will be buoyed up, or held in suspension, as long as the velocity of the turbulent water is equal to, or greater than, the terminal velocity of the grain.

Terminal velocity increases with the size of the particle if its general shape and density remain the same. The bigger a particle, the more turbulent the flow needed to keep it in suspension. And because turbulence increases when the velocity of stream flow increases, it follows that the greatest amount of material is moved during flood time, when velocities and turbulence are highest. The graph in Figure 13.16 shows how the suspended load of a stream increases as the discharge increases. In just a few hours or a few days during flood time, a stream transports more material than it does during the much longer periods of low or normal flow. Observations of the area drained by Coon Creek, at Coon Valley, Wisconsin, over a period of 450 days showed that 90 percent of the stream's total suspended load was carried during an interval of 10 days, slightly over 2 percent of total time.

Silt and clay-sized particles are distributed fairly evenly through the depth of a stream, but coarser particles in the sand-sized range are carried in greater amounts lower down in the current, in the zone of greatest turbulence.

Bed Load Materials in movement along a stream bottom constitute the stream's **bed load,** in contrast to its suspended load and solution load. Because it is difficult to observe and measure the movement of the bed load, we have few data on the subject. Measurements on the Niobrara River near Cody, Nebraska, however, have shown that at discharges between about 6 and 30 m³/s, the bed load averaged about 50 percent of the total load. Particles in the bed load move along in three ways: saltation, rolling, and sliding.

The term **saltation** has nothing to do with salt. It is derived from the Latin *saltare,* "to jump." A particle moving by saltation jumps from one point on the stream bed to another. First it is shaken loose from the bed of the stream by a surge of turbulent water and carried upward into the forward-moving current. When the particle reaches a level

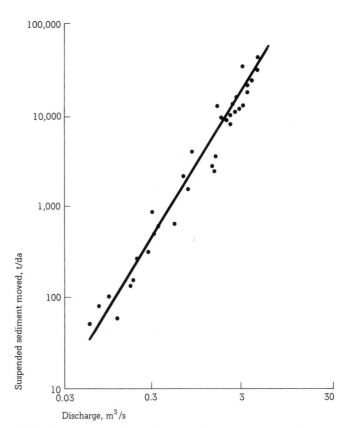

13.16 The suspended load of a stream increases very rapidly during floods, as illustrated by this graph based on measurements in the Rio Puerco near Cabezon, New Mexico. [After Luna B. Leopold and John P. Miller, "Ephemeral Streams — Hydraulic Factors and Their Relation to the Drainage Net," *U.S. Geol. Surv. Prof. Paper 282-A,* p. 11, 1956.]

in the stream at which its terminal velocity is greater than the upward motion of the turbulent water, it drops back to the bed of the stream. In this process of rising and falling, the particle is carried forward by the stream to a new resting place downstream from its original position. Movement by saltation, then, is transitional between movement by continuous suspension and movement by continuous contact between the particle and the floor of the channel. In the latter situation particles are too heavy to be picked up, even momentarily, by water currents, but they may be pushed along the stream bed and, depending on their shape, move forward by either *rolling* or *sliding.*

EROSION

People have not always been convinced that most of the world's stream valleys were fashioned by the streams that flow in them, and by the processes that streams encouraged. Today, however, we know enough about running water to confirm this. In the next few paragraphs we look at the various ways a stream may remove material from its channel and its banks. But you may wish, as you continue this chapter, to keep in mind the larger question and consider what added evidence we might cite to demonstrate that a stream is responsible for its own valley.

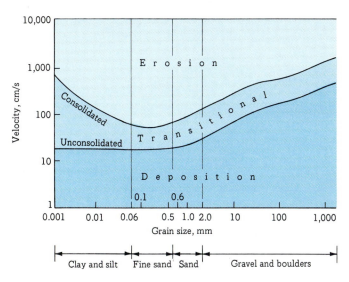

13.17 The central band in this diagram shows the velocity range at which turbulent water will lift particles of differing size off the stream bed. The width is affected by the shape, density, and consolidation of the particles. [After Ake Sundborg, "The River Klarälven, a Study in Fluvial Processes," *Geografis. Ann.*, vol. 38, p. 197, 1956.]

13.18 The solid material carried by a stream may erode the bedrock of the stream channel. Abrasion has fashioned an irregular channel in the bedrock of this small stream north of Rome, Italy.

Direct Lifting   In turbulent flow, as we have seen, water travels along paths that are not parallel to the bed. The water eddies and whirls, and if an eddy is powerful enough, it dislodges particles from the stream channel and lifts them into the stream. Whether or not this will happen in a given situation depends on a number of variables that are difficult to measure, but if we assume that the bed of a stream is composed of particles of uniform size, then the graph in Figure 13.17 gives us the approximate stream velocities that are needed to erode particles of various sizes, such as clay, silt, sand, granules, and pebbles. A stream bed composed of fine-sized sand grains, for example, can be eroded by a stream with a velocity of between 18 and 50 cm/s, depending on how firmly the sand grains are packed. As the fragments become larger and larger, ranging from coarse sand to granules to pebbles, increasingly higher velocities are required to move them, as we should expect.

But what we might *not* expect is that as the particles become smaller than about 0.06 mm in diameter, they do not become more easily picked up by the stream. In fact, if the clay and silt are firmly consolidated, then increased velocities will be needed to erode the particles. The reason for this is generally that the smaller the particles, the more firmly packed the deposit is and thus the more resistant to erosion. Moreover, the individual particles may be so small that they do not project sufficiently high into the stream to be swept up by the turbulent water.

Abrasion, Impact, and Solution   The solid particles carried by a stream may themselves act as erosive agents, for they are capable of abrading (wearing down) the bedrock itself or larger fragments in the bed of the stream. When the bedrock is worn by abrasion, it usually develops a series of smooth, curving surfaces, either convex or concave, or a series of holes—potholes—drilled by pebbles swirled in

the turbulent stream (Figures 13.18 and 5.14). Individual cobbles or pebbles on a stream bottom are sometimes moved and rolled about by the force of the current, and as they rub together, they become both rounder and smoother.

The impact of large particles against the bedrock or against other particles knocks off fragments, which are added to the load of the stream.

Some erosion also results from the solution of channel debris and bedrock in the water of the stream. Most of the dissolved matter carried by a stream, however, is probably contributed by the underground water that drains into it.

DEPOSITION

As soon as the velocity of a stream falls below the point necessary to hold material in suspension, the stream begins to deposit its suspended load. Deposition is a selective process. First, the coarsest material is dropped; then, as the velocity (and hence the energy) continues to slacken, finer and finer material settles out. We shall reconsider stream deposits later.

13.6
FEATURES OF STREAM VALLEYS

CROSS-VALLEY PROFILES

Earlier in this chapter we mentioned the longitudinal profile of a stream. Now let us turn to a discussion of the **cross-valley profile,** that is, the profile of a cross section at right angles to the trend of the stream's valley. In Figure

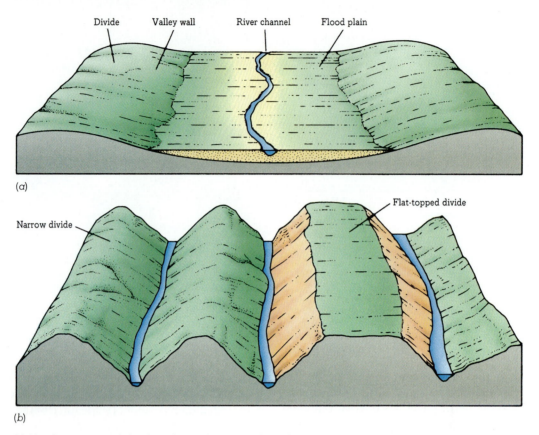

Divide   Valley wall   River channel   Flood plain

(a)

Flat-topped divide

Narrow divide

(b)

13.19 Cross-sectional sketches of typical stream valleys. The major features of valleys in cross section include divides, valley walls, river channel, and in some instances a floodplain. Divides may be flat-topped, broadly rounded, or narrow.

13.19(a) notice that the channel of the river runs along a broad, relatively flat **floodplain.** During flood time, when the channel can no longer accommodate the increased discharge, the stream overflows its banks and floods this area. On either side of the floodplain **valley walls** rise to crests called **divides,** separations between a valley and other valleys on either side of it. In Figure 13.19(b) there is no floodplain because the valley walls descend directly to the banks of the river. This diagram also illustrates two different shapes of divide. One is broad and flat; the other is narrow, almost knife-edged. Both are in contrast to the broadly convex divides shown in Figure 13.19(a).

DRAINAGE BASINS, NETWORKS, AND PATTERNS

A **drainage basin** is the entire area from which a stream and its tributaries receive their water. The Mississippi River and its tributaries drain a tremendous section of the central United States, reaching from the Rockies to the Appalachian Mountains, and each tributary of the Mississippi has its own drainage area, which forms a part of the larger basin. Every stream, even the smallest brook, has its own drainage basin (Figure 13.20). These basins are shaped differently from stream to stream but are characteristically pear-shaped, with the main stream emerging from the narrow end (Figure 13.21).

Individual streams and their valleys are joined together into **networks.** In any single network the streams prove to have a definite geometric relationship in a way first detailed in 1945 by the United States hydraulic engineer Robert Horton. We can devise a demonstration of this relationship by first ranking the streams in a hierarchy. This ranking, shown in Figure 13.22, lists small headwater streams without tributaries as belonging to the first order in the hierarchy. Two or more first-order streams join to form a second-order stream; then two or more second-order streams join to form a third-order stream; and so on. In other words, a stream segment of any given order is formed by the junction of at least two stream segments of the next lower order. The main segment of the system always has the highest-order number in the network, and the number of this stream is assigned to describe the drainage basin. Thus the main stream in the basin shown in Figure 13.22 is a fourth-order stream, and the basin is a fourth-order basin.

It is apparent from Figure 13.22 that the number of stream segments of different order decreases with increasing order. When we plot the number of streams of a given order against order, the points define a straight line on semilogarithmic paper; see Figure 13.23(a). Likewise, if we plot order against the average length of streams of a given order or plot order against areas of drainage basins of a particular order, we find well-defined relationships; see Figure 13.23(b) and (c). Using these relationships, we can de-

13.20 Big or small, wet or dry, every stream channel has its own drainage basin. Here, northwest of St. Anthony, Idaho, these upland slopes and channels form a network of drainage basins. [Austin Post, University of Washington.]

13.21 The stream's drainage basin is the area from which the stream and its tributaries receive water. This basin displays a pattern reminiscent of a tree leaf and its veins.

13.22 Method of designating stream orders. [After A. N. Strahler, "Quantitative Geomorphology in Erosional Landscapes," *Proc. Intern. Geol. Congr.*, sec. 13, pt. 3, p. 344, 1954.]

Boundaries of secondary drainage basins

Boundary of main drainage basin

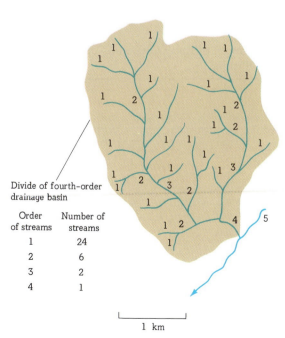

Divide of fourth-order drainage basin

Order of streams	Number of streams
1	24
2	6
3	2
4	1

1 km

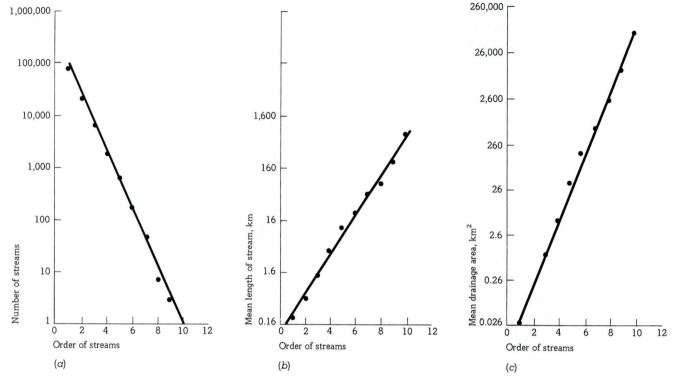

13.23 Relation between stream order and the number of streams, the mean length of streams, and the mean drainage area of streams of a location in central Pennsylvania. The system used to designate stream orders is that proposed by Robert Horton in 1945 and differs slightly from that illustrated in Figure 13.22. [After Lucien Brush, Jr., *U.S. Geol. Surv. Prof. Paper 282-F*, 1961.]

scribe the river channels and basins of the United States as in Table 13.2.

We have seen that when streams are arranged in a hierarchical fashion, the arrangement bears a definite relation to the number and lengths of the stream segments in each order and to the size of the basins of the stream segments of different orders. Beyond this, we find that streams

develop patterns not so easily quantified as are the relationships of stream segments and stream basins based on hierarchical arrangements. We describe some of these below.

The overall pattern developed by a system of streams and tributaries depends partly on the nature of the underlying rocks and partly on the history of the streams (Figure 13.24). Almost all streams follow a branching pattern in the

TABLE 13.2
Number and Length of River Channels of Various Sizes in the United States[a]

Order	Number	Av. length, km	Total length, km	Mean drainage area, incl. tributaries, km²	River representative of size
1[b]	1,570,000	1.6	2,512,000	2.6	
2	350,000	3.7	1,295,000	12	
3	80,000	8.5	680,000	59	
4	18,000	19	342,000	283	
5	4,200	45	189,000	1,348	Charles
6	950	102	97,000	6,400	Raritan
7	200	219	44,000	30,300	Allegheny
8	41	541	22,000	144,000	Gila
9	8	1,243	10,000	683,000	Columbia
10	1	2,880	2,880	3,238,000	Mississippi

[a] From Luna B. Leopold, "Rivers," *Am. Sci.*, vol. 50, p. 512, 1962.
[b] The size of the order 1 channel depends on the scale of the maps used; these order numbers are based on the determination of the smallest order using maps of scale 1 : 62,500.

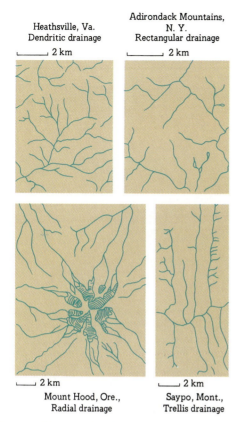

Heathsville, Va.
Dendritic drainage

⊢——⊣ 2 km

Adirondack Mountains,
N. Y.
Rectangular drainage

⊢——⊣ 2 km

⊢——⊣ 2 km
Mount Hood, Ore.,
Radial drainage

⊢——⊣ 2 km
Saypo, Mont.,
Trellis drainage

13.24 The overall pattern developed by a stream system depends in part on the nature of the bedrock and in part on the history of the area. See the text for discussion.

sense that they receive tributaries, and the tributaries, in turn, are joined by still smaller tributaries; but the manner of branching varies widely.

A stream that resembles the branching habit of a maple, oak, or similar deciduous tree is called **dendritic,** "treelike." A dendritic pattern develops when the underly-

ing bedrock is uniform in its resistance to erosion and exercises no control over the direction of valley growth. This situation occurs when the bedrock is composed either of flat-lying sedimentary rocks or of massive igneous or metamorphic rocks. The streams can cut as easily in one place as another; thus the dendritic pattern is, in a sense, the result of the random orientation of the streams.

Another type of stream pattern is **radial:** Streams radiate outward in all directions from a high central zone. Such a pattern is likely to develop on the flanks of a newly formed volcano where the streams and their valleys radiate outward and downward from various points around the cone.

A **rectangular** pattern occurs when the underlying bedrock is crisscrossed by fractures that form zones of weakness particularly vulnerable to erosion. The master stream and its tributaries then follow courses marked by nearly right-angle bends.

Some streams, particularly in a belt of the Appalachian Mountains running from New York to Alabama, follow what is known as a **trellis** pattern. This pattern, like the rectangular one, is caused by zones in the bedrock that differ in their resistance to erosion. The trellis pattern usually, though not always, indicates that the region is underlain by alternate bands of resistant and nonresistant rock.

Water Gaps   Particularly intriguing features of some valleys are short, narrow segments walled by steep, rocky slopes or cliffs. These are called **water gaps.** The river, in effect, flows through a narrow notch in a ridge or mountain that lies across its course, as shown in Figure 13.25.

Here is one way water gaps can form: Picture an area of hills and valleys carved by differential erosion from rocks of varying resistance. Then imagine sediments covering this landscape so that the valleys and hills are buried beneath the debris. Any streams that flow over the surface of this cover will establish their own courses across the buried hills and

13.25 The notch in the ridge on the skyline is the Delaware Water Gap. It was cut by the Delaware river which here flows along the New Jersey – Pennsylvania border.

BOX 13.2   Captured and Beheaded by Pirate

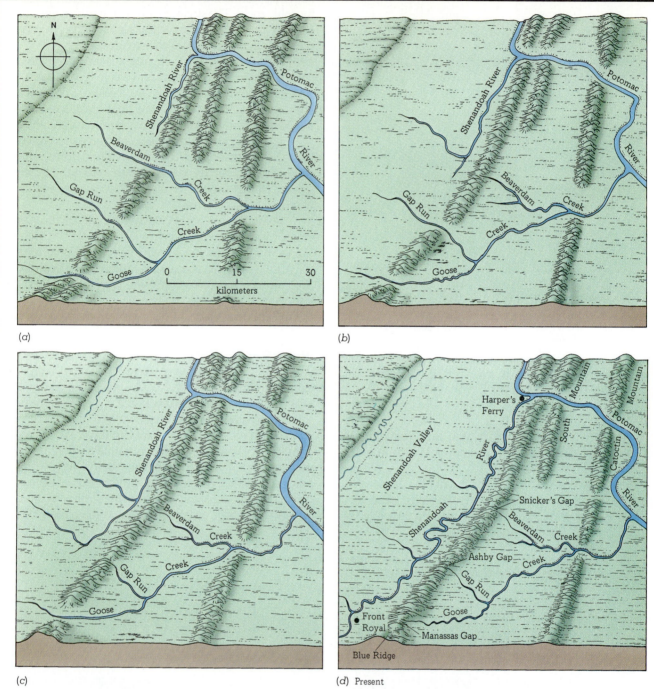

(a)

(b)

(c)

(d) Present

B13.2.1   The Shenandoah River has expanded its drainage basin at the expense of several other streams through a process of stream piracy. Along the way some old water gaps have become wind gaps.

Though our box title sounds like a sensational newspaper headline, it describes a geological process known as **stream piracy,** in which one stream captures a part of an adjacent stream and its drainage basin. Here is an example from the Appalachian Mountains of Virginia.

The Potomac River cuts through the Blue Ridge in a water gap where Virginia, West Virginia, and Maryland meet near Harpers Ferry. Before it does so, it is joined by a large tributary, the Shenandoah River, flowing northeastward. At some time in the past the Shenandoah was a very much shorter

river. Then other streams to the south of Harpers Ferry also crossed the Blue Ridge in their water gaps, as did the Potomac River of the time (see Figure B13.2.1). The Shenandoah flowed parallel to the strike of the rocks within a belt of nonresistant limestone, dolomite, and shale. At the same time the other streams flowed across the general strike of the rocks, which included the resistant rocks of the Blue Ridge in which they had fashioned their water gaps. This set the stage for piracy by the Shenandoah.

The Shenandoah's channel lay in the easily eroded belt of nonresistant rock; the channels of the streams destined for capture crossed the resistant rocks of the Blue Ridge. These rocks served as a base level for the upper parts of the drainages of Beaverdam Creek, Gap Run, and Goose Creek. The ancestral Shenandoah was able first to lower its valley floor below that of the upper Beaverdam, and then to erode head-ward until it intercepted the headwaters of the Beaverdam and diverted them into its own drainage. We can say, therefore, that the **pirate stream,** the Shenandoah, has **captured** the upper drainage of the Beaverdam and that the remaining drainage of the Beaverdam has been **beheaded.** This process was repeated as the Shenandoah captured, in succession, the headwaters of Gap Run and Goose Creek.

The water gaps that originally carried the channels of the three drainages across the Blue Ridge were abandoned during this process of piracy. Now they stand as gaps along the Blue Ridge and are known locally, from northeast to southwest, as Snickers Gap, Ashby Gap, and Manassas Gap. These had previously been the water gaps for the Beaverdam, Gap Run, and Goose Creek. Inasmuch as today wind flows through these abandoned water gaps they are known as **wind gaps** (Figure B13.2.1).

From the human point of view, the three gaps carry more than just wind. Four-lane highways convey traffic across the Blue Ridge from and to the Shenandoah valley: Virginia State Highway 7 (Snickers Gap), U.S. Highways 17 and 50 (Ashby Gap); and Interstate 66 (Manassas Gap).

---

valleys. When these streams erode downward, some of them may encounter an old ridge crest. If such a stream has sufficient erosional energy, it may cut down through the resistant rock of the hill. The course of the stream is thus superimposed across the old hill, and we call the stream a **superimposed stream.** Differential erosion may excavate the sedimentary fill from the old valleys, but the main superimposed stream flows through the hill in a new, narrow gorge, or water gap. One example of such a gap is that followed by the Big Horn River through the northern end of the Big Horn Mountains in Montana; another is that followed by the Wind River across the Owl Creek Mountains in Wyoming (Figure 13.26).

Stream Piracy   An intriguing history that is recorded by some streams results from the enlargement of one stream valley at the expense of another. An example is given in Box 13.2.

ENLARGEMENT OF VALLEYS

We cannot say with assurance how running water first fashioned the great valleys and drainage basins of the continents because the record has been lost in time. But we do know that certain processes are now at work in widening and deepening valleys, and it seems safe to assume that they also operated in the past.

13.26   The Wind River has been superimposed onto the Owl Creek Mountains of Wyoming and has cut the Wind River Canyon.

13.27 This gully in Wisconsin has been formed by the downward cutting of the stream combined with the slope wash and mass movement on the gully walls. [U.S. Forest Service.]

If a stream were left to itself in its attempt to reach base level, it would erode its bed straight downward, forming a vertically-walled chasm in the process. But because the stream is not the only agent at work in valley formation, the walls of most valleys slope upward and outward from the valley floor. In time, the cliffs of even the steepest gorge will be angled away from the axis of its valley.

As a stream cuts downward and lowers its channel into the land surface, weathering, slope wash, and mass movement come into play, constantly wearing away the valley walls and pushing them farther back (Figure 13.27). Under the influence of gravity, material is carried down from the valley walls and dumped into the stream, to be moved toward the sea. The result is a valley whose walls flare outward and upward from the stream in a typical **cross-valley profile** (Figure 13.28).

The rate at which valley walls are reduced and the angles that they assume depend on several factors. If the walls are made up of unconsolidated material that is vulnerable to erosion and mass movement, the rate will be rapid; but if the walls are composed of resistant rock, the rate of erosion will be very slow, and the walls may rise almost

vertically from the valley floor (Figure 13.29).

In addition to cutting downward into its channel, a stream also cuts from side to side, or laterally, into its banks. In the early stages of valley enlargement, when the stream is still far above its base level, downward erosion is dominant. Later, as the stream cuts its channel closer and closer to base level, downward erosion becomes progressively less important. Now a larger percentage of the stream's energy is directed toward eroding its banks. As the stream meanders back and forth, it forms an ever-widening floodplain on the valley floor, and the valley itself broadens.

How fast do these processes of valley formation proceed? In some instances we can measure the ages of gully formation in years and even months, but for large valleys we must be satisfied with rough approximations—and usually with plain guesses. For instance, one place where we can make an approximation of the rate of valley formation is the Grand Canyon of the Colorado in Arizona. Several lines of evidence suggest that it has taken something on the order of 10 million years for the Colorado to cut its channel downward about 1,800 m and expand its valley walls 6 to 20 km.

13.28 If a stream of water were the only agent in valley formation, we might expect a vertically walled valley no wider than the stream channel, as suggested by the colored lines in (a) and (b). Mass movement and slope wash, however, are constantly wearing away the valley walls, carving slopes that flare upward and away from the stream channel, as shown in the diagrams.

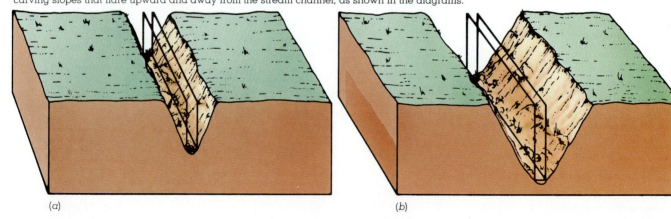

(a)   (b)

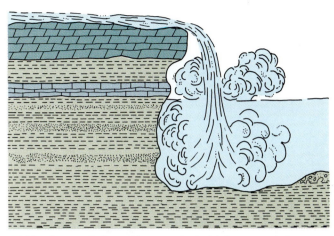

13.30 Niagara Falls tumble over a bed of dolomite, underlain chiefly by shale. As the less-resistant bed is eroded, the undermined ledge of dolomite breaks off, and the lip of the falls retreats.

13.29 Erosion by the Aude River in southwestern France has fashioned this limestone-walled gorge. Slope processes have only just begun to lower the angle of the valley sides.

FEATURES OF NARROW VALLEYS

Waterfalls are among the most fascinating spectacles of the landscape. Thunderous and powerful as they are, however, they are actually short-lived features in the history of a stream. They owe their existence to some sudden drop in the river's longitudinal profile — a drop that may be eliminated with the passage of time.

Waterfalls are caused by many different conditions. Niagara Falls, for instance, are held up by a relatively resistant bed of dolomite underlain by beds of nonresistant shale (Figure 13.30). This shale is easily undermined by the swirling waters of the Niagara River as they plunge over the lip of the falls. When the undermining has progressed far enough, blocks of the dolomite collapse and tumble to the base of the falls. The same process is repeated over and over again as time passes, and the falls slowly retreat upstream. Historical records suggest that the Horseshoe, or Canadian, Falls (by far the larger of the two falls at Niagara) have been

retreating at a rate of 1.2 to 1.5 m/year, whereas the smaller American falls have been eroding away 5 to 6 cm/year. The 11 km of gorge between the foot of the falls and Lake Ontario are evidence of the headward retreat of the falls through time.

Yosemite Falls in Yosemite National Park, California, plunge 770 m over the Upper Falls, down an intermediate zone of cascades, and then over the Lower Falls. The falls leap from the mouth of a small valley high above the main valley of the Yosemite. The Upper Falls alone measure 430 m, nine times the height of Niagara. During the Ice Age glaciers scoured the main valley much deeper than they did the unglaciated side valley. Then, when the glacier ice melted, the main valley was left far below its tributary, which now joins it after a drop of nearly three quarters of a kilometer.

Rapids, like waterfalls, occur at a sudden increase in the gradient of the stream's channel. Although rapids do not plunge straight down as waterfalls do, the underlying cause of their formation is often the same. In fact, many rapids have developed directly from preexisting waterfalls (Figure 13.31).

Even without waterfalls and rapids narrow valleys usually have **steep gradients** as compared with those of broad valleys.

FEATURES OF BROAD VALLEYS

If conditions permit, the various agents working to enlarge a valley ultimately produce a broad valley with a wide level floor. During periods of normal or low water the river running through the valley is confined to its channel, but during high water it overflows its banks and spreads out over the floodplain.

A floodplain that has been created by the lateral erosion and the gradual retreat of the valley walls is called an **erosional floodplain,** and has a thin cover of gravel, sand,

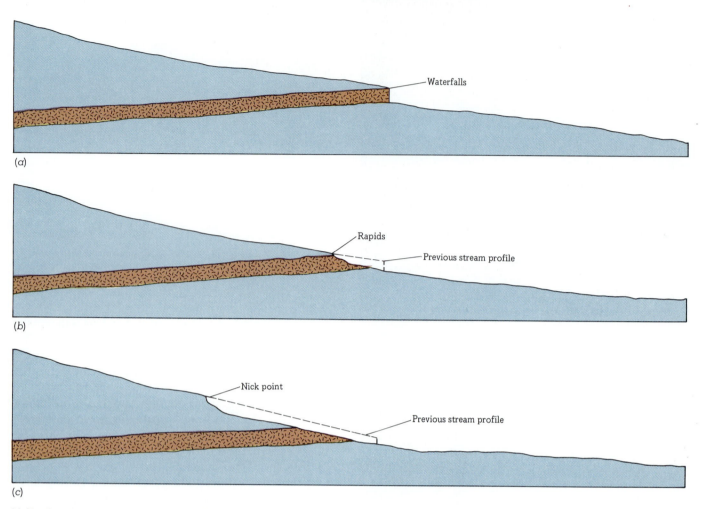

(a)

Waterfalls

(b)

Rapids

Previous stream profile

(c)

Nick point

Previous stream profile

13.31   Rapids may represent a stage in the destruction of waterfalls, as suggested in this diagram.

and silt a few meters or a few tens of meters in thickness. On the other hand, the floors of many broad valleys are underlain by deposits of gravel, sand, and silt scores of meters thick. These thick deposits are laid down as changing conditions force the river to drop its load across the valley floor. Such a floodplain, formed by the building up of the valley floor, or aggradation, is called an **aggradational floodplain.** Aggradational floodplains are much more common than erosional floodplains, and are found in the lower reaches of the Mississippi, Nile, Rhône, and Yellow Rivers, to name but a few areas.

Both erosional floodplains and aggradational floodplains exhibit the following characteristics.

Meanders   The channel of the Menderes River in western Turkey curves back on itself in a series of broad hairpin bends, and today all such bends are called **meanders** (Figure 13.32).

Both erosion and deposition are involved in the formation of a meander. First, some obstruction swings the current of a stream against one of the banks, and the current is deflected over to the opposite bank. Next, erosion takes place on the outside of each bend, where the turbulence is greatest. The material detached from the banks is then

moved downstream, there to be deposited in zones of decreased turbulence — either along the center of the channel or on the inside of the next bend. As the river swings from side to side, the meander continues to grow by erosion on the outside of the bends and by deposition on the inside.

We have already seen that a stream erodes on the outside of a bend. The erosion has both a lateral, cross-valley component and a down-valley component. The result is that the bends of a sinuous stream move both across and down the valley. In a meandering stream, with its hairpin bends, this downstream sweep of the meanders produces features called **cutoffs, meander scars,** and **oxbow lakes.**

In their down-valley migration different meanders as well as different parts of the same meander may move at different rates. The result is that the narrow neck of a meander may be cut through and the stream, taking a new course, will abandon the old bend (see Box 13.3). The new channel is called a **neck cutoff.** The abandoned meander — the meander scar — is called an **oxbow** because it has a shape similar to the U-shape of an ox's collar. Usually both ends of the oxbow are gradually silted in, and the old meander becomes completely isolated from the new channel. If the abandoned meander fills up with water, an **oxbow lake** results.

13.32 The Laramie River in southeastern Wyoming meanders across its floodplain. Continuing migration of the meanders has left behind the scroll-like patterns marking former positions of the stream channel. [James R. Balsley, U.S. Geological Survey.]

BOX 13.3   Time and the River

Mark Twain, the pen name adopted by Samuel Langhorne Clemens, was one of the most successful of nineteenth-century American authors. He started out as a printer at the age of 11, took to the Mississippi in 1857 at the age of 22, became a licensed river pilot by 1859, and left the river when it closed at the outbreak of the Civil War. *Life on the Mississippi* (1883) is based in part on these early experiences and in part on a trip Twain made on the Mississippi from New Orleans to St. Paul over 20 years later. Here, from that book, are some of Twain's conclusions about the past and future of the Mississippi and about the fascination of science, all based on his extrapolations from a few simple observations about cutoffs of meanders.

Since my own day on the Mississippi, cut-offs have been made at Hurricane Island, at Island 100, at Napoleon, Arkansas, at Walnut Bend, and at Council Bend. These short-ened the river in the aggregate, sixty-seven miles. In my own time a cut-off was made at

American Bend, which shortened the river ten miles or more.

Therefore the Mississippi between Cairo and New Orleans was twelve hundred and fifteen miles long, one hundred and seventy-six years ago. It was eleven hundred and eighty after the cut-off of 1722. It has lost sixty-seven miles since. Consequently, its length is only nine hundred and seventy-three miles at present.

Now, if I wanted to be one of those ponderous scientific people, and "let on" to prove what had occurred in the remote past by what had occurred in a given time in the recent past, or what will occur in the far future by what has occurred in late years, what an opportunity is here! Geology never had such a chance, nor such exact data to argue from! Nor "development of species," either! Glacial epochs are great things, but they are vague—vague. Please observe:

In the space of one hundred and seventy-six years the Lower Mississippi has shortened itself two hundred and forty-two miles. That is an average of a trifle over one mile and a third per year. Therefore, any calm person, who is not blind or idiotic, can see that in the Old Oölitic Silurian Period, just a million years ago next November, the Lower Mississippi River was upward of one million three hundred thousand miles long, and stuck out over the Gulf of Mexico like a fishing rod. And by the same token any person can see that seven hundred and forty-two years from now the Lower Mississippi will be only a mile and three-quarters long, and Cairo and New Orleans will have joined their streets together, and be plodding comfortably along under a single mayor and a mutual board of aldermen. There is something fascinating about science. One gets such whole-some returns of conjecture out of such a trifling investment of fact.

Although a cutoff will eliminate a particular meander, the stream's tendency toward meandering still exists, and before long the entire process begins to repeat itself.

We found that a meander grows and migrates by erosion on the outside and downstream side of the bend and by deposition on the inside. This deposition on the inside leaves behind a series of low ridges and troughs, collectively known as point bars (discussed later in this section). Swamps often form in the troughs, and during flood time

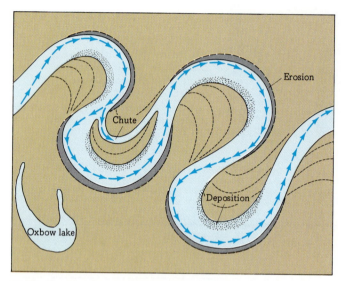

13.33 Erosion takes place on the outside of a meander bend, whereas deposition is most marked on the inside. If the neck of a meander is eroded through, an oxbow forms. A chute originates along the inside of a meander where irregular deposition creates ridges and troughs as the meander migrates.

the river may develop an alternate channel through one of the troughs. Such a channel is called a **chute cutoff**, or simply a **chute** (Figure 13.33).

The meandering river demonstrates a unity in ways other than the balance of erosion and deposition. The length of a meander, for example, is proportional to the width of the river, and this is true regardless of the size of the river. It holds for channels a few meters wide as well as those as large as that of the Mississippi River. As shown in Figure 13.34(*a*), this principle is also true of the Gulf Stream, even though this "river" is unconfined by solid banks. A similar relationship holds between the length of the meander and the radius of curvature of the meander (Figure 13.34[*b*]). These relationships seem to be controlled largely by the discharge and the sediments carried. Increasing discharge causes an increase in width, length, and wavelength. For a given discharge the wavelength tends to be greater for streams carrying a high proportion of sand and gravel in their loads than for those transporting mainly fine sand.

Braided Streams On some floodplains, particularly where large amounts of debris are dropped rapidly, a stream may build up a complex tangle of converging and diverging channels separated by sandbars or islands (Figure 13.35). A stream of this sort is called **braided.** The pattern generally occurs when the discharge is highly variable and the banks are easily eroded to supply a heavy load. It is characteristic of alluvial fans, glacial-outwash deposits, and certain heavily laden rivers.

In general, the gradient of a braided stream is higher than that for a meandering stream of the same discharge, as

13.34 (*a*) Length of the meander increases with the widening, meandering stream. (*b*) A similar orderly relationship exists between length of the meander and the mean radius of curvature of the meander. [After Luna B. Leopold and M. Gordon Wolman, "River Meanders," *Geol. Soc. Am. Bull. 71*, p. ;773, 1960.]

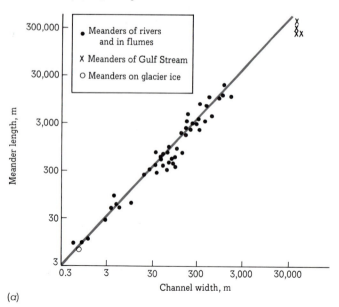

(a)

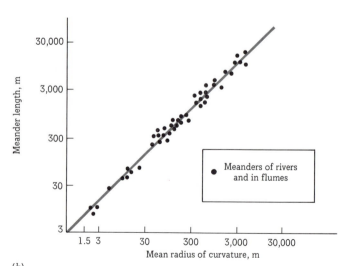

(b)

13.35 This low aerial photograph shows the complex braided pattern of the Platte River near Grand Island, Nebraska. [Kevin Crowley.]

indicated in Figure 13.36. This is apparently indicative of a tendency of a stream to increase its efficiency in order to transport proportionally larger loads.

Natural Levees   In many floodplains the water surface of the stream is held above the level of the valley floor by banks of sand and silt known as **natural levees,** a name derived from the French *lever,* "to raise." These banks slope gently, almost imperceptibly, away from their crest along

the river toward the valley wall. The low-lying floodplain adjoining a natural levee may contain marshy areas known as **back swamps.** Levees are built up during flood, when the water spills over the river banks onto the floodplain. Because the muddy water rising over the stream bank is no longer confined by the channel, its velocity and turbulence drop immediately, and much of the suspended load is deposited close to the river; but some is carried farther along to be deposited across the floodplain. The deposit of one flood

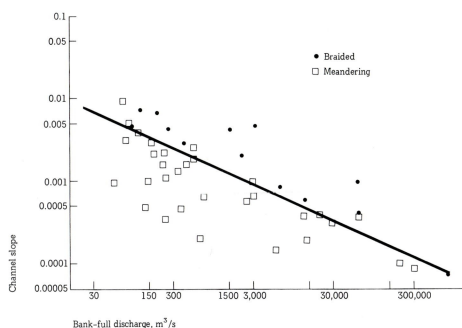

13.36 Meandering streams have lower gradients than do braided streams of the same discharge. In this diagram each symbol represents measurements at a single point on a stream. [Adapted from Luna B. Leopold, M. Gordon Wolman, and John P. Miller, *Fluvial Processes in Geomorphology,* Fig. 7-39, W. H. Freeman and Co., New York, 1964.]

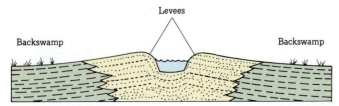

Levees

Backswamp   Backswamp

13.37   Natural levees characterize many aggrading streams. They build up during periods of flood as coarser material is deposited closest to the stream channel to form the levee and finer material is deposited in the back swamps. As the banks build up, the floor of the channel also rises.

is a thin wedge tapering away from the river, but over many years the cumulative effect produces a natural levee that is considerably higher beside the river bank than away from it (Figure 13.37). On the Mississippi delta, for instance, the levees stand 5 to 6 m above the back swamps. Although natural levees tend to confine a stream within its channel, each time the levees are raised slightly, the bed of the river is also raised. In time, the level of the bed is raised above the level of the surrounding floodplain. A stream may cut a new channel through a confining levee and assume a new channel across the lowest parts of the floodplain toward the back swamps.

A tributary stream entering a river valley with high levees may be unable to find its way directly into the main channel, so it will flow down the back-swamp zone and may run parallel to the main stream for many kilometers before finding an entrance. Because the Yazoo River typifies this situation by running 320 km parallel to the Mississippi, all rivers following similar courses are known as **yazoo-type** rivers.

Floodplains and Their Deposits   Most streams are bordered by floodplains. These may range from very narrow (a few meters wide) to the floodplains whose widths are measured in kilometers, as is that of the lower Mississippi River. The floodplain appears very flat, particularly when viewed in relation to the slopes of the valley walls that flank it. But the plain is not without its ups and downs. A large floodplain may have differences of relief of several meters and be marked by such features as natural levees, meander scars, and oxbow lakes.

The floodplain is made up of stream-carried sediments. Two general categories of deposit are common. One of these is made up of fine silts, clays, and sands, which may be spread across the plain by a river that overflows its banks during flood. These are called **overbank deposits.** The other group of deposits is made of coarse material, gravel and sand, and is related directly to the channel of the stream. These deposits include chiefly the material that is deposited in the slack water on the inside of the bends of a winding or meandering river. These bars of sand and gravel are called **point bars.**

We found in discussing meanders that the loops of a channel migrate laterally so that the river swings both across the valley and down-valley. As the river migrates laterally, it leaves behind coarse deposits in point bars and in the deeper parts of the abandoned channel. While erosion takes place on one bank of the river, deposition of sediments from upstream takes place on the point bars and, during flood, by overbank deposits—the river thus builds its floodplain in some places and simultaneously destroys it in others. The floodplain becomes a temporary storage place for sediments.

We can view the floodplain, then, as a depositional feature more or less in equilibrium. This equilibrium may be disturbed in several ways. The stream may lose some of its ability to erode and carry sediment so that a net gain of deposits occurs and the floodplain builds up. On the other hand, the flow of water or the gradient may increase or the supply of sediments may decrease, and there will be a net loss of sediments as erosion begins to destroy the previously developed floodplain.

Deltas and Alluvial Fans   For thousands of years the Nile River has been depositing sediments as it empties into the Mediterranean Sea and thus forming a great triangular plain with its apex upstream. In the fifth century B.C. the historian Herodotus introduced the term **delta** for this plain because of the similarity between its plan and the outline of the Greek letter Δ (Figures 13.38 and 13.39).

Whenever a stream flows into a body of standing water, its velocity and transporting power are quickly stemmed. If it carries enough debris and if conditions in the body of standing water are favorable, a delta will gradually form. The drainage across a delta is by a series of channels diverging from the apex. These branching channels, or **distributary channels,** build the delta. As sediment is deposited along a distributary channel, its bed and banks build up above the surrounding area. In time, the distributary stream escapes its old channel and forms a new one. The process continues to repeat itself on the ever-growing delta (Figure 13.40).

At the outer end of a distributary channel sediments may be arranged in a definite pattern. The coarse material is dumped first, forming a series of dipping beds called **foreset beds.** But the finer material is swept farther along to settle across the sea or lake floor as **bottomset beds.** As the delta extends farther and farther out into the water body, the stream must extend its channel to the edge of the delta. As it does so, it covers the delta with **topset beds,** which lie across the top of the foreset beds (Figure 13.41).

Very few deltas, however, show either the perfect delta shape or this regular sequence of sediments. Many factors, including shore currents, varying rates of deposition, the compaction of sediments, and the downwarping of the Earth's crust, conspire to modify the typical deltaic form and sequence.

Deltas are characteristic of many of the larger rivers of the world, including the Nile, Mississippi, Ganges, Rhine, and Rhône. On the other hand, many rivers have no deltas, either because the deposited material is swept away as soon

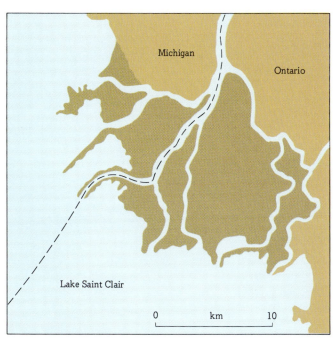

13.38 The delta (darker area) of the Saint Clair River in Lake Saint Clair has the classic shape as well as its distributary channels.

13.39 In this picture taken from *Gemini 4* the delta of the Nile stands out because it is well vegetated and thus contrasts with the desert country that borders it. [NASA.]

13.40 The Mississippi delta is called a "birdfoot delta" for its similarity to the clawed foot of a bird. Here we see the pattern of the main distributary channel of the Mississippi and its smaller ones. Note also the muddiness of the waters around the mouths of the distributaries. [NASA]

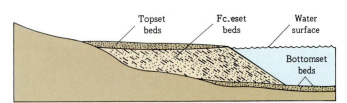

13.41 The ideal arrangement of sediments beneath a delta. Some material deposited in a lake or sea is laid on the bottom as bottomset beds. Other material is dumped in inclined foreset beds, built farther into the water and partly covering the bottomset beds. Over the foreset beds the stream lays down topset beds.

13.42 Alluvial fans in Death Valley, California. The varying sizes of fans relate to the varying sizes of the drainage basins providing material to the fans. The larger the drainage basin, the larger the fan. The largest fan shown here has a radius of about 2.5 km. [Fairchild Aerial Photograph Collection, Whittier College.]

as it is dumped or because the streams do not carry sufficient detrital material to build up a delta.

An **alluvial fan** is the land counterpart of a delta (Figures 13.42 and 13.43). These fans are typical of arid and semiarid climates, but they may form in almost any climate if conditions are right. A fan marks a sudden decrease in the carrying power of a stream as it descends from a steep gradient to a flatter one—for example, when the stream flows down a steep mountain slope onto a plain. As the velocity is checked, the stream rapidly begins to deposit the sediments it carries. Eventually the fan is established and its gradient is close to that of the stream as it flows from the mountain. Deposition on the fan continues, however, for two reasons. First, as the stream emerges from the mountain, it loses water into the permeable sediments of the fan and thereby loses its ability to transport material. Second, as the stream comes onto the fan at its apex, it is able to widen its channel in the unconsolidated sediments of the fan, and this decreases the velocity and leads to deposition.

13.43 This alluvial fan on the shore of the Aachen See in the Austrian Alps has a typically triangular profile, as seen in frontal view.

As the stream deposits material, it builds up its channel, often constructing small natural levees along its banks. Eventually, as the levees continue to grow, the stream may flow above the general level. Then during a time of flood it seeks a lower level and shifts its channel to begin deposition elsewhere. As this process of shifting continues, the alluvial fan continues to grow.

Stream Terraces   A **stream terrace** is a relatively flat surface running along a valley, with a steep bank separating it either from the floodplain or from a lower terrace. It is a remnant of an old, higher floodplain of a stream that now has succeeded in cutting its way down to a lower level (Figure 13.44).

The so-called **cut-and-fill terrace** is created when a stream first clogs a valley with sediments and then cuts its way down to a lower level (Figures 13.45 and 13.46). The initial aggradation may be caused by a change in climate that leads either to an increase in the stream's load or to a decrease in its discharge. Or the base level of the stream may rise, reducing the gradient and causing deposition. In any event, the stream chokes the valley with sediment, and the floodplain gradually rises. Now, if the equilibrium is upset and the stream begins to erode, it will cut a channel down through the deposits it has already laid down. The level of flow will be lower than the old floodplain, and at this lower level the stream will begin to carve out a new floodplain. As time passes, remnants of the old floodplain may be left standing as terraces on either side of the new one. Terraces that face each other across the stream at the same elevation are referred to as **paired terraces.** Sometimes the downward erosion by streams creates **unpaired terraces:** If the stream swings back and forth across the valley, slowly eroding as it moves, it may encounter resistant rock beneath the unconsolidated deposits; the exposed rock will then deflect the stream and prevent further erosion, and a single terrace will

13.44   Well-developed terraces along Cave Creek, 65 km northwest of Christchurch, New Zealand. These terraces are formed from sand and gravels washed from melting glaciers in the mountains in the distance.

13.45   The terraces' levels are matched in elevation across this intermittent stream, La Plaza, 15 km north of Gela in southern Sicily. The top of the terrace marks the elevation of a past floodplain. Since then the stream has cut downward through its previously deposited sediments to form the matching terraces. (See also Figure 13.46.)

13.46   One example of the formation of paired terraces. (a) The stream has partially filled its valley and has created a broad floodplain. (b) Some change in conditions has caused the stream to erode into its own deposits; the remnants of the old floodplain stand above the new river level as terraces of equal height. This particular example is referred to as a cut-and-fill terrace. (See also Figure 13.45.)

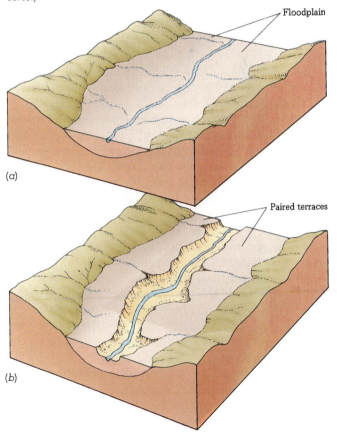

be left behind, with no corresponding terrace on the other side of the stream.

Terraces, either paired or unpaired, may be cut into bedrock as well. A thin layer of sand and gravel usually rests on the beveled bedrock of these terraces.

13.7

OF RIVERS AND CONTINENTS

So far we have been discussing some of the details of streams and their action. In concluding this chapter we look at some global aspects of running water.

CYCLE OF EROSION

Of all the agents at work in moving sediments from the continents to the ocean basins, the most important is running water. Like great conveyor belts, the rivers of the world move weathered waste from the land to the oceans. They annually carry to the oceans an estimated 10 billion t of Earth material, material delivered to them down slopes by mass movement. This rate of erosion could in a few million years wear down the continents close to base level and leave very little to erode and provide new sediments to the oceans. But geologic studies tell us that sediments and sedimentary rocks have been continually forming for billions of years. We therefore conclude that continuous or repeated uplift of the Earth's surface has for eons provided material for erosion. Thus the tectonics associated with plate motions have kept erosion from flattening the continents. Nonetheless, it is startling to observe on a world scale how close erosion comes to beveling the continents and reducing them toward sea level.

How far can erosion reduce continents toward sea level? It is clear that as long as orogenic activity continues there will be mountain belts standing high above the ultimate base level, the ocean. But, given a long enough time and the absence of orogeny, may not a mountain range be worn down to a plain of low relief sloping gently to the coast?

Indeed, over a century ago, the American geographer, geologist, and meteorologist William Morris Davis suggested that subcontinental segments of the Earth's surface, mountains and all, could be reduced by erosion to just such a plain. He coined the word **peneplain** (Latin *paene,* "almost," plus "plain") to describe such a geographic feature. The requisites to produce this regional surface of low relief were long erosion and equally long cessation of the mountain-building process.

Davis pictured a region as passing through a series of stages that he termed **youth, maturity,** and **old age.** Areas of newly formed mountains such as the Andes and the Himalaya are in the youthful stage of what Davis called the **cycle**

of erosion. The Appalachians are in the mature stage of the cycle. Erosion, Davis reasoned, would be very rapid during youth and become progressively less so as the cycle progressed toward old age, by which time it would be extremely slow.

Davis felt that the Appalachians, where he formulated many of his concepts of landscape development, had indeed been reduced to a peneplain in the past, and that today the peneplain's remnants still existed in the more or less level ridge tops of the mountains. Subsequent to this peneplanation, he argued, the Appalachians were lifted upward. This epeirogenic movement had the effect of rejuvenating the sluggish streams of the old-age stage and initiating a new period or cycle of erosion.

Where should we look for an extensive erosional surface of low relief today? Certainly not in the low-lying coastal plain of eastern United States, for that is basically a geologically recent depositional feature, not an erosional one. But turn to the stable interior sections of the continents, the cratons and more specifically the shields. There, unobscured by younger strata, Precambrian igneous and metamorphic rocks, the products of ancient orogenies, immediately underlie vast areas of low relief. Thus the surface of most of the Canadian Shield is well below a kilometer in elevation, as is much of the shield area of the other continents. The shields have lain protected from the orogenies focused on the continental margins. Streams have been able to remove the products of erosion, however slowly, without tectonic forces periodically putting new mountain barriers in their paths. We are tempted to see in such zones the end product of long-continued erosion in tectonically quiet areas — the peneplain.

How much time is needed to create a peneplain? The last major orogeny in the Appalachians was about 250 million years ago, and the Appalachians are still with us, although greatly reduced in elevation. The last major orogenies in the low-lying Canadian Shield occurred perhaps a billion years ago. So as a first approximation we suggest that the period necessary to reduce a full-fledged mountain system to an erosional surface of low relief is between a quarter of a billion and one billion years.

RIVERS AND CONTINENTAL ELEVATION

The mean elevation of the continents is less than a kilometer — 840 m, to be more precise. But if we look at the average elevation above sea level of individual continents, we find some interesting relationships. The greater the area of a continent, the greater its average elevation, as shown in Figure 13.47. Note also the aberrant elevation of Antarctica as compared with other continents. This is because a large part of the continent — some $27 \times 10^6 km^3$ — is ice, which is very low in density compared to normal rocks. Therefore, because of its relatively low mass per unit volume, Antarctica stands relatively high. The Antarctic icecap, which has

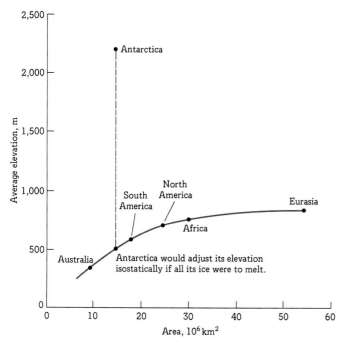

13.47 As the area of a continent increases, so does its elevation. Antarctica's elevation is aberrantly high because of the great mass of ice that covers it. If the ice, which has a relatively low density, were to be removed, the continent would, after isostatic adjustment, have the lower elevation indicated.

an average thickness of about 2.3 km, bends downward the underlying bedrock surface of the continent. Were we to remove the ice from Antarctica, the down-warped bedrock would rise isostatically to an average elevation of about 500 m above sea level, as suggested in Figure 13.47.

There is no immediately demonstrable reason why the elevation of continents increases with their areal extent.

One reasonable suggestion is that the continents of greater area have deeper roots than those of smaller area. Another suggestion relates the explanation in part to the drainage systems of continents. Any stream must have a gradient on which to flow. The longer the drainage network, the higher above sea level must be the interior parts of the drainage. The larger landmasses, on average, must of necessity have longer drainage systems, and that is why they stand higher than continents of smaller area.

MAJOR DRAINAGE DIVIDES

This leads us to a final observation about some of the major features of the world's drainage systems. If we look at a world map, we find that many of the major divides are related to plate boundaries. South America provides a spectacular example. There the Andes, formed by the collision of the Nazca and American plates, form the continental divide. Most of the continent drains to the east into the Atlantic Ocean, and only a narrow area sheds water to the west, into the Pacific Ocean. The great mountain masses of the Himalaya, the Tibetan Plateau, and the mountains of Afghanistan, Iran, the Caucasus, and on to the Alps represent a complex convergence of the great Eurasian plate with several other plates along its southern border. The continental divide of North America, from southern Mexico to Alaska, is associated with plate movements, although their details are still to be worked out. We also note that ancient mountains, such as the Appalachians of eastern North America and the Urals of Eastern Europe, mark old lines of plate convergence, and they still serve as significant drainage divides hundreds of millions of years later.

<hr>

Summary

The hydrologic cycle is the path of water circulation from oceans to atmosphere to land to oceans.

The equation for the hydrologic cycle is as follows:
   Precipitation = runoff + evapo-transpiration
   Flow is either **turbulent** or **laminar,** but turbulent is the rule.
   Water in a stream has **velocity,** flows down a **gradient,** and is measured in terms of **discharge.**

The economy of a stream is part of a dynamic equilibrium that characterizes earth processes.
   It is clear that in a stream **discharge = width × depth × velocity.** The change of one characteristic will affect one or more of the others. In adjusting to an increased discharge downstream, the gradient decreases, and width, depth, and velocity increase.
   Floods are a normal, but relatively rare, stage of stream flow. Flood-recurrence intervals are statistically predictable.
   Human activity can affect a stream's velocity, discharge, and channel characteristics.
   Base level of a stream is the point below which it cannot erode. Lowering of the base level produces erosion, and raising of the base level produces deposition.

Work of running water includes transportation, erosion, and deposition.

Transportation by a stream involves its **load** (the amount it carries at any one time), its **capacity** (the total amount it can carry under given conditions), and its **competence** (the maximum-sized particle it can move under given conditions). Material is moved in **solution**, in **suspension**, and as **bed load**.

Erosion by a stream involves **direct lifting** of material, **abrasion**, **impact**, and **solution**.

Features of valleys include those of the valley bottom, the drainage basin, and the river channel.

A **cross-valley profile** of a typical valley shows **floodplains**, **valley walls**, and **divides**.

A **drainage basin** is the area drained by a river and its tributaries. Individual streams and their valleys are joined in definite geometrical relationships. The patterns developed by stream systems include **dendritic, radial, rectangular,** and **trellis.** Some streams form **water gaps. Enlargement of valleys** is accomplished through downward and lateral erosion by the stream and by mass movement and water erosion on the valley walls.

Features of narrow valleys include **waterfalls, rapids,** and **steep gradients.**

Broad valleys have **meanders, braided streams, natural levees, floodplains, deltas, alluvial fans,** and **stream terraces.**

Rivers are the most important agent in removing materials from the continents. Plate tectonics keep raising the land surface to provide new materials for erosion. A **peneplain** may form as the end product of a region that undergoes long-continued erosion in the **cycle of erosion.**

Questions

1. With the aid of a diagram, describe the hydrologic cycle. You can make your answer quantitative by adding data from the text or Appendix C.

2. You will get a better grasp of the hydrologic cycle by putting some numbers on it. Try this. If the precipitation in a drainage basin equals 10^9 m^3/year and the sum of the runoff and infiltration is 4×10^8 m^3/year, how much water is returned annually to the atmosphere by evaporation and transpiration? What is this in terms of percentage of the total precipitation?

(We got our answer for total evaporation and transpiration as 6×10^8 m^3/year by substituting in the formula):

Evaporation + transpiration = precipitation
− (runoff + infiltration)

3. What is the difference between laminar and turbulent flow? Which is more important in streams?

4. Draw and label a stream hydrograph. Explain how human intervention in a stream's natural regime may modify its hydrograph.

5. Referring to Figure B13.1.1, what is a reasonable suggestion for the recurrence interval of a flood of magnitude of flow of about 6,250 m^3/s?

(We got about 200 years.)

6. Base level is an important concept in understanding the behavior of a stream. What is the base level of a stream? What is the ultimate base level? Is it constant or changing, and why do you answer as you do? Give two examples of temporary base level. (Here is another place where a simple sketch can sharpen your answer.)

7. What happens to the activity of a stream when its base level rises? When it falls?

8. Distinguish between load, capacity, and competence in a stream.

9. What are the ways by which material is carried by a stream?

10. How does land use affect the availability of sediments to a stream?

11. Stream order is one way of describing the organization of streams, stream networks, and their drainage basins. Referring to Figure 13.23(b) and (c), what would be the approximate length of an 11th-order stream? What would be the approximate size of the drainage basin of this stream?

(We assumed that we could extend the graph in both instances. We got an 11th-order stream of about 1,600 km in length, with a drainage basin of about 1.1×10^6 km^2. Remember that the vertical axes on the graphs are marked in logarithmic scales.)

12. Nature provides a number of different stream patterns when seen on a map or from the air. List two different patterns and discuss what each means in terms of the geology of the country on which they have developed.

13. Many different features distinguish streams and valleys. You should be familiar enough with them to define and discuss the following: waterfalls, rapids, floodplain, meander, oxbow, cutoff, braided stream, natural levee, point bar, delta, alluvial fan, stream terrace.

14. For reasons best known to yourself, you find yourself in the midst of a discussion on the origin of stream valleys. The person with whom you are having this discussion says, "I don't believe that streams are responsible for the valleys in which they run. I think that streams run in valleys because the valleys were there in the first place."

What evidence can you muster to substantiate your belief that most valleys were formed by the streams that flow in them?

Or take the other side: What is the evidence that streams do not form their own valleys, but rather the valleys are there for other reasons and streams just flow along the bottoms of these preexisting valleys?

15. If an agency designed a spillway for the Cumberland River at Nashville that would carry 5,750 m³/s of water, claiming that the design should be safe for 200 years because 5,750 m³/s is the largest known flood that occurred at that site, would you believe it? See Figure B13.1.1 and Box 13.1 for data.

16. Evaluate the remarks of Mark Twain about the Mississippi River in Box 13.3.

17. What are the conditions under which a mountain range can be reduced to a low-lying plain?

18. Name one or more geographic areas that might be considered to be in each of the three stages of the cycle of erosion.

Supplementary Readings

BLOOM, ARTHUR L. *Geomorphology,* Prentice Hall, Englewood Cliffs, N.J., 1978.
Four chapters, 9 through 12, emphasize the development of landforms by the fluvial process.

CHORLEY, RICHARD J., STANLEY A. SCHUMM, AND DAVID E. SUGDEN *Geomorphology,* Methuen & Co., New York, 1985.
An extensive volume in which Chapters 13, 14, and 15 address the subject of running water.

GREGORY, K. J., AND D. E. WALLING *Drainage Basin Form and Process,* Cambridge University Press, New York, 1973.
A very good advanced text devoted to the forms of drainage basins and the processes that go on in them.

MORISAWA, MARIE *Streams: Their Dynamics and Morphology,* McGraw-Hill Book Co., New York, 1968.

A short book of intermediate level, which is a readable summary of the hydrology and work of streams.

RICHARD, KEITH *Rivers: Form and Process in Alluvial Channels,* Methuen & Co., New York, 1984.
A good intermediate text on the subject.

RITTER, DALE F. *Process Geomorphology,* 2nd ed., William C. Brown Co., Dubuque, Iowa, 1986.
Three chapters deal with drainage basins, fluvial processes, and fluvial landforms in more detail than is possible in this volume.

SCHUMM, STANLEY A. *The Fluvial System,* John Wiley & Sons, New York, 1977.
An excellent, although somewhat specialized, treatment.

Groundwater gushes out of permeable rock to form Roaring Springs on the north wall of the Grand Canyon of the Colorado. [C. Sharyn Magee.]

CHAPTER ▪ 14

UNDERGROUND WATER

". . . we must infer a subterranean circulation connected with that of the surface. . . ."

Ferencz Pošepný, Transactions of the American Institute of Mining Engineers, 1893

In discussing the hydrologic cycle in the last chapter, we found that a large percentage of the moisture falling on the surface of the Earth infiltrated through the soil zone into the underground. In this chapter we take up this connection between surface and subsurface and follow water in its subterranean journey. As we do this, we will seek to understand how and why groundwater moves and what results it may have both at and below the surface.

We will find that water seeps from the surface through a *zone of aeration* to the *zone of saturation*, where all pore spaces are filled with water. This water is called *groundwater* and its upper surface is the *groundwater table*, or, more simply, the *water table*.

Groundwater flows through *aquifers* and most flow is laminar. We will see that the direction and velocity of movement depend on characteristics of rock material called *porosity* and *permeability*, on *gravity*, on the *viscosity* of the water, and on the *hydraulic gradient* of the water table. This is expressed in a relationship known as *Darcy's law*.

Simple *springs* and *wells* occur at the intersections of the surface with the groundwater table. An *artesian well* or *spring* flows under pressure great enough to raise the water above the level of the *aquifer*. *Thermal springs* are heated by cooling igneous rocks or by the earth's normal thermal gradient. *Geysers* are thermal springs that eject water intermittently and forcefully.

Caves develop most commonly in limestone. Slightly acidic water reacts with the calcite of limestone to dissolve large amounts of rock. Following the formation of a cave, decoration occurs as *stalactites* grow down from cave ceilings and *stalagmites* up from cave floors. Solution of limestone also creates a landscape called *karst*.

The human use and misuse of groundwater create a number of problems. These include *pollution* by landfills, sites of toxic waste, and agriculture, as well as by *saltwater invasion;* and overpumping, which has caused *land subsidence* in some areas.

14.1

BASIC DISTRIBUTION

Beneath the surface, hidden from our sight, is a vast supply of water. In fact, for the Earth as a whole, subsurface water exceeds by more than 66 times the amount of water in streams and freshwater lakes (see Appendix C, Table C.1). In the United States each year streams discharge an estimated 30,000 km³ of water into the oceans. In contrast, an estimated 7,575,000 km³ of water lie beneath the surface.

Underground water, groundwater, subsurface water, and **subterranean water** are all general terms used to refer to water in the pore spaces, cracks, tubes, and crevices of the consolidated and unconsolidated material beneath our feet. The study of underground water is largely an investigation of these openings and of what happens to the water that finds its way into them.

ZONES OF AERATION AND SATURATION

Some of the water that moves down from the surface is caught by rock and earth materials and is checked in its downward progress. The zone in which this water is held is known as the **zone of aeration,** and the water itself is called **suspended water.** The spaces separating particles in this zone are filled partly with water and partly with air. Two forces operate to prevent suspended water from moving deeper into the earth: (1) the molecular attraction exerted on the water by the rock and earth materials; and (2) the attraction exerted by the water particles on one another (Figures 14.1 and 14.2).

14.1 A drop of water held between two fingers illustrates the molecular attraction that prevents downward movement. Water is similarly suspended within the pore spaces of the zone of aeration shown in Figure 14.2

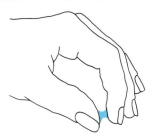

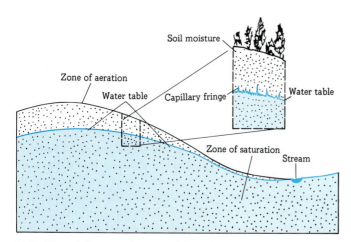

14.2 Water infiltrating the ground from the surface passes through a zone of aeration in which some of the pore space is filled with air. Part of this water may be held in the soil and used by plants. The water that passes downward eventually reaches the zone of saturation, where all pore spaces are water-filled. The upper surface of this zone is the groundwater table, or simply the water table. This may be capped by a capillary fringe.

The zone of aeration can be subdivided into three belts: **belt of soil moisture, intermediate belt,** and **capillary fringe.** Some of the water that enters the belt of soil moisture from the surface is used by plants, and some is evaporated back into the atmosphere. But some water also passes down to the intermediate belt, where it may be held by molecular attraction (as suspended water). Little movement occurs in the intermediate belt except when rain or melting snow sends a new wave of moisture down from above. In some areas the intermediate belt is missing, and the belt of soil moisture lies directly above the third belt, the capillary fringe. Water rises into the capillary fringe from below, to a height ranging from a few centimeters to 2 or 3 m.

Beneath the zone of aeration lies the **zone of saturation.** Here the openings in the rock and earth materials are completely filled with **groundwater,** and the surface between the zone of saturation and the zone of aeration is called the **groundwater table,** or simply the **water table.** The level of the water table fluctuates with variations in the supply of water coming down from the zone of aeration, with variations in the rate of discharge in the area, and with variations in the amount of groundwater that is drawn off by plants and human beings.

It is the water below the water table, within the zone of saturation, that we shall focus on in this chapter.

THE WATER TABLE

The water table is an irregular surface of contact between the zone of saturation and the zone of aeration. Below the water table lies the groundwater; above it lies the suspended water. The thickness of the zone of aeration differs from one place to another, and the level of the water table fluctuates

accordingly. In general, the water table tends to follow the irregularities of the ground surface, reaching its highest elevation beneath hills and its lowest elevation beneath valleys. Although the water table reflects variations in the ground surface, the irregularities in the water table are less pronounced.

In looking at the topography of the water table, let us consider an ideal situation. Figure 14.3 shows a hill under-

14.3 Ideally, the water table is a subdued reflection of the surface of the ground. In (a) and (b) the water table rises as a horizontal plane until it reaches the level of the valley bottoms on either side of the hill. Thereafter, as more water soaks into the ground, it seeks an outlet toward the valleys. Were the movement of the water not slowed down by the material making up the hill, it would remain essentially horizontal. The friction caused by the water passing through the material (and even to some extent, the internal friction of the water itself) results in a piling up of water beneath the hill; the bulge is highest beneath the crest and lowest toward the valleys (c). The shape of the water table may be altered by pumping water from a well (d). The water flows to this new outlet and forms a cone of depression.

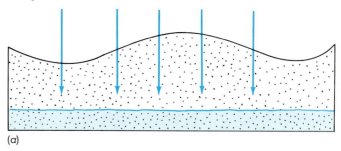

(a)

(b)

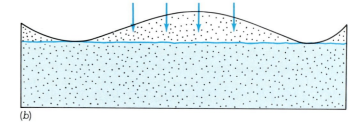

(c)

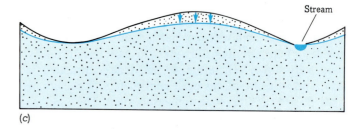

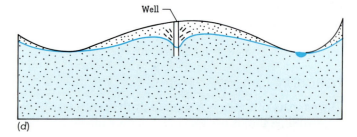

(d)

lain by completely homogeneous material. Assume that, initially, this material contains no water at all. Then a heavy rainfall comes along, and the water soaks slowly downward, filling the interstices at depth. In other words, a zone of saturation begins to develop. As more and more water seeps down, the upper limit of this zone continues to rise. The water table remains horizontal until it just reaches the level of the two valley bottoms on either side of the hill. Then, as additional water seeps down to the water table, some of it seeks an outlet into the valleys. But this added water is "supported" by the material through which it flows, and the water table is prevented from maintaining its flat surface. The water is slowed by the friction of its movement through the interstices and even, to some degree, by its own internal friction. Consequently more and more water is piled up beneath the hill, and the water table begins to reflect the shape of the hill. The water flows away most rapidly along the steeper slope of the water table near the valleys and most slowly on its gentler slope beneath the hill crest.

We can modify the shape of the groundwater surface by providing an artificial outlet for the water. For example, we can drill a well on the hill crest and extend it down into the saturated zone. Then, if we pump out the groundwater that flows into the well, we will create a dimple in the water table. The more we pump, the more pronounced the depression—a **cone of depression**—will become.

Returning to our ideal situation, we find that if the supply of water from the surface were to be completely stopped, the water table under the hill would slowly flatten out as water discharged into the valleys. Eventually it would almost reach the level of the water table under the valley bottoms; then the flow would stop. This condition is common in desert areas where the rainfall is sparse.

14.2

MOVEMENT OF UNDERGROUND WATER

The preceding chapter stated that the flow of water in streams on the surface could be measured in terms of so many meters per second. But in dealing with the flow of underground water, we must change our scale of measurement; for here, although the water moves, it usually does so very, very slowly. Therefore we find that centimeters per day, and in some places even centimeters or meters per year, provide a better scale of measurement. The main reason for this slow rate of flow is that the water must travel through small, confined passages if it is to move at all. It will therefore be worthwhile for us to consider the porosity and permeability of Earth materials.

POROSITY

The **porosity** of a rock is measured by the percentage of its total volume that is occupied by voids or interstices. The more porous a rock is, the greater the amount of open space it contains.

Porosity differs from one material to another. What is the porosity of a rock made up of particles and grains derived from preexisting rocks? Here porosity is determined largely by the shape, size, and assortment of these rock-building units. A sand deposit composed of round quartz grains with fairly uniform size has a high porosity. But if mineral matter enters the deposit and cements the grains into a sandstone, the porosity is reduced by an amount equal to the volume of the cementing agent. A deposit of sand poorly sorted with finer particles of silt and clay mixed in has low porosity because the smaller particles fill up much of the space between the larger particles.

Even a compact massive rock may become porous as a result of fracturing. And a soluble massive rock, such as limestone, may have its original planes of weakness enlarged by solution.

Clearly, then, the range of porosity in earth materials is extremely great. Recently deposited muds (called **slurries**) may hold up to 90 percent by volume of water, whereas unweathered igneous rocks such as granite, gabbro, or obsidian may hold only a fraction of 1 percent. Unconsolidated deposits of clay, silt, sand, and gravel have porosities ranging from about 20 to as much as 50 percent. But when these deposits are consolidated into sedimentary rocks by cementation or compaction, their porosity is sharply reduced. Average porosity values for individual rock types have little meaning because of the extreme variations within each type. In general, however, a porosity of less than 5 percent is considered low; from 5 to 15 percent represents medium porosity; and over 15 percent is considered high.

PERMEABILITY

Whether or not we find a usable supply of fresh groundwater in a given area depends on the ability of the Earth materials to transmit water as well as on their ability to contain it. The ability to transmit underground water is termed **permeability.**

The rate at which a rock transmits water depends not only on its total porosity but also on the size of the interconnections between its openings. For example, although a clay may have a higher porosity than a sand, the particles that make up the clay are minute flakes and the interstices between them are very small. Therefore water passes more readily through the sand than through the more porous clay simply because the molecular attraction on the water is much stronger in the tiny openings of the clay. The water moves more freely through the sand because the passageways between particles are relatively large and the molecular attraction on the water is relatively low. Of course, no matter how large the interstices of a material are, there must be connections between them if water is to pass through. If

they are not interconnected, as in pumice or scoria, the material is impermeable.

A permeable material that actually carries underground water is called an **aquifer,** from the Latin for "water" and "to bear." Perhaps the most effective aquifers are unconsolidated sand and gravel, sandstone, and some limestones. The permeability of limestone is usually due to solution that has enlarged the fractures and bedding planes into open passageways. The fractured zones of some of the denser rocks such as granite, basalt, and gabbro also act as aquifers, although the permeability of such zones decreases rapidly with depth.

Clay, shale, and most metamorphic and igneous crystalline rocks are generally poor aquifers and water passes through them very slowly, if at all. These rocks of low permeability are called **aquicludes,** from the Latin words for "water" and "close" or "shut," in reference to their ability to close off the passage of water.

Because the flow of underground water is usually very slow, it is largely laminar; in contrast, the flow of surface water is largely turbulent. (There is one exception, however: the turbulent flow of water in large underground passageways formed in such rocks as cavernous limestone.) In laminar flow the water near the walls of interstices is presumably held motionless by the molecular attraction of the walls. Water particles farther away from the walls move more rapidly, in smooth, threadlike patterns because the resistance to motion decreases toward the center of an opening. The most rapid flow is reached at the very center.

Even as a stream needs a gradient on which to flow, so also does groundwater. For groundwater this depends on the slope of the water table and is called the **hydraulic gradient.** We measure it by the difference in the elevation of the water table between two points divided by the distance between the two points. Therefore hydraulic gradient is measured by

$$\frac{h_2 - h_1}{l}$$

where h_2 is the greater of two elevations on the slope of the water table, h_1 the lesser, and l the distance between the two points.

An equation to express the velocity of water movement through rock was proposed by the French engineer Henri Darcy in 1856. What is now known as **Darcy's law** may be expressed as follows:

$$\text{Velocity} = \frac{K(h_2 - h_1)}{l}$$

where K is the *hydraulic conductivity,* a coefficient that depends on the permeability of the material, the acceleration of gravity, and the viscosity of the water. It measures the rate of groundwater flow. But because $(h_2 - h_1)/1$ is simply a way of expressing the hydraulic gradient, we may say that in a rock of constant permeability, the velocity of water will increase as the hydraulic gradient increases. Remembering that the hydraulic gradient depends on the

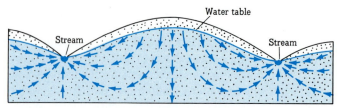

14.4 The flow of groundwater through uniformly permeable material is suggested here. Movement is not primarily along the groundwater table; rather, particles of water define broadly looping paths that converge toward the outlet and may approach it from below.

slope of the groundwater table, we may also say that the velocity of groundwater varies with the slope of the water table. Other things being equal, the steeper the slope of the water table, the more rapid the flow. In ordinary aquifers the rate of water flow has been estimated as not faster than 1.5 m/day and not slower than 1.5 m/year, although rates of over 120 m/day and as low as a few centimeters per year have occasionally been recorded.

Gravity exerts a directly downward pull on the groundwater. But below, the water is diverted laterally in the direction of the slope of the water table. In a completely homogeneous aquifer the course of groundwater is not directly toward the stream but follows broad, looping curves, as suggested in Figure 14.4. The explanation for this is not difficult to visualize. The column of groundwater beneath the hill crest is higher than that below the valley bottom, resulting in a greater pressure within the groundwater beneath the hill at any elevation than at a similar elevation beneath the stream bed. Flow is, then, from high- to low-pressure zones, the flow curves reflecting direction of pressure-gradient vectors.

14.3
WELLS, SPRINGS, AND GEYSERS

So far we have assumed that groundwater is free to move on indefinitely through a uniformly permeable material of unlimited extent. Subsurface conditions actually fall far short of this ideal situation. Some layers of rock material are more permeable than others, and the water tends to move rapidly through these beds in a direction more or less parallel to bedding planes. Even in a rock that is essentially homogeneous, the groundwater tends to move in some preferred direction.

SIMPLE SPRINGS AND WELLS

Underground water moves freely downward from the surface until it reaches an impermeable layer of rock or until it arrives at the water table. Then it begins to move laterally.

Sooner or later it may flow out again at the surface of the ground in an opening called a **spring.**

Springs have attracted human attention throughout history. In early days they were regarded with superstitious awe and were sometimes selected as sites for temples and oracles. To this day many people feel that spring water possesses special medicinal and therapeutic values. Water from "mineral springs" contains salts in solution that were picked up by the water as it percolated through the ground. The same water pumped up out of a well would be regarded as merely hard and undesirable for general purposes.

Springs range from intermittent flows that disappear when the water table recedes during a dry season, through tiny trickles, to an effluence of 3.8 billion liters daily—the abundant discharge of springs along a 16-km stretch of Fall River, California.

This wide variety of types of springs is the result of subterranean conditions that vary greatly from one place to another. As a general rule, however, a spring results wherever the flow of groundwater is diverted to a discharge zone at the surface (see chapter opening figure and Figure 14.5). For example, a hill made up largely of permeable rock may contain a zone of impermeable material, as shown in Figure 14.6. Some of the water percolating downward will be blocked by this impermeable rock, and a small saturated zone will be built up. Because the local water table here is actually above the main water table, it is called a **perched water table.** The water that flows laterally along this impermeable rock may emerge at the surface as a spring. Springs are not confined to points where a perched water table flows from the surface, and it is clear that if the main water table intersects the surface at a point along a slope, a spring will form.

Even in impermeable rocks permeable zones may develop as a result of fractures or solution channels. If these openings fill with water and their lower parts are intersected by the ground surface, the water will issue forth as a spring.

A **spring** is the result of a **natural** intersection of the ground surface and the water table, but a **well** is an **artificial** opening cut down from the surface into the zone of saturation. A well is productive only if it is drilled into permeable material and penetrates below the water table. The greater the demands that are made on a well, the deeper it must be

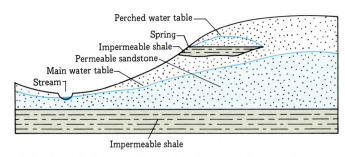

14.6 A perched water table results when groundwater collects over an impermeable zone and is separated from the main water table.

drilled below the water table. Continuous pumping creates the cone of depression previously described, which distorts the water table and may reduce the flow of groundwater into the well (Figure 14.7).

The type of bedrock that makes up an aquifer will affect both the yield and the quality of groundwater. For instance, wells drilled into fractured crystalline rock, such as granite, may produce an adequate supply of water at relatively shallow depths, but we cannot increase the yield of such wells appreciably by deepening them because the number and size of the fractures commonly decrease the farther down into the earth we go (Figure 14.8). It is clear, also, that a highly permeable aquifer such as an unconsolidated gravel of an old stream bed will, all other conditions being equal, yield larger supplies of water than a tightly cemented sandstone.

The composition of an aquifer affects the composition of underground water. Water flowing through rock rich in calcite will acquire calcium ions in solution. Such water does not lather readily with soap and forms a scale when evaporated in a container. It is familiar to us as **hard water.** In contrast, aquifers made up of insoluble silicate minerals carry very little material in solution, and essentially no calcium. The water they carry lathers well with soap; we call it **soft water.** Should an aquifer contain such

14.5 Nature seldom, if ever, provides uniformly permeable material. In this diagram a hill is capped by permeable sandstone that overlies impermeable shale. Water soaking into the sandstone from the surface is diverted laterally by the impermeable beds. Springs result where the water table intersects the surface at the contact of the shale and sandstone.

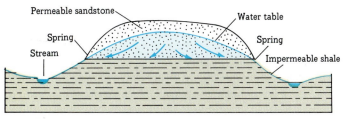

14.7 To provide a reliable water source a well must penetrate deep into the zone of saturation. In this diagram Well 1 reaches only deep enough to tap the groundwater during periods of high water table; a seasonal drop of this surface will dry up the well. Well 2 reaches to the low water table, but continued pumping may produce a cone of depression that will reduce effective flow. Well 3 is deep enough to produce reliable amounts of water, even with continued pumping during low water-table stages.

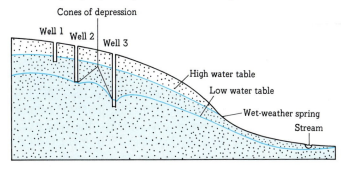

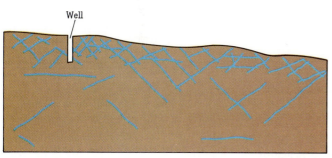

14.8 Wells may produce water from a fractured zone of impermeable rocks, such as granite. The supply of water, however, is likely to be limited, not only because the size and number of fractures decrease with depth, but also because the fractures do not interconnect.

minerals as halite or gypsum, the water it carries may well be unpotable.

The quantity of water passing through a given section in a unit of time is determined by the area of the section and by the velocity of flow. Therefore

Quantity = area of cross section of flow × velocity

or

$$Q = A \times V$$

which rearranged is

$$V = \frac{Q}{A}$$

But in Darcy's law

$$V = \frac{K(h_2 - h_1)}{l}$$

Therefore by substitution for V in $Q = A \times V$ thus

$$Q = \frac{AK(h_2 - h_1)}{l}$$

we have a statement of Darcy's laws that gives quantity of water in terms of hydraulic gradient, area of cross section of

flow, and a coefficient that includes permeability, gravity, and viscosity. It is most usefully applied to sandstone and rocks with similar characteristics of permeability.

ARTESIAN WATER

Contrary to common opinion, artesian water does not necessarily come from great depths. But other definite conditions characterize an artesian water system: (1) The water is contained in a permeable layer, the aquifer, inclined so that one end is exposed to receive water at the surface. (2) The aquifer is capped by an impermeable layer. (3) The water in the aquifer is prevented from escaping either downward or off to the sides. (4) There is enough head or pressure to force the water above the aquifer wherever it is tapped. Sometimes, if the head is sufficiently great, the water will flow out to the surface either as a well or a spring (Figure 14.9). The term **artesian** is derived from the name of a French town, Artois (originally called **Artesium** by the Romans), where this type of well was first studied.

THE SEARCH FOR UNDERGROUND WATER

Geologists are guided in the search for water in the underground by the principles outlined in the preceding sections. Others, known as *dowsers* or *water witches,* use a different approach, as discussed in Box 14.1.

THERMAL SPRINGS

Springs that bring warm or hot water to the surface are called **thermal springs, hot springs,** or **warm springs** (Figures 14.10 and 14.11). A spring is usually regarded as a thermal spring if the temperature of the water is about 6.5°C higher than the mean temperature of the air. There

14.9 The wells in the diagram meet the conditions that characterize an artesian system: (1) an inclined aquifer, (2) capped by an impermeable layer, (3) with water prevented from escaping either downward or laterally, and (4) sufficient head to force the water above the aquifer wherever it is tapped. In the well at the right of the diagram the head is great enough to force water to appear at the surface.

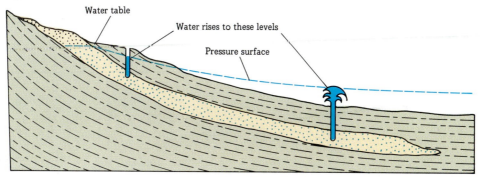

BOX 14.1   Water Witching

Water witching, often called **divining** or **dowsing,** is the search for underground water using, generally, a forked stick. The dowser grasps the arms of the stick, usually with a palms-upward grip, points it up and forward at about a 45° angle (see Figure B14.1.1), and walks the area to be surveyed. At some place during the survey the stick may twist downward with a force that the dowser's grip cannot control. People who believe in water witching are convinced that underground water in plentiful supply lies beneath the point at which the stick points.

Water witching is an ancient practice. Many believers claim Moses as the first water witch, citing Numbers 20:11, wherein Moses "struck the rock with his rod twice; and water came forth abundantly, and the congregation drank, and their cattle." Others think that he had merely developed a good geologic sense.

The first detailed description of the use of the divining rod is found in Agricola's *De Re Metallica,* published in 1556. This early treatise reports that German miners used the forked stick to prospect for minerals. Most historians agree that divining spread throughout Western Europe, was adapted to the search for water, and was carried by migrants to other continents. In the United States today there are an estimated 25,000 water witches and countless believers in the efficacy of their practice.

The questions, Does it work? and How does it work? get us into a controversy that has raged at least since Agricola gave us our first good description of divining over four centuries ago.

Those who "have the gift" and those who believe in them point to seemingly endless examples in which water has been found by divining. Many of these

examples record situations in which those needing the water had failed to find it with other methods.

Still, many remain unconvinced and call the practice nonsense. Scientists consider the evidence in favor of dowsing too anecdotal to be acceptable.

14B.1.1 A forked willow stick is held with a palms-upward grip. This is the type of stick often used by dowsers or water witches in their search for water. [Robert P. Matthews.]

In the few controlled tests that have been reported, in fact, the success rate is no better than what would be expected by pure chance. The believers counter that the art of dowsing is so sensitive that the very act of testing interferes with the process and invalidates the experiments.

When asked why the rod behaves as it does, a dowser may well shrug off the question with "I don't know, it just does." Some skeptics suggest that the way the rod is held may explain its behavior. As the stick is grasped and the hands twist the two branches (Figure B14.1.1), stresses are set up within the rod. Because the stick is in very delicate balance, any of several small muscle movements in hands or wrists can upset the balance. The rod then twists downward and the hands cannot control it. The dowsers protest that they do not move their hands or wrists. The skeptics charge that the movements are involuntary and unnoticed, and offer several possibilities as to how this might happen.

Plainly both believers and skeptics remain unshaken in their positions. Furthermore, both are sincere in their beliefs. For the record, the authors of this volume are skeptics, though they do not believe the dowsers are frauds (and not simply because the mother of one of us was a firm supporter of water witching).

You won't find much on water witching in treatises on groundwater or in the textbooks. There is, however, a fairly specialized literature, both pro and con. Here are some volumes you might wish to look up: *Water Witching,* by Evon Z. Vogt and Ray Hyman, 2nd edition, published by the University of Chicago Press, 1979, presents the skeptic's view in measured manner and contains an extensive bibliography. Kenneth Roberts, a well-known historical novelist, wrote in his lifetime three widely read books in support of water witching: *Henry Gross and His Dousing Rod* (1951); *The Seventh Sense* (1953); and *Water Unlimited* (1957), all published by Doubleday & Co., New York.

14.10 Boiling Springs, California, are hot springs in the Long Valley caldera at the foot of the eastern escarpment of the Sierra Nevada. This is an area that is being closely monitored for warnings of possible explosive volcanic activity. Heat is probably supplied indirectly from a magma chamber thought to exist below the caldera. [Pamela Hemphill.]

14.11 Mammoth Hot Springs, Yellowstone National Park, are thermal springs that have built these terraces by depositing travertine. [Jane Grigger.]

14.12 Old Faithful Geyser, Yellowstone National Park, Wyoming. The periodic eruption of a geyser is due to the particular pattern of its plumbing and its proximity to a liberal supply of heat and groundwater. [William E. Bonini.]

are over 1,000 thermal springs in the western mountain regions of the United States, 46 in the Appalachian Highlands of the East, 6 in the Ouachita area in Arkansas, and 3 in the Black Hills of South Dakota.

Most of the western thermal springs derive their heat from masses of magma that have pushed their way into the crust almost to the surface and are now cooling. In the eastern group, however, the circulation of the groundwater carries it so deep that it is warmed by the normal increase in earth heat (see **thermal gradient** in the Glossary).

The well-known spring at Warm Springs, Georgia, is heated in just this way. Long before the Civil War this spring was used as a health and bathing resort by the people of the region. Then, with the establishment of the Georgia Warm Springs Foundation for the treatment of victims of infantile paralysis, the facilities were greatly improved. Rain falling on Pine Mountain, about 3 km south of Warm Springs, enters a rock formation known as the Hollis. At the start of its journey downward, the average temperature of the water is about 16.5°C. It percolates through the Hollis Formation northward under Warm Springs at a depth of around 100 m, and then follows the rock as it plunges into the Earth to a depth of 1,140 m, 1.6 km farther north. Normal rock temperatures in the region increase about 1.8°C/100 m of depth, and the water is warmed as it descends along the bottom of the Hollis bed. At 1,140 m, the bed has been broken and shoved against an impervious layer that turns the water back. This water is now hotter than the water coming down from above, and it moves upward along the top of the Hollis Formation, cooling somewhat as it rises. Finally it emerges in a spring at a temperature of 36.5°C.

Less than 1 km away is Cold Spring, whose water comes from the same rainfall on Pine Mountain. A freak of circulation, however, causes the water at Cold Spring to emerge before it can be conducted to the depths and be warmed. Its temperature is only about 16.5°C.

GEYSERS

A **geyser** is a special type of thermal spring that ejects water intermittently and with considerable force (Figure 14.12). The word comes from the Icelandic name of a spring of this type, *geysir*, probably based on the verb *gjosa*, "to rush forth."

In general, a geyser's behavior is caused by its plumbing and a good supply of heat. Here is what probably hap-

14.13 Whether water exists as liquid, solid, or vapor depends upon both its temperature and pressure. Consider ice with the temperature and pressure of point A. If the temperature is held constant but the pressure increased, the ice will convert to liquid at point A'. Water with the temperature of point B will convert to vapor at point B' if the temperature is held constant but the pressure reduced. The transformation between water and vapor with changing temperature and pressure helps to explain the activity of geysers. See text for discussion.

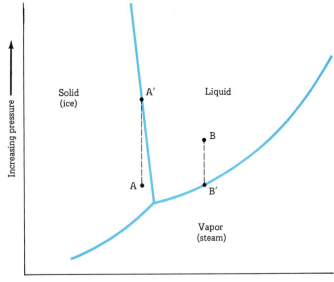

pens: Groundwater fills a natural pipe, or conduit, that opens to the surface. Hot igneous rocks, or gases given off by them, heat the water column in the pipe and raise its temperature toward the boiling point. Now, the higher the pressure on water, the higher its boiling point (Figure 14.13). Because the pressure in the water column increases downward, so does the boiling point. Eventually the column of water becomes so hot that it is close to the boiling or vapor phase throughout. At this stage a small increase in temperature or decrease in pressure will bring the water to a boil. More probably it is a small increase in temperature at some point in the pipe that brings about boiling. This, in turn, causes expansion, which pushes the overlying column of water upward until some of the water spills out at the surface. The consequent reduction in the amount of water in the conduit reduces pressure all along the water column. As a result, more of the heated column boils and steam and water gush upward to the surface. The process continues until the energy is spent. Then the pipe fills again with water and the process begins over again.

14.4
RECHARGE OF GROUNDWATER

As we have seen, the ultimate source of underground water is precipitation that finds its way below the surface of the land.

Some of the water from precipitation seeps into the ground, reaches the zone of saturation, and raises the water table. Continuous measurements over long periods of time at many places throughout the United States show an intimate connection between the level of the water table and rainfall (Figure 14.14). Because water moves relatively slowly in the zone of aeration and the zone of saturation, fluctuations in the water table usually lag a little behind fluctuations in rainfall.

14.14 Relationship between the water level in an observation well near Antigo, Wisconsin, and precipitation, as shown by records from 1945 to 1952. The water table reflects the changes in precipitation. The graphs represent 3-year running monthly averages. For example, 5.8 cm of precipitation for May means that precipitation averaged 5.8 cm/month from May 1947 to May 1950, inclusive. [After A. H. Harder and William J. Drescher, "Ground Water Conditions in Southwestern Langlade County, Wisconsin," U.S. Geol. Surv. Water Supply Paper 1294, 1954.]

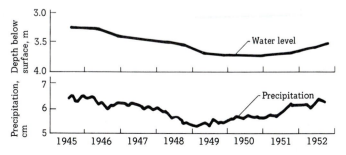

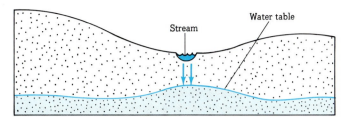

14.15 The groundwater may be recharged by water from a surface stream leaking into the underground.

Several factors control the amount of water that actually reaches the zone of saturation. For example, rain that falls during the growing season must first replenish moisture used up by plants or passed off through evaporation. If these demands are great, very little water will find its way down to **recharge** the zone of saturation. Then, too, during a very rapid, heavy rainfall most of the water may run off directly into the streams instead of soaking down into the ground. A slow, steady rain is much more effective than a heavy, violent rain in replenishing the supply of groundwater. High slopes, lack of vegetation, or the presence of impermeable rock near the surface may promote runoff and reduce the amount of water that reaches the zone of saturation. It is true, however, that some streams are themselves sources for the recharge of underground water. Water from the streams leaks into the zone of saturation, sometimes through a zone of aeration (Figure 14.15).

In many localities the natural recharge of the underground water cannot keep pace with human demands for water. Consequently attempts are sometimes made to recharge these supplies artificially. On Long Island, New York, for example, water that has been pumped out for air-conditioning purposes is returned to the ground through special recharging wells or, in winter, through idle wells that are used in summer for air conditioning. In the San Fernando Valley, California, water from the Owens Valley aqueduct is fed into the underground in an attempt to keep the local water table at a high level.

14.5
CAVES AND RELATED FEATURES

Caves are probably the most spectacular examples of the handiwork of underground water. In dissolving great quantities of solid rock in its downward and lateral course, the water fashions large rooms, galleries, and systems of underground streams as the years pass. In many caves the water deposits calcium carbonate as it drips off the ceilings and walls, building up fantastic shapes of **dripstone**. Certainly the exploration of caves can provide a great experience (Box 14.2).

Caves of all sizes tend to develop in highly soluble rocks such as limestone, $CaCO_3$, and small ones occur in

BOX 14.2   Cave Exploration

Cave exploration is exciting. It can also be dangerous to life and limb and destructive to caves. Here are a few observations for those intent upon the adventure.

The first rule is that you should never go caving alone. The second is that you should not start out without the proper equipment and instruction on its use.

You should wear a hard hat. You obviously need light. This is usually battery-powered and may be mounted on your hat. Always carry a backup.

Many caves have very steep to vertical drops. They are the sites of most serious cave injuries and deaths. Here mountain-climbing equipment is needed. In fact, some aspects of cave exploration are like mountain or rock climbing in reverse: You generally start your cave climbing by going down and usually finish by going up.

Make sure that your equipment is sound and in good working order. Don't use ropes or ladders left by pre-vious visitors because they may well have deteriorated in the cave environment even though they appear perfectly sound.

There are some very real differences between climbing in a cave and climbing in the open air. In the open air you see the entire panorama. In the underground you see only a small part of the scene at any one time, and that not always too clearly. You will be surprised also by how rapidly you lose track of time without the changing light of the upper world as an approximate guide. A real help, then, is a watch calibrated into 24 hours.

Parts of some caves are filled with water. Here further exploration is not only underground, it is underwater. Therefore it demands underwater equipment—and training.

Once in a cave, remember that the cave environment is fragile. The stalactites and stalagmites grow very slowly; certainly they do not replace themselves within a human lifetime. So do not collect them for collection's sake or even for display. Scientific investigation sometimes justifies some type of collection, but it should be done very carefully, economically, and under the supervision of the responsible scientist engaged in a legitimate scientific investigation. It is even good practice to leave undisturbed any broken pieces you may find, if only to avoid tempting others to begin unnecessary collecting when they see what you have brought back. The same caveats hold for the rare and sparse fauna found in caves.

Modern graffiti have no place in caves. And you shouldn't put survey or directional marks on walls. Not only do these disfigure the cave, but they may also confuse later visitors. Trash should leave the cave with you.

Cave etiquette, then, dictates that you leave things as you found them and take out only what you take in. The exception, of course, is the memories you will take out of an exciting experience.

---

the sedimentary rock dolomite, $CaMg(CO_3)_2$. Rock salt, $NaCl$, gypsum, $CaSO_4 \cdot 2H_2O$, and similar rock types are the victims of such rapid solution that underground caverns usually collapse under the weight of overlying rocks before erosion exposes them.

Calcite, the main component of limestone, is very insoluble in pure water. But when the mineral is attacked by water containing small amounts of carbonic acid, it undergoes rapid chemical weathering; most natural water contains carbonic acid, H_2CO_3—the combination of water, H_2O, with carbon dioxide, CO_2. The carbonic acid reacts with the calcite to form calcium bicarbonate, $Ca(HCO_3)_2$, a soluble substance that is then removed in solution. If not redeposited, it eventually reaches the ocean.

Let us look more closely at underground water as it brings about the decay of calcite. Calcite contains the complex carbonate ion, $(CO_3)^{2-}$ (built by packing three oxygen atoms around a carbon atom), and the calcium ion, Ca^{2+}. These two ions are combined in much the same way as sodium and chlorine in forming salt, or halite. The weathering or solution of calcite takes place when a hydrogen ion, H^+, approaches $(CO_3)^{2-}$. Because the attraction of hydrogen for oxygen is stronger than the attraction of carbon for oxygen, the hydrogen ion pulls away one of the oxygen atoms of the carbonate ion and, with another hydrogen ion, forms water. The two other oxygen atoms remain with the carbon atom as carbon dioxide gas, CO_2. The calcium ion, Ca^{2+}, now joins with two negative bicarbonate ions $(HCO_3)^-$, to form $Ca(HCO_3)_2$ in solution. We can express these activities as follows:

Two parts water *plus* two parts carbon dioxide *yield* two parts carbonic acid
$$2H_2O + 2CO_2 \rightleftharpoons 2H_2CO_3$$

Two parts carbonic acid *yield* two hydrogen ions *plus* two bicarbonate ions
$$2H_2CO_3 \rightleftharpoons 2H^+ + 2(HCO_3)^-$$

Two hydrogen ions *plus* two bicarbonate ions *plus* calcite *yield*
$$2H^+ + 2(HCO_3)^- + CaCO_3 \rightleftharpoons$$

water *plus* carbon dioxide *plus* calcium bicarbonate in solution
$$H_2O + CO_2 + Ca^{2+} + 2(HCO_3)^-$$

The principal chemical reaction in the formation of caves is given in the above equations. However, even though the reaction can be demonstrated in the laboratory and occurs in nature, its extension to the formation of caves is not as direct as it might first appear.

There is a difference of opinion as to whether caves

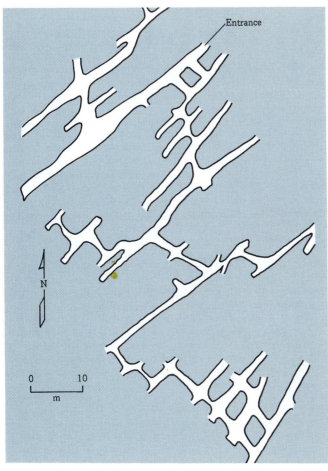

14.16 The rectangular plan of the passages in the Frieze Cave near Eddyville, Pope County, Illinois, reflects the jointing along which chemical solution has been concentrated. [From J. Harlan Bretz and S. E. Harris, Jr., *Caves of Illinois*, p. 43, Fig. 9, Report of Investigations, 215, Illinois State Geological Survey, Urbana, Ill., 1961.]

form above, at (or just below), or deep beneath the water table. Regardless of how caves form, however, we can be certain of two things. First, formation of caves is closely controlled by the fractures and bedding planes in carbonate rocks, as suggested in Figure 14.16. Second, there are two main phases in the formation of caves: the initial excavation of the caves itself, followed by a later phase of decoration of ceilings, walls, and floors.

The bizarre forms so characteristic of caves are composed of calcite deposited by underground water that has seeped down through the zone of aeration. They develop either as **stalactites,** which look like stony icicles hanging from the cave roof (from the Greek *stalaktos* "oozing out in drops"), or as **stalagmites,** heavy posts growing up from the floor (from the Greek *stalagmos,* "dripping"). When a stalactite and stalagmite meet, a **column** is formed (Figure 14.17). A stalactite forms as water charged with calcium bicarbonate in solution seeps through the cave roof. Drop after drop forms on the ceiling and then falls to the floor. But during the few moments that each drop clings to the ceiling, a small amount of evaporation takes place; some carbon dioxide is lost, and a small amount of calcium carbonate is deposited. Over the centuries a large stalactite may gradually develop. Part of the water that falls to the cave floor runs off, and part is evaporated. This evaporation again causes the deposition of calcite, and a stalagmite begins to grow upward to meet the stalactite.

On the ground surface above soluble rock, depressions sometimes develop that reflect areas where the underlying rock has been carried away in solution. These depressions, called **sinkholes** or merely **sinks**, may form in one of two different ways. In one example the limestone immediately below the soil may be dissolved by the seepage of waters downward. The process of solution may be focused

14.17 Stalactites grow downward from the cave ceiling and stalagmites grow upward from the cave floor. Temple of the Sun in the Big Room of Carlsbad Caverns, New Mexico. [National Park Service.]

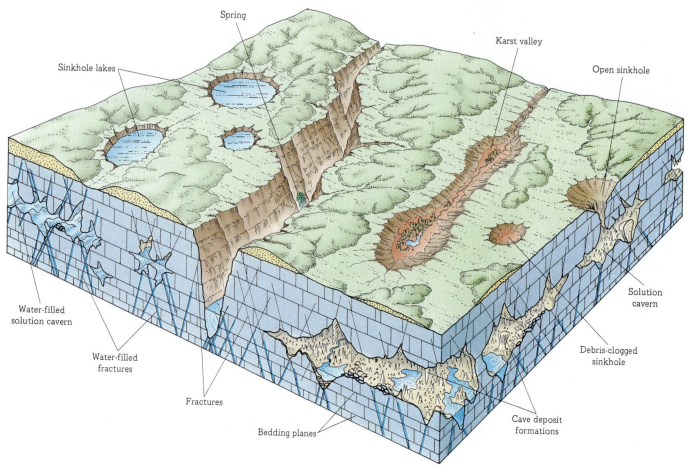

Spring

Sinkhole lakes

Karst valley

Open sinkhole

Water-filled
solution cavern

Water-filled
fractures

Fractures

Bedding planes

Cave deposit
formations

Debris-clogged
sinkhole

Solution
cavern

14.18 Solution of limestone by groundwater produces a number of different features, some of them suggested in this figure. Caverns are formed in the subsurface. Some of these may be completely filled with water. Others may be drained or partially drained and contain stalactites, stalagmites and related forms. At the surface various types of sinkholes may occur as may karst valleys whose streams disappear into the underground.

14.19 The "December Giant," a sinkhole near Montevallo in central Alabama, is believed to have formed December 2, 1972. It is about 45 m deep and measures about 100 m by 130 m at the surface. [U.S. Geological Survey.]

14.20 Solution of limestone has produced these steep-sided hills along the Li River in southern China. They are the residuals of a once-continuous layer of limestone. [Richard J. Cross.]

by local factors, such as a more abundant water supply or greater solubility of the limestone. Eventually a depression —a sink—evolves. It is probably more common for sinks to form when the surface collapses into a large cavity below. In either event, surface water may drain through the sinkholes to the underground; or if the sinks' subterranean outlets become clogged, they may fill with water to form lakes. An area with numerous sinkholes is said to display **karst topography,** from the name of a plateau in Yugoslavia and northeastern Italy where this type of landscape is well developed (see Figures 14.18 and 14.19). In some places solution has gone so far that only limestone residuals remain, standing like great steep-sided monuments (Figure 14.20).

14.6
SOME GROUNDWATER PROBLEMS CAUSED BY HUMAN USE

POLLUTION

Water underground, like water running at the surface, is subject to contamination. There are, however, differences.

Sources of contamination of surface water are usually more easily identified than those for underground water, precisely because they are at the surface where they can be spotted more readily. Furthermore, once the origin of pollution of stream water is eliminated, the stream cleanses itself fairly rapidly because of the high velocity of stream flow.

The sources of contamination of underground water are usually landfills, toxic waste dumps, and areas of extensive use of septic tanks. These sites must be sealed off from the groundwater beneath lest pollution seep slowly into it. Once in the groundwater, pollution will spread with it at a rate equal to the rate of the flow of the groundwater. It may be years before the contaminated water reaches a point of use, such as a well. Assuming that the source of pollution can be identified and stemmed, it will take many more years for the contaminated water to be flushed out of the system.

There is a second problem in correcting pollution of groundwater. The direction and flow of groundwater are determined by the characteristics of the aquifer, and there are many possible variations in permeability and structure of the aquifer. These variations can usually be determined

BOX 14.3   The Case of the "Mottled Tooth"

In 1901 the first medical reference appeared to what the dental profession later came to call "chronic endemic dental fluorosis." Translated, this means that many people in particular localities suffered a long-lasting mottling or staining of their teeth. This disease was informally known as "mottled enamel" or "Colorado brown stain."

Between the two World Wars the evidence mounted that the problem arose when children's teeth, particularly permanent teeth, erupted. More and more regions where the problem was widespread were discovered until it was being reported from all continents. In the United States it was most prevalent west of the Mississippi River, with extensive occurrence in Texas, Colorado, the Dakotas, and Iowa.

By the mid-1930s the disease was firmly linked to drinking water that contained several parts per million of fluorides. In virtually all instances groundwater was identified as the culprit. The source of fluorides in the groundwater seemed to be beds of volcanic ash to which the groundwater has

access. The happy, straightforward solution was either to change the water source to one low in fluoride or to remove the fluoride from contaminated supplies.

This, however, was by no means the end of the story, for by 1940 another and unexpected result of the extensive study of mottled teeth had become apparent. In those areas where the incidence of mottled teeth was high, the incidence of tooth decay was impressively low. Accumulating evidence convinced most investigators that the fluorides that caused mottled teeth also deterred the development of cavities (known to your dentist as *caries*). On the basis of data from laboratory and field, it appeared that small amounts of fluoride (about 1 ppm) in drinking and cooking water drastically reduced the number of cavities but, happily, did not affect the coloration of teeth.

In 1945 two cities—Grand Rapids, Michigan, and Newburgh, New York—became the first communities to add fluoride to their water. Today approximately half the people in the United States drink fluoridated water.

Also, routine fluoride treatment of children's teeth is a procedure used by many dentists. Beyond this, the toothpaste industry has been reoriented so that most dentifrices on the market today contain fluoride.

Exactly how fluorides help prevent tooth decay is still unknown, but since they do seem to, most dentists are pro-fluoride and fluoridation has the support of the American Dental Association and the American Medical Association. People who advocate fluoridation argue that it has been proved safe, can prevent two-thirds of all cavities, and reduces dental expenses accordingly.

Despite wide acceptance, fluoridation has its opponents. These people argue that fluoridation of public water supplies is compulsory medication and prevents the individual from controlling fluoride intake. They suggest that those who wish to use fluorides to prevent tooth decay can do so by individually adding fluorides to their liquids and by using fluoridated dental products.

only with extensive programs of drilling and monitoring. So it can take a long time and extensive effort to demonstrate beyond reasonable doubt the connection between, for instance, a contaminated well and the source of the trouble.

Of course, once the source is identified, there are usually political, economic, and technological problems that delay the solution of the problem. The same types of difficulties also attend the cleanup of the sources of contamination of surface water.

Groundwater may also have undesirable qualities owing to natural causes as well as to those of human origin. These depend upon the nature of the materials through which the groundwater moves. Box 14.3 describes an example of the natural pollution of groundwater and its unexpected result.

Underground water may be subject to thermal pollution as well as microbial or chemical pollution. The very process of recycling may give it the added undesirable quality of increased temperature. For instance, water used for air conditioning is withdrawn from the ground at one tem-

perature, used in cooling, and returned to the ground at a higher temperature.

SALTWATER INVASION

A particular type of pollution to which underground freshwater supplies are subject is the **invasion of salt water.** This may be salt water of an adjacent ocean environment or it may be salt water trapped in rocks lying beneath the freshwater aquifers.

Fresh water has a lower specific gravity (1.000) than normal salt water (about 1.025). Therefore fresh water will float on top of salt water. If there is little or no subsurface flow, this can lead to an equilibrium in which a body of fresh water is buoyed on salt water. In a groundwater situation in which fresh water and salt water are juxtaposed, we might have a situation such as that shown in Figure 14.21. Here an island (it could as well be a peninsula) is shown underlain by a homogeneous aquifer that extends under the adjacent ocean. Fresh water falling on the land has built up

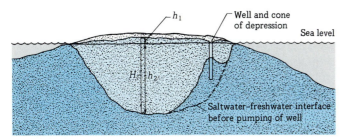

14.21 Cross section through an oceanic island (or a peninsula) underlain by homogeneous, permeable material. The fresh water forms a prism floating in hydrostatic balance with the neighboring salt water. Reduction of the groundwater table on the island will cause a change in the position of the saltwater-freshwater interface. See the text for discussion.

a prism of fresh water in hydrostatic equilibrium with the salt water that surrounds it. The height of a column of fresh water is balanced by an equal mass of salt water. In the example given, the column of fresh water (H) is equal to h_1, the height of the water table above sea level, plus h_2, the thickness of the groundwater below the sea level. A column of salt water equivalent in length to h_2 will balance a column of fresh water of length H, or $h_1 + h_2$. Working this out, we

find that $h_2 = 40h_1$. This means that if the water table is 10 m above sea level, then the freshwater-saltwater contact will be 400 m below sea level. It also means that as we lower the water table by whatever means, we raise the elevation of the saltwater-freshwater interface at the rate of 40 m of rise for each meter of lowering of the water table. It is very possible, then, that a well that originally bottomed in fresh water could, by a lowering of the water table, turn into a saltwater well, as suggested in Figure 14.21.

LAND SUBSIDENCE

Excess pumping may deplete supplies of groundwater and in certain places bring about saltwater invasion of wells. But another bothersome problem connected with heavy pumping of groundwater is **land subsidence.** It expresses itself at the surface either as broad, gentle depressions of the land or as catastrophic collapse and the formation of sinkholes.

Zones of broad subsidence have been created in California over large areas of the San Joaquin valley and secondarily in the Santa Clara valley south of Oakland and San Francisco (Figure 14.22). The phenomenon is related to the

14.22 Principal areas in California where withdrawal of groundwater has caused subsidence of the land surface. The darker shade represents greater subsidence and the lighter shade less subsidence. [After B. E. Lofgren and R. L. Klausing. "Land Subsidence Due to Groundwater Withdrawal, Tulare-Wasco Area, California," *U.S. Geol. Surv. Prof. Paper 437-B*, p. 2, 1969.]

pumping of groundwater and the resulting compaction of sediments. The Tulare-Wasco area of the San Joaquin valley is a good example. In 1905 intensive pumping for irrigation began in the area. By 1962 more than 20,000 km² of irrigable land had undergone subsidence of more than 0.3 m, and as much as 4 m of subsidence had occurred in some places. Subsidence was slow and hardly noticeable to residents. But the casings of many wells have been damaged, the problems of surveying and construction have been complicated, and the operation of irrigation districts has been endangered.

More recently excessive withdrawal of groundwater, largely for urban and suburban use, has caused subsidence over large areas in the Galveston-Houston area of Texas. A continent away in Italy, Venice has been slowly sinking because of land subsidence caused by pumping groundwater from the deltaic sediments that underlie the ancient city.

Sometimes just the weight of a building will cause subsidence by squeezing water out of the underlying sediments and thus compacting the deposits. The opera house in Mexico City is a famous example. So also is the Leaning Tower of Pisa (Figure 14.23).

14.23 The Leaning Tower of Pisa, begun in 1173 and completed in 1372, stands on deltaic sediments near the mouth of the River Arno. It started to subside differentially during the construction of the first three floors, at which time the builders made an adjustment and the next four floors tilt at a slightly smaller angle. Tilting continued and another adjustment was made for the topmost floor, which houses the bells. By the mid 1980s the tower was over 5 m out of line and continued to tilt a little over a millimeter a year.

Summary

Underground water is over 66 times as abundant as the water in streams and lakes.

Distribution is in the **zone of aeration** and in an underlying **zone of saturation.** The irregular surface that separates these zones is called the **groundwater table.**

Movement of underground water is usually by **laminar flow.**
  Porosity is the total percentage of void space in a given volume of Earth material.
  Permeability is the ability of Earth material to transmit water. Flow is driven by pressure gradients according to
  Darcy's law: $V = K(h_2 - h_1)/l$.

Groundwater in nature moves in aquifers. It comes to the surface as springs and wells.

A **spring** occurs when the ground surface intersects the water table. **Simple wells** draw water from the zone of saturation.
Artesian wells are those in which pressure drives the water to elevations above the top of the aquifer.
Thermal springs derive their heat from the cooling of igneous rocks or by a normal increase of the Earth's heat with depth.
Geysers are thermal springs marked by periodic, violent eruptions and are controlled by a particular arrangement of underground passages.

Recharge of groundwater is either natural or artificial.

Caves are usually created by the solution of limestone. The chemical reaction involves carbonic acid (formed by the combination of water and carbon dioxide) with limestone.

Formation of caves apparently takes place in two stages: (1) the creation of chambers, galleries, and tunnels, and (2) the decoration of these large openings by forms growing from the ceilings **(stalactites)** and forms growing from the floor **(stalagmites).** Solution of limestone develops an irregular "pitted" topography of depressions with no exterior surface drainage. This is called **karst topography** and is marked by numerous **sinkholes.**

Misuse of groundwater can cause numerous problems including **pollution, saltwater invasion,** and **land subsidence.**

Questions

1. The streams of the conterminous United States discharge approximately 30,000 km^3 of water into the oceans each year. How long would it take them to carry a volume of water equal to that stored as underground water in the 48 states? You will find the second number you need for the answer in the first paragraph of Section 14.1.

(We got 252.5 years.)

2. Define water table. In a landscape underlain by a uniformly permeable material, what is the shape of the water table? Draw a labeled diagram through two hills separated by a stream valley to show this relationship.

3. Distinguish between porosity and permeability in earth materials.

4. What is Darcy's law?

5. How is the slope of the groundwater table defined?

6. Draw a labeled diagram to show the nature of flow of groundwater through a uniformly permeable material. The diagram you used in Question 2 would be a good place to start.

7. What is a perched water table? Diagram and label an example.

8. Myth has it that a well must be very deep to qualify as an artesian well. Although no particular depth defines an artesian well, there are several conditions that must be met before we can say we have an artesian well. What are they? Here again, a neat, simple, concisely labeled diagram will help to give precision to your answer.

9. You should understand the chemistry involved in the solution of limestone caves and in the deposition of stalactites and stalagmites that decorate them. Do you?

10. What is karst topography?

11. You find that the water table in your island well is 5 m above sea level. How far down will fresh water persist before salt water is encountered? Your well is 50 m deep, and drawdown due to pumping produces a cone of depression of 4-m depth. What problems do you face?

(We got a depth of 200 m of fresh water beneath sea level — 5m × 40. With a 4-m cone of depression, salt water will rise 160 m and thus will be 5 m above the bottom of the well, and that is a problem!)

12. Refer to Figure 14.13. What would happen if ice with temperature and pressure of point A were to undergo decreasing pressure at a constant temperature? Give two ways in which the vapor with temperature and pressure of point B^1 could be converted to liquid.

Supplementary Readings

FETTER, C. W., JR. *Applied Hydrogeology,* Charles E. Merrill Publishing Co., Columbus, Ohio, 1980.
A textbook designed for an undergraduate course on underground water.

FREEZE, R. ALLEN, AND JOHN A. CHERRY *Groundwater,* Prentice Hall, Englewood Cliffs, N.J., 1979.
A fairly advanced text on the subject.

HOLZER, THOMAS L., ED. *Man-Induced Land Subsidence: GSA Reviews in Engineering Geology VI,* Geological Society of America, Boulder, Colo., 1985.
A series of nine papers dealing with subsidence of the ground surface due to human activity.

MOORE, GEORGE W., AND G. NICHOLAS SULLIVAN *Speleology: The Study of Caves,* 3rd ed., Zephyrus Press, Teaneck, N.J., 1978.
A short book but it covers a variety of subjects, including cave formation, cave atmosphere, and cave life from microorganisms to man.

RINEHART, JOHN S. *Geysers and Geothermal Energy,* Springer-Verlag, New York, 1980.
A comprehensive treatment of geysers with an overview of geothermal energy.

SWEETING, MARJORIE M. *Karst Landforms,* Columbia University Press, New York, 1973.
An excellent volume on caves, karst, and associated landscape.

TODD, DAVID K. *Groundwater Hydrology,* 2nd. ed., John Wiley & Sons, New York, 1980.
A basic text on the subject.

UNESCO *Proceedings of the International Symposium on Land Subsidence,* Tokyo, 1979, UNESCO, Paris, 1970.
Contains case histories and discussions of many examples of different types of land subsidence, including those due to groundwater.

Rock Creek Canyon is in the Beartooth Mountains of Montana and Wyoming just northeast of Yellowstone National Park. Its broad, open U-shaped profile testifies to the past presence of mountain glaciers.

GLACIATION

"For only a glacier produces all these features at once."

Louis Agassiz, Etudes sur les glaciers, *1840*

The Swiss naturalist Louis Agassiz did more than any other person to establish the reality of the "Great Ice Age." The method he used to do this was simple. He studied modern glaciers. He described their erosional effects on bedrock and on the fragments enclosed within the ice. He noted the deposits of glaciers and their characteristic forms. He argued that "only a glacier produces all these features at once." Then, when he found similar features beyond the limits of the present ice, he concluded that the glaciers had had a much vaster extent in the past—that there had been in times gone by a "Great Ice Age," a term he coined.

In this chapter we examine the formation of *glacier ice* and the different *types of glaciers* it forms. We look at a glacier's *zones of nourishment and wastage* and find that a glacier may be either *warm* or *cold*.

Spectacular forms of *erosion* include such features as *cirques, horns, arêtes, U-shaped valleys, hanging valleys,* and *fiords*. On a smaller scale, glacier ice produces *striations, grooves,* and *polish*.

Glaciers deposit directly an unsorted material, *till*, which we find in *moraines* and the less common streamlined hills, *drumlins*. The water from melting glaciers lays down stratified material called *outwash*, which forms *outwash plains, kames, kame terraces,* and *eskers* among other features.

The *proof of former glaciations* lies in applying the doctrine of *uniformitarianism* even as Louis Agassiz did in the 1830s. We will find that major glaciations occurred several times from the Precambrian to the Pleistocene.

As we look for the causes of glaciation we will combine long-term changes of climate, made possible by the *drift of continents* into polar positions, with short-term changes caused by variations of the *orbital motions* of the Earth in its revolution around the Sun.

It is risky to predict the future, but if history is a reliable guide, we can expect to have a *major future glaciation*.

15.1

GLACIERS

A **glacier** is a mass of ice, formed by the recrystallization of snow, that flows forward—or has flowed at some time in the past—under the influence of gravity. This definition eliminates the pack ice formed from seawater in polar latitudes and, by convention, the icebergs, even though they are large fragments broken from the seaward end of glaciers.

Like streams on the surface and water underground, glaciers depend on the oceans for their nourishment. Some of the water drawn up from the oceans by evaporation falls on the land in the form of snow. If the climate is right, part of the snow may last through the summer without melting. Gradually, as the years pass, the accumulation may grow deeper and deeper until at last a glacier is born. In areas where the winter snowfall exceeds the amount of snow that melts away during the summer, stretches of perennial snow known as **snowfields** cover the landscape. At the lower limit of a snowfield lies the **snow line**. Above the snow line glacier ice may collect in the more sheltered areas of the snowfields. The exact position of the snow line varies from one climatic region to another. In polar regions, for example, it reaches down to sea level, but near the equator it recedes to the mountaintops. In the high mountains of East Africa it ranges from elevations of 4,500 to 5,400 m. The highest snow lines in the world are in the dry regions known as the "horse latitudes," between 20° and 30° north and south of the equator. Here the snow line reaches higher than 6,000 m.

Fresh snow falls as a feathery aggregate of complex and beautiful crystals with a great variety of patterns. All the crystals are basically hexagonal, however, and all reflect their internal arrangement of hydrogen and oxygen atoms (Figure 15.1). Snow is not frozen rain; rather it forms from the crystallization of water vapor at temperatures below the freezing point. This is called **sublimation,** the process by which material in the gaseous state passes directly to the solid state without first becoming liquid.

After snow has been lying on the ground for some time, it changes from a light fluffy mass to a heavier granular material called **firn** or **névé.** *Firn* derives from a German adjective meaning "of last year," and *névé* is a French word from the Latin for "snow." Solid remnants of large snowbanks, those tiresome vestiges of winter, are largely firn. Several processes are at work in the transformation of snow into firn. The first is sublimation—only this time solid material passes to the gaseous from the solid state rather than in the opposite direction, as in the formation of snow. Molecules of water vapor escape from the snow, particu-

15.1 Snowflakes exhibit a wide variety of patterns, all hexagonal and all reflecting the internal arrangement of hydrogen and oxygen. It is from snowflakes that glacier ice eventually forms.

larly from the edges of the flakes. Some of the molecules attach themselves to the center of the flakes, where they adapt themselves to the structure of the crystals of snow. Then, as time passes, one snowfall follows another, and the granules that have already begun to grow as a result of sublimation are packed tighter and tighter together under the pressure of the overlying snow.

Water has the rare property of increasing in volume when it freezes; conversely, it decreases in volume as the ice melts. But the cause and effect may be interchanged: If added pressure on the ice squeezes the molecules closer together, the ice may melt, as we saw earlier in Figure 14.13. This, for example, makes ice skating possible. As the blade of the skate presses down on the ice, a small amount of ice melts and the film of water that results allows the skate to glide smoothly over the ice. This is called **pressure melting** and it plays an important role in the formation of glacier ice. At those points where individual particles of firn are in contact melting can take place as pressure increases because of the weight of overlying firn and snow. The resulting meltwater trickles down and refreezes on still lower granules at points where they are not yet in contact. And all through this process the basic hexagonal structure of the original crystals of snow is maintained.

Gradually, then, a layer of granules of firn, ranging from a fraction of a millimeter to approximately 3 or 4 mm in diameter, is built up. The thickness of this layer varies, but the average seems to be 30 m on many mountain glaciers.

The firn itself undergoes further change as continued pressure forces out most of the air between the granules, reduces the space between them, and finally transforms the firn into **glacier ice,** a true solid composed of interlocking crystals. In large blocks it is usually opaque and blue-gray because of the air and the fine dirt it contains.

The ice crystals that make up glacier ice are minerals; the mass of glacier ice, made up of many interlocking crystals, is a metamorphic rock because it has been transformed from snow into firn and eventually into glacier ice.

Later in this section, as we study the movement of glaciers, we will find that new ice moves deep into the glacier from its place of formation near the surface. When it reappears at the glacier's snout, it has undergone additional transformations: It has been more extensively metamorphosed. The ice crystals, originally a few millimeters in size, have grown to several centimeters in maximum dimension. Furthermore, the original, generally equidimensional texture has changed to one in which the crystals may be elongated. This traveled ice is also marked by bands that reflect planes of shear, and indeed recrystallization of the ice has taken place along these planes of shearing. Truly, glacier ice represents several stages of metamorphism. The stages resemble those of the metamorphic rocks we studied in Chapter 7, except of course glacier ice melts at a much lower temperature.

CLASSIFICATION OF GLACIERS

The glaciers of the world fall into four principal classifications: valley glaciers, piedmont glaciers, ice sheets, and continental glaciers.

Valley glaciers are streams of ice that flow down the valleys of mountainous areas. Like streams of running water, they vary in width, depth, and length. A branch of the Hubbard Glacier in Alaska is 120 km long, whereas some of the valley glaciers that dot the higher reaches of our western mountains are only a few hundred meters in length. Valley glaciers that are nourished on the flanks of high mountains and that flow down the mountainsides are sometimes called **mountain glaciers** or **Alpine glaciers** (Figures 15.2 and 15.3).

Piedmont glaciers form when two or more glaciers emerge from their valleys and coalesce to form an apron of moving ice on the plains below.

Ice sheets are broad, moundlike masses of glacier ice that tend to spread radially under their own weight. The Vatna Glacier of Iceland is a small ice sheet measuring about 120 by 160 km and 225 m in thickness. A localized sheet of this sort is sometimes called an **icecap.** The term **continental glacier** is usually reserved for great ice sheets that obscure the mountains and plains of large sections of a continent, such as those of Greenland and Antarctica. On Greenland ice exceeds 3,000 m in thickness near the center of the icecap. Ice in Antarctica averages about 2,300 m in thickness, and in some places this ice is over 4,000 m thick.

15.2 The collecting ground for the Eldridge Glacier is high on the flanks of Mt. McKinley, at 6,195 m the highest peak in North America. Here several smaller glaciers converge to form the main stream of ice. [U.S. Air Force Cambridge Research Laboratories, photograph no. 244 by T/Sgt Roland E. Hudson, October 4, 1965. courtesy Richard S. Williams, Jr., U.S. Geological Survey.]

15.3 The Tasman Glacier in the Mt. Cook National Park, New Zealand, flows down its U-shaped valley. The slight pinkish color of the snow in the foreground is caused by a blue-green algae that finds this habitat congenial. [C. Sharyn Mcgee.]

DISTRIBUTION OF MODERN GLACIERS

Modern glaciers cover approximately 10 percent of the land area of the world. They are found in widely scattered locations in North and South America, Europe, Asia, Africa, Antarctica, and Greenland, and on many of the north polar islands and the Pacific islands of New Guinea and New Zealand. A few valley glaciers are located almost on the equator. Mount Kenya in East Africa, for instance, only 0.5° from the equator, rises over 5,100 m into the tropical skies and supports at least ten valley glaciers.

The total land area covered by existing glaciers is estimated as 17.9 million km², of which the Greenland and Antarctic ice sheets account for about 96 percent. The Antarctic ice sheet covers approximately 15.3 million km², and the Greenland sheet covers about 1.8 million km². Small icecaps and numerous mountain glaciers scattered around the world account for the remaining 4 percent.

NOURISHMENT AND WASTAGE OF GLACIERS

When the weight of a mass of snow, firn, and ice above the snow line becomes great enough, movement begins and a glacier is created. The moving stream flows downward across the snow line until it reaches an area where the loss through evaporation and melting is so large that the forward edge of the glacier can push no further. A glacier, then, can be divided into two zones: a **zone of accumulation** and a **zone of wastage.** Both are illustrated in Figure 15.4.

The position of the front of a glacier depends on the relationship between the glacier's rate of nourishment and its rate of wastage. When nourishment just balances wastage, the front becomes stationary, and the glacier is said to be in equilibrium. This balance seldom lasts for long, however, for a slight change in either nourishment or wastage will cause the front to advance or recede.

15.4 A glacier is marked by a zone of accumulation and a zone of wastage. Within a glacier ice may lie either in the zone of fracture or deeper in the zone of flow. A valley glacier originates in a basin, the *cirque,* and it is separated from the headwall of the cirque by a large crevasse known as the *bergschrund.*

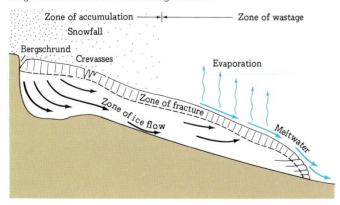

Today most of the glaciers of the world are receding. With only a few exceptions this process has been going on since the latter part of the nineteenth century, although at varying rates. A striking feature of modern glaciers is that they follow the same general pattern of growth and wastage the world over and serve as indicators of widespread climatic changes.

Valley glaciers are nourished not only by snow falling directly on them but also by masses of snow that avalanche down steep slopes along their course. In fact, according to one interpretation, avalanches caused by earthquakes have enabled certain glaciers to advance in a single month as far as they would have if fed by the normal snowfall of several years.

A particular form of wastage occurs if a glacier terminates in a body of water. Here great blocks of ice can break off and float away in a process called **calving.** This is the action that produces the **icebergs** of the polar seas.

GLACIER MOVEMENT

Except in rare cases glaciers move only a few centimeters or at most a few meters per day. That they actually move, however, can be demonstrated in several ways. The most conclusive test is to measure the movement directly, by emplacing a row of stakes across a valley glacier. As time passes, the stakes move down-valley with the advancing ice, the center stakes more rapidly than those near the valley walls. A second source of evidence is provided by the distribution of rock material on the surface of a glacier. When we examine the boulders and cobbles lying along a valley glacier, we find that many of them could not have come from the walls immediately above and that the only possible source lies up-valley. We can infer, then, that the boulders must have been carried to their present position on the back of the glacier. Another indication of glacier movement is that when a glacier melts, it often exposes a rock floor that has been polished, scratched, and grooved. The simplest explanation of this surface is to assume that the glacier actually moved across the floor, using embedded debris to polish, scratch, and groove it.

Clearly, then, a glacier moves (Figure 15.5). In fact, different parts of it move at different rates. But although we know a good bit about how a glacier flows forward, certain phases are not yet clearly understood. In any event, we can distinguish two zones of movement: (1) an upper zone, between 30 and 60 m thick, which reacts like a brittle substance—that is, it breaks sharply rather than undergoing gradual, permanent distortion; and (2) a lower zone, which because of the pressure of the overlying ice behaves like a plastic substance. The first is the **zone of fracture;** the second is the **zone of flow** (Figure 15.4).

As plastic deformation takes place in the zone of flow, the brittle ice above is carried along. But the zone of flow moves forward at different rates—faster in some parts,

15.5 The flow patterns in the Malaspina Glacier, Alaska, are emphasized by the dark bands of ice-carried debris. [Austin Post, University of Washington.]

more slowly in others—and the rigid ice in the zone of fracture is unable to adjust itself to this irregular advance. Consequently the upper part of the glacier cracks and shatters, giving rise to a series of deep, treacherous **crevasses** (Figures 15.4 and 15.6).

One student of the Earth has drawn a tempting analogy between glacier flow and plate tectonics. He sees the glacier's zone of fracture, where crevasses develop, as comparable to the Earth's lithosphere. The zone of flow in a glacier behaves plastically, as does the asthenosphere. As the ice in the zone of flow moves, it carries along the zone of brittle crevasses just as the lithosphere is moved on the underlying asthenosphere.

A glacier attains its greatest velocity in midstream, for the sides and bottom are retarded by friction against the valley walls. In this respect, the movement of an ice stream resembles that of a stream of water. The mechanics of ice flow, however, are still under study—a study made difficult by the fact that we cannot actually observe the zone of flow because it lies concealed within the glacier. Yet the ice from the zone of flow eventually emerges at the snout of the glacier, and there it can be studied. By the time it has emerged, it is brittle but bears the imprint of its passage through the glacier, as we pointed out earlier in this section when we identified glacier ice as a metamorphic rock.

Measurements of the direction of flow of ice in modern glaciers show that in the zone of wastage the flow of ice is upward toward the ice surface at a low angle. In the zone of accumulation the direction of movement is downward from the ice surface at a low angle (Figure 15.4).

No matter how the glacier ice moves in the interior of a glacier, there is also a movement involving the entire ice tongue or sheet. The glacier literally slips along its base, moving across the underlying ground surface. This **basal slip** is added to the internal flow of the glacier to give the total movement of the ice body. Although glaciers generally move very slowly, some surge forward at remarkably high rates, as described in Box 15.1.

TEMPERATURES OF GLACIERS

All glaciers are cold, but some are colder than others. Therefore students of glaciers speak of "temperate" and "cold" glaciers. A **cold glacier** is one in which no surface melting occurs during the summer months; its temperature is always below freezing. A **temperate glacier** is one that reaches the melting temperature throughout its thickness during the summer season.

Obviously no glacier can exist above the melting point of ice; yet there are glaciers that hover at this temperature. All glaciers must form at a subfreezing temperature, and therefore we must ask how a glacier warms from its formation temperature to the melting point. The Sun's heat cannot penetrate more than a few meters into the poorly conducting ice, so some other mechanism must operate to warm a glacier. One such mechanism involves the downward movement of water from the surface of the glacier. There the heat of the Sun melts surface and near-surface ice, and the meltwater percolates downward into the glacier. When eventually it freezes, it gives up heat at the rate of 80 cal/g (gram) of water. (Because each gram of ice requires 0.5 cal to rise 1°C, these 80 cal increase the temperature of 160 g by 1°C.) In addition, there is the normal heat flow from the Earth to the surface.

In a temperate glacier meltwater may be present in large amounts both within the ice and at its base, and facili-

15.6 Crevassed surface of the Columbia glacier, Chugach Mountains, Alaska. [Austin Post, University of Washington.]

BOX 15.1   Surging Glaciers

Most well-behaved glaciers move fairly slowly, a few meters to a hundred or somewhat more meters per year. But as more and more observations are made, we are finding that some glaciers that have for many years flowed with a certain slow sedateness suddenly move very rapidly. We call such glaciers **surging glaciers** (Figure B15.1.1).

An early report by a villager in the Karakoram Mountains in what is now Pakistan describes a surge there.

> One day . . . we noticed that the water in the irrigation channels was very muddy and was coming down in larger quantities than usual. We went up the nullah [Hindi for "ravine"] to see what had happened and saw the glacier advancing. It came down like a snake, quite steadily: we could see it moving. . . . At the same time water and mud gushed out from the ice while it was still advancing and flooded . . . some fields. . . . The ice continued to move for eight days and eight nights and came to a stop about forty yards from the Hispar River.

The largest reported surge occurred on the Bråsellbreen Icecap in Spitzbergen. The ice advanced 20 km along a 21-km front sometime between 1935 and 1938, as shown by aerial photographs. In the summer of 1986 the Hubbard Glacier in Alaska surged at 12 m/day, cutting off the Russell fiord from the sea and turning it into a lake, at least temporarily. The most rapid surge known was 110 m/day on the Kutiah Glacier, northern India.

Glaciers surge because of an instability

B15.1.1   Rapid surges along the Yanert glacier in the Alaska Range have displaced morainal loops down-valley. The photograph was taken in 1964, and the latest surge at that time had been in 1942. [Austin Post, University of Washington.]

that allows gravity to tear them loose from the ground surface on which they lie. The cause of this instability is not yet completely understood, but it involves the accumulation of an excess of water at the glacier's base. One way this water may accumulate is through a thickening of the ice and thus an increase of pressure in the glacier. At the base this increase may be sufficient to allow some melting there.

Surges are as yet unrecorded from the Antarctic icecap. Nevertheless, it has been estimated that should a surge occur there, it might well involve the movement into the Antarctic Ocean of enough ice to raise the world's sea level between approximately 10 and 30 m during a period of a few years or a few tens of years. The results would make some science fiction tame reading by comparison.

tate glacier movement. The result is that a temperate glacier has a greater velocity than a cold glacier and is a more effective agent of erosion. The presence of excess water at the base of a glacier may be one reason why some glaciers surge (see Box 15.1).

15.2
RESULTS OF GLACIATION

EROSIONAL PROCESSES

Glaciers have special ways of eroding, transporting, and depositing Earth materials. A valley glacier, for example, acquires debris by means of **frost action, landsliding,** and **avalanching.** Fragments pried loose by frost action clatter down from neighboring peaks and come to rest on the back of the glacier. And great snowbanks, unable to maintain themselves on the steep slopes of the mountainsides, avalanche downward to the glacier, carrying along quantities of rock debris and rubble. This material is either buried beneath fresh snow or avalanches, or tumbles into gaping crevasses in the zone of fracture and is carried along by the glacier.

When a glacier flows across a fractured or jointed stretch of bedrock, it may lift up large blocks of stone and move them off. This process is known as **plucking,** or **quarrying.** The force of the flow of ice itself may suffice to pick up the blocks, and the action may be helped along by the great pressures that operate at the bottom of a glacier. Suppose the moving ice encounters a projection or rock jutting up from the valley floor. As the glacier ice forces itself over and around the projection, the pressure on the ice is increased, and some of the ice around the rock may melt. This meltwater trickles toward a place of lower pressure, perhaps into a crack in the rock itself. There it refreezes, forming a strong bond between the glacier and the rock. Continued movement by the glacier may then tear the block out of the valley floor.

At the heads of valley glaciers plucking and frost action sometimes work together to pry rock material loose. Along the back walls of the collection basins of mountain glaciers, great hollows called **cirques** develop in the mountainside. As the glacier begins its movement downslope, it pulls slightly away from the back wall, forming a crevasse known as a **bergschrund.** One wall of the bergschrund is formed by the glacier ice; the other is formed by the nearly vertical cliff of bedrock. During the day meltwater pours into the bergschrund and fills the openings in the rock. At night the water freezes, producing pressures great enough to loosen blocks of rock from the cliff. Eventually these blocks are incorporated into the glacier and are moved away from the headwall of the cirque. The cirque is thus enlarged by headward erosion.

The streams that drain from the front of a melting glacier are charged with **rock flour,** very fine particles of pulverized rock. So great is the volume of this material that it gives the water a characteristically grayish-blue color similar to that of skim milk. Here, then, is further evidence of the grinding power of the glacier mill.

Glaciers also pick up rock by means of abrasion. As the moving ice drags rocks, boulders, pebbles, sand, and silt across the glacier floor, the bedrock is cut away as though by a great rasp or file. And the cutting tools themselves are abraded. It is this mutual abrasion that produces rock flour and gives a high **polish** to many of the rock surfaces across which a glacier has ridden. But abrasion sometimes produces scratches, or **striations,** on both the bedrock floor and on the grinding tools carried by the ice (Figure 15.7). More extensive abrasion creates deep gouges, or **grooves,** in the bedrock (Figure 15.8). The striations and grooves along a surface of bedrock show the direction of the glacier's movement. At Kelleys Island, in Lake Erie north of Sandusky, Ohio, the bedrock is marked by grooves 0.3 to 0.6 m deep and 0.6 to 1 m wide. In the Mackenzie valley, west of Great Bear lake in Canada, grooves as wide as 45 m have been described, with an average depth of 15 m and lengths from several hundred meters to over 1 km.

FEATURES OF EROSION

The erosional effects of glaciers are not limited to the fine polish and striations mentioned above. Glaciers also operate on a much grander scale, producing spectacularly sculptured peaks and valleys in the mountainous areas of the world (Figure 15.9).

15.7 Glacial polish and striations on basalt at the Devil's Post Pile, California. The striations are oriented across the rock exposure from left to right. The rough, dull areas represent postglacial weathering of the polished surface. Diameter of an individual basalt prism is about 40 cm. [William C. Bradley.]

15.8 Grooves with striations near the Parliament Buildings, Victoria, British Columbia. Glacier flow is away from observer.

Cirques As we have seen, a **cirque** is the basin from which a mountain glacier flows, the focal point for the glacier's nourishment. After the glacier has disappeared and all its ice has melted away, the cirque is revealed as a great amphitheater, or bowl, with one side partially cut away.

The headwall rises a few scores of meters to over 900 m above the floor, often as an almost vertical cliff. The floor of a cirque lies below the level of the low ridge separating it from the valley of the glacier's descent. A lake that may fill the bedrock basin of the cirque is called a **tarn** (Figure 15.10).

A cirque begins with an irregularity in the mountainside formed either by preglacial erosion or by a process called **nivation,** a term that refers to erosion beneath and around the edges of a snowbank. Nivation works in the following way: When seasonal thaws melt some of the snow, the meltwater seeps down to the bedrock and trickles along the margin of the snowbank. Some of the water works its way into cracks in the bedrock, where it freezes again, producing pressures that loosen and pry out fragments of the rock. These fragments are moved off by solifluction, forming a shallow basin. As this basin gradually grows deeper, a cirque eventually develops. Continued accumulation of snow leads to the formation of firn; if the basin becomes deep enough, the firn is transformed into ice. Finally the ice begins to flow out of the cirque into the valley below, and a small glacier is born.

The actual mechanism by which a cirque is enlarged is still a matter of dispute. Some observers claim that frost action and plucking on the cirque wall within the bergschrund are sufficient to produce precipitous walls hundreds of meters in height. Others, however, point out that the bergschrund, like all glacier crevasses, remains open only in the zone of fracture, 60 m at most. Below that depth pressures cause the ice to deform plastically, closing the bergschrund.

15.9 Erosion by glaciers has deepened and widened valleys and sharpened ridges and peaks here in the Grand Teton Mountains of Wyoming. [Franklyn B. Van Houten.]

15.10 A cirque with tarn near the Goodnews River in Southwestern Alaska.

This debate has led to the development of the so-called **meltwater hypothesis** to explain erosion along the headwalls below the base of the bergschrund. The proponents of this theory explain that meltwater periodically descends the headwalls of cirques, melts its way down behind the ice and into crevices in the rock, and there freezes at night and during cold spells. The material thus broken loose is then removed by the glacier, and erosion of the cirque proceeds mainly by this form of recession of the headwall.

Horns, Arêtes, and Cols   A **horn** is a spire of rock formed by the headward erosion of a ring of cirques around a single high mountain. When the glaciers originating in these cirques finally disappear, they leave a steep, pyramidal mountain outlined by the headwalls of the cirques (Figure 15.11). The classic example of a horn is the famous Matterhorn of Switzerland.

An **arête** (French for "fishbone," "ridge," or "sharp edge") is formed when a number of cirques gnaw into a ridge from opposite sides. The ridge becomes knife-edged, jagged, and serrated (Figure 15.11).

A **col** (from the Latin *collum,* "neck"), or pass, is fashioned when two cirques erode headward into a ridge from opposite sides. When their headwalls meet, they cut a sharp-edged gap in the ridge (Figure 15.11).

Glaciated Valleys   Instead of fashioning their own valleys, glaciers probably follow the course of preexisting valleys, modifying them in a variety of ways; usually the resulting valleys have a **broad U-shaped cross profile,** whereas mountain valleys created exclusively by streams have **narrow V-shaped cross profiles.** Because the tongue of an advancing glacier is relatively broad, it tends to broaden and deepen the V-shaped stream valleys, transforming them

15.11   The progressive development of cirques, horns, arêtes, and cols. (a) Valley glaciers have produced cirques; but since erosion has been moderate, much of the original mountain surface has been unaffected by the ice. The result of more extensive glacial erosion is shown in (b). In (c) glacial erosion has affected the entire mass and has produced not only cirques but also a horn, jagged, knife-edged arêtes, and cols. [After William Morris Davis, "The Colorado Front Range," *Ann. Assoc. Am. Geo.,* vol. 1, p. 57, 1911.]

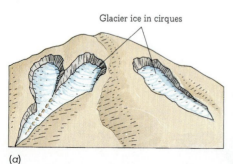

(a)

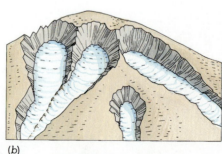

(b)

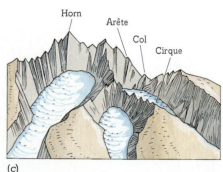

(c)

15.12 A mountainous area before, during, and after glaciation. (a) The unglaciated mountains have V-shaped valleys; smoothly rounded divides and ridge spurs overlap as they come to the valley bottoms. (c) With the disappearance of the ice we find U-shaped valleys, hanging valleys, and truncated spurs, as well as horns, arêtes, cols, and cirques. [Redrawn from William Morris Davis, "The Sculpture of Mountains by Glaciers," *Scot. Geog. Mag.*, vol. 22, pp. 80–83, 1906.]

into broad U-shaped troughs (Figures 15.12 and 15.13). And because the moving body of ice has difficulty manipulating the curves of a stream valley, it tends to straighten and simplify its course. In this process of straightening, the ice snubs off any spurs of land that extend into it from either side. The cliffs thus formed are shaped like large triangles, or flatirons, with their apex upward, and are called **truncated spurs** (Figure 15.12).

A glacier also gives a mountain valley a characteristic longitudinal profile from the cirque downward. The course of a glaciated valley is marked by a series of **rock basins,** probably formed by plucking in areas where the bedrock was shattered or closely jointed. Between the basins are relatively flat stretches of rock that was more resistant to plucking. As time passes, the rock basins may fill up with water, producing a string of lakes that are sometimes re-

15.13 Glaciated landscape in the Himalaya Mountains of Bhutan. The U-shaped valley in the foreground lies at an elevation of about 4,100 m. About 3 km distant from the observer it is joined from the left by a valley in which the snout of a valley glacier is visible. Out beyond the glacier a lake is held in by a terminal moraine and just toward the observer are the remnants of an older terminal moraine. The morainal dam holding in the lake was partially breached in the early 1950s (the outlet channel is visible as a white V-shaped notch across the moraine). The lake was rapidly lowered to its present level and flood water caused widespread loss of life and damage downstream at the religious center of Punaka. The gray strip along the river is a "trim" line below which vegetation was swept away during the flood and has yet to recover. Up the main valley beyond is the nose of another glacier. In front of it are high lateral moraines. The unvegetated sides of the moraines, facing inward toward the axis of the valley, are white because they are made up of the debris resulting from the mechanical weathering of white granite, a dominant rock type in the region. [Lincoln Hollister.]

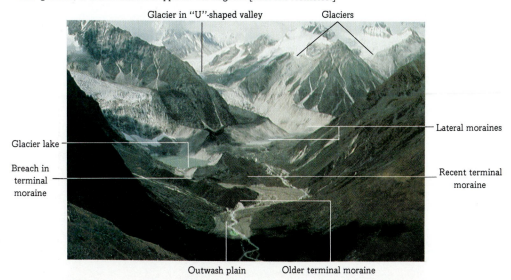

ferred to as **paternoster lakes** because they resemble a string of beads.

Hanging valleys are another characteristic of mountainous areas that have undergone glaciation. The mouth of a hanging valley is left stranded high above the main valley through which a glacier has passed. As a result, streams from hanging valleys plummet into the main valley in a series of falls and plunges (Figure 15.14). Hanging valleys may be formed by processes other than glaciation, but they are almost always present in mountainous areas that formerly supported glaciers and are thus very characteristic of past valley glaciation. What has happened to leave these valleys stranded high above the main valley floor? During the time when glaciers still moved down the mountains, the greatest accumulation of ice would tend to travel along the central valley. Consequently the erosion there would be greater than in the tributary valleys with their relatively small glaciers, and the main valley floor would be cut correspondingly deeper. This action would be even more pronounced where the main valley was underlain by rock that was more susceptible to erosion than was the rock under the tributary valleys. Finally, some hanging valleys were probably created by the straightening and widening action of a glacier on the main valley. In any event, the difference in level between the tributary valleys and the main valley does not become apparent until the glacier has melted.

Cutting deep into the coasts of Alaska, Norway, western Canada, Greenland, Labrador, and New Zealand are deep, narrow arms of the sea — **fiords** (Figure 15.15). Actually these inlets are stream valleys that were modified by glacier erosion and then were partially filled by the sea. The deepest known fiord, Vanderford in Vincennes Bay, Antarctica, has a maximum depth of 2,287 m.

15.15   The ocean has flooded partway up this glaciated valley on the southwest coast of Norway to create the Hardanger Fiord, which extends inland 180 km.

Some valleys have been modified by continental glaciers rather than by the valley glaciers that we have been discussing so far. The valleys occupied by the Finger Lakes of central New York State are good examples. These long, narrow lakes lie in basins that were carved out by the ice of a continental glacier. As the great sheet of ice moved down from the north, its progress seems to have been checked by

15.14   Bridal Veil Falls tumbles from a hanging valley into "U"-shaped Yosemite Valley.

15.16   The smooth, gentle slope of Lembert Dome in the Tuolumne Meadows area of Yosemite National Park has been produced by glacial abrasion and the steep slope by plucking. Ice moved from right to left. [Pamela Hemphill.]

the northern scarp of the Appalachian Plateau. But some of the ice moved up into the valleys that had previously drained the plateau. The energy concentrated in the valleys was so great that the ice was able to scoop out the basins that are now filled by the Finger Lakes.

Asymmetric Rock Knobs and Hills   In many places glacial erosion of bedrock produces small, rounded, asymmetric hills with gentle, striated, and polished slopes on one side and steeper slopes lacking polish and striations on the opposite side. The now-gentle slope faced the advancing glacier and was eroded by abrasion. The opposite slope was steepened by the plucking action of the ice as it rode over the knob (Figure 15.16).

Large individual hills have the same asymmetric profiles as the smaller hills. Here, too, the gentle slope faced the moving ice.

TYPES OF GLACIAL DEPOSITS

The debris carried along by a glacier is eventually deposited either because the ice that holds it melts or, less commonly, because moving ice plasters the debris across the land surface.

The general term **drift** is applied to all deposits that are laid down directly by glaciers or that, as a result of glacial activity, are laid down in lakes, oceans, or streams. The term dates from the days when geologists thought that the unconsolidated cover of sand and gravel blanketing much of Europe and America had been made to drift into its present position either by the sea or by icebergs. Drift can be

divided into two general categories: **unstratified** and **stratified**.

Deposits of Unstratified Drift   Unstratified drift laid down directly by glacier ice is called **till**. It is composed of rock fragments ranging in size from boulders weighing several tonnes to tiny clay and colloid particles, and all are mixed together (Figures 15.17 and 15.18). Many of the large pieces are striated, polished, broken, and faceted as a

15.17   The variation in the size of particles composing a till is very large. In this exposure of till near Netcong, New Jersey, boulders of varying sizes are mixed with finer particles that, when examined closely, turn out to range downward in size from pebbles through grains of sand and silt to clay-sized particles.

15.18   This moraine composed of till lies along the East Rosebud River and records the last major advance of glaciers from the Beartooth Mountains, Montana. The largest boulders measure 3 m in maximum dimension.

15.19   The differences between pebbles transported by glaciers and those transported by streams are shown here. The glacier-transported stones (a) are striated. The upper black pebble is polished as well, and the two lower stones are faceted and have broken irregular margins. The stream-carried pebbles (b) are, by contrast, well-rounded and unstriated and have a matte surface. [John Simpson.]

(a)

(b)

result of the wear they underwent while transported by the glaciers. Their form is in contrast to that of the familiar well-rounded stream pebbles (Figure 15.19). Some of the material picked up along the way was smeared across the landscape during the glacier's progress, but most of it was dropped when the rate of wastage began to exceed the rate of nourishment and the glacier gradually melted away.

The type of till varies from one glacier to another. Some tills, for instance, are known as **clay tills** because clay-sized particles predominate, with only a scattering of larger units. Many of the most recent tills in northeastern and eastern Wisconsin are of this type, but in many parts of New England the tills are composed mostly of large rock fragments and boulders. Deposits of this sort are known as **boulder,** or **stony, tills.**

Some tills seem to have been worked over by melt-water. The sediments have begun to be sorted out according to size, and some of the finer particles may even have been washed away. This is the sort of winnowing action we should expect to find near the nose of a melting glacier, where floods of meltwater wash down through the deposits.

Till is deposited by glaciers in a great variety of topographic forms, including moraines, drumlins, erratics, and boulder trains.

Moraines.   *Moraine* is a general term used to describe many of the landforms that are composed largely of till.

A **terminal moraine,** or **end moraine,** is a ridge of till that marks the utmost limit of a glacier's advance. These ridges vary in size from ramparts scores of meters high to very low, interrupted walls of debris. A terminal moraine forms when a glacier reaches the critical point of equilibrium — the point at which it wastes away at exactly the same rate as it is nourished. Although the front of the glacier is now stable, ice continues to push forward, so it will deliver a continuous supply of rock debris. As the ice melts in the zone of wastage, the debris is released and the terminal moraine grows. At the same time water from the melting ice pours down over the till and sweeps part of it out in a broad flat fan that butts against the forward edge of the moraine like a giant ramp (Figure 15.20).

The terminal moraine of a mountain glacier is crescent-shaped, with the convex side extending down-valley. The terminal moraine of a continental ice sheet is a broad loop or series of loops traceable for many miles across the countryside.

Behind the terminal moraine, and at varying distances from it, a series of smaller ridges known as **recessional moraines** may build up. These ridges mark the position where the glacier front was stabilized temporarily during the retreat of the glacier.

Not all the rock debris carried by a glacier finds its way to the terminal and recessional moraines, however. A great deal of till is laid down as the main body of the glacier melts to form gently rolling plains across the valley floor. Till in this form, called a **ground moraine,** may be a thin veneer

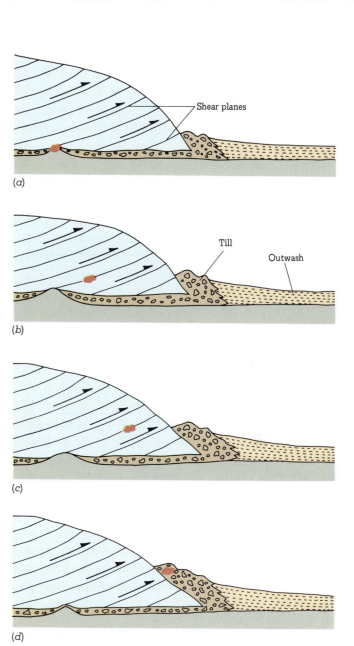

15.20   A sequence of diagrams to suggest the growth of a terminal moraine at the edge of a stable ice front. The progressive movement of a single particle is shown. In (a) it is moved by the ice from the bedrock floor. Forward motion of ice along a shear plane carries it ever closer to the stabilized ice margin, where it is finally deposited as part of the moraine in (d). Diagram (e) represents the relation of the terminal moraine, ground moraine, and outwash after the final melting of the glacier.

lying on the bedrock, or it may be scores of meters thick, partially or completely clogging preglacial valleys.

Finally, valley glaciers produce two special types of moraine. While a valley glacier is still active, large amounts

15.21 A drumlin near Elbridge in central New York State. The blunt end of the asymmetric long profile faces north, the direction from which the glacier advanced.

of rubble keep tumbling down from the valley walls, collecting along the side of the glacier. When the ice melts, this debris is stranded as a ridge along each side of the valley, forming a **lateral moraine.** At its down-valley end the lateral moraine may grade into a terminal moraine. The other special type of deposit produced by valley glaciers is a **medial moraine,** created when two valley glaciers join to form a single stream of ice; material formerly carried along on the edges of the separate glaciers is combined in a single moraine near the center of the enlarged glacier. A streak of this kind builds up whenever a tributary glacier joins a larger glacier in the main valley (Figures 15.5 and B15.1.1). Although medial moraines are very characteristic of living glaciers, they are seldom preserved as topographic features after the disappearance of the ice.

Drumlins.   Drumlins are smooth, elongated hills composed largely of till. The ideal drumlin shape has an asymmetric profile, with a blunt nose pointing in the direction from which the vanished glacier advanced and a gentler, longer slope pointing in the opposite direction (Figure 15.21). Drumlins range from about 8 to 60 m in height, the average being about 30 m. Most drumlins are between 0.5 and 1 km in length and are usually several times longer than they are wide.

In most areas drumlins occur in clusters, or **drumlin fields.** In the United States these are most spectacularly developed in New England, particularly around Boston; in eastern Wisconsin; in west-central New York State, particularly around Syracuse; in Michigan (Figure 15.22); and in certain sections of Minnesota. In Canada extensive drumlin fields are located in western Nova Scotia and in northern Manitoba and Saskatchewan; Figure 15.23 shows a drumlin field in British Columbia.

Just how drumlins were formed is still not clear. Because their shape is a nearly perfect example of streamlin-

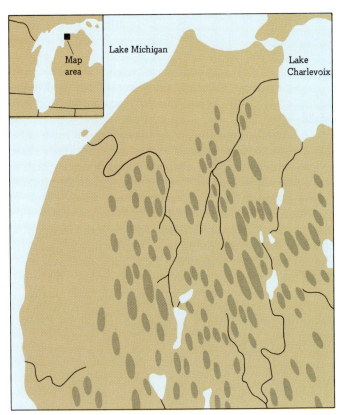

15.22 Map of part of the drumlin-field area south and east of Charlevoix, Michigan. Ice moved toward the south-southeast. [After Frank Leverett and F. B. Taylor, "The Pleistocene of Indiana and Michigan and the History of the Great Lakes," *U.S. Geol. Surv. Monog. 53,* p. 311, 1915.]

ing, it seems probable that they were formed deep within active glaciers in the zone of plastic flow.

Erratics and boulder trains.   A stone or a boulder that has been carried from its place of origin by a glacier and

15.23 This drumlin field shows the streamlined form of these ice-molded features. Ice movement was from upper right to lower left. Near Carp Lake, north-central British Columbia; width of view is about 10 km. [U.S. Army Air Force.]

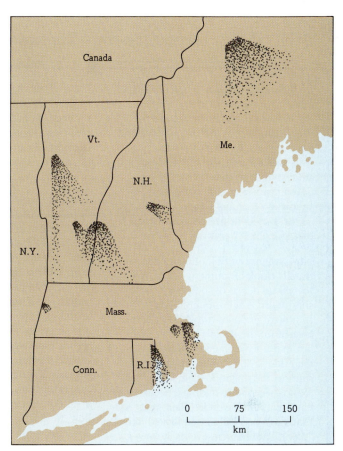

15.24 The boulder trains plotted on this map indicate the general direction of ice movement across New England. The apex of the fan indicates the area from which the boulders were derived; the fan itself covers the area across which they were deposited. [After J. W. Goldthwait in R. F. Flint, "Glacial Map of North America," *Geol. Soc. Am. Spec. Paper 60*, 1945.]

left stranded on bedrock of different composition is called an **erratic.** The term is used whether the stone is embedded in a till deposit or rests directly on the bedrock. Some erratics weigh several tonnes, and a few are even larger. Near Conway, New Hampshire, there is a granite erratic 27 m in maximum dimension, weighing close to 9,000 t. Although most erratics have traveled only a limited distance, many have been carried along by the glacier for hundreds of kilometers. Chunks of native copper torn from the Upper Peninsula of Michigan, for example, have been transported as far as southeastern Iowa, a distance of nearly 800 km, and to southern Illinois, over 960 km.

Boulder trains consist of a series of erratics that have come from the same source; they usually have some characteristic that makes it easy to recognize their common origin. The trains appear either as a line of erratics stretching down-valley from their source or in a fan-shaped pattern with the apex near the place of origin. By mapping boulder trains that have been left behind by continental ice sheets, we can obtain an excellent indication of the direction of the ice flow (Figure 15.24).

Deposits of Stratified Drift   Stratified drift is ice-transported sediment that has been washed and sorted by glacial meltwaters according to particle size. Because water is a much more selective sorting agent than ice, deposits of stratified drift are laid down in recognizable layers, unlike the random arrangements of particles typical of till. Stratified drift occurs in outwash plains, kettles, eskers, kames, and varves—all discussed below.

Outwash sand and gravel.   The sand and gravel that are carried outward by meltwater from the front of a

glacier are referred to as **outwash.** As a glacier melts, streams of water heavily loaded with reworked till or with material washed directly from the ice weave a complex, braided pattern of channels across the land in front of the glacier. These streams, choked with clay, silt, sand, and gravel, rapidly lose their velocity and deposit their load of debris as they flow away from the ice sheet. In time, a vast apron of bedded sand and gravel is built up, which may extend for kilometers beyond the ice front. If the zone of wastage happens to be located in a valley, the outwash deposits are confined to the lower valley and compose a **valley train.** But along the front of a continental ice sheet the outwash deposits stretch out for kilometers, forming what is called an **outwash plain** (Figure 15.25).

Deposits of outwash can be very valuable economically as a source of aggregate for concrete. An unexpected discovery of economic value, however, is reported in Box 15.2.

Kettles.   Sometimes a block of stagnant ice becomes isolated from the receding glacier during wastage and is partially or completely buried in till or outwash before it finally melts. When it disappears, it leaves a **kettle,** a pit or

In 1876 Mrs. Clarissa Wood engaged a laborer to sink a well on property in the town of Eagle, in southeastern Wisconsin, where she and her husband were tenants. In the process the well-digger found a small, peculiar stone, which he gave to Mrs. Wood. Seven years later the pebble, pale yellow in color and weighing over 15 carats, was identified as a diamond. (A carat weighs 0.2 grams.)

Now, when we think of diamonds, we usually think of southern Africa, not the glacial debris of southern Wisconsin. Indeed, Eagle, about 30 km southwest of downtown Milwaukee, is located on the glacial outwash of the last major ice advance. Finding a diamond in such a locality—even assuming one might be there to be found—is akin to finding a needle in a haystack. Small wonder, then, that Clarissa Wood's diamond looked like a "plant." Subsequent events, however, proved otherwise.

Since that first recorded diamond find in the glacial deposits of the upper Midwest, an unexpectedly large number of diamonds have been discovered there. Today we have well-authenticated finds, not only from Wisconsin, but also from Minnesota, Illinois, Michigan, Indiana, and Ohio.

The tantalizing question is, Where did the diamonds come from? Their distribution, associated with the till and outwash of the last major glaciation, places some constraint on our speculation. They must have been brought to the upper Midwest by glaciers. Such indicators of glacier flow as striations, drumlins, lithology of glacial deposits, and shapes of moraines allow us to reconstruct the path of the glacier that carried the diamonds. The evidence suggests an origin for the stones in the James Bay–southern Hudson Bay area. But precisely where, and even how, to look for commercial accumulations is another question. Still, this has not stopped periodic searches, some of which have been on a concerted and sophisticated scale.

And what about that diamond found in 1876 in Eagle, Wisconsin? Seven years after its discovery Mrs. Wood, thinking it a topaz, sold it for $1.00 to Colonel S. B. Boynton, a jeweler in Milwaukee. Later, after it was established that the stone was actually a diamond, Mrs. Wood offered to buy it back for $1.10, an offer that was refused. The ensuing dispute ended up in litigation. Finally, the state supreme court ruled against Mrs. Wood on the ground that Colonel Boynton had not known the pebble's true nature when he bought it. The stone was sold for $850 to Tiffany and Company in New York, where it joined the jeweler's collection of uncut stones. Later it was sold to J. P. Morgan, who gave it to the American Museum of Natural History in New York City. In 1964 it was stolen from the museum along with other valuable gemstones and has not yet been recovered.

depression, in the drift (Figure 15.26). These depressions range from a few meters to several kilometers in diameter and from a few meters to over 30 m in depth. Many outwash plains are pockmarked with kettles and are referred to

15.25   These sand and gravel deposits represent the outwash from a now-vanished continental glacier. North of Otis Lake, Wisconsin. [W. C. Alden, U.S. Geological Survey.]

as **pitted outwash plains.** Water sometimes fills the kettles to form lakes or swamps, features found through much of Canada and the northern United States (Figure 15.27).

Eskers and crevasse fillings.   Winding, steep-walled ridges of stratified gravel and sand, sometimes branching and often discontinuous, are called **eskers** (Figure 15.28). They usually vary in height from about 3 to 15 m, although a few are over 30 m high. Eskers range from a fraction of a kilometer to over 160 km in length, but they are only a few meters wide. Most investigators believe that eskers were formed by the deposits of streams running through tunnels beneath stagnant ice. Then, when the body of the glacier finally disappeared, the old stream deposits were left standing as a ridge.

Crevasse fillings are similar to eskers in height, width, and cross profile, but unlike the sinuous and branching pattern of eskers, they run in straight ridges. As their name suggests, they were probably formed by the filling of a crevasse in stagnant ice.

Kames and kame terraces.   In many areas stratified drift has built up low, relatively steep-sided hills called **kames,** either as isolated mounds or in clusters. Unlike

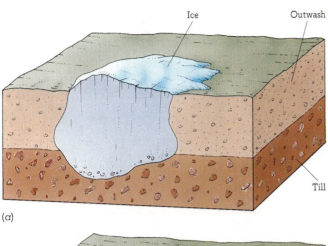

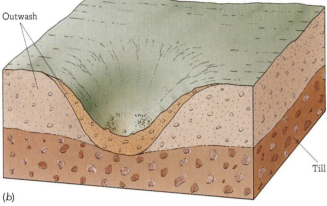

15.26 Kettle formation: (a) Stagnant ice is almost buried by outwash. (b) Eventual melting forms a depression.

15.27 A small kettle-hole lake near Rochester, New York. This depression was formed by the melting of a block of stagnant glacier ice. [H. L. Fairchild.]

15.28 An aerial view of the MacKinnon esker east of the Salmon River in north-central British Columbia. The feature is about 250 m wide at its base and branches at both ends. [U.S. Army Air Force.]

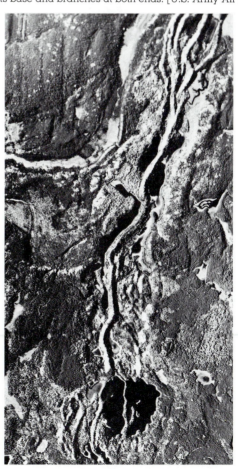

drumlins, kames are of random shape and the deposits that compose them are stratified. They were formed by the material that collected in openings or small lakes in stagnant ice. In this sense they are similar to crevasse fillings, but without the linear pattern.

A **kame terrace** is a deposit of stratified sand and gravel that has been laid down between a wasting glacier and an adjacent valley wall. When the glacier disappears, the deposit stands as a terrace along the side of the valley (Figure 15.29).

Varves. A **varve** is a pair of thin sedimentary beds, one of coarse and one of fine sediment (Figure 8.9). This couplet of beds is usually interpreted as representing the deposits of a single year and is thought to form in the following way: During the period of summer thaw waters from a melting glacier carry large amounts of clay, fine sand, and silt out into lakes along the ice margin. The coarser particles sink fairly rapidly and blanket the lake floor with a thin layer of silt and silty sand. But as long as the lake is unfrozen, the wind creates currents strong enough to keep the finer clay particles in suspension. When the lake freezes over in the winter, these wind-generated currents cease, and the fine particles sink through the quiet water to the bottom,

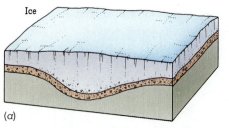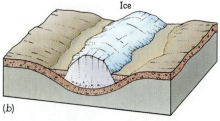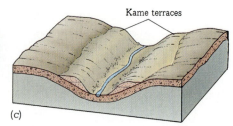

15.29 The sequence in the development of a kame terrace. (a) Ice wasting from an irregular topography lingers longest in the valleys. (b) While the ice still partially fills one of these valleys, outwash may be deposited between it and the valley walls. (c) The final disappearance of the ice leaves the outwash in the form of terraces along the sides of the valley.

covering the somewhat coarser summer layer. A varve is usually a few millimeters to 1 cm thick, although thicknesses of 5 to 8 cm are not uncommon. There are rare instances of varves 30 cm or more thick.

COMPARISON OF FEATURES OF VALLEY AND CONTINENTAL GLACIATION

Some of the glacial features that we have been discussing are more common in areas that have undergone valley glaciation; others usually occur only in regions that have been overridden by ice sheets. Many other features, however, are found in both types of area. Table 15.1 lists and compares the features that are characteristic of the two types.

TABLE 15.1
Features of Valley and of Continental Glaciation Compared

Features	Valley glacier	Continental ice sheet
Striations, polish, etc.	Common	Common
Cirques	Common	Absent
Horns, arêtes, cols	Common	Absent
U-shaped valleys, truncated spurs, hanging valleys	Common	Rare
Fiords	Common	Absent
Till	Common	Common
Terminal moraine	Common	Common
Recessional moraine	Common	Common
Ground moraine	Common	Common
Lateral moraine	Common	Absent
Medial moraine	Common, easily destroyed	Absent
Drumlins	Rare or absent	Locally common
Erratics	Common	Common
Stratified drift	Common	Common
Kettles	Common	Common
Eskers, crevasse fillings	Rare	Common
Kames	Common	Common
Kame terraces	Common	Present in hilly country

INDIRECT EFFECTS OF GLACIATION

The glaciers that diverted rivers, carved mountains, and covered half a continent with debris also gave rise to a variety of indirect effects that were felt far beyond the glaciers' immediate margins. Sea level fell with the accumulation of glacier ice and rose as glaciers melted (Section 16.7). Desert and near-desert areas were better watered during maximum glaciation. In now-temperate latitudes beyond the margin of the ice, the climate also changed, the wind laid down vast blankets of dust called **loess** (Section 17.2), and the erosional processes on slopes were speeded by solifluction (Section 12.3).

15.3
DEVELOPMENT OF THE GLACIAL THEORY

Geologists have made extensive studies of the behavior of modern glaciers and have carefully interpreted the traces left by glaciers that disappeared thousands of years ago. On the basis of their studies they have developed the **glacial theory**: In the past great ice sheets covered large sections of the earth where no ice now exists, and many existing glaciers once extended far beyond their present limits.

THE BEGINNINGS

The glacial theory took many years to evolve, years of trying to explain the occurrence of erratics and the vast expanses of drift strewn across northern Europe, the British Isles, Switzerland, and adjoining areas. The exact time when inquisitive minds first began to seek an explanation of these deposits is shrouded in the past, but by the beginning of the eighteenth century explanations of what we now know to be glacial deposits and features were finding their way into print. According to the most popular early hypothesis, a great inundation had swept these deposits across the face of the land with cataclysmic suddenness or else had drifted them in by means of floating icebergs. Then, when the flood receded, the material was left where it is now.

By the turn of the nineteenth century a new theory was in the air — the theory of ice transport. We do not know who first stated the idea or when it was first proposed, but it seems quite clear that it was not hailed immediately as a great truth. As the years passed, however, more and more observers became intrigued with the idea. The greatest impetus came from Switzerland, where the activity of living glaciers could be studied on every hand.

In 1821 J. Venetz, a Swiss engineer delivering a paper before the Helvetic Society, presented the argument that Swiss glaciers had once expanded on a great scale. It has since been established that from about 1600 to the middle of the eighteenth century there was actually a time of moderate but persistent expansion of glaciers in many localities, a time known as the **Little Ice Age.** Abundant evidence in the Alps, Scandinavia, and Iceland indicates that the climate was milder during the Middle Ages than it is at present, that communities existed and farming was carried on in places later invaded by advancing glaciers or devastated by glacier-fed streams. We know, for example, that a silver mine in the valley of Chamonix was being worked during the Middle Ages and that it was subsequently buried by an advancing glacier, and lies buried to this day. And the village of St. Jean de Perthuis has been buried under the Brenva Glacier since about 1600.

Although Venetz's idea did not take hold immediately, by 1834 Jean de Charpentier was arguing in its support before the same Helvetic Society. Yet the theory continued to have more opponents than defenders. It was actually one of the skeptics, Jean Louis Rodolphe Agassiz (1807 – 1873), who did more than anyone else to develop the glacial theory and bring about its acceptance.

AGASSIZ

As a young zoologist, Louis Agassiz had listened to Charpentier's arguments and emphatically disagreed. In 1836 Agassiz planned fieldwork in the area of Bex along the upper Rhône valley, where he would search for fossil fish. Indeed he got to the upper Rhône, but not to the fossil fish. Instead he, Charpentier, and Venetz set out to examine modern glaciers. One supposes that Agassiz was bent on demonstrating to his two colleagues the errors of their way, but as it turned out, it was Agassiz who recanted. In 1837, speaking before the Helvetic Society, he championed the glacial theory and suggested that during a "Great Ice Age" not only the Alps but much of northern Europe and the British Isles were overrun by a sea of ice.

Agassiz's statement of the glacial theory was not accepted immediately, but in 1840 he visited England and won the support of leading British geologists. In 1846 he arrived in the United States, where in the following year he became professor of zoology at Harvard College and later founded the Museum of Comparative Zoology. In this country he convinced geologists of the validity of the glacial

theory. By the third quarter of the nineteenth century the theory was firmly entrenched; the last opposition died with the turn of the twentieth century.

PROOF OF THE GLACIAL THEORY

What proof is there that the glacial theory is valid? The most important evidence is that certain features produced by glacier ice are produced by no other known process. Thus Agassiz and his colleagues found isolated stones and boulders quite alien to their present surroundings. They noticed too that boulders were actually being transported from their original location by modern ice. Some of the boulders they observed were so large that rivers could not possibly have moved them, and others were perched on high places that a river could have reached only by flowing uphill. They also noticed that when modern ice melted, it revealed a polished and striated pavement unlike the surface fashioned by any other known process. To explain the occurrence of these features in areas where no modern glaciers exist, they postulated that the ice once extended far beyond its present limits.

Notice that the development of this theory sprang from the concept of uniformitarianism that we mentioned in Section 1.2: "The present is the key to the past." The proof of glaciation lies not in the authority of the textbook or of the lecture. It lies in observing modern glacial activity directly and in comparing the results of this activity with features and deposits found beyond the present extent of the ice.

THEORY OF MULTIPLE GLACIATION

Even before universal acceptance of the glacial theory, which spoke of a single great Ice Age, some investigators were coming to the conclusion that the ice had advanced and retreated not just once but **several times in the recent geological past.** By the early twentieth century a broad fourfold division of the Ice Age, or Pleistocene, had been demonstrated in this country and in Europe. According to this theory, each major glacial advance was followed by a glacial retreat and a return to climates that were sometimes even warmer than those of the present. Early in this century four stages of major continental glaciation were discovered in the United States; originally they were called, from oldest to youngest, **Nebraskan, Kansan, Illinoian,** and **Wisconsin,** for the midwestern states where deposits of a particular period were first studied or where they are well exposed.

This fourfold division of the Ice Age eventually proved to be oversimplified. First of all, it has long been known that the Wisconsin glaciation was multiple. Furthermore, we have discovered that the Kansan and Nebraskan glaciations included not two, but several glaciations. So now we generally use the term *Pre-Illinoian* to refer to these

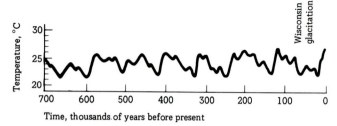

15.30 Analysis of deep-sea sediments shows that the average surface temperature of the ocean fluctuated many times from warm to cold in the recent geologic past. Here is the record for the surface waters of the central Caribbean during the last 700,000 years; the most recent cooling is labeled to represent the last major glaciation in North America. [After Cesare Emeliani and Nicholas J. Shackleton, "The Brunhes Epoch: Isotopic Paleotemperatures and Geochronology," *Science*, vol. 183, p. 513, 1974.]

early advances and retreats of ice in the Midwest. Second, in central Europe successive sheets of loess suggest at least 17 periods of Pleistocene glaciation before the last major advance. Finally, analysis of cores of sediments taken from the deep oceans reveals repeated cooling and warming of the oceans during the last 2 million years (Figure 15.30).

About 20 of these cycles have been identified, and we believe that they represent the climatic changes accompanying repeated glaciation and deglaciation on land.

15.4
EXTENT OF PLEISTOCENE GLACIATION

During the maximum advance of the glaciers of the Wisconsin age (the last of the four great ice advances during the Pleistocene), 39 million km² of the Earth's surface—about 27 percent of the present land area—were probably buried by ice. Approximately 15 million km² of North America were covered. Greenland was also under a great mass of ice, as it is now. In Europe an ice sheet spread southward from Scandinavia across the Baltic Sea and into Germany and Poland, and the Alps and the British Isles supported their own icecaps. Eastward, in Asia, the northern plains of Russia were covered by glaciers, as were large sections of Siberia and of the Kamchatka Peninsula and the high plateaus of Central Asia (Figure 15.31).

15.31 Extent of Pleistocene glaciation (darker areas) in the Northern Hemisphere. [After Ernst Antevs, "Maps of the Pleistocene Glaciations," *Bull. Geol. Soc. Am., 40*, p. 636, 1929.]

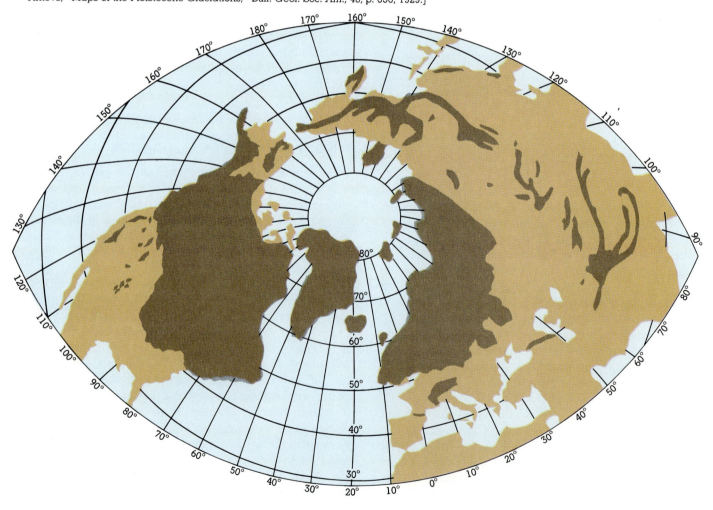

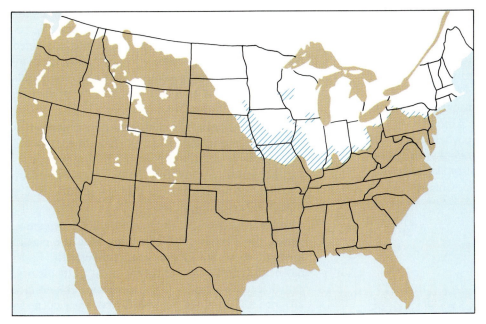

15.32 Extent of Pleistocene glaciation in the United States. White zones indicate areas covered at various times during the Wisconsin glaciation. Diagonal lines represent area glaciated during pre-Wisconsin stages but not covered by the later Wisconsin ice. Tint area unglaciated. [After R. F. Flint, "Glacial Map of North America," *Geol. Soc. Am. Spec. Paper 60*, 1945.]

In eastern North America ice moved southward out of Canada to New Jersey, and in the Midwest it reached as far south as Saint Louis. The western mountains were heavily glaciated by small icecaps and valley glaciers. The southernmost glaciation in the United States was in the Sierra Blanca of south-central New Mexico. The maximum extent of the Wisconsin glaciation in the United States and the maximum limit of glaciation during the Pleistocene are shown in Figure 15.32.

15.5

PRE-PLEISTOCENE GLACIATIONS

So far we have discussed only the glaciers that exist today and those that moved within the last few million years, the Pleistocene. Geologists have found evidence that glaciers

appeared and disappeared in other periods as well. The record is fragmentary, as we should expect, for time tends to conceal, jumble, and destroy the various manifested effects of glaciation.

Recent evidence indicates that glaciation and deglaciation, so characteristic of the Pleistocene, reached back into the late Miocene in Alaska. Glacial deposits in Antarctica have been dated by the potassium-argon method as 10 million years in age. Farther back in the geologic-time scale, it seems certain that there were other extensive glaciations. About 230 to 290 million years ago (toward the end of the Paleozoic Era) glaciation was widespread in what is now South Africa, South America, India, Australia (Figure 15.33), and Antarctica. There is evidence of glaciation during the Silurian and Devonian in South America; and earlier, in the Ordovician, ice spread across what is now the Sahara Desert. In the late Precambrian, a little over 600 million years ago, glaciation affected various landmasses

15.33 In Permian time, toward the end of the Paleozoic Era, glaciation affected the ancient supercontinent of Gondwanaland. The continent subsequently fragmented and its pieces then drifted into their present positions, so that traces of the Permian glaciation are now found in South America, South Africa, India, Antarctica, and here in southern Australia. Glacier ice smoothed and striated the slightly metamorphosed siltstone. The knife is parallel to the major set of striations and its blade points in the direction of ice movement. These striations are at a high angle to an earlier, less obvious set of striations whose direction is indicated by the metal pencil at the top edge of the exposure. [William C. Bradley.]

now in the Northern Hemisphere. Still earlier, perhaps 850 million years ago, large parts of the ancient continent of Gondwanaland (see Chapter 11) were ice-covered. Some Russian geologists believe that sections of northwestern Siberia were glaciated some 1.2 billion years ago. Extensive glaciation, dating from 2.2 billion years ago, is recorded in what is now south-central Ontario, Canada. At about the same time glaciation occurred in South Africa.

Much of this glaciation took place where no ice can exist today. The geologic evidence tells us that this is because the landmasses have since drifted from latitudes hospitable to glaciation to new locations (see Chapter 11).

15.6
CAUSES OF GLACIATION

As Agassiz did over 150 years ago, we can travel about the world today and observe modern glaciers at work, and we can reason convincingly that glaciers were more extensive in the past than they are at present. We can even make out a good case for the belief that glacier ice advanced and receded many times in the immediate and more remote geological past. But can we explain why glaciation takes place? The answer to that question is probably yes.

The geologic record has contributed some basic data that any theory of the causes of glaciation must take into account. Among these are the following:

1. **Glaciation is related to the elevation and arrangement of continents.** During much of geologic time the continents were lower than they are today, and shallow seas flooded across their margins. Such conditions were unfavorable for widespread glaciation. But for several million years the continents have been increasing in elevation, until now, in the Pleistocene, they stand on the average an estimated 450 m higher than they did

in the mid-Cenozoic. We have already found that the last great Ice Age came with the Pleistocene. We believe too that the other great glaciations coincided with high continents. In addition to having adequate elevations, to allow glaciation to begin, continents must be located in high latitudes, either over the pole of rotation or clustered about it.

2. **Glaciation is not due to a slow, long-term cooling off of the Earth since its formation.** We have already found that extensive glaciation occurred several times during the geological past. But these glacial periods are unusual; for during most of geologic history the climate has been nonglacial.

3. **There was a cooling of the Earth's climate beginning during the Tertiary and climaxing with the glacial fluctuations of the Pleistocene.** Although the Earth has been generally warm during most of its history, evidence now shows that its mean temperature dropped an estimated 8 to 10°C from the Eocene to the end of the Pliocene, and that the glacial and interglacial epochs of the Pleistocene were short-term fluctuations at the end of the long-term cooling (Figure 15.34).

4. **The advance and retreat of glaciers were probably broadly simultaneous throughout the world.** For instance, dating by means of radioactive carbon has demonstrated that geologically recent fluctuations of the continental glaciers in North America occurred at approximately the same time as similar fluctuations in Europe. Furthermore, observations indicate that the general retreat of mountain glaciers now recorded in North America and Europe was duplicated in South America.

One way to approach the problem is to consider that we are dealing with two great classes of causes: One is long-term and has caused the general decline of temperatures since the Eocene; the other is shorter-range and accounts

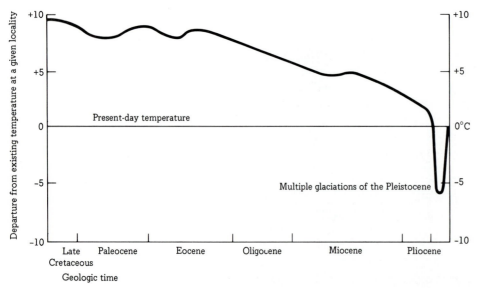

15.34 Temperatures trended downward from the Late Cretaceous to the Pliocene. This general decline was followed by the several glacial and interglacial episodes, the details of which cannot be shown at the scale of this diagram. The values in degrees Celsius represent approximate departures from today's temperatures of localities in the middle latitudes. [Data in part from Erling Dorf.]

for the comings and goings of the ice sheets during the Pleistocene. One operates on a scale of approximately 100 million years, and the other on the scale of perhaps 100,000 years. In considering these two different types of cause, we can envisage the long-term cause as cooling the climate to the point where one or more other causes of shorter wavelength and lower amplitude can trigger glaciation and deglaciation.

LONG-TERM CHANGES IN CLIMATE

Geologic evidence shows that the Earth's "normal" climate has been more equable than that of the present. Through most of the Earth's history the climate has been warmer than it is now and landmasses presently in the poleward zones enjoyed average annual temperatures about 10°C higher than today. At some time during the early Tertiary, probably in the Eocene, average temperatures began to drop.

Causes of Long-Term Climatic Fluctuation What could have caused the decline of temperatures during the last 50 to 60 million years? We can make a number of suggestions.

Increasing continentality.   Today the average height of continents is computed to be about 840 m, as we have stated, about 450 m higher than it was in the beginning of the Eocene. Rising landmasses would result in a general drop in temperature. It is estimated that about one-third of the 10°C decline in average temperatures might be accounted for by the increase in the height of continents since the Eocene.

The rise of continents not only would lower global temperatures but could also interfere with the patterns of worldwide atmospheric circulation. It is conceivable that the transfer of heat from equatorial latitudes toward the poles would be hindered and that the poleward latitudes would become colder. A corollary to this would be an interruption and change in the oceanic circulation and a lessened efficiency in the transfer of heat poleward.

Continental drift.   Glaciers need a landmass on which to form and grow. The landmass has to be in a polar position. Thus Antarctica, positioned on the South Pole, is in an ideal spot to support glacier ice—and it does. In the Northern Hemisphere the landmasses are clustered around the North Pole and restrict circulation of warm waters into the Arctic Ocean. In this situation we have Greenland glaciated at present, and other, adjacent lands that have recently been glaciated. We conclude that glaciation occurs when continental drift has positioned landmasses in polar areas where climate is conducive to the inception of glaciation.

We know, however, that the rate of continental drift, while measurable, is slow. Certainly continents did not

move in and out of polar positions rapidly enough to account for the short, 100,000-year cycle of the various glacial epochs of the Pleistocene. It seems, then, that the role of continental drift in causing glaciation lies in bringing the continents slowly into polar latitudes and causing a general cooling of climate. At some point this cooling climate triggers another set of factors that push the climate back and forth between the shorter cycles of glaciation and deglaciation.

CLIMATIC CHANGES OF SHORTER DURATION

We now ask what might be the factors that superimposed the short-term climatic fluctuations—the glaciations and deglaciations of the Pleistocene—on the longer and slower cooling of climate since the early Tertiary. A wide variety of suggestions has been put forward over the years. Recently, however, a growing body of evidence has brought about a near-consensus in favor of one of these suggestions. It involves the changing geometry of the Earth's orbit around the Sun.

The changes in the Earth's orbit produce variations in the geographical and seasonal distribution of solar energy received on the Earth. We refer to the hypothesis as the **orbital hypothesis,** though it is sometimes called the **Milankovitch hypothesis** after the Yugoslavian astronomer instrumental in its development.

Three factors enter into the changing position of the Earth in relation to the Sun, each of which can be measured and its rate of change determined. One is the **eccentricity** of the Earth's orbit around the Sun. The Earth is somewhat closer to the Sun at some epochs than others. This motion has a period of about 93,000 years. A second factor is the **obliquity of the ecliptic,** which produces seasons and is determined by the angle that the Earth's axis makes with the plane in which the Earth circles the Sun. This angle, or obliquity, changes about 3° every 41,000 years. Finally, there is the **precession of the equinoxes,** which merely means that the axis of the Earth wobbles because of the gravitational effect of the Sun, the Moon, and the planets. Like a giant top, the Earth completes one wobble every 21,000 years. All these factors affect the relation of the Earth to the Sun—hence the amount of heat received at different places on the Earth at any one time. The variation can be calculated backward in time to any desired date for any latitude. These calculations have been made, and a curve showing maxima and minima of heat received at the Earth during the Pleistocene is the result. The periods of low heat are said to coincide with, and be the cause of, the various glaciations.

The orbital hypothesis has been around in increasingly sophisticated versions for over 100 years. During the last 40 years more and more bits of geologic history have accumulated and now suggest a tantalizing fit between the astronomical prediction and the actual Earth record. Many workers have been involved, but here we cite only three:

James D. Hayes (Columbia University), John Imbrie (Brown University), and N. J. Shackleton (Cambridge University) have recently marshalled the climatic evidence found in deep-sea cores extending back for nearly 0.5 million years, and from these they have constructed a curve not unlike that of Figure 15.30. The correlation of this curve with the climatic fluctuations predicted from the orbital motions of the Earth is striking. And their conclusion that "changes in the Earth's orbital geometry are the fundamental cause of" Pleistocene ice ages is extremely persuasive.

15.7

IMPLICATIONS FOR THE FUTURE

If geology teaches us nothing else, it demonstrates that our globe is in constant change—that the face of the Earth is mobile. Mountains rise only to be laid low by erosion; seas lap over the continents; entire regions progress through various stages; and glaciers come and go. We still live in the Pleistocene Ice Age, a pinpoint in time that has been preceded, and will be followed, by extensive climatic changes.

Of course, we cannot predict with any surety anything but change. On the other hand, some feel that it is reasonable to extrapolate the Earth's orbital variations into the future and to assume that these will control major climatic fluctuations, as they appear to have controlled them during the Pleistocene. If we accept these premises, then the long-term outlook is toward widespread glaciation in the Northern Hemisphere.

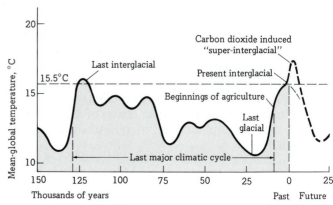

15.35   The mean global temperature for the last 150,000 years is shown. If the orbital variations of the Earth control the climate of the future as many believe they controlled the climate of the past, then we will enter a full-fledged glacial period in about 23,000 years. See text for discussion of "superinterglacial." [After J. Murray Mitchell, Jr., "Carbon Dioxide and Future Climate," *Environmental Data Service*, p. 8, March 1977.]

Figure 15.35 is one such prediction. It shows the climatic fluctuations in temperature for the last 150,000 years and projects present temperature along a cooling course that reaches a full-fledged glacial stage in 23,000 years. The author of this particular prediction also suggests that we may inadvertently extend the present interglacial by continued burning of coal, oil, and gas. This would increase the carbon dioxide in the atmosphere, create a global "greenhouse" effect, and perhaps produce a **superinterglacial** period warmer than today and warmer than previous interglacials. Among other effects, this might extend the present interglacial for some 2,000 years, before the decline in temperatures leads to the next glaciation.

Summary

Glacier formation is a low-temperature, metamorphic process that converts snow to firn to ice.
> The principal classifications of glaciers are valley glaciers, piedmont glaciers, ice sheets, and continental glaciers.
> Nourishment of glaciers occurs in the zone of accumulation. The zone of wastage lies below the snow line, where melting and evaporation exceed snowfall.
> Glacier movement is usually a few centimeters per day, but some glaciers move rapidly in surges. Below a 60-m brittle zone glacier ice is in a zone of flow. The glacier moves by slipping along its base, by recrystallization of individual crystals, and by shearing.
> Temperatures of glaciers vary. A temperate glacier reaches the pressure melting point throughout the summer. In a cold glacier no melting occurs even in summer.

Results of glaciation are both erosional and depositional.
> Erosion of the ground surface takes place by plucking or by abrasion. Features formed include cirques, horns, arêtes, cols, U-shaped valleys, hanging valleys, fiords, striations, grooves, and polish.
> Glacial deposits include unstratified material called till, which is found in moraines and drumlins. Boulder trains are special depositional features.
> Stratified deposits include outwash of sand and gravel, which is usually found in outwash plains, eskers, crevasse fillings, kames, and kame terraces. Varves usually consist of clay and silt, and form in glacial lakes.

The glacial theory was born in Switzerland and was first clearly stated by Louis Agassiz. Proof of the glacial theory rests on the principle of uniformitarianism.

The Pleistocene is marked by multiple glaciations, including several in the central United States.

Pre-Pleistocene glaciations are known to have occurred several times in the Paleozoic and the Precambrian.

Causes of glaciation in the Pleistocene seem to involve a long-term cooling of climate that began in the early Tertiary and was brought on as **continental drift** moved large landmasses into, or close to, polar positions. Most geologists now believe that short-term climatic fluctuations caused by **orbital variations** of the Earth were superimposed on the long-term cooling and triggered the glaciations and deglaciations of the Pleistocene.

The implication for the future is that major climatic change is certain.

Questions

1. Define a glacier and describe how snow is converted to glacier ice.

2. With reference to the zones of accumulation and wastage, explain why a glacier's snout may move forward, retreat, or stand still.

3. Glaciers move in several ways. What are they?

4. Erosion by glaciers produces a variety of features and you should be familiar with them. Briefly and succinctly, what are the following: cirque, tarn, striation, horn, arête, col, U-shaped valley, hanging valley, fiord, asymmetric knob or hill?

5. Glacial deposits are both stratified and unstratified. What is the term applied to unstratified material? Why is it unstratified?

6. Unstratified glacial material accumulates in certain characteristic forms. What are some of them and how do they form?

7. What are each of the following: drumlin, erratic, boulder train?

8. Why are stratified glacial deposits stratified? Describe the nature and origin of the following, all of which are made of stratified material: outwash plain, valley train, esker, crevasse filling, kame, kame terrace, varve.

9. In what way is the development of the glacial theory related to the doctrine of uniformitarianism?

10. What are some of the proofs of multiple glaciation?

11. Discuss the causes of Pleistocene glaciations.

12. The albedo (the ratio of the amount of light reflected by a surface to that received by it) in part determines the effectiveness of the Sun in warming that surface. Decrease the whiteness—and hence the reflectiveness of the surface—and the temperature increases. An international effort to melt at least some of the ice of Antarctica and thus make that continent more habitable has been proposed. This is to be achieved by spreading a layer of black dust over the ice and thus increasing the effectiveness of the Sun in melting the underlying ice. As a consultant to this project, you are asked for an opinion. What would you recommend and why?

13. You are prospecting for lead and zinc in a glaciated region and discover that local farmers regularly turn up in their fields fragments of galena and sphalerite, common ore-forming minerals of lead and zinc, respectively. Using your knowledge of glacial geology, how would you go about trying to locate the bedrock source of these minerals?

Supplementary Readings

AGASSIZ, LOUIS *Studies on Glaciers,* Neuchâtel, 1840; trans. and ed. A. V. Carozzi, Harper & Row, Publishers, New York, 1967.
No better exposition of the proof of the former extent of glaciers can be found.

CAROZZI, ALBERT V. "Glaciology and the Ice Age," *Jour. Geol. Education,* vol. 32. pp. 158–170, 1984.
An excellent account of the history of glaciology beginning with sixteenth-century observations on the Swiss glaciers.

CHORLEY, RICHARD J., STANLEY A. SCHUMM, AND DAVID E. SUGDEN *Geomorphology,* Methuen & Co., New York, 1985.
Chapters 17 and 19 of this textbook on landforms are devoted to glaciers and their effects.

EMBLETON, CLIFFORD, AND CUCHLAINE A. M. KING *Glacial Geomorphology,* 2nd ed., John Wiley & Sons, New York, 1975.
A very good volume stressing glacier ice, erosion, and deposition.

FLINT, RICHARD FOSTER *Glacial and Quaternary Geology,* John Wiley & Sons, New York, 1971.
A standard text in the field and a very good one.

IMBRIE, JOHN, AND KATHERINE P. IMBRIE *Ice Ages: Solving the Mystery,* Enslow Publishers, Short Hills, N.J., 1979. Reissued in paperback, Harvard University Press, Cambridge, Mass., 1986.
A very readable account of the ice ages and, as the preface states, "what they were like, why they occurred and when the next one is due." It tells the story of the search for a solution to the "ice-age mystery" that has engaged so many people over the last 150 years.

PATERSON, W. S. B. *The Physics of Glaciers,* 2nd ed., Pergamon Press, London, 1981.
If you are interested in physics, this is the best summary available on its application to glacier ice.

RITTER, DALE F. *Process Geomorphology,* 2nd ed., William C. Brown Co., Dubuque, Iowa, 1986.
Two chapters treat glacier and glacier mechanics, glacial erosion, deposition, and landforms.

SUGDEN, DAVID E., AND BRIAN S. JOHN *Glaciers and Landscape,* John Wiley & Sons, New York, 1976.
Discusses (in intelligible style) glacier movement and distribution and the effects of glacial erosion and deposition. Well illustrated; good bibliography.

Individual seas and oceans connect up to form a single large world ocean that accounts for over 70 percent of the area of the Earth.

THE OCEANS

"Of Neptune's empire let us sing . . ."

Thomas Campion, from "In Praise of Neptune"

OVERVIEW

Over 70 percent of the Earth's surface is submerged beneath the waters of the world's ocean. It is not surprising that our knowledge of this difficult-to-explore realm has lagged behind our knowledge of the continents. In the last several decades, however, data on the oceanic basins have expanded enormously. Samples from dredge and drill, topographic sounding of the ocean's floor, geophysical investigation of the Earth below the ocean, studies of the circulation of seawater, and an ever-increasing understanding of marine plants and animals (both past and present) have all contributed to the emerging portrait of the ocean.

In this chapter we look first at the origin and nature of seawater and then at its circulation. Next we briefly examine the interconnecting basins that hold the seawater.

We find that we can divide the topography of the sea floor into a few major zones that include the *continental margins,* the *abyssal plains,* and the *oceanic ridges,* and that other major topographic features include towering *seamounts* and *trenches.*

The oceans provide the ultimate collecting ground for the sediments produced by weathering and erosion of the continents. Here lies a blanket of sediments, ranging in thickness from several kilometers down to a feather edge. As we shall see, these sediments differ. Some are silts, some coarse sands and gravels, others fine deposits made up of the hard parts of microscopic organisms or far-traveled volcanic ash or desert dust.

It is, reasonably enough, the shoreline of the oceans that has been most intensively studied and that is best understood. In the section of the chapter that discusses shoreline features we study the nature and behavior of *waves* and how they affect the shoreline. More specifically we examine the *depositional* and *erosional* features of a shoreline. Finally, we look at some of the factors that produce a constantly *changing sea level.*

16.1

SEAWATER

DISTRIBUTION

We refer to the Northern Hemisphere as the **land hemisphere** because north of the equator oceans and seas cover only about 60 percent of the Earth's surface, in contrast with the Southern Hemisphere, 80 percent of which is flooded by marine waters. Between 45° and 70°N, the ocean occupies only 38 percent of the surface, whereas 98 percent of the Earth's surface is covered by oceans between 35° and 65°S (Figure 16.1).

The greatest oceanic depth so far recorded is in the Pacific Ocean near the southern end of the Mariana Trench, between Guam and Yap. Here the ocean bottom lies more than 11 km below the surface, a distance considerably greater than the height of Mt. Everest, which is not quite 9 km above sea level. The average oceanic depth is about 3,800 m, while the mean elevation of the continents is only 840 m (Figure 16.2).

ORIGIN OF SEAWATER

Some minerals contain water as part of their crystal structure, hornblende being an example. It has been suggested that, over millions of years, weathering of rock containing hydrous minerals may have been the origin of seawater. But a closer examination of this suggestion shows this process could not account for the volume of oceanic water, nor for the chlorine in the dissolved salt (NaCl) we now see. Two other possibilities seem more probable.

One of these possibilities is that volcanic eruptions have been continuously adding water and other elements from great depths below the surface, depths that far exceed the depths of weathering. This process, called **continuous degassing** of the Earth, has the drawback that apparently most of the water now coming to the surface through volcanic eruptions is recycled and is not being added to the hydrologic cycle for the first time.

The other possibility assigns the accumulation of most oceanic water to an early stage of Earth history, a period for which we have little tangible evidence. In this suggestion the excess hydrogen combined with oxygen to

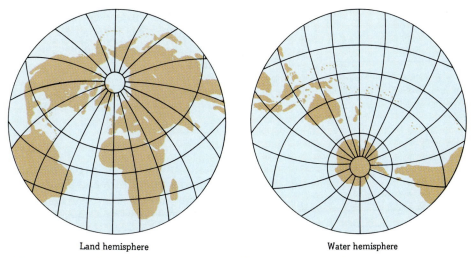

Land hemisphere   Water hemisphere

16.1 On the land hemisphere map, centered on Western Europe, land and sea are about evenly divided. But an indisputable predominance of the seas is seen on the water hemisphere map, centered on New Zealand.

form water and bubbled to the surface during the first few hundred million years of Earth time, when temperatures were high and the present gross stratification of the Earth (see Section 10.11) was still incomplete. The excess chlorine came to the surface at the same time.

At present we do not have sufficient evidence to choose between these two possibilities. In fact, both processes may have contributed to the creation of oceanic waters.

NATURE OF SEAWATER

Most of the naturally occurring elements have been identified in dissolved form in the ocean. On the average, 1,000 g of water contain 35 g of dissolved inorganic salts. The salinity of the ocean is thus frequently expressed as 35 ‰, although it actually varies between 30 ‰ and 38 ‰. Chlorine and sodium together account for over 86 percent (or about

16.2 The relative distribution of land and sea. Note that the mean ocean depth is 3,800 m, whereas the mean elevation of the land is only 840 m. [After H. U. Sverdrup, Martin W. Johnson, and Richard H. Fleming, *The Oceans*, p. 19, Prentice Hall, Englewood Cliffs, N.J., 1942.]

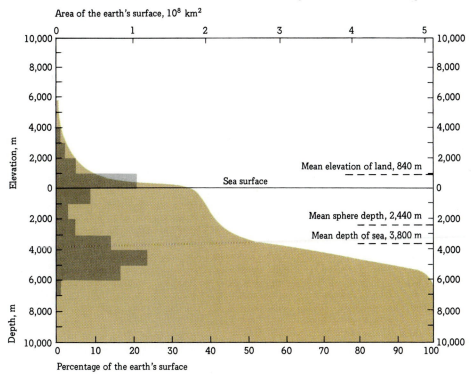

TABLE 16.1
The Major Constituents Dissolved in
Seawater[a]

Ion		Percentage of all dissolved material
Chlorine	Cl^-	55.04
Sodium	Na^+	30.61
Sulfate	SO_4^{2-}	7.68
Magnesium	Mg^{2+}	3.69
Calcium	Ca^{2+}	1.16
Potassium	K^+	1.10
Bicarbonate	HCO_3^-	.41
		99.69

[a] From H. U. Sverdrup, Martin W. Johnson, and Richard H. Fleming, *The Oceans*, p. 166, Prentice Hall, Englewood Cliffs, N.J., 1942.

30 parts per thousand) of the dissolved material in seawater (Table 16.1).

The salinity of the ocean varies from place to place, although the proportion of the major components of the salinity remains constant. High rainfall in equatorial climates reduces salinity. In the subtropical belts north and south of the equator, low rainfall and high evaporation tend to increase salinity (Figure 16.3). In the Arctic and Antarctic areas, on the other hand, the melting of glacier ice reduces the saltiness of the sea.

16.2
CURRENTS

Seawater is in constant movement, in some places horizontally, in others downward, and in still others upward. Its rate of movement varies from spot to spot, but it has been estimated that there is a complete mixing of all water of the

16.3 In general, the total salinity of seawater varies inversely with precipitation and directly with evaporation. [After M. Grant Gross, *Oceanography*, 4th ed., Fig. 7.5, p. 160, Prentice Hall, Englewood Cliffs, N.J., 1987.]

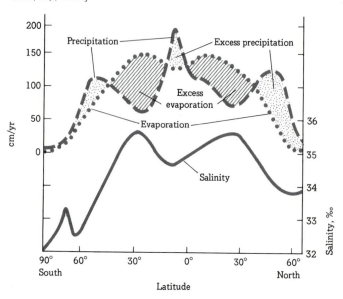

oceans about once every 2,000 years. The currents of the ocean are caused by tides, by the differences in density of seawater from place to place, and by the Earth's system of winds.

TIDAL CURRENTS

The gravitational forces that operate between the Sun, the Moon, and the Earth set the waters of the ocean in motion to produce tidal currents (Figure 16.4). The speed of these currents may reach several kilometers per hour if local conditions are favorable. Velocities of 20 km/hour develop during extreme tides in Seymour Narrows, between Vancouver Island and the mainland of British Columbia. Tidal currents of half this velocity are not uncommon.

Tidal currents close to shore, often confined by the topographic configuration of the shore and the near-shore topography, are stronger than those in the open ocean. Constricted as they are, the near-shore currents are **reversing,** flowing landward with the rising tide and seaward with the falling tide. In the open waters of the ocean the currents are much weaker, generally not more than 1 km/hour, and have a rotary motion, clockwise in the Northern Hemisphere.

DENSITY CURRENTS

The density of seawater varies from place to place with changes in salinity, temperature, and the amount of material held in suspension. These changes in density cause movement of seawater. Thus water of high salinity is heavier than water of low salinity and sinks beneath it; cold, heavy water sinks below warmer and lighter water; and heavy, muddy water sinks beneath light, clear water. An

16.4 Twice a day tides cause the withdrawal of ocean water and the exposure of this muddy bottom along a Maine inlet.

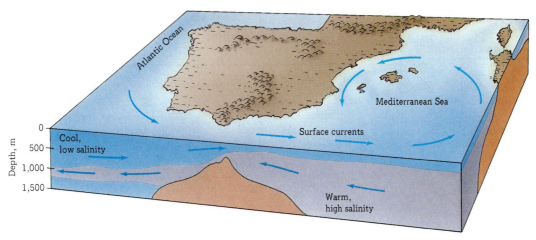

16.5 A density current flows from the Mediterranean Sea through the Straits of Gibraltar and spreads out into the Atlantic Ocean. High evaporation and low rainfall in the Mediterranean area produce a more saline and hence heavier water than the water of the neighboring Atlantic Ocean. As a result, Mediterranean water flows out through the lower portion of the straits, and lighter Atlantic water moves above it and in the opposite direction to replace it. The higher temperature of the Mediterranean water is more than counteracted by its greater salinity, and the water sinks to a level in the Atlantic somewhat lower than 1,000 m below the surface.

example of each of the three types of density currents is given in the following paragraphs.

Water flowing through the Straits of Gibraltar illustrates density currents caused by differences in salinity. A pair of currents, one above the other, flow in opposite directions between the Mediterranean Sea and the Atlantic Ocean. They originate as follows: The Mediterranean, lying in a warm, dry climatic belt, loses about 1.5 m of water every year through evaporation. Consequently the saltier, heavier water of the Mediterranean moves outward along the bottom of the straits and sinks downward into the less salty, lighter water of the Atlantic. At the same time the lighter surface water of the Atlantic moves into the Mediterranean basin. The water flowing from the Mediterranean settles to a depth of about 1,000 m in the Atlantic and then spreads slowly outward beyond the equator on the south, the Azores on the west, and Ireland on the north. It has been estimated that as a result of this activity the water of the Mediterranean basin is changed once every 75 years (Figure 16.5). What might happen if the straits of Gibraltar were closed is the subject of Box 16.1.

Differences in temperature are responsible for major circulation of water through much of the deep ocean. For example, as a result of such variations, water from the cold Arctic and Antarctic regions creeps slowly toward the warmer environment near the equator. The cold, relatively dense water from the Arctic sinks near Greenland in the North Atlantic and can be traced to the equator and beyond as far as 60°S. This water is called **North Atlantic Deep Water.** Colder and denser water **(Antarctic Bottom Water)** moves downward to the sea floor off Antarctica and creeps northward, pushing beneath the North Atlantic Deep Water. In fact, the Antarctic Bottom Water reaches well north of the equator before it loses its identity. Lighter than these two types of water is the **Antarctic Intermediate Water,** which sinks in the area of 50°S and flows northward above the North Atlantic Deep Water until it loses its identity about 30°N in the area where the Mediterranean water flows from east to west. These relationships are shown in Figure 16.6.

A third type of density current, known as a **turbidity current,** is caused by the fact that turbid or muddy water has

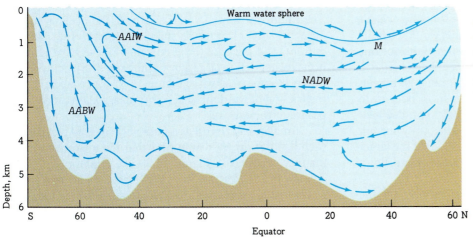

16.6 Circulation of water in the Atlantic Ocean shown along a north-south profile as determined by temperature and salinity. *NADW*, North Atlantic Deep Water; *AAIW*, Antarctic Intermediate Water; *AABW*, Antarctic Bottom Water; and *M*, Mediterranean water.

BOX 16.1   The Mediterranean Was a Desert

It is hard to conceive that the basin of the Mediterranean Sea, with its sparkling water of turquoise blue, was, not long ago geologically, a desert—a giant Death Valley—whose floor in places was over 3,000 m below today's shoreline. But so it was.

Twenty million years ago, in the early Miocene, a seaway stretched eastward from the Atlantic through the Mediterranean and linked up with the Indian Ocean. As the Miocene wore on, the circulation between the Mediterranean and the Atlantic on the west and the Indian Ocean on the east slackened. Eventually rising land barriers eliminated all connection between the Mediterranean basin and the Atlantic and Indian Oceans. As explained in Section 16.2, today there is less water draining from the lands to the Mediterranean than is lost by evaporation. So it was in the Miocene. Only then there was no way to make up the deficit, as there is today with water from the Atlantic via the Straits of Gibraltar. The result was inevitable. The level of

the landlocked Mediterranean began to fall. Its water became progressively more saline. Eventually calcium carbonate began to precipitate. As amounts of dissolved material increased, gypsum was deposited, followed in sequence by halite and then a complex of potassium and magnesium salts (see Section 6.2). River water reaching the basin either evaporated completely or in some places formed, at best, shallow temporary salt lakes. The Mediterranean basin had become a desert.

Then about 5 million years ago, as the Miocene drew to a close, the dam between the Mediterranean and the Atlantic broke at what is now the Straits of Gibraltar. Water rushed in from the Atlantic Ocean over a gigantic waterfall, greater than any we know today: 40,000 km³ of water cascaded into the Mediterranean basin until, a little over 100 years later, the Mediterranean was again a sea.

The facts now used to support this history had been accumulating over the

years. But it was not until 1970 that the research vessel *Glomar Challenger,* with an international team of scientists on board, brought up the samples that demonstrated the validity of the Miocene Mediterranean desert. The story of this cruise of the *Glomar Challenger* (Leg 13 of the Deep Sea Drilling Program sponsored by the National Science Foundation) appeared 13 years later.

Kenneth S. Hsu, a geologist at the Swiss Federal Institute of Technology in Zurich, has told the story of this cruise in his book *The Mediterranean Was a Desert: A Voyage of the Glomar Challenger.* This is no dry-bones account, but one of those infrequent readable and informative books of scientific reporting by a scientist deeply involved in the investigations reported. Hsu, a co-chief scientist for the cruise, recounts with verve and color the frustrations, the difficulties, the personality conflicts, and above all the thrill of discovery as a scientific idea develops from vague hypothesis to established fact.

a greater density than clear water and therefore sinks beneath it. We can demonstrate how these currents operate on a small scale in the laboratory, and they have actually been observed in freshwater lakes and reservoirs. Turbidity currents explain several different phenomena in the present-day oceans as well as in the oceans of the geological past.

Turbidity currents offer the most plausible explanation of certain deposits that have been studied from the deep oceans. Samples of sediments from these basins contain thin layers of sand that could not have been carried so far from shore by the slow drift of water along the ocean's floor, but rapidly moving turbidity currents would be quite capable of moving the sand down the slopes of the oceanic basins to the deep floors.

Analysis of other samples from the ocean floor reveals that particles become progressively finer from bottom to top of the deposit. The best explanation of this **graded bedding** is that the deposit was laid down by a turbidity current carrying particles of many different sizes and that the larger particles were the first to be dropped on the ocean floor. Such deposits are called **turbidites.**

Turbidity currents may be set in motion by the

slumping and sliding of material along the slopes of the ocean basin under the influence of gravity, or aided by the jarring of an earthquake. Or the currents may be created by the churning up of bottom sediments under the influence of violent storms.

MAJOR SURFACE CURRENTS

The major movements of water near the ocean's surface occur in such currents as the Gulf Stream, the Japanese current, and the equatorial currents. These great rivers of ocean water define large gyres, some moving clockwise and others counterclockwise (Figure 16.7). They are driven by the friction of wind moving over water, so let us look at the nature of these winds.

Imagine our Earth as completely covered by water. In such a situation we would find that the air moved in the manner depicted in Figure 16.8. The equator, where the greatest heating by the Sun takes place, would be a belt of warm, rising air marked by low pressure, as it is today. At

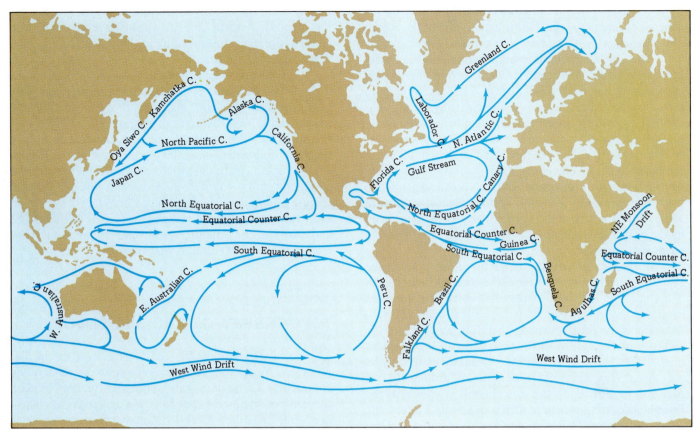

16.7   Major surface currents of the world.

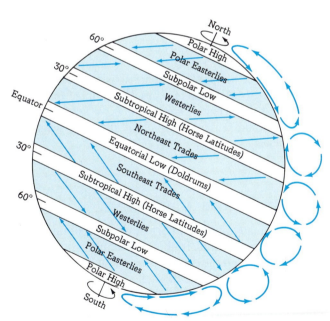

16.8   Idealized circulation of the atmosphere on a rotating globe completely covered with water. See text for discussion and relation to the major surface currents of the Earth.

the poles colder air would cause the air to settle and create a high-pressure zone. Other belts of high pressure would lie at the southern pole and in the areas of descending air located about 30° north and south of the equator. Near 60° north

and south would be two more low-pressure zones. At the Earth's surface the air would move from zones of high pressure to zones of low pressure.

These winds do not blow directly north or south because of the **Coriolis effect** caused by the rotation of the Earth. The result of the Coriolis effect is that moving things, whether they be bullets, planes, winds, or ocean currents, appear to veer to the right in the Northern Hemisphere and to the left in the Southern Hemisphere. This explains the direction of the trade winds as well as that of the westerlies and of the polar easterlies, as shown in Figure 16.8.

Turn again to the surface currents of the ocean in Figure 16.7 and examine by way of illustration the surficial currents of the Atlantic Ocean in both Northern and Southern Hemispheres. The equatorial currents lie on each side of the equator and they move almost due west. They derive their energy from the trade winds blowing constantly toward the equator from the northeast in the Northern Hemisphere and from the southeast in the Southern Hemisphere. The westerly direction of the currents is explained by the Coriolis effect. As the water driven by the trade winds moves toward the equator, it is deflected to the right (toward the west) in the Northern Hemisphere and to the left (also toward the west) in the Southern Hemisphere. As a result, the north and south equatorial currents are formed. As these currents approach South America, one is deflected north and the other south. This deflection is caused largely by the shape of the ocean basins, but it is aided by the Coriolis effect and by the slightly higher level of the oceans along the equator where rainfall is greater than elsewhere.

The north equatorial current moves into the Caribbean waters and then northeastward, first as the Florida current and then as the Gulf Stream. The Gulf Stream, in turn, is deflected to the east (to the right) by the Coriolis effect. This easterly movement is strengthened by prevailing westerly winds between 35° and 45°N, where it becomes the North Atlantic current. As it approaches Europe, the North Atlantic current splits. Part of it moves northward as a warm current past the British Isles and parallel to the Norwegian coast. The other part is deflected southward as the cool Canary current and eventually is caught up again by the northeast trade winds, which drive it into the north equatorial current.

In the South Atlantic the picture is very much the same — a kind of mirror image of the currents in the North Atlantic. After the south equatorial current is deflected southward, it travels parallel to the eastern coast of South America as the Brazil current. Then it is bent back to the east (toward the left) by the Coriolis effect and is driven by prevailing westerly winds toward Africa. This easterly moving current veers more and more to the left until finally, off Africa, it is moving northward. There it is known as the Buenguela current, which in turn is caught up by the trade winds and is turned back into the south equatorial current.

The cold surface water from the Antarctic regions moves along a fairly simple course, uncomplicated by large landmasses. It is driven in an easterly direction by the prevailing winds from the west. In the Northern Hemisphere, however, the picture is complicated by continental masses. Arctic water emerges from the polar seas through the straits on either side of Greenland to form the Labrador current on the west and the Greenland current on the east. Both currents subsequently join the North Atlantic current and are deflected easterly and northeasterly.

We need not examine in detail the surface currents of the Pacific and Indian Oceans. We note, however, that the surface currents of the Pacific follow the same general patterns as those of the Atlantic. Furthermore, the surface currents of the Indian Ocean differ only in detail from those of the South Atlantic (Figure 16.7).

16.3
THE OCEANIC BASINS

Most of the seawater surrounding the continents is held in one great basin which girdles the Southern Hemisphere and branches northward into the Atlantic, Pacific, and Indian Oceans. The Atlantic and Pacific Oceans, in turn, are connected with the Arctic Ocean through narrow straits.

As we saw in Sections 1.4 and 10.11, the floors of the oceans are underlain by a layer of basaltic material approxi-

mately 5 km thick. This is covered by a layer of sedimentary deposits, which varies in thickness from 2 km to a feather edge near the spreading centers of the midoceanic ridges.

The oceanic basins have not always been as we see them today. In Sections 1.4 and 11.1 we found that they have changed in shape and size through time. These changes have been the result of sea-floor spreading as new oceanic crust formed along the active oceanic ridges and old crust was consumed along the margins of colliding plates. The geologic record allows us to reconstruct the locations of continents and oceanic basins at various times in the past (Figure 16.9) and even to predict where they might be at some future time. Figure 16.9 thus shows the changing size and shape of the Atlantic Ocean and its bordering land masses through 200 million years of time.

16.4
TOPOGRAPHY OF THE SEA FLOOR

The more we learn about the sea floor, the more spectacular we find its topography to be, even more so than that of the continents. To be sure, the landmasses present us with some impressive landscapes, but most dry land is less than a kilometer above sea level. Continuous erosion has worn down most of the continental masses close to sea level. In contrast, volcanism and tectonic forces have created large, bold features on the oceanic floor, features that have been unaffected by the vigorous erosion almost everywhere active above sea level. Only where sedimentation has blanketed the oceanic topography can the sea floor be said to be nearly featureless.

The three major topographic divisions of the oceanic realm are the **continental margins,** the **floors of the oceanic basins,** and the **oceanic ridges,** which we describe below. Two additional striking forms bear mention: **seamounts** and **trenches.**

THE CONTINENTAL MARGINS

We have already noted that ocean waters spill up over the edges of the continents. In this zone of transition between the continents and the deep oceans, the subsea topography is commonly a sequence of shelf-slope-rise from land to oceanic deep. The **continental shelf,** which accounts for about 7 percent of the ocean's area, slopes gently from the shore to an average depth of about 130 m, its very gentle gradient approximating 1 : 1,700. It varies greatly in width. For instance, off the western coast of South America the shelf is measured in a few hundred meters, whereas off the eastern coast the shelf measures hundreds of kilometers in width up to a maximum of 560 km off the Rio de la Plata.

At the outer edge of the shelf the sea floor steepens and becomes the **continental slope.** The slope leads downward

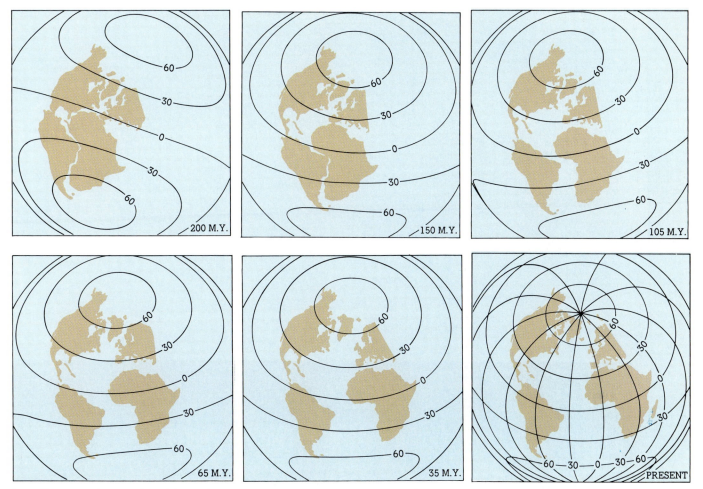

16.9 The size and shape of the Atlantic Ocean basin at various moments in time from 200 million years ago to the present. Lines of ancient latitude (paleolatitudes) are shown in the first five frames. Modern latitude and longitude are shown in the map of the present. [After J. D. Phillips and D. Forsythe, "Plate Tectonics, Paleomagnetism, and the Opening of the Atlantic," *Bull. Geol. Soc. Amer.,* vol. 83, p. 1579, Fig. 1, 1972.]

on a gradient of about 1 : 300 to the **continental rise.** The rise, with a gradient similar to that of the shelf, represents sediments that have moved down the continental slope to accumulate in vast fans at the edge of the deep oceanic basins. Somewhere, at a depth between 2,500 and 3,000 m, lies the true edge of the continent, if we define that boundary as the limit of continental crust.

The types of continental margins relate, in a general way, to the various types of plate boundaries: converging, diverging, and transform.

The western margin of the South American continent marks the convergence of the westward-moving American plate and the eastward-moving Nazca plate. The Nazca oceanic plate is being subducted beneath South America, and the continental margin is backed by the towering Andean mountain range. From the coast the surface of the continental block plunges precipitously beneath the sea. The continental shelf, if it exists at all, is very narrow and the rise is missing. The continental slope leads down directly into a deep oceanic trench formed as the oceanic crust of the Nazca plate pushes beneath the continental

plate (Figure 16.10). We can refer to the western edge of South America as a **leading continental margin,** for it is the front edge of the moving continent.

Diverging continental margins present a very much different picture. Consider the margins of the Red Sea as examples of the borders of an incipient ocean forming between diverging plates. In our discussion of plate tectonics in Chapter 11 we found that divergent boundaries were zones of high heat flow. As a result, thermal expansion raises the crust and produces a topographic high. At the same time lateral transfer of the crust results in tension along the boundary between the two plates and rifting occurs. This activity then produces the down-dropped zone along the axis of the thermally raised ridge and is flanked on either side by steep, high margins facing each other across the central rift.

The Red Sea and its adjacent borders in Arabia and Africa illustrate what happens when a spreading center develops within a continental area. Figure 16.11 shows the abrupt escarpments facing each other across the Red Sea rift. These bordering highlands slope away from their crests

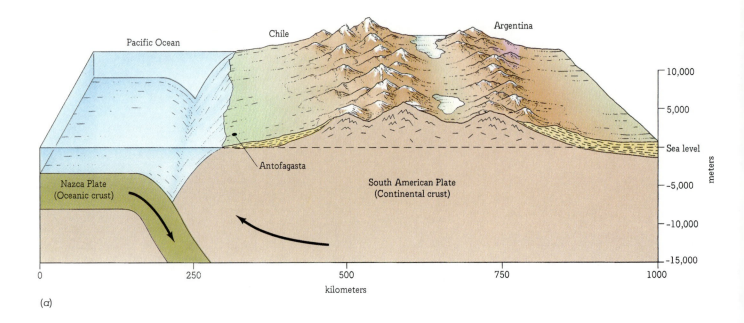

(a)

(b)

16.10   The western coast of South America is a converging border between a continental plate
and an oceanic plate. (a) A geologic and topographic sketch at 23½°S. (b) The same area as seen
looking toward the Chilean coast from Earth orbit. Clouds over the Andes form as the mountains
force moist, warm air upward to cooler elevations. [Photo by NASA.]

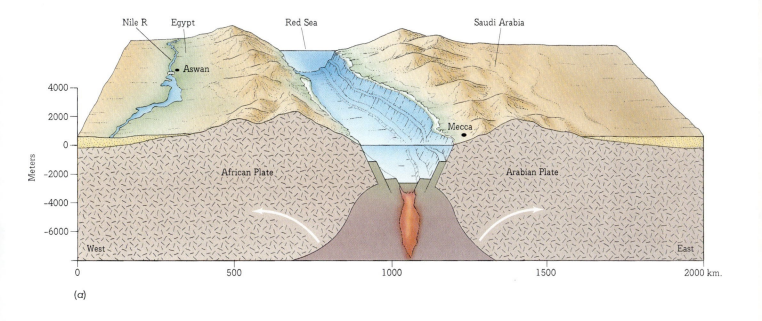

(a)

(b)

16.11   An early stage in the rifting of a continent is represented by the present-day Red Sea. The continued widening of the sea floor from a similar rift produced our modern Atlantic Ocean with its Midatlantic Ridge, as shown in Figure 16.12. (a) A geologic and topographic section looking north toward Aswan, Egypt, and Mecca, Saudi Arabia. (b) A photograph taken from Earth orbit looking south. The Gulf of Aqaba (left) and the Gulf of Suez (center) open into the Red Sea. The Nile River (right) flows toward viewer. [NASA.]

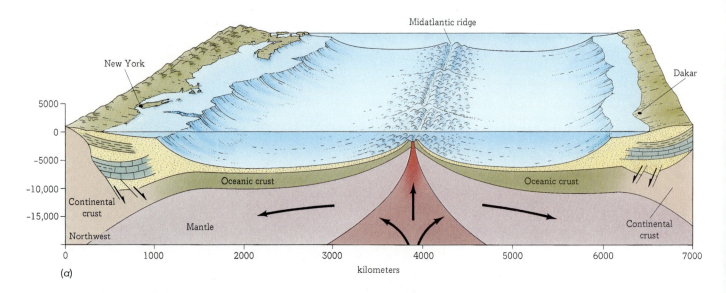

(a)

(b)

16.12 The basin of the Atlantic Ocean developed through sea-floor spreading to its present size from a narrow sea similar to today's Red Sea shown in Figure 16.11. (a) The Midatlantic Ridge marks the approximate position of the original rift. The continental margins of eastern North America and western Africa are called "trailing continental margins" because they are on the back or trailing sides of the continents as they move away from the spreading center, here the Midatlantic Ridge. By contrast, the western edge of South America (Figure 16.10) is termed a "leading continental margin" because it is the forward edge of a moving plate. (b) A small segment of the trailing edge of North America is seen in this satellite photo of the New York City area. The bright strips along the coasts of Long Island and northeastern New Jersey are barrier beaches, typical of the North American Coastal Plain from Cape Cod to Mexico. Offshore water depths over the continental shelf are shallow, as suggested by the hazy patterns in the water. This is a false-color infrared image in which the bright red patches are cultivated land and the browns are vegetation, generally trees. Densely populated areas, particularly those around New York City, are in blues. The white spots in the New Jersey area are clouds. [U.S. Geological Survey, Eros Data Center.]

toward the continental interiors. The highlands underwent extensive erosion as they rose, erosion that stripped away the cover of the nearly flat-lying sedimentary rocks that once covered them and are still to be seen in the lower regions farther away from the Red Sea. The margins of the separating landmasses are continental margins in the early stage of evolution. The continental shelf-slope-rise sequence described earlier is yet to develop. Only the steep slopes of the facing escarpments are present.

As diverging plates continue to separate, the continental margin is extensively modified. The North American margin south of New England can serve as an example. Some 200 million years ago the North American continent split from larger Gondwanaland along what is now the Midatlantic Ridge. The incipient ocean was similar to the Red Sea area of today. Its western margin (the eastern edge of the newly formed North American continent) was bold and steep and without a continental shelf. At least four processes contributed to the evolution of the eastern margin of North America as it appears today: thinning of the continental crust by rifting caused by tensional stress; additional thinning of the continental margin by erosion of the escarpment; contraction and subsidence of the continental margin by cooling as it moved away from the spreading center; and deposition of sediments on the sinking continental margin (Figure 16.12). These processes of thinning, subsidence, and deposition produced the shelf, slope, and rise we see today. We refer to such a margin as a **trailing continental margin,** for it forms on the back side of a continent as it moves away from the spreading center.

In those places where the edge of a landmass is located along a transform plate border and plates slip by each other side by side, a third general type of continental margin develops. Western United States from San Diego to north of San Francisco provides an example. The continental margin is complexly faulted (Figure 16.13). The San Andreas transverse fault is the longest and most familiar, but other such faults occur as well and give the Pacific borderlands here a topographic texture generally parallel to the plate border. This texture is interrupted by movements along both thrust and normal faults attendant upon the strike-slip motion. The result is a series of highlands (such as the Coast, Transverse, and Peninsular Ranges) and valleys and basins (such as the San Francisco Bay area, the Salinas valley, and the Los Angeles basin). These structures and resulting topography continue offshore. Off southern California, where their highest parts project above sea level, they form the Channel Islands (Figure 16.13).

Drowned valleys occur on the continental shelves. Many of these seem to be extensions of large valleys on the adjacent land. One of the best known of these valleys is the submerged extension of the Hudson River valley off the eastern coast of North America. This valley is cut some 60 m into the gently sloping shelf and widens down-valley from about 5 km to 24 km at its seaward end. It is generally agreed that this and similar valleys were cut during periods

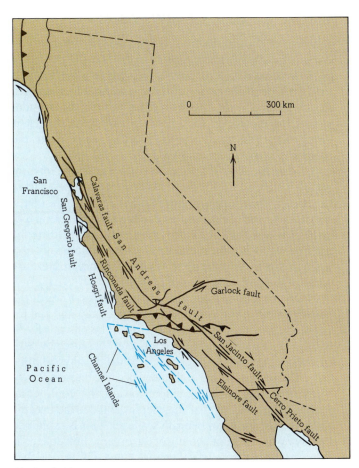

16.13 California is characteristic of the margin of a continent that is formed by the side-by-side motion between two plates moving in opposite directions. This figure shows the major strike-slip faults in the area. In general, the topography, both onshore and off, follows the direction of the faulting and produces a linearity to the topography. [Modified from John Suppe, *Structural Geology*, Prentice Hall, Englewood Cliffs, N.J., 1985.]

of Pleistocene glaciation when the sea level was 100 m or more lower than it is at present.

More difficult to explain than these valleys on the continental shelves are the deep submarine canyons along the continental slopes. Some canyons, such as those on the slopes off the Hudson and Congo Rivers, appear to be extensions of valleys on land or on the continental shelves. Others appear to have no association with such valleys. In any event, these canyons have V-shaped cross profiles and gradients that decrease along the lower sections of the continental slopes. Their slightly winding courses may extend 235 km out to sea, as does the Congo canyon. Some cut down more than 1 km below the surface of the continental slopes. Turbidity currents moving down the continental slope are most generally cited as the erosional agents responsible for these canyons.

FLOORS OF THE OCEANIC BASINS

Seaward of the continental margins lie the floors of the oceanic basins with a mean depth of 3,800 m. The creation of these floors is directly linked to diverging-plate boundaries, for here at the crests of the spreading ridges new sea floor is formed by the basaltic volcanism along the boundaries of the plates. As each new strip of oceanic crust is emplaced, older crust moves laterally away from the spreading center (Section 11.4).

Near the foot of the continental rise lie the **abyssal plains,** which, except for the trenches, are the deepest part of the oceans. Here turbidity currents flowing down the continental slope and rise deposit beds of sand and silt. Here, too, settle clay-sized particles carried in suspension from the continents, as do the remains of microscopic plants and animals that lived in the upper waters of the ocean. Here are some of the flattest regions of the Earth.

As we move from the abyssal plains toward the spreading ridge, the ocean shallows gradually and the topography becomes more pronounced. The effects of volcanism and fracturing are not completely obscured by sediments, and the hills found here are called the **abyssal hills.**

The regional slope of the deep oceanic floors is downward from the spreading ridges toward the continental margins. This means that in general the older the oceanic floor, the deeper it lies beneath sea level. This is explained largely by the effect of cooling of the oceanic lithosphere as it moves farther from its spreading center. This subsidence is proportional to the square root of the age of the oceanic crust, as illustrated in Figure 16.14.

16.14 The depth of the sea floor varies in a regular way with its age: The deeper the floor below sea level, the older it is. [John Suppe, *Structural Geology*, Fig. 1.10, Prentice Hall, Englewood Cliffs, N.J., 1985.]

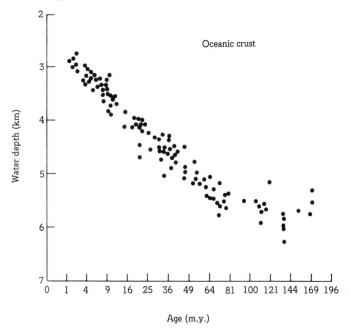

OCEANIC RIDGES

A network of broad ridges segments the deep oceanic basins. There are two general types of ridges. One is the **spreading center ridge,** at which new oceanic crust is being created. The second type is called an **aseismic ridge** because, in contrast to the seismically active spreading-center ridges, this type exhibits little or no earthquake activity.

System of Spreading-Center Ridges   The spreading centers of the oceans make up a continuous system of oceanic ridges. That section of the system that we know as the Midatlantic Ridge is characteristic. It bisects the Atlantic Ocean, generally paralleling the continental margins to the east and west. It rises toward the oceanic surface and in some places—for example, in Iceland and the Azores—emerges above sea level. It is between 1,500 and 2,000 km wide, and it slopes gently away from its central axis. Shallow-focus earthquakes and volcanic eruptions of basalt mark the crest, as does a trench or rift valley 25 to 50 km wide and 1 to 2 km deep. Transform faults offset and segment the ridge and its axial valley. In previous paragraphs we found that the plates of the Atlantic basin diverge east and west from the summit of this restless ridge.

The Midatlantic Ridge is continuous with the rest of the Earth's spreading-center ridge system, although other segments are not all so perfectly centered in the oceanic basins (Figure 11.20). The entire system, however, is a long, narrow heat bulge along which thermal energy and new crustal material are transferred upward from the mantle.

Aseismic Ridges   The second type of ridge, the **aseismic ridge,** is also a thermal feature. Unlike the spreading-center ridge, however, where heat is applied all along the ridge, volcanic activity takes place only at the extreme tip of the aseismic ridge, which is also the highest point along the ridge.

The Hawaiian Islands can serve as an example (see Section 4.1). This chain of islands, built on top of an aseismic ridge, begins with the active volcanic island of Hawaii. We can follow the ridge northwestward past extinct volcanic islands such as Molokai and Oahu, and finally, 2,000 km from Hawaii, to the coral reef islands of Midway and Kure. The ridge becomes submerged and continues another 500 km beyond Kure, at which point it turns almost due north. The ridge then all but disappears, but its trend is marked by a series of submarine peaks that can be traced into the North Pacific, where the most northerly peaks are being consumed, along with adjacent sea floor, in the Kurile trench off southern Kamchatka.

The ridge, its associated islands, and submarine peaks trace the direction of plate movement across a stationary plume or hot spot. In this case, the active volcanoes of Hawaii are the surface expressions of a plume. The hot, rising mantle of the plume fuels the volcanic activity and bulges up the adjacent sea floor in a domal feature

1,000 km in diameter. As the plate moves over the plume, it takes the earlier-formed heat bulge and volcanoes farther and farther from the heat source (Figure 4.15). With continued cooling, the ridge and its volcanoes subside, and eventually the volcanoes disappear below sea level. The Hawaiian Ridge, its islands, and submerged peaks mark, then, the direction of movement of the Pacific plate over tens of millions of years. The hot spot trace tells us that the plate has been moving northwestward and that at an earlier time it moved northward.

Numerous aseismic ridges characterize the ocean basins and we think their origin was similar to that of the Hawaiian Ridge.

SEAMOUNTS AND TRENCHES

Seamounts are found on the oceanic ridges and across the oceanic floors. **Trenches** occur on the oceanic floor and along the margins of some continents.

Seamounts   We have referred to seamounts, thousands of which occur on the sea floor. Some have the characteristic conical form we associate with volcanoes, as indeed they are. Others are flat-topped cones. These latter, called **Guyots** (pronounced *gee*—hard "g" as in *geese*—*yoz*), are named after Arnold Guyot, the nineteenth-century Swiss-American geologist-geographer. They were once volcanic islands truncated by wave action before they sank beneath the oceanic surface. The subsidence of both types of seamounts was caused by cooling of the oceanic crust as their distance from a spreading center or a hot-spot plume increased with time.

Trenches   The greatest oceanic depths are found in the trenches. Most of these lie around the margins of the Pacific Ocean. Usually arcuate in plan, they are as much as 200 km wide and up to 2,400 km in length. They reach 2 to 3 km below the adjacent oceanic floor. These deeps, as we have seen, mark the convergence of plates where the oceanic crust plunges beneath continental masses or beneath other oceanic crust.

16.5

SEDIMENTS OF THE OCEANS

In earlier chapters we spoke from time to time of the processes by which rocks are weathered, eroded, transported, and finally deposited to be transformed into sedimentary rocks. The great oceanic basins of the world constitute the ultimate collection area for the sediments and dissolved material that are carried from the land, and the great bulk of the sedimentary rocks found on our modern landmasses were once deposits on the ocean floors of the past. In this section we shall speak briefly of the sediments being laid down in the modern oceans—the sediments destined to become the sedimentary rocks of the future.

DEPOSITS ON THE CONTINENTAL MARGINS

Theoretically, when particles of solid material are carried out and deposited in a body of water, the largest particles should fall out nearest the shore and the finest particles farthest away from the shore, in a neatly graduated pattern. But there are a great many exceptions to this generalization. Many deposits on the continental shelves show little tendency to grade from coarse to fine away from the shoreline. We would expect to find sand only close to the shore, but actually it shows up from place to place all along the typical continental shelf right up to the lip of the continental slope. It is particularly common in areas of low relief on the shelf. In fact, on glaciated continental shelves sand mixed with gravel and cobbles constitutes a large part of the total amount of material deposited.

Also common on the continental shelves are deposits of mud, especially off the mouths of large rivers and along the courses of currents that sweep across the river-laid deposits. Mud also tends to collect in shallow depressions across the surface of the shelves, in lagoons, sheltered bays, and gulfs.

Where neither sand nor mud collects on the shelves, the surface is often covered with fragments of rock and gravel. This is commonly the case on open stretches of a shelf where strong currents can winnow out the finer material, and off rocky points and exposed stretches of rocky shoreline. In narrow straits running between islands or giving access to bays, the energy of tide and current is often so effectively concentrated that the bottom is scoured clean and the underlying, solid rock is exposed.

The sedimentary patterns of the shelves are also complicated by the fact that the shelves have been dry land during the low stands of sea level accompanying glacial stages and are flooded during interglacial stages, as is the situation today. This alternation between marine and nonmarine conditions is bound to have affected depositional patterns on the shelf.

Sediments not trapped on the shelves continue on out beyond the break in shelf. Most of these sediments are deposited on the slopes and rises. Here clay, silt, sand, and even gravel accumulate. In general, however, the sediments are finer than on the shelf. Nevertheless, fully as much, if not more, sedimentation occurs on slopes and rises as on the shelves. The continental margins, considered as a unit, receive over 90 percent of the land-derived sediments. The rest of it goes on to the deep oceanic basins.

DEPOSITS ON THE DEEP-SEA FLOORS

The deposits that are found across the floors of the deep sea are generally much finer than those on the slopes, rises, and shelves, although occasional beds of sand have been found even in the deeps. One would expect this. One would also expect that the rates of sedimentation would be slower than on the continental margins, and indeed they are. The rates on the continental margins are from 20 to 200 times that of the deep sea.

Deposits of the deep sea may be derived from the landmasses. For instance, windborne volcanic ash may settle to the ocean surface and, eventually, reach the sea floor. In the polar regions silts and sands from glacier ice make up much of the bottom deposit, and around the margins of the continents we often find deep-sea mud deposits of silt and clay washed down from the landmasses. Much of the oceanic bottom, especially in the Pacific, is covered by an extremely fine-grained deposit known as **brown clay,** which may have originated on the continents and then drifted out into the open ocean. Some of the clay in the deep seas also represents the insoluble residue remaining after the solution of carbonate matter. Thus when a one-celled animal with a calcitic shell dies in near-surface waters, the test settles toward the sea floor. But once it reaches a certain depth, the chemistry of the water column is such that the calcite dissolves. The depth at which solution occurs is known as the **carbonate compensation depth,** or **CCD** (see Section 6.2).

In the Antarctic and Arctic Oceans some of the deep oceanic sediments are believed to have been ice-rafted from the land to the oceans by icebergs. These **glacial marine deposits** are chiefly silt containing coarse fragments. Near the landmasses are extensive layers of sand and coarse silt deposited by turbidity currents, the turbidites, discussed previously.

Extensive areas of the deep-sea floors are covered with deposits of mud called **oozes.** At least 30 percent of these deposits is made up of the skeletal remains of small plants and animals that lived in the upper waters of the oceans.

The balance is a very fine clay. The oozes may be either calcareous or siliceous, depending upon the composition of the skeletal material.

The calcareous oozes are often referred to as **foraminiferal oozes** because a significant percentage of their volume is made up of the tests of one-celled animals called *foraminifera.*

Siliceous oozes include **radiolarian oozes,** made up largely of the delicate and complex hard parts of minute marine protozoa called *radiolaria,* and **diatomaceous oozes,** made up of the siliceous cell walls of one-celled marine algae known as *diatoms.* Radiolarian mud predominates in a long east-west belt in the Pacific Ocean just north of the equator, and the greatest concentration of diatomaceous mud occurs in the North Pacific and in the Antarctic Ocean.

More spectacular deposits are the **manganese nodules** (Figures 16.15 and 16.16). These average 10 cm in diameter and when cut open exhibit the patterns of concentric growth. Composed largely of oxides of manganese and iron, they contain in addition the oxides of many of the rarer elements. They were originally thought to be inorganic in origin, but there is now evidence that iron-depositing bacteria play an important role in their formation. It is estimated that manganese nodules are abundant across some 20 percent of the floor of the Pacific. Their composition and abundance represent a commercial resource, not only as an ore of manganese, but also as an ore of copper and of such rarer elements as cobalt, titanium, and zirconium.

16.6
SHORELINES

Not all of us have the occasion to study firsthand the currents of the oceans or the forms and processes of the ocean floor. But most of us have the opportunity from time to time to observe the activity of water along the shorelines of oceans or lakes.

16.15   Manganese nodules 5 to 10 cm in diameter cover the sea floor in the D.O.M.E.S. (Deep Ocean Mining Environmental Studies) area of the east-central Pacific Ocean; photographed by the New Zealand Oceanographic Institution, 1974. (Photo from the National Geophysical Data Center)

16.16 This section through a manganese nodule shows that the concentric layers deposited as the nodule grew. The nodule, 6 cm across, was dredged from a depth of 5,400 m off the southeast coast of South Africa. [Raymond C. Gutschick.]

THE PROCESSES

Wind-Formed Waves   Most waves in the ocean have a much different origin from the earthquake-induced tsunami discussed in Section 10.9. They are produced by the shear stress between air and water as the air moves across the surface of the water. The harder the wind blows, the higher the water is piled up into long waves and intervening troughs at right angles to the direction of the wind. The distance between two successive waves is the **wavelength,** and the vertical distance between the **wave crest** and the bottom of an adjacent trough is the **wave height** (Figure 16.17). When the wind is blowing, the waves it generates are called a **sea.** But wind-formed waves persist even after the

wind that formed them dies. Such waves, or **swells,** may travel for hundreds or even thousands of kilometers from their zone of origin.

We are concerned with both the movement of the waveform and the motion of the particles of water in the path of the wave. Obviously the waveform itself moves forward at a measurable rate. But in deep water the particles of water in the path of the wave describe a circular orbit: Any given particle moves forward on the crest of the wave, sinks as the following trough approaches, moves backward under the trough, and rises as the next crest advances. Such a motion can best be visualized by imagining a cork bobbing up and down on the water surface as successive wave crests and troughs pass by. The cork itself makes only very slight forward progress under the influence of the wind. Wave motion extends downward until at a depth equal to about one-half the wavelength it is virtually negligible. But between this level and the surface, particles of water move forward under the crest and backward under the trough of each wave in orbits that decrease in diameter with depth (Figure 16.18).

As the wave approaches a shoreline and the water becomes more shallow, definite changes take place in the motion of the particles and in the form of the wave itself. When the depth of water is about half the wavelength, a depth called **wave base,** the bottom begins to interfere with the motion of the particles of water in the path of the wave, and their orbits become increasingly elliptical. As a result, the wavelength decreases, as does the velocity of the wave, and the front of the wave becomes steeper. When the water becomes shallow enough and the front of the wave steep enough, the wave crest falls forward as a breaker, producing what we call **surf** (Figure 16.19). At this moment the particles of water within the wave are thrown forward against the shoreline. The energy thus developed is then available to erode the shoreline or to set up currents along the shore that are able to transport the sediments produced by erosion.

16.18   The motion of water particles relative to wave motion in deep water. Water particles move forward under the crest and backward under the trough, in orbits that decrease in diameter with depth. Such motion extends downward to a distance of about one-half the wavelength. [After *U.S. Hydrog. Off. Pub. 604,* 1951.]

16.17   Diagrammatic explanation of terms used to describe water waves.

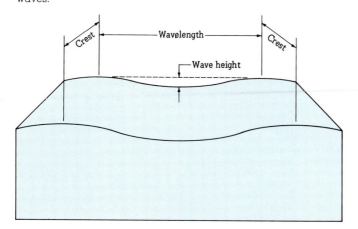

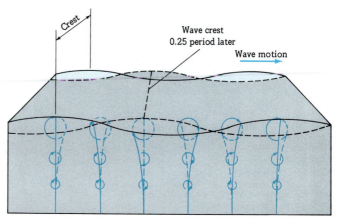

16.19 A wave steepens as it moves into shallow water and eventually its crest may fall forward to form a breaker, as in this example off Long Beach Island, New Jersey. It is this forward motion that generates most of the energy for erosion and transportation along a shore.

16.20 Wave crests that advance at an angle on a straight shoreline and across a bottom that shallows at a uniform rate are bent shoreward, as suggested in this diagram. Refraction is caused by the increasing interference of the bottom with the orbits of water-particle motion within the wave.

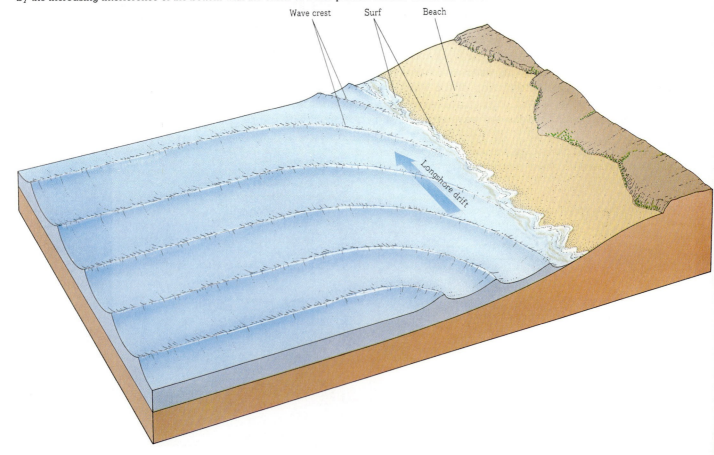

Wave crest   Surf   Beach

Longshore drift

Wave Refraction and Coastal Currents   Although most waves advance obliquely toward a shoreline, the influence of the sea floor tends to bend or **refract** them until they approach the shore nearly head-on.

Let us assume that we have a relatively straight stretch of shoreline with waves approaching it obliquely over an even bottom that grows shallower at a constant rate. As a wave crest nears the shore, the section closest to land feels the effect of the shelving bottom first and is retarded, while the seaward part continues along at its original speed. The effect is to swing the wave around and to change the direction of its approach to the shore, as shown in Figure 16.20.

As a wave breaks, not all its energy is expended on the erosion of the shoreline. Some of the water thrown forward is deflected and moves laterally, parallel to the shore. The energy of this water movement is partly used up by friction along the bottom and partly by the transportation of sediment.

Refraction also helps explain why, on an irregular shoreline, the greatest energy is usually concentrated on the headland and the least along the bays. Figure 16.21 shows a bay separating two promontories and a series of wave crests sweeping into the shore across a bottom that is shallow off the headland and deep off the mouth of the bay. Where the depth of the water is greater than one-half the wavelength, the crest of the advancing wave is relatively straight. Closer to shore, off the headlands, however, the depth becomes less than half the wavelength, and the wave begins to slow down. In the deeper water of the bay the wave continues to move rapidly shoreward until there, too, the water grows shallow and the wave crest slows. This differential bending of the wave tends to make it conform in a general way to the shoreline. In so doing, the energy of the wave is concentrated on the headland and dispersed around the bay, as suggested in Figure 16.21.

SHORELINE FEATURES

Erosion and deposition work hand in hand to produce most of the features of the shoreline. An exception to this generalization is an offshore island that is merely the top of a hill or a ridge that was completely surrounded by water as the sea rose in relation to the land. But even islands formed in this way are modified by erosion and deposition.

A composite profile of a shoreline from a point above high tide seaward to some point below low tide reveals features that change constantly as they are influenced by the kinds of waves and currents along the shore. Not all features are present on all shorelines, but several are present in most shore profiles. The **offshore** section extends seaward from low tide. The **shore**, or **beach**, section reaches from low tide to the foot of the **sea cliff** and is divided into two segments: In front of the sea cliff is the **backshore**, characterized by one or more **berms**, resembling small terraces with low ridges on their seaward edges built up by storm waves; seaward from the berms to low tide is the **foreshore**. Inland from the shore lies the **coast**. Deposits of the shore may veneer a surface cut by the waves on bedrock and known as the **wave-cut terrace**. In the offshore section, too, there may be an accumulation of unconsolidated deposits composing a **wave-built terrace** (Figure 16.22).

The shoreline profile is ever-changing. During great storms the surf may pound in directly against the sea cliff, eroding it back and at the same time scouring down through the beach to abrade the wave-cut terrace. As the storm (and

16.21   Refraction of waves on an irregular shoreline. It is assumed that the water is deeper off the bay than off the headlands. Consider that the original wave is divided into three equal segments, *AB*, *BC*, and *CD*. Each segment has the same potential energy. But observe that by the time the wave reaches the shore, the energy of *AB* and *CD* has been concentrated along the short shoreline of headlands *A'B'* and *C'D'*, whereas the energy of *BC* has been dispersed over a greater front (*B'C'*) around the bay. Energy for erosion per unit of shoreline is therefore greater on the headlands than it is along the bay.

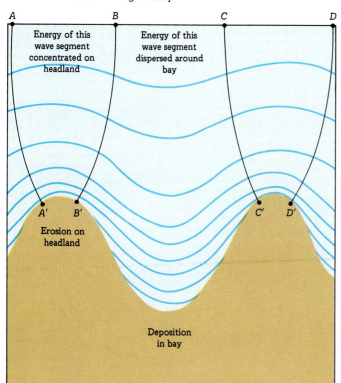

16.22   Some of the features along a shoreline and the nomenclature used in referring to them. [After F. P. Shepard, *Submarine Geology*, 2nd ed., p. 168, Harper & Row, Publishers, New York, 1963.]

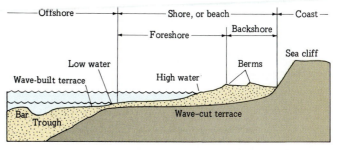

16.23   This platform, a meter or two above sea level, was cut by wave action when the land stood lower in relation to the sea along the California coast near Santa Cruz.

16.24   The sea has cut arches through this promontory on Monterey Bay, California.

hence the available energy) subsides, new deposits build up on the beach out in front of the sea cliff. The profile of a shoreline at any one time, then, is an expression of the available energy: It changes as the energy varies. This relation between profile and available energy is similar to the changing of a stream's gradient and channel as the discharge (and therefore the energy) of the stream varies (see Chapter 13).

Features Caused by Erosion   **Wave-cut cliffs** are common erosional features along a shore, particularly where it slopes steeply down beneath the sea. Here waves can break directly on the shoreline and thus can expend the greatest part of their energy in eroding the land. Wave erosion pushes the wave-cut cliff steadily back, producing a wave-cut terrace or platform at its foot (Figure 16.23). Because the surging water of the breaking waves must cross this terrace before reaching the cliff, it loses a certain amount of energy through turbulence and friction. Therefore the farther the cliff retreats and the wider the terrace becomes, the less effective are the waves in eroding the cliff. If sea level remains constant, the retreat of the cliffs becomes slower and slower.

Waves pounding against a wave-cut cliff produce various features as a result of the differential erosion of the weaker sections of the rock. Wave action may hollow out cavities, or **sea caves,** in the cliff, and if this erosion should cut through a headland, a **sea arch** is formed. The collapse of the roof of a sea arch leaves a mass of rock, a **stack,** isolated in front of the cliff (Figure 16.24).

Features Caused by Deposition   Features of deposition along a shore are built of material eroded by the waves from the headlands and of material brought down by the rivers that carry the products of weathering and erosion from the landmasses. For example, part of the material eroded from a headland may be drifted by currents into the protection of a neighboring bay, where it is deposited to form a sandy beach.

The coastline of northeastern New Jersey (Figure 16.25) illustrates some of the features caused by deposition. Notice that the Asbury Park–Long Branch section of the coastline is a zone of erosion formed by the destruction of a broad headland. Erosion still goes on along this part of the coast, where the soft sedimentary rocks are easily cut by the waves of the Atlantic. The material eroded from this section is moved both north and south along the coastline. Much additional sediment is derived from just offshore. Storm waves move sediment shoreward until it is added to the supply of sand moving either northward or southward. Sand swept northward is deposited in Raritan Bay and forms a long sandy beach projecting northward, a **spit** known as Sandy Hook.

Just south of Sandy Hook the flooded valleys of the Navesink River and of the Shrewsbury River are bays that have been almost completely cut off from the open ocean

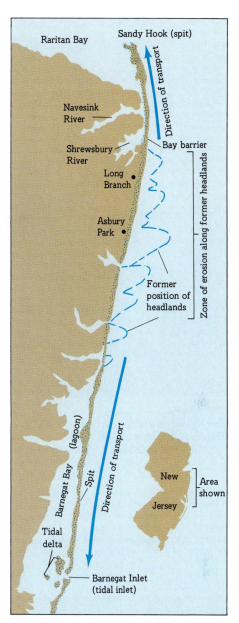

16.25   Erosion by the sea has pushed back the New Jersey coastline as indicated on this map. Some of the material eroded from the headlands has been moved northward along the coast to form Sandy Hook, a spit. To the south, a similar but longer feature encloses Barnegat Bay, a lagoon with access to the open ocean through a tidal inlet. [After an unpublished map by Paul MacClintock.]

by sandy beaches built up across their mouths. These beaches are called **bay barriers.**

Sand moved southward from the zone of erosion has built up another sand spit. Behind it lies a shallow lagoon, Barnegat Bay, that receives water from the sea through a **tidal inlet,** Barnegat Inlet. This passage through the spit was probably first opened by a violent storm, presumably a hurricane. Just inside the inlet a delta has been formed of sediment deposited partly by the original breakthrough of the bar and partly by continued tidal currents entering the lagoon.

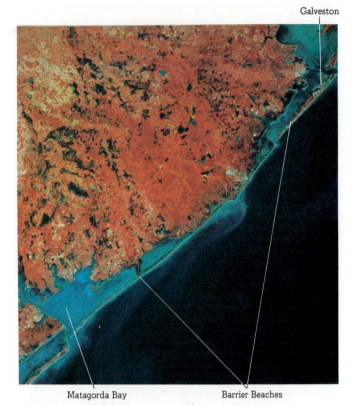

Long stretches of the shoreline from Long Island to Florida and from Florida westward around the Gulf Coast to Mexico are marked by shallow, often marshy lagoons separated from the open sea by narrow sandy beaches. Many of these beaches are similar to those that enclose Barnegat Bay, apparently elongated spits attached to broad headlands. Others, such as those that enclose Pamlico Sound at Cape Hatteras, North Carolina, have no connection with the mainland. These sandy beaches are best termed **barrier islands.** (See Figure 16.26.)

Human activity can affect processes of the shoreline. Figure 16.27 illustrates deposition caused by the Santa Monica pier and breakwater in Los Angeles.

16.7

THE CHANGING SEA LEVEL

Sea level is not constant. We find evidence of marine animals in the rocks of our highest mountains, and deep drilling has recorded shallow marine sediments hundreds and even thousands of meters below the modern sea level. In Section 10.9 we found that large earthquakes can change the relative position of land and sea in local areas. In the following paragraphs we look at some geologically recent changes of sea level felt over wide areas.

Isostasy accompanying continental glaciation and deglaciation has led to some changes in shoreline. The weight of a continental glacier is so great as to cause a

16.26 A system of barrier beaches backed by tidal lagoons extends from Cape Cod to Mexico. This satellite image shows a 240-km section of the system along the Texas coast. At the upper right Galveston is seen at the end of the island on the southern side of the entrance to Galveston Bay. Matagorda Bay lies behind the barrier beach at the lower left. This is an infrared image and vegetation shows in tones of red. [NASA.]

16.27 (a) Santa Monica pier, Los Angeles, California, 1931. Construction of the pier has caused increased deposition of sand on either side of its base. (b) Santa Monica pier with breakwater, 1949. Construction of the breakwater has caused growth of the beach seaward on the north side of the pier. [Fairchild Aerial Photography Collection, Whittier College.]

(a)

(b)

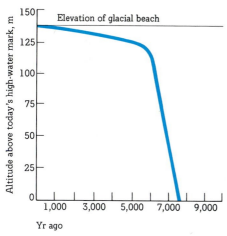

16.28 Change between land and sea in northern Ungava, Quebec, Canada. Land has risen over 125 m since the formation of the last glacial beach in the area more than 7,000 years ago. This change was very rapid until about 6,000 years ago, when it slowed abruptly and continued at a much lower rate to the present. The change in the relationship between land and sea is due to the isostatic recovery of the land after the disappearance of the ice sheet that had depressed it. [After B. Matthews, "Late Quaternary Land Emergence in Northern Ungava, Quebec," *Arctic*, vol. 20, no. 3, p. 186, 1967.]

downward warping of the land it covers. When the glacier melts and the load of ice is removed, the land slowly recovers and achieves its original balance (Figure 16.28).

We have good evidence of the recoil of land following glacial retreat along the shores of the Gulf of Bothnia and the Baltic Sea. Accurate measurements show that between 1800 and 1918 the land rose at rates ranging from 0.0 cm/year at the southern end to 1.1 cm/year at the northern end. Studies indicate further that the land has been rising at a comparable rate for 5,000 years. Areas that were obviously sea beaches are now elevated from a few meters to 240 m above sea level. A comparison of precise measurements along railroads and highways in Finland from 1892 to 1908 with measurements from 1935 to 1950 shows that uplift there has proceeded at the rate of from 0.3 to 0.9 m/century. The greatest change has been on the Gulf of Bothnia, at about 64°N. Most of this movement has been caused by the recoil of the land following the retreat of the Scandinavian icecap.

Similar histories of warping and recoil have been established in other regions, including the Great Lakes area in the United States. At one stage in the final retreat of the North American ice sheet, the present Lakes Superior, Michigan, and Huron had higher levels than they do now and were joined together to form glacial Lake Algonquin. Beaches around the borders of Lake Algonquin were horizontal at the time of their formation. Today they are still horizontal south of Green Bay, Wisconsin, and Manistee, Michigan, but 290 km north, at Sault Sainte Marie, Michigan, the oldest Algonquin beach is 108 m higher than it is at Manistee.

The water that is now locked up in the ice of glaciers originally came from the oceans. It was transferred landward by evaporation and winds, precipitated as snow, and finally converted to firn and ice. If all this ice were suddenly to melt, it would find its way back to the ocean basins and would raise the sea level an estimated 60 m. A rise of this magnitude would transform the outline of the Earth's landmasses and would submerge towns and cities along the coasts. For the last several thousand years melting glaciers have been raising the sea level (Figure 16.29), and modern records of sea level indicate that the sea is still rising at about 1 mm/year (Box 16.2).

16.29 Sea level has been rising for the last several thousand years. Here radiocarbon dates on samples of wood, shell, and peat originally deposited at or close to sea level have been combined to produce these curves of sea-level rise at various points along the eastern coast of the United States. [After David Scholl and Minze Stuiver, "Recent Submergence of Southern Florida," *Bull. Geol. Soc. Am.*, vol. 78, p. 448, 1967.]

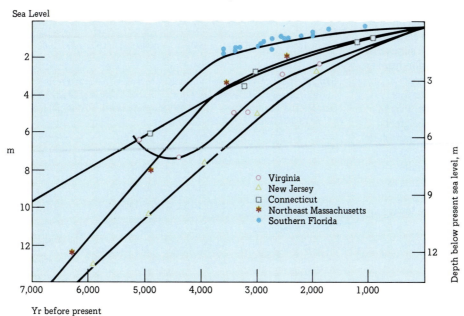

BOX 16.2   Sea Level, Present and Future

During the last 100 years the global rise in sea level has been between 10 and 15 cm, as shown in Figure B.16.2.1. Part of this rise has resulted from the melting of glaciers and the return to the oceans of water locked up in them. Another part has come from the thermal expansion of the ocean because of the warming atmosphere. Most authorities agree that this rise has been caused by a worldwide rise of surface-air temperature, as shown in Figure B.16.2.2.

The worldwide rise in temperature appears to be due to the **greenhouse effect** of several different gases released into the atmosphere by human activity. These greenhouse gases, which include carbon dioxide, methane, nitrous oxide, and the chlorofluorocarbons, have been increasing rapidly with the advance of industrial technology. Carbon dioxide, for instance, has increased 20 percent since the mid-nineteenth century as the result of the burning of coal, oil, and natural gas. Current measurements indicate that it is accumulating at an increasing rate in the atmosphere, as are the other greenhouse gases.

The phenomenon takes its name from the familiar horticultural greenhouse, in which solar energy enters through glass walls. Short-wave energy that heats the walls, floors, and contents of the greenhouse is reradiated as long-wave energy, but is partly blocked by the glass and is therefore available to warm the air inside the greenhouse. Here is how the greenhouse gases warm the atmosphere. The Earth's surface is heated by short-wave radiation from the Sun. Some of this energy is reradiated into the atmosphere as long-wave infrared radiation. Part of this energy is absorbed by the greenhouse gases and warms the atmosphere rather than escaping to space. Were it not for the presence of naturally occurring greenhouse gases, it is estimated that the mean global temperature would be 30°C or more lower than it is.

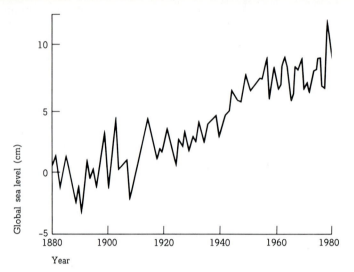

B16.2.1   Global mean sea-level trend based on tidal-gauge records. [Adapted from V. Gornitz et al., "Global Sea Level Trend in the Past Century," *Science*, vol. 215, Fig. 2, p. 1613, 1982.]

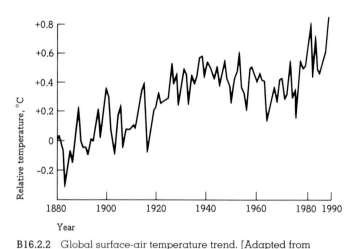

B16.2.2   Global surface-air temperature trend. [Adapted from James Hansen and Sergej Lebedeff, "Global Trends of Measured Surface Air Temperature," *Jour. Geophys Res.*, vol. 92, no. D11, Fig. 15, p. 13,370, 1988.]

Current estimates suggest that the amount of greenhouse gases will double sometime between the years 2030 and 2050. As greater amounts of humanly produced greenhouse gases are added to the atmosphere, we can expect an increasing rate of atmospheric warming. Studies by the National Academy of Sciences conclude that a doubling of the greenhouse gases could raise the mean global temperature between 1.5° and 4.5°C.

The estimated effect of an increasing temperature rise on sea level is shown in Figure B.16.2.3. The actual rise presumably will lie somewhere between the two extremes depicted in the figure. Some of the expected consequences of such a rise in sea level are:

- Increased beach erosion
- A decrease in wetlands areas
- Damage to structures on low-lying shores
- Saltwater invasion of groundwater supplies (see Figure 14.21)
- Saltwater invasion of coastal rivers and pollution of the freshwater supplies of adjoining cities

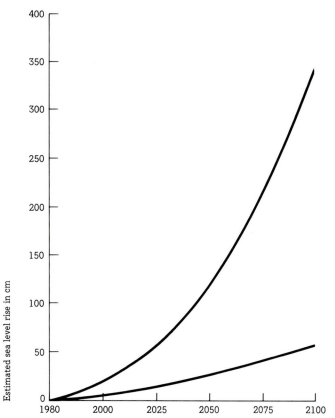

B16.2.3   Estimates of future sea-level rise. The lower curve is the minimum estimate and the upper curve the maximum. Eventual reality is expected to fall between the two extremes. [Adapted from James G. Titus, ed., *Greenhouse effect, Sea Level Rise and Coastal Wetlands*, Fig. 1–4, p. 9, Office of Policy Analysis, U.S. Environmental Protection Agency, Washington, D.C., 1988.]

During most of the glacial periods of the past water impounded on the land in the form of ice was more extensive than it is at present. Consequently the sea level must have been lower than it is now. It has, in fact, been estimated that during the maximum extent of Pleistocene glaciation the sea level was from 105 to 120 m lower. Most geologists accept this estimate, although some feel that it is too conservative. During the height of the Wisconsin glaciation, the last of the major ice advances, the sea level is usually estimated as having been between 70 and 100 m lower than at present. On the other hand, during the interglacial periods, when glaciers were somewhat less widespread than they are today, the sea level must have been higher.

Summary

Oceans cover more than 70 percent of the Earth's surface. **Origin of seawater** was probably by "degassing" of the Earth's interior. This may have been fairly continuous by volcanic eruptions throughout Earth history, or may have been confined to an early thermal stage.
Dissolved material gives oceans their saltiness. The material is moved through the rock cycle from lands to oceans and back to lands.

Currents of the ocean mix seawater once about every 2,000 years.
Tidal currents, surface currents (wind-driven), and **density currents** are the three major types of currents.

One worldwide **oceanic basin** contains the world's oceans and branches northward as the Pacific, Atlantic, and Indian Oceans.

Topography of the oceanic floor is divisible into three major units.

The **continental margins** include the **continental shelves, slopes,** and **rises.** The characteristics of the margins reflect the kind of plate boundaries involved — **converging, diverging,** or **transform.**

The **floors of the oceanic basins** lie on the average 3,800 m below the surface. Aside from the trenches, the **abyssal plains** are the deepest parts of the basins. The older the sea floor, the farther it lies below sea level.

Oceanic ridges segment the ocean basins. New oceanic crust is created at a **system of spreading-center ridges. Aseismic ridges** mark the direction of plate movement as it crosses a plume or hot spot fixed in the mantle.

Seamounts are drowned volcanic peaks. **Guyots** are volcanoes that were truncated by wave action and then subsided below sea level. **Trenches** mark convergence of plates where one is subducted beneath the other.

Sediments in the oceans are most rapidly accumulated on the continental margins. They thin to a feather edge close to the spreading ridges.

Shorelines acquire characteristics generated by the energy of wind-driven waves.

Shoreline features include those formed by **erosion** and those formed by **deposition.** Erosional features include **cliffs, caves, arches, stacks,** and **tidal inlets.** Depositional features include **beaches, spits, bay barriers,** and **barrier islands.**

Sea level is not constant. Isostasy attendant upon glaciation and deglaciation changes sea level. The sea level also rises and falls as glaciers melt or expand. Humanly-produced **greenhouse** gases appear to be responsible for increasing global surface air temperatures and a continuing rise of sea level.

Questions

1. What is degassing of the Earth and what bearing does it have on the origin of seawater?

2. Most of the known elements have been identified in dissolved form in seawater. Which are the volumetrically most important?

3. What are the three types of density currents found in the ocean? Give an example of each.

4. How is water moved in the major currents of the ocean?

5. Referring to the bar graph in Figure 16.2, what is a reasonable explanation for the doubly peaked distribution of the Earth's surface below and above sea level?

6. By now you should be familiar with the major topographic subdivisions of the oceanic floor, so write a brief description of each.

7. Explain why the Hawaiian Islands and their drowned continuation to the northwest are cited as an indication of the direction of movement of the Pacific plate.

8. Here's a question the answer to which none of us will live long enough to prove right or wrong. Sediments eroded from the continents are being added to the ocean's floor at a rate of about 10^{10} t/year. How long will it be before erosion wears the continents down to nearly level surfaces approximating sea level?

Here are some added data: From Figure 16.2, the average elevation of the continents is 840 m, and from Appendix Table C.8, the land area of the world is 149×10^6 km². We can use 2.65 for the average density of sediments.

(We got 32 million years, as follows):

$$\text{Years to flatten continents} = \frac{\text{land above sea level (tonnes)}}{\text{tonnes eroded per year}}$$

$840 \text{ m} \times 149 \times 10^6 \text{ km}^2 \times 10^6 \text{ m}^2 \times 2.65 \text{ t/m}^3 = 332 \times 10^{15} \text{ t}$

$$\frac{332 \times 10^{15} \text{ t}}{10^{10} \text{ t}} = 32 \times 10^6 \text{ y}$$

(Now go back over this question and answer and see if you can think of some reasons why this answer might not bear too much relation to reality.)

9. How do sediments of the deep ocean differ from those of the continental margins? Why?

10. With a diagram, show how the refraction of waves advancing on an irregular shoreline concentrates more energy per unit length of shoreline on the headlands than in the bays.

11. How many meters would sea level rise if the ice on Antarctica were to melt? From Appendix Table C.8, the area of the oceans is 361×10^6 km². The volume of ice on Antarctica is variously estimated; we use 22.5×10^6 km³.

(Our answer was 62 m of sea-level rise. Because an answer in meters was asked for, we first converted km to m. We then divided ice volume by ocean area to get sea-level rise):

$$\frac{22.5 \times 10^6 \text{ km}^3 \times 10^9 \text{ m}^3}{361 \times 10^6 \text{ km}^2 \times 10^6 \text{ m}^2} = 62 \text{ m}$$

(This answer looks impressively precise, but it has its failings. Again, can you think of some factors that should be considered to refine the answer we got?)

Supplementary Readings

BIRD, ERIC C. F. *Coasts,* The M.I.T. Press, Cambridge, Mass., 1968.
A good, brief, introductory volume on the subject.

BIRD, ERIC C. F., and MARTIN L. SCHWARTZ, ed. *The World's Coastline,* Van Nostrand Reinhold Co., New York, 1985.
Truly a worldwide survey in which 129 authors each produce a thumbnail sketch (with references) of a particular segment of the earth's coast.

GROSS, M. GRANT *Oceanography,* 5th ed., Prentice Hall, Englewood Cliffs, N.J., 1990.
A very good general reference that covers oceanic circulation, geology, marine life, and shorelines.

HSU, KENNETH S. *The Mediterranean Was a Desert: A Voyage of the Glomar Challenger,* Princeton University Press, Princeton, N.J., 1983. See Box 16.1.

ROSS, DAVID A. *Introduction to Oceanography,* 4th ed., Prentice Hall, Englewood Cliffs, N.J., 1988.

A sound text written to be understandable to the nonscientist.

SHEPARD, FRANCIS P., and HAROLD R. WANLESS *Our Changing Shorelines,* McGraw-Hill Book Co., New York, 1971.
A wonderfully illustrated and extensive study of the coastlines of the United States. It should convince even the casual browser that the shoreline has been, is, and will continue to be in constant change.

TITUS, JAMES G. (ED.) *Greenhouse Effect, Sea Level Rise and Coastal Wetlands,* Office of Policy Analysis, Environmental Protection Agency, Washington D.C., 1988.
Specifically a study of potential loss of wetlands, this short volume also provides an excellent overview of the greenhouse effect on modern climate.

TUREKIAN, KARL K. *Oceans,* 2nd ed., Prentice Hall, Englewood Cliffs, N.J., 1976.
Brief, authoritative, and a good place to begin further reading.

Wind-formed ripples of sand, Algodones dune field, New Mexico. [Franklyn B. Van Houten.]

WIND AND DESERTS

"Who has seen the wind?
  Neither you nor I:
But when the trees bow down their heads,
The wind is passing by."

Christina Rossetti, from "Who Has Seen the Wind?"

Here we examine some processes operating in the desert, of which the work of the wind is most characteristic. As Christina Rossetti suggests, we don't really see the wind. We infer its presence, present or past, by its results.

Because the effects of the wind are most clearly demonstrated in the desert, our first goal is to examine the *distribution and causes of deserts.* Next we turn to the way wind moves sediment and find that this process varies depending upon whether *sand-sized or dust-sized particles* are involved. As an agent of erosion, wind leaves its marks through two processes: *abrasion* and *deflation.* As an agent of deposition, wind is responsible for blankets of dust known as *loess.* When sand-sized particles are deposited by wind, they accumulate in a variety of topographic forms, all of which fall into the general category of *dunes.*

Although deserts provide wind with its most effective environment for work, other processes are also at work in deserts. Weathering proceeds but slowly, with *mechanical weathering* dominant over *chemical weathering.* And though *water* is scarce in deserts, it is probably, as it is in more humid zones, the single most important agent of erosion, transportation, and deposition.

17.1
DISTRIBUTION OF DESERTS

Although there is no generally accepted definition of a **desert,** we can at least say that a desert is characterized by a lack of moisture, leading to—among other things—a restriction on the number of living things that can exist there. There may be too little initial moisture, or the moisture that occurs may be evaporated by extremely high temperatures or locked up in ice by extreme cold. We are not concerned here with polar deserts; we shall consider only those in the hotter climates. The distribution of middle- and low-latitude deserts is shown in Figure 17.1; they fall into two general groups.

The first are the so-called **topographic deserts,** deficient in rainfall either because they are located toward the center of continents, far from the oceans, or more commonly because they are cut off from rain-bearing winds by high mountains. Takla Makan, north of Tibet and Kashmir in extreme western China, is an example of a desert located deep inside a continental landmass. The desert of large sections of Nevada, Utah, Arizona, and Colorado, on the other hand, is caused by the Sierra Nevada range of California, which cuts off the rain-bearing winds blowing in from the Pacific. A similar, though smaller, desert in western Argentina has been created by the Andes.

Much more extensive than the topographic deserts are the **subtropical deserts** lying in zones that range 5° to 30° north and south of the equator. Their origin is best explained by the general circulation of the Earth's atmosphere (Figure 16.8). Air heated at the equator rises and is diverted north and south. At about 30° on either side of the equator, this air begins to pile up and form permanent atmospheric high-pressure zones. In these zones, known as the **subtropical high-pressure zones,** the air settles.

As the air settles, it warms. The warmer the air becomes, the more moisture it can hold and the less likely it is to release moisture as precipitation. Furthermore, this descending and warming air is generally clear and without clouds. Therefore the Sun is all the more effective in warming the Earth and the air that comes in contact with it.

At the Earth's surface the descending air moves laterally away from the high-pressure zones. Poleward, the winds become the **prevailing westerlies.** Toward the equator, the winds are the **northeast trades** (in the Northern Hemisphere) and the **southeast trades** (in the Southern Hemisphere). These trade winds blow toward the equator, where they become even more heated. They rise and cool, thus giving up moisture. Eventually, in the upper atmosphere, they are diverted north and south toward the subtropical high-pressure zones, to begin the path all over again.

It is continuing circulation of the air that creates the great deserts on either side of the equator. These include the Sahara Desert of North Africa; the Arabian Desert of the Middle East; the Victoria Desert of Australia; the Kalahari Desert of South-West Africa and Botswana; the Sonora Desert of northwestern Mexico, southern Arizona, and California; the Atacama Desert of Peru and Chile; and the deserts of Afghanistan, Pakistan, and northwestern India.

Some of the smaller deserts along the tropical coastlines have been created by the influence of oceanic currents bordering the continents. Winds blow in across the cool water of the ocean and suddenly strike the hot tropical landmass. There the air is heated and its ability to hold water is increased. The resulting lack of precipitation gives rise to deserts. The coast of southern Peru and northern Chile, where the cool Humboldt current flows north toward the equator, is an example.

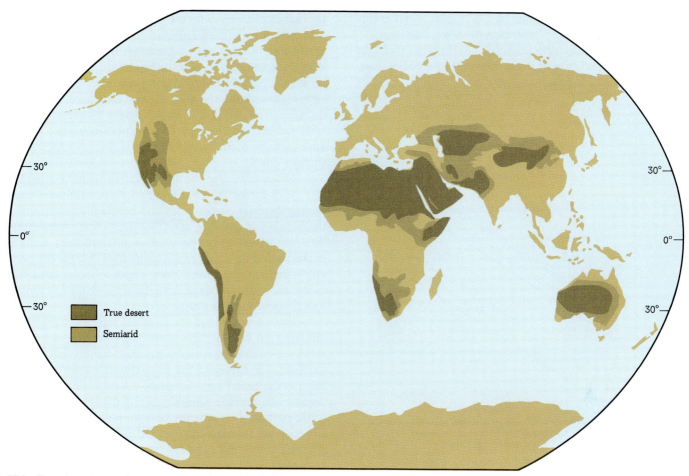

17.1 Deserts and near-deserts cover nearly one-third of the land surface of the earth. Middle- and low-latitude deserts, but not polar deserts, are shown here.

17.2

WORK OF THE WIND

In humid lands a virtually continuous vegetative cover blankets the Earth's surface, but in deserts the lack of moisture ensures that vegetation is absent or sparse, so the wind can directly affect the surface. Therefore it is not surprising that we find good examples of erosion, transportation, and deposition by wind in deserts. It is also true, however, that we see the effects of wind in other environments wherever the wind has access to sand and dust. Thus sand dunes are common near beaches from which sand is blown inland, where it is trapped by vegetation.

MOVEMENT OF SEDIMENT

Wind velocity increases rapidly with height above the ground, just as the velocity of running water increases at levels above its channel. Furthermore, like running water, most air moves in turbulent flow. But wind velocity increases at a greater rate than that of water, and the maximum velocity attained is much higher.

The general movement of wind is forward, across the surface of the land; but within this general movement the air is moving upward, downward, and from side to side. In the zone about 1 m above the ground the average velocity of upward motion in an air eddy is approximately one-fifth the average forward velocity of the wind. This upward movement greatly affects the wind's ability to transport small particles, as we shall see.

Right along the surface of the ground there is a thin but definite zone where the air moves very little or not at all. Field and laboratory studies have shown that the depth of this zone depends on the size of the particles that cover the surface. On the average, the depth of this **zone of no movement** is about one-thirtieth the average diameter of the surface grains (Fig. 17.2). Thus, over a surface of evenly distributed pebbles with an average diameter of 30 mm, the zone of no movement would be about 1 mm deep. This fact, too, has a bearing on the wind's ability to transport sediment.

Dust Storms and Sandstorms Material blown by the wind usually falls into two groups based on size: **sand** and **dust.** The diameter of wind-driven sand grains averages between 0.15 and 0.30 mm, and a few grains as fine as

0.06 mm. All particles smaller than 0.06 mm, whether silt-sized or clay-sized, are classified as dust (see Table 6.1).

In a true **dust storm** (Figure 17.3) the wind picks up fine particles and sweeps them upward hundreds or even thousands of meters into the air, forming a great cloud that may blot out the Sun and darken the sky. In contrast, a **sandstorm** is a low-moving blanket of wind-driven sand with an upper surface 1 m or less above the ground. The greatest concentration of moving sand is usually just a few centimeters above the ground, and individual grains seldom rise even as high as 2 m. Above the blanket of moving sand the air is quite clear, and a person on the ground appears to be partially submerged, as though standing in a shallow pond.

Often, of course, the dust and sand are mixed together in a wind-driven storm (Figure 17.4); but the wind soon sweeps the finer particles off, and eventually the air above the blanket of moving sand becomes clear. Apparently, then, the wind handles particles of different sizes in different ways. A dust-sized grain is swept high into the air, and a sand-sized grain is driven along closer to the ground. The difference arises from the strength of the wind and the terminal velocity of the grain.

Earlier we defined the **terminal velocity** of a grain as the constant rate of fall attained by the grain when the acceleration due to gravity is balanced by the resistance of the fluid — in this case, the air — through which the grain falls (see Section 13.5). Terminal velocity varies only with the size of a particle when shape and density are constant. As the particle size increases, both the pull of gravity and the air resistance increase too. But the pull of gravity increases at a faster rate than the air resistance: A particle with a diameter of 0.01 mm has a terminal velocity in air of about 0.01 m/s; a particle with a 0.2 mm diameter has a terminal velocity of about 1 m/s; and a particle with a diameter of 1 mm has a terminal velocity of about 8 m/s.

To be carried upward by an eddy of turbulent air, a particle must have a terminal velocity that is less than the upward velocity of the eddy. Close to the ground, where the upward currents are particularly strong, dust particles are swept up into the air and are carried in suspension. Sand grains, however, have terminal velocities greater than the velocity of the upward-moving air; they rise for a moment

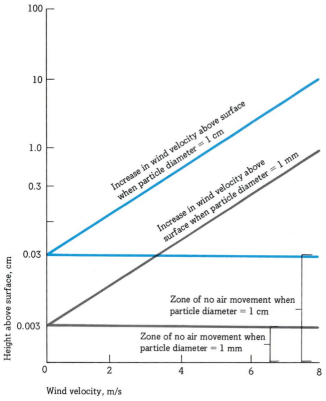

17.2 In a thin zone close to the ground there is little or no air movement, regardless of the wind velocity immediately above. This zone is approximately one-thirtieth the average diameter of surface particles. Two zones are shown in the graph: one for surfaces on which the particles average 1 mm in diameter and one for surfaces with 1-cm particles. Diagonal lines represent the increase in velocity of a wind of given intensity blowing over surfaces covered with particles of 1-mm and 1-cm average diameter. [By permission, after R. A. Bagnold, *The Physics of Blown Sand and Desert Dunes*, p. 54, Methuen & Co., London, 1941.]

17.3 Turbulent air sweeps silt-sized particles high into the air in this dust storm on the Great Plains. [National Oceanographic and Atmospheric Administration.]

17.4 Sand drifts across this railroad in Texas, while finer particles are winnowed out and are carried upward, partially obscuring the sky. [U.S. Department of Agriculture.]

and then fall back to the ground. But how does a sand grain get lifted into the air at all if the eddies of turbulent air are unable to support it?

Movement of sand grains.   Careful observations, both in the laboratory and on open deserts, show sand grains moving forward in a series of jumps, in a process known as **saltation.** We used the same term to describe the motion of particles along a stream bed, but there is a difference: An eddy of water can actually lift individual particles into the main current, whereas wind by itself cannot pick up sand particles from the ground.

Sand particles are thrown into the air only under the impact of other particles. When the wind reaches a critical velocity, grains of sand begin to roll forward along the surface. Suddenly one rolling grain collides with another; the impact may lift either the second particle or the first into the air.

Once in the air, the sand grain is subjected to two forces. First, gravity tends to pull it down to earth again, and eventually it succeeds. But even as the grain falls, the horizontal velocity of the wind drives it forward. The resulting course of the sand grain is parabolic from the point where it was first thrown into the air to the point where it finally hits the ground. The angle of impact varies between 10° and 16° (Figure 17.5).

When the grain strikes the surface, it may either bounce off a large particle and be driven forward once again by the wind, or it may bury itself in the loose sand, perhaps throwing other grains into the air by its impact.

In any event, it is through the general process of saltation that a sand cloud is kept in motion. Countless grains are thrown into the air by impact and are driven along by the wind until they fall back to the ground. Then they either bounce back into the air again or else pop other grains upward by impact. The initial energy that lifts each grain into the air comes from the impact of another grain, and the

wind contributes additional energy to keep it moving. When the wind dies, all the individual particles in the sand cloud settle to earth.

Some sand grains, particularly the large ones, never rise into the air at all, even under the impact of other grains. They simply roll forward along the ground, very much like the rolling and sliding of particles along the bed of a stream of water. It has been estimated that between one-fifth and one-quarter of the material carried along in a sandstorm travels by rolling and the rest by means of saltation.

Notice that after the wind has started the sand grains moving along the surface, initiating saltation, it no longer acts to keep them rolling. The cloud of saltating grains obstructs the wind and shields the ground from its force; thus, as soon as saltation begins, the velocity of near-surface wind drops rapidly. Saltation continues only because the impact of the grains continues. The stronger the winds blow during saltation, the heavier will be the blanket of sand and

17.5   A sand grain is too heavy to be picked up by the wind but may be put into the air by saltation. Here a single grain is rolled forward by the wind until it bounces off a second grain. Once in the air, it is driven forward by the wind and is then pulled to the ground by gravity. It follows a parabolic path, hitting the ground at an angle between 10° and 16°.

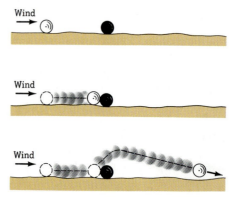

the less the possibility that surface grains will be rolled by the wind.

Movement of dust.   As we have seen, dust particles are small enough and have low enough terminal velocities to be lifted aloft by currents of turbulent air and to be carried along in suspension. But just how does the wind lift these tiny particles in the first place?

Laboratory experiments show that under ordinary conditions, particles smaller than 0.03 mm in diameter cannot be swept up by the wind after they have settled to the ground. In dry country, for example, dust may lie undisturbed on the ground even though a brisk wind is blowing, but if a flock of sheep passes by and kicks loose some of the dust, a dust plume will rise into the air and move along with the wind.

The explanation for this seeming reluctance of dust particles to be disturbed lies in the nature of air movement. The small dust grains lie within the thin zone of negligible air movement at the surface. They are so small that they do not create local eddies and disturbances in the air, and the wind passes them by — or the particles may be shielded by larger particles against the action of the wind.

Some agent other than the wind must set dust particles in motion and lift them into a zone of turbulent air — perhaps the impact of larger particles or sudden downdrafts in the air movement. Irregularities in a plowed field or in a recently exposed stream bed may help the wind begin its work by creating local turbulence at the surface. Also, vertical downdrafts of chilled air during a thunderstorm sometimes strike the ground with velocities of 40 to 80 km/h and churn up great swaths of dust.

Lest this discussion of windborne dust seem remote, we need only recall the Dust Bowl that brought so much suffering to the Great Plains in the 1930s (see Box 17.1).

EROSION

Erosion by the wind is accomplished through two processes: **abrasion** and **deflation.**

Abrasion   Like the particles carried by a stream of running water, saltating grains of sand driven by the wind are highly effective abrasive agents in eroding rock surfaces. As we have seen, wind-driven sand seldom rises more than 1 m above the surface of the Earth, and measurements show that most of the grains are concentrated in the 0.5 m closest to the ground. In this layer their abrasive power is greatest.

Although evidence of abrasion by sand grains is rather meager, there is enough to indicate that this erosive process takes place. For example, we sometimes find fence posts and telephone poles abraded at ground level and bedrock cliffs with a small notch along their base. In desert areas the evidence is more impressive, for here the wind-driven sand has in some places cut troughs or furrows in the softer rocks.

The cross profile of one of these troughs is not unlike that of a glaciated mountain valley in miniature, the troughs ranging from a few centimeters to perhaps 8 m in depth. They run in the usual direction of the wind, and their deepening by sand abrasion has actually been observed during sandstorms.

The most common products of abrasion are certain pebbles, cobbles, and even boulders that have been eroded in a particular way. These pieces of rock are called **ventifacts,** from the Latin for "wind" and "made." They are found not only on deserts but also along modern beaches — in fact, wherever the wind blows sand grains against rock surfaces (Figures 17.6 and 17.7). Their surface is characterized by a relatively high gloss, or sheen, and by facets, pits, gouges, and ridges.

The surface of an individual ventifact may display one facet, or 20 facets or more — sometimes flat but more commonly curved. Where two facets meet, they often form a well-defined ridge, and the intersection of three or more facets gives the ventifact the appearance of a small pyramid. Apparently the surface becomes pitted when it lies across the direction of wind movement at an angle of 55° or more, but becomes grooved when it lies at angles of less than 55°

Deflation   Deflation (from the Latin for "to blow away") is the erosive process of the wind carrying off unconsolidated sediment. The process creates several recognizable features in the landscape. For example, it often scoops out basins in soft, unconsolidated deposits ranging from a few meters to several kilometers in diameter. These basins are known as **blowouts,** for obvious reasons. Even in relatively consolidated sediment the wind will excavate sizable basins if some other agency is at work loosening the material. We find such depressions in the almost featureless High Plains of eastern New Mexico and western Texas, where the bedrock is loosely cemented by calcium carbonate. Several times during the Pleistocene the climate in this area shifted back and forth between moist and dry. During

17.6 The facets on these stones are characteristic of abrasion by wind-driven sand. The maximum dimension of the largest ventifact is 10 cm. [John Simpson.]

BOX 17.1   Dust Bowl

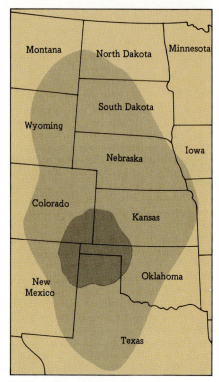

B17.1.1 The greatest damage during the Dust Bowl of the 1930s occurred where Colorado, Kansas, Oklahoma, Texas, and New Mexico come together, shown here in darker color. The effect of the wind was also felt beyond this zone over much of the Great Plains, as suggested by the lighter color on the map. In the 1950s dust storms and wind damage again visited the Great Plains, but although their effect was more widespread than in the 1930s, it was less disastrous.

The Great Plains of North America stretch the length of the continent from the Canadian Arctic to Texas. From Alberta, Saskatchawan, and Manitoba southward to northern Texas is the center of the continent's wheat production. And the American section has the dubious honor of being the site of some of the continent's most extensive wind activity, both ancient and modern. We know the area of most intense recent wind work as the **Dust Bowl,** centered where the six states of Nebraska, Kansas, Oklahoma, Texas, New Mexico, and Colorado cluster (Figure B17.1.1).

The name was earned in the 1930s when strong winds stripped the soil from plowed lands and dumped their dusty burden on parched and struggling young shoots of winter wheat planted in preceding autumns. Towering walls of swirling dust—"black rollers"—turned day to night and sent people scurrying for what shelter they could find. The economic disaster and human suffering of the "dirty thirties" inspired John Steinbeck's Pulitzer Prize–winning novel *The Grapes of Wrath,* a record of the westward trek of displaced Oklahomans (the "Okies") to California along old U.S. Highway 66, now converted to Interstate 40. And Woody Guthrie is said to have begun his song "So Long, It's Been Good to Know You" as, on April 14, 1935, a particularly bad dust storm seemed to herald the world's end.

The causes of the Dust Bowl were several. The grasses that ranchers and then farmers found as they moved west onto the plains had established a thick, tough sod on unconsolidated silts and sands. Much of this sediment had been deposited by windstorms long before the arrival of the white man. First came the ranchers. Then came the farmers with their plows. Wheat cultivation removed the protective cover of the grasses and laid bare the loose soil. As long as winter snow and summer rains were adequate, the soil remained relatively stable. But in the 1930s rainfall was well under the average, which made the plowed land vulnerable to wind damage (Figure B17.1.2.) Gale-velocity winds combined with unprotected soil made the Dust Bowl almost inevitable. Crops were ruined and farmers went bankrupt. This natural disaster was magnified by the financial problems of the 1930s, particularly the plummeting price of grain.

In 1939 a series of wet years set in and the wheatland began to recover. At the same time farming practices designed to reduce the susceptibility of the soil to wind damage were introduced and hastened the recovery. Through the 1940s a few dust storms occurred but damage was minimal.

Then from 1955 to 1957 wind and dust returned to the Dust Bowl states as another series of low-rainfall years took its toll. The area affected was more extensive than before, but the damage was not as great, in large part because of the introduction of new farming techniques and effective soil conservation methods.

Since the 1950s the storms have not been as numerous. They have occurred, however—a reminder that given dry years and unprotected soil, the Dust Bowl can return.

B17.1.2 This photo of an abandoned plow in a deserted field was taken during the Dust Bowl days of the 1930s. [Bettmann Archive.]

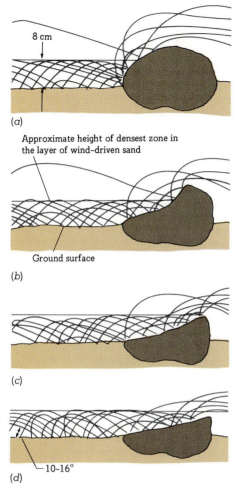

(a)

Approximate height of densest zone in
the layer of wind-driven sand

Ground surface

(b)

(c)

10-16°

(d)

17.7 A facet on a ventifact is cut by the impact of grains of wind-driven sand. [After Robert P. Sharp, "Pleistocene Ventifacts East of the Big Horn Mountains, Wyoming," *J. Geol.*, vol. 57, p. 182, 1949.]

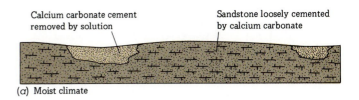

Calcium carbonate cement removed by solution

Sandstone loosely cemented by calcium carbonate

(a) Moist climate

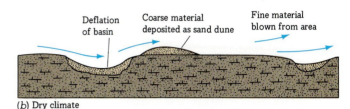

Deflation of basin

Coarse material deposited as sand dune

Fine material blown from area

(b) Dry climate

17.8 The High Plains of eastern New Mexico and western Texas are pockmarked with broad, shallow depressions fashioned in loosely consolidated sandstone. In this instance, wind deflation has created blowouts, but only after the calcite cement of the sandstone was destroyed by downward-percolating waters. Destruction of the cement took place during moist periods in the Pleistocene, and deflation occurred in intervening dry periods. [After Sheldon Judson, "Geology of the San Jon Site, Eastern New Mexico," *Smithsonian Misc. Coll.*, vol. 121, no. 1, p. 13, 1953.]

17.9 Stone-littered surfaces are common in the desert, as illustrated by this example from the Sahara. [Franklyn B. Van Houten.]

the moist periods water dissolved some of the calcium carbonate and left the sandstone particles lying on the surface. Then during the dry periods the wind came along and removed the loosened sediment. Today we find the larger particles piled up on sand hills on the leeward side of the basin excavated by the wind. The smaller dust particles were carried farther along and spread in a blanket across the plains to the east (Figure 17.8).

Deflation removes only the sand and dust particles from a deposit; it leaves behind the larger particles of pebble or cobble size. As time passes, these stones form a surficial cover, a **desert pavement,** that cuts off further deflation (see Figure 17.9).

DEPOSITION

Whenever the wind loses its velocity, and hence its ability to transport the sand and dust particles it has picked up from the surface, it drops them back to the ground. The features formed by wind-deposited materials are of various types, depending on the size of the particles, the presence or ab-

sence of vegetation, the constancy of the wind's direction, and the amount of sediment available for movement by the wind. We still have a great deal to learn about this sort of deposit, but certain generalizations seem valid.

Loess Loess is a buff unstratified deposit composed of small, angular mineral fragments. Deposits of loess range in thickness from a few centimeters to 10 m or more in the central United States to over 100 m in parts of China. A large part of the surficial deposits across some 0.5 million

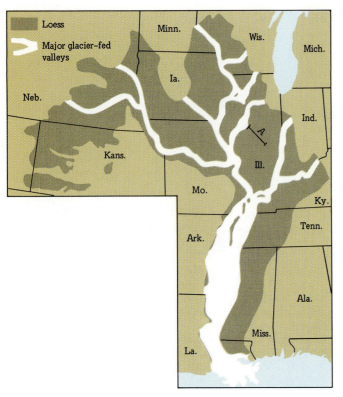

17.10 The great bulk of the loess in the central United States is intimately related to the major glacier-fed valleys of the area and was probably derived from the floodplains of these valleys. In Kansas and parts of Nebraska, however, the loess is probably nonglacial in origin and has presumably been derived from local sources and the more arid regions to the west. The line of section marked *A* in Illinois refers to Figure 17.12.

17.11 In many places unweathered till is overlain by loess on which a soil zone has developed. The lack of a weathering zone on the till beneath loess often indicates rapid deposition of the loess immediately after the disappearance of the glacier ice and before weathering processes could affect the till. Not until loess deposition has slowed or halted is there time available to allow weathering and organic activity capable of producing a soil.

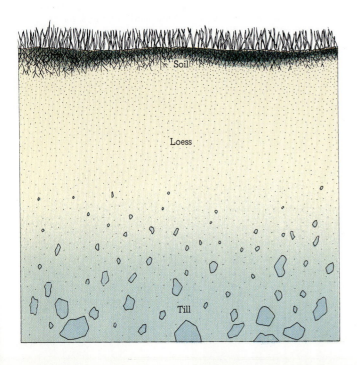

km² of the Mississippi River basin is made up of loess, and this material has produced the modern fertile soils of several midwestern states, particularly Iowa, Illinois, and Missouri (Figure 17.10).

Most geologists believe loess to be material originally deposited by the wind. They base their conclusion on several facts. The individual particles in a loess deposit are very small, strikingly like the particles of dust carried about by the wind today. Moreover, loess deposits stretch over hill slopes, valleys, and plains alike, an indication that the material settled from the air. And the shells of air-breathing snails present in loess strongly impugn the possibility that the deposits were actually laid down by water.

Many exposures in the north-central United States reveal that loess there is intimately associated with till and outwash deposited during the great Ice Age. Because the loess lies directly on top of the glacial deposits in many areas, it seems likely that it was deposited by the wind during periods when glaciation was at its height rather than during interglacial intervals. Also, because there is no visible zone of weathering on the till and outwash, the loess probably was laid down on the newly formed glacial deposits before any soil could develop on them (Figure 17.11).

Certain relationships between the loess deposits in the Midwest and the streams that drain the ancient glacial areas serve to strengthen the conclusion that there is a close connection between glaciation and the deposit of wind-borne materials. Figure 17.10, for example, shows that the major glacial streams cut across the belt of loess. Furthermore, as seen in Figure 17.12, the thickness of the loess decreases toward the east and away from the banks of the streams. Moreover, the mean size of the particles decreases away from the glacial streams. These facts can best be explained as follows: We know that loess is not forming in this area at

17.12 Loess related to the major glacier-fed rivers in the Midwest shows a decrease in thickness away from the rivers and a decrease in the size of individual particles. An example is shown in this diagram, based on data gathered along line *A* in Figure 17.10. [After G. D. Smith, "Illinois Loess — Variations in Its Properties and Distributions," *Univ. Ill. Agric. Expt. Sta. Bull. 490*, 139–184, 1942.]

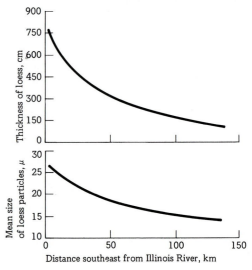

present, so we must look for more favorable conditions in the past. During the great Ice Age of the Pleistocene the rivers of the Midwest carried large amounts of debris-laden meltwater from the glaciers. Consequently the floodplains of these rivers built up at a rapid rate and were broader than they are today. During periods of low water the floodplains were wide expanses of gravel, sand, silt, and clay exposed to strong westerly winds. These winds whipped the dust-sized material from the floodplains, moved it eastward, and laid down the thickest and coarsest of it closest to the rivers.

Not all loess is derived from glacial deposits, however. For example, in one of the earliest studies of loess it was shown that the Gobi Desert provided the source material for the vast stretches of yellow loess that blankets much of northern China and gives the characteristic color to the Yellow River and the Yellow Sea. Also much of the land used for growing cotton in the eastern Sudan of Africa is thought to be made up of particles blown from the Sahara Desert to the west. We have already seen that finely divided mineral fragments are swept up in suspension during sandstorms and are carried by the wind far beyond the confines of the desert. Clearly, then, the large amounts of very fine material in most deserts would make an excellent source of loess.

Deposits of Sand   Unlike deposits of loess, which form featureless blankets of wind-deposited material over wide areas, deposits of sand assume characteristic and recognizable shapes. These range from sand ripples a few centimeters in height and wavelength to forms a kilometer or more in width, 100 km or more in length, and over 100 m in height. Some deserts are veritable "sand seas" whose surfaces are marked by various complex forms. In other places well-defined isolated forms are the rule.

In Chapter 13 we found that as the velocity of a stream falls, so does the energy available for the transportation of material; consequently deposition of sediment takes place. The same relationship between decreasing energy and increasing deposition applies to the wind. But in dealing with wind-deposited sand we need to examine the relationship more closely and to explain why sand is deposited in the form of dunes rather than as a regular, continuous blanket.

The wind shadow.   Any obstacle — large or small — across the path of the wind will divert moving air and create a **wind shadow** to the leeward, as well as a smaller shadow to the windward immediately in front of the obstacle. Within each wind shadow the air moves in eddies, with an average motion less than that of the wind sweeping by outside. The boundary between the two zones of air moving at different velocities is called the **surface of discontinuity** (Figure 17.13).

When particles of sand driven by the wind strike an obstacle, they settle in the wind shadow immediately in front of it. Because the velocity of the wind (hence energy) is low in this wind shadow, deposition takes place, and gradually a small mound of sand builds up. Other particles move

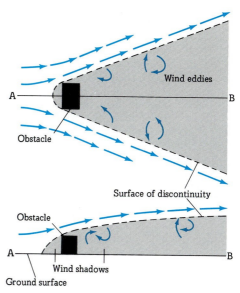

17.13   The shaded area indicates the wind shadow created by an obstacle. The wind is diverted over and around the obstacle. Within the wind shadow wind velocity is low and air movement is marked by eddies. A surface of discontinuity separates the air within the wind shadow from the air outside. [By permission, after R. A. Bagnold, *The Physics of Blown Sand and Desert Dunes*, p. 190, Methuen & Co., London, 1941.]

past the obstacle and cross through the surface of discontinuity into the leeward wind shadow behind the barrier. Here again, the velocities are low, deposition takes place, and a mound of sand (a dune) builds up—a process aided by eddying air that tends to sweep the sand in toward the center of the wind shadow (Figures 17.14 and 17.15). This process is familiar in dry sandy regions, where sand piles up before and in the lee of obstacles, but it is also seen in snowy climates. The principle is illustrated by snow fences (de-

17.14   Because of its momentum, the sand in the more rapidly moving air outside the wind shadow either passes through the surface of discontinuity to settle in the wind shadow behind the obstacle or strikes the obstacle and falls in the wind shadow in front of the obstacle. [By permission, after R. A. Bagnold, *The Physics of Blown Sand and Desert Dunes*, p. 190, Methuen & Co., London, 1941.]

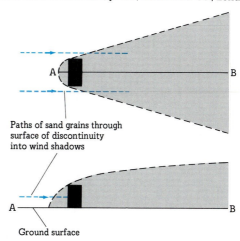

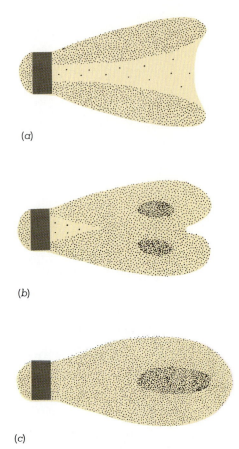

(a)

(b)

(c)

17.15 Sand falling in the wind shadow tends to be gathered by wind eddies within the shadow to form a shadow dune, as shown in this sequence of diagrams. [By permission, after R. A. Bagnold, *The Physics of Blown Sand and Desert Dunes*, p. 190, Methuen & Co., London, 1941.]

vices that have also been used effectively against wind-blown sand).

Wind shadow of a dune. A sand dune itself acts as a barrier to the wind; and by disrupting the flow of air, it may cause the continued deposition of sand. A profile through a dune in the direction toward which the wind blows shows a gentle slope facing the wind and a steep slope to the leeward. A wind shadow exists next to the leeward slope, and it is here that deposition is active. The wind drives the sand up the gentle windward slope to the dune crest and then drops it into the wind shadow. The leeward slope is called the **slip face** of the dune because of the small slides of sand that take place there.

The slip face is necessary for the existence of a true wind shadow. Here is how the slip face is formed: A mound of sand affects the flow of air across it, as shown in the topmost diagram of Figure 17.16. Notice that the wind flows over the mound in streamlined patterns. These lines of flow tend to converge toward the top of the mound and diverge to the leeward. In the zone of diverging air flow velocities are less than in the zone of converging flow. Consequently sand tends to be deposited on the leeward slope just over the top of the mound where the velocity begins to slacken. This slope steepens because of deposition, and eventually the sand slumps under the influence of gravity. The slump usually takes place at an angle of about 34° from the horizontal. A slip face is thus produced, steep enough to create a wind shadow in its lee (Figure 17.17). Within this shadow grains of sand fall like snow through quiet air. Continued deposition and periodic slumping along the slip face account for the slow growth or movement of the dune in the direction toward which the wind blows.

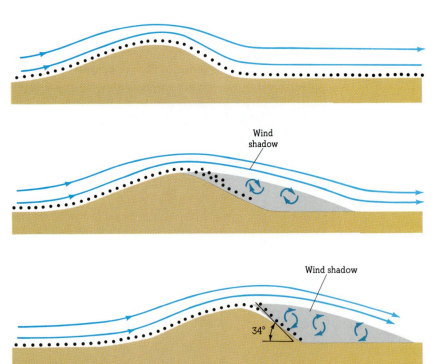

Wind shadow

Wind shadow

34°

17.16 The development of a slip face on a dune. Wind converges on the windward side of the dune and over its crest and diverges to the lee of the dune. The eventual result is the creation of a wind shadow in the lee. In this wind shadow sand falls until a critical angle of slope (about 34°) is reached. Then a small landslide occurs, and the slip face is formed. [By permission, after R. A. Bagnold, *The Physics of Blown Sand and Desert Dunes*, p. 202, Methuen & Co., London, 1941.]

17.17   The steep slope facing the viewer is the slip face on the lee side of a New Mexican dune. The dune moves forward toward the viewer as sand is pushed up its windward side and is deposited on the lee side. Continuing small slips of loose sand keep the slip face steep. [Franklyn B. Van Houten.]

Shoreline dunes.   Not all dunes are found in deserts. Along the shores of the ocean and of large lakes ridges of windblown sand called **foredunes** are built up even in humid climates. They are well developed along the southern and eastern shores of Lake Michigan, along the Atlantic coast from Massachusetts southward, along the southern coast of California, and at various fonts along the coasts of Oregon and Washington (Figure 17.18).

These foredunes are fashioned by the influence of strong onshore winds acting on the sand of the beach. On most coasts the vegetation is dense enough to check the inland movement of the dunes (Figure 17.19), and they are concentrated in a narrow belt that parallels the shoreline. These dunes usually have an irregular surface, sometimes pockmarked by blowouts (see the subsection on deflation under "Erosion" above).

Sometimes, however, in areas where vegetation is scanty, the sand moves inland in a series of ridges at right angles to the wind. These **transverse dunes** exhibit the gentle windward slope and the steep leeward slope characteristic of other dunes. Transverse dunes are also common in arid and semiarid country where sand is abundant and vegetation scarce. In the Sahara Desert some transverse

17.18   These shoreline dunes form complex patterns behind the beach at Coos Bay, Oregon. The beach serves as a source of sand, and this source is continuously renewed by longshore currents of ocean water. Onshore winds (from the left) drive the beach sand inland. The photograph shows about 450 m of shoreline.

17.19 Dunes along the shore of Tusket Island near Yarmouth, Nova Scotia, are partially anchored by grass and bushes. [Information Canada Photothèque.]

forms measure over 100 km in length and 1 to 3 km in width. They may carry on their crests smaller forms of varying shapes (Figure 17.20).

Barchans.   Barchans are sand dunes shaped like a crescent, with their horns pointing downwind. They move slowly with the wind, the smaller ones at a rate of about 15 m/year, the larger ones about 7.5 m/year. The maximum height obtained by barchans is about 30 m, and their maximum spread from horn to horn is about 300 m (Figure 17.21).

Just what leads to the formation of a barchan is still a matter of dispute. However, certain conditions seem essential, including a wind that blows from a fixed direction, a relatively flat surface of hard ground, a limited supply of sand, and a lack of vegetation.

17.20   This transverse dune in the Sahara Desert south of the Atlas Mountains is nearly 1 km wide. Its surface exhibits a complex pattern characteristic of many smaller dunes. [U.S. Army Air Corps.]

17.21   These barchans are moving across the Pampa de Islay, Peru, in the direction in which their horns point. [Aerial Explorations, Inc.]

17.22 Longitudinal dunes in the Simpson Desert, Northern Territory, Australia. They are oriented parallel with the direction of the wind: here we are looking in the downward direction. [William C. Bradley.]

Parabolic dunes.   Long and scoop-shaped, **parabolic dunes** look rather like barchans in reverse: that is, their horns point upwind rather than downwind. They are usually covered with sparse vegetation, which permits limited movement of the sand. Parabolic dunes are quite common in coastal areas and in various places throughout the southwestern states. Ancient parabolic dunes, no longer active, exist in the upper Mississippi valley and in Central Europe.

Longitudinal dunes.   Longitudinal dunes are long ridges of sand running in the general direction of wind movement (Figure 17.22). The smaller types are about 3 m high and about 60 m long. In the Libyan and Arabian Deserts, however, they commonly reach a height of 100 m and may extend for 100 km across the country. There they are known as **seif dunes,** from the Arabic word for "sword." As with large transverse dunes, the large longitudinal forms may be decorated by smaller dunes along their crests.

Stratification in dunes.   The layers of sand within a dune are usually inclined. The layers along the slip face have an angle of about 34°, whereas the layers along the windward slope have a gentler inclination.

Because the steeper beds along the slip face are analogous to the **foreset beds** in a delta (see Figure 13.41), we can use the same term in referring to them in dunes (Section 6.3). These beds develop if there is a continuous deposition of sand on the leeward side of the dune, as in barchans and actively moving transverse dunes.

Backset beds develop on the gentler slope to the windward. These beds constitute a large part of the total volume of a dune, especially if there is enough vegetation to trap most of the sand before it can cross over to the slip face. **Topset beds** are nearly horizontal beds laid down on top of the inclined foreset or backset beds.

17.3
OTHER PROCESSES OF THE DESERT

WEATHERING AND SOILS

Because of the lack of moisture in the desert, the rate of weathering, both chemical and mechanical, is extremely slow. As most of the weathered material consists of chemically unaltered rock and mineral fragments, we can conclude that mechanical weathering predominates.

Some **mechanical weathering** is simply the result of gravity, such as the shattering of rock material when it falls from a cliff. Wind-driven sand can bring about some degree of mechanical weathering. Sudden flooding of a desert by a cloudburst also moves material to lower elevations, reducing the size of the rock fragments and scouring the bedrock surface in the process. And in almost every desert in the world temperatures fall low enough to permit frost action. But here again, the deficiency of moisture slows the process. Finally, the wide temperature variations characteristic of deserts cause rock materials to expand and contract and may produce some mechanical weathering.

This low rate of weathering is reflected in the soils of the desert. Seldom do we find extensive areas of residual soil, for the lack of protective vegetation permits the winds and occasional floods to strip away the soil-producing minerals before they can develop into true soils. Even so, soils sometimes develop in local areas, but they lack the humus of the soils in moister climates, and they contain concentrations of such soluble substances as calcite, gypsum, and even halite because there is insufficient water to carry these minerals away in solution. In the deserts of Australia rock-like concentrations of calcite, iron oxide, and even silica sometimes form a crust on the surface.

One minor weathering feature of the desert is of special interest because it seems to have its counterpart on the surface of Mars. A thin, shiny, reddish-brown to blackish coating called **desert varnish** occurs on some stones. The coating is a thin film—a fraction of a millimeter thick—made up largely of iron and manganese oxides. Interestingly enough, there has long been disagreement as to how this varnish forms on Earth. Some suggest that it is the result of entirely inorganic processes. Other workers believe that some organic activity is also involved.

WORK OF WATER

Although rainfall is extremely sparse in deserts, there is still enough water to act as an important agent of erosion, transportation, and deposition. In fact, water is probably more effective than even the driving winds in molding desert landscapes.

Very few streams flowing through desert regions ever find their way to the sea, and the few that do, such as the Colorado River in the United States and the Nile of Egypt, originate in areas from which they receive sufficient water to sustain them through their long courses across the desert. Exploitation of such rivers for agricultural, domestic, and industrial uses can drastically affect their flow. An example is the Colorado River. Diversion of water to human use in California, Arizona, Nevada, and Colorado reduces that river to but a trickle by the time it reaches the Gulf of California. Also the flow of the Nile has been modified by the construction of Lake Nasser just above the first cataract and by the diversion of water to the Fayum depression west of Cairo.

Stream beds in most deserts, however, are dry over long periods of time and carry water only during floods. Even then, the flow is short-lived, for the water either evaporates rapidly or vanishes into the highly permeable rubble and debris of the desert. In some places, however, such as the western United States, broad, dry plains slope down from the mountain ranges toward central basins, called **playas,** where surface runoff collects from time to time. But the **playa lakes** formed in these basins usually dry up in a short time or at best exist as shallow, salty lakes, of which Great Salt Lake is the best-known example.

Floods in the desert are unlike floods in humid areas. The typical flood in the desert, like the rain that produces it, is local in extent and of short duration. In moist regions most floods arise from a general rain falling over a relatively long period of time; consequently these floods affect large areas. Because of the widespread vegetative cover, the floods tend to rise and fall slowly. But on the bare ground of the desert the runoff moves swiftly and floods rise and fall with great rapidity. These **flash floods** give little warning. The experienced traveler has a healthy respect for them and will never pitch camp on a dry stream floor, even though the stream banks offer protection from the wind. At any moment a surging wall of debris-laden water may sweep down the stream bed, destroying everything in its path.

The amount of moisture in deserts does change. Thus during the Ice Age, when glaciers lay across Canada and the northern United States and draped the flanks of most of the higher mountains of the West, the climate of the arid and semiarid areas of the western United States was quite different from what it is today. Glaciations produced **pluvial periods** (from the Latin *pluvia,* "rain"), when the climate was undoubtedly not so moist as that of the eastern United States today, but was certainly more hospitable than what we now find in the arid regions. During any single pluvial period rainfall was greater, temperature lower, rate of evaporation less, and vegetation more extensive. At such a time large sections of the southwestern states were dotted with lakes known as **pluvial lakes** (Figure 17.23).

17.23 During glaciation many basins of the western United States held lakes where only intermittent water bodies exist today; modern saltwater lakes are remnants of basins formerly filled with freshwater lakes. Modern lakes (b) are compared with those during the glaciation of higher mountains and the western United States (a). After O. E. Meinzer, "Map of the Pleistocene Lakes of the Basin-and-Range Province and Its Significance," *Bull. Geol. Soc. Am.,* vol. 33, pp. 543, 545, 1922.]

(a)

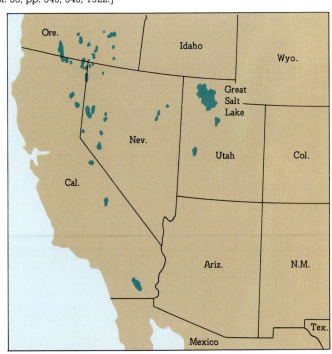

(b)

What was responsible for this pluvial climate? The presence of continental glaciers in the north is thought to have modified the general atmospheric circulation of the globe so that the belt of rain-bearing winds moved to the south and the temperatures fell. Consequently the rates of evaporation decreased, and at the same time the amount of precipitation increased. When the ice receded, the climate again became very much what it is today.

Summary

Deserts cover one-third of the Earth's surface. A **topographic** desert owes its aridity to the distance from a source of moisture or to protection from rain-bearing winds by a mountain mass. A **subtropical** desert lies between 5° and 30° north or south of the equator in one of two zones of subtropical high pressure developed by planetary atmospheric circulation.

Work of the wind is more effective in the desert than anywhere else on the lands. Even here, however, it is subordinate to the work of running water in shaping the landscape.
  Movement of material by wind depends in part on the size of the particle: Dust-sized particles move differently than do sand-sized particles. Dust particles are carried high into the atmosphere in **suspension.** Sand moves along in continuous contact with the ground or moves a few centimeters above the surface in **saltation.**
  Erosion by wind consists of **abrasion,** of which **ventifacts** are the most common example, and **deflation,** of which **blowouts** are examples.

Deposition consists of dust (**loess**) and sand (usually **dunes**). Deposition in sand dunes begins in the **wind shadow** in the lee of an obstacle. Once established, the dune provides its own wind shadow. In the dune sand is moved up the windward slope and is deposited on the steep **slip face** in the dune's wind shadow. Dune types include **foredunes, transverse dunes, barchans, parabolic dunes,** and **longitudinal dunes.**

Other desert processes are also important.
  Weathering, both mechanical and chemical, is slow because of lack of water.
  Soils are rubbly and sometimes marked by crustlike accumulations of calcite, iron oxide, and silica.
  Water, despite its scarcity, is abundant enough to create streamways in even the driest desert. Flow of streams is short-lived but often catastrophic. During the Pleistocene **pluvial** periods occurred in desert and semiarid regions when glaciers advanced elsewhere.

Questions

1. What are some of the different types of deserts and their causes?
2. Distinguish between dust storms and sandstorms.
3. How does wind move sand-sized grains?
4. What is a ventifact and how is it formed?
5. Silt- and clay-sized particles dominate deposits of loess. What are two different origins for such deposits?
6. What is a wind shadow and what function does it play in the deposition of sand?
7. On a dune, what is a slip face and what is its function in the dune-building process?
8. There are several different forms of dunes. Write a brief, informative paragraph about each of the following: foredune, barchan, parabolic dune, longitudinal dune, and transverse dune.
9. What is a pluvial climate? What significance would it have in western United States were one to occur there?

Supplementary Readings

BAGNOLD, R. A. *The Physics of Blown Sand and Desert Dunes,* Methuen & Co., London, 1941.
  Although this book appeared nearly five decades ago, it is yet to be replaced by a better one on the subject. The study of the movement of sand and dust by wind and of sand dunes begins here.
BLOOM, ARTHUR L. *Geomorphology,* Prentice Hall, Englewood Cliffs, N.J., 1978.

Chapter 13 is a discussion of arid, semiarid, and savannah landscapes. Chapter 14 deals with the work of the wind and the forms that result from it.

HUNT, R. DOUGLAS *The Dust Bowl,* Nelson-Hall, Chicago, 1981.
A history of the American Dust Bowl.

MABBUTT, J. A. *Desert Landforms,* The M.I.T. Press, Cambridge, Mass., 1977.
A short but comprehensive volume of intermediate level that covers the various desert processes and landforms, including, of course, wind action and its results.

PETROV, M. P. *Deserts of the World,* John Wiley & Sons, New York, 1976.
Translated from the Russian, this volume surveys the deserts of the world and considers as well the physical and biological processes of the desert.

The United States imports over 40 percent of its petroleum today, and the percentage will rise as we become more and more dependent upon foreign oil fields. At this offshore platform petroleum is being pumped from the Saladin field off western Australia. [Courtesy of WAPET/Western Image Pty. Ltd., Perth Western Australia.]

CHAPTER ▪ 18

ENERGY

"Energy is the sine qua non of a modern society's ability to do the things it wants to do."

Report to the President of the United States by Dixie Lee Ray, December 1973

By about 1750 the extensive coal deposits in England began to be fully exploited, displacing water, wind, and wood, the traditional sources of energy. The plentiful British coal fueled the beginnings of the Industrial Revolution, for industrial society is dependent on the availability of very large supplies of energy. Industrialism spread rapidly throughout the Western world, then to some Eastern nations, and today is moving onto the less developed countries.

This momentous change in social organization has been driven by energy, at first in the form of coal and later as petroleum and natural gas (see Chapter Opening figure). Now, as we approach the end of the twentieth century, industrial society seeks ways to maintain reliable and economically feasible supplies of energy. We make no attempt here to offer solutions to this question, but as a basis for dealing with solutions, we look at the various forms of energy that are available on Earth. Our energy, both actual and potential, has three major sources: the Sun, the nuclei of some elements, and gravity.

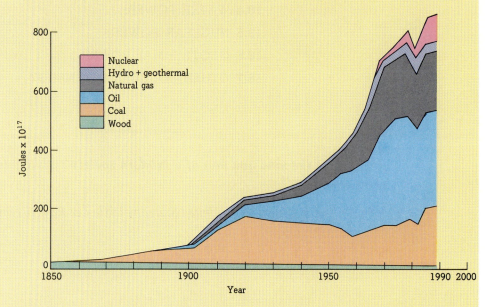

18.1 Energy consumption in the United States has expanded dramatically from 1850 to 1985. During this time the major source of energy has shifted from wood to coal to oil and gas. The slight decline in energy consumption in the early 1980s was due to fuel-conservation measures. Since then, the demand for energy has increased again. [Data from various sources.]

The *fossil fuels* (coal, petroleum, and natural gas) represent the remains of plants or animals that gathered their energy indirectly from the Sun millions of years ago. Energy from the Sun drives the hydrologic cycle and, combined with gravity, provides us with *water power*. The Sun heats the atmosphere differentially, which in turn causes its movement and produces *wind power*. The nuclei of some elements give off energy by splitting, or *fission*, as does, for example, the radioactive form of uranium. The combination, or *fusion*, of the nuclei of some forms of hydrogen also produces energy. Both fission and fusion can be used to create *nuclear power*. Much energy from the Earth—*geothermal energy*—is derived from the heat created by the long-continued decay of radioactive elements in the Earth's crust. Gravity, as represented by the rising and falling of the ocean surface, produces *tidal power*.

By far the most important sources of energy, however, are still the fossil fuels. In the United States they account for about 90 percent of the energy consumed. The other 10 percent is predominantly hydroelectric and nuclear power. Past and present major sources of energy used in the United States are shown in Figure 18.1.

18.1

NATURAL RESOURCES

Before proceeding to a consideration of energy sources, we should pause briefly to make a few general observations about natural resources. The world's natural resources are concentrated in relatively small areas and are unevenly distributed throughout the crust; deposits that have been located may soon be exhausted unless the natural reserves are huge. These natural resources have been concentrated by the geologic processes we have been studying, such as weathering, running water, igneous activity, and metamorphism; but only a few geologic environments have favored their formation. And after they are used, they are gone forever. Unlike our practice with plant resources, we cannot grow another crop of natural resources.

To put this another way, most of the natural resources that we exploit for energy and minerals are **nonrenewable.** Once they have been used up, they cannot be replenished — at least in our lifetime. Coal and oil, when burned, vanish into water and carbon dioxide.

Estimating the amount of a nonrenewable resource available to us is at best a difficult and often imprecise exercise. Most of our Earth resources lie hidden beneath the surface. Direct evidence of their abundance comes from surface occurrences, existing mines, and exploratory drilling. Most estimates, however, are heavily dependent on indirect geological, geophysical, and geochemical data. Today's estimates may be modified by new discoveries. New technology may make possible extraction of resources that are today impractical or impossible to bring to the surface. Or new uses for an Earth material may raise a once-useless deposit to the rank of a valuable resource. Despite these considerations, many of our natural resources are finite. Changing estimates can only change the useful life of the resource.

18.2

COAL

Coal, as we found in Section 6.2, is a biochemically formed sedimentary rock often considered metamorphic because it goes through various stages. It is formed from plants that grew in ancient swamps. Conditions favored preservation, which means that the remains of the plants accumulated in a nonoxidizing environment and were eventually buried by other sediments, usually sand or mud, which are now the sandstone and shale typically associated with coal beds (Figure 18.2).

The coal deposits started off as accumulations of organic materials made up chiefly of carbon, oxygen, and hydrogen. With rising temperature and pressure, due to the burial of the deposits, the hydrogen and oxygen were gradually driven off. The higher the percentage of carbon in the

18.2   In northeastern New Mexico this coal of Cretaceous age crops out along the highway.

remaining deposit, the higher the rank of the coal. We find, therefore, that the lowest grade of coal, **lignite,** has about 70 percent carbon; that **bituminous,** or **soft,** coal has about 80 percent carbon; and that the highest rank of coal, **anthracite,** contains from 90 to 95 percent carbon (Figure 18.3).

It takes about 2 m of peat, the organic deposits of swamps, to produce 1 m of soft coal or 0.5 m of anthracite. Some coal beds, or **seams,** reach 100 m in thickness. The

18.3   Increasing calorific value of coal with increasing rank. [After Brian J. Skinner, *Earth Resources*, p. 34, Prentice Hall, Englewood Cliffs, N.J., 1986.]

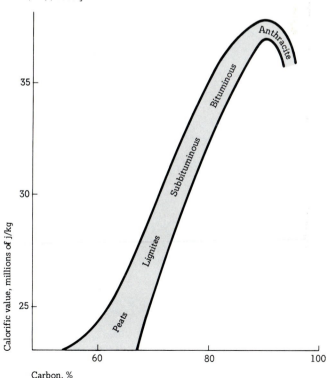

thickest coal bed known is a bituminous deposit in Australia, which is nearly 200 m thick. But the average coal seam is less than a meter in thickness, and seams greater than 3 m are rare.

The geology of coal is well understood, and the search for coal is governed by the relatively simple characteristics of flat-lying to moderately folded sedimentary rocks. Coal occurs widely but unevenly around the world, with the United States being particularly well endowed (Figures 18.4 and 18.5). Most of the major coal-bearing basins have been identified, and we have a pretty good idea of both the known reserves and potential resources yet to be proved. Estimates of the total recoverable coal in the world vary, but something like 7.1×10^{12} t seems to be a reasonable figure. For the United States, total recoverable coal is about 1.5×10^{12} t. These are impressive figures and represent a resource that can last for a few hundred years, a much longer time than we can count on for oil and natural gas.

We usually think of coal as a solid fuel, but for well over a century methane gas (CH_4) has been produced from coal. Today natural gas has generally replaced coal gas because it is more economical to produce. But gasification of coal remains an alternative. Indeed, the increasing costs of natural gas are making the gasification of coal seem more and more attractive.

Not only can gas be produced from coal but gasification can serve as an intermediate step to the production of oil. Direct liquefaction of coal is also possible, and commercial installations producing oil directly from coal may not be far off.

Despite coal's versatility, its use as a fuel has created problems that were virtually unrecognized three decades ago. In addition to its principal constituents, carbon, oxygen, and hydrogen, coal contains many other elements in small amounts. One, sulfur, has made some coal a dangerous pollutant of air and water. It enters the air as sulfur dioxide gas, SO_2, which can be harmful to life if it reaches certain concentrations. Furthermore, SO_2, when combined with the moisture of the atmosphere, forms sulfuric acid, H_2SO_4. This in turn, when washed out of the atmosphere, produces a rain of higher than normal acidity. **Acid rain** has become a problem for certain areas downwind from regions burning large amounts of **both coal and oil.** It has made some lakes so acid that fish life is adversely affected. Furthermore, it has had deleterious effects on some forests. Affected zones include eastern Canada, northern New York State, New England, higher elevations in parts of the Rocky Mountains, Scandinavia, and Germany.

In addition, the drainage from some coal mines can acidify streams. **Acid mine drainage** is more locally concentrated than acid rain, but where it occurs, the H_2SO_4 carried in the streams makes them unsuitable for many fish species.

18.4 Known and assumed occurrences of coal in the world. [After Günter B. Fettweis, *World Coal Resources*, Fig. J, p. 415, Elsevier Scientific Publishing Co., New York, 1979.]

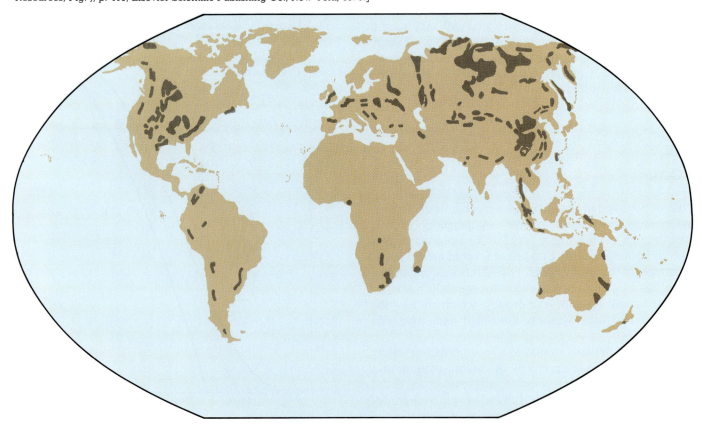

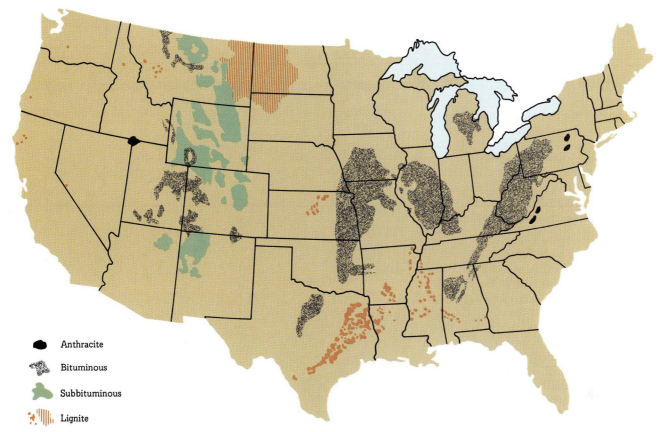

18.5 The coal fields of the United States are large. Extensions of deposits in the western United States are found beneath the plains of Canada, and coals typical of the eastern United States are mined in Nova Scotia. Deposits of lignitic through bituminous rank occur widely in Alaska. ["Strippable Reserves of Bituminous Coal and Lignite in the United States," *U.S. Bur. Mines Info. Cir. 8531, 1971.*]

Legend:
- Anthracite
- Bituminous
- Subbituminous
- Lignite

18.3

OIL AND NATURAL GAS

Although petroleum from natural seeps had been used for thousands of years, the petroleum industry as we know it had its birth in the decade of the 1860s, following the successful completion of the first drilled oil well in 1859 near Titusville, Pennsylvania (see Box 18.1).

Oil and natural gas have proved to be economical, efficient, and relatively clean fuels. As a result, by 1945 they had overtaken coal as the primary sources of energy in the United States, and today nearly three times as much energy is produced in this country by burning oil and natural gas as is produced from coal (see Figure 18.1).

Like coal, the burning of oil and gas has environmental consequences. Along with coal it contributes greenhouse gases to the atmosphere, thus increasing the potential for global warmings (See Box 16.2). Another type of environmental hazard connected to oil is described in Box 18.2.

ORIGIN

With a few exceptions, **petroleum and natural gas are associated with sedimentary rocks of marine origin.** Petroleum and natural gas are mixtures of hydrocarbon compounds (composed largely of hydrogen and carbon) with minor amounts of sulfur, nitrogen, and oxygen. They were derived originally from organic remains—both plant and animal. The hydrocarbons found in our oil and gas fields, however, differ somewhat from those we know in living things. So some changes must take place between the stage of organic remains and the end product.

The first step is the accumulation of an oceanic sediment rich in remains of plant and animal life. This is not so simple as it sounds, for most sea environments provide oxygen to destroy the organic remains before they can be buried by subsequent sediments. However, we have some examples of basins in which the circulation of water is so slow that the bottom waters become depleted of oxygen, and organic-rich sediments can accumulate. But even this oxygen-poor environment may not suffice to protect the

BOX 18.1   Drake's Folly

"Drake's Folly" marked the real beginnings of the petroleum industry. The exact date was Sunday, August 28, 1859, and the place was northwestern Pennsylvania near Titusville, along Oil Creek, a tributary of the Allegheny River. The event was the completion of the first drilled oil well. "Colonel" Edwin L. Drake (his military title was gratuitous) was the chief protagonist.

Drake's well was by no means fortuitous. The site had been chosen on the basis of persuasive evidence. Along Oil Creek natural seeps of crude oil had long been known and collected by immersing blankets in oil-bearing springs along the creek and then wringing out the oil. The product was variously known as "rock oil" (Figure B18.1.1), "Seneca Oil" (after the Iroquois Indian tribe that had exploited the source long before the white man), and "snake oil." Its chief use was medicinal—it was applied both internally and externally for a wide variety of ills.

A number of investors from New York City and New Haven, Connecticut,

18B.1.2   A reconstruction of the pump house and derrick housing for Colonel Edwin L. Drake's first oil well at Titusville, Pennsylvania, in 1859. [Jane Grigger.]

18B.1.1   Thomas Gale lived on Oil Creek near the spot where, in the summer of 1859, "Colonel" Drake drilled the first oil well in the United States. He lost no time in writing a short treatise on petroleum—or, as many then called it, rock oil.

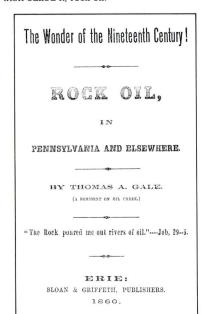

The Wonder of the Nineteenth Century!

ROCK OIL,

IN

PENNSYLVANIA AND ELSEWHERE.

BY THOMAS A. GALE.
(A RESIDENT ON OIL CREEK.)

"The Rock poured me out rivers of oil."---Job, 29--5.

ERIE:
SLOAN & GRIFFETH, PUBLISHERS.
1860.

sponsored Drake's effort. They were convinced, in no small measure by a report written by Yale Professor Benjamin Silliman, Jr., that oil had great economic potential if it could be obtained in large enough quantities.

One of the promoters of the new enterprise, George H. Bissell, had once examined a bottle of "Kier's Petroleum, or Rock Oil" in a New York drugstore and been struck by the fact that the oil had come from a salt well and was associated with the brine that was pumped. The well had been a drilled well, not a shallow, hand-dug one. If you could drill for water, he thought, why not for oil? It is very possible that Bissell's analogy was decisive in the early history of the oil industry.

Drake was dispatched from New Haven in December 1857 to oversee the venture. On his way west he visited salt wells at Syracuse in central New York State, and when he got to Titusville, he hired an old-time salt-well driller to take charge of the operation. Twenty months later Drake brought in the discovery well (Figure B18.1.2), much to the amazement of those who had labeled the enterprise a great folly.

Drake's strike started an oil boom in northwestern Pennsylvania as frantic as the Gold Rush in California a decade earlier. Drilling went on virtually everywhere in the area. In the two score years following Drake's discovery well the use of rock oil (petroleum) for illumination and lubrication grew rapidly. Simultaneously transportation facilities graduated from barrels in horse-drawn wagons to pipelines, tank cars, and tankers. The oil industry was well on its way to becoming the twentieth-century giant it still is.

BOX 18.2   Oil Spills and Blowouts at Sea

Our continuing demand for power derived from petroleum products, to say nothing of the mountains of plastic we use that are produced from petroleum, initiates a complicated network of environmental hazards. One of these involves oil spills at sea.

On March 18, 1967, the 295 meter tanker Torrey Canyon awakened us to the problem of oil pollution at sea. The tanker, carrying 117,000 tons of crude oil from Kuwait to English refineries went aground off Land's End at the southwestern tip of England. Over a two week period the stranded hulk gave up its cargo of petroleum, sending a black, oily contamination to the French and English shores along the English Channel and either directly or indirectly caused the death of untold numbers of sea birds and marine organisms.

More than a decade later, in March 1978, the Amoco Cadiz, broke up off the coast of Brittany and lost an estimated 200,000 tons of petroleum which was spread by high tides and gale winds along 300 km of the French coast. Again, the loss of wild life was extensive as was the damage to commercial fisheries.

The largest tanker spill in American waters also happened in the month of March, this time in 1989 in Prince William Sound, off southern Alaska. On the 24th of the month the tanker, Exxon Valdez, leaving Valdez, the southern terminal for the Alaskan pipeline, hit submerged rocks. An amount variously estimated as between 31,000 and 35,000 tons of the 165,000 tons of crude oil on board was spilled. The effect on marine life, particularly along the western portion of Prince William Sound, was catastrophic.

It is not just petroleum in transport that can cause problems. We found this out in 1969 when, on January 28, an oil well 'blew' while being drilled off the coast of Santa Barbara, California. We are not sure how much oil escaped before the blowout could be contained. One estimate suggests that over 1,000 tons of oil covered an ocean area of about 800 km². Ten years later, in 1979, an even larger amount of oil was lost from an off-shore Mexican pumping platform in Campeche Bay off southeastern Mexico.

All oil spills are a threat to marine and bird life. But those the size of the ones described are disastrous. The best protection is obviously prevention because once a major spill occurs it is very difficult to control, and its harmful effects linger for months and perhaps years.

The first order of business, once a spill has occurred, is to contain mechanically as much of the oil as possible. This is usually done with floating barriers which concentrate the oil so that it can be pumped into barges or other tankers. Oil that escapes is handled in different ways. For instance, the French in the case of both the Torrey Canyon and the Amoco Cadiz spread a dust of chalk and sodium stearate on the spill. Oil adhered to the particles and sank with it to the bottom, although the effect of this on bottom life is not fully known. The English tried spreading detergents to emulsify and break up the slick and thus hasten evaporation. The detergents, however, proved as fatal to marine life along the beaches as did the oil. In the Exxon Valdez spill, oil was removed from the shore by cleaning with the help of steam hoses. In that same spill attempts to burn the oil failed, as did similar attempts by the French to torch the oil spilled from the Amoco Cadiz.

Crude oil spilled on the sea surface moves as a coherent mass 5 to 60 cm thick and parallel with the direction, and at a speed of about 3.4 percent, of the wind velocity. Therefore knowing the direction and velocity of the wind we should be able to predict the direction of movement of an oil spill and its estimated time of arrival at a particular location.

Natural processes help eliminate the spill. Evaporation is effective and about 25 percent of an oil spill will evaporate in a few days. Rough seas accelerate the process by breaking up the oil mass. After three months 85 percent will have volatilized. This process is slowed in cold waters, such as those involved with the Exxon Valdez. Some bacteria degrade crude oil and photochemical reactions play a role in eliminating the oil. In addition oil can adhere to sedimentary particles and sink toward the bottom.

---

organic sediments because some bacteria can utilize the sulfates of seawater to oxidize organic matter. Therefore we need to have a marine basin that accumulates organic sediments in an oxygen- and sulfate-deficient environment.

The sediments destined to provide a source of petroleum start off as muds and silts, which eventually end up as shales, siltstones, and fine-grained limestones. Their organic content varies between 0.5 and 5 percent, and averages about 1.5 percent. A small amount of this may be in the form of hydrocarbons from the beginning, but the bulk of hydrocarbons—over 90 percent—form after the initial deposition of sediments.

Formation of hydrocarbons takes place in two stages. First, biological, chemical, and physical processes begin to break down the organic matter into what is called **kerogen,** a precursor of oil and gas. The second stage is marked by the thermal alteration of kerogen to hydrocarbons as the deposit is buried deeper and deeper by younger, overlying sediments. The production of hydrocarbons begins at a temperature of about 50° to 60°C and a depth of 2 to 2.5 km. Hydrocarbon formation continues to depths of 6 to 7 km and temperatures of 200° to 250°C. Formation of oil dominates in the lower range of temperature and burial, and gas in the higher range, as suggested for the oil and gas

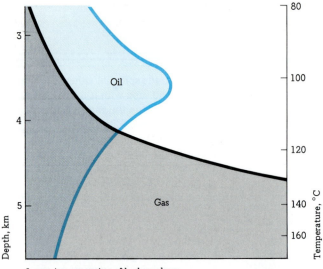

3

Oil

4

Gas

5

Depth, km

Temperature, °C

80

100

120

140

160

Increasing generation of hydrocarbons

18.6 Data from an oil and gas field in the Aquitaine basin in southwestern France show that the maximum generation of oil occurs at a shallower depth and a lower temperature than does that of gas. In this example oil generation peaks at a depth of 3.5 km and a temperature of 100°C. This is followed by the intense generation of gas at greater depths and temperatures. Note that the rate of temperature rise increases with depth. [Generalized from K. Le Tran, "Geochemical Study of Hydrogen Sulfide Sorbed on Sediments," in *Advances in Organic Geochemistry*, Fig. 2, p. 722, ed. H. R. Gaertner and H. Werner, Pergamon Press, Elmsford, N.Y., 1971.]

field diagrammed in Figure 18.6. Whether the hydrocarbons, once formed, will gather into an oil and/or gas field depends on a number of additional factors, discussed in the following paragraphs.

ACCUMULATION

We have found that a sedimentary unit rich in organic material and buried to depths between 2 and 7 km is needed for the creation of oil and gas. This rock becomes the **source rock** for petroleum, defined as a sedimentary rock that has naturally generated enough hydrocarbons to form a commercial accumulation of oil and gas. Several processes move the hydrocarbons in the source bed to a more permeable rock. First, the weight of the overlying rocks tends to squeeze the oil and gas through pores and cracks in the rock. Second, the formation of gas during the breakdown of the hydrocarbons generates a pressure that helps to drive the oil from the source rock. Third, there is usually water in the source bed as well as in adjacent rocks, and if the pores are large enough, the low-density oil and gas tend to move upward through the water-saturated layers. Eventually the oil and gas reach the surface and are lost there, unless something intervenes to collect and hold them in the underground.

An impermeable rock unit will block or divert oil and gas in their migration. If the impermeable unit is properly

positioned, it will halt the hydrocarbons on their way toward the surface and keep them underground. Associated with this impermeable **capping bed** is a permeable rock unit into which the hydrocarbons have moved. We call this the **reservoir rock** because it contains the accumulated oil and gas. The capping bed and related reservoir rock together form a **trap** and serve as a natural tank. The traps can be of several types, as illustrated in Figure 18.7.

Once the seal provided by the capping bed is broken by the drill, the oil and natural gas move from the pores of the reservoir rock to the drill hole and then can be brought to the surface for processing and distribution. Of course, the trap may be broken by natural means as well. For instance, Earth movements after the formation of the trap may so disturb it that fractures develop, and the hydrocarbons will escape either to another trap or to the surface. Or surface erosion may eventually reach and destroy the trap. Therefore one would expect that the older the rock, the less chance there is that it still contains oil, and in a general way this is true. The greatest amount of production comes from rocks of Cenozoic age, followed by fields of oil and gas in Mesozoic and then Paleozoic rocks. There is essentially no production from Precambrian rocks.

THE SEARCH FOR PETROLEUM

The search for petroleum is carried on largely by indirect methods. Most of the geologic structures that contained oil and gave surficial indication of their presence have long since been drilled. So now we must resort to means other than surface observation to discover potential traps of oil and gas buried beneath kilometers of sedimentary rocks and contained in the beds submerged beneath the oceans along the continental margins. One exploration technique is shown in Figures 18.8 and 18.9.

Figure 18.10 shows the areas in the world favorable to oil and natural gas exploration, both onshore and offshore. It also shows the producing areas. The conterminous United States has been very effectively explored, as have the Canadian provinces of Alberta and Saskatchewan. Probably half the oil and gas in these areas has been found and used, and almost certainly all the giant fields have been discovered. The new petroleum frontiers for the North American continent lie in the arctic lands of Canada and Alaska and along the offshore margins of the continent (Figure 18.11).

How long will the supply of oil and natural gas in the United States last? That is a question we cannot answer with precision. We know that there is a finite amount of these liquid and gaseous hydrocarbons in the ground, and we also know that they are not renewable resources like trees or rivers. Several estimates of the remaining U.S. supply duration have reached roughly the same conclusion: namely, that early in the twenty-first century we shall run out of oil and natural gas. Two estimates for crude-oil pro-

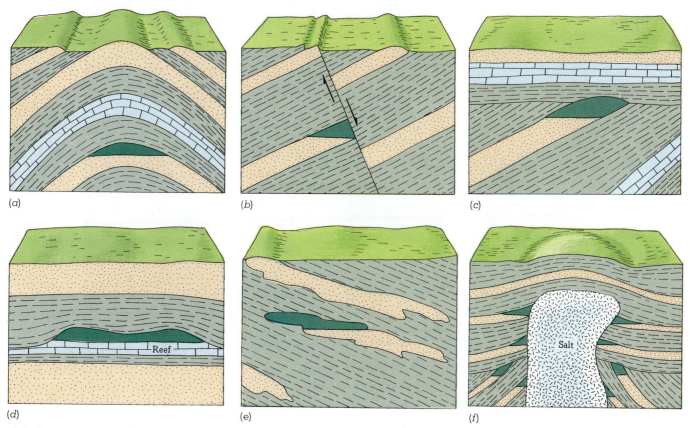

18.7   The common types of oil traps, drawn here in cross section, include (a) anticlines, (b) faults, (c) unconformities, (d) reefs, (e) sand lenses, and (f) salt domes. Oil accumulation is shown in dark green.

18.8   The operation of a seismic-exploration effort on land is shown in this diagram. First a drill prepares a hole about 50 m deep. A shooting truck then loads the bottom of the hole with explosives, plugs the hole, and explodes the charge electrically on a command from the recording truck. Sound waves from the explosion follow paths shown by the colored lines from the explosion to a reflecting layer at depth and back up to miniature seismometers connected by cable to the recording truck. Computer processing of the seismic records produces a diagram that resembles a geologic cross section. [After Sheldon Judson, Kenneth S. Deffeyes, and Robert B. Hargraves, *Physical Geology*, p. 443, Prentice Hall, Englewood Cliffs, N.J., 1976.]

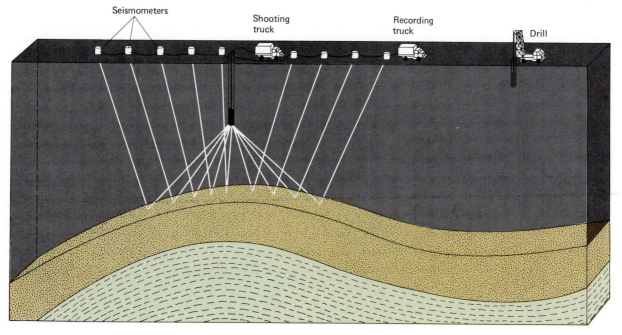

18.9 Off northern California the seismic ship *Indian Seal* trails air guns, the energy from which is reflected off different rock layers beneath the sea floor. As with land-based seismic exploration, this shipboard operation will produce a cross section that shows the geologic structures. [James Sugar/Black Star.]

18.10 Areas favorable to the accumulation of petroleum both on the continents and subsea. Only small portions of the areas shown as favorable actually contain producible accumulations. [After V. E. McKelvey et al., "World Subsea Mineral Resources," *U.S. Geol. Surv. Misc. Geol. Invest. Map I-632,* sheet 3, 1969.]

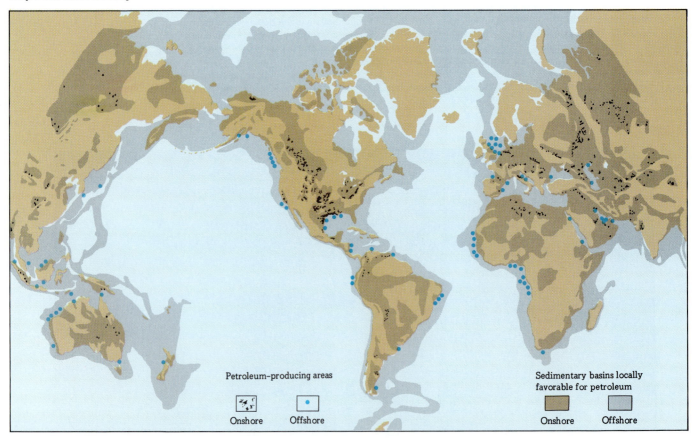

18.11 Petroleum and natural gas are the major sources of the world's energy supply today. In recent years explorations for oil and gas have expanded from dry land to the submerged borders of the continents. Offshore reserves, which are increasingly important, are tapped by wells drilled from platforms set at sea like this one in the Gulf of Mexico off Louisiana. [Amoco Corporation.]

duction in the United States are given in Figure 18.12, one made in 1956 and the other in 1989. The 1956 estimate suggested that the United States would reach its peak production of crude oil between 1960 and 1970. By 1989, experience told us that the United States production of crude oil had peaked in 1967. Furthermore, it became apparent that by the year 2000 we shall have produced 80 percent of all the crude oil available to us in the United States, exclusive of the state of Alaska.

On a worldwide scale, for both oil and gas, the picture is similar to that in the United States, although throughout the world there is a larger percentage of the original hydrocarbons yet to be found than there is in the extensively explored United States (Figure 18.13). (The richest, most concentrated reserves of crude oil known are those of Saudi Arabia, the leading member of OPEC, the subject of Box 18.3). Nevertheless, when we consider the "big three" of

our present sources of energy — coal, oil, and natural gas — we find that the world supply will be consumed over a very short time span in terms of human history (Figure 18.14). So the fossil fuels as we currently know them will become exhausted. What will replace them?

OIL SHALE AND TAR SANDS

Oil shale and tar sands are potential sources of additional fossil fuels. In fact, the potential resources exceed the known and postulated supplies of natural gas and oil.

Oil Shale Most fine-grained sedimentary rocks contain some organic matter. Given the proper conditions of burial, these rocks may produce commercial amounts of oil and

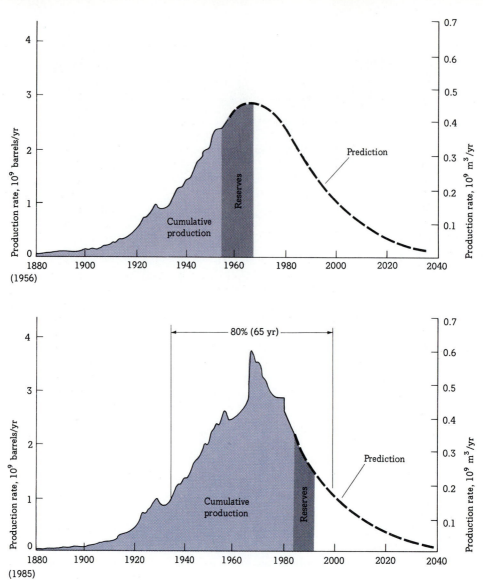

18.12 Crude-oil production histories and predictions for the United States and adjacent continental shelves (exclusive of Alaska) made in 1956 and 1989. Past production is shown in tint color, reserves in gray, and oil predicted to be discovered in white. [Earlier diagram after M. King Hubbert, "Nuclear Energy and the Fossil Fuels," Publication 95, p. 23, Shell Development Company, Houston, 1956; 1989 diagram updated from M. King Hubbert, "Energy Resources," in *Resources and Man*, W. H. Freeman & Co., Publishers, San Francisco, 1969. Copyright National Academy of Sciences.]

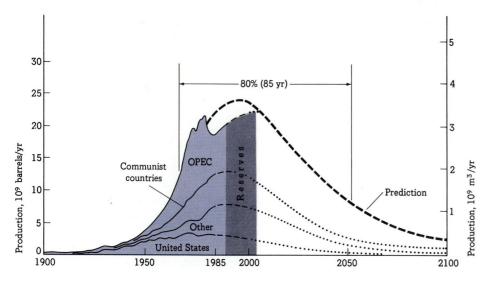

18.13 Figure 18.12 shows that crude-oil production in the United States peaked in 1967. This figure shows one estimate of when peak production for the world will be reached —early in the twenty-first century. It also suggests that crude-oil production worldwide will satisfy demand through the year 2000. Thereafter an ever-increasing deficit of supply in relation to demand will have to be met by energy sources other than conventionally produced petroleum. The figure shows the role of the United States becoming less and less important throughout this period and the world supply of oil being dominated by OPEC. Past production is shown in color, reserves in gray, and oil predicted to be discovered in white. [Data from Chevron Corporation.]

BOX 18.3   OPEC

The Organization of Petroleum Exporting Countries (OPEC), founded in 1960 by five petroleum-rich countries, is now composed of 13 countries: Algeria, Ecuador, Gabon, Indonesia, Iran, Iraq, Kuwait, Libya, Nigeria, Qatar, Saudi Arabia, the United Arab Emirates, and Venezuela. Among them these nations control about 70 percent of the world's recoverable oil reserves. They work together within OPEC to obtain maximum income from the exportation of their petroleum.

The early development of petroleum production in the OPEC countries was carried on by American and European oil companies, which paid the host countries income taxes and royalties based on the price they charged for oil in the world market. By the end of the 1950s the world supply of oil was exceeding demand, so the oil compa-

nies had to reduce the price they charged for oil. This, of course, reduced the income received by the host countries. The latter formed OPEC with the aim of establishing greater control over the price of oil in the world market.

OPEC had little effect on prices until the early 1970s, when world demand for oil began to outrun supply. OPEC rapidly pushed up the price of oil, and the results were dramatic. In the United States the average price per gallon of gasoline at the pump went from about 36 cents in 1972 to $1.31 in 1981. The retail price for heating oil increased from less than 20 cents per gallon to nearly $1.25 during the same period. The economic and social effects of such price increases were profound. OPEC's action forced the United States and European countries into extensive conservation efforts, an intensified

search for new sources of petroleum, and a variety of programs to establish alternative sources of energy. By the mid-1980s these measures had produced strong results: The supply of crude oil was outstripping demand, dependence on OPEC oil had been reduced, and the price of oil on the world market had fallen sharply.

All this was cause for some satisfaction. But the source of future energy supplies was yet to be solved. Thus by 1989 energy demand in the United States was being satisfied approximately as follows: oil, 42 percent; natural gas, 25 percent; coal, 23 percent; hydro and geothermal, 5 percent, and nuclear, 5 percent. The country was still overwhelmingly dependent on fossil fuels. The observation of Dixie Lee Ray quoted at the opening of this chapter has the same implications that it did in 1973.

---

gas. But some organic-rich rocks have not been deeply enough buried to allow the conversion of the kerogen they contain into hydrocarbons. One can, however, produce oil and gas from these deposits by heating them. We thus utilize an "immature" source rock to make synthetic petroleum. So far the production of petroleum from shale has

been on a small scale, with some production in China, Estonia, and other parts of the Soviet Union.

The richest known deposits of high-grade oil shale are in the United States. During the Eocene Epoch large freshwater lakes spread across portions of what are now Colorado, Utah, and Wyoming (Figure 18.15). In them accumulated great masses of organic-rich sediments, now known as the Green River oil shale. Some shale is rich enough to produce well over a 42-gallon barrel of oil per ton, and the potential reserves are large. Just how large is not presently clear, but they certainly are equal to all proven and predicted reserves of oil and natural gas still left in the United States. For many years pilot plants have been experimenting with the technology of processing the Green River shale. Large-scale exploitation of the resource has stalled, however, in part because the oil from processed shale is more expensive than the traditional oil-field petroleum and in part because of the environmental problems associated with extensive shale-oil operations. For instance, as oil shale is heated to extract oil, the shale expands, so that spent shale is 20 percent greater in volume than it was before treatment. Furthermore, immense volumes of shale must be used in any practical exploitation of the oil shale. Therefore the problem remains of what to do with the leftover shale once the oil is produced. Beyond this, the extraction process itself demands large amounts of water, already in short supply in the western states.

18.14 The period during which most of the fossil fuels will be consumed is short compared to human history and is an instant in terms of geologic time. This graph, showing the use of fossil fuels as a pimplelike bump on the scale of Earth's history, has become known as "Hubbert's pimple" after M. King Hubbert, who originally published this graph in 1962. [After M. K. Hubbert, "Energy Resources: A Report to the Committee on Natural Resources," *Natl. Res. Council Pub. 1000-D*, National Academy of Sciences, Washington, 1962.]

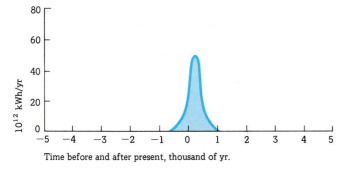

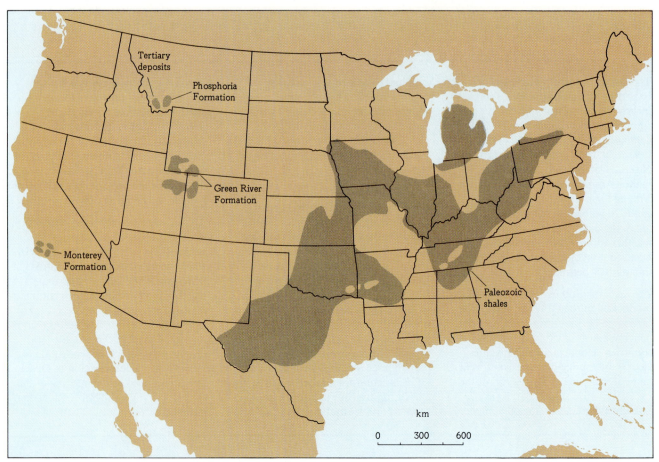

18.15 Organic-rich shales can be a source of oil and gas. The richest in the United States are found in the Green River Formation of Colorado, Utah, and Wyoming. Additional shales, accessible to mining and capable of yielding 10 gallons/ton or more, are widespread in the United States. [Data from U.S. Geological Survey.]

Besides the Green River shale, the United States possesses vast amounts of other organic-rich shales, although their concentration of organic matter by no means compares with that in the Green River shale. Those shales close enough to the surface for mining are shown in Figure 18.15. However, present technology is far from making these deposits useful to us as an energy source, even though they may be exploited eventually.

Tar Sands   Earlier in this section we pointed out that hydrocarbons formed in the underground work their way upward and, if not contained by a trap, will eventually reach the surface and there be lost. But in some situations certain hydrocarbons may be brought to, or close to, the surface and survive there for a considerable time. These hydrocarbons are the dark, heavy, low-volatility, high-viscosity ones. We know them as **tar, asphalt,** or **bitumin.** And in some localities they are preserved as if we were dealing with a reservoir bed from which all the flowing oil had been removed and in which only the heavy tar remained.

Three very large deposits of tar sands are known, of which the Canadian are probably the largest, although the other two are probably comparable in size. One of the latter is located in Venezuela along the Orinoco River; the other is in the Soviet Union. When the three largest deposits are considered together with several known small ones, tar sands appear to have the potential of yielding as much energy in the form of crude oil as all the oil and natural gas originally available to us in the traditional petroleum fields of the world. Estimates of recoverable oil from the Canadian deposits alone range between 250 and 500 billion barrels.

Extraction of the bitumen from the sands is an obvious problem in the production from tar sands. Another problem is transportation. Furthermore, not many refineries are equipped to process this very heavy hydrocarbon. Nonetheless, the Canadian deposits produced 175,000 barrels per day in 1985, most of it from the Fort McMurray area (Figure 18.16), where the sands are mined and the asphalt is separated from the host rock. Some 250 km to the south, at Cold Lake and Wolf Lake, the tar is softened by steam injected into the well and then extracted by pumping. Another 80 km south, at the settlements of Lindbergh and Elk Point, similar operations have been established.

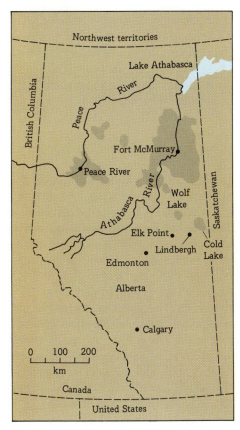

18.16 The areas of Alberta known to be underlain by tar sands. Some production now comes from deposits near Fort McMurray, Wolf Lake, Cold Lake, Lindbergh, and Elk Point.

SOME CONCLUSIONS

Before leaving this discussion of oil and gas, it may be worthwhile to draw some conclusions about the future of fossil fuels in general. As we have stated, we know that all fossil fuels are nonrenewable; that is, once they are used up, they cannot be replaced. We also know that the traditional sources of oil and natural gas are very limited and will be essentially gone in a few decades. Coal, being more abundant, will extend the life of fossil fuels as an energy source, and so may oil shales and tar sands. But whatever we do, the prediction in Figure 18.14 will not be modified much if we continue to use fossil fuels at the ever-increasing rates we have grown accustomed to. Somewhere 100 to 200 years down the line, we shall have used up half the fossil fuels that were originally available to us.

What, then, are the alternatives to our dependence on fossil fuels? We shall look at some of them in the following sections.

18.4
WATER POWER

Water power has been used as a source of energy for at least 2,000 years and perhaps longer. Water has the advantage of being a renewable resource: The Sun pumps it by evapora-

tion from the sea, and the atmosphere delivers it over the landmasses, where it falls as precipitation and flows off in rivers and streams back to the sea. And water power is clean and relatively cheap. But there are some drawbacks to its use. The dams generally necessary for the storage of water have limited lifetimes and usefulness because sediments silt up the lakes formed behind them. There is also the objection that damming floods out a great deal of prized scenery by drowning valleys and their associated rivers.

About 25 percent of the water-power capacity of the United States has already been developed. Worldwide, about 6 percent of water-power capacity has been developed. It is estimated that if the world's water power were fully developed, it would produce energy equal to only about 30 percent of the total energy now consumed. Africa and South America have the greatest water-power potential; thus far very little, about 1 percent, has been developed on these continents.

18.5
NUCLEAR POWER

The nucleus of an atom is, as we discussed in Chapter 2, made up of protons or protons and neutrons. It is the binding energy, holding these particles together in the nucleus, that forms the basis for nuclear power. This energy can be released either by breaking the nucleus of an atom into nuclei of lighter-weight elements, a process known as **fission,** or by joining the nuclei of two elements together to form the nucleus of a heavier element, a process called **fusion.** Both of these processes have been achieved at the explosive level in the atomic bomb (fission) and the hydrogen bomb (fusion). We have been able to develop technology that controls fission so that the slow release of the energy can be converted to the generation of electric power. We cannot yet control the fusion process. So far we have been able to tap only a small fraction of the atom's potential energy. At present we are using only that available from the heavy element uranium. It is not the common isotope of **uranium,** with an atomic weight of **238,** that fissions under neutron bombardment, but rather the lighter isotope, **uranium 235.** This is found in nature intermixed with ^{238}U in the proportion of about 1 part in 140. Because the two isotopes are chemically identical, the ^{235}U can be separated only by a physical process capitalizing on the slight difference in mass. The process of making uranium into a gaseous compound and then pumping it against plates pierced with billions of tiny holes causes the lighter ^{235}U to move through the holes to concentrate a product more than 90 percent ^{235}U.

In nuclear fission a slow-moving neutron smashes into the nucleus of the ^{235}U atom. It splits the nucleus into two new atomic nuclei of less mass, plus several neutrons. Because the mass of the fission products is less than that of the original atom, some of the mass has been converted into energy. For nuclear fission to take place spontaneously, the piece of material must exceed a certain critical mass. That is, there must be a sufficient number of nuclei so that neutrons bombarding the material cannot fail to hit a nucleus and split it, producing more active neutrons. After the process is started, it thus proceeds by a chain reaction. This reaction, so explosive in the atomic bomb, is controllable and is used in generating electrical energy. By 1988 about 575 nuclear reactors were operating throughout the world on a commercial basis.

The fuel used in the nuclear plant is the ^{235}U extracted from the more abundant ^{238}U. It is a nonrenewable resource, so that question arises as to how much ^{238}U (and hence ^{235}U) is available and how inexpensively it can be obtained. The evidence points to a limited supply of fissionable uranium. If this is true, we are facing the same difficulty with respect to nuclear power similar to that we face with the fossil fuels.

There is, however, a particular situation that may help to solve the problem of declining reserves of uranium. We can, in effect, replace ^{235}U with artificially produced fissionable atoms. The fairly abundant ^{238}U and the ^{232}Th (thorium), though not naturally fissionable, can be converted into fissionable elements. Thus ^{238}U may absorb a neutron in its nucleus and convert into ^{239}Pu (plutonium), which is fissionable and can sustain the chain reaction necessary to our purposes. In the same way, neutrons will convert ^{232}Th into fissionable ^{236}U. We then need to use the neutrons of decaying ^{235}U to create radioactive plutonium or ^{236}U at a rate equal to, or faster than, the fissioning of ^{235}U. This creation of new radioactive isotopes is called **breeding,** and the reactors in which they are used are called **breeder reactors.**

In the United States the richest uranium deposits occur in sedimentary rocks in New Mexico, Wyoming, and Texas. Lower-grade deposits are found with phosphate deposits, lignite beds, or shales.

18.6

GEOTHERMAL POWER

The volcanoes that dot sections of our Earth are expressions of geothermal energy, as are the geysers and hot springs in such areas as Yellowstone National Park, Iceland, and New Zealand. It would be useful if this energy could be extensively harnessed. Our use of geothermal energy is already a reality on a small scale. For instance, most of the space heat for Reykjavik, the capital of Iceland, comes from hot springs on the island; The Geysers, a hot steam field north of San Francisco, generates electricity on a commercial scale; and the geothermal field of Larderello in central Italy has been producing power since before World War II.

The decay of radioactive elements within the Earth continues to produce heat and to support a slow heat flow

toward the surface (Section 3.1). In trying to harness this energy, we run into a number of problems, the first of which is that the heat flow in the Earth is so diffuse that general exploitation is impractical. We look, therefore, for those areas where heat flow is unusually high. These areas are of two kinds: the plate boundaries and the hot spots away from the plate boundaries. Within such areas we can consider two possibilities in the use of geothermal power to generate electricity.

The first possibility, where the source of energy is pools of geothermally produced hot water and steam, is already being exploited. The energy comes from hot igneous rocks in still-active or quiescent igneous areas. The geothermal pool is not unlike an oil trap in some of its characteristics, but instead of hydrocarbons, we are dealing with heated water. The permeable-rock units containing the water are sealed off, at least in part, from the surface by impermeable rocks. The water in the aquifer is heated by the igneous rocks below. Some of the heated water may leak to the surface in the form of hot springs or geysers (Figure 18.17). That which remains trapped may be tapped by drill holes and brought to the surface to produce electrical power. Commercial installations are of this type, and the bulk of them rely on geothermal steam as the source of energy.

The second possibility is still in the experimental stage. It may be that there are hot igneous rocks at shallow depth, but the overlying rocks lack water, usually because they are too impermeable (Figure 18.18). In this situation it may be possible to drill to some depth where the rocks are hot enough and then create an artificial permeability by **hydrofraction,** a process of fracturing subsurface rock with

18.17   A geothermal system involves the heating of water by hot igneous rocks at some depth below the surface. The system shown here involves a down-faulted basin in which sediments have accumulated. Surface water percolates downward as groundwater in the coarse sediments at the margins of the basin and is held in by finer, impermeable sediments in the center of the basin. Some heated water is shown as escaping back to the surface as a hot spring. Wells drilled into the basin would be able to tap the heated water at depth and bring it to the surface as an energy source.

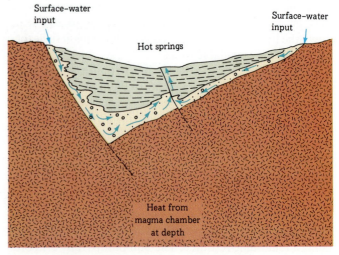

Surface-water input

Hot springs

Surface-water input

Heat from magma chamber at depth

One drawback to geothermal power is that its use will be restricted to those areas where geothermal flow is particularly high. A major technological problem involves the mineral-forming elements dissolved in the hot water or steam. These can precipitate out in the generating installation, thereby rapidly reducing effectiveness and increasing maintenance costs. Also, a considerable amount of noise is generated in the process of using geothermal steam. And finally, geothermal pools of steam are not a renewable resource, or at least we have so far been unable to develop a renewal process.

18.7
OTHER SOURCES OF ENERGY

We should not close this chapter without mentioning some other possibilities.

Wind continues to have a small place in the energy-supply picture. Windmills on land can supply small energy needs of individual homes or farms. On a larger scale, **wind farms,** each with many individual windmills, have been established in some places, including New Mexico and California. It is even conceivable that we can occasionally return to sail for supplementary transportation over the oceans.

The **tides** of the world represent the gravitational pull of Moon and Sun, the Moon exerting the greater influence. Using the rise and fall of tides to generate electricity has often seemed attractive. The Rance River estuary in northern France is producing tidally generated electricity, as is a plant in the northern Soviet Union along the coast of the White Sea. The Passamaquoddy area in the Bay of Fundy on the Maine-Canadian border has long been cited as a possible place to install a tidal plant, although thus far it has not proved economically or politically feasible.

The **geothermal gradient** in the ocean has been suggested by some as a large-scale source of energy. A difference of about 20°C, for instance, is not uncommon between the cold bottom waters and the warm surface waters. Theoretically this difference in temperature is convertible into power, though the development of this source of energy is still in the future.

18.18   Drill rig at the site of the Hot Dry Rock Geothermal Project in the Jemez Mountains of north-central New Mexico.

water introduced into the drill hole under very high pressure. The next step is to introduce water through the drill hole into the fractured zone where the water is to be heated. A recovery hole is drilled, and the heated water is brought to the surface to power electric generators. In such a system the water, after being used for power generation, is recycled into the ground, heated again, and once more used to produce electrical power.

Summary

Fossil fuels (oil, gas, and coal) are presently the most important sources of energy in the industrialized world.

Coal is a sedimentary rock that represents ancient swamps and associated peat deposits. Coal increases in carbon content from **lignite** through **bituminous coal** to **anthracite.**

Oil and natural gas are today the primary sources of energy. The origin of oil and gas involves organic-rich, fine-grained sediments. First the organic matter is transformed to **kerogen.** This turns to oil and gas as the sediment is buried to depths between 2 and 7 km and reaches temperatures between 50° and 250°C.

Accumulation of oil and natural gas in the underground demands a **source bed,** a permeable **reservoir rock,** an impermeable **capping bed,** and a **trap** to contain the hydrocarbons.

The **search** for petroleum relies largely on indirect methods of geophysics and geology. In the conterminous United States we appear to have found and used over half of the hydrocarbons originally available. Worldwide, the supplies of oil and natural gas will be close to exhaustion in the early part of the twenty-first century.

Oil shale and **tar sands** are potential sources of large amounts of hydrocarbons and can extend somewhat the supply from conventional petroleum fields.

Water power is renewable, clean, and relatively inexpensive. One drawback is that the dams needed to impound water have a limited useful life because of siltation.

Nuclear power involves either the splitting of an atomic nucleus **(fission)** or the joining of two nuclei **(fusion).** Controlled fission is in commercial use. We do not yet know if controlled fusion is feasible. Artificial breeding of fissionable material (plutonium) could alleviate the supply problem for fission but does not answer questions of health, environment, or politics.

Geothermal energy can locally supply some energy needs by the use of steam and hot water generated at shallow depths by still-hot igneous rocks.

Other sources of energy include the **wind** (another indirect effect of the Sun), **tides,** and the **thermal gradient in the oceans.**

Questions

1. Distinguish between renewable and nonrenewable natural resources.

2. Explain why the various grades of coal might be called members in a metamorphic series.

3. Review the steps involved in the formation of oil and gas in nature.

4. You are involved in deep drilling for hydrocarbons and you strike them at a depth of 6,500 m. Is it more likely that you have hit oil or gas? Why?

5. To show that even a small pool of crude oil has real value, try the following evaluation. Let us suppose that you have discovered an oil pool that is roughly circular and has a diameter of 200 m. The reservoir rock is 20 m thick, has a 20 percent porosity, and is saturated with oil. You can expect to recover 25 percent of the oil. What is the gross value of this oil to you if the price per barrel is $15 (A cubic meter of oil equals 6.3 barrels.)

(We got just under $3 million as follows: The area of a circle is πr^2. To get the volume of the reservoir rock, first multiply 3.1416 by 100 m^2 to determine area, and then multiply by 20 m to get volume in m^3. It all comes out thus:)

$$3.1416 \times 10,000 \times 20 \times .2 \times .25 \times 6.3(\text{bbls}) \times \$15 =$$
$$\$2,968,812$$

What happens to the value if the price drops to $10 per barrel? If it rises to $30 per barrel?

6. Even though nature has produced much oil and gas, it does not always provide a place to accumulate them so that they can be drawn on for energy. Draw well-labeled diagrams to illustrate three different types of geologic structures in which oil and gas are known to have accumulated.

7. If liquid petroleum has a limited future among our supplies of energy, then what role might be played by oil shales and tar sands? Discuss them as sources of energy.

8. What are the advantages and disadvantages of hydroelectric power?

9. How does a nuclear fission plant produce energy?

10. Discuss geothermal power.

11. We expect that the recoverable crude oil reserves in the United States will be approaching exhaustion in the first quarter of the twenty-first century. Most of you reading this will by then be less than 60 years old and, we hope, in good health. Consider what this declining availability of petroleum will mean to you and your children, who will average perhaps 30 years of age. What will you be using as sources of energy for heating, cooling, transportation, light, and appliances? Do you have any speculations about the economics of this energy? Do you have any other thoughts about your future in relation to the future of sources of energy?

Supplementary Readings

ARMSTEAD, H. CHRISTOPHER *Geothermal Energy,* John Wiley & Sons, New York, 1978.
A comprehensive view of geothermal energy discussing sources of energy, exploration methods, and environmental and engineering aspects.

FETTWEIS, GÜNTER B. *World Coal Resources,* Elsevier Scientific Publishing Co., New York, 1979.
This is a translation from the original German edition of 1976. There is a wealth of information in this book, although the lack of an index sometimes makes it a bit hard to find.

GIDDENS, PAUL H. *The Birth of the Oil Industry,* The Macmillan Co., New York, 1938.
The emphasis is on the 10 years following the drilling of the Drake well near Titusville, Pennsylvania, in 1859. Ida Tarbell, American author and editor and one-time resident of Titusville, provides a lengthy introduction that takes us from Colonial times to 1859.

HUNT, JOHN M. *Petroleum Geochemistry and Geology,* W. H. Freeman & Co., San Francisco, 1979.
This is an excellent text that integrates the geochemistry of oil and gas with geology. The serious reader can learn a great deal from this book.

McLAREN, D. J., AND BRIAN J. SKINNER *Resources and World Development,* John Wiley & Sons, New York, 1987.
Contains useful analyses of world energy resources.

ORBAN, JOHN, III *Money in the Ground: Oil and Gas Investments Explained,* Meridian Press, Oklahoma City, 1985.
Here, in clear language and simple illustration, is the explanation of oil and gas investments from prospect development to tax computations.

SKINNER, BRIAN J. *Earth Resources,* 3d ed., Prentice Hall, Englewood Cliffs, N.J., 1986.
Two of the chapters of this excellent and concise book focus on energy resources.

TISSOT, B. P., AND D. H. WELTE *Petroleum Formation and Occurrence,* Springer-Verlag, New York, 1978.
As one advocate of the book writes: "A clear, practical, readable account by two of the world's leading researchers in the field, of the formation, migration, and accumulation of petroleum, primarily from the geochemical viewpoint, but adequately seasoned with geology."

Sulfur, produced by Amoco Canada Petroleum, Ltd., for use in the chemical industry is loaded on board ship here in Vancouver. [Amoco Corporation.]

CHAPTER ■ 19

USEFUL MATERIALS

"And since the art [of mining] is one of the most ancient, the most necessary and the most profitable to mankind, I considered that I ought not to neglect it."

Georgius Agricola, from the preface to De Re Metallica, *1556*

Mining has gone on for thousands of years. The oldest known mine is said to date back further than 40,000 years, when Stone Age man dug for red ochre (hematite), seeking for funeral rites the blood-red deposits at Bomvu Ridge in what is now Swaziland, in southern Africa. Now we extract from the Earth a wide variety of minerals and rocks out of which we construct the vast number of things so characteristic of modern civilization. Roads and buildings, cameras and silicon chips, automobiles and planes, belt buckles and pins—an almost unending list of things depends on our ability to extract raw materials from the ground.

The processes by which nature has produced *minerals* useful to us are not unique; they are the same processes involved in all other geologic phenomena. For example, the sand shoveled from a sand deposit to mix with cement to form concrete was laid down by some agent such as a stream or the seawater sloshing on a beach. The gold of a brooch may have concentrated in veins of quartz during the final stages of igneous activity. The ore that provided the aluminum for your soft-drink can may have been mined from a deposit formed by long-continued weathering that produced a soil-related deposit high in bauxite, a hydrated oxide of aluminum. What is special about these deposits is that we find them of value.

In the following pages we look at the *geologic processes* that concentrate valuable deposits, all of them processes that we studied in earlier chapters of this book.Then we examine *metallic deposits,* including their *temperature* and *process of formation,* their *ore minerals,* and their *relation to plate tectonics.* Next we look at the *nonmetallic deposits,* including those of importance as fertilizers, building materials, and resources for the chemical industry, and close the section with a brief look at gems. Finally, we add a note on *earth materials and society.*

19.1

MINERAL DEPOSITS

We can consider **mineral deposits** as either **metallic** or **nonmetallic.** The metallic deposits include those mined for such common elements as iron and aluminum, as well as for the rarer elements such as chromium, tin, uranium, silver, mercury, and gold, to name but a few. The numerous nonmetallic products we take from the Earth include salt, building stone, sand and gravel, phosphorous, and sulfur.

MINING

Mining is the removal of economically useful materials from the Earth, a process first described in detail by Agricola in the sixteenth century (Box 19.1). The operation may go on either above or below ground. For instance, the familiar surface process of **quarrying** is used to extract building stone and rock aggregate. **Open-pit,** or **strip, mining** is also carried on from the surface (Figure 19.1). The procedure often involves the removal of some amount of valueless material so that the sought-for material can be reached. Much of the coal mining done in this country is of this type. **Placer mining** usually involves the separation of valuable metals or nonmetals from unconsolidated, near-surface, or surface deposits of sand and gravel. Some diamond mining is of this type, and some gold, tin, and titanium come from placers (Figure 19.2).

When we think of mining, however, we usually think of **underground mining** (Figure 19.3), a process that has gone on for more than 3,000 years. Underground mines may be a few tens of meters into a hillside or may extend to 2 or more kilometers below the surface.

Whether a mineral deposit is mined or not depends on economics. Is it worthwhile to mine? If it is, then the product is called an **ore,** a mineral deposit that can be worked at a profit. If you think about this definition, you will recognize that factors other than the geology of the deposit, factors such as technology and politics, can affect whether or not a deposit is classed as an ore. Here is an example. The metal titanium occurs chiefly in two minerals, ilmenite ($FeTiO_3$) and rutile (TiO_2), and until World War II had little economic value. Two developments turned titanium-bearing deposits into ores. First, titanium and its alloys are lightweight and exhibit very high tensile strength at high temperatures, characteristics that established titanium as an important metal in the postwar aircraft and space industries. Second, titanium in the oxide form came to be extensively used in the paint industry to replace lead, which is toxic.

19.1   The Chino mine near Santa Rita in southern New Mexico is an open-pit operation mining a porphyry for its copper content.

19.2   Placer mining for diamonds along the coast of Namibia in Africa. In this case, rivers have transported diamonds to the shoreline, where wave action has distributed them in gravels along the beach. Later a cover of sand buried the diamond-bearing deposits. Here the covering sands are being removed from the gravels. [De Beers Consolidated Mines.]

BOX 19.1 *De Re Metallica*

De Re Metallica (Concerning Things Metallic), written by Georgius Agricola (the latinized name for George Bauer, *Bauer* being the German for "peasant") and published in 1556, stands as a landmark in the development of our knowledge of the Earth, particularly of mining (Figure B19.1.1.)

As a physician, scientist, and public servant, Agricola had access to the works of the ancient writers at a time when they were just being rediscovered. But the bases for his treatise were his own observations and investigations.

De Re Metallica is a very complete record of mining practices in Agricola's day. He describes the occurrence of metalliferous veins, the mineralogy of the deposits, the extraction of ores, and their treatment. For nearly two centuries Agricola's treatise was the basic authority on the subject, the one on which later students built.

De Re Metallica appeared a year after Agricola's death. The first edition was in Latin, as were several subsequent printings. Later the book was translated into German and Italian, but it was not until 1912 that an English translation appeared. That edition was translated by the mining engineer, geologist, and future president of the United States Herbert Hoover, in collaboration with his wife, Lou Henry Hoover.

B19.1.1 The title page for Agricola's treatise of mining, *De Re Metallica.*

19.3 More than 750 m beneath the surface of the Grace mine in eastern Pennsylvania, a miner operates a load-haul-dump vehicle to transport iron ore. Predominantly magnetite, the ore is the result of contact metamorphism caused by a gabbroic intrusion into limestone. [Bethlehem Steel Corporation.]

Even as a mineral deposit may be elevated by circumstances to the status of an ore deposit, so may an ore lose its economic value and be downgraded to a mineral deposit. Titanium again serves as an example. Deposits of ilmenite concentrated by river and shoreline processes were mined in New Jersey in the years immediately following World War II. These ilmenite operations were later closed down because American purchasers found it more economical to import the mineral than to buy it domestically. Thus the New Jersey deposits lost their status as ore.

GEOLOGIC PROCESSES AND MINERAL DEPOSITS

A convenient way of thinking about mineral deposits is to consider the **geologic processes** that formed them.

Igneous Activity You will remember that in Chapter 3 we discussed the formation of igneous rocks from a mixture of elements in a solution called a **magma.** We found that minerals crystallize from the magma in a sequence. Some minerals, forming early in the cooling of the magma and being heavier than the remaining melt, may settle toward the bottom of the magma chamber. This process of **fractionation,** known also as **magmatic segregation,** is responsible for the formation of some ore deposits. In the famous iron-mining district of Kiruna in northern Sweden, for example, one of the early-formed minerals in the cooling magma was magnetite (Fe_3O_4). Being denser than the remaining melt, the magnetite settled to the floor of the magma chamber. The result was an accumulation of high-grade iron ore (over 60 percent iron on average) that had segregated out from the main igneous mass.

In the final stages of cooling of a granitic magma the remaining fluids may give rise to the coarse-textured rock called a pegmatite, as described in Section 3.3. Most pegmatites are simply large crystals of orthoclase and quartz with some mica. But a few pegmatites contain in addition some rare minerals, and from them come such metals as tantalum, niobium, and lithium.

Hydrothermal Activity Igneous activity often gives rise to the circulation of heated fluids that migrate beyond the immediate area of igneous activity. These **hydrothermal solutions** may carry metallic ions into surrounding rocks whether they be igneous, sedimentary, or metamorphic.

During the crystallization of a magma the volatiles contained in the melt become more and more concentrated as minerals form. Eventually, in the final stages of crystallization, these volatiles are driven out into the surrounding rock as hydrothermal solutions. In this new and cooler environment they may deposit as minerals the elements that are carried in solution.

Many gold and silver deposits have formed from hydrothermal solutions and represent the final stages of igneous activity. Box 19.2 discusses an historically famous hydrothermal deposit. The great copper porphyry deposits of the western states (Section 19.2 and Figures 19.1 and 19.8) are also examples. Characteristically the ore minerals in these deposits are very finely disseminated and a great deal of rock must be processed to extract a relatively small amount of metal. The Chino mine (Figure 19.1), for instance, mines rock that contains about 0.05 percent copper. Some gold-mining properties have a significantly lower percentage of metal.

Exploration of the sea floor in recent years has shown that the spreading ridges are in places the location of great hot springs called **black smokers** (Figure 19.4). These springs are rich in sulfur and carry also metallic ions in solution. From the waters of these springs rich deposits of sulfides of zinc, lead, and copper are precipitated in what are known as **massive sulfides.** The process is different from the mineralization by solutions coming directly from the magmatic source. In this instance seawaters are thought to circulate through the pile of hot lavas forming at the spreading centers. There these waters pick up ions and then, as they rise to the cooler environment of the sea floor, they deposit sulfide minerals.

Sedimentation In some instances valuable minerals have been formed directly by the process of sedimentation. A case in point is a sedimentary deposit called **banded iron formation,** known also as **taconite.** The rock, the result of chemical sedimentation in Precambrian seas, is characteristically thinly bedded or laminated so that iron-rich bands alternate with bands of silica — usually chert. The rock type is widely distributed and is mined in the Lake Superior district (Minnesota, Wisconsin, and Michigan) as well as in eastern Canada, Venezuela, Brazil, western Australia, the Soviet Union, and the west African countries of Gabon, Liberia, and Mauritania.

A second type of sedimentary iron ore has long been important in England, Germany, France, and Belgium,

---

BOX 19.2   The Mines of Laurium

The city-state of ancient Athens attained its greatest glory during its so-called Golden Age in the fifth century B.C., when it was the artistic and intellectual center of the Western world. Golden was the age, but it was made possible in no small part by the silver extracted from the state-owned mines at Laurium, some 40 km southeast of Athens.

Rich ores of silver-bearing galena

contained in limestone had been worked for nearly 1,000 years previously. Under Athenian ownership the mines were leased by private citizens, who paid the state a royalty of about 4 percent of the value of the metal produced. From these mines came the wealth that built much of the combined Greek fleet that defeated the Persians at Salamis in 480 B.C. And Athenian silver did much to establish the silver standard that dominated

Mediterranean trade of the times.

Strife among the Greek city-states led to the replacement of Athens by Sparta as the dominant Greek state in the late fifth century B.C., and then by Thebes in the early fourth century. By this time the mines of Laurium, although still producing, were becoming exhausted. In the second century B.C. only the old mine dumps were being worked.

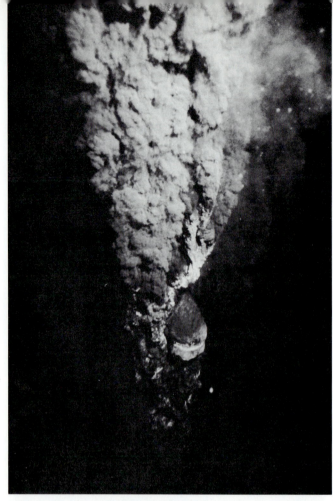

19.4 In the rifts separating diverging oceanic plates hydrothermal springs may discharge into the bottom waters of the ocean, as here along the East Pacific Rise at latitude 29° North. Called "black smokers," they carry metals and sulfur from magmatic sources at depth. On their way upward to the sea floor they may deposit metallic minerals in the newly forming oceanic crust. [Robert D. Ballard, Woods Hole Oceanographic Institution.]

and more recently in Canada, the United States, and Egypt. It also is marine, but the iron-bearing minerals have accumulated generally in **oölites** (see Section 6.2) and are associated with sandstone and shale, often calcareous.

A particular type of sedimentation is represented by the **marine evaporites.** Seawater, as we found in Section 16.1, contains dissolved salts in the proportion of about 3.5 per 1,000. As seawater evaporates, salts begin to crystallize out of the fluid and do so in a particular sequence that is inverse to their solubility. Calcium carbonate ($CaCO_3$), being the least soluble, precipitates first. It is followed by gypsum ($CaSO_4 \cdot 2H_2O$), by halite (NaCl), and finally, when the original solution reaches about 4 percent of its original volume, by a complex mixture of potassium and magnesium salts (Figure 19.5). These last-deposited minerals furnish potassium to the fertilizer industry. Gypsum is used in the construction industry to make plasterboard, and halite is in demand in the chemical industry.

Groundwater   Some ore minerals appear to have been deposited by groundwater as it moved through permeable rocks. An example is the brilliant canary-yellow mineral carnotite, $K_2(UO_2)_2(VO_4)_2$. Organic matter within the rock was involved in the deposition of carnotite. In some places the powdery mineral is found replacing the remains of trees buried in the rock.

Weathering   The process of weathering (see Chapter 5) has been responsible for the creation of some of our most important mineral deposits. The process operates in two ways: It may leach out undesirable elements from the weathering zone and leave the desirable elements concentrated in the upper soil zones; or it may move the desirable elements in low concentration near the surface to deeper

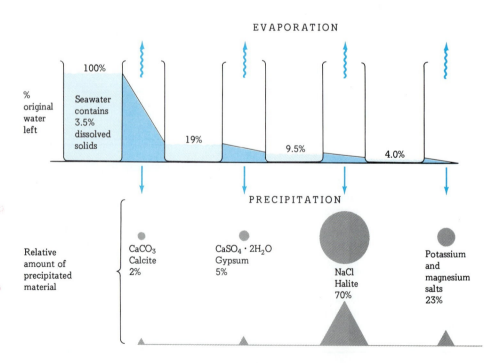

19.5 The sequence of solids precipitated with the continuing evaporation of seawater. [After Brian J. Skinner, *Earth Resources*, 3rd ed., Fig. 7.2, p. 133, Prentice Hall, Englewood Cliffs, N.J., 1986.]

zones, where they are redeposited in a zone enriched in a metal-bearing mineral.

Bauxite ($Al_2O_3 \cdot nH_2O$), the chief ore of aluminum, illustrates a mineral deposit formed by weathering. In Section 5.5 we found bauxite to be the residuum of long-continued weathering of aluminum silicate minerals in tropical climates. The process removes the silica and concentrates the hydrous aluminum oxide in the soil as an ore.

Placer Formation   Some minerals freed by weathering are dense enough and resistant enough to abrasion that they can be concentrated in favorable localites by the action of water in a stream or on the beach. Being denser than the usual sand and gravel, the minerals may collect in pockets, many of which prove rich enough to mine. Such a deposit is called a **placer** (pronounced *plasser*).

Placer deposits of gold are famous. The Gold Rush of 1849 was set off by the discovery of gold in stream gravels at Sutter's Mill in California. In 1898 the discovery of gold in beach sands on the Seward Peninsula near the present site of Nome triggered the Alaskan Gold Rush. Not all placers are modern. The world's richest deposits of gold are found in the Witwatersrand of South Africa, where they occur as native gold in Precambrian conglomerates of wide extent. The deposits give every indication of being an ancient placer.

Rock Formation   In many instances the mineral deposits constitute the whole rock rather than just a small enriched fraction of it. The processes that create, for example, limestone or marble or granite are creating an often-useful material, the rock itself (Figure 19.6). Examples abound in which the rock is used directly as it comes from the earth, being modified only in size and shape for immediate practical purposes. Building stone and crushed rock for aggregate are illustrative.

19.2
METALLIC DEPOSITS

In the previous section we reviewed the ways in which mineral deposits may form. Here we examine metallic **mineral** deposits in greater detail.

TEMPERATURE AND PROCESS

Many metal-bearing minerals and the processes that form them are associated with particular ranges of temperature. Figure 19.7 shows this for two dozen different metals. The figure is worth a close look.

In a way the diagram reflects the rock cycle that we first discussed in the opening chapter of this book. The temperatures cover the range characteristic of the Earth's crust. The processes duplicate many of those of the rock cycle, from igneous activity through weathering, sedimentation, groundwater, and stream and shore activity.

Still referring to Figure 19.7, note that at high temperatures (above 1,000°C) magmatic segregation may form minerals that can serve as the ores of platinum, chromium, iron, nickel, and titanium. With the exception of chromium, however, all these metals are also concentrated by other processes at lower temperatures. So it is with the other metals. Some are found in ore-grade concentrations in a particular temperature range and in minerals formed by a particular process.

METALS AND THEIR ORE MINERALS

The metals listed in Figure 19.7 are repeated in alphabetical order in Table 19.1, along with their major ore minerals, process of formation, chief producing areas, and some comments on their uses.

Even a casual perusal of the table reveals some interesting facts. Some metals, for instance, have more than one mineral as an ore. Often the minerals represent different geologic processes of formation. Most of the important ore minerals, and certainly those mined in large volume, are either sulfides or oxides. But the native elements and, in general, the pegmatitic minerals do not follow this rule. The table also makes clear that the major sources of economi-

19.6   Some rock material, as in this Vermont granite quarry, is used as it comes from the ground, needing only to be cut to the desired dimensions. [Jane Grigger.]

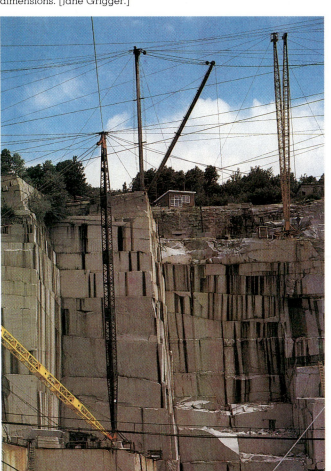

Temperature, °C

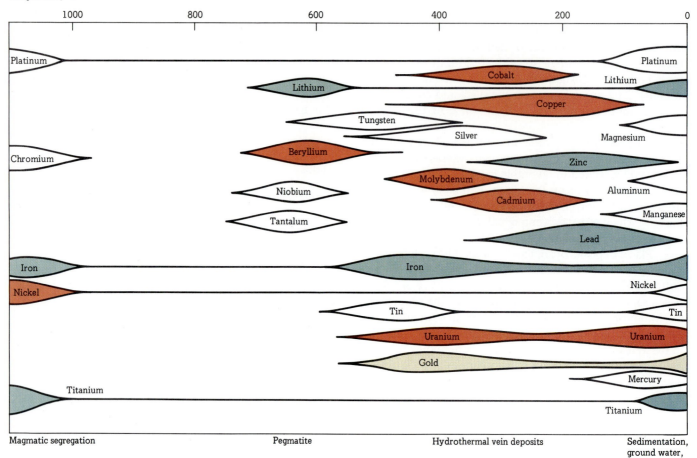

19.7 Ores of metals can be arranged according to their temperatures of accumulation. These range from temperatures of over 1000°C during magmatic segregation to the comparatively cool temperatures under which placers form at the Earth's surface. [From Sheldon Judson, Kenneth S. Deffeyes, Robert B. Hargraves, *Physical Geology*, Fig. 17.16, p. 424. Prentice Hall, Englewood Cliffs, N.J., 1976.]

TABLE 19.1
Major Metals

Metal	Major ore mineral(s)	Chief process(es) of ore accumulation	Percent of world production, 1983[a]	Chief uses
Aluminum	Bauxite, $Al_2O_3 \cdot nH_2O$	Weathering	Australia, 32; Guinea, 15; USSR, 9; Jamaica, 9; Yugoslavia, 8	Aircraft; electrical equipment; cans
Berylium	Beryl, $Be_3Al_2Si_6O_{18}$	Pegmatitic, hydrothermal	Exclusive of U.S. production:[b] USSR, 65; Brazil, 28	Copper alloys; electronic equipment
Cadmium[c]	By-product of smelting sphalerite (ZnS), which carries traces of Cd	Hydrothermal	USSR, 17; Japan, 13; USA, 6; Canada, 6; Belgium, 6; West Germany, 6; Australia, 6	Electroplating steel; batteries; pigments
Chromium[c]	Chromite, $FeCr_2O_4$	Magmatic segregation	USSR, 30; South Africa, 28; Albania, 11; Turkey, 5; Zimbabwe, 5; Phillipines, 4; India, 4; Finland, 4	Heat and corrosion-resistant alloys with steel
Cobalt[c]	Siegenite, $(CoNi)_3S_4$ Cobaltite, CoAsS by-product of Ni-Cu sulfide ores[4]	Hydrothermal	Zaire, 47; Zambia, 13; USSR, 10; Australia, 8; Cuba, 7; Canada, 7; Finland, 4	Petroleum refining; pigments; paint driers; aerospace alloys
Copper	Chalcopyrite, $CuFeS_2$ Bornite, Cu_5FeS_4 Chalcocite, Cu_2S Malachite, $Cu_2CO_3(OH)_2$	Hydrothermal Sedimentation Weathering	Chile, 15; U.S., 13; USSR, 12; Canada, 8; Zambia, 7; Zaire, 7; Poland, 5	Alloys with tin (bronze) and zinc (brass); electrical equipment; construction

TABLE 19.1 (*Continued*)

Metal	Major ore mineral(s)	Chief process(es) of ore accumulation	Percent of world production, 1983[a]	Chief uses
Gold	Native element, Au	Hydrothermal, placer	South Africa, 49; USSR, 19; Canada, 5; U.S., 4; China, 4	Bullion; dentistry, jewelry
Iron	Magnetite, Fe_3O_4 Hematite, Fe_2O_3 Goethite, $Fe_2O_3 \cdot H_2O$ Pyrite, FeS_2	Magmatic segregation Sedimentary Weathering Hydrothermal	USSR, 33; Brazil, 11; China, 10; Australia, 9; U.S., 5; Canada, 5; India, 5	Steelmaking
Lead	Galena, PbS	Hydrothermal	U.S., 14; Australia, 14; USSR, 13; Canada, 8; Peru, 6	Storage batteries; gasoline; construction
Lithium	Lepidolite, $K_1Li_2Al(Si_4O_{10})(OH)_2$ Spodumene, $LiAlSi_2O_6$	Pegmatitic	Exclusive of U.S. production:[b] USSR, 69; China, 17; Zimbabwe, 6; Brazil, 4	Aluminum industry; ceramics; glass; nuclear energy & chemical industries
Magnesium	Seawater } Magnesite, $MgCO_3$ Dolomite, $CaMg(CO_3)_2$	Weathering Sedimentation	U.S., 40; USSR, 31; Norway, 13	Steel manufacture; cement, glass, and copper industries
Manganese[c]	Pyrolusite, MnO_2 Psilomelane, $(Ba_1H_2O)_2Mn_5O_{10}$	Sedimentation Weathering	USSR, 46; South Africa, 13; Brazil, 9; Gabon, 8; India, 7; Australia, 6	Steel manufacture
Mercury	Cinnabar, HgS	Hydrothermal	USSR, 34; Spain, 26; U.S., 13; China, 11; Algeria, 5	Chemical industry
Molybdenum	Molybdenite, MoS_2	Hydrothermal	U.S., 25; Chile, 24; USSR, 18; Canada, 17; Mexico, 9	Steel industry
Nickel[c]	Pentlandite, $(FeNi)_9S_8$ Garnierite, $(Mg,Ni)_6Si_4O_{10}(OH)_8$	Magmatic segregation Weathering	USSR, 25; Canada, 18; Australia, 13; New Caledonia, 9; Cuba, 5	Stainless steel; heat- and corrosion-resistant steel
Niobium[c] (formerly called Columbium)	Pyrochlore, $(NaCa)_2(NbTa)_2O_6(O,OH,F)$ Columbite, $(FeMn)(Nb,Ta)_2O_6$	Pegmatitic	Brazil, 82; Canada, 15	High-strength steel; superconducting alloys
Platinum[c]	In minerals of Fe, Ni, Cr Native element	Magmatic segregation Placers	USSR, 54; South Africa, 40; Canada, 3	Chemical and metallurgical industries; jewelry; laboratory equipment
Silver[c]	By-product of Pb, Cu, Zn sulfide ores	Hydrothermal	Mexico, 16; Peru, 14; USSR, 12; U.S., 11; Canada, 9; Bolivia, 8	Photography, electrical and electric equipment; sterling ware
Tantalum[c]	Pyrochlore, $(NaCa)_2(NbTa)_2O_6(O,OH,F)$ Tantalite, $(FeMn)(TaNb_2O_6)$	Pegmatitic	Australia, 22; Brazil, 13	Electronics, metallurgical, and chemical industries
Tin[c]	Cassiterite, SnO_2	Hydrothermal, placers	Malaysia, 20; USSR, 18; Indonesia, 13; Bolivia, 12; Thailand, 9; China, 7	Tinplating, solder; alloys with copper (bronze)
Titanium	Ilmenite, $FeTiO_3$ Rutile, TiO_2	Magmatic segregation, placers	Australia, 27; Canada, 16; Norway, 14; USSR, 11; India, 5; Malaysia, 5	Paint pigment; lightweight metal
Tungsten	Wolframite, $(Fe,Mn)WO_4$ Scheelite, $CuWO_4$	Hydrothermal, placers	China, 32; USSR, 23; Bolivia, 6; South Korea, 6; Australia, 5; Portugal, 4	Hard steel; cutting and wear-resistant alloys
Uranium	Uraninite, UO_2 Carnotite, $K_2(UO_2)_2(VO_4)_2 \cdot 3H_2O$	Hydrothermal Groundwater	Main producers: U.S., USSR, South Africa, Canada	Nuclear fuel and explosives
Zinc	Sphalerite, ZnS	Hydrothermal	Canada, 17; USSR, 13; Australia, 11; Peru, 9; U.S., 4	Galvanizing iron and steel; alloys with copper (brass)

[a] From U.S. Bureau of Mines, *Minerals Yearbook, Metals and Minerals*, vol. I, 1984.
[b] U.S. production withheld for proprietary reasons.
[c] Included among the strategic metals. See Section 19.4.

cally useful deposits are unevenly distributed geographically. No single country dominates all others in its supply of useful minerals, although certain areas of the world are more abundantly endowed than others, an observation to which we will return in Section 19.4.

PLATE TECTONICS AND MINERAL DEPOSITS

Most igneous activity is concentrated along plate boundaries, particularly along diverging and converging boundaries. Many mineral deposits are related to igneous activity and associated hydrothermal activity. So it should come as no surprise that there is a relation in time and space between the boundaries of the Earth plates and many metallic mineral deposits. Let us look at some examples.

Oceanic spreading centers along diverging boundaries are marked not only by volcanism and shallow earthquakes but also by thermal springs (see Figure 19.4). These springs represent recirculation of seawater and may carry such metals as lead, copper, and zinc picked up below the ocean floor. Investigations in the Red Sea demonstrate that

19.8 A belt of copper-bearing porphyry is located along the western margins of the North American and South American continents. These deposits seem to be related to the convergence of the eastern Pacific plates and the American landmasses during Cenozoic mountain-building activity. [After Brian J. Skinner, *Earth Resources*, 3rd ed., Fig. 6.11, p. 109, Prentice Hall, Englewood Cliffs, N.J., 1986.]

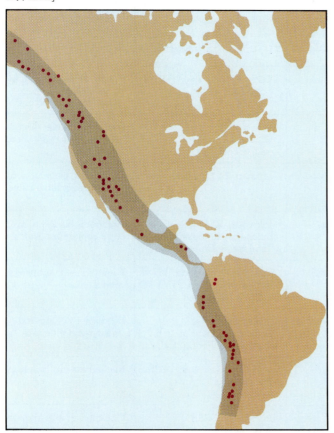

in some places modern and very recent sediments are being mineralized by hydrothermal solutions. Beneath these sediments we think that sulfide minerals are being deposited in the cracks and fissures in the basalts as well as in the more mafic mantle rocks below the sediments.

One can picture what might happen if these mineralized sediments and igneous rock were to be carried laterally in an ever-widening ocean basin by the process of sea-floor spreading. The rich, long-worked mineral deposits of Cyprus, for instance, are found in the basalts and peridotites associated with an ancient spreading center. In this case the mineralized rock has been arched upward along a converging boundary.

A different situation is represented by the **porphyry copper deposits** of western North and South America. These extend in a great belt from Alaska to Chile (Figure 19.8). The host rock is generally a diorite or a granodiorite. The ore is a low-grade copper sulfide deposit ranging from about 0.05 percent to 2 percent copper. The mineralization is finely disseminated through the country rock that is taken in bulk from open-pit mines. The emplacement of the host rock and its mineralization appear coincident with the convergence of the plates of the eastern Pacific Ocean with the North and South American continents and the accompanying igneous and hydrothermal activity above the zone of subduction. Similar deposits are found on the opposite side of the Pacific along converging boundaries in Malaysia, the Philippines, and New Guinea.

19.3
NONMETALLIC DEPOSITS

When we think of mining, it is usually the metals that first come to mind. But the nonmetallic deposits are equally important to us. We include here minerals mined for use in agriculture, in the chemical industry, and in building and construction. Then, of course, there is that exciting group of minerals we know as gems.

FERTILIZERS

The increased productivity of agricultural land during the last 50 years has been partly due to the enormous increase in the use of chemical fertilizers. The most important elements for the fertilizer industry are nitrogen, potassium, and phosphorous.

Nitrogen   Most of the naturally occurring nitrates of commercial grade have been mined out. But nitrogen is the most abundant element in the atmosphere, although plants are unable to use it directly. It must be **fixed** from the atmosphere (that is, combined with other elements) before it can be used in the life process. The manufacture of nitrogenous fertilizers involves the artificial fixation of nitrogen in compounds of ammonium or as nitrates of calcium or potassium. Artificial fixation demands large amounts of energy, and this means that the cost of nitrogenous fertilizers is closely tied to the cost of energy.

Potassium   There is a great abundance of potassium in the silicate minerals of the Earth's crust, but it is unusable in organic processes until turned into soluble form. We obtain the potassium for fertilizer, then, not directly from such minerals as orthoclase and potassic plagioclase, but from the group of minerals known as the **marine evaporites,** discussed in Section 19.1. The evaporite minerals supplying the bulk of our resources of potassium are sylvite (KCl) and carnalite ($KCl \cdot MgCl_2 \cdot 6H_2O$), two of the end members of the marine evaporite series (see Figure 19.5). These minerals were deposited in ancient, restricted shallow-water marine basins at various times in the past. The largest deposits are in the two Germanies and the Soviet Union. In the United States extensive deposits occur in New Mexico, Utah, Colorado, North Dakota, and Montana; in Canada they are in Saskatchewan.

Phosphorus   Our source of phosphorus is the mineral apatite with the composition $Ca_5(PO_4)_3OH$. Our supplies come from certain marine sedimentary rocks where apatite has been concentrated by the sedimentary process. Minable deposits in the United States are found in Montana, Idaho, Wyoming, Florida, and North Carolina.

Sulfur   Sulfur is heavily used in the fertilizer industry to produce **superphosphate** and **ammonium sulfate.** Some deposits are in the form of the native element, but more quantitatively important is the sulfur in deposits of gypsum ($CaSO_4 \cdot H_2O$) associated with salt domes (see Section 18.3). Some of this gypsum has been reduced to sulfur by bacteria that use the oxygen in the sulfate. Sulfur is also recovered in the roasting of sulfide ores, in the burning of some natural gas, and from the burning of coal.

MINERALS FOR THE CHEMICAL INDUSTRY

The chemical industry depends on raw materials from the Earth. All of the minerals we have discussed in this chapter (as well as the many more that have gone unmentioned) provide elements for the myriad of chemicals used in the industry. We will not pursue this further here except to point again to the marine evaporites, particularly to one, halite (NaCl). Halite not only provides salt for our table but also serves as the raw material from which we manufacture such diverse chemical products as chlorine gas, soda ash (Na_2CO_3), and sodium sulfate (Na_2SO_4). The sources are both from geologic deposits, the remnant of ancient seas, and from the modern ocean. The supply is unlimited.

BUILDING MATERIALS

The crust of the Earth provides us with a tremendous variety of materials for construction. Some material is used as it comes from the Earth, modified only in shape and size. Other rock goes through extensive change as it is converted into essentially new material.

Building Stone   Until the widespread use of concrete and steel in construction, building stone was used as the structural support in buildings. Today steel and concrete form the framework for most structures and we use stone as an exterior facing and trim, as well as for interior paneling, steps, and decorative purposes. Most widely used are the sedimentary rocks limestone and sandstone, the igneous rock granite, and the metamorphic rocks marble and slate. But these by no means exhaust the list of stones in buildings. Builders have found uses for gneiss, anorthosite, basalt, indurated volcanic ash, conglomerate, peridotite, diabase, and many others.

Crushed Rock, Sand, and Gravel   The dollar value of crushed rock, sand, and gravel produced in this country annually is exceeded only by the value of the petroleum and natural gas extracted from the country's wells. The primary use of crushed rock is as an aggregate for concrete and as a base for streets and roads. Most crushed rock is limestone, but other common rocks such as granite, diabase, and well-cemented sandstone are also quarried for this purpose.

Sand and gravel are taken directly from stream beds, from the terraces of old streams, from the glacial outwash of now-vanished glaciers (Figure 19.9), from beaches, and even from the continental shelves. Sand and gravel are used chiefly as an aggregate for concrete and as a base for streets and roads.

Cement   Cement was discovered by the Romans, a discovery that made possible the creation of many of their monumental buildings. The Roman cement was a mix of quicklime (CaO), produced by heating limestone and a particular volcanic ash called **pozzuolana,** named after the town of Pozzuoli near Naples. When water was added to this mix, the result was a hard, strong product — **concrete.** Pozzuolana is still quarried from the volcanic slopes near Rome and Naples and is used in the manufacture of concrete. The use of cement declined along with the Roman Empire and was eventually forgotten. It was rediscovered in the mid-eighteenth century, this time in England.

Today's cement is made from a mixture of limestone and clay or shale, or from an impure limestone with a high clay content. This material is ground, heated almost to fusion, and then ground again to a fine powder. Mixed with water, it hardens and forms the binder or cement that holds together the aggregate of crushed rock, sand, or gravel that makes concrete. Recently a strong, quick-curing, chemically-resistant cement has been developed. It is made of the standard limestone and clay plus certain additives. Among these is pozzuolana.

Clay   Clay, molded into the desired form and heated to the proper temperature, produces a variety of products from bricks to fine porcelains, from tiles to ceramic pipes. It is used in the manufacture of paper, rubber, and plastics, for oil-drilling muds, and in the chemical industry.

We have found that **clay** is a term that can apply both to a specific group of minerals (the clay minerals) and to any mineral of a certain size (less than 0.004 mm). The clay minerals are generally sought for fine ceramics, chemical industries, and manufacturing. Minerals of many types, ground to clay size, are used in coarse pottery ware and in brick manufacture.

Clay forms from the weathering of rocks and their minerals and is a large component of the most abundant of the sedimentary rocks, shale (see Sections 5.2 and 6.3).

Plaster   Plaster is used mainly as a wall-surfacing material, most commonly now in prefabricated sheets of plaster board.

Gypsum ($CaSO_4 \cdot 2H_2O$) is the raw material for plaster. Gypsum is heated, that is, calcined, thus driving off some of the water and changing the composition to $2CaSO_4 \cdot H_2O$. The treated gypsum is then ground to a powder. When water is readded, the powder reverts back to gypsum, and a fine-grained, smooth-surfaced product results.

We have already discussed gypsum as a marine evaporite. It has been recorded in many places within the geologic column. It is very widespread indeed, so we need have no concern about its future supplies.

Glass   We need not detail the many uses of glass. We will note only that recent technology has increased the use of glass tremendously, including as a structural material.

We found in Section 3.3 that natural glass, obsidian, comes from a rock melt that cooled so rapidly that the atoms were unable to assemble into mineral crystals. Artificial glass is essentially the same, although quartz is the material used. Sources include both sandstone and unconsolidated sand deposits. It needs to be very pure quartz to be used as a glass sand. There is no problem with supply.

Asbestos   Asbestos is used in electrical and thermal insulation and is in addition very resistant to corrosion. Some forms of asbestos are so fibrous that the threads can be woven into fabrics. Its uses range from brake linings to fire-protective suits.

The major commercial source of asbestos is the mineral chrysotile ($Mg_3Si_2O_5(OH)_4$), which forms from the minerals in the rock peridotite. The largest reserves are in Quebec just across the border from Vermont, in the Union of South Africa, and in the Soviet Union. These reserves should be large enough to meet demand for many years.

Much concern has arisen recently about the use of asbestos because some forms of it are closely linked to certain lung cancers.

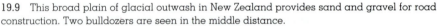

19.9   This broad plain of glacial outwash in New Zealand provides sand and gravel for road construction. Two bulldozers are seen in the middle distance.

GEMS

The qualities sought in **gems** are beauty, durability, and rarity. A gem's beauty depends upon its optical properties, including **color, luster** (how it looks in reflected light), and **fire** (how it transmits light). Durability is a function of the mineral's **hardness** (its scratchability as measured by the Mohs scale) and of its **lack of easy cleavage and fracture.** We usually separate gems into **precious stones** and **semiprecious stones.**

Precious Stones   **Diamonds** are considered the most precious of all stones. They are a form of pure carbon (C) and the hardest of all known natural substances. They are formed in nature under very high pressures in the upper mantle and are brought to the surface in conduits called **pipes** that carry a particular type of peridotite called **kimberlite.** The South African kimberlites are the best known and longest exploited, having been discovered in 1870 (Figure 19.10). Other kimberlites occur in various places around the world, some containing diamonds and others barren.

We had long believed that the diamonds were formed in the kimberlite pipes. Recently, however, it has been discovered that the kimberlites in South Africa are much younger than the diamonds found in them. This indicates that, at least in South Africa, the diamonds formed in the mantle and at a much later time were picked up by the rising kimberlite. What caused the kimberlite pipes is still obscure, but one theory suggests that they are the traces of hot spots or plumes as continental crust passed over them.

Although diamonds occur in kimberlites, it is weathering that eventually sets them free. Because they are heavier than the normal sand grain or pebble and also very tough, they collect in placers. Most production today is from placer deposits, both stream and beach.

Artificial diamonds are commercially available. They are usually considered suitable only for industrial purposes, but some gem-quality stones have been made.

Sapphires and rubies are varieties of the mineral corundum, (Al_2O_3), whose hardness of 9 on the Mohs scale is second only to that of diamond. The blue of sapphire is due to small amounts of iron and titanium substituting for aluminum. Natural stones are found chiefly in Burma, Thailand, Sri Lanka, India, and Australia, as well as in Montana. Ruby gets its red color from traces of chromium substituting for aluminum. Rubies are found primarily in Burma, Thailand, and Sri Lanka.

Corundum is made artificially for use as an abrasive. Artificial rubies and sapphires are also common.

Emerald is the gem variety of beryl ($Be_3Al_2Si_6O_{18}$) and has a hardness between 7 and 9 on the Mohs scale. The color is due to small amounts of chromium. Natural stones are associated with pegmatitic rocks, hydrothermal deposits, and biotite schists. The most famous emeralds have come from Colombia. Synthetic stones can be made.

Although not strictly a mineral because of its organic origin, a fine **pearl** is usually considered a precious stone.

19.10   The "Big Hole" at Kimberley, South Africa, was opened in the 1870s. The mine followed a "pipe" of kimberlite, an igneous intrusion that carried the diamonds. Before this now-flooded mine was closed in 1914, it had produced over 14.5 million carats of diamonds. [De Beers Consolidated Mines.]

Semiprecious Stones   Among the semiprecious stones are **aquamarine** (a blue variety of beryl), **amethyst** (purple quartz), **topaz, garnet, tourmaline, chrysoberyl, quartz, opal, moonstone,** and **jade.**

19.4

EARTH MATERIALS AND SOCIETY

We have already pointed out that no nation is completely self-sufficient in its mineral resources. This geographically patchy distribution of minerals leads us to consider the subject of **strategic minerals.** We define a strategic mineral as one that is vital to the security of a nation but is in short supply within the national boundaries, so that domestic production would be unable to meet the country's requirement in time of war. For instance, a number of elements can be classed as strategic for the United States. They include tin, chromium, manganese, and tungsten.

There are several ways in which the need for strategic minerals can be met. Substitutes can be found for the element involved. Or an intensive search can be carried on for

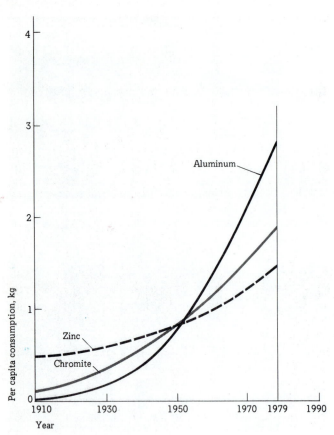

19.11 An example of the rapid increase in the per capita consumption of three different mineral products. [Data from the U.S. Bureau of Mines.]

the mineral within national boundaries, as the United States did for uranium minerals in World War II. Minerals can be stockpiled during times of peace, when excess supplies are obtainable from countries that have them in abundance. And of course conquest, either through colonies or military occupation, has been used.

The use of mineral resources has increased dramatically during the present century (Figure 19.11). In fact, we now recognize that some of these raw materials are being used up at such a rapid rate that they may not be available to us in the near future, or if they are, only at a very high cost. This raises a set of social, economic, and political problems.

Recently other questions have been raised about the exploitation of natural resources. Many of these are environmental questions that range from the detrimental effect of strip mining for coal to the health hazards related to the mining and use of asbestos.

Although the use of natural resources has increased worldwide, there are large discrepancies in the use and availability of these resources. Industrial, developing, and underdeveloped nations do not share equally in the resources on a per capita basis. In fact, many raw materials abundant in the less developed countries are largely consumed in the developed countries.

The solutions to the problems raised by these facts are not easily come by. But certainly any adequate solutions will demand a recognition and understanding of the origin and the geologic and geographic distribution of natural resources, which this book has attempted to provide.

Summary

Mineral deposits are generally classed as **metallic** or **nonmetallic** deposits.

Mining involves the removal of economically useful minerals from the Earth. An **ore** is an earth material that can be mined at a profit.

Geologic processes that concentrate mineral deposits include **igneous activity, hydrothermal activity, sedimentation, groundwater, weathering, placer formation,** and **rock formation.**

Metallic deposits of different types vary in their **temperature of formation.** Metallic ore minerals are generally sulfides or oxides. Many metallic deposits are associated with **plate tectonics,** particularly along **plate boundaries.**

Nonmetallic deposits include minerals used in agriculture, in the chemical and building industries, and as gems.

Fertilizers use soluble compounds of nitrogen, potassium, and phosphorous. Sulfur is used in superphosphates and ammonium sulfates.

The **chemical industry** relies on minerals and fuels for most of its raw materials.

Building stone, crushed rock, and **sand and gravel** are used in construction more or less as they come from the ground, except for some modification in shape and size.

Cement, clay, plaster, glass, and **asbestos** are building materials that have been extensively modified from nonmetallic Earth materials.

Gems are classed on the basis of their beauty, durability, and rarity. **Precious stones** include diamonds, sapphires, rubies, and emeralds. **Semiprecious stones** include aquamarine, amethyst, topaz, garnet, tourmaline, chrysoberyl, garnet, opal, moonstone, and jade.

Earth materials involve societal problems, the answers to which must take into account the origin and the geologic and geographic distribution of Earth fuels and minerals.

Questions

1. What is an ore? How may a mineral deposit be raised to the rank of an ore deposit? How may an ore deposit cease to be classed as an ore deposit?

2. What are the various ways in which mineral deposits may be formed? Give an example of each.

3. Discuss the relation between plate tectonics and metallic deposits.

4. What is cement? Where do we get it?

5. What are the three most important elements in the manufacture of fertilizer? What is the chief source of each?

For what is sulfur used in making fertilizers? What are some of its sources?

6. What is plaster and what is the source of the material from which it is made?

7. Look around the room. What different things that contain material derived from rocks and/or minerals can you identify?

8. What is a strategic mineral? Name some strategic minerals for the United States.

Supplementary Readings

AGRICOLA, GEORGIUS *De Re Metallica,* trans. by Herbert Clark Hoover and Lou Henry Hoover from the first Latin edition of 1556, Dover Publications, New York, 1950.
Facsimile edition with introduction, annotations, and appendices by the translators. A scholarly rendition of the Renaissance classic on mining.

BROBST, D. A., AND W. P. PRATT (EDS.) "United States Mineral Resources," *U.S. Geol. Surv. Prof. Paper 820,* 1973.
An authoritative compendium of what we know we have and what we hope we have in the way of mineral resources.

COMMITTEE ON MINERAL RESOURCES AND THE ENVIRONMENT *Mineral Resources and the Environment,* National Academy of Sciences–National Research Council, Washington, 1975.
A thoughtful, informative, and readable discussion of the subject.

DENNEN, WILLIAM H. *Mineral Resources: Geology Exploration and Development,* Taylor and Francis, Philadelphia, 1989.
The nature of ore deposits, the role of the geologist in their discovery and the intricacies of the mineral industry are all examined.

DOUGLAS, R. J. W. (ED.) *Geology and Economic Minerals of Canada,* Economic Geology Report 1, Geologic Survey of Canada, Ottawa, 1970.
A good place to start for the Canadian picture.

EDWARDS, RICHARD, AND KEITH ATKINSON "Ore Deposit Geology and Its Influence on Exploration," Chapman and Hall, London, 1986.
Discusses and illustrates the various types of deposits from magmatic ores to placers.

GUILBERT, JOHN M., AND CHARLES F. PARK, JR. *The Geology of Ore Deposits,* W. H. Freeman and Co., San Francisco, 1986.
A standard textbook that deals with the principles of deposition of metallic ores and provides detailed descriptions of various mines representative of the different types of deposits.

HARBEN, PETER W., AND ROBERT L. BATES *Geology of the Nonmetallics,* Metal Bulletin, New York, 1980.
A worldwide geological inventory of 52 nonmetallic commodi-

ties, each with its own bibliography. This is the first place to turn if you need to know more about the nonmetals.

HUTCHISON, CHARLES S. *Economic Deposits and Their Tectonic Setting,* John Wiley & Sons, New York, 1983.
The emphasis is on minerals but one of the 12 chapters considers fuels. An advanced text, but usable by a student who has absorbed introductory geology.

MCLAREN, D. J., AND BRIAN J. SKINNER *Resources and World Development,* John Wiley & Sons, New York, 1987.
We cited this volume in the chapter on energy, and it is equally useful here. It is particularly valuable for readers concerned with the societal aspects of natural resources.

SAWKINS F. J. *Metal Deposits in Relation to Plate Tectonics,* Springer-Verlag, Berlin, 1984.
The title is properly descriptive: The author examines metallic deposits in the various settings of plate tectonics. An upper-level text.

SKINNER, BRIAN J. *Earth Resources,* 3rd ed., Prentice Hall, Englewood Cliffs, N.J., 1986.
An excellent compact review of the subject.

U.S. BUREAU OF MINES, *Minerals Yearbook, Vols. I and II: Metals, Minerals, and Fuels,* Washington, D.C. (issued annually).
Here are the details on the current United States situation.

WEBSTER, ROBERT *Gems: Their Sources, Descriptions and Identification,* 4th ed., rev. by B. W. Anderson, Butterworth and Co., Boston, 1983.
Here is a complete, lengthy, authoritative reference on precious and semiprecious stones, including synthetic and imitation gems.

YOUNG, OTIS E., JR. *Western Mining: An Informal Account of Precious-Metals Prospecting, Placering, Lode Mining, and Milling on the American Frontier from Spanish Times to 1893,* University of Oklahoma Press, Norman, 1970.
A well-written and illustrated book outlining the methods by which gold and silver were obtained in the American West.

Saturn as seen from *Voyager 1* on October 18, 1980. Three separate images, taken through ultraviolet, green, and violet filters, were used to construct this color composite of the planet. [NASA, *Voyager 1*.]

CHAPTER ▪ 20

PLANETS, MOONS, AND METEORITES

"But it does move!"

Abbé Irailh, Querelles litteraires, 1761, vol. III, p. 49. Attributed to Galileo Galilei (1564–1642) after he was forced to recant before the Inquisition in 1633 his belief that the Earth moved around the Sun.

Galileo's remark, apocryphal though it may be, is fitting, for the Earth does move around the Sun, and can in no way be considered the center of the universe.

In this chapter we look beyond Earth at our *Sun* and *Moon*, at the other *planets* and their moons, at *asteroids*, and at *meteorites* and *comets*.

As we examine the *inner, terrestrial planets,* we will find that the geologic principles we learned by observation of the Earth apply also to these members of the solar system. We will find similarities among the planets as well as real differences, but all will be generally explicable in the familiar context of the Earth.

The *giant outer planets* beyond the belt of asteroids are very different from the inner, smaller planets and are not explicable in earthly terms. Their moons, however, are solid bodies and our experience on Earth again applies, at least in part, as we examine the processes and histories of these distant satellites.

Meteorites, most of them probably from the asteroid belt, bring us samples from beyond our own planet. Because these samples date from the inception of the planetary system, they give us clues to the chemical and physical conditions attending the birth of the planetary system.

20.1

OUR SUN AND SOLAR SYSTEM

We are physically able to exist because of heat and light from our own star, the Sun. As we observe it from Earth, the Sun rises in the east to bring daylight and sets in the west to bring night. This phenomenon anciently led to the assumption that the Sun moves around a stationary Earth. Likewise, the stars appear to rise in the east and set in the west. Therefore, for a long segment of human history the Earth was believed to be at the center of the universe, with everything else revolving around it.

This belief, supported by noted Greek scientists who were dominated by the teachings of Aristotle, was widely held until about 3 centuries ago. In the third century B.C. one Greek, Aristarchus, had the temerity to suggest that the planets, including the Earth, circle about the Sun and that the Earth rotates on its axis, giving us night and day. But he failed to convince most of the people of his time, and his explanation for the movements of the planets around the Sun was not to be firmly established until 1543 — by Copernicus. Still another century passed, however, before there was universal acceptance of Aristarchus' idea.

We now know that the Sun is the center of our physical existence. Everything within the Sun's gravitational control constitutes what we call our **solar system** (Figure 20.1). The solar system includes nine planets, over three score moons, thousands of asteroids, millions of meteors, and billions of comets. All these revolve about the Sun, forming a system with a diameter that may exceed 14×10^{12} km.

The Sun is a star. Like all other stars, it is a hot, self-luminous ball of gas. Located 150 million km from the Earth, it has a diameter of about 1,391,000 km and a mass 332,000 times that of the Earth. It is the source of the solar

system's light, heat, and charged particles. Its gravitation holds the system together.

The bulk of the Sun is composed of two elements: Its volume is 81.76 percent hydrogen and 18.17 percent helium. All other elements total only 0.07 percent. The Sun's energy comes from the conversion of hydrogen to helium — the process of fusion that we are currently attempting to develop with the hydrogen of this planet into a source of energy for our earthly uses. The Sun is "burning" hydrogen at an estimated rate of 4 million t per second, producing an "ash" of helium. At this rate, the Sun has an estimated life of about 30 billion years. The continuous output of solar energy is estimated at 30×10^{16} W (watts).

The Sun is thought to have originated from the accumulation of material in interstellar space. In the beginning stages an early Sun, perhaps a part of what is called a **dust cloud,** grew by the gravitational attraction of the dust and gas surrounding it. The dust cloud assumed a more and more disklike shape that, when taken with the evolving Sun, we can refer to as a **nebula.**

The planets formed from the material in the disk in ways that are not yet completely understood. Be that as it may, the planets had assumed their primordial shapes by about 4.6 billion years ago. We are much less certain of the age of the universe, but most theories call for an origin between 12 and 15 billion years ago.

PLANETS AND SATELLITES

The name **planet** was given to certain celestial bodies that appear to wander about the sky, in contrast to the seemingly fixed stars. It came from the Greek *planēs,* meaning "wanderer." Our Earth is a planet, one of nine that circle the Sun. All travel around it from west to east in nearly the same plane, the **ecliptic.** And, with the exception of Venus,

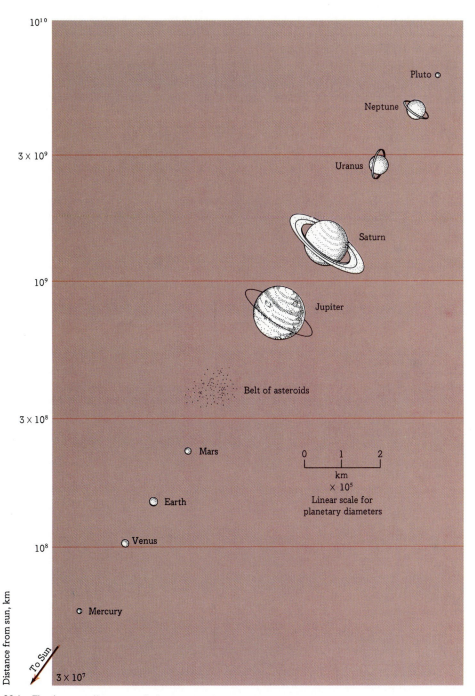

Distance from sun, km

10^{10}

Pluto ○

Neptune

3×10^9

Uranus

Saturn

10^9

Jupiter

Belt of asteroids

3×10^8

○ Mars

```
0   1   2
|---|---|
   km
  × 10⁵
Linear scale for
planetary diameters
```

○ Earth

10^8

○ Venus

○ Mercury

To Sun

3×10^7

20.1 The four small terrestrial planets are closest to the sun. Beyond the asteroid belt lie the giant planets and finally the terrestrial-like planet of Pluto. In this diagram distances from the Sun are given by a logarithmic scale and the diameters of the planets by an arithmetic scale.

Uranus, and Pluto, they rotate on their respective axes in a "forward" sense, or from west to east.

In order of distance from the Sun the first four planets are Mercury, Venus, Earth, and Mars. These planets are about the same size and fairly dense, as though they were made of iron and stone. They are called the **terrestrial planets** because of their similarity to the Earth. Next in order of distance from the Sun are Jupiter, Saturn, Uranus, Neptune, and Pluto. The first four of these are of relatively large size and low density. Pluto is solid and thought to be made of frozen methane (CH_4) and to have a very thin

atmosphere of the same composition. Its tilted and very elliptical orbit is such that sometimes the planet is inside the orbit of Neptune, as it was in 1979 and will be through 1999. A uniform pattern of spacing outward from the Sun is broken between Mars and Jupiter. In the "gap" are the thousands of asteroids, ranging in size from 2 km to about 771 km in diameter. These asteroids are believed to be either the remains of a planet that exploded or matter that never completed the planet-forming process.

Tables 20.1 and 20.2 list some of the characteristics of the planets.

TABLE 20.1
Orbital, Rotational, Dimensional, and Mass Characteristics of the Planets, Moon, and Sun[a]

Object	Mean distance to sun, km × 10^6	Diameter at equator, km	Mass, g	Mean density, (g/cm³)	Rotational Period, days	Obliquity, degrees	Revolution period, years (except moon)	Escape velocity, km/s
Sun	—	1,391,400	1.987×10^{33}	1.4	25.5	7.25	—	
Mercury	57.9	4,878	3.3×10^{26}	5.44	58.6	~2	0.2408	4.2
Venus	108.2	12,100	4.87×10^{27}	5.25	243 R	~179	0.6152	10
Earth	149.6	12,756	5.98×10^{27}	5.52	1.00	23.5	1.000	11.2
Mars	227.7	6,786	6.44×10^{26}	3.94	1.02	25.0	1.881	5.0
Jupiter	778.1	142,796	1.90×10^{30}	1.33	0.41	3.1	11.86	61
Saturn	1428.3	120,000	5.69×10^{29}	0.70	0.43	26.7	29.46	37
Uranus	2872.7	50,800	8.66×10^{28}	1.30	0.7 R	97.9	84.1	22
Neptune	4498.1	46,300	1.03×10^{29}	1.76	0.7	28.8	164.8	25
Pluto	5914.3	2,100	1.5×10^{25}	1.0?	6.4 R	118?	284.5	?
Moon	0.38[b]	3,476	7.35×10^{25}	3.34	27.3	6.7	23.3 days	2.4

[a] Compiled from various sources.
[b] Distance from Earth
R = retrograde motion.

TABLE 20.2
Some Characteristics of the Planets[a]

Planet	Number of satellites	Diameter of largest satellites, km	Magnetic field	Surface pressure in bars	Surface temp., °C	Major atmospheric gases, percent	Interior composition	Albedo	Acceleration of gravity, cm/s²
Mercury	0	—	Present	2×10^{-15}	350 d[b] −170 n	He 98 H 2	Silicates Fe core	0.07	3.95×10^2
Venus	0	—	Present	90	457	CO_2 96 N_2 3.5	Silicates Fe-Ni core	0.7	8.8×10^3
Earth	1	3,476	00.3–0.7 gauss	1	25	N_2 77 O_2 21 H_2O ~ 1[c] Ar 0.93	Silicates Fe-Ni core	~0.39[c]	9.78×10^2
Mars	2	18	Weak	0.007	−60	CO_2 95 N_2 2.7 Ar 1.6	Silicates Fe-FeS core	0.15	3.73×10^2
Jupiter	16	5,276	Strong	>> 100[d]	−140	H_2 ~ 89[c] He ~ 11[c]	H_2 He Rocky core	0.51	2.32×10^3
Saturn	23	5,140	Present	>> 100[d]	−175	H_2 ~ 87[c] He ~ 13[c]	Similar to Jupiter	~0.50[c]	8.77×10^2
Uranus	15	1,620	Present	>> 100[d]	−225	H_2 ~ 85[c] He ~ 15[c]	Similar to Jupiter	~0.66[c]	9.46×10^2
Neptune	8	2,700	?	>> 100[d]	−220	Similar to other large planets	Similar to Jupiter	~0.62[c]	1.37×10^3
Pluto	1	1,200?	?	?	−230?	CH_4?	CH_4	~0.16[c]	?

[a] Compiled from various sources.
[b] d = day; n = night.
[c] ~ = approximately.
[d] >> = much greater than.

20.2
MERCURY

Hermes was the messenger of the gods and the god of dawn and twilight in the Greek Pantheon; he was Sabuku to the ancient Egyptians; Woden, chief god of the hunt, to pagan Anglo-Saxons (hence Woden's day — our Wednesday); and Mercury to the Romans, their messenger of the Gods.

Mercury is the planet closest to the Sun, and the smallest of the inner terrestrial planets, being intermediate in size between our Moon and Mars. Appropriate to its name, it has the quickest orbit around the Sun of all the planets, just under 88 days.

The surface of Mercury is very heavily cratered (Figure 20.2) and in this respect looks very much like the surface of our own Moon. The largest of these craters, the Caloris Basin, is 1,300 km in diameter.

20.2 The cratered surface of Mercury is similar to that of our Moon and to some of the moons of Jupiter and Saturn. This detail of a portion of the planet also shows a cliff extending for over 300 km from lower left to upper right. [NASA, *Mariner 10*.]

Unlike the Moon, Mercury has a magnetic field, albeit weak, which strongly suggests that the planet has—or once had—a liquid core. Various lines of evidence indicate that Mercury contains a very high percent by weight of metal, most probably iron. Most of this is presumably in the core, which accounts for over 40 percent of the planet's volume. By comparison, the Earth's core is only 16 percent of our planet's volume.

The active events in the history of Mercury are confined largely to the first billion years or so of the planet. Robert G. Strom of the University of Arizona has suggested the following sequence of events. Just after its formation Mercury differentiated into a heavy core, a mantle of silicates, and a lighter-weight lithosphere, the last subjected to heavy bombardment by meteorites. The heat from the very large core melted the mantle and caused expansion of the planet. This placed the lithosphere under tensional stress and it fractured. The fractures thus formed allowed volcanic eruptions, which took place along with the continued heavy meteoritic bombardment of the surface. With time the cooling of the mantle allowed planetary contraction, which in turn placed the lithosphere under compressional stress. This caused thrust faulting and turned off the flow of lava. The large-impact crater, the Caloris Basin, formed about this time and very possibly caused new outpourings of lava. As the planet continued to cool and contract, these last volcanic fissures were sealed off. The period of heavy bombardment came to a close and the rest of Mercury's history has been marked by an ever-decreasing rate of bombardment by meteorites.

20.3

VENUS

Named for the Roman goddess of love and beauty, Venus, like Mercury, also appears as a morning and evening "star." It lies between Earth and Mercury and is the Earth's nearest planetary neighbor. It has been called the Earth's "twin" because of its almost identical size. Space probes, however, have shown us how different the two planets are.

The planet is shrouded in a thick blanket of clouds (Figure 20.3). This represents a heavy atmosphere composed chiefly of CO_2 (96 percent) and N_2 (about 3.5 percent), with minor amounts of H_2, SO_2, Ar, Ne, CO, HCl, HF, and some aerosols of sulfuric acid.

At the surface of Venus the atmosphere exerts a pressure more than 90 times that at the Earth's surface. There the temperatures are extremely high. On the side facing the Sun temperatures exceed 500°C; on the dark side of the planet they drop to something over 300°C. These elevated surface temperatures, the highest of all the planets, result from the "greenhouse effect" of the CO_2 in the Venusian atmosphere. Short-wave solar radiation passes through the atmosphere to the surface, where it is reradiated as long-wave radiation. This is in part absorbed by the CO_2, thus heating the atmosphere. Eventually a temperature balance is reached as incoming and outgoing radiation equalize. The end results are the very high surface and atmospheric temperatures.

The permanent cover of clouds has made it impossible to photograph the surface of Venus with conventional Earth-based or spacecraft cameras. But radar imagery from orbiting spacecraft has provided a tantalizing view of the planet's surface, and we have a few surface views from the Russian *Venera* landers.

The surface pictures of Venus show a litter of slabby, angular blocks separated by deposits of finer sandlike material. This is very similar to the Martian surface, discussed in Section 20.6.

Radar imagery from orbiting spacecraft has provided a blurred look at some of the major surface features of Venus, features known only vaguely from Earth-based observations. We seem to be seeing, among other things, a high plateau standing 5 to 10 km above the surrounding terrain. Features that include volcanoes, large-impact

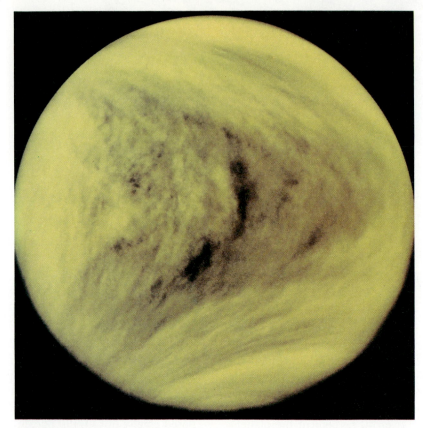

20.3 This computer-enhanced photomosaic showing Venus veiled in yellow clouds was compiled from images obtained February 19, 1979. The horizontal Y-shaped pattern is dimly visible in Earth-based telescopes. The mottled patterns in the center and to the left of center are believed to be convection cells in the Venusian atmosphere driven by the Sun's heat. [NASA, *Pioneer Venus Spacecraft.*]

craters, long ridges, and rift valleys are all tentatively identified. Positive resolution of these and other features will come only when and if radar-imaging equipment better-suited to the task is orbited around the planet. With imagery of sharp definition we may very well able to settle a major controversy about the planet: Has the process of plate tectonics been active on Venus?

20.4
THE EARTH

In the first 19 chapters of this book we examined in some detail the materials and processes of our own particular planet, Earth. Indeed we now know a great deal about how the Earth operates today and have been able to reconstruct its past with what would have seemed amazing precision to a nineteenth-century geologist. But the farther back in time we go, the more incomplete the record and the less secure we are in our reconstructions. In fact, for the very early history of the Earth we are dependent upon what we have been able to find out about the early histories of Mercury, Mars, and our Moon. It is to the last of these bodies that we now turn.

20.5
EARTH'S MOON

We probably know more about the Moon than any of the other bodies in the planetary system with the exception of Earth. The first man to see the Moon with other than the naked eye was the Italian scientist Galileo Galilei (1564–1642). A Dutch lens grinder had just constructed the first telescope, and Galileo, imitating it, built his own instrument. Early in 1609 Galileo, then a mathematics professor at the University of Padua, turned his telescope on, among other celestial objects, the Moon. What he saw is published in a treatise, *Sidereus nuncius (The Starry Messenger),* one of the great landmarks of scientific discovery. For the first time man found that, as Galileo wrote, the "Moon is not robed in a smooth and polished surface, but is rough and uneven" (Figure 20.4). In the last three and one-half centuries better telescopes, the camera, spacecraft and their instruments, and direct observation have given us a very good idea of what the Moon is like and how it got that way (Figure 20.5).

The Moon's diameter is 3,476 km and its mass is 7.35×10^{25} kg, two orders of magnitude less than that of the Earth. Its mean density is 3.34.

The major surface divisions of the Moon are the dark,

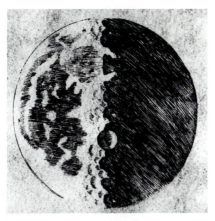

20.4 These illustrations from *The Starry Messenger* show some of what Galileo saw through his telescope. [Willard Starks.]

topographically lower **maria** or "seas" and the higher, lighter-toned **highlands.** A heavily cratered surface is the trademark of the moonscape. These craters range in size from a few micrometers to the gigantic Mare Imbrium, over 1,000 km in diameter. There are two types of craters: (1) the **primary craters,** formed by the original impact of meteorites; and (2) the **secondary,** or **satellite, craters** made by falling fragments ejected from a primary crater.

Copernicus, a well-known, fairly large primary crater, shows many of the features characteristic of Moon craters.

It measures about 90 km in diameter and over 3 km in depth. The relatively level floor has a raised center portion, a feature that is more pronounced as a central peak in somewhat smaller craters. The walls facing into the crater are terraced or stepped and represent the slumping of large slices of the crater walls into the central depression. The rim of the crater is about 1 km high and slopes away from the crest at the crater's edge, to feather out at the edge of an irregular zone up to 150 km wide and concentric to the crater.

20.5 The Moon as seen through a present-day telescope. North is to the top, the last quarter is to the left, and the first quarter is to the right. The lighter zones are the highlands of the Moon, and the darker zones are the topographically lower, younger maria. [Lick Observatory.]

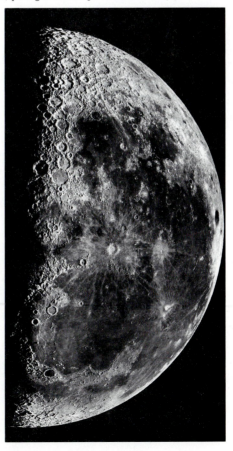

The material that forms the rim is zoned into three rings. The interior zone is very hummocky, with ridges concentric to the crater. Outward beyond this is a less hummocky intermediate zone, and the topography is one of branching ridges more or less radial to the central crater. The outermost zone is still less hummocky, but the radial ridges of the intermediate zone persist in subdued form. This outermost zone is pocked by many small satellite craters thought to have been formed by fragments thrown out of the primary crater. Superimposed on the rim material and spreading far beyond it to a distance of 500 km from the crater are light-colored, discontinuous streaks, the **rays** of Copernicus (Figure 20.6).

Copernicus is interpreted by most observers as an **impact crater.** The central rise on the floor of the crater is thought to result from upward adjustment following impact. The rim is thought to have been pushed upward and outward as a result of the impact explosion, and the hummocky ridges of the interior zone of the rim material are thought to stem from the same cause. Outward from this zone the areas of radial ridges are interpreted as material ejected from the crater and arranged in ridges subparallel to the direction of blast. The ray material is believed to be finely crushed matter thrown out from the primary crater and, to a lesser extent, similar material kicked up during formation of the satellite craters. Copernicus is one of a

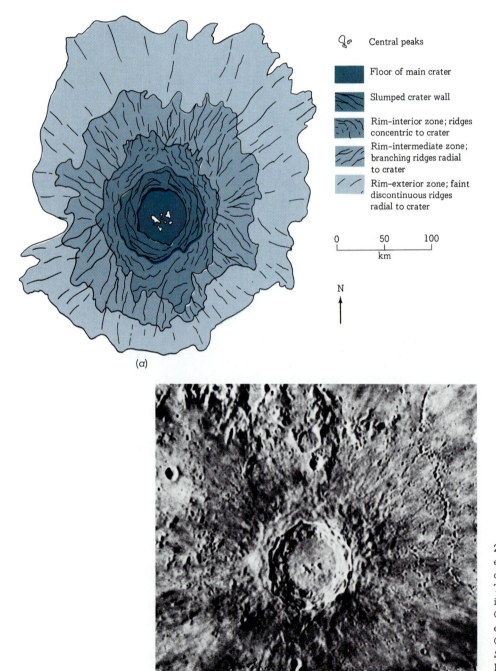

Central peaks

Floor of main crater

Slumped crater wall

Rim-interior zone; ridges concentric to crater

Rim-intermediate zone; branching ridges radial to crater

Rim-exterior zone; faint discontinuous ridges radial to crater

0   50   100
km

N

(a)

(b)

20.6 (a) Map of Copernicus, showing the extent of the major zones of the central crater. (b) Telescopic view of Copernicus. The rays and smaller craters can be seen in the photograph. [Mount Wilson Observatory. Map based on H. H. Schmitt et al., "Geologic Map of the Copernicus Quadrangle of the Moon," *U.S. Geol. Surv. Atlas of the Moon,* map I-515, LAC-58, 1967.]

number of **ray craters** on the Moon. Most craters, however, lack rays. Their absence is attributed to the greater age—hence longer weathering—of the craters.

Some craters show only a low lip for a rim. In general, these are old craters all but buried by material ejected from younger craters or by lava flows that filled the maria.

MOON ROCKS

The several hundred kilograms of material returned to the Earth from the Moon by the *Apollo* missions have shown us that four general rock materials dominate the lunar surface.

Basaltic rocks are common and seem to form the greater part of the filling of the many maria. These basalts differ slightly from those we find on Earth. They contain, for instance, more titanium than our earthly basalts and are somewhat lower in silica, alumina, and sodium. Like all lunar samples, they contain essentially no water.

The early *Apollo* missions brought samples of basalt and its plutonic equivalent, **gabbro.** In addition, they brought us a third rock type, a **breccia.** These breccias contain fragments of basalt, gabbro, and a fourth rock type, **anorthosite.** We theorize that these rocks, composed of the fragments of other rocks, were consolidated by compaction arising from meteoritic impact.

The discovery of anorthosite fragments early in the *Apollo* program led to the speculation that there were large masses of this material on the Moon. Anorthosite, a rock composed largely of the feldspar mineral anorthite, $Ca(Al_2Si_2O_8)$, usually contains a small amount of olivine and pyroxene. Anorthosite has a somewhat lower specific gravity than the basalt-gabbro group. Various lines of evidence lead us to believe that the highland areas of the Moon are composed of anorthosite.

In addition to the rocks already mentioned, there is a lunar "soil," which covers much of the Moon's surface. This is not a soil as we think of soil on Earth. It is made up of fragments of crystalline rocks, small glassy objects, and some meteoritic material. The assemblage appears to be the result of debris created by meteoritic impact.

LUNAR HISTORY

Even though we cannot yet determine the origin and earliest history of the Moon to the satisfaction of most students of that body, we have a fairly good idea of a great deal of the Moon's history. Much of this story has been worked out by geologists of the United States Geological Survey working from photographs taken through Earth-based telescopes and cameras carried in space probes. Rock units have been mapped on the basis of relative age, surface characteristics, and genetic types. Ages are established by cross-cutting relationships and superposition, stratigraphic principles discussed in Chapter 8 on geologic time.

Age determinations on lunar materials now give us some absolute ages that can be fitted into the relative chronology of events established by these other methods. The first samples returned to Earth were of maria material, which turned out to have an age of about 3.6 billion years, at that time older than any known Earth rock. Subsequent dating has produced ages from about 3.2 billion years back to 4.6 billion years for different materials and events.

As we put together what we know about lunar history, we can construct the following thumbnail biography of the Moon:

The Moon formed some 4.6 billion years ago, along with the Earth and the other planets. The details of the Moon's birth are still argued. Some think that it was born, as was the Earth, from a portion of the solar dust cloud. Others think the evidence points to a very early, giant impact event that spread material from a newly created Earth and that some of this matter coagulated into our Moon. Once formed, however, the Moon heated up quickly and differentiation began. A crust of anorthosite rose to the surface, and heavier material formed a core that, according to the testimony provided by remanent magnetism, was molten. Massive bombardment by meteorites characterized the first 1.5 billion years. The great basins—the maria—were excavated by the impact of enormous meteorites. The undestroyed highland sections of anorthosite crust were deeply pocked by smaller meteorites. The large-impact basins were eventually filled by basaltic materials, the bulk of the eruptions being between 3.2 and 3.8 billion years ago. Thereafter major volcanism ceased, and most students of the question believe that this marks the cooling of the Moon and the solidification of a small liquid core. This cooling also marks (but did not cause) a decrease in the numbers and sizes of meteorites impacting the Moon. True, impressive craters have formed since then (as, for example, Copernicus and Tycho), but the period of mare-producing impacts had ceased.

Today the Moon is a very quiet celestial body. There are few moonquakes (these are tidally induced or represent small meteoritic impacts), little if any volcanism, no mountain building, no atmosphere, no water, no life.

20.6
MARS

The red planet was named for Mars, the Roman god of war, who gave his name also to our month of March. Mars is the outermost of the terrestrial planets.

Telescopic observation of Mars in the nineteenth century led some to believe in the presence there of long, straight "canals" that some unknown civilization had built to bring water from the poles to the arid regions of lower latitudes. Although twentieth-century spacecraft explorations have laid this thought to rest, they have confirmed the presence of polar icecaps and of planet-obscuring dust storms, which earlier telescopic work had reported. The biological experiments conducted by Martian landers have not identified past or present life.

Our first spacecraft pictures of Mars came from *Mariner 4*, which photographed the red planet in a flyby mission in 1965. By November 1971 *Mariner 9* had been inserted into orbit around Mars, and after the subsidence of a violent dust storm it relayed to us images that revealed a much more varied planet than we had expected. More information arrived from *Viking 1* and *2* orbiters and from their landing vehicles, which were successfully set down in July and September 1976.

THE MARTIAN LANDSCAPE

We can divide Mars into two hemispheres: one of old, heavily cratered plains and the other of younger, lower, and lightly cratered plains (Figure 20.7). The boundary between the two is inclined at about 50° to the Martian equator. By extrapolating from cratered surfaces of known age on the Moon, we believe that the heavily cratered surface of Mars is about 3.5 billion years old. The lightly cratered plains are estimated to be generally younger than 1.5 billion years.

Before spacecraft had circled Mars, the telescope had already told us of the existence of polar icecaps that waxed and waned with the seasons. At that time they were thought to be made of solid CO_2 (dry ice). So one surprise provided by the *Viking* orbiters was that the northern cap is H_2O ice, and only the southern cap may contain CO_2 ice.

The major activity on the Martian surface today is wind action. The atmosphere, although very thin, does transport dust in suspension and moves sand grains by saltation. In many places today we are able to identify large fields of sand dunes (Figure 20.8).

In the past a number of other processes had major roles in shaping the Martian surface. These included cratering by the impact of meteorites, the record being particularly extensive in the uplands.

Sometime after the major period of meteorite bombardment extensive volcanism took place. The largest volcano, Olympus Mons, measures more than 500 km across its base and stands about 27 km in height (Figure 20.9). As such, it is several times larger than any volcano on Earth. And Olympus Mons is but one of nearly two score volcanic centers known on Mars. Olympus Mons and several other large volcanoes are located on a broad topographic swell or dome known as Tharsis Montes, a feature that seems to indicate unwarping, perhaps similar to (but larger than) those over the hot spots known on Earth. Around the margins of this uplift are fractures and faults. But we see no evidence of plate movement or of the associated mountain belts that we have on Earth.

The volcanic slopes are comparatively fresh, and the sparseness of craters suggests that the volcanism can be as young as 100 to 200 million years.

Among the most puzzling and startling features on Mars is the evidence of running water on the planet's surface. Associated with the boundary between the lower sparsely cratered plains and the higher heavily cratered plains are great canyons and sinuous channels.

20.7   Generalized physiographic map of Mars. [Adapted from a more detailed map by T. A. Mutch and R. S. Saunders.]

☐ Lightly cratered plains

▨ Old, highly cratered plains

▧ Young volcaneous

Permanent ice

Olympus Mons
Ascraeus Mons
Pavonis Mons
Tharsis Montes
Arsia Mons
Valles Marineris

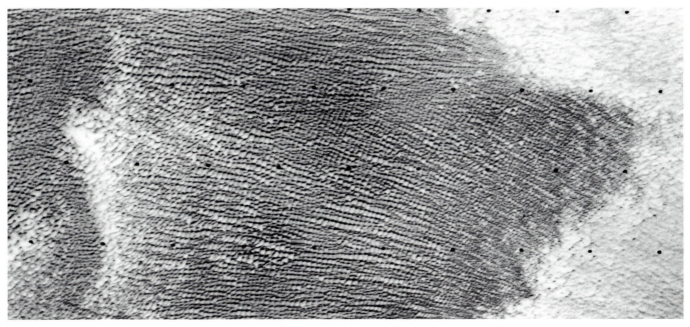

20.8 Particles moved by the wind can totally obscure Mars and in some places large areas of dunes occur on its surface. The dark linear pattern in this image of Mars records extensive sand dunes in the high northern latitudes of the planets near the edge of the polar icecap. The area is about 60 km across. [NASA, *Viking 1.*]

20.9 Olympus Mons is the largest known volcano in our planetary system, measuring over 500 km across its base and standing about 27 km above the surrounding plains. Its complex, multiple crater measures 40 km in diameter. [NASA, *Viking Orbiter.*]

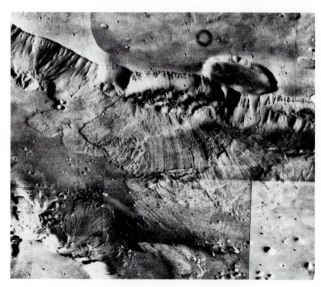

20.10 A view across the Valles Marineris shows a series of slump blocks along the canyon walls. Below the slumps stretch aprons of material that seem to have behaved as gigantic mudflows. At the bottom of the canyon obstacles have created wind shadows in which wind-driven material has accumulated. In the cliffs, which measure several hundreds of meters in height, differential erosion has etched out nearly horizontal layers of rock. The valley is about 130 km wide. [NASA, *Viking 1.*]

The canyons, such as Coprates Canyon and Valles Marineris, are huge. Valles Marineris (Figure 20.10) is 125 km wide and about 6 km deep and stretches for nearly 5,000 km. Such a feature seems to be a gigantic rift and points to vertical movements of the Martian crust. Dendritic canyons feed into the main valley, as shown in Figure 20.11. In addition, sinuous, even meandering, channels are also found, as well as wide, braided streamways. These features, taken together, indicate that running water has played an important role in fashioning the Martian surface at one or more times in the history of the planet.

Various lines of evidence point to the presence of ground ice beneath the Martian surface. Its melting has produced such features as karstlike topography and debris flows, and may have provided a source of stream water.

All in all, Mars turns out to have been a very dynamic planet: not as dynamic as Earth, but certainly more dynamic than Mercury. From the imagery sent back to us by various spacecraft we can suggest something about the past processes that have affected the planet, as shown in Figure 20.12.

20.7
OVERVIEW OF OUR MOON AND THE TERRESTRIAL PLANETS

We should not be surprised that the terrestrial planets have some history and processes in common and some that are unique to each planet.

IMPACT

We have found that the Moon, Mercury, and Mars all experienced extremely heavy bombardment early in their history. Because Earth and Venus lie between Mars and Mercury in the planetary sequence, we can be certain that they had a similar early history. Extrapolating from the dated sequence of events on the Moon, we can say that this period of intense early bombardment ended about 3.8 billion years ago.

The record of impact cratering continues but at a declining rate on the Moon, Mercury, and Mars. We can assume that a similar decline took place on the Earth and Venus. In the areas of Precambrian rocks of Earth we find the blurred scars of old cratering events. We find also that meteorites are still impacting the Earth, although most of

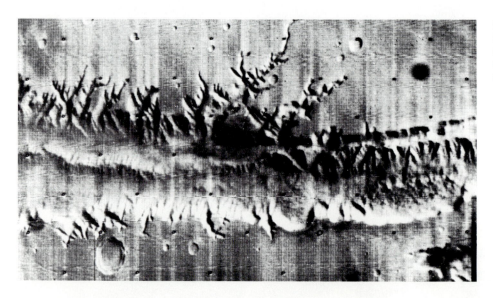

20.11 Part of Valles Marineris. Its width in this area is about 110 km. Note the dendritic canyons, particularly in the upper center of the area, which feed into the main valley. [NASA, *Mariner 9.*]

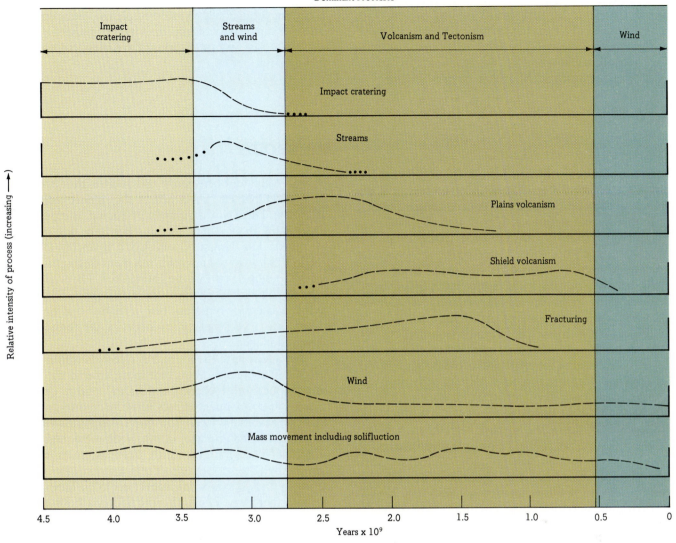

20.12 The intensities of various geologic processes on Mars have varied with time, as suggested in this diagram. The dominant processes at various stages in Martian history are listed across the top. Individual processes are shown below. Intensity scales are relative only. [Modified from a diagram by Lisa Rossbacher.]

the more recent ones have been small compared to those of early planetary time. We know too little about the surface of Venus to say much about its cratering record, although radar imaging reveals circular patterns that may well be the record of impact craters.

WEATHERING AND EROSION

The processes of weathering and erosion so familiar to us depend upon an atmosphere—in the case of Earth, an atmosphere rich in oxygen and water. Mercury and the moon have essentially no atmosphere. Mars has a very thin atmosphere. The atmosphere of Venus is rich in carbon dioxide.

On Earth the processes of weathering very rapidly break down the minerals of the crustal rocks, converting them into unconsolidated sediments and, most usually, into new minerals. Gravity and running water move these sediments through the rivers to the settling basis of the ocean. Here they continue their course through the rock cycle, emerging again at some future time in the zone of weathering (Figure 1.9). The Earth is unique in the rapidity of these processes of weathering and transportation. But it is also unique in that plate tectonics continuously raise rock material above sea level into the zones of weathering and transportation.

The rapidity of this recycling of rock material is characteristic of Earth. Mercury and the Moon are essentially without surface weathering and transportation today. On Mars the only surficial activities we can document as occurring today are wind action and the waxing and waning of the polar icecaps. We have found, however, that water was active on Mars in the past, and even today water probably exists as ground ice in permanently frozen ground. The atmospheric veil of Venus has so far denied clear views of that planet. When we finally penetrate the Venusian atmosphere, we will probably find that the intensity of its surface activity is closer to that of Mars than to that of Earth.

TECTONICS

The nature of tectonic movement seems to be different for each of the inner planets. On Earth the process of plate tectonics is dominant. We have given considerable attention to this style of tectonic activity in previous chapters, particularly Chapter 11. A worldwide system of plates forms the Earth's crust. These plates are defined by active boundaries in which compression dominates along converging boundaries, tension along diverging boundaries, and shear along transform margins. Here is where most earthquake and volcanic activity is concentrated.

When we turn to Mercury we find that tectonic activity has been confined to the planet's early history and seems to have been related to the very large core that forms such a large percent of Mercury's volume. Tectonic activity on the Moon also was early in that body's history; it seems to have been connected to the giant outpourings of lava. On Mars identifiable deformation is associated with the giant bulge of Tharsis Montes and is marked by features produced by both tensional and compressional stresses. Through the atmosphere of Venus we can dimly see craters, ridges, highlands, and lowlands, but we cannot yet make out enough detail to reconstruct either the style or history of the planet's tectonism.

VOLCANIC ACTIVITY

The Moon, Mars, Mercury, Earth, and presumably Venus all experienced volcanic activity in the past, but only Earth can undeniably be shown to have present-day volcanism.

On Mercury volcanism was associated with the heat generated by its very large core, and later by the creation of the large Caloris impact basin. Presumably the lavas were basaltic in composition. On the Moon the volcanism was of the flood basalt variety, the great bulk of the flows occurring between 3.8 and 3.2 billion years ago, probably about the same time as the volcanism on Mercury. But on the Moon, in contrast to Mercury, volcanism continued on down to between 1 and 2 billion years ago. On Mars large shield volcanoes have developed and Olympus Mons extends nearly 27 km above its surroundings. The great size of these volcanoes is in part explained by the absence of plate tectonics on the planet. On Mars volcanoes are stationary over their sources rather than located on a plate that moves over them, as is true on Earth. In addition to the shield volcanoes

there have been periods of basaltic flows, particularly in the northern, younger, lightly cratered plains. Venus undoubtedly has supported volcanic activity, but present data are too few to say very much about the timing and nature of that activity.

20.8
ASTEROIDS

Beyond Mars and between it and Jupiter, the first of the giant outer planets, lies a belt of tiny planets called the **asteroids.** There are thousands of known asteroids, and presumably many times this number have not yet been identified. The largest, Ceres, has a diameter of about 770 km. Some of the asteroids represent fragments resulting from collisions within the asteroid belt. The origin of the asteroid belt itself is still uncertain but it is generally thought that the asteroids represent bits of matter that never accreted into a single planetary mass, or the remains of a shattered planet.

In addition to the myriad of asteroids within the asteroid belt there are others that move around the Sun in very elliptical orbits that cross the Earth's orbit. One such asteroid had an unexpectedly close encounter with the Earth on March 23, 1989. The object, designated 1989 FC, had an estimated diameter of over a kilometer and passed within 765,000 km of the Earth. These asteroids orbit the Sun annually so we can expect that asteroid 1989 FC will return in future years, sometimes closer to the Earth, sometimes farther away. The further prediction is that sometime in the far distant future it will impact Mars or Earth.

This brings us to the subject of meteorites. They have their origin as asteroids or, perhaps in some cases, as comets. We discuss these visitors from space in some detail in Section 20.13.

20.9
JUPITER

In Roman mythology Jupiter was the son of Saturn and king of all the gods. In our planetary system Jupiter is one of the four giant outer planets, the others being Saturn, Uranus, and Neptune.

Layers of dense clouds surround the planet (Figures 20.13 and 20.14). These are made up primarily of hydrogen gas and secondarily of helium. The patterns seen at the top of these clouds reflect a complex system of atmospheric circulation. We have never penetrated to the surface beneath this cloudy shroud. Indeed, there is very probably no surface similar to that of the terrestrial planets. Rather, most of the planet could be liquid hydrogen with a core of iron-rich rock.

20.13 Jupiter as seen from a distance of 28.4 million km through the cameras of *Voyager 1* on February 5, 1979. The photo shows a swirling, turbulent atmosphere composed largely of hydrogen. The large, eyelike feature in the lower left—the Great Red Spot of Jupiter— has been known from ground-based observations for over 300 years. Jupiter's 30-kilometer thick ring is not visible in this view but two of its four largest moons are. Io can be seen against Jupiter's disk and Europa is the small bright sphere to the right of the planet. [NASA, *Voyager 1*.]

20.14 The photo of Jupiter's Great Red Spot (upper right) was taken on March 1, 1979, when the spacecraft was 5 million km from the planet. It shows some of the complexity of the turbulence in the Jovian atmosphere. [NASA, *Voyager 1*.]

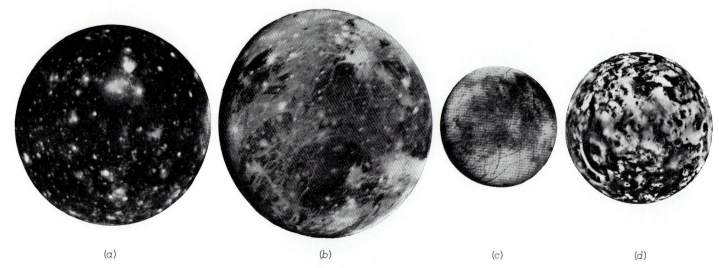

(a)   (b)   (c)   (d)

20.15   Jupiter has 16 moons. The four largest are known as the Galilean moons, after Galileo, who discovered them in 1610. Here they are shown in their correct relative size. Callisto (a) is farthest from Jupiter. Ganymede (b), Europa (c), and Io (d) are progressively closer to the planet. Ganymede and Callisto are larger than the planet Mercury, and Europa and Io are about the size of our Moon. Both Ganymede and Callisto appear to have a crust of ice and rock. Europa's crust is an intricately fractured rind of ice. Io, volcanically the most active object yet known in the planetary system, has long-lived volcanoes that are apparently fueled by sulfur. [NASA, *Voyagers 1 and 2.*]

Although we have not seen beneath Jupiter's cloud cover, spacecraft have provided us with excellent views of Jupiter's moons. The four largest satellites of the planet are known as the **Galilean moons** because of their discovery by Galileo in 1610. In order of decreasing size, they are Ganymede, Callisto, Io, and Europa (Figure 20.15).

Ganymede, 5,276 km in diameter, exhibits a cratered surface imposed on a mixture of rock and ice, a puzzling grooved terrain, and evidence that portions of the surface are cut by transverse faults (Figure 20.16).

Callisto, 4,820 km in diameter, like Ganymede, seems to have an outer crust of rock and ice. The surface is heavily cratered and very possibly represents one of the oldest surfaces in the planetary system.

Io, the innermost of the Galilean moons, is 3,630 km in diameter and is marked by a bright yellow-orange equatorial band and reddish-brown polar zones. This remarkable body is the only one in the planetary system (with the exception of our Earth) to have present-day volcanic activity (Figures 20.17 and 20.18). But the volcanism on Io is very different from that on Earth. Individual eruptions may last for several months or more, and appear to involve burning sulfur. The energy needed to initiate eruptions appears to originate in the stretching of the moon under the gravitational attraction of the Sun and Neptune. The result is a surface that is young when compared with Ganymede or Callisto or with our Moon or Mercury.

Europa, 3,126 km in diameter, exhibits a very young surface crisscrossed by what appear to be long fractures (Figure 20.19). A few impact craters are present, but again, their scarcity suggests a young surface. It is speculated that the outer 100 km or so of the planet is an icy rind over a

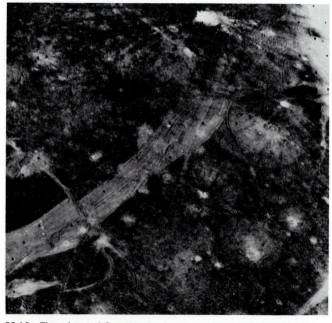

20.16   This photo of Ganymede shows a bright zone of grooved terrain about 150 km wide. It is offset as if by faulting. Craters and their ejecta exhibit variations in tone from dark to light, which may reflect variations in layering of surface material. [NASA, *Voyager 2.*]

silicate crust. If this is true, the reticulated pattern may be due to fractures in the rind. Furthermore, old crater scars may have been erased by flow of the ice substrate, thereby accounting for their low incidence.

The *Voyager* missions also discovered that Jupiter has a ring, about 30 km thick, in its equatorial plane.

20.17 On March 4, 1979, a few hours before its closest approach to Io, *Voyager 1* took this picture of a volcanic eruption on the innermost of the Galilean moons. The plume of the eruption can be seen on the bright edge or limb of Io and towers over 150 km above its surface. This eruption was still active 4 months later when *Voyager 2* passed by. It was one of 6 such eruptions identified by the *Voyager* missions. Many of the patterns on the surface of Io represent volcanic activity. [NASA, *Voyager 1*.]

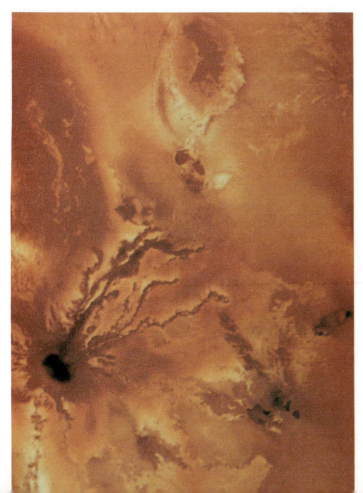

20.18 This picture of the surface of Io was taken March 5, 1979. The spiderlike pattern with a dark center is thought to be a volcanic crater with lava flows radiating from it. Evidence points to the fact that sulfur is intimately associated with the volcanism on Io. The width pictured is about 1,000 km. [NASA, *Voyager 1*.]

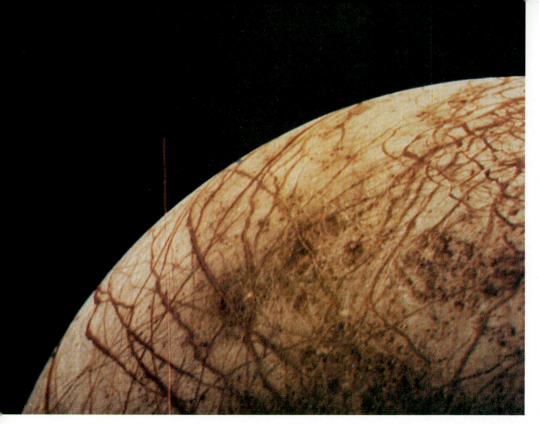

20.19 Europa's surface is thought to consist predominantly of water ice. Here we see a view of its strangely marked surface of intersecting, cracklike lineations. Adjacent dark-speckled zones may represent cratered terrain. But large craters are absent, suggesting that the surface of Europa is young when compared with the surfaces of Ganymede and Callisto. [NASA, *Voyager 2.*]

20.10

SATURN

Saturn was named for an early Roman god of fertility, a mythological figure from whom we also have the word **saturnalia** to designate an event of exuberant revelry. Space-craft flybys in 1979 *(Pioneer 11)* and in 1980 and 1981 *(Voyagers 1* and *2)* have added a wealth of information to our understanding of the ringed planet.

Like Jupiter, Saturn has a thick and turbulent atmosphere composed primarily of hydrogen. This atmosphere, some 1,000 km thick, covers a surface of liquid helium, which at greater depths becomes metallic. This in turn may envelop a rocky core.

The moons and rings of Saturn, long known from Earth-based observations, provided spectacular shows for the *Voyager* encounters. The three rings of icy particles, first identified from Earth, are now seen to be made of 500 to 1,000 or more rings (Figure 20.20). The moons of Sa-

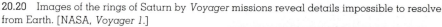

20.20 Images of the rings of Saturn by *Voyager* missions reveal details impossible to resolve from Earth. [NASA, *Voyager 1.*]

20.21 In this montage the artist has shown Saturn, its rings, and its moons, which together resemble a miniature solar system. Ringed Saturn is partially obscured by the moon Dione in the foreground. Tethys and Mimas fade away to the right, and Enceladus and Rhea to the left. Titan is above and beyond Saturn to the right. [NASA.]

20.22 Rhea is the most heavily cratered of Saturn's moons. Its surface may be the oldest in the planetary system. [NASA, *Voyager 1*.]

turn, from giant Titan to tiny Phoebe, give Saturn the look of a miniature planetary system, even more so if one considers the rings an analogue to the belt of asteroids (Figure 20.21). In this regard Saturn mimics Jupiter, although on a smaller scale.

The moons of Saturn have surfaces of ice mixed with rock. Cratered landscape is common (Figure 20.22) and reminds us of our own Moon, as well as of Mercury, portions of Mars, and the outer moons of Jupiter. Only Titan's surface is obscured from us. It is large enough to maintain an atmosphere of nitrogen with about 1 percent methane.

20.11

URANUS

Uranus was discovered in 1781 by the English astronomer and musician William Herschel, who named it the Georgian Star in honor of his royal patron, King George III. Later, following the convention of naming planets after mythological figures, the name Uranus was assigned. Uranus, the god of the heavens, and Gaea, goddess of Earth, were the two original deities of Greek mythology. Cronus (in Latin, Saturn) was their son and Zeus (in Latin, Jupiter) was their grandson and chief among the 12 Olympian gods.

The spacecraft *Voyager 2,* nearly 8½ years after its launching, had its closest encounter with Uranus on January 24, 1986. The data returned by this only planned spacecraft visit to the planet revealed a startling number of new facts. Uranus apparently has a core of rocky material about the size of Earth. But most of the planet is a cloud of hydrogen (about 87 percent) and helium (about 13 percent). A zone of liquid hydrogen may separate this gaseous envelope from the rocky core.

Uranus turns out to have a magnetic field almost as strong as that of Earth. Surprisingly, this field is offset from the rotational pole by approximately 55°, an apparently unique situation among the planets.

Voyager 2 showed that the 5 moons of Uranus previously known from ground-based observations are accompanied by 10 smaller companions. The larger moons resemble the icy satellites of Jupiter and Saturn and are probably a mixture of water ice and rock. Their surfaces reveal a wide variety of features, including the familiar patterns of impact craters and fault scarps bordering linear rift basins. The strangest topography is found on Miranda (Figure 20.23).

20.12

NEPTUNE AND BEYOND

Beyond Uranus lie the planets of Neptune and Pluto. *Voyager 2* received a gravitational boost from Uranus that sent it on its way to Neptune for a close encounter on August 24, 1989, 12 years and 4 days after its departure from Earth. This gave us our first close look of that planet and its moons. See Box 20.1.

With the Neptune encounter, spacecraft have visited all of the planets in the solar system, except Pluto, in the program of space exploration that began with the launching of a Russian earth-orbiting satellite on October 4, 1957.

Beyond Neptune, *Voyager 2* is scheduled to leave the plane of the solar system and join *Pioneers 10* and *11* and *Voyager 1* in a search for the **heliopause,** the limit of the sun's magnetic field. We estimate that one of these spacecrafts will cross the heliopause between 2005 and 2015.

20.13

METEORITES

Those streaks of light that we often call "falling stars" are bits of rock that enter our atmosphere and there burn as they are ignited by friction. These are properly called **meteors.** If the rock is large enough (probably over 1,000 kg), some of it will survive passage through our atmosphere and impact our Earth. Those that do are called **meteorites.** The very large ones will explode upon impact and leave a crater

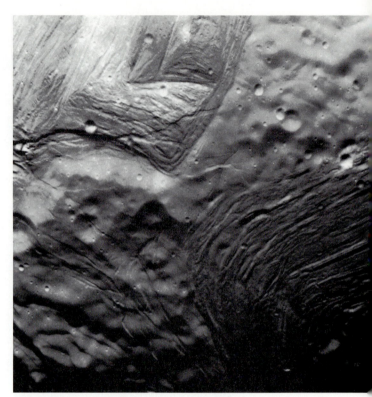

20.23   The surface of Miranda, one of the moons of Uranus, has a puzzling topography unknown elsewhere in the solar system. Chevron patterns of ridges and valleys (upper left and lower right) are separated by a zone of rolling terrain. Relatively fresh craters are superimposed on both types of terrain, as are long, generally linear fault scarps. The picture is about 220 km across. [NASA, *Voyager 2.*]

as the primary evidence of their arrival. Such craters are called **astroblemes** (from the Greek for "star" and the Middle English for "wound" or "disfigurement").

The space-exploration program that began in the mid-twentieth century has demonstrated the importance of meteoritic impact on the history and surfaces of the terrestrial planets and set us to studying the evidence for meteoritic impact on Earth. Actually the study of meteorite craters on Earth began at the end of the last century with an examination of Arizona's Meteor Crater, also known as Barringer Crater, after an early student of the feature, and as Coon Butte Crater and Canon Diablo Crater. It was a long time before it was accepted that Meteor Crater was formed by the impact of an extraterrestrial object. Once it was accepted, however, additional impact craters were soon reported. We have now identified several score craters that are definitely or probably due to impact.

METEOR CRATER

Meteor Crater is a circular depression 1.2 km in diameter and 180 m deep located about 150 km southeast of the

BOX 20.1   Voyager 2 Views Neptune

Neptune, named for the Roman god of the sea, is the fourth largest planet in the solar system. Its existence was first predicted on the basis of anomalies in the orbit of Uranus and telescopic confirmation came in 1845.

Like the other giant planets Neptune is enveloped in an atmosphere of primarily hydrogen and helium. A small amount of methane gives it a blue color and an informal name, The Blue Planet. The Neptunian atmosphere is very active with—among other features—scattered, bright cirrus clouds, some of them crowning a large, dark turbulent zone near its equator called the Great Dark Spot for its superficial similarity to the Great Red Spot of Jupiter.

Voyager 2 showed that the partial rings or arcs earlier predicted from telescopic observations are actually complete around the planet to the number of at least three and possibly four. The outer most ring is composed in part of moonlets.

Six new moons were discovered during the Voyager 2 visit which brought to eight the total number for the planet. Triton, the largest of the two previously

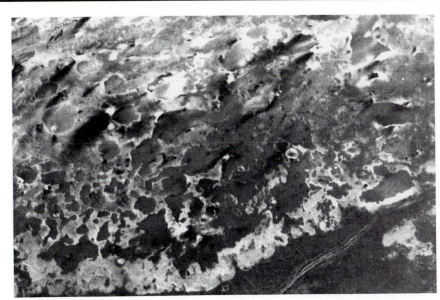

B20.1.1   A portion of Triton's southern polar region shows dark streaks across a mottled surface. See text for discussion. [NASA Voyager 2.]

known moons, provided startling images for the spacecraft's sensors. Its moonscape has developed across an icy crust of nitrogen and methane. In places, dark streaks, some up to 200 km in length, are splashed across the surface (Figure B20.1.1). These have been interpreted as the results of ice volcanoes driven by liquid nitrogen.

Barely frozen at Triton's surface temperatures, nitrogen can liquefy beneath an overburden of only a few tens of meters. This liquid nitrogen is then thought to have exploded through a fractured ice cover into Triton's thin atmosphere before falling back as frozen crystals of nitrogen which appear dark against a bright surface.

Grand Canyon (Figure 20.24). Within a radius of 5 km of the crater thousands of small fragments of iron meteorites have been found and are part of a field of such fragments oriented northeast-southwest over a distance of 500 km. It was the presence of these meteorites along with the explosive appearance of the feature and the analogy with lunar craters that persuaded people of the impact origin of the Meteor Crater.

20.24   Aerial view of Meteor Crater, Coconino County, Arizona. This broadly circular bowl is 200 m deep and over 1250 m in diameter. The ridge or rim surrounding the crater is 30 to 60 m above the plateau and is composed of angular debris, that ranges from dust-sized particles to blocks nearly 30 m across. [Warren B. Hamilton, USGS.]

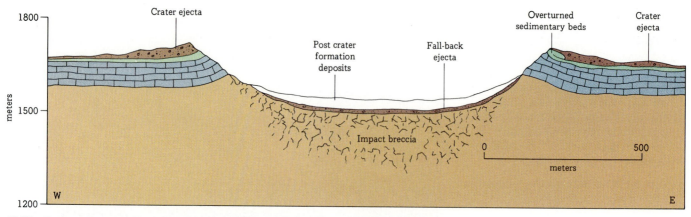

20.25 Geologic section through Meteor Crater. [After E. M. Shoemaker, "Penetration Mechanics of High-Velocity Meteorites, Illustrated by Meteor Crater, Arizona," *21st Int. Geol. Congress, Norden,* vol. 18, pp. 418–434, 1960.]

There are other features, however, that are found at Meteor Crater and point to a high-energy impact for its formation. The material around the outside of the hole appears to have been "splashed" out as if some great force blew it from the crater site. Furthermore the sedimentary layers at the rim have been turned upward in places and folded back flaplike from the crater rim so that older material lies on younger (Figure 20.25). Drilling into the bottom of the crater revealed the rock to be shattered and broken to a depth of 300 m before solid fresh bedrock was encountered.

On a smaller scale a diagnostic feature of high-velocity impact is the presence of **shatter cones** (Figure 20.26). As their name implies, shatter cones have a conical shape. They average in height from 6 to 12 cm, although many are smaller and some have been reported up to 2 m in size. The angle of the apex is about 90°. The conical areas are marked by fine striations that radiate from the apex toward the base. When in place the cones point upward and inward toward the center of the crater. Shatter cones of a very small scale can be duplicated in the laboratory in high-velocity impact experiments.

Finally we mention two forms of quartz, known to be formed at the high pressures associated with impact, that are found at Meteor Crater. These are **coesite** and **stishovite,** which have densities of about 2.92 and 4.3, respectively, as compared with 2.65 for ordinary quartz. Both forms were first discovered in the laboratory, and so far have been found outside the laboratory only in terrestrial impact craters.

20.26 Shatter cones in late Precambrian sandstone, Island Butte, southwestern Montana. Maximum height is 35 cm. [Robert B. Hargraves]

OTHER TERRESTRIAL CRATERS

Few impact craters on Earth are as clear and well-defined as Meteor Crater. This is not particularly surprising given the low incidence of meteoritic impact on Earth at present and the dynamic nature of our planet. Weathering and erosion soon obliterate the scars of impact. It has been estimated that all vestiges of a crater the size of Meteor Crater would be erased by erosion in 10 million years. Some craters, however, were buried before they could be destroyed by erosion. Three examples have been found in the course of drilling for oil in the Williston basin of Montana (two proved to be petroleum reservoirs). Other craters that were long buried were exhumed by erosion. Still others were most likely caught up and destroyed in the mountain-building process.

Nevertheless a significant number of features can be confidently assigned to impact. For instance, the Manicouagen Crater in Quebec, Canada measures 65 km across, and was formed by a meteoritic impact over two-hundred million years ago (Figure 20.27). The Gosses Bluff Crater in the Macdonnell Ranges of central Australia measures 20 km across and by all signs is geologically recent (Figure 20.28).

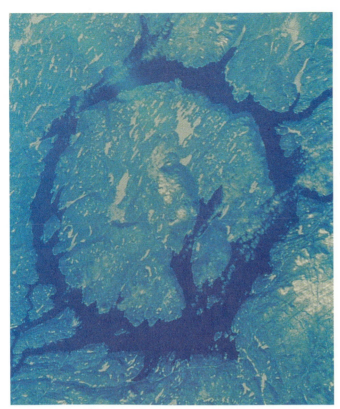

20.27 Manicouagen Crater, Quebec, Canada. This circular feature was produced over 200 million years ago by meteoritic impact. It measures 65 km in diameter and is one of the largest known impact craters on Earth. [NASA.]

RATE OF METEORITE IMPACT

We are not sure what the rate of meteoritic impact on the Earth is. Certainly it is small. One estimate says that it is a rate that has produced 100,000 craters larger than 1 km in diameter during the last 2 billion years, and that it was much greater in the earliest history of Earth. We would expect that from the earliest time to about 3.5 billion years ago meteoritic bombardment was intense, just as it was, for instance, on the Moon and Mercury and, by extension, on Mars (Figure 20.12).

Types of Meteorites   We have already suggested in Section 20.8 that meteorites originate from asteroids or perhaps from comets. We can divide these meteorites into two main groups: the **iron meteorites** and the **stony meteorites.**
    Box 20.2 describes the discovery of an unexpectedly large supply of meteorites.

Iron Meteorites   The **iron meteorites** are primarily an alloy of iron with approximately 10 percent nickel. Their texture indicates that they formed by slow cooling and crystallization. We believe that this took place in a small planet or moon that, on the basis of radiometric dating, broke up about a billion years ago, probably as the result of collision with another planetary object. It is now thought that iron meteorites — and probably others as well — came from asteroids with radii between 70 and 200 km.

20.28   Gosses Bluff Crater, Macdonnell Ranges, central Australia, is the circular feature in the lower part of the picture. The light-colored rim is 3 km in diameter and represents the central uplift in the crater. Overall the crater measures about 20 km in diameter. Its outer portion is marked by the circular zone of discontinuous darker material outside of the inner ring. [NASA, STS-41D, August–September 1984, Picture #14-41-028.]

Twenty years ago the world's collection of catalogued meteorites numbered about 2,000, and it had taken over 200 years to find them. Then, in 1969, a Japanese expedition to Antarctica stumbled on an unexpected and very rich source of meteorites: glacier ice on the Antarctic continent. In 15 years Japanese and American scientists had collected over 6,000 specimens of meteorites from locations separated by 3,000 km. How many individual meteorites these represent is yet to be determined, but by any standard the Antarctic icecap has provided a bonanza of extraterrestrial samples.

The concentration of the meteorites in-volves the nature of glacier flow. The meteorites, having fallen on the ice, are carried deep into the glacier as more and more new ice is added from above. The flow of ice is also lateral toward the continental margins. But the route is not always direct, and in some places partially ice-buried mountain ranges direct glacier flow back up toward the surface. It has been in such areas, where ice continues to rise toward the surface, there to be eroded, that the meteorites have been concentrated.

The Antarctic collections contain examples of meteorites already known from other localities, including the very rare types. In addition, new varieties of meteorites have been discovered.

One almost unbelievable discovery is a meteorite that gives every indication of having come from the Moon. How this could have happened is yet to be demonstrated. Even more of a puzzle are two Antarctic meteorites that some think have come from Mars. These are not the first meteorites discovered that are thought to be from Mars, a handful having been collected on other continents. So far there is no plausible explanation of how Martian rock could survive a liftoff that would send it on a collision path with Earth.

Stony Meteorites   The **stony meteorites** are composed chiefly of silicate minerals and don't look much different from ordinary rocks on earth. They provide us with much more information about our planetary system than do the iron meteorites, partly because there is a much larger variety of stony meteorites, and partly because they haven't gone through the high temperatures associated with the melting and crystallization characteristic of the iron meteorites.

Most stony meteorites are called **chondrites** because of their unusual texture, which consists of small round silicate grains—**chondrules** (from the Greek for "grain" plus the diminutive *ule*). Some 10 percent of the stony meteorites are without chondrules; they are called **achondrites.**

The chondrules are a few millimeters in diameter, composed in most examples of pyroxene or olivine, and set in a fine-grained groundmass of the same composition plus, in some chondrites, metallic iron. The chondrites have radiometric ages of between 4.5 and 4.6 billion years and must represent conditions that existed during the early history of the formation of the planets.

A small number of the chondrites, the **carbonaceous chondrites,** contain carbon, hydrocarbons, and even amino acids. It is generally agreed that these constituents are inorganic in origin, but it is intriguing to know that many of the building blocks of life were present when the planetary system began to form.

The carbonaceous chondrites have well-defined chondrules of sharp outline set in a fine-grained earthy matrix of low-temperature minerals. The chondrules themselves are high-temperature minerals. Matrix and chondrules, then, are not in equilibrium with each other and represent two different environments of formation. In addition, inclusions of Ca-Al-Ti minerals are found in some carbonaceous chondrites, and on theoretical grounds these minerals are thought to have condensed first in the cooling of a primitive solar nebula. Finally, the carbonaceous chondrites resemble the Sun in composition except that they contain much less hydrogen, carbon, nitrogen, and the noble gases. The absence of these elements is not unexpected, for they are not likely to have condensed in the inner portion of the solar system because of their high volatility.

All these characteristics, along with their age of 4.6 billion years, suggest that the carbonaceous chondrites are representative of the material from which planets formed.

About 8 percent of the stony meteorites are **achondritic.** Some are breccias and probably represent the lithified outer rind of an asteroid. Others appear to be igneous rocks that formed by the processes of melting, differentiation, and crystallization with which we are familiar here on Earth.

20.14
COMETS

Comets are generally thought to be accumulations of dust and ice. They are believed to be present in large numbers (10^{12}) in the far reaches of the solar system in the Oort cloud, named after its discoverer, the Dutch astronomer J. H. Oort. Some of these comets have been diverted into the inner solar system as their orbits have been perturbed by passing stars.

The comets we see are of two types: Periodic and nonperiodic. The vast majority of them are **nonperiodic,** making a single pass through the solar system, never to return—at least on a predictable schedule. The **periodic** comets, in contrast, have predictable orbits—as, for instance, does Halley's Comet, which reappears every 77 years, the last nearest approach being in February 1986.

As a comet approaches the Sun, the Sun's heat begins to evaporate the outer portion of the icy body. Dust and gas released from the nucleus of the comet form a hazy cloud around the icy core or nucleus of the comet. The smaller particles of dust and gas thus freed from the nucleus then string away from the Sun, pushed by the solar wind. This is the comet's tail, and its light is in part reflected sunlight and in part the release of energy absorbed from the Sun by gas molecules in the tail.

Summary

Our **solar system** centers on, and moves around, the Sun. It includes **planets and their satellites, meteors** and **meteorites,** and **comets.** The system is thought to have originated from a **dust cloud** 4.6 billion years ago.
   The inner, terrestrial planets are **Mercury, Venus, Earth,** and **Mars.** The giant outer planets are **Jupiter, Saturn, Uranus, Neptune,** and **Pluto;** they are separated from the inner planets by a belt of **asteroids.**

Mercury, the smallest planet, has an ancient, heavily cratered surface marked also by ancient lava flows.

Venus is shrouded in a thick atmosphere of CO_2 and has a complex surface only dimly delineated by radar.

Earth is the most active of the inner planets.

Earth's Moon looks very much like Mercury.
   Lunar rocks brought back to Earth by the astronauts include **basalt, gabbro, anorthosite,** and **breccia.**

Mars shows evidence of impact cratering, volcanism, running water, permanently frozen ground, wind action, and landsliding.

Comparison of the Moon and the individual terrestrial planets can be made in the areas of **impact craters, weathering and erosion, tectonics,** and **volcanic activity.**

The **belt of asteroids** is the source of most of our meteorites.

Jupiter, Saturn, Uranus, and **Neptune** have thick, turbulent atmospheres made up chiefly of hydrogen. Saturn has many **rings;** Uranus, several; Jupiter, one. All have a number of moons. These planets all record impact cratering, tectonism, and vulcanism, and several have icy crusts.

Meteoritic impact craters on Earth are rare, but exist.

Most **meteorites** come from the asteroid belt, particularly from the **Amor asteroids.** They are either **iron meteorites** or **stony meteorites.** The stony group contains both **chondrites** and **achondrites.** Among the former are a few **carbonaceous chondrites** that seem to be samples of the dust cloud from which the planetary system formed.

Comets are **periodic** or **nonperiodic** in their passage through the inner planetary system.

Questions

1. What similarities exist among Moon, Mercury, Venus, Earth, and Mars?
2. In what ways does Earth differ from the Moon and the other terrestrial planets?
3. What are the differences between iron and stony meteorites?
4. Why do we think that carbonaceous chondrites represent a sample of the dust cloud from which our planetary system formed?
5. How does volcanism on Earth differ from that on Io, a moon of Jupiter?

6. How old will you be when Halley's comet makes its next approach to the Sun?
7. In what ways are comets different from meteorites?
8. Why are surface temperatures so high on Venus and so low on Mars?
9. Why are seasonal shifts in climate less extreme on Earth than on Mars?
10. Why are meteoritic impact craters rare on Earth?
11. What are some of the characteristic features of meteoritic impact craters on Earth?

Supplementary Readings

CARR, MICHAEL H. *The Surface of Mars,* Yale University Press, New Haven, Conn., 1984.
A well-illustrated discussion of the landforms of Mars and the processes that fashioned them. Picks up where the volume by Mutch et al., cited below, leaves off.

CARR, MICHAEL H., R. STEPHEN SAUNDERS, ROBERT G. STROM, AND DON E. WILHELMS *The Geology of the Terrestrial Planets,* National Aeronautics and Space Administration, Washington, D.C., 1984.
An excellent, fairly detailed overview of the inner planets. Also included is a short chapter on asteroids, comets, and the formation of planets.

FISHER, DAVID E. *The Birth of the Earth: A wanderlied through space, time, and the human imagination.* Columbia University Press, New York, 1987.
For the general reader, the story of the birth of our planet as told by an historian of science.

GLASS, BILLY P. *Introduction to Planetary Geology,* Cambridge University Press, Cambridge, 1982.
An excellent textbook with thorough coverage of all aspects of the subject.

HAWKING, STEPHEN W. *A Brief History of Time: From the Big Bang to Black Holes,* Bantam Books, New York, 1988.
A best seller by the distinguished English theoretical physicist. As the author tells us it addresses in words rather than mathematics such questions as "Where did the universe come from? How and why did it begin? Will it come to an end, and if so why?".

MUTCH, THOMAS A., R. E. ARVIDSON, JAMES W. HEAD, K. L. JONES, AND R. STEPHEN SAUNDERS *The Geology of Mars,* Princeton University Press, Princeton, N.J., 1976.
An authoritative review of the geology of the planet as known up to the time of the Viking *missions and their landers, which put down on the planet in 1976.*

SOLAR SYSTEM EXPLORATION COMMITTEE OF THE NASA ADVISORY COUNCIL *Planetary Exploration through Year 2000,* U.S. Government Printing Office, Washington, D.C., 1988.
Sets out the scientific rationale for planetary exploration in the decade of the 1990's.

APPENDIXES

APPENDIX A

MINERALS

Many of the most common minerals may be identified in hand specimens by physical properties. Among the characteristics useful for this purpose are those discussed in Chapter 2, including crystal form, cleavage, striations, hardness, specific gravity, color, and streak. Still other properties are found to be useful in identifying certain groups of minerals. Some of these properties are summarized in the text and tables that follow.

A.1
MINERAL CHARACTERISTICS

CRYSTAL SYMMETRY AND SYSTEMS

As noted in Chapter 2, when a mineral grows without interference, it is bounded by plane surfaces symmetrically arranged, which give it a characteristic crystal form. This form is the external expression of its definite internal crystalline structure. The faces of crystals are defined by surface layers of atoms.

Every crystal consists of atoms arranged in a three-dimensional pattern that repeats itself regularly. Even in irregular mineral grains the atoms are arranged according to their typical crystalline structure.

Crystals are classified in six different systems according to the **symmetry** of their faces and the arrangement of their **axes of symmetry.** (An axis is an imaginary straight line that is drawn from the center of a face to the center of the opposite face.) The systems are described and illustrated in Table A.1.

HARDNESS

We can determine the hardness of a mineral by scratching its smooth surface with the edge of another. We must be sure that the mineral tested is actually scratched. Some-

TABLE A.1
Basic Crystal Systems[a]

Unit-cell shape[b]	Name: description	Typical forms of common-mineral crystals
	Isometric: three equal-length axes—all at right angles	Garnet · Magnetite · Halite, pyrite
	Tetragonal: two equal axes and a third either longer or shorter—all at right angles	Zircon
	Hexagonal: three equal axes in same plane intersecting at 60°, a fourth perpendicular to other three	Quartz · Calcite, dolomite · Hematite
	Orthorhombic: three unequal axes—all at right angles	Olivine · Aragonite
	Monoclinic: three unequal axes—two at right angles, a third perpendicular to one but oblique to other	Pyroxene (augite) · Mica, clay minerals · Orthoclase · Gypsum
	Triclinic: three unequal axes—all at oblique angles	Plagioclase · Calcium aluminum silicate

[a] From Sheldon Judson, Kenneth S. Deffeyes, and Robert B. Hargraves, *Physical Geology*, Prentice Hall, Englewood Cliffs, N.J., 1976.
[b] Symmetry axes in color.

TABLE A.2
Mohs Scale of Hardness

Scale	Mineral	Test
1	Talc	[Softest]
2	Gypsum	
2.5		Fingernail
3	Calcite	Copper coin
4	Fluorite	
5	Apatite	
5.5–6		Knife blade or plate glass
6	Orthoclase	
6.5–7		Steel file
7	Quartz	
8	Topaz	
9	Corundum	
10	Diamond	[Hardest]

times particles simply rub off the specimen, suggesting that it has been scratched when it has not.

In Table A.2 ten common minerals have been arranged as examples of the degrees of the Mohs scale of relative hardness. Each of these minerals will scratch all those lower in number on the scale and will be scratched by all those higher. In other words, this is a **relative scale.** In terms of absolute hardness, the steps are approximately uniform up to 9; that is, number 7 is 7 times as hard as 1, and 9 is 9 times as hard as 1. But 10 is about 40 times as hard as 1. A more extensive listing is in Tables A.3 and A.8).

MAGNETISM

Minerals that in their natural state are attracted to a magnet are said to be **magnetic.** Magnetite, Fe_3O_4, and pyrrhotite, $Fe_{1-x}S$, with x between 0 and 0.2, are the only common magnetic minerals, although many others containing iron are drawn to a sufficiently powerful electromagnet.

PYROELECTRICITY

Pyroelectricity is the simultaneous development of positive and negative charges of electricity on different parts of the same crystal under the proper conditions of temperature change. Quartz is a good example. If it is heated to about 100°C, it will, on cooling, develop positive electric charges at three alternate prismatic edges and negative charges at the other three edges.

PIEZOELECTRICITY

Piezoelectricity is the charge developed in a crystallized body by pressure. Quartz is probably the most important piezoelectric mineral; an extremely slight pressure parallel to its "electric axis" can be detected by the electric charge set up. It is used in specially oriented plates in radio equipment and in sonic sounders.

LUSTER

Luster is the way a mineral looks in reflected light. There are several kinds of luster:

- *Metallic.* Of metals.
- *Adamantine.* Of diamonds.
- *Vitreous.* Of a broken edge of glass.
- *Resinous.* Of yellow resin.
- *Pearly.* Of pearl.
- *Silky.* Of silk.
- *Greasy.* Of grease.
- *Waxy.* Of wax.
- *Earthy.* Of earth.

TABLE A.3
Minerals Arranged According to Hardness

Hardness	Mineral	Hardness	Mineral	Hardness	Mineral	Hardness	Mineral
1	Talc	3.5–4	Chalcopyrite	5–6	Tremolite	6.5–7	Jadeite
1–2	Graphite	3.5–4	Dolomite	5.5	Chromite	6.5–7	Olivine
1–3	Bauxite	3.5–4	Siderite	5.5	Enstatite	6.5–7	Spodumene
2	Gypsum	3.5–4	Sphalerite	5.5	Uraninite	6.5–7.5	Almandite
2	Stibnite	4	Azurite	5.5–6	Anthophyllite	6.5–7.5	Garnet
2–2.5	Chlorite	4	Fluorite	5.5–6.5	Hematite	7	Kyanite (across crystal)
2–2.5	Kaolinite	5	Apatite	6	Albite	7	Quartz
2–2.5	Muscovite	5	Kyanite (along crystal)	6	Anorthite	7–7.5	Staurolite
2–5	Serpentine	5–5.5	Goethite	6	Arfvedsonite	7–7.5	Tourmaline
2.5	Galena	5–5.5	Limonite	6	Magnetite	7.5	Andalusite
2.5	Halite	5–5.5	Wollastonite	6	Orthoclase	7.5	Zircon
2.5–3	Biotite	5–6	Actinolite	6–6.5	Aegirite	8	Spinel
2.5–3	Chalcocite	5–6	Augite	6–6.5	Pyrite	8	Topaz
3	Bornite	5–6	Diopside	6–7	Cassiterite	9	Corundum
3	Calcite	5–6	Hornblende	6–7	Epidote	10	Diamond
3–3.5	Anhydrite	5–6	Opal	6–7	Sillimanite		

TABLE A.4
Minerals Arranged According to Specific Gravity

Specific gravity	Mineral	Specific gravity	Mineral	Specific gravity	Mineral	Specific gravity	Mineral
1.9–2.2	Opal	2.8–3.2	Biotite	3.3–3.5	Jadeite	4.02	Corundum
2.0–3.0	Bauxite	2.85	Dolomite	3.3–4.37	Goethite	4.1–4.3	Chalcopyrite
2.16	Halite	2.85–3.2	Anthophyllite	3.35–3.45	Epidote	4.25	Almandite
2.2–2.65	Serpentine	2.89–2.98	Anhydrite	3.4–3.55	Aegirite	4.52–4.62	Stibnite
2.3	Graphite	3.0–3.25	Tourmaline	3.4–3.6	Topaz	4.6	Chromite
2.32	Gypsum	3.0–3.3	Actinolite	3.45	Arfvedsonite	4.68	Zircon
2.57	Orthoclase	3.0–3.3	Tremolite	3.5	Diamond	5.02	Pyrite
2.6	Kaolinite	3.15–3.2	Apatite	3.5–4.1	Spinel	5.06–5.08	Bornite
2.6–2.9	Chlorite	3.15–3.2	Spodumene	3.5–4.3	Garnet	5.18	Magnetite
2.62	Albite	3.16	Andalusite	3.56–3.66	Kyanite	5.26	Hematite
2.65	Quartz	3.18	Fluorite	3.6–4.0	Limonite	5.5–5.8	Chalcocite
2.7–2.8	Talc	3.2	Hornblende	3.65–3.75	Staurolite	6.8–7.1	Cassiterite
2.72	Calcite	3.2–3.4	Augite	3.77	Azurite	7.4–7.6	Galena
2.76	Anorthite	3.2–3.5	Enstatite	3.85	Siderite	9.0–9.7	Uraninite
2.76–3.1	Muscovite	3.23	Sillimanite	3.9–4.1	Sphalerite		
2.8–2.9	Wollastonite	3.27–3.37	Olivine	4.0	Carnotite		

SPECIFIC GRAVITY

The ratio of the mass of a mineral to the mass of an equal volume of water at a temperature of 4°C, is known as the **specific gravity** of that mineral. Table A.4 lists the most common minerals in order of increasing specific gravity.

FLUORESCENCE AND PHOSPHORESCENCE

Minerals that become luminescent during exposure to ultraviolet light, X rays, or cathode rays are **fluorescent.** If the luminescence continues after the exciting rays are shut off, the mineral is said to be **phosphorescent.**

FUSIBILITY

Minerals can be divided into those fusible and those infusible in a blowpipe flame. Seven minerals showing different degrees of fusibility have been used as a scale to which fusible minerals can be referred. They are listed in Table A.5.

SOLUBILITY

Concentrated hydrochloric acid, HCl, diluted with three parts of water, is commonly used for the solution of minerals being tested. Other wet reagents are used for special tests to help identify minerals.

TABLE A.5
Scale of Fusibility

Scale	Mineral	Approx. fusing point, °C	Remarks
1	Stibnite	525	Easily fusible in candle flame
2	Chalcopyrite	800	Small fragment easily fusible in Bunsen burner flame
3	Garnet (almandite)	1050	Infusible in Bunsen flame but easily fusible in blowpipe flame
4	Actinolite	1200	Sharp-pointed splinter fuses with little difficulty in blowpipe flame
5	Orthoclase	1300	Fragment edges rounded with difficulty in blowpipe flame
6	Bronzite	1400	Only fine splinter ends rounded in blowpipe flame
7	Quartz	1470	Infusible in blowpipe flame

FRACTURE

Many minerals that do not exhibit cleavage (Chapter 2) break or fracture in a distinctive manner. Some types of fracture are:

- *Conchoidal.* Along smooth curved surfaces like the surface of a shell (conch); commonly observed in glass and quartz.
- *Fibrous or splintery.* Along surfaces roughened by fibers or splinters.
- *Uneven or irregular.* Along rough, irregular surfaces.
- *Hackly.* Along a jagged, irregular surface with sharp edges.

TENACITY

A mineral's cohesiveness, as shown by its resistance to breaking, crushing, bending, or tearing, is known as its **tenacity.** Various kinds of tenacity in minerals include the following:

- *Brittle.* Breaks or powders easily.
- *Malleable.* Can be hammered into thin sheets.
- *Sectile.* Can be cut by a knife into thin shavings.
- *Ductile.* Can be drawn into wire.
- *Flexible.* Bends but does not return to its original shape when pressure is removed.
- *Elastic.* After being bent, will resume its original position upon release of pressure.

A.2
IMPORTANT MINERALS

SILICATES

More than 90 percent of rock-forming minerals are silicates, with structures based on the $(SiO_4)^{4-}$ tetrahedron. Important classes are listed in Table 2.2.

PLAGIOCLASE FELDSPARS

The plagioclase feldspars, also called the **soda-lime feldspars,** form a complete solid-solution series from pure albite to pure anorthite. Sodium substitutes for calcium in all proportions, with accompanying substitution of silicon for aluminum. The series is divided into the six arbitrary species names listed in Table A.6 (also see Table A.8).

PYROXENES AND AMPHIBOLES

The pyroxene family of minerals and the amphibole family of minerals are chain silicates that parallel each other. The amphiboles contain OH^-. The pyroxenes crystallize at higher temperatures than their amphibole analogs. Qualities of the two minerals are shown in Tables A.7 and A.8.

TABLE A.6
Plagioclase Feldspars

Species	Albite, %	Anorthite, %
Albite, $Na(AlSi_3O_8)$	90	10
Oligoclase	70	30
Andesine	50	50
Labradorite	30	70
Bytownite	10	90
Anorthite, $Ca(Al_2Si_2O_8)$		

TABLE A.7
Ions in Common Pyroxenes and Amphiboles

Ions		Pyroxenes[a]	Amphiboles[b]
X	Y		
Mg	Mg	Enstatite	Anthophyllite
Ca	Mg	Diopside	Tremolite
Li	Al	Spodumene	
Na	Al	Jadeite	Glaucophane
Na	Fe^{3+}	Aegirite	Arfvedsonite
Ca, Na	Mg, Fe, Mn, Al, Fe^{3+}, Ti	Augite	Hornblende

[a] Basic structure: single chain, SiO_3; formula, $XY(Si_2O_6)$.

[b] Basic structure: double chains, Si_4O_{11}; formula, $X_{0-7}Y_{7-14}Si_{16}O_{44}(OH)_4$.

TABLE A.8
Physical Characteristics of Minerals

Mineral	Chemical name and/or composition	Form	Cleavage and/or fracture	Hardness	Specific gravity
Actinolite (an *asbestos;* an *amphibole*)	Calcium iron silicate, $Ca_2(Mg, Fe)_5Si_8O_{22}(OH)_2$ (*tremolite* with more than 2% iron)	Slender crystals, usually fibrous	See *amphibole*	5–6	3.0–3.3
Aegirite (a *pyroxene*)	$NaFe^{3+}(Si_2O_6)$	Slender prismatic crystals	Imperfect cleavage at 87° and 93°	6–6.5	3.40–3.55
Albite	Sodic plagioclase *feldspar*, $Na(AlSi_3O_8)$	Tabular crystals; striations caused by twinning	Good in 2 directions at 93°34′	6	2.62
Almandite	$Fe_3Al_2(SiO_4)_3$	12- or 24-sided; massive or granular		6.5–7.5	4.25
Amphibole family	See *anthophyllite, arfvedsonite, hornblende, tremolite*		Perfect prismatic at 56° and 124°, often yielding splintery surface		
Andalusite	Aluminum silicate, Al_2SiO_5	Usually in coarse, nearly square prisms; cross section may show black cross	Not prominent	7.5	3.16
Andesine	A *plagioclase feldspar* 50–70% albite				
Anhydrite	Anhydrous calcium sulfate, $CaSO_4$	Commonly in massive fine aggregates not showing cleavage; crystals rare	3 directions at right angles to form rectangular blocks	3–3.5	2.89–2.98
Ankerite	Dolomite in which ferrous iron replaces more than 50% magnesium				
Anorthite	Calcic *plagioclase feldspar*, $Ca(Al_2Si_2O_8)$	Striations caused by twinning; lathlike or platy grains	Good in 2 directions at 94°12′	6	2.76
Anthophyllite (an *amphibole*)	$(Mg, Fe)_7(Si_8O_{22})(OH)_2$	Lamellar or fibrous	See *amphibole*	5.5–6	2.85–3.2
Apatite	Calcium fluorophosphate, $Ca_5(F, Cl)(PO_4)_3$	Massive, granular	Poor in 1 direction; conchoidal fracture	5	3.15–3.2
Arfvedsonite (a sodium-rich *amphibole*)	$Na_3Mg_4Al(Si_8O_{22})(OH, F)_2$	Long prismatic crystals	See *amphibole*	6	3.45
Asbestos	See *actinolite, chrysotile, serpentine*				
Augite (a *pyroxene*)	Ferromagnesian silicate, $Ca(Mg, Fe, Al)(Al, Si_2O_6)$	Short, stubby crystals with 4- or 8-sided cross section; often in granular, crystalline masses	Perfect prismatic along 2 planes at nearly right angles, often yielding splintery surface	5–6	3.2–3.4
Azurite	Blue copper carbonate, $Cu_3(CO_3)_2(OH)_2$	Crystals complex in habit and distorted; sometimes in radiating spherical groups	Fibrous	4	3.77
Bauxite	Hydrous aluminum oxides of indefinite composition; not a mineral	In rounded grains or earthy, claylike masses	Uneven fracture	1–3	2–3
Beryl	$Be_3Al_2Si_6O_{18}$	Prismatic crystals, some several feet long; also massive	Poor in 1 direction	8	2.66–2.92

Color	Streak	Luster	"Transparency"	Other properties
White to light green	Colorless	Vitreous	Transparent to translucent	A common ferromagnesian metamorphic mineral; a common component of greenschists
Brown or green		Vitreous	Translucent	Rare rock former, chiefly in rocks rich in soda and poor in silica
Colorless, white, or gray	Colorless	Vitreous to pearly	Transparent to translucent	Opalescent variety, *moonstone*
Deep red	White	Vitreous to resinous	Transparent to translucent	A garnet used to define one of the zones of middle-grade metamorphism; striking in schists
				A group of silicates with tetrahedra in double chains; *hornblende* is the most important; contrast with *pyroxene*
Flesh red, reddish brown, or olive green	Colorless	Vitreous	Transparent to translucent	Found in schists formed by middle-grade metamorphism of aluminous shales and slates; the variety chiastolite has carbonaceous inclusions in the pattern of a cross
				As grains in igneous rock; chief *feldspar* in andesite lavas of the Andes Mountains
White; may have faint gray, blue, or red tinge	Colorless	Vitreous or pearly	Transparent to translucent	Found in limestones and in beds associated with salt deposits; heavier than *calcite*, harder than *gypsum*
				Formed in low-grade regional metamorphism from conversion of *calcite* or *dolomite* or both between 80 and 120°C
Colorless, white, gray, green, yellow, or red	Colorless	Vitreous to pearly	Transparent to translucent	Occurs in many igneous rocks
Gray to various shades of green and brown		Vitreous		From metamorphism of *olivine*
Green, brown, or red	White	Glassy	Translucent to transparent	Widely disseminated as an accessory mineral in all types of rocks; an important source of fertilizer; a transparent variety is a gem, but too soft for general use
Deep green to black		Vitreous	Translucent	Rock-forming mineral in rocks poor in silica
				General term for certain fibrous minerals with similar physical characteristics though different composition; *chrysotile* is most common
Dark green to black	Greenish gray	Vitreous	Translucent only on thin edges	An important igneous rock-forming mineral found chiefly in simatic rocks
Intense azure blue	Pale blue	Vitreous to dull or earthy	Opaque	An ore of copper; a gem mineral; effervesces with HCl
Yellow, brown, gray, or white	Colorless	Dull to earthy	Opaque	An ore of aluminum; produced under subtropical to tropical climatic conditions by prolonged weathering of aluminum-bearing rocks; a component of *laterites;* clay odor when wet
Pale green, white, or yellow; blue = aquamarine, dark green = emerald	White	Vitreous	Transparent to translucent	Only commercial source of beryllium; usually occurs in granite pegmatites; gem

Mineral	Chemical name and/or composition	Form	Cleavage and/or fracture	Hardness	Specific gravity
Biotite (black *mica*)	Ferromagnesian silicate, $K(Mg, Fe)_3AlSi_3O_{10}(OH)_2$	Usually in irregular foliated masses; crystals rare	Perfect in 1 direction into thin, elastic, transparent, smoky sheets	2.5–3	2.8–3.2
Bornite (peacock ore; purple copper ore)	Copper iron sulfide, Cu_5FeS_4	Usually massive; rarely in rough cubic crystals	Uneven fracture	3	5.06–5.08
Bronzite (a *pyroxene*)	*Enstatite* with 5–13% FeO				
Bytownite	A *plagioclase feldspar*, 10–30% albite				
Calcite	Calcium carbonate, $CaCO_3$	Usually in crystals or coarse to fine granular aggregates; also compact, earthy; crystals extemely varied—over 300 different forms	Perfect in 3 directions at 75° to form unique rhombohedral fragments	3	2.72
Carnotite	Potassium uranyl vanadate, $K_2(UO_2)_2(VO_4)_2$	Earthy powder	Uneven fracture	Very soft	4
Cassiterite (tinstone)	Tin oxide, SnO_2	Commonly massive granular	Conchoidal fracture	6–7	6.8–7.1
Chalcocite (copper glance)	Copper sulfide, Cu_2S	Commonly aphanitic and massive; crystals rare but small, tabular with hexagonal outline	Conchoidal fracture	2.5–3	5.5–5.8
Chalcopyrite (copper *pyrites*; yellow copper ore; fool's gold)	Copper iron sulfide, $CuFeS_2$	Usually massive	Uneven fracture	3.5–4	4.1–4.3
Chlorite	Hydrous ferromagnesian aluminum silicate, $(Mg, Fe^{2+})_5 (Al, Fe^{3+})_2Si_3O_{10}(OH)_8$	Foliated massive or in aggregates of minute scales	Perfect in 1 direction, like *micas*, but into inelastic flakes	2–2.5	2.6–2.9
Chromite	Iron chromium oxide, $FeCr_2O_4$	Massive, granular to compact	Uneven fracture	5.5	4.6
Chrysotile (*serpentine asbestos*)	See *serpentine*				
Cinnabar	Mercuric sulfide, HgS	Usually massive, granular	1 perfect cleavage	2.5	8.1
Clay	See *illite, kaolinite, montmorillonite*				
Cobaltite	Cobalt-arseno sulfide, CoAsS	Commonly in cubes, sometimes granular	1 perfect cleavage	6	6.3
Columbite	$(Fe, Mn)Nb_2O_6$	Short prismatic crystals; often massive	1 distinct cleavage, 1 less distinct	6	5.2
Corundum (ruby, sapphire)	Aluminum oxide, Al_2O_3	Barrel-shaped crystals; sometimes deep horizontal striations; coarse or fine granular	Basal or rhombohedral parting	9	4.02

Color	Streak	Luster	"Transparency"	Other properties
Black, brown or dark green	Colorless	Pearly or glassy	Transparent or translucent	Constructed around tetrahedral sheets; a common and important rock-forming mineral in both igneous and metamorphic rocks
Brownish bronze on fresh fracture; tarnishes to variegated purple and blue, then black	Grayish black	Metallic	Opaque	An important ore of copper
				Primarily as grains in igneous rocks
Usually white or colorless; may be tinted gray, red, green, blue, or yellow	Colorless	Vitreous	Transparent to opaque	A very common rock mineral, occurring in masses as limestone and marble; effervesces freely in cold dilute HCl
Brilliant canary yellow		Earthy	Opaque	An ore of vanadium and uranium
Brown or black; rarely yellow or white	White to light brown	Adamantine to submetallic and dull	Translucent; rarely transparent	Principal ore of tin
Shiny lead gray; tarnishes to dull black	Grayish black	Metallic	Opaque	One of the most important ore minerals of copper; occurs principally as a result of secondary sulfide enrichment
Brass yellow; tarnishes to bronze or iridescence, but more slowly than bornite or chalcocite	Greenish black; also greenish powder when scratched	Metallic	Opaque	An ore of copper; distinguished from *pyrite* by being softer than steel, distinguished from gold by being brittle; like pyrite, known as "fool's gold"
Green of various shades	Colorless	Vitreous to pearly	Transparent to translucent	A common metamorphic mineral characteristic of low-grade metamorphism
Iron black to brownish black	Dark brown	Metallic to submetallic or pitchy	Subtranslucent	The only ore of chromium; a common constituent of perido-tites and *serpentines* derived from them; one of the first minerals to crystallize from cooling magma
Red	Red	Adamantine	Transparent	Chief ore of mercury; usually related to volcanic activity
Silver white, steel gray	Gray-black	Metallic	Opaque	An ore of cobalt
Iron black to brownish black	Dark red to black	Submetallic	Transparent in small slivers	Ore of niobium; continuous series of solid solutions with tantalite
Brown, pink, or blue; may be white, gray, green, ruby red, or sapphire blue	Colorless	Adamantine to vitreous	Transparent to translucent	Common as an accessory mineral in metamorphic rocks such as marble, mica schist, gneiss; occurs in gem form as *ruby* and *sapphire*; the abrasive emery is black granular corundum mixed with *magnetite*, *hematite*, or the magnesian aluminum oxide *spinel*

Mineral	Chemical name and/or composition	Form	Cleavage and/or fracture	Hardness	Specific gravity
Diamond	Carbon, C	Octahedral crystals, flattened, elongated, with curved faces	Octahedral	10	3.5
Diopside (a *pyroxene*)	$CaMg(Si_2O_6)$	Prismatic crystals; also granular massive	Poor prismatic	5–6	3.2
Dolomite	Calcium magnesium carbonate, $CaMg(CO_3)_2$	Rhombohedral crystals with curved faces; granular cleavable masses or aphanitic compact	Perfect in 3 directions at 73°45'	3.5–4	2.85
Emery	Black granular *corundum* intimately mixed with *magnetite*, *hematite*, or iron *spinel*				
Enstatite (a *pyroxene*)	Magnesium inosilicate, $Mg_2(Si_2O_6)$	Usually massive	Good at 87° and 93°	5.5	3.2–3.5
Epidote	Hydrous calcium aluminum iron silicate, $Ca_2(Al, Fe)Al_2O(SiO_4)(Si_2O_7)(OH)$	Prismatic crystals striated parallel to length; usually coarse to fine granular; also fibrous	Good in 1 direction	6–7	3.35–3.45
Fayalite (an *olivine*)	$Fe_2(SiO_4)$ (see *olivine*)				
Feldspars	Aluminosilicates		Good in 2 directions at or near 90°	6	2.55–2.75
Fluorite	Calcium fluoride, CaF_2	Well-formed interlocking cubes; also massive, coarse or fine grains	Good in 4 directions parallel to faces of an octahedron	4	3.18
Forsterite (an *olivine*)	$Mg_2(SiO_4)$ (see *olivine*)				
Galena	Lead sulfide, PbS	Cube-shaped crystals; also in granular masses	Good in 3 directions parallel to faces of a cube	2.5	7.4–7.6
Garnet	$R_3''R_2'''(SiO_4)_3$ (R'' may be Ca, Mg, Fe, or Mn; R''' may be Al, Fe, Ti, or Cr)	Usually in 12- or 24-sided crystals; also massive granular, coarse or fine	Uneven fracture	6.5–7.5	3.5–4.3
Garnierite	$(Mg, Ni)_6Si_4O_{10}(OH)_8$	Massive, compact or fibrous	No cleavage	2.5	2.5–2.6
Glaucophane (a sodium-rich *amphibole*)	Variety of *arfvedsonite*				
Goethite (bog-iron ore)	$HFeO_2$	Massive, in radiating fibrous aggregates; foliated	Perfect 010	5–5.5	3.3–4.37
Gold	Au	Seldom in crystal form; usually plates, scales, or masses	None	2.5–3	19.3 when pure
Graphite (plumbago; black lead)	Carbon, C	Foliated or scaly masses common; may be radiated or granular	Good in 1 direction, folia flexible but not elastic	1–2	2.3

Color	Streak	Luster	"Transparency"	Other properties
Colorless or pale yellow; may be red, orange, green, blue, or black	Colorless	Adamantine or greasy	Transparent	Gem and abrasive; 95% of natural diamond production is from South Africa; abrasive diamonds have been made in commerical quantities in the laboratory in the United States
White to light green		Vitreous	Transparent to translucent	Contact metamorphic mineral in crystaline limestones
Pinkish; may be white, gray, green, brown, or black	Colorless	Vitreous or pearly	Transparent to opaque	Occurs chiefly in rock masses of dolomitic limestone and marble or as the principal constituent of the rock named for it; distinguished from limestone by its less vigorous action with cold HCl (the powder dissolves with effervescence, large pieces in hot acid)
Grayish, yellowish, or greenish white to olive green and brown		Vitreous	Translucent	Common in pyroxenites, peridotites, gabbros, and basalts; also in both stony and metallic meteorites
Pistachio green or yellowish to blackish green	Colorless	Vitreous	Transparent to translucent	A metamorphic mineral often associated with *chlorite*; derived from metamorphism of impure limestone; characteristic of contact metamorphic zones in limestone
				The most common igneous rock-forming group of minerals; chemically weather to clay minerals
Variable; light green, yellow, bluish green, purple, etc.	Colorless	Vitreous	Transparent to translucent	A common, widely distributed mineral in *dolomite* and limestone; an accessory mineral in igneous rocks; used as a flux in making steel; some varieties fluoresce
Lead gray	Lead gray	Metallic	Opaque	The principal ore of lead; so commonly associated with silver that it is also an ore of silver
Red, brown, yellow, white, green, or black	Colorless	Vitreous to resinous	Transparent to translucent	Common and widely distributed, particularly in metamorphic rocks; brownish-red variety, *almandite*
Apple green	White	Greasy	Translucent	Product of weathering nickel-bearing ultramafic rocks
Yellowish brown to dark brown	Yellowish brown	Adamantine to dull	Subtranslucent	An ore of iron; one of the commonest minerals formed under oxidizing conditions as a chemical weathering product of iron-bearing minerals
Yellow	Yellow	Metallic	Opaque	Malleable
Black to steel gray	Black	Metallic or earthy	Opaque	Feels greasy; common in metamorphic rocks such as marble, schists, and gneisses

Mineral	Chemical name and/or composition	Form	Cleavage and/or fracture	Hardness	Specific gravity
Gypsum	Hydrous calcium sulfate, $CaSO_4 \cdot 2H_2O$	Crystals prismatic, tabular, diamond-shaped; also in granular, fibrous, or earthy masses	Good in 1 direction, yielding flexible but inelastic flakes; fibrous fracture in another direction; conchoidal fracture in a third direction	2	2.32
Halite (rock salt; common salt)	Sodium chloride, NaCl	Cubic crystals; massive granular	Perfect cubic	2.5	2.16
Hematite	Iron oxide, Fe_2O_3	Crystals tabular; botryoidal; micaceous and foliated; massive	Uneven fracture	5.5–6.5	5.26
Hornblende (an *amphibole*)	Complex ferromagnesian silicate of Ca, Na, Mg, Ti, and Al	Long, prismatic crystals; fibrous; coarse- to fine-grained masses	Perfect prismatic at 56° and 124°	5–6	3.2
Hypersthene	*Enstatite* with more than 13% FeO				
Illite *(clay)*					
Ilmenite	$FeTiO_3$	Tabular crystals; massive, granular compact	No cleavage	5.5–6	4.7
Jadeite (a *pyroxene*)	$NaAl(Si_2O_6)$	Fibrous in compact massive aggregates	87° and 93°	6.5–7	3.3–3.5
Kaolinite *(clay)*	Hydrous aluminum silicate, $Al_2Si_2O_5(OH)_4$	Claylike masses	None	2–2.5	2.6
Kyanite	Aluminum silicate, Al_2SiO_5	In bladed aggregates	Good in 1 direction	5 along, 7 across, crystals	3.56–3.66
Labradorite	A *plagioclase feldspar*, 30–50% *albite*				
Lepidolite	Lithia mica, $KLi_2Al(Si_4O_{10})(OH)_2$	In fine- to coarse-grained scaly aggregates	Perfect in 1 direction	2.5	3.6–4
Limonite (brown *hematite*; bog-iron ore; rust)	Hydrous iron oxides; not a mineral	Amorphous; mammillary to stalactitic masses; concretionary, nodular, or earthy	None	5–5.5 (finely divided, apparent H as low as 1)	3.6–4
Magnesite	$MgCO_3$	Commonly massive	1 perfect cleavage	3.5–5	3.0–3.2
Magnetite	Iron oxide, Fe_3O_4	Usually massive granular, granular or aphanitic	Some octahedral parting	6	5.18
Malachite	Copper carbonate, $Cu_2CO_3(OH)_2$	Usually in radiating crystals forming botryoidal masses	1 perfect rarely seen	3.5–4	3.9–4
Mica	See *biotite*, *muscovite*				

Color	Streak	Luster	"Transparency"	Other properties
Colorless, white, or gray; with impurities, yellow, red, or brown	Colorless	Vitreous, pearly, or silky	Transparent to translucent	A common mineral widely distributed in sedimentary rocks, often as thick beds; *satin spar* is a fibrous gypsum with silky luster; *selenite* is a variety that yields broad, colorless, transparent folia; *alabaster* is a fine-grained massive variety
Colorless or white; with impurities, yellow, red, blue, or purple	Colorless	Glassy to dull	Transparent to translucent	Salty taste; permits ready passage of heat rays; a very common mineral in sedimentary rocks; interstratified in rocks of all ages to form a true rock mass
Reddish brown to black	Light to dark blood-red; blackens on heating	Metallic	Opaque	The most important ore of iron; red earthy variety known as *red ocher*; botryoidal form known as *kidney ore*; micaceous form, *specular*; widely distributed in rocks of all types and ages
Dark green to black	Colorless	Vitreous; fibrous variety often silky	Translucent on thin edges	Distinguished from *augite* by cleavage; a common rock-forming mineral that occurs in both igneous and metamorphic rocks
				A general term for clay minerals that resemble *micas;* the chief constituent in many shales
Iron black	Black to brownish red	Metallic, submetallic	Opaque	Often associated with magnetite; magmatic segregation and placers
Apple green, emerald green, or white		Vitreous		Occurs in large masses in *serpentine* by metamorphism of a nepheline-*albite* rock
White	Colorless	Dull earthy	Opaque	Usually unctuous and plastic; other clay minerals similar in composition and physical properties but different in atomic structure are *illite* and *montmorillonite;* derived from weathering of *feldspars*
Blue; may be white, gray, green, or streaked	Colorless	Vitreous to pearly	Transparent to translucent	Characteristic of middle-grade metamorphism; compare with *andalusite*, which has same composition but has different crystal habit; contrast with *sillimanite*, which has same composition but different crystal habit and forms at highest metamorphic temperatures
				Widespread as a rock mineral; the only important constituent in large masses of rocks called *anorthosite*
Commonly pale lilac	Colorless	Vitreous to pearly	Transparent	Fairly rare; occurs in pegmatities; ore of lithium
Dark brown to black	Yellow brown	Vitreous	Opaque	Always of secondary origin from alteration or solution of iron minerals; mixed with fine clay, it is a pigment, *yellow ocher*
White, gray, yellow, brown	Colorless	Vitreous	Transparent to translucent	Often in association with serpentine, of which it is an alteration product
Iron black	Black	Metallic	Opaque	Strongly magnetic; may act as a natural magnet, known as *lodestone;* an important ore of iron; found in black sands on the seashore; mixed with *corundum*, it is a component of *emery*
Green	Colorless	Silky in fibrous form	Translucent	A minor ore of copper; found in limestones in which copper deposits have been oxidized

Mineral	Chemical name and/or composition	Form	Cleavage and/or fracture	Hardness	Specific gravity
Molybdenite	Molybdenum sulfide, MoS_2	Usually foliated, massive or in scales	1 perfect	1–1.5	4.6–4.7
Montmorillonite *(clay)*	Hydrous aluminum silicate				
Muscovite (white *mica*; potassium *mica*; common *mica*)	Nonferromagnesian silicate, $KAl_3Si_3O_{10}(OH)_2$	Mostly in thin flakes	Good in 1 direction, giving thin, very flexible, and elastic folia	2–2.5	2.76–3.1
Oligoclase	A *plagioclase feldspar*, 70–90% *albite*				
Olivine (peridot)	Ferromagnesian silicate, $(Mg, Fe)_2SiO_4$	Usually in embedded grains or granular masses	Conchoidal fracture	6.5–7	3.27–3.37
Opal	$SiO_2 \cdot nH_2O$; a mineraloid	Amorphous; massive; often botryoidal or stalactitic	Conchoidal fracture	5–6	1.9–2.2
Orthoclase	Potassium *feldspar*, $K(AlSi_3O_8)$	Prismatic crystals; most abundantly in rocks as grains of irregular form	Good in 2 directions at or near 90°	6	2.57
Pentlandite	$(Fe,Ni)_9S_8$	Massive	No cleavage	4	4.6–5
Plagioclase	Soda-lime *feldspar*				
Platinum	Native element, Pt	As grains in places, sometimes nugget size	No cleavage	4–4.5	14–19
Psilomelane	Barium-manganese oxide, $(Ba,H_2O)_2Mn_5O_{10}$	Typically in botryoidal crusts; also earthy	None	5–6	3.7–4.7
Pyrite (iron pyrites; fool's gold)	Iron sulfide, FeS_2	Cubic crystals with striated faces; also massive	Uneven fracture	6–6.5	5.02
Pyrochlore	$(Na,Ca)_2(Nb,Ta)_2O_6(O,OH,F)$	Crystals (octohedral) or in masses	1 fair cleavage direction	5–5.5	4.2–6.4
Pyrolusite	MnO_2	Crystals rare; usually massive, fibrous, or columnar	1 perfect cleavage	6–6.5 crystals, 2–6 massive etc.	4.4–5
Pyroxene family	Inosilicates; see *aegirite, augite, diopside, enstatite, jadeite, spodumene*				

Color	Streak	Luster	"Transparency"	Other properties
Lead gray	Grayish black	Metallic	Opaque	Feels greasy; laminae flexible but not elastic; chief ore of molybdenum; most commonly associated with granitic porphyry
				Unique capacity for absorbing water and expanding
Thin, colorless; thick, light yellow, brown, green, or red	Colorless	Vitreous, silky, or pearly	Thin, transparent; thick translucent	Widespread and very common rock-forming mineral; characteristic of sialic rocks; also very common in metamorphic rocks such as gneiss and schist; the principal component of some mica schists; sometimes used for stove doors, lanterns, etc., as transparent *isinglass;* used chiefly as an insulating material
				Found in various localities in Norway with inclusions of *hematite,* which give it a golden shimmer and sparkle; this is called *aventurine* oligoclase, or *sunstone*
Olive to grayish green or brown	Pale green or white	Vitreous	Transparent to translucent	A common rock-forming mineral found primarily in simatic rocks; the principal component of peridotite; actually a series grading from *forsterite* to *fayalite;* the most common olivines are richer in magnesium than in iron; the clear-green variety *peridot* is sometimes used as a gem
Colorless; white; pale yellow, red, brown, green, gray, or blue; opalescent		Vitreous or resinous	Transparent to translucent	Many varieties; lines and fills cavities in igneous and sedimentary rocks, where it was deposited by hot waters
White, gray, or pink	White	Vitreous	Translucent to opaque	Characteristic of sialic rocks
Light bronze-yellow	Light bronze-brown	Metallic	Opaque	Ore of nickel associated with pyrrhotite; formed by magmatic segregation
				A continuous series varying in composition from pure *albite* to pure *anorthite;* important rock-forming minerals; characteristic of simatic rocks
Steel-gray	Colorless	Bright metallic	Opaque	Malleable, ductile; originates in ultramafic rocks; often mined from placers
Black	Black	Submetallic	Opaque	Formed by weathering in surface or near-surface environments
Brass yellow	Greenish or brownish black	Metallic	Opaque	The most common of the sulfides; used as a source of sulfur in the manufacture of sulfuric acid; distinguished from *chalcopyrite* by its paler color and greater hardness, from gold by its brittleness and hardness
Brown with reddish or yellowish shades to black	Colorless	Vitreous	Opaque to translucent	An ore of niobium, usually associated with pegmatites
Steel to iron gray	Black	Metallic	Opaque	An ore of manganese
				A group of silicates with tetrahedra in single chains; *augite* is the most important; contrast with *amphibole*

Mineral	Chemical name and/or composition	Form	Cleavage and/or fracture	Hardness	Specific gravity
Quartz (silica)	Silicon oxide, SiO_2, but structurally a silicate with tetrahedra sharing oxygens in 3 dimensions	Prismatic crystals with faces striated at right angles to long dimension; also massive forms of great variety	Conchoidal fracture	7	2.65
Rutile	Titanium oxide, TiO_2	Prismatic crystals, granular massive	1 distinct cleavage; 1 less distinct	6–6.5	4.25
Scheelite	Calcium tungstate, $CaWO_4$	Commonly massive, granular	1 distinct cleavage	4.5–5	6.12
Serpentine	Hydrous magnesium silicate, $Mg_3Si_2O_5(OH)_4$	Platy or fibrous	Conchoidal fracture	2–5	2.2–2.65
Siderite (spathic iron; chalybite)	Iron carbonate, $FeCO_3$	Granular, compact, or earthy	Perfect rhombohedral	3.5–4	3.85
Sillimanite (fibrolite)	Aluminum silicate, Al_2SiO_5	Long, slender crystals without distinct terminations; often in parallel groups; frequently fibrous	Good in 1 direction	6–7	3.23
Sphalerite (zinc blende; blackjack)	Zinc sulfide, ZnS	Usually massive; crystals many-sided, distorted	Perfect in 6 directions at 120°	3.5–4	3.9–4.1
Spinel	$MgAl_2O_4$	Octahedral crystals	Conchoidal to subconchoidal fracture	8	3.5–4.1

Spinel group	(XY_2O_4)				
	X	Y = AL	Y = Fe	Y = Cr	
	Mg	Spinel, $MgAl_2O_4$	Magnesioferrite, $MgFe_2O_4$	Magnesiochromite, $MgCr_2O_4$	
	Fe	Hercynite, $FeAl_2O_4$	Magnetite, $FeFe_2O_4$	Chromite, $FeCr_2O_4$	
	Zn	Gahnite, $ZnAl_2O_4$	Franklinite, $ZnFe_2O_4$		
	Mn	Galaxite, $MnAl_2O_4$	Jacobsite, $MnFe_2O_4$		

Mineral	Chemical name and/or composition	Form	Cleavage and/or fracture	Hardness	Specific gravity
Spodumene (a pyroxene)	Lithium aluminum inosilicate, $LiAl(Si_2O_6)$	Prismatic crystals; coarse, some large	Perfect at 87° and 93°	6.5–7	3.15–3.20
Staurolite	Iron aluminum silicate, $Fe^{+2}Al_5Si_2O_{12}(OH)$	Usually in crystals, prismatic, twinned to form a cross; rarely massive	Not prominent	7–7.5	3.65–3.75
Stibnite	Antimony trisulfide, Sb_2S_3	Slender prismatic habit; often in radiating groups	Perfect in 1 direction	2	4.52–4.62
Taconite	Not a mineral				
Talc (soapstone; steatite)	Hydrous magnesium silicate, $Mg_3Si_4O_{10}(OH)_2$	Foliated, massive	Good cleavage in 1 direction, thin, flexible, but inelastic folia	1	2.7–2.8
Tantalite	$(FeMn)Ta_2O_6$	Short prismatic crystals, massive	1 distinct cleavage, 1 less distinct	6–6.5	7.9

Color	Streak	Luster	"Transparency"	Other properties
Colorless or white; with impurities, any color	Colorless	Vitreous or greasy	Transparent to translucent	An important constituent of sialic rocks; coarsely crystalline varieties; *rock crystal, amethyst* (purple), *rose quartz, smoky quartz, citrine* (yellow), *milky quartz, cat's eye;* cryptocrystalline varieties: *chalcedony, carnelian* (red chalcedony), *chrysoprase* (apple-green chalcedony), *heliotrope,* or *bloodstone* (green chalcedony with small red spots), *agate* (alternating layers of chalcedony and opal); granular varieties: *flint* (dull to dark brown), *chert* (like flint but lighter in color), *jasper* (red from hematite inclusions), *prase* (like jasper but dull green)
Generally reddish brown	Pale brown to grayish black	Adamantine to metallic	Transparent in thin slices	A high-temperature mineral that is commonly concentrated in placer deposits
Colorless to white, pale yellow, or brownish	Colorless	Vitreous	Translucent	Ore of tungsten
Variegated shades of green	Colorless	Greasy, waxy, or silky	Translucent	Platy variety, *antigorite;* fibrous variety, *chrysotile,* an asbestos; an alteration product of magnesium silicates such as *olivine, augite,* and *hornblende;* common and widely distributed
Light to dark brown	Colorless	Vitreous	Transparent to translucent	An ore of iron; an accessory mineral in *taconite*
Brown, pale green, or white	Colorless	Vitreous	Transparent to translucent	Relatively rare but important as a mineral characteristic of high-grade metamorphism; contrast with *andalusite* and *kyanite,* which have the same composition but form under conditions of middle-grade metamorphism
White or green; with iron, yellow to brown and black; red	White to yellow and brown	Resinous	Transparent to translucent	A common mineral; the most important ore of zinc; the red variety is called *ruby zinc;* streak lighter than corresponding mineral color
White, red, lavender, blue, green, brown, or black	White	Vitreous	Usually translucent; may be clear and transparent	A common metamorphic mineral imbedded in crystalline limestone, gneisses, and *serpentine;* when transparent and finely colored, it is a gem; the red is spinel ruby, or *balas ruby;* some are blue
White, gray, pink, yellow, or green	White	Vitreous	Transparent to translucent	A source of lithium, which improves lubricating properties of greases; some gem varieties
Red brown to brownish black	Colorless	Fresh, resinous or vitreous; altered, dull to earthy	Translucent	A common accessory mineral in schists and slates; characteristic of middle-grade metamorphism; associated with *garnet, kyanite, sillimanite, tourmaline*
Lead gray to black	Lead gray to black	Metallic	Opaque	The chief ore of antimony, which is used in various alloys
				Unleached iron formation in the Lake Superior district, consists of chert (see *quartz*) with *hematite, magnetite, siderite,* and hydrous iron silicates; an ore of iron
Gray, white, silver white, or apple green	White	Pearly to greasy	Translucent	Of secondary origin, formed by the alteration of magnesium silicates such as *olvine, augite,* and *hornblende;* most characteristically found in metamorphic rocks
Iron black to reddish black	Dark red to black	Submetallic		Ore of tantalite, associated with columbite; is one end of a continuous solid-solution series, columbite is the other

Mineral	Chemical name and/or composition	Form	Cleavage and/or fracture	Hardness	Specific gravity
Topaz	Aluminum fluorosilicate, $Al_2SiO_4(F, OH)_2$	Usually in prismatic crystals, often with striations in direction of greatest length	Good in 1 direction	8	3.4–3.6
Tourmaline	Complex boron-aluminum silicate with Na, Ca, F, Fe, Li, or Mg	Usually in crystals; common with cross section of spherical triangle	Not prominent; black variety fractures like coal	7–7.5	3–3.25
Tremolite (an *amphibole*)	$Ca_2Mg_5(Si_8O_{22})(OH)_2$	Often bladed or in radiating columnar aggregate	Good in 1 direction	5–6	3.0–3.3
Uraninite (pitch-blende)	Complex uranium oxide with small amounts of Pb, Ra, Th, Y, N, He, and A	Usually massive and botryoidal	Not prominent	5.5	9–9.7
Wollastonite	Calcium silicate, $CaSiO_3$	Commonly massive, fibrous, or compact	Good in 2 directions at 84° and 96°	5–5.5	2.8–2.9
Wolframite	$(Fe, Mn)WO_4$	Crystals short prismatic, commonly granular massive	1 perfect cleavage	4–4.5	7.1–7.5
Zircon	Zirconium nesosilicate, $Zr(SiO_4)$	Tetragonal prism and dipyramid	Conchoidal fracture	7.5	4.68

Color	Streak	Luster	"Transparency"	Other properties
Straw yellow, wine yellow, pink, bluish, or greenish	Colorless	Vitreous	Transparent to translucent	Represents 8 on Mohs scale of hardness; a gemstone
Varied: black or brown; red, pink, green, blue or yellow	Colorless	Vitreous to resinous	Translucent	Gemstone; an accessory mineral in pegmatites, also in metamorphic rocks such as gneisses, schists, marbles
White to light green	White	Vitreous	Transparent to translucent	Frequently in impure, crystalline *dolomitic* limestones where it formed on recrystallization during metamorphism; also in *talc* schists
Black	Brownish black	Submetallic or pitchy	Opaque	An ore of uranium and radium; the mineral in which helium and radium were discovered
Colorless, white, or gray	Colorless	Vitreous or pearly on cleavage surfaces	Translucent	A common contact-metamorphic mineral in limestones
Brownish black to iron black	Reddish brown to brownish black	Submetallic	Opaque	Ore of tungsten
Brown; also gray, green, red, or colorless	Colorless	Adamantine	Translucent; sometimes transparent	Transparent variety is a gemstone; a source of zirconium metal, which is used in the construction of nuclear reactors

APPENDIX B

PERIODIC CHART

Legend (key to each element cell):

- Atomic number and symbol
- Mass compared to carbon 12 (approximate in parentheses)
- Most common electric charge / Ion radius ($\times 10^{-10}$ m) with that charge
- Next most common charge / Ion radius with that charge

Example:

16 S		
32.0		
−2	1.84	
+6	0.30	
sulfur		

Shading key:
- Element with only radioactive species in the earth's crust
- Radioactive element not found in the crust

Elements (atomic number, symbol, name, mass, charge / ion radius):

Z	Symbol	Name	Mass	Charge / radius	Next charge / radius
1	H	hydrogen	1.01	+1 / 0.01	
2	He	helium	4.00	0 / 1.08	
3	Li	lithium	6.94	+1 / 0.68	
4	Be	beryllium	9.01	+2 / 0.35	
5	B	boron	10.8	+3 / 0.23	
6	C	carbon	12.0	+4 / 0.16	
7	N	nitrogen	14.0	+5 / 0.13	−3 / 1.71
8	O	oxygen	16.0	−2 / 1.32	
9	F	fluorine	19.0	−1 / 1.33	
10	Ne	neon	20.2	0	
11	Na	sodium	23.0	+1 / 0.97	
12	Mg	magnesium	24.3	+2 / 0.66	
13	Al	aluminum	27.0	+3 / 0.51	
14	Si	silicon	28.1	+4 / 0.42	
15	P	phosphorus	31.0	+5 / 0.35	
16	S	sulfur	32.0	−2 / 1.84	+6 / 0.30
17	Cl	chlorine	35.5	−1 / 1.81	
18	Ar	argon	40.0	0	
19	K	potassium	39.1	+1 / 1.33	
20	Ca	calcium	40.1	+2 / 0.99	
21	Sc	scandium	45.0	+3 / 0.73	
22	Ti	titanium	47.9	+4 / 0.68	
23	V	vanadium	50.9	+5 / 0.59	
24	Cr	chromium	52.0	+3 / 0.63	+6 / 0.52
25	Mn	manganese	54.9	+4 / 0.60	+2 / 0.80
26	Fe	iron	55.8	+2 / 0.74	+3 / 0.64
27	Co	cobalt	58.9	+2 / 0.72	+3 / 0.63
28	Ni	nickel	58.7	+2 / 0.72	
29	Cu	copper	63.5	+2 / 0.72	+1 / 0.96
30	Zn	zinc	65.4	+2 / 0.74	
31	Ga	gallium	69.7	+3 / 0.62	
32	Ge	germanium	72.6	+4 / 0.53	
33	As	arsenic	74.9	+3 / 0.58	
34	Se	selenium	79.0	−2 / 1.91	
35	Br	bromine	79.9	−1 / 1.96	
36	Kr	krypton	83.8		
37	Rb	rubidium	85.5	+1 / 1.47	
38	Sr	strontium	87.6	+2 / 1.12	
39	Y	yttrium	88.9	+3 / 0.89	
40	Zr	zirconium	91.2	+4 / 0.79	
41	Nb	niobium	92.9	+5 / 0.69	
42	Mo	molybdenum	95.9	+4 / 0.70	+6 / 0.62
43	Tc	technetium	(99)		
44	Ru	ruthenium	101.	0 / 1.32	
45	Rh	rhodium	103.	0 / 1.34	
46	Pd	palladium	106.	0 / 1.37	
47	Ag	silver	108.	+1 / 1.26	
48	Cd	cadmium	112.	+2 / 0.97	
49	In	indium	115.	+3 / 0.81	
50	Sn	tin	119.	+4 / 0.71	
51	Sb	antimony	122.	+3 / 0.76	
52	Te	tellurium	128.	−2 / 2.11	
53	I	iodine	127.	−1 / 2.20	
54	Xe	xenon	131.		
55	Cs	cesium	133.	+1 / 1.67	
56	Ba	barium	137.	+2 / 1.34	
57	La	lanthanum	139.	+3 / 1.02	
58	Ce	cerium	140.	+3 / 1.03	
59	Pr	praseodymium	141.	+3 / 1.01	
60	Nd	neodymium	144.	+3 / 0.99	
61	Pm	promethium	(147)		
62	Sm	samarium	150.	+3 / 0.96	
63	Eu	europium	152.	+3 / 0.95	+2 / 1.09
64	Gd	gadolinium	157.	+3 / 0.94	
65	Tb	terbium	159.	+3 / 0.92	
66	Dy	dysprosium	162.	+3 / 0.91	
67	Ho	holmium	165.	+3 / 0.89	
68	Er	erbium	167.	+3 / 0.88	
69	Tm	thulium	169.	+3 / 0.87	
70	Yb	ytterbium	173.	+3 / 0.86	
71	Lu	lutetium	175.	+3 / 0.85	
72	Hf	hafnium	178.	+4 / 0.78	
73	Ta	tantalum	181.	+5 / 0.68	
74	W	tungsten	184.	+6 / 0.62	
75	Re	rhenium	186.	0 / 1.37	
76	Os	osmium	190.	0 / 1.33	
77	Ir	iridium	192.	0 / 1.36	
78	Pt	platinum	195.	0 / 1.37	
79	Au	gold	197.	0 / 1.44	
80	Hg	mercury	201.	+2 / 1.10	
81	Tl	thallium	204.	+3 / 0.95	
82	Pb	lead	207.	+2 / 1.20	
83	Bi	bismuth	209.	+3 / 0.96	
84	Po	polonium	(210)		
85	At	astatine	(210)		
86	Rn	radon	(222)		
87	Fr	francium	(223)		
88	Ra	radium	(226)		
89	Ac	actinium	(227)		
90	Th	thorium	232.	+4 / 1.02	
91	Pa	protactinium	(231)		
92	U	uranium	238.	+4 / 0.97	+6 / 0.80
93	Np	neptunium	(237)		
94	Pu	plutonium	(242)		
95	Am	americium	(243)		
96	Cm	curium	(247)		
97	Bk	berkelium	(247)		
98	Cf	californium	(249)		
99	Es	einsteinium	(254)		
100	Fm	fermium	(253)		
101	Md	mendelevium	(256)		
102	No	nobelium	(254)		
103	Lw	lawrencium	(257)		

[a] Lanthanides, or "rare earths"

[b] Actinides

EARTH DATA

TABLE C.1
Distribution of World's Estimated Supply of Water[a]

	Area, thousands of		Volume, thousands of		
	km²	mi²	km³	mi³	Total volume, %
World (total area)	510,000	197,000	—	—	—
Land area	149,000	57,500	—	—	—
Water in land areas:					
Freshwater lakes	850	330	125	30	0.009
Saline lakes and inland seas	700	270	104	25	0.008
Rivers (average instantaneous volume)	—	—	1.25	0.3	0.0001
Soil moisture and vadose water	—	—	67	16	0.005
Groundwater to depth of 4,000 m	—	—	8,350	2,000	0.61
Icecaps and glaciers	19,400	7,500	29,200	7,000	2.14
Atmospheric moisture	—	—	13	3.1	0.001
World ocean	361,000	139,500	1,320,000	317,000	97.3
Total water volume (rounded)			1,360,000	326,000	100

[a] In part after R. L. Nace, *U.S. Geol. Surv. Circ.* 536, Table 1, 1967.

TABLE C.2
Composition of Seawater at 35 Parts per Thousand Salinity[a]

Element	µg/l	Element	µg/l	Element	µg/l
Hydrogen	1.10×10^8	Nickel	6.6	Praesodymium	0.00064
Helium	0.0072	Copper	23	Neodymium	0.0023
Lithium	170	Zinc	11	Samarium	0.00042
Beryllium	0.0006	Gallium	0.03	Europium	0.000114
Boron	4,450	Germanium	0.06	Gadolinium	0.0006
Carbon (inorganic)	28,000	Arsenic	2.6	Terbium	0.0009
(dissolved organic)	2,000	Selenium	0.090	Dysprosium	0.00073
Nitrogen (dissolved N_2)	15,500	Bromine	6.73×10^4	Holmium	0.00022
(as NO_3^-, NO_2^-, NH_4^+)	670	Krypton	0.21	Erbium	0.00061
Oxygen (dissolved O_2)	6,000	Rubidium	120	Thulium	0.00013
(as H_2O)	8.83×10^8	Strontium	8,100	Ytterbium	0.00052
Fluorine	1,300	Yttrium	0.003	Lutetium	0.00012
Neon	0.120	Zirconium	0.026	Hafnium	<0.008
Sodium	1.08×10^7	Niobium	0.015	Tantalum	<0.0025
Magnesium	1.29×10^6	Molybdenum	10	Tungsten	<0.001
Aluminum	1	Ruthenium	—	Rhenium	—
Silicon	2,900	Rhodium	—	Osmium	—
Phosphorus	88	Palladium	—	Iridium	—
Sulfur	9.04×10^5	Silver	0.28	Platinum	—
Chlorine	1.94×10^7	Cadmium	0.11	Gold	0.011
Argon	450	Indium	—	Mercury	0.15
Potassium	3.92×10^5	Tin	0.81	Thallium	—
Calcium	4.11×10^5	Antimony	0.33	Lead	0.03
Scandium	<0.004	Tellurium	—	Bismuth	0.02
Titanium	1	Iodine	64	Radium	1×10^{-13}
Vanadium	1.9	Xenon	0.047	Thorium	0.0015
Chromium	0.2	Cesium	0.30	Protactinium	2×10^{-10}
Manganese	1.9	Barium	21	Uranium	3.3
Iron	3.4	Lanthanum	0.0029		
Cobalt	0.39	Cerium	0.0013		

[a] Adapted from Karl K. Turekian, *Oceans*, p. 92, Prentice Hall, Englewood Cliffs, N.J., 1968.

TABLE C.3
Runoff from the Continents[a]

| | Area, millions of | | Annual runoff | | | |
| | | | Total, $10^{15} \times$ | | Depth/unit area | |
	km²	mi²	l	gal	cm	in
Asia	46.6	18.0	11.1	3.0	23.8	9.4
Africa	29.8	11.5	5.9	1.6	19.8	7.8
North America	21.2	8.2	4.5	1.2	21.1	8.3
South America	19.6	7.6	8.0	2.1	41.4	16.3
Europe	10.9	4.2	2.5	0.6	23.1	9.1
Australia	7.8	3.0	0.4	0.1	2.5	1.0
Total or (mean)	135.9	52.5	32.4	8.6	(24.9)	(9.8)

[a] Calculated from Data from D. A. Livingstone, *U.S. Geol. Surv. Prof. Paper* 440-G, 1963.

TABLE C.4
Composition of River Waters of the World[a]

Substance	Ppm
HCO_3	58.4
Ca	15
SiO_2	13.1
SO_4	11.2
Cl	7.8
Na	6.3
Mg	4.1
K	2.3
NO_3	1
Fe	0.67

[a] From Daniel A. Livingstone, "Data of Geochemistry," *U.S. Geol. Surv. Prof. Paper* 440-G, p. G-41, 1963.

TABLE C.5
Earth Volume, Density, and Mass

	Av. thickness or radius, km	Volume, millions of km³	Mean density, g/cm³	Mass, $\times 10^{24}$ g
Total Earth	6,371	1,083,230	5.52	5,976
Oceans and seas	3.8	1,370	1.03	1.41
Glaciers	1.6	25	0.9	0.023
Continental crust	35	6,210	2.8	17.39
Oceanic crust	8	2,660	2.9	7.71
Mantle	2,883	899,000	4.5	4,068
Core	3,471	175,500	10.71	1,881

TABLE C.6
Average Composition of the Crust[a]

	Av. igneous rock, %	Av. shale, %	Av. sandstone, %	Av. limestone, %	Weighted-av. crust,[b] %
SiO_2	59.12	58.11	78.31	5.19	59.07
TiO_2	1.05	0.65	0.25	0.06	1.03
Al_2O_3	15.34	15.40	4.76	0.81	15.22
Fe_2O_3	3.08	4.02	1.08 ⎫	0.54	⎧ 3.10
FeO	3.80	2.45	0.30 ⎭		⎩ 3.71
MgO	3.49	2.44	1.16	7.89	3.45
CaO	5.08	3.10	5.50	42.57	5.10
Na_2O	3.84	1.30	0.45	0.05	3.71
K_2O	3.13	3.24	1.32	0.33	3.11
H_2O	1.15	4.99	1.63	0.77	1.30
CO_2	0.10	2.63	5.04	41.54	0.35
ZrO_2	0.04	—	—	—	0.04
P_2O_5	0.30	0.17	0.08	0.04	0.30
Cl	0.05	—	Tr[c]	0.02	0.05
F	0.03	—	—	—	0.03
SO_3	—	0.65	0.07	0.05	—
S	0.05	—	—	0.09	0.06
$(Ce, Y)_2O_3$	0.02	—	—	—	0.02
Cr_2O_3	0.06	—	—	—	0.05
V_2O_3	0.03	—	—	—	0.03
MnO	0.12	Tr[c]	Tr[c]	0.05	0.11
NiO	0.03	—	—	—	0.03
BaO	0.05	0.05	0.05	0.00	0.05
SrO	0.02	0.00	0.00	0.00	0.02
Li_2O	0.01	Tr[c]	Tr[c]	Tr[c]	0.01
Cu	0.01	—	—	—	0.01
C	0.00	0.80	—	—	0.04
Total	100.00	100.00	100.00	100.00	100.00

[a] After F. W. Clarke and H. S. Washington, "The Composition of the Earth's Crust," *U.S. Geol. Surv. Prof. Paper* 127, p. 32, 1924.
[b] Weighted average: igneous rock, 95%; shale, 4%; sandstone, 0.75%; limestone, 0.25%.
[c] Trace.

TABLE C.7
Earth Size

	Thousands of km
Equatorial radius	6.378
Polar radius	6.357
Mean radius[a]	6.371
Polar circumference	40.009
Equatorial circumference	40.077
Ellipticity [(equatorial radius − polar radius)/equatorial radius], 1/297	

[a] Term used by geophysicists to designate radius of a sphere of equal volume.

TABLE C.8
Earth Areas

	Millions of km^2
Total area	510
Land (29.22% of total)	149
Oceans and seas (70.78% of total)	361
Glacier ice	15.6
Continental shelves	28.4

TOPOGRAPHIC
AND GEOLOGIC MAPS

Topography refers to the shape of the physical features of the land. A **topographic map** is a representation of the shape, size, position, and relation of the physical features of an area. In addition to mountains, hills, valleys, and rivers most topographic maps show the culture of a region, that is, political boundaries, towns, houses, roads, and similar features.

D.1
TOPOGRAPHIC MAPS

Topographic maps are used in the laboratory for the observation and analysis of the geological processes that are constantly changing the face of the Earth.

DEFINITIONS

Elevation, or Altitude   The vertical distance between a given point and the datum plane.

Datum Plane   The reference surface from which all altitudes on a map are measured. This is usually mean sea level.

Height   The vertical difference in elevation between an object and its immediate surroundings.

Relief   The difference in elevation of an area between tops of hills and bottoms of valleys.

Bench Mark   A point of known elevation and position —usually indicated on a map by the letters B.M., with the altitude given (on American maps) to the nearest foot. (Some maps now use meters.)

Contour Line   A map line connecting points representing places on the Earth's surface that have the same eleva-

tion. It thus locates the intersection with the Earth's surface of a plane at any arbitrary elevation parallel to the datum plane. Contours represent the vertical, or third, dimension on a map, which has only two dimensions. They show the shape and size of physical features such as hills and valleys. A depression is indicated by an ordinary contour line except that hachures, or short dashes, are used on one side and point toward the center of the depression.

Contour Interval   The difference in elevation represented by adjacent contour lines.

Scale   The ratio on a map of the distance between two points on the ground and the same two points on the map. It may be expressed in three ways:

Fractional Scale.   If two points are 1 km apart in the field, they may be represented on the map as separated by some fraction of that distance, say, 1 cm. In this instance the scale is 1 cm to the kilometer. There are 100,000 cm in 1 km; so this scale can be expressed as the fraction, or ratio, 1:100,000. Many topographic maps of the United States Geological Survey have a scale of 1:62,500; and many recent maps have a scale of 1:31,250, and others of 1:24,000.

Graphic Scale.   This scale is a line printed on the map and divided into units that are equivalent to some distance, as much as 1 km or 1 mi.

Verbal Scale.   This is an expression in common speech, such as "four centimeters to the kilometer," "an inch to a mile," or "two miles to the inch."

CONVENTIONAL SYMBOLS

An explanation of the symbols used on topographic maps is printed on the back of each topographic sheet, along the margin; or, for newer maps, on a separate legend sheet. In general, culture (artificial works) is shown in black. All water features, such as streams, swamps, and glaciers, are shown in blue. Relief is shown by contours in brown. Red may be used to indicate main highways, and green overprints may be used to designate areas of woods, orchards, vineyards, or scrub.

The United States Geological Survey distributes free of charge a single sheet, entitled "Topographic Maps," that includes an illustrated summary of topographic-map symbols. (Apply to Map Distribution Center, U.S. Geological Survey, Federal Center, Building 41, Box 25286, Denver, Colorado 80225.)

LOCATING POINTS

Any particular point or area may be located in several ways on a topographic map. The three most commonly used are:

In Relation to Prominent Features   A point may be referred to as being so many kilometers in a given direction from a city, mountain, lake, river mouth, or other easily located feature on the map.

By Latitude and Longitude   Topographic maps of the Geological Survey are bounded on the north and south by parallels of latitude and on the east and west by meridians of longitude. These intersecting lines form the grid into which the earth has been divided. Latitude is measured north and south from the equator, and longitude is measured east and west from the prime meridian that passes through Greenwich, England. Thus maps of the United States are within north latitude and west longitude.

By Township and Range   Much of the area of the United States has been subdivided by a system of land survey in which a square 6 mi on a side is the basic unit, called a **township** when measured north and south of a given base line, and called a **range** when measured east and west of a given principal meridian. Not included in this system are all the states along the eastern seaboard (with the exception of Florida), West Virginia, Kentucky, Tennessee, Texas, and parts of Ohio. Townships are laid off north and south from a base line and east and west from a principal meridian. Each township is divided into 36 sections, usually 1 mi on a side. Each section may be further subdivided into half sections, quarter sections, or sixteenth sections. Thus in Figure D.1 the point x is located in the northeast quarter of the northwest quarter of section 3, township 9 north, range 5 west, abbreviated as NE¼NW¼Sec 3, T9N, R5W, or NE NW Sec 3–9N-5W.

CONTOUR SKETCHING

Many contour maps are now made from aerial photographs. Before this can be done, however, the position and location of a number of reference points, or bench marks,

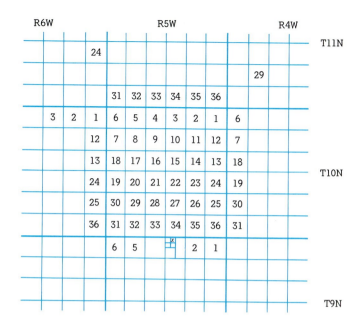

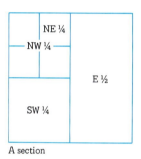

A section

D.1   Subdivision by township and range (see text for discussion).

must be determined in the field. If the topographic map is prepared from field surveys rather than from aerial photographs, the topographer first determines the location and elevation of bench marks and a large number of other points selected for their critical position. They may be on hilltops, on the lowest point in a saddle between hills, along streams, or at places where there is a significant change in slope. On the basis of these points, contours may be sketched through loci of equal elevation. The contours are preferably sketched in the field in order to include minor irregularities that are visible to the topographer.

Because contours are not ordinary lines, certain requirements must be met in drawing them to satisfy their definition. These are listed below:

1. All points on one contour line have the same elevation.
2. Contours separate all points of higher elevation than the contour from all points of lower elevation.
3. The elevation represented by a contour line is always a simple multiple of the contour interval. Every contour line that is a multiple of five times the contour interval is heavier than the others. (An exception is

25-unit contours, in which every multiple of four times the interval is heavier.)

4. Contours never cross or intersect one another.

5. A vertical cliff is represented by coincident contours.

6. Every contour closes on itself either within or beyond the limits of the map. In the latter case, the contours will end at the edge of the map.

7. Contour lines never split.

8. Uniformly spaced contour lines represent a uniform slope.

9. Closely spaced contour lines represent a steep slope.

10. Contour lines spaced far apart represent a gentle slope.

11. A contour line that closes within the limits of the map indicates a hill.

12. A hachured contour line represents a depression. The short dashes, or hachures, point into the depression.

13. Contour lines curve up a valley but cross a stream at right angles to its course.

14. Maximum ridge and minimum valley contours always go in pairs: that is, no single lower contour can lie between two higher ones and vice versa.

TOPOGRAPHIC PROFILES

A topographic profile is a cross section of the Earth's surface along a given line. The upper line of this section is irregular and shows the shape of the land along the line of profile, or section.

Profiles are most easily constructed with graph paper. A horizontal scale, usually the map scale, is chosen. Then a vertical scale sufficient to bring out the features of the surface is chosen. The vertical scale is usually several times larger than the horizontal—that is, it is exaggerated. The steps in the construction of a profile are:

1. Select a base (one of the horizontal lines on the graph paper). This may be sea level or any other convenient datum.

2. On the graph paper number each fourth or fifth line above the base, according to the vertical scale chosen.

3. Place the graph paper along the line of profile.

4. With the vertically ruled lines as guides, plot the elevation of each contour line that crosses the line of profile.

5. If great accuracy is not important, plot only every heavy contour and the tops of hills and the bottoms of valleys.

6. Connect the points.

7. Label necessary points along the profile.

8. Give the vertical and horizontal scales.

9. State the vertical exaggeration.

10. Title the profile.

Vertical Exaggeration   The profile represents both vertical and horizontal dimensions. These dimensions are not usually on the same scale because the vertical needs to be greater than the horizontal to give a clear presentation of changes in level. Thus if the vertical scale is 500 m to the centimeter and the horizontal scale is 5,000 m to the centimeter, the vertical exaggeration is 10 times, written 10×. This is obtained by dividing the horizontal scale by the vertical scale. Note that both horizontal and vertical scales must be expressed in the same unit (in this instance meters to the centimeter) before division.

D.2
GEOLOGIC MAPS

Geologic maps show the distribution of Earth materials on the surface. In addition they indicate the relative age of these materials and suggest their arrangement beneath the surface.

DEFINITIONS

Formation   The units illustrated upon a geologic map are usually referred to as **formations.** We define a formation as a rock unit with upper and lower boundaries that can be recognized easily in the field and that is large enough to be shown on the map. A formation receives a distinctive designation made up of two parts: The first part is geographic and refers to the place or general area where the formation is first described; the second refers to the nature of the rock. Thus **Trenton Limestone** is a formation dominantly composed of limestone and is named after Trenton Falls in central New York, where it was first formally described. **Wausau Granite** designates a body of granite in the Wausau, Wisconsin, area. If the lithology is so variable that no single lithologic distinction is appropriate, the word *formation* may be used. For instance, the **Raritan Formation** is named for the area of the Raritan River and Raritan Bay in New Jersey, and its lithology includes both sand and clay.

Dip and Strike   The dip and strike of a rock layer refer to its orientation in relation to a horizontal plane. In Chapter 9 we found that the dip is the acute angle that a tilted rock layer makes with an imaginary plane. We also found that the strike is the compass direction of a line formed by the intersection of the dipping surface with an imaginary horizontal plane. The direction of strike is always at right angles to the direction of dip. The dip-and-strike symbol used on a geologic map is in the form of a top-heavy tee. The crossbar represents the direction of the strike of the bed. The short upright represents the direction of the dip of the bed. Very often the angle of dip is indicated alongside the symbol. Here is an example (the top of the page is considered to be north):

$\top$ 30 = strike E-W; dip 30°S

$\swarrow$ 25 = strike N45°E; dip 25°SE

Contact A contact is the plane separating two rock units. It is shown on the geologic map as a line that is the intersection of the plane between the rock units and the surface of the ground.

Outcrop An outcrop is rock material that is exposed at the surface through the cover of soil and weathered material. In areas of abundant rainfall soil and vegetation obscure the underlying rock and only a small fraction of 1 percent of the surface may be in outcrop. In dry climates, where soils are shallow or absent and the plant cover is discontinuous, bedrock usually crops out much more widely.

Legend and Symbols A legend is an explanation of the various symbols used on the map. There is no universally accepted set of standard symbols, but some that are widely used are given in Figure D.2. In addition to the graphic symbols in Figure D.2, letter symbols are sometimes used to designate rock units. Such a symbol contains a letter or letters referring to the geologic column, followed by a letter or letters referring to the specific name of the rock unit. Thus in the symbol Ot the O stands for Ordovician and the t for the Trenton Limestone of central New York. The letters or abbreviations generally used for the geologic column are given in Table D.1.

Sometimes different colors are used to indicate different rock systems. There is no standardized color scheme, but many of the geologic maps of the Geological Survey use the colors given in Table D.1, combined with varying pat-

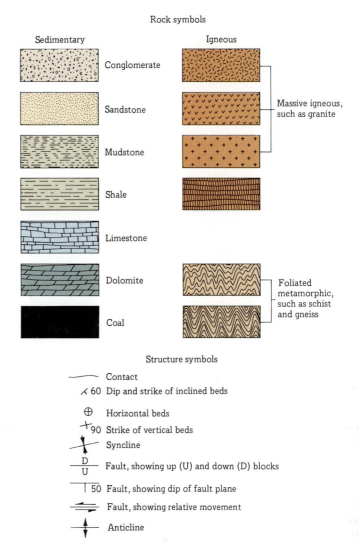

D.2 Symbols commonly used on geologic maps.

TABLE D.1
Letter Symbols and Colors Commonly Used to Designate Units in the Geologic Column

Period	Symbol	Color
Pleistocene	Q	Yellow and gray
Pliocene	Tpl	Yellow ocher
Miocene	Tm	Yellow ocher
Oligocene	To	Yellow ocher
Eocene	Te	Yellow ocher
Paleocene	Tp	Yellow ocher
Cretaceous	K	Olive green
Jurassic	J	Blue green
Triassic	T_R	Light peacock blue or bluish gray green
Permian	P	Blue
Pennsylvanian	Cp	Blue
Mississippian	Cm	Blue
Devonian	D	Gray purple
Silurian	S	Purple
Ordovician	O	Red purple
Cambrian	$\mathcal{C}$	Brick red
Precambrian	P_C	Terracotta and gray brown

terns, for systems of sedimentary rocks. No specific colors are designated for igneous rocks, but when colors are used, they are usually purer and more brilliant than those used for sedimentary rocks.

CONSTRUCTION OF A GEOLOGIC MAP

The basic idea of geologic mapping is simple. We are interested first in showing the distribution of the rocks at the Earth's surface. Theoretically all we need to do is plot the occurrence of the different rocks on a base map, and then we have a geologic map. Unfortunately the process is not quite so simple.

In most areas the bedrock is more or less obscured in one way or another, and only a small amount of outcrop is available for observation, study, and sampling. From the few exposures available, the geologist must extrapolate the general distribution of rock types. In this extrapolation the field data are obviously of prime importance. But one will also be guided by changes in soil, vegetation, and landscape as well as by patterns that can be detected on aerial photo-

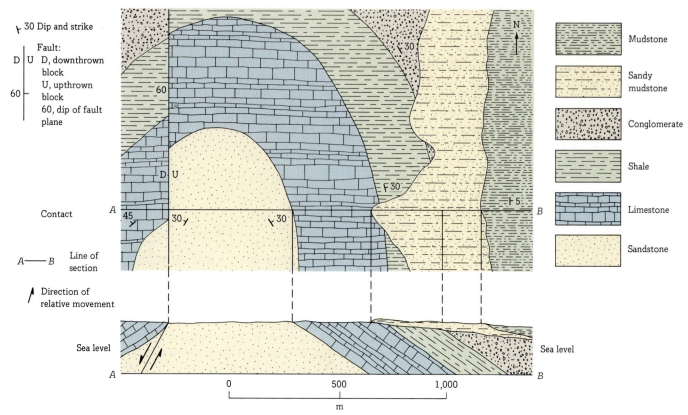

Legend labels within figure:

30 Dip and strike

Fault:
D | U   D, downthrown block
        U, upthrown block
60   60, dip of fault plane

Contact

A —— B   Line of section

Direction of relative movement

Mudstone

Sandy mudstone

Conglomerate

Shale

Limestone

Sandstone

D.3 Construction of a geologic cross section from a geologic map.

graphs. Furthermore, one may be aided by laboratory examination of field samples and by the records of both deep and shallow wells. One may also have available geophysical data that help determine the nature of obscured bedrock. Eventually, when one has marshaled as many data as possible, one draws the boundaries delineating the various rock types.

In addition to the distribution of rock types, the geologist is also concerned with depicting, as accurately as one can, the ages of the various rocks and their arrangement beneath the surface. These goals will also be realized in part through direct observations in the field and in part through other lines of evidence. The preparation of an accurate, meaningful, geologic map demands experience, patience, and judgment.

GEOLOGIC CROSS SECTIONS

A geologic map tells us something of how rocks are arranged in the underground. Often, to show these relations more clearly, we find it convenient to draw geologic cross sections. Such a section is really a diagram showing a side view of a block of the Earth's crust as it would look if we could lift it up to view. We have used cross sections in many illustrations throughout this book.

A geologic cross section is drawn, insofar as possible, at right angles to the general strike of the rocks. The general manner in which a geologic cross section is projected from a geologic map is shown in Figure D.3. If the projection is made onto a topographic profile in which the vertical scale has been exaggerated, then the angle of dip of the rocks should be exaggerated accordingly.

DIRECTORY OF EARTH SCIENCE ORGANIZATIONS AND GOVERNMENT GEOLOGIC SURVEYS IN MEXICO, UNITED STATES AND CANADA*

E.1

EARTH SCIENCE ORGANIZATIONS

Abilene Geological Society
Box 974, Abilene, Tex., 79604

Alabama Geological Society
Box 6184, University, Ala., 35486

Alaska Geological Society
Box 101288, Anchorage, Alaska, 99510

Albuquerque Geological Society
Box 26884, Albuquerque, N.M., 87125

Alumni Association of the San Diego State University Geology Department
Geology Department, San Diego State University, San Diego, Calif., 92182

American Association for the Advancement of Science
1333 H St. NW, Washington, D.C., 20005

American Association for Crystal Growth
Solar Energy Research Institute, 1617 Cole Blvd., Golden, Colo., 80401

* Compiled by the American Geological Institute, 4220 King Street, Alexandria, Virginia 22302-1507. For organizations and government surveys outside of North America see *Geotimes,* Vol. 34, No.10, October 1989.

American Association of Petroleum Geologists
Box 979, Tulsa, Okla., 74101
918/584-2555

American Association of Stratigraphic Palynologists
Gordon D. Wood, Amoco Production Co., Box 3092, Houston, Tex., 77253

American Astronautical Society
6212 Old Keene Mill Court, Springfield, Va., 22152

American Astronomical Society
Louisiana State University, Box BK, LSU OBS, Baton Rouge, La., 70803

American Astronomical Society, Division for Planetary Sciences
Lowell Observatory, 1400 W. Mars Hill Road, Flagstaff, Ariz., 86001

American Congress on Surveying & Mapping
210 Little Falls St., Falls Church, Va., 22046

American Crystallographic Association
Box 96 Ellicott Station, Buffalo, N.Y., 14205-0096

American Gas Association
1515 Wilson Blvd., Arlington, Va., 22209

American Geographical Society
156 5th Ave., Suite 600, New York, N.Y., 10010

American Geographical Society Collection
University of Wisconsin Library, Box 399, Milwaukee, Wis., 53201

American Geological Institute
4220 King St., Alexandria, Va., 22302-1507
703/379-2480

American Geomorphological Field Group
Department of Geology & Geophysics, University of California, Berkeley, Calif., 94720

American Geophysical Union
2000 Florida Ave. NW, Washington, D.C., 20009

American Institute of Aeronautics & Astronautics
1633 Broadway, New York, N.Y., 10019

American Institute of Hydrology
3416 University Ave. SE, Suite 200, Minneapolis, Minn., 55414

American Institute of Mining, Metallurgical & Petroleum Engineers
345 E. 47th St., New York, N.Y., 10017

American Institute of Professional Geologists
7828 Vance Drive, Suite 103, Arvada, Colo., 80003
303/431-0831

American Meteorological Society

45 Beacon St., Boston, Mass., 02108

American Mining Congress
1920 N St. NW, Washington, D.C., 20036

American Nuclear Society
555 N. Kensington Ave., La Grange Park, Ill., 60525

American Petroleum Institute
1220 L St. NW, Washington D.C., 20005

American Quaternary Association
Department of Geological Sciences, University of Michigan, Ann Arbor, Mich., 48109-1063

American Society for Photogrammetry & Remote Sensing
210 Little Falls St., Falls Church, Va., 22046

American Society of Agronomy
667 S. Segoe Road, Madison, Wis., 53711

American Society of Civil Engineers
Geotechnical Engineering Division, 345 E. 47th St., New York, N.Y., 10017

American Society of Limnology & Oceanography
Virginia Institute of Marine Science, College of William & Mary, Gloucester Point, Va., 23062

American Water Resources Association
5410 Grosvenor Lane, Suite 220, Bethesda, Md., 20814

American Water Works Association
6666 W. Quincy Ave., Denver, Colo., 80235

Appalachian Geological Society
Box 2605, Charleston, W.Va., 25329

Ardmore Geological Society
Box 1552, Ardmore, Okla., 73402

Arizona Geological Society
Box 40952, Tucson, Ariz., 85717

Asociación Mexicana de Geofísicos de Exploración
Apartado Postal 57275, C.P. 06501, D.F., Mexico, D.F.

Asociación Mexicana de Geólogos Petroleros
Apartado Postal 57275, C.P. 06500, Mexico, D.F.

Association for Women Geoscientists
Box 1005, Menlo Park, Calif., 94026
303/422-8527

Association of American Geographers
1710 6th St. NW, Washington, D.C., 20009-3198

Association of American State Geologists
American Geological Institute, 4220 King St., Alexandria, Va., 22302
703/379-2480

Association of Black Geoscientists
Howard University, Department of Geology & Geography, Box 1098, Washington, D.C., 20059

Association of Earth Science Editors
USGS, Mail Stop 903, National Center, Reston, Va., 22092
703/648-4332

Association of Engineering Geologists
62 King Philip, Sudbury, Mass., 01776
615/377-3578

Association of Exploration Geochemists
Box 523, Rexdale, Ont., M9W 5L4, Canada

Association of Missouri Geologists
Box 250, Rolla, Mo., 65401

Association of Professional Engineers, Geologists & Geophysicists of Alberta
Scotia Place, 15th Floor, Tower 1, 10060 Jasper Ave., Edmonton, Alberta, T5J 4A2, Canada

Association Québécoise pour l'Etude du Quaternaire
Département de Géographie, Université de Sherbrooke, Sherbrooke, Quebec, J1K 2R1, Canada

Astronomical Society of the Pacific
390 Ashton Ave., San Francisco, Calif., 94112

Atlantic Coastal Plain Geological Association
Delaware Geological Survey, University of Delaware, Newark, Del., 19716

Atlantic Geoscience Society
Atlantic Geoscience Centre, Bedford Institute of Oceanography, Box 1006, Dartmouth, N.S., B2Y 4A2, Canada

Austin Geological Society
Box 1302, Austin, Tex., 78767

Baton Rouge Geological Society
Box 19151, University Station, Baton Rouge, La., 70893

Bay Area Geophysical Society
625 Post St., Suite 417, San Francisco, Calif., 94109

Bay Area Mineralogists
Menlo College, 1000 El Camino Real, Atherton, Calif., 94025-4185

Baylor Geological Society
CSB 367, Baylor University, Waco, Tex., 76798

Bemidji State Geologic Association
Sattgast Hall, Bemidji State University, Bemidji, Minn., 56601

Billings Geological Society
Box 3481, Billings, Mont., 59103

Botanical Society, Paleobotany Section
Department of Botany, University of Alberta, Edmonton, Alberta, T6G 2E9, Canada

Buffalo Association of Professional Geologists, Inc.
Box 1254, Ellicott Station, Buffalo, N.Y., 14205-1254

Byrd Polar Research Center
125 S. Oval Mall, Ohio State University, Columbus, Ohio, 43210

California Earthquake Society
Box 686, San Marcos, Calif., 92069

Canadian Exploration Geophysical Society
Allan Spector & Associates, 24 Strathallon Blvd., Toronto, Ont., M5N 1S7, Canada

Canadian Geoscience Council
Department of Earth Sciences, University of Waterloo, Waterloo, Ont., N2L 3G1, Canada

Canadian Geotechnical Society
602, 170 Attwell Drive, Rexdale, Ont., M9W 5Z5, Canada

Canadian Geothermal Resources Association
Box 5059, Vancouver, B.C., V6B 4A9, Canada

Canadian Institute of Mining & Metallurgy
400-1130 Sherbrooke St. West, Montreal, Quebec, H3A 2M8, Canada

Canadian Society of Exploration Geophysicists
501, 206 7th Ave. SW, Calgary, Alberta, T2P 0W7, Canada

Canadian Society of Petroleum Geologists
505, 206 7th Ave. SW, Calgary, Alberta, T2P 0W7, Canada

Canadian Society of Soil Science
Guelph Agriculture Centre, 52 Royal Road, Guelph, Ontario, N1H 1G3, Canada

Canadian Well Logging Society
229, 640 5th Ave. SW, Calgary, Alberta, T2P 0M6, Canada

Carolina Geological Society
Department of Geology, Duke University, Durham, N.C., 27706

Casper Geophysical Society
UNOCAL, Box 2620, Casper, Wyo., 82602

Central Carolina Borehole Grotto
3302 Starmount, Greensboro, N.C., 27403

Charles Camsell Geological Society
Geological Surveys, Bag 9100, Yellowknife, Northwest Territories, X1A 2R3, Canada

Circum-Pacific Council for Energy & Mineral Resources
Michel T. Halbouty, 5100 Westheimer Road, Houston, Tex., 77056

Circum-Pacific Jurassic Research Group
(IGCP Project 171) Department of Geology, McMaster University, Hamilton, Ont., L8S 4M1, Canada

Clay Minerals Society
Box 2295, Bloomington, Ind., 47402

Club de Minéralogie de Montréal
C.P. 305, Succursale St-Michel, Montréal, Québec, H2A 3M1, Canada

Coastal Bend Geophysical Society
Box 2741, Corpus Christi, Tex., 78403-2741

Colorado River Association
417 South Hill St., Suite 1024, Los Angeles, Calif., 90013

Colorado Scientific Society
Box 150495, Lakewood, Colo., 80215

Computer Oriented Geological Society
Carol Petersen, Box 1317, Denver, Colo., 80201

Corpus Christi Geological Society
Box 1068, Corpus Christi, Tex., 78403

Cushman Foundation for Foraminiferal Research
Room E-206, MRC NHB 121, National Museum of Natural History, Washington, D.C., 20560

Dallas Geological Society
One Energy Square, Suite 100, Dallas, Tex., 75206

Denver Region Exploration Geologists Society
5025 Ward Road, Suite 508, Wheat Ridge, Colo., 80033

Dibblee Geological Foundation
Box 60560, Santa Barbara, Calif., 93160

Earthquake Engineering Research Institute
6431 Fairmount Ave., #7, El Cerrito, Calif., 94530

East Texas State University Geological Society
Department of Earth Sciences, East Texas State University, Commerce, Tex., 75428

Edmonton Geological Society
15th Floor, Scotia Place, Tower 1, 10060 Jasper Ave., Edmonton, Alberta, T57 4A2, Canada

El Paso Geological Society
Department of Geological Sciences, University of Texas, El Paso, Tex., 79968

Fine Particle Society
University of Iowa, Iowa City, Iowa, 52242

Florida Paleontological Society
Florida State Museum, University of Florida, Gainesville, Fla., 32611

Fort Worth Geological Society
900 Oil & Gas Bldg., Fort Worth, Tex., 76102

Four Corners Geological Society
Box 1501, Durango, Colo., 81302

Franklin-Ogdensburg Mineralogical Society
Box 146, Franklin, N.J., 07416

Friends of Mineralogy
1173 W. Lake Ave., Guilford, Ct., 06437

Friends of Mineralogy, Colorado Chapter
Department of Geology, Denver Museum of Natural History, City Park, Denver, Colo., 80205

Friends of Mineralogy, Pacific Northwest Chapter
Box 85, Mailroom, Seattle University, Seattle, Wash., 98122

Friends of Sherlock Holmes

3900 Tunlaw Road #119, Washington, D.C., 20007-4830

Gemological Institute of America
1660 Stewart St., Santa Monica, Calif., 90404

Geochemical Society
Department of Geological Sciences, Wright State University, Dayton, Ohio, 45435

Geological Association of Canada
Department of Earth Sciences, Memorial University of Newfoundland, St. John's, Newfoundland, A1B 3X5, Canada

Geological Society of America
3300 Penrose Place, Box 9140, Boulder, Colo., 80301
303/477-2020

Geological Society of Iowa
Iowa Department of Natural Resources, Geological Survey Bureau, 123 N. Capitol St., Iowa City, Iowa, 52242

Geological Society of Kentucky
228 Mining & Mineral Resources Bldg., University of Kentucky, Lexington, Ky., 40506-0107

Geological Society of the Oregon Country
Box 8579, Portland, Ore., 97207

Geophysical Society of Tulsa
Phillips Petroleum, 2919 Ridge Court, Bartlesville, Okla., 74006

Geophysical Society of Washington
U.S. Geological Survey, MS 911, Reston, Va., 22092

Geophysical Society of Edmonton
Imperial Oil Enterprises Ltd., 10025 Jasper Ave., Edmonton, Alberta, T5J 0J0, Canada

Georgia Geological Society
Department of Geology, Georgia State University, University Plaza, Atlanta, Ga., 30303-3083

Geoscience Information Society
American Geological Institute, 4220 King St., Alexandria, Va., 22302
703/379-2480

Geothermal Resources Council
Box 1350, Davis, Calif., 95617

Grand Canyon Natural History Association
Box 129, Grand Canyon, Ariz., 86023

Grand Junction Geological Society
William L. Chenoweth, 707 Brassie Drive, Grand Junction, Colo., 81506-3911

Gulf Coast Association of Geological Societies
Earth Enterprises Inc., Box 672, Austin, Tex., 78767

Harrisburg Area Geological Society
Pennsylvania Geological Survey, Box 2357, Harrisburg, Pa., 17120

Hells Canyon Geological Society
Department of Geology, University of Wisconsin, Eau Claire, Wis., 54702-4004

Herrick Society
Department of Geology & Geography, Denison University, Granville, Ohio, 42023

History of the Earth Sciences Society
Kennard B. Bork, Department of Geology & Geography, Denison University, Granville, Ohio, 43023

Houston Geological Society
6916 Ashcroft, Houston, Tex., 77081

Idaho Association of Professional Geologists
Box 7584, Boise, Idaho, 83707

Illinois Groundwater Association
Illinois State Water Survey, 2204 Griffith Drive, Champaign, Ill., 61820

Independent Petroleum Association of America
1101 16th St. NW, Washington, D.C., 20036

Inland Geological Society
San Bernardino County Museum, 2024 Orange Tree Lane, Redlands, Calif., 92374

Institute of Environmental Sciences
940 E. Northwest Highway, Mount Prospect, Ill., 60056

Institute of the Expanding Earth
Richard W. Guy, Box 144, Kingston 5, Jamaica

Institute on Lake Superior Geology
Department of Geological Engineering, Geology & Geophysics, Michigan Technological University, Houghton, Mich., 49931

Instituto de Geología U.N.A.M.
Cesar Jacques-Ayala, Estación Regional del Noroeste, Apartado Postal 1039,

Hermosillo Sonora, Mexico, 83000

Instrument Society of America
Instrument Division, Bourns Inc., 6135 Magnolia Ave., Riverside, Calif., 92506

International Association for Mathematical Geology
U.S. Geological Survey, National Center, Mail Stop 920, Reston, Va., 22092

International Association of Geochemistry & Cosmochemistry
Brian Hitchon, Alberta Research Council, Box 8330, Station F, Edmonton, Alberta, T6H 5X2, Canada

International Association on the Genesis of Ore Deposits
Geological Survey of Canada, 601 Booth St., Ottawa, Ont., K1A 0E8, Canada

International Landslide Research Group
William Cotton & Associates, 330 Village Lane, Los Gatos, Calif., 95030

International Mountain Society
Box 3128, Boulder, Colo., 80307

International Palaeontological Association
U.S. Geological Survey, E-305 Natural History Building, Smithsonian Institution, Washington, D.C., 20560

International Permafrost Association
J. Ross Mackay, Department of Geography, 217-1984 West Mall, University of British Columbia, Vancouver, B.C., V6T 1W5, Canada

International Stop Continental Drift Society
Star Route 38, Winthrop, Wash., 98862

International Union of Geodesy & Geophysics
U.S. National Committee, American Geophysical Union, 2000 Florida Ave. NW, Washington, D.C., 20009

Interstate Natural Gas Association of America
1660 L St. NW, Washington, D.C., 20036

Jackson Geophysical Society
Box 52, Jackson, Miss., 39205

Kansas Geological Society
212 North Market, Suite 100, Wichita, Kan., 67202

Kentucky Geological Society, Lexington Branch
Department of Geology, University of

Kentucky, Lexington, Ky., 40506

Lafayette Geological Society
Box 51896, Oil Center Station, Lafayette, La., 70505

Lamar University Geological Society
Box 10031, L.U. Station, Beaumont, Tex., 77710

Los Angeles Basin Geological Society
Box 1072, Bakersfield, Calif., 93302

Marine Technology Society
2000 Florida Ave. NW, Suite 500, Washington, D.C., 20009

Meteoritical Society
Geology Department, University of Tennessee, Knoxville, Tenn., 37996-1410

Mexican Geological Society
Museo de Geología, Calle Jaime Torres Bodet 176, Col. Santa María La Ribera, Mexico 06400, D.F. Mexico

Miami Geological Society
Box 144333, Coral Gables, Fla., 33114

Michigan Basin Geological Society
Department of Geological Sciences, Michigan State University, 206 Natural Science Building, East Lansing, Mich., 48824-1115

Michigan Earth Science Teachers Association
Science Department, Lansing Community College, Lansing, Mich., 48901

Microbeam Analysis Society
Institute for Materials Research, National Bureau of Standards, Washington, D.C., 20234

Mid-American Paleontology Society
2623 34th Ave. Court, Rock Island, Ill., 61201

Mineralogical Association of Canada
Department of Mineralogy, Royal Ontario Museum, 100 Queen's Park, Toronto, Ont., M5S 2C6, Canada

Mineralogical Society of America
Susan L. Myers, 1625 I St. NW, Suite 414, Washington, D.C., 20006
202/775-4344

Mining & Metallurgical Society of America
160 Sansome St., 16th floor, San Francisco, Calif., 94104

Montana Geological Society
Box 844, Billings, Mont., 59103

Monterey Bay Geological Society
Moss Landing Marine Laboratories, Box 450, Moss Landing, Calif., 95039

National Academy of Sciences
2101 Constitution Ave. NW, Washington, D.C., 20418

National Association of Black Geologists & Geophysicists
Box 720157, Houston, Tex., 77272
913/843-1235

National Association of Geology Teachers
Box 368, Lawrence, Kan., 66044

National Earth Science Teachers Association
Department of Geological Sciences, Michigan State University, East Lansing, Mich., 48824

National Geothermal Association
Box 1350, Davis, Calif., 95617

National Petroleum Council
1625 K St. NW, Washington, D.C., 20006

National Science Teachers Association
1742 Connecticut Ave. NW, Washington, D.C., 20009

National Speleological Society
1 Cave Ave., Huntsville, Ala., 35810

National Water Well Association
6375 Riverside Drive, Dublin, Ohio, 43017

New Orleans Geological Society
Box 52172, New Orleans, La., 70152-9989

New York City Department of General Services
Subsurface Exploration Section, Rm. 2214, 1 Centre St., New York, N.Y., 10007

New York State Geological Association
Department of Earth Sciences, SUNY College, Plattsburgh, N.Y., 12901

North American Cartographic Information Society
6010 Executive Blvd., Suite 100, Rockville, Md., 20852

North Dakota Geological Society
Box 82, Bismarck, N.D., 58501

North Texas Geological Society
Box 1671, Wichita Falls, Tex., 76307

Northern California Geological Society
Lawrence Livermore National Lab., Box 808 (L-279), Livermore, Calif., 94550

Northern Illinois University Geophysical Society
Department of Geology, Northern Illinois University, DeKalb, Ill., 60115

Northern Ohio Geological Society
Department of Geological Sciences, Case Western Reserve University, Cleveland, Ohio, 44106

Northwest Geology Society
19344 11 Ave. NW, Seattle, Wash., 98177

Northwest Mining Association
414 Peyton Bldg., Spokane, Wash., 99201

Offshore Technology Conference
Box 833868, Richardson, Tex., 75083

Ohio Geological Society
Box 14322, Beechwold Station, Columbus, Ohio, 43214-0322

Oklahoma City Geological Society
1020 Cravens Bldg., Oklahoma City, Okla., 73102

Pacific Science Association
Box 17801, Honolulu, Hawaii, 96817

Paleontological Society
John Pojeta Jr., 1492 Dunster Lane, Rockville, Md., 20854

Panamerican Institute of Geography & History
Commission of Geophysics, R.R. #1, Union, Ont., N0L 2L0, Canada

Pander Society
Indiana Geological Survey, 611 N. Walnut Grove, Bloomington, Ind., 47405

Peninsula Geological Society
Dept. of Geophysics, Stanford University, Stanford, Calif., 94305

Permian Basin Geophysical Society
Box 361, Midland, Tex., 79702

Petroleos Mexicanos Pemex
Alfredo Eduardo Guzman, A.P. Postal 1260, Chihuahua, Chih., Mexico

Petroleum Philatelic Society International
(with Les Amis du Pétrole) 2808 Baylor St., Bakersfield, Calif., 93305

Pittsburgh Geological Society
Box 3432, Pittsburgh, Pa., 15230

Planetary Society
65 N. Catalina Ave., Pasadena, Calif., 91106

Rocky Mountain Association of Geologists
1531 Stout St., Suite 210, Denver, Colo., 80202

Rocky Mountain Mineral Law Foundation
Porter Administration Building, 3d floor, 7039 East 18th Ave., Denver, Colo., 80220

Rocky Mountain Oil & Gas Association
1860 Lincoln St., #404, Denver, Colo., 80295-0001

Roswell Geological Society
Box 1171, Roswell, N.M., 88201

Royal Society of Canada
344 Wellington St., Ottawa, Ont., K1A 0N4, Canada

San Angelo Geological Society
Box 2568, San Angelo, Tex., 76902

San Diego Association of Geologists
2926 Harris Drive, Vista, Calif., 92084

San Diego Society of Natural History
Box 1390, San Diego, Calif., 92112

San Diego State University Geology Alumni
Department of Geological Sciences, San Diego State University, San Diego, Calif., 92182

San Joaquin Geological Society
Box 1056, Bakersfield, Calif., 93302

Seismological Society of America
El Cerrito Plaza Professional Building, Suite 201, El Cerrito, Calif., 94530
415/525-5474

Shreveport Geological Society
Box 750, Shreveport, La., 71162

Sigma Gamma Epsilon
Oklahoma Geological Survey, University of Oklahoma, 830 Van Vleet Oval, Norman, Okla., 73019

Society of Economic Geologists
Box 571, Golden, Colo., 80402
303/236-5538

Society of Exploration Geophysicists
Box 702740, Tulsa, Okla., 74170-2740

Society of Independent Professional Earth Scientists
4925 Greenville Ave., Suite 170, Dallas, Tex., 75206
214/363-1780

Society of Mining Engineers
8307 Shaffer Parkway, Box 625002, Littleton, Colo., 80162-5002
303/973-9550

Society of Petroleum Engineers
Box 833836, Richardson, Tex., 75083

Society of Professional Well Log Analysts
6001 Gulf Freeway, Suite C129, Houston, Tex., 77023
213/744-3445

Society of Vertebrate Paleontology
Los Angeles County Museum of Natural History, 900 Exposition Blvd., Los Angeles, Calif., 90007

Soil Science Society of America
677 South Segoe Road, Madison, Wis., 53711

South Coast Geological Society
Box 10244, Santa Ana, Calif., 92711

South Texas Geological Society
D-100 Petroleum Center, San Antonio, Tex., 78209

Southern Geological Society
University of Southern Mississippi, Box 5044, Southern Station, Hattiesburg, Miss., 39406

Texas A&M Geological Society
Department of Geology, Texas A&M University, College Station, Tex., 77843

Tobacco Root Geological Society
Box 470, Dillon, Mont., 59725

Toronto Geological Discussion Group
130 Adelaide St. West, Suite 1400, Toronto, Ont., M5H 1T8, Canada

Tulsa Geological Society
Box 4508, Tulsa, Okla., 74159-0508

Twin Cities Geologists
Minnesota Geological Survey, University of Minnesota, Minneapolis, 2642 University Ave., St. Paul, Minn., 55114

Union Geofisica Mexicana
(Mexican Geophysical Union) Apartado Postal 142, 024, Mexico 16100, D.F., Mexico

U.S. National Committee on Geology
U.S. Geological Survey, MS 917, Reston, Va., 22092

U.S. National Committee on History of Geology
Department of the History of Science,

University of Oklahoma, Norman, Okla., 73019

Utah Geological Association
Box 11334, Salt Lake City, Utah, 84147

Vermont Geological Society
Box 304, Montpelier, Vt., 05602

Virginia Geological Field Conference
S.O. Bird, 2201 West Broad St., Richmond, Va., 23220

Walker Mineralogical Club
100 Queen's Park, Toronto, Ont., M5S 2C6, Canada

West Texas Geological Society
Box 1595, Midland, Tex., 79702

Wyoming Association of Petroleum Landmen
Box 1012, Casper, Wyo., 82602

Wyoming Geological Association
Box 545, Casper, Wyo., 82602

Yellowstone-Bighorn Research Association
Box 638, Red Lodge, Mont., 59068

E.2

GEOLOGICAL SURVEY OF CANADA

GSC Sector
580 Booth St., Room 2064, Ottawa, Ont., K1A OE4
613/992-5910

Continental Geoscience & Mineral Resources Branch
601 Booth St., Room 213, Ottawa, Ont., K1A 0E8
613/995-4093

Lithosphere & Canadian Shield
613/995-4314

Mineral Resources
613/995-4093

Geophysics & Terrain Sciences Branch
GSC Sector, 580 Booth St., Room 2048, Ottawa, Ont., K1A OE8
613/995-0623

Geophysics
1 Observatory Crescent, Ottawa, Ont., K1A 0Y3
613/995-5484

Polar Continental Shelf Project
344 Wellington St., Room 6130, Ot-

tawa, Ont., K1A 0E4
613/990-6987

Programs, Planning & Services Branch
601 Booth St., Room 212, Ottawa, Ont., K1A 0E8
613/995-4482

Sedimentary & Marine Geoscience Branch
Atlantic Geoscience, Bedford Institute of Oceanography, Box 1006, Dartmouth, N.S., B2Y 4A2
902/426-2367

Cordilleran & Pacific Geoscience Division
100 W. Pender St., Vancouver, B.C., V6B 1R8
604/666-0529

Institute of Sedimentary & Petroleum Geology
3303-33d St. NW, Calgary, Alberta, T2L 2A7
403/284-0110

E.3

CANADIAN PROVINCIAL GEOLOGICAL SURVEYS

Alberta
Alberta Geological Survey, Alberta Research Council, Box 8330 Postal Station F, Edmonton, Alberta, T6H 5X2
403/438-7615

Mineral Resources Division, Department of Energy & Natural Resources, Petroleum Plaza, South Tower 9915-1083, Edmonton, Alberta, T5K 2C9
403/427-7749

British Columbia
Geological Survey Branch, Mineral Resources Division, Ministry of Energy, Mines & Petroleum Resources, Parliament Buildings, Victoria, B.C. V8V 1X4
604/387-0687

Manitoba
Geological Services Branch, Manitoba Energy & Mines, 555-330 Graham Ave., Winnipeg, Manitoba, R3C 4E3
204/945-6559

New Brunswick
Geological Surveys Branch, New Brunswick Department of Natural Resources, Box 6000, Fredericton, N.B., E3B 5H1
506/453-2206

Newfoundland
Mineral Development Division, Department of Mines & Energy, Box 4750, St. John's, Newfoundland, A1C 5T7
709/576-2763

Northwest Territories
Geology Division, Northern Affairs Program, Box 1500, Yellowknife, Northwest Territories, X1A 2R3
403/920-8212

Nova Scotia
Nova Scotia Department of Mines & Energy, Box 1087, Halifax, N.S., B3J 2X1
902/424-4700

Ontario
Ontario Geological Survey, Mines & Minerals Division, Ministry of Northern Development & Mines, 77 Grenville St., Room 1121, Toronto, Ont., M7A 1W4
416/965-1283

Prince Edward Island
Department of Energy & Minerals, Box 2000, Charlottetown, Prince Edward Island, C1A 7N8
902/368-5010

Quebec
Direction générale, Exploration géologique et minérale (Mines), Ministère de l'Energie et des Resources, Gouvernement du Québec, 1620 Boul. de l'Entente, Quebec, G1S 4N6
418/643-4617

Saskatchewan
Geology & Mines Division, Saskatchewan Energy & Mines, Toronto Dominion Bank Bldg., 1914 Hamilton St., Regina, Sask., S4P 4V4
306/787-2613

Yukon Territory
Department of Indian Affairs & Northern Development, Exploration & Geological Services Division, 200 Range Road, Whitehorse, Yukon Territory, Y1A 3V1

E.4
UNITED STATES GEOLOGICAL SURVEY

Headquarters, Virginia
USGS, National Center, Mail Stop 101, 12201 Sunrise Valley Drive, Reston, Va., 22092
703/648-4000

Alaska
USGS, 4230 University Drive, Anchorage, Alaska, 99508-4664
907/271-4398

Arizona
USGS, 2255 N. Gemini Drive, Flagstaff, Ariz., 86001
602/527-7000

California
USGS, 345 Middlefield Road, Menlo Park, Calif., 94025
415/853-8300

Denver
USGS, Denver Federal Center, Box 25046, Mail Stop 911, Lakewood, Colo., 80225
303/236-5438

Hawaii
USGS, Hawaiian Volcano Observatory, Box 51, Hawaii National Park, Hawaii, 96718
808/967-7328

Massachusetts
USGS, Quissett Campus, Woods Hole, Mass., 02543
617/548-8700

Washington
USGS, Cascades Volcano Observatory, 5400 MacArthur Blvd., Vancouver, Wash., 98661
206/696-7693
94025

E.5
UNITED STATES STATE GEOLOGICAL SURVEYS

Geological Survey of Alabama
Box O, Tuscaloosa, Ala., 35486-9780
205/349-2852

Alaska Division of Geological & Geophysical Surveys
794 University Ave., Suite 200, Fairbanks, Alaska, 99709
907/479-7625

Arizona Geological Survey
845 North Park Ave., Suite 100, Tucson, Ariz., 85719
602/621-7906

Arkansas Geological Commission
Vardelle Parham Geology Center, 3815 W. Roosevelt Road, Little Rock, Ark., 72204
501/371-1488

California Division of Mines & Geology
1416 Ninth St., Room 1341, Sacramento, Calif., 95814
916/445-1923

Colorado Geological Survey
1313 Sherman St., Room 715, Denver, Colo., 80203
303/866-2611

Connecticut Geological & Natural History Survey
State Office Building, 165 Capitol Ave., Room 553, Hartford, Conn., 06106
203/566-3540

Delaware Geological Survey
University of Delaware, 101 Penny Hall, Newark, Del., 19716
302/451-2833

Florida Geological Survey
903 W. Tennessee St., Tallahassee, Fla., 32304-7795
904/488-4191

Georgia Geologic Survey
Department of Natural Resources, Room 400, 19 Martin Luther King Jr. Drive SW, Atlanta, Ga., 30334
404/656-3214

Hawaii Division of Water & Land Development
Box 373, Honolulu, Hawaii, 96809
808/548-7533

Idaho Geological Survey
Morrill Hall, Room 332, University of Idaho, Moscow, Idaho, 83843
208/885-6195

Illinois Geological Survey
Natural Resources Building, 615 E. Peabody Drive, Champaign, Ill., 61820
217/333-4747

Indiana Geological Survey
611 N. Walnut Grove, Bloomington, Ind., 47405
812/855-9350

Iowa Geological Bureau
Iowa Department of Natural Resources, 123 N. Capitol St., Iowa City, Iowa, 52242
319/335-1575

Kansas Geological Survey
1930 Constant Ave., West Campus, University of Kansas, Lawrence, Kan., 66046

913/864-3965

Kentucky Geological Survey
228 Mining & Mineral Resources Building, University of Kentucky, Lexington, Ky., 40506-0107
606/257-5500

Louisiana Geological Survey
Box G, University Station, Baton Rouge, La., 70893
504/388-5320

Maine Geological Survey
Department of Conservation, State House Station 22, Augusta, Maine, 04333
207/289-2801

Maryland Geological Survey
2300 St. Paul St., Baltimore, Md., 21218
301/554-5503

Massachusetts Department of Environmental Quality Engineering
Executive Office of Environmental Affairs, 100 Cambridge St., 20th Floor, Boston, Mass., 02202
617/727-9800

Michigan Geological Survey Division
Box 30028, Lansing, Mich., 48909
517/334-6923

Minnesota Geological Survey
2642 University Ave., St. Paul, Minn., 55114-1057
612/627-4780

Mississippi Bureau of Geology
Box 5348, Jackson, Miss., 39216
601/354-6228

Missouri Division of Geology & Land Survey
Box 250, Rolla, Mo., 65401
314/364-1752

Montana Bureau of Mines & Geology
Montana College of Mineral Science & Technology, Butte, Mont., 59701
406/496-4180

Nebraska Conservation & Survey Division
113 Nebraska Hall, University of Nebraska, Lincoln, Neb., 68588-0517
402/472-3471

Nevada Bureau of Mines & Geology
University of Nevada, Reno, Nev.,

89557-0088
702/784-6691

New Hampshire Department of Environmental Services
117 James Hall, University of New Hampshire, Durham, N.H., 03824
603/862-3160

New Jersey Geological Survey
CN-029, Trenton, N.J., 08625
609/292-1185

New Mexico Bureau of Mines & Mineral Resources
Campus Station, Socorro, N.M., 87801
505/835-5420

New York State Geological Survey
3136 Cultural Education Center, Empire State Plaza, Albany, N.Y., 12230
518/474-5816

North Carolina Department of Natural Resources & Community Development
Division of Land Resources, Box 27687, Raleigh, N.C., 27611
919/733-3833

North Dakota Geological Survey
University Station, Grand Forks, N.D., 58202-8156
701/777-2231

Ohio Division of Geological Survey
Fountain Square, Building B, Columbus, Ohio, 43224
614/265-6605

Oklahoma Geological Survey
830 Van Vleet Oval, Room 163, Norman, Okla., 73019
405/325-3031

Oregon Department of Geology & Mineral Industries
910 State Office Bldg., 1400 SW 5th Ave., Portland, Ore., 97201-5528
503/229-5580

Pennsylvania Bureau of Topographic & Geologic Survey
Department of Environmental Resources, Box 2357, Harrisburg, Pa., 17120
717/787-2169

Puerto Rico Department of Natural Resources
Geological Survey Division, Box 5887, Puerta de Tierra, San Juan, P.R., 00906

809/724-8774

Rhode Island State Geologist
Department of Geology, University of Rhode Island, Kingston., R.I., 02881
401/792-2265

South Carolina Geological Survey
5 Geology Road, Columbia, S.C., 29210
803/737-9440

South Dakota Geological Survey
Science Center, University of South Dakota, Vermillion, S.D., 57069-2390
605/677-5227

Tennessee Division of Geology
Custom's House, 701 Broadway, Nashville, Tenn., 37219-5237
615/742-6691

Texas Bureau of Economic Geology
University of Texas, Box X, University Station, Austin, Tex., 78712-7508
512/471-1534

Utah Geological & Mineral Survey
606 Black Hawk Way, Salt Lake City, Utah, 84108-1280
801/581-6831

Vermont Geological Survey
Agency of Natural Resources, 103 South Main St., Waterbury, Vt., 05676
802/244-5164

Virginia Division of Mineral Resources
Box 3667, Charlottesville, Va., 22903
804/293-5121

Washington Division of Geology & Earth Resources
Department of Natural Resources, Olympia, Wash., 98504
206/459-6372

West Virginia Geological & Economic Survey
Mont Chateau Research Center, Box 879, Morgantown, W.Va., 26507-0879
304/594-2331

Wisconsin Geological & Natural History Survey
3817 Mineral Point Road, Madison, Wis., 53705
608/262-1705

Wyoming Geological Survey
Box 3008, University Station, University of Wyoming, Laramie, Wyo., 82071
307/742-2054

GLOSSARY

In the definitions here words in *italics* are often separately defined in their alphabetic place.

Å   Abbreviation for *angstrom,* a unit of length (10^{-8} cm).

aa lava   *Lava* whose surface is covered with random masses of angular jagged blocks.

ablation   As applied to glacier ice, process by which ice below snow line is wasted by evaporation and melting.

abrasion   Erosion of rock material by friction of solid particles moved by gravity, water, ice, or wind.

absolute time   Geologic time measured in years. Compare with *relative time.*

achondrite   Stony meteorite without chondrules.

acid rain   Rain higher in acid than normal. Forms from sulfur dioxide (SO_2) from the burning of fossil fuels; combines with water to form sulfuric acid.

aftershock   Earthquake that follows a larger earthquake and originates at or near focus of larger earthquake. Major shallow earthquakes are generally followed by many aftershocks, which decrease in number as time goes on but may continue for days or even months.

agate   Variety of *chalcedony* with alternating layers of chalcedony and *opal.*

aggradational flood plain   Flood plain formed by building up of valley floor by sedimentation.

A horizon   Soil zone immediately below surface, from which soluble material and fine-grained particles have been moved downward by water seeping into soil. Varying amounts of organic matter give *A* horizon color ranging from gray to black.

Airy hypothesis   Explains *isostasy* by assuming earth's crust has same density everywhere and differences in elevation result from differences in thickness of outer layer.

albite   Feldspar in which diagnostic positive ion is Na$^+$; sodic feldspar, Na(AlSi$_3$O$_8$). One of plagioclase feldspars.

alkali rocks   Igneous rocks in which abundance of alkalies is unusually high, generally indicated by soda pyroxenes, soda amphiboles, and/or feldspathoids.

alluvial fan   Land counterpart of a delta: an assemblage of sediments marking place where a stream moves from a steep gradient to a flatter gradient and suddenly loses transporting power. Typical of arid and semiarid climates but not confined to them.

almandite   A deep-red garnet of iron and aluminum formed during regional metamorphism.

alpha decay   Radioactive decay taking place by loss of an alpha particle from nucleus. Mass of element decreases by 4, and atomic number decreases by 2.

alpha particle   A helium atom lacking electrons and therefore having a double positive charge.

Alpine, or mountain, glacier   See *valley glacier.*

Amor asteroids   Those asteroids having paths whose minimum orbit lies between Earth and Sun.

amorphous   A state of matter in which there is no orderly arrangement of atoms.

amphibole group   Ferromagnesian silicates with a double chain of silicon-oxygen tetrahedra. Common example: hornblende. Contrast with *pyroxene group.*

amphibolite   A faintly foliated metamorphic rock developed during regional metamorphism of *simatic* rocks. Composed mainly of hornblende and plagioclase feldspars.

amphibolite facies   An assemblage of minerals formed during regional metamorphism at moderate to high pressures between 450 and 700°C.

andalusite   A silicate of aluminum built around independent tetrahedra, Al$_2$SiO$_5$. Characteristic of middle-grade metamorphism. Compare with *kyanite,* which has same composition but forms under higher pressures and has different crystal habit. Contrast with *sillimanite,* which has same composition but different crystal habit and forms at highest metamorphic temperatures.

andesite   A fine-grained igneous rock with no quartz or orthoclase, composed of about 75 percent plagioclase feldspars, balance ferromagnesian silicates. Important as lavas; possibly derived by fractional crystallization from basaltic magma. Widely characteristic of mountain-making processes around borders of Pacific Ocean. Confined to continental sectors.

angstrom   A unit of length equal to one hundred-millionth of a centimeter (10^{-8} cm). Abbreviation, Å.

angular unconformity   An *unconformity* in which older strata dip at different angle from that of younger strata.

anhydrite   Mineral calcium sulfate, CaSO$_4$, which is *gypsum* without water.

anion   A negatively charged atom or group of atoms produced by gain of electrons. Compare *cation.*

anorthite   Feldspar in which diagnostic positive ion is Ca^{2+}; calcic feldspar, Ca(Al$_2$Si$_2$O$_8$). One of plagioclase feldspars.

anorthosite   A plutonic igneous rock composed of 90 percent or more of feldspar mineral anorthite. Pyroxene and some olivine usually make up balance of the rock.

Antarctic bottom water   Seawater that sinks to ocean floor off Antarctica and flows equatorward beneath North Atlantic deep water.

Antarctic intermediate water   Seawater that sinks at about 50°S and flows northward above North Atlantic deep water.

anthracite   Metamorphosed bituminous coal of about 95 to 98 percent carbon.

anticline   A configuration of folded, stratified rocks in which rocks dip in two directions away from a crest, as principal rafters of a common gable roof dip away from ridgepole. Reverse of *syncline.* The "ridgepole," or crest, is called *axis.*

aphanitic texture   Individual minerals present but in particles so small that they cannot be identified without a microscope.

aquiclude   Rock of low permeability that will not transmit water, or if it does, too slowly to provide usable amounts of water.

aquifer   A permeable material through which groundwater moves.

arête   A narrow, saw-toothed ridge developed between cirques.

arkose   A detrital sedimentary rock formed by cementation of individual grains of sand size and predominantly composed of quartz and feldspar. Derived from disintegration of granite.

artesian water   Water under pressure when tapped by a well and able to rise above level at which first encountered. It may or may not flow out at ground level.

asbestos   A general term applied to certain fibrous minerals displaying similar physical characteristics although differing in composition. Some asbestos has fibers long enough to be spun into fabrics with great heat resistance, such as those for automobile brake linings. Types with shorter fibers are compressed into insulating boards, shingles, etc. Most common asbestos mineral (95 percent of United States production) is *chrysotile,* a variety of *serpentine,* a metamorphic mineral.

aseismic ridge   An oceanic ridge without extensive seismic activity as, for example, the Hawaiian Island-Emperor Seamount ridge in the Pacific Ocean.

ash   Volcanic fragments of sharply angular glass particles, smaller than cinders.

asphalt   A brown to black, solid or semisolid bituminous substance. Occurs in nature but is also obtained as residue from refining of certain hydrocarbons ("artificial asphalt").

asteroids   Orbiting small bodies believed to be either fragments of a disintegrated planet or of matter that never completed planet-forming

497

process.

asthenosphere A zone within Earth's *mantle* where plastic movements occur to permit *isostatic* adjustments. Begins 70 to 100 km below surface and extends perhaps to 500 km.

astrobleme The scar left by the impact of a meteorite. Usually a circular depression.

asymmetric fold A fold in which one limb dips more steeply than the other.

Atlantic Ridge See **Midatlantic Ridge.**

atoll A ring of low coral islands arranged around a central lagoon.

atom A building block of matter; combination of protons, neutrons, and electrons, of which 103 kinds are now known.

atomic energy Energy associated with nucleus of an atom. It is released when nucleus is split.

atomic mass The nucleus of an atom contains 99.95 percent of its mass. Total number of protons and neutrons in nucleus is called *mass number.*

atomic number Number of protons in nucleus.

atomic reactor A huge apparatus in which a radioactive core heats water under pressure and passes it to a heat exchanger.

atomic size Radius of an atom (average distance from center to outermost electron of neutral atom). Commonly expressed in angstroms.

augite A rock-forming ferromagnesian silicate mineral built around single chains of silicon-oxygen tetrahedra.

aureole A zone in which contact metamorphism has taken place.

axial plane A plane through a rock fold that includes the axis and divides the fold as symmetrically as possible.

axis Ridge or place of sharpest folding of an anticline or syncline.

backset beds Inclined layers of sand developed on gentler dune slope to windward. These beds may constitute a large part of total volume of a dune, especially if there is sufficient vegetation to trap most sand before it can cross over to slip face.

banded iron formation Stratified deposit with marked banding, generally of iron-rich minerals and chert or fine-grained quartz.

bank-full stage Stage of flow at which a stream fills its channel up to level of its bank. Recurrence interval averages 1.5 to 2 years.

barchan A crescent-shaped dune with wings, or horns, pointing downwind. Has gentle windward slope and steep lee slope inside horns; about 30 m high and 300 m wide from horn to horn. Moves with wind about 15 m/year across flat, hard surface where limited supply of sand is available.

barrier island A low, sandy island near shore and parallel to it, on a gently sloping offshore bottom.

barrier reef A reef separated from a landmass by a lagoon of varying width and depth opening to sea through passes in reef.

basal slip Movement of an entire glacier over underlying ground surface.

basalt A fine-grained igneous rock dominated by dark-colored minerals, consisting of high calcium plagioclase feldspars and ferromagnesian silicates. Basalts and andesites represent about 98 percent of all extrusive rocks.

base level Level below which a stream cannot erode. There may be temporary base levels along a stream's course, such as those established by lakes or resistant layers of rock. Ultimate base level is sea level.

basement complex Undifferentiated rocks underlying oldest identifiable rocks in any region. Usually sialic, crystalline, metamorphosed. Often, but not necessarily, Precambrian.

basin See *dome.*

batholith A *discordant pluton* that increases in size downward, has no determinable floor, and shows an area of surface exposure exceeding 100 km².

bauxite Chief ore of commercial aluminum. A *mineraloid* mixture of hydrous aluminum oxides.

bay barrier A sandy beach, built up across mouth of a bay so that it is no longer connected to main body of water.

bedding (1) A collective term used to signify existence of beds, or layers, in sedimentary rocks. (2) Sometimes synonymous with *bedding plane.*

bedding plane Surface separating layers of sedimentary rocks. Each bedding plane marks termination of one deposit and beginning of another of different character, such as surface separating a sand bed from a shale layer. Rock tends to separate, or break, readily along bedding planes.

bed load Material in movement along stream bottom or, if wind is moving agency, along surface. Contrast with material carried in *suspension* or solution.

beheaded stream The diminished lower part of a stream whose headwaters have been captured by another stream.

belt of soil moisture Subdivision of *zone of aeration.* Belt from which water may be used by plants or withdrawn by evaporation. Some water passes down into intermediate belt, where it may be held by molecular attraction against influence of gravity.

bench mark See Appendix D.

Benioff zone A seismic zone dipping beneath a continental margin and having a deep-sea trench as surface expression.

bergschrund Gap, or crevasse, between glacier ice and headwall of a cirque.

berms In coastline terminology berms are storm-built beach features that resemble small terraces; on seaward edges berms are low ridges built up by storm waves.

beta decay Radioactive decay taking place by loss of a beta particle (electron) from a neutron in nucleus. Mass of element remains same, but atomic number increases by 1.

B horizon Soil zone of accumulation below *A* horizon. Here is deposited some material moved down from *A* horizon.

big-bang theory Theory that presently expanding universe originated as primeval cosmic fireball in very short period of time 10 to 20 billion years ago. Compare *steady-state theory.*

binding energy Amount of energy that must be supplied to break an atomic nucleus into its component fundamental particles. It is equivalent to mass that disappears when fundamental particles combine to form a nucleus.

biochemical rock A sedimentary rock made up of deposits resulting directly or indirectly from life processes of organisms.

biotite "Black mica," ranging in color from dark brown to green. Rock-forming ferromagnesian silicate mineral with tetrahedra in sheets.

bituminous coal Soft coal, containing about 80 percent carbon.

blowout A basin scooped out of soft, unconsolidated deposits by process of deflation. Ranges from a few meters to several kilometers in diameter. Also refers to an oil well which gushes oil and gas to the surface before it can be capped.

body wave *Push-pull* or *shake* earthquake wave traveling through body of a medium; distinguished from waves traveling along free surface.

bornite A mineral Cu_5FeS_4; an important ore of copper.

bottomset bed Layer of fine sediment deposited in a body of standing water beyond advancing edge of a growing delta, which eventually builds up on bottomset beds.

boulder size Volume greater than that of a sphere with diameter of 256 mm.

boulder train Series of glacier erratics from the same bedrock source, usually with some property that permits easy identification. Arranged across country in a fan, with apex at source and widening in direction of glacier movement.

Bowen's reaction series Series of minerals for which any early-formed phase tends to react with melt that remains to yield a new mineral further along in the series. Thus early-formed crystals of olivine react with remaining liquids to form augite crystals; these in turn may further react with liquid then remaining to form hornblende. See also *continuous reaction series* and *discontinuous reaction series.*

braided stream Complex tangle of converging and diverging stream channels separated by sandbars or islands. Characteristic of flood plains where amount of debris is large in relation to discharge.

breccia Clastic rock made up of angular fragments of such size that an appreciable percentage of rock volume consists of particles of granule size or larger.

brittle A property of material whereby strength in *tension* is greatly different from strength in *compression.* Substance ruptures easily with little or no flow.

brown clay An extremely fine-grained deposit found on the deep oceanic floors, most abundantly in the Pacific Ocean. Derived from the continents and drifted to the open ocean, where the clay-sized particles settle.

burial metamorphism Changes resulting from pressures and temperatures in rocks buried to depth of several kilometers.

calcic feldspar Anorthite, $Ca(Al_2Si_2O_8)$.

calcite   A mineral composed of calcium carbonate, $CaCO_3$.

caldera   Roughly circular, steep-sided volcanic basin with diameter at least three or four times depth. Commonly at summit of a volcano. Contrast with *crater*.

caliche   Whitish accumulation of calcium carbonate in soil profile.

capillary fringe   Belt above *zone of saturation*, in which underground water is lifted against gravity by surface tension in passages of capillary size.

capillary size   "Hairlike," or very small, such as tubes from 0.0025 to 0.25 cm in diameter.

carbohydrate   Compound of carbon, hydrogen, and oxygen. Carbohydrates are chief products of life process in plants.

carbonaceous chondrite   Group of friable, dull-black chondritic stony meteorites, characterized by presence of great variety of organic compounds thought to be of extraterrestrial origin.

carbonate mineral   Mineral formed by combination of complex ion $(CO_3)^{2-}$ with a positive ion. Common example: calcite, $CaCO_3$.

carbon cycle   Process in Sun's deep interior by which radiant energy is generated in formation of helium from hydrogen.

carbon 14   Radioactive isotope of carbon, $^{14}_{6}C$, which has half-life of 5,730 years. Used to date events up to about 50,000 years ago.

carbon ratio   A number obtained by dividing amount of fixed carbon in coal by sum of fixed carbon and volatile matter and multiplying by 100. This is same as percentage of fixed carbon, assuming no moisture or ash.

cassiterite   A mineral, tin dioxide, SnO_2. Ore of tin with specific gravity 7; nearly 75 percent of world's tin production is from placer deposits, mostly from cassiterite.

cataclastic metamorphism   Textural changes in rocks in which brittle minerals and rocks are broken and flattened as a result of intense folding or faulting; produces fragmentation of rocks as coarse-grained breccias and fine-grained mylonites.

cation   A positively charged atom or group of atoms produced by loss of electrons. Compare *anion*.

cementation   Process by which a binding agent is precipitated in spaces among individual particles of an unconsolidated deposit. Most common cementing agents are calcite, dolomite, and quartz; others include iron oxide, opal, chalcedony, anhydrite, and pyrite.

cement rock   Clayey limestone used in manufacturing hydraulic cement. Contains lime, silica, and alumina in varying proportions.

chalcedony   General name applied to fibrous cryptocrystalline silica with waxy luster. Deposited from aqueous solutions and frequently found lining or filling cavities in rocks. *Agate* is a variety with alternating layers of chalcedony and opal.

chalcocite   A mineral, copper sulfide, Cu_2S; sometimes called *copper glance;* one of most important ore minerals of copper.

chalcopyrite   A mineral, a sulfide of copper and iron, $CuFeS_2$; sometimes called *copper pyrite* or *yellow copper ore.*

chalk   Variety of limestone made up in part of biochemically derived calcite, in form of skeletons or skeletal fragments of microscopic oceanic plants and animals mixed with very fine-grained calcite deposits of biochemical or inorganic-chemical origin.

chemical rock   In terminology of sedimentary rocks, chemical rock is composed chiefly of material deposited by chemical precipitation, whether organic or inorganic (compare with *detrital sedimentary rock*). Chemical sedimentary rocks may have either clastic or nonclastic (usually crystalline) texture.

chemical weathering   Weathering of rock material by chemical processes that transform original material into new chemical combinations. Thus chemical weathering of orthoclase produces clay, some silica, and a soluble salt of potassium.

chert   Granular cryptocrystalline silica, similar to *flint* but usually light in color. Occurs as compact massive rock or as nodules.

chlorite   Family of tetrahedral sheet silicates of iron, magnesium, and aluminum, characteristic of low-grade metamorphism. Green color, with cleavage like mica except that chlorite small scales are not elastic.

chondrite   Stony meteorite containing *chondrules* embedded in fine-grained matrix of pyroxene, olivine, and nickel-iron.

chondrule   Spheroidal granule, about 1 mm in diameter, consisting primarily of olivine and pyroxene, embedded in stony meteorite.

C **horizon**   Soil zone that contains partially disintegrated and decomposed parent material. Lies directly under *B* horizon and grades downward into unweathered material.

chromite   Mineral oxide of iron and chromium, $FeCr_2O_4$, only ore of chromium. One of first minerals to crystallize from magma; is concentrated within the magma.

chrysotile   Metamorphic mineral; an asbestos, fibrous variety of *serpentine*. Silicate of magnesium, with tetrahedra arranged in sheets.

chute or chute cutoff   Applied to stream flow, *chute* refers to new route taken by a stream when main flow is diverted to inside of a bend, along trough between low ridges formed by deposition on inside of bend, where water velocity is reduced. Compare with *neck cutoff.*

cinder cone   Structure built exclusively or predominantly of pyroclastic ejecta dominated by cinders. Parasitic to a major volcano, seldom exceeds 500 m in height. Slopes up 30° to 40°. Example: Parícutin.

cinders   Volcanic fragments; small, slaglike, solidified pieces of magma 0.5 to 2.5 cm across.

cirque   A steep-walled hollow in a mountainside at high elevation, formed by ice-plucking and frost action and shaped like a half bowl or half amphitheater. Serves as principal gathering ground for ice of a valley glacier.

clastic texture   Texture shown by sedimentary rocks from deposits of mineral and rock fragments.

clay minerals   Finely crystalline, hydrous silicates formed from weathering of such silicate minerals as feldspar, pyroxene, and amphibole. Most common clay minerals belong to *kaolinite, montmorillonite,* and *illite* groups.

clay size   Volume less than that of a sphere with diameter of 1/256 mm (0.004 mm).

cleavage   (1) *Mineral cleavage:* property possessed by many minerals of breaking in certain preferred directions along smooth plane surfaces. Planes of cleavage are governed by atomic pattern and represent directions in which atomic bonds are relatively weak. (2) *Rock cleavage:* property possessed by certain rocks of breaking with relative ease along parallel planes or nearly parallel surfaces. Rock cleavage is designated as *slaty, phyllitic, schistose,* and *gneissic.* See also *foliation.*

coal   Sedimentary rock composed of combustible matter derived from partial decomposition and alteration of plant cellulose and lignin.

cobble size   Volume greater than that of a sphere with diameter of 64 mm and less than that of a sphere with diameter of 256 mm.

coesite   High-pressure form of quartz, with density of 2.92. Associated with impact craters and cryptovolcanic structures.

col   Pass through a mountain ridge. Created by enlargement of two cirques on opposite sides of ridge until headwalls meet and are broken down.

cold glacier   One in which no surface melting occurs during summer months and whose temperature is always below freezing.

colloidal size   Between 0.2 and 1 μm (0.0002 to 0.001 mm).

column   Column, or post, of dripstone joining floor and roof of a cave; result of joining of stalactite and stalagmite.

columnar jointing   Pattern of jointing that blocks out columns of rock. Characteristic of tabular basalt flows or sills.

comet   Celestial body consisting of bright nucleus of dust, ice crystals, and gas, and a long tail that points away from the sun; its orbit varies between nearly round and parabolic.

compaction   Reduction in pore space between individual grains from pressure of overlying sediments or pressures of earth movements.

composite volcanic cone   Composed of interbedded lava flows and pyroclastic material and characterized by slopes of close to 30° at summit, progressively reducing to 5° near base. Example: Mayon.

compression   Squeezing stress that tends to decrease volume of a material.

concentric folding   Elastic bending of an originally horizontal sheet with all internal movements parallel to a basal plane (lower boundary of the fold).

conchoidal fracture   A mineral's habit of fracturing to produce curved surfaces like interior of a shell *(conch).* Typical of glass and quartz.

concordant pluton   An intrusive igneous body with contacts parallel to layering or foliation surfaces of rocks into which it has intruded.

concretion   An accumulation of mineral matter formed around a center, or axis, of deposition after a sedimentary deposit has been laid down. Cementation consolidates the deposit as a whole, but the concretion is a body within host rock that represents local concentration of cementing material: enclosing rock is less firmly cemented than the concretion. Commonly spheroidal or disk-shaped and composed of such cementing agents as calcite, dolomite, iron oxide, or silica.

cone of depression   A dimple in the water table, which forms as water is pumped from a well.

cone sheet   A dike, part of a concentric set dipping inward, like an inverted cone.

conglomerate   Detrital sedimentary rock made up of more or less rounded fragments of such size that an appreciable percentage of volume of rock consists of particles of granule size or larger.

contact metamorphism   Metamorphism at or very near contact between magma and rock during intrusion.

continental drift   Slow, lateral movement of continents; involves rigid plates that may carry both continental and oceanic areas as they move.

continental glacier   An ice sheet that obscures mountains and plains of a large section of a continent. Continental glaciers exist on Greenland and Antarctica.

continental margin   Continental rise, continental slope, and continental shelf together make up the continental margin, particularly on the trailing edge of a continent.

continental platform   The portion of the *craton* in which the old, Precambrian rocks are covered by a veneer of generally flat-lying sedimentary rocks.

continental rise   In some places base of continental slope is marked by somewhat gentler continental rise, which leads downward to deep ocean floor.

continental shelf   Shallow, gradually sloping zone extending from sea margin to a depth at which there is marked or rather steep descent into ocean depths down continental slope. Seaward boundary of shelf averages about 130 m in depth.

continental shield   The portion of the *craton* in which the old, Precambrian rocks are exposed at the surface.

continental slope   Portion of ocean floor extending downward from seaward edge of continental shelves. In some places, such as south of Aleutian Islands, slopes descend directly to ocean deeps. In other places, such as off eastern North America, they grade into somewhat gentler continental rises, which in turn lead to deep ocean floors.

continuous reaction series   Branch of Bowen's reaction series comprising plagioclase feldspars, in which reaction of early-formed crystals with later liquids takes place continuously —that is, without abrupt phase changes.

contour interval   See Appendix D.

contour line   See Appendix D.

convection   Mechanism by which material moves because its density differs from that of surrounding material. Density differences are often brought about by heating.

convection cell   Pair of *convection currents* adjacent to each other.

convection current   Closed circulation of material sometimes developed during convection. Convection currents normally develop in pairs, each pair called *convection cell*.

convergent plate boundary   Boundary between two plates moving toward each other. Compare *divergent plate boundary*.

coquina   A coarse-grained, porous, friable variety of clastic limestone made up chiefly of fragments of shells.

core   Innermost zone of Earth, surrounded by *mantle*.

Coriolis effect   Tendency of any moving body, on or starting from surface of Earth, to continue in direction in which Earth's rotation propels it. Direction in which the body moves because of this tendency combined with direction in which it is aimed determines ultimate course of the body relative to Earth's surface. In the Northern Hemisphere Coriolis effect causes a moving body to veer or try to veer to right of its direction of forward motion; in the Southern Hemisphere, to left. Magnitude of effect is proportional to velocity of a body's motion. This effect causes cyclonic storm-wind circulation to be counterclockwise in the Northern Hemisphere and clockwise in the Southern Hemisphere and determines final course of ocean currents relative to trade winds.

correlation   Process of establishing contemporaneity of rocks or events in one area with rocks or events in another area.

covalent bond   Bond in which atoms combine by sharing their electrons.

crater   Roughly circular, steep-sided volcanic basin with diameter less than three times depth. Commonly at summit of a volcano (contrast with *caldera*). Applied also to depressions caused by meteorites.

craton   The stable portion of a continent that has escaped *orogeny* for the last billion years or so.

creep   Applied to soils and surficial material, slow downward plastic movement. As applied to elastic solids, slow permanent yielding to stresses less than yield point if applied for a short time only.

crevasse   (1) Deep crevice, or fissure, in glacier ice. (2) Breach in a natural levee.

cross bedding   See *false bedding*.

crosscutting relationships, law of   A rock is younger than any rock across which it cuts. (Similarly for faults.)

crust   Outermost shell of the Earth. Continental crust averages 35 km thick, density 2.6-2.8 t/m³; oceanic crust, about 5 km thick, density 3 t/m³.

cryptocrystalline   State of matter in which there is actually orderly arrangement of atoms characteristic of crystals but in units so small (material is so fine grained) that crystalline nature cannot be determined with an ordinary microscope.

crystal   Solid with orderly atomic arrangement, which may or may not develop external faces that give it crystal form.

crystal habit   Geometrical form taken by a mineral, giving external expression to orderly internal atomic arrangement.

crystalline structure   Orderly arrangement of atoms in a crystal. Also called *crystal structure*.

crystallization   Process through which crystals separate from fluid, viscous, or dispersed state.

cumulates   Rocks that result from the accumulation of early formed crystals segregated from a *magma* by floating or sinking.

Curie temperature   Temperature above which ordinarily magnetic material loses magnetism. On cooling below this temperature, it regains magnetism. Example: Iron loses magnetism above 760°C and regains it while cooling below this temperature. Thus 760°C is the Curie temperature of iron.

current bedding   See *false bedding*.

cutoff   See *chute cutoff, neck cutoff*.

cyclosilicate   Mineral with crystal structure containing silicon-oxygen, $(SiO_4)^{4-}$, tetrahedra arranged as rings.

datum plane   See Appendix D.

debris avalanche   See *rock slide*.

debris flow   Rapid movement of well-mixed mass of water, soil, sand and rock. Similar to a *mudflow* but contains coarser material.

debris slide   Downslope movement of moderate velocity involving comparatively dry and unconsolidated material. With addition of water becomes an *earthflow*.

decomposition   Synonymous with *chemical weathering*.

deep focus   Earthquake focus deeper than 300 km. Greatest depth of focus known is 700 km.

deflation   Erosive process in which wind carries off unconsolidated material.

deformation of rocks   Any change in original shape or volume of rock masses; produced by mountain-building forces. *Folding, faulting,* and *plastic flow* are common modes of rock deformation.

delta   Plain underlain by an assemblage of sediments accumulated where a stream flows into a body of standing water, its velocity and transporting power suddenly reduced. Originally named after Greek letter *delta* (Δ) because many are roughly triangular in plan, with apex pointing upstream.

dendritic pattern   An arrangement of stream courses that, on a map or viewed from the air, resemble branching habit of certain trees, such as oaks or maples.

density   Measure of concentration of matter, expressed as mass per unit volume.

density current   Current due to differences in density of water from place to place caused by changes in temperature and variations in salinity or amount of material held in suspension.

depositional remanent magnetism   Magnetism resulting from tendency of magnetic particles such as hematite to orient themselves in Earth's magnetic field as they are deposited. Orientation is maintained as soft sediments are lithified and thus records Earth's field when particles were laid down. Abbreviation, DRM.

desert varnish   Thin, shiny layer of iron and

manganese oxides that coats some desert-rock surfaces.

detrital sedimentary rock   Rock formed from accumulation of minerals and rocks derived from erosion of previously existing rocks or from weathered products of these rocks.

diabase   Rock of basaltic composition, essentially labradorite and pyroxene, characterized by *ophitic* texture.

diagenesis   Changes that take place in sedimentary deposits after deposition and before metamorphism.

diamond   A mineral composed of element carbon; hardest substance known. Used as a gem and, in industry, for cutting tools.

differential erosion   Process by which different rock masses or different parts of same rock erode at different rates.

differential weathering   Process by which different rock masses or different parts of same rock weather at different rates.

dike   Tabular *discordant pluton.*

dike swarm   Group of approximately parallel dikes.

dilatancy   Tendency of rocks to expand along minute fractures immediately prior to failure; stress may be from Earth movements or from controlled laboratory experiments.

diorite   Coarse-grained igneous rock with composition of andesite (no quartz or orthoclase), composed of about 75 percent plagioclase feldspars and balance ferromagnesian silicates.

dip   (1) Acute angle that a rock surface makes with a horizontal plane. Direction of dip is always perpendicular to *strike.* (2) See *magnetic declination.*

dipole   Any object oppositely charged at two points. Most commonly refers to molecule that has concentrations of positive or negative charge at two different points.

dip pole   See *magnetic pole.*

dip-slip fault   Fault in which displacement is in direction of fault's dip.

discharge   With reference to stream flow, quantity of water that passes a given point in unit time. Measured in cubic meters per second or, often, cubic feet per second (abbreviation, cfs).

disconformity   *Unconformity* in which beds on opposite sides are parallel.

discontinuity   Within Earth's interior, sudden or rapid changes with depth in one or more of physical properties of materials constituting the Earth, as evidenced by seismic data.

discontinuous reaction series   Branch of Bowen's reaction series including minerals olivine, augite, hornblende, and biotite, for which each series change represents abrupt structural change.

discordant pluton   An intrusive igneous body with boundaries that cut across surfaces of layering or foliation in rocks into which it has intruded.

disintegration   Synonymous with *mechanical weathering.*

divergent plate boundary   Boundary between two plates moving apart. New oceanic-type lithosphere is created at the opening. Compare *convergent plate boundary.*

divide   Line separating two drainage basins.

divining rod   A forked wooden stick or similar object used in *dowsing;* it supposedly dips downward sharply when held over the object being sought.

dolomite   Mineral composed of carbonate of calcium and magnesium, $CaMg(CO_3)_2$. Also used as rock name for formations composed largely of mineral dolomite.

dome   Anticlinal fold without clearly developed linearity of crest so that beds involved dip in all directions from a central area, like an inverted but usually distorted cup. Reverse of *basin.*

drainage basin   Area from which a given stream and its tributaries receive water.

drift   Any material laid down directly by ice or deposited in lakes, oceans, or streams as result of glacial activity. Unstratified glacial drift is called *till* and forms *moraines;* stratified forms *outwash plains, eskers, kames,* and *varves.*

dripstone   Calcium carbonate deposited from solution by underground water entering a cave in *zone of aeration.* Sometimes called *travertine.*

dowsing   The practice of locating underground water, mineral deposits, or other objects by means of a *divining rod* or a pendulum.

DRM   See *depositional remanent magnetism.*

drumlin   Smooth, streamlined hill composed of *till.* Long axis oriented in direction of ice movement: blunt nose points upstream, and gentler slope tails off downstream. In height drumlins range from 8 to 60 m, with average somewhat less than 30 m. Most drumlins are between 0.5 and 1 km in length, the length commonly several times width. Diagnostic characteristics are shape and composition of unstratified glacial drift, in contrast to *kames,* of stratified glacial drift and random shapes.

dune   Mound or ridge of sand piled by wind.

dust-cloud hypothesis   Hypothesis that solar system was formed from condensation of interstellar dust clouds.

dust size   Volume less than that of sphere with diameter 0.06 mm; used in reference to particles carried in suspension by wind.

earthflow   Plastic movement of moderate velocity involving unconsolidated soil and weathered rock. With addition of more water grades into *mudflow.*

earth waves   Mechanism for transmitting energy from earthquake focus.

ecliptic   Apparent path of the Sun in the heavens; plane of most planets' orbits.

eclogite facies   Metamorphic rocks of *gabbroic* composition, consisting primarily of *pyroxene* and *garnet.*

elastic deformation   Nonpermanent deformation, after which body returns to original shape or volume when deforming force removed.

elastic energy   Energy stored within a solid during elastic deformation and released during elastic rebound.

elastic limit   Maximum *stress* that produces only *elastic deformation.*

elastic rebound   Recovery of elastic *strain*

when material breaks or deforming force is removed.

elastic solid   A solid that yields to applied force by changing shape, volume, or both but returns to original condition when force is removed. Amount of yield is proportional to force.

elasticity   A property of materials that defines extent to which they resist small deformations, from which they recover completely when deforming force is removed. Elasticity equals *stress* divided by *strain.*

electric charge   Property of matter, resulting from imbalance between number of *protons* and number of *electrons* in given piece of matter. The electron has negative charge; the proton, positive charge. Like charges repel each other; unlike charges attract each other.

electric current   Flow of electrons.

electrical energy   Energy of moving *electrons.*

electron   Fundamental particle of matter, the most elementary unit of negative electrical charge. Mass, 0.00055 u (*atomic mass* unit).

electron shell   Imaginary spherical surface representing all possible paths of electrons with same average distance from nucleus and with approximately same energy.

electron-capture decay   Radioactive decay that takes place as an orbital electron is captured by a proton in nucleus. Mass of element remains constant, but atomic number decreases by 1.

element   Unique combinations of *protons, neutrons,* and *electrons* that cannot be broken down by ordinary chemical methods. Fundamental properties of an element are determined by number of protons, each element assigned a number corresponding to its number of protons. Combinations containing from 1 through 103 protons are now known.

end moraine   Ridge or belt of *till* marking farthest advance of a glacier. Sometimes called *terminal moraine.*

energy   Capacity for producing motion. Energy holds matter together and can become *mass* or be derived from mass. Takes such forms as kinetic, potential, heat, chemical, electrical, and atomic energy; one form of energy can be changed to another.

epicenter   Area on surface directly above *focus* of earthquake.

epidote-amphibolite facies   Assemblage of minerals formed between 250 and 450° C during *regional metamorphism.*

equilibrium crystallization   Process by which a *magma* cools at a rate that allows early-formed crystals to react completely with the residual melt.

erosion   Movement of material from one place to another on Earth's surface. Agents of movement include gravity, water, ice, and wind.

erosional flood plain   Flood plain created by lateral erosion and gradual retreat of valley walls.

erratic   In terminology of glaciation a stone or boulder carried by ice to a place where it rests on or near bedrock of different composition.

escape velocity   Minimum velocity an object must have to escape from gravitational field. For moon this is about 2.38 km/s and for earth about 11.2 km/s.

esker Winding ridge of stratified glacial *drift*, steep-sided, 3 to 15 m high, and from a fraction of a kilometer to over 160 km long.

eugeosyncline Part of *geosyncline* in which volcanism is associated with clastic sedimentation, generally located away from *craton*.

eustatic change Change in sea level produced entirely by increase or decrease in amount of water in oceans; hence of worldwide proportions.

evaporation Process by which liquid becomes vapor at temperature below boiling point.

evaporite Rock composed of minerals precipitated from solutions concentrated by evaporation of solvents. Examples: *rock salt, gypsum, anhydrite.*

exfoliation Process by which rock plates are stripped from larger rock mass by physical forces.

exfoliation dome Large, rounded domal feature produced in homogeneous coarse-grained igneous rocks and sometimes in conglomerates by process of exfoliation.

exotic terranes Small fragments of continental masses that have moved great distances and have been accreted to other larger continents.

external magnetic field Component of Earth's field originating from activity above earth's surface. Small when compared with dipole and nondipole components of field, which originate beneath surface.

extrusive rock Rock solidified from mass of magma poured or blown out upon earth's surface.

facies Assemblage of mineral, rock, or fossil features reflecting environment in which rock was formed. See *sedimentary facies, metamorphic facies.*

false bedding Bedding laid down at angle to horizontal. Also referred to as *cross bedding* or *current bedding.*

fault Surface of rock rupture along which has been differential movement.

feldspars Silicate minerals composed of silicon-oxygen and aluminum-oxygen tetrahedra linked together in three-dimensional networks with positive ions fitted into interstices of negatively charged framework of tetrahedra. Classed as aluminosilicates. When positive ion is K^+, mineral is *orthoclase;* when Na^+, mineral is *albite;* when Ca^{2+}, mineral is *anorthite.*

felsite General term for light-colored, fine-grained igneous rocks.

ferromagnesian silicate Silicate in which positive ions are dominated by iron, magnesium, or both.

fibrous fracture Mineral habit of breaking into splinters or fibers.

fiery cloud (pyroclastic flows) Avalanche of incandescent *pyroclastic debris* mixed with steam and other gases, heavier than air, and projected down a volcano's side. Also called *nuée ardente.*

filter pressing Process by which early-formed crystals are squeezed out of a *magma.*

fiord Glacially deepened valley now flooded by the sea to form long, narrow, steep-walled inlet.

firn Granular ice formed by recrystallization of snow. Intermediate between snow and glacier ice. Sometimes called *névé.*

fission Process by which atomic nucleus breaks down to form nuclei of lighter atoms.

fissure eruption Extrusion of lava from fissure in Earth's crust.

flashy stream Stream with high flood peak of short duration, which may be caused by urbanization.

flint Granular cryptocrystalline silica, usually dull and dark. Often occurs as lumps or nodules in calcareous rocks, such as Cretaceous chalk beds of southern England.

flood basalt Basalt poured out from fissures in floods that tend to form great plateaus. Sometimes called *plateau basalt.*

flood frequency Time within which a flood of a given size can be expected to occur.

flood plain Area bordering a stream, over which water spreads in time of flood.

fluid Material that offers little or no resistance to forces tending to change its shape.

focus Source of given set of earthquake waves.

fold Bend, flexure, or wrinkle in rock produced when rock was in a plastic state.

folded mountains Mountains consisting primarily of elevated, folded sedimentary rocks.

foliation Layering in some rocks caused by parallel alignment of minerals; textural feature of some metamorphic rocks. Produces rock cleavage.

footwall One of blocks of rock involved in fault movement. One that would be under feet of person standing in tunnel along or across fault; opposite *hanging wall.*

foredune *Dune* immediately behind shoreline of ocean or large lake.

foreset beds Inclined layers of sediment deposited on advancing edge of growing *delta* or along lee slope of advancing sand *dune.*

foreshock Relatively small earthquake that precedes larger earthquake by a few days or weeks and originates at or near *focus* of larger earthquake.

fossil Evidence of past life, such as dinosaur bones, ancient clam shell, footprint of long-extinct animal, or impression of leaf in rock.

fossil fuels Organic remains used to produce heat or power by combustion. Include *coal, petroleum,* and *natural gas.*

fractional distillation Recovery—one or more at a time—of fractions of complex liquid, each of which has different density.

fractionation or fractional crystallization Process whereby crystals that formed early from magma have time to settle appreciably before temperature drops much further. They are thus effectively removed from environment in which they formed.

fracture As mineral characteristic, way in which mineral breaks when it does not have cleavage. May be *conchoidal* (shell-shaped), *fibrous, hackly,* or *uneven.*

fracture cleavage System of joints spaced fraction of centimeter apart.

frost action Process of mechanical weathering caused by repeated cycles of freezing and thawing. Expansion of water during freezing cycle provides energy for process.

fumarole Vent for volcanic steam and gases.

fundamental particles *Protons, neutrons,* and *electrons,* which combine to form atoms. Each particle is defined in terms of its *mass* and its *electric charge.*

fusion Process by which nuclei of lighter atoms join to form nuclei of heavier atoms.

gabbro Coarse-grained igneous rock with composition of basalt.

galena A mineral; lead sulfide, PbS. Principal ore of lead.

gangue Commercially valueless material remaining after ore-mineral extraction from rock.

garnet Family of silicates of iron, magnesium, aluminum, calcium, manganese, and chromium, which are built around independent tetrahedra and appear commonly as distinctive 12-sided, fully developed crystals. Characteristic of *metamorphic rocks;* generally cannot be distinguished from one another without chemical analysis.

gas (1) State of matter that has neither independent shape nor volume, can be compressed readily, and tends to expand indefinitely. (2) In geology "gas" is sometimes used to refer to *natural gas,* gaseous hydrocarbons that occur in rocks, dominated by methane. Compare "oil," referring to *petroleum.*

geode Roughly spherical, hollow or partially hollow accumulation of mineral matter from a few centimeters to nearly 0.5 m in diameter. Outer layer of chalcedony lined with crystals that project toward hollow center. Crystals, often perfectly formed, usually quartz although calcite and dolomite also found and—more rarely—other minerals. Geodes most commonly found in limestone and more rarely in shale.

geographic poles Points on Earth's surface marked by ends of axis of rotation.

geologic column Chronologic arrangement of rock units in columnar form, with oldest units at bottom and youngest at top.

geologic-time scale Chronologic sequence of units of Earth time.

geology Organized body of knowledge about the Earth, including *physical geology* and *historical geology,* among others.

geophysical prospecting Mapping rock structures by methods of experimental physics. Includes measuring magnetic fields, force of gravity, electrical properties, seismic-wave paths and velocities, radioactivity, and heat flow.

geophysics Physics of the Earth.

geosyncline Literally "Earth syncline." Term now refers, however, to a *basin* in which thousands of meters of sediments have accumulated, with accompanying progressive sinking of basin floor explained only in part by load of sediments. Common usage includes both accumulated sediments themselves and geometrical form of basin in which they are deposited. All folded mountain ranges were built from geosynclines, but not all geosynclines have become mountain ranges.

geothermal field Area where wells drilled to obtain elements contained in solution in hot brines and to tap heat energy.

geothermal gradient See *thermal gradient.*

geyser Special type of thermal spring that intermittently ejects its water with considerable force.

glacier A mass of ice, formed by recrystallization of snow, that flows forward or has flowed at some past time under influence of gravity. By convention we exclude icebergs even though they are large fragments broken from seaward end of glaciers.

glacier ice Unique form of ice developed by compression and recrystallization of snow and consisting of interlocking crystals.

glass Form of matter that exhibits properties of a solid but has atomic arrangements, or lack of order, of a liquid.

Glossopteris **flora** A late-Paleozoic assemblage of fossil plants named for seed fern *Glossopteris,* a plant in the flora. Widespread in South America, South Africa, Australia, India, and Antarctica.

gneiss Metamorphic rock with *gneissic cleavage.* Commonly formed by metamorphism of granite.

gneissic cleavage Rock cleavage where surface may be a few hundredths of a millimeter to a centimeter or more apart.

goethite Hydrous iron oxide, FeO(OH).

Gondwanaland Hypothetical continent thought to have broken up in Mesozoic. Resulting fragments are postulated to form present-day South America, Africa, Australia, India, and Antarctica.

graben Elongated, trenchlike structural form bounded by parallel normal faults created when block that forms trench floor moves downward relative to blocks that form sides.

grade Term used to designate extent to which *metamorphism* has advanced. Found in such combinations as "high-grade" or "low-grade metamorphism." Compare with *rank.*

graded bedding Type of bedding shown by sedimentary deposit when particles become progressively finer from bottom to top.

gradient Slope of stream bed or land surface; expressed in percent, as feet per mile, or meters per kilometer, or in degrees.

granite Coarse-grained *igneous rock* dominated by light-colored minerals, consisting of about 40 percent *orthoclase,* 25 percent *quartz,* and balance of plagioclase *feldspars* and *ferromagnesian silicates.* Granites and *granodiorites* comprise 95 percent of all intrusive rocks.

granitization Special type of *metasomatism* by which solutions of magmatic origin move through solid rocks, change ions with them, and convert them into rocks that achieve granitic character without having passed through magmatic stage.

granodiorite Coarse-grained igneous rock intermediate in composition between *granite* and *diorite.*

granular texture Composed of mineral grains large enough to be seen by unaided eye; also called *phaneritic texture.*

granulite facies *Gneissic* rocks produced by deep-seated high-grade *regional metamorphism.*

graphic structure Intimate intergrowth of *potassic feldspar* and *quartz.* Quartz part is dark and feldspar is light in color; so pattern suggests Egyptian hieroglyphs. Commonly found in *pegmatites.*

graphite "Black lead." A mineral composed entirely of carbon. Very soft because of crystalline structure; diamond, in contrast, has same composition but is hardest substance known.

gravity anomaly Difference between observed and computed values of gravity.

gravity fault Fault in which *hanging wall* appears to have moved downward relative to *footwall.* Also called *normal fault.*

gravity meter An instrument for measuring force of gravity. Also called *gravimeter.*

gravity prospecting Mapping force of gravity at different places to determine differences in *specific gravity* of rock masses and, through this, distribution of masses of different specific gravity. Done with *gravity meter* (gravimeter).

graywacke A variety of *sandstone* generally characterized by hardness, dark color, and angular grains of *quartz, feldspar,* and small rock fragments set in matrix of clay-sized particles. Also called *lithic sandstone.*

greenschist Schist characterized by green color. Product of *regional metamorphism* of *simatic rocks.* (Green color is imparted by mineral *chlorite.*)

greenschist facies Assemblage of minerals formed between 150 and 250°C during *regional metamorphism.*

groundmass Finely crystalline or glassy portion of porphyry.

ground moraine *Till* deposited from a glacier as veneer over landscape and forming gently rolling surface.

groundwater Undergroud water within *zone of saturation.*

groundwater table Upper surface of *zone of saturation* for underground water. An irregular surface with slope or shape determined by quantity of groundwater and permeability of earth materials. In general, highest beneath hills and lowest beneath valleys. Also referred to as *water table.*

guyot Flat-topped *seamount* rising from ocean floor like a volcano but planed off on top and covered by appreciable water depth. Synonymous with *tablemount.*

gypsum Hydrous calcium sulfate, $CaSO_4 \cdot 2H_2O$. A soft, common mineral in sedimentary rocks, where it sometimes occurs in thick beds interstratified with limestones and shales. Sometimes occurs as layer under bed of *rock salt* since it is one of first minerals to crystallize on evaporation of seawater. Alabaster is a fine-grained massive variety of gypsum.

H Symbol for mineral *hardness.*

hackly fracture Mineral's habit of breaking to produce jagged, irregular surfaces with sharp edges.

half-life Time needed for one-half of nuclei in sample of radioactive element to decay.

halide Compound made from a halogen, such as chlorine, iodine, bromine, or fluorine.

halite A mineral; *rock salt* or common salt, NaCl. Occurs widely disseminated or in extensive beds and irregular masses precipitated from seawater and interstratified with rocks of other types as true sedimentary rock.

hanging valley A valley that has greater elevation than the valley to which it is tributary, at point of junction. Often (but not always) created by deepening of main valley by a glacier. Hanging valley itself may or may not be glaciated.

hanging wall One of blocks involved in fault movement. One that would be hanging overhead for person standing in tunnel along or across fault; opposite *footwall.*

hardness Mineral's resistance to scratching on a smooth surface. Mohs scale of relative hardness consists of 10 minerals, each scratching all those below it in scale and being scratched by all those above it: (1) talc, (2) gypsum, (3) calcite, (4) fluorite, (5) apatite, (6) orthoclase, (7) quartz, (8) topaz, (9) corundum, (10) diamond.

head Difference in elevation between intake and discharge points for a liquid. In geology most commonly of interest in connection with movement of *underground water.*

heat flow Product of thermal gradient and thermal conductivity of earth materials. Average over whole Earth, 1.2 ± 0.15 $\mu cal/cm^2$-s.

height see Appendix D.

hematite Iron oxide, Fe_2O_3. Principal ore mineral for about 90 percent of iron produced in United States. Characteristic red color when powdered.

historical geology Branch of *geology* that deals with history of the Earth, including record of life on Earth as well as physical changes in Earth itself.

horn Spire of bedrock left where *cirques* have eaten into a mountain from more than two sides around a central area. Example: Matterhorn of the Swiss Alps.

hornblende A rock-forming *ferromagnesian silicate* mineral with double chains of silicon-oxygen tetrahedra. An *amphibole.*

hornfels Dense metamorphic rock. Since this term is commonly applied to metamorphic equivalent of any fine-grained rock, composition is variable.

hornfels facies Assemblage of minerals formed at temperatures greater than 700°C during *contact metamorphism.*

horst Elongated block bounded by parallel *normal faults* in such a way that it stands above blocks on both sides.

hot spot Localized melting region in mantle near base of lithosphere, a few hundred kilometers in diameter and persistent over tens of millions of years. Existence of heat is assumed from volcanic activity at surface.

hot spring *Thermal spring* that brings hot water to surface. Water temperature usually 6.5°C or more above mean air temperature.

hydration Process by which water combines

chemically with other molecules.

hydraulic gradient *Head* of underground water divided by distance of travel between two points: If head is 10 m for two points 100 m apart, hydraulic gradient is 0.1, or 10 percent. When head and distance of flow are same, hydraulic gradient is 100 percent.

hydraulic mining Use of strong water jet to move deposits of sand and gravel from original site to separating equipment, where sought-for mineral is extracted.

hydrocarbon Compound of hydrogen and carbon that burns in air to form water and oxides of carbon. There are many hydrocarbons. The simplest, methane, is chief component of natural gas. Petroleum is a complex mixture of hydrocarbons.

hydroelectric power Conversion of energy to electricity by free fall of water. This method supplies about 4 percent of world's electrical energy.

hydrograph Graph of variation of stream flow over time.

hydrologic cycle General pattern of water movement by evaporation from sea to atmosphere, by precipitation onto land, and by return to sea under influence of gravity.

hydrothermal solution Hot, watery solution that usually emanates from *magma* in late stages of cooling. Frequently contains, and deposits in economically workable concentrations, minor elements that, because of incommensurate *ionic radii* or electronic charges, have not been able to fit into atomic structures of common minerals of igneous rocks.

icecap Localized *ice sheet.*

ice sheet Broad, moundlike mass of glacier ice of considerable extent that has tendency to spread radially under own weight. Localized ice sheets are sometimes called *icecaps.*

igneous rock Aggregate of interlocking silicate minerals formed by cooling and solidification of magma.

illite Clay mineral family of hydrous aluminous silicates. Structure is similar to that of *montmorillonite,* but aluminum replaces 10 to 15 percent of silicon, which destroys montmorillonite's property of expanding with addition of water because weak bonds replaced by strong potassium-ion links. Structurally illite is intermediate between montmorillonite and *muscovite.* Montmorillonite converts to illite in sediments; illite, to muscovite under conditions of low-grade metamorphism. Illite is the commonest clay mineral in clayey rocks and recent marine sediments and is present in many soils.

ilmenite Iron titanium oxide. Accounts for much of unique abundance of titanium on moon.

index minerals Those that are useful indicators of particular facies: *chlorite,* low-grade metamorphism; *almandite,* middle-grade metamorphism; *sillimanite,* high-grade metamorphism.

induced magnetism In terminology of rock magnetism one of components of rock's *natural remanent magnetism.* It is parallel to earth's

present field and results from it.

inertia member Central element of a seismograph, consisting of weight suspended by wire or spring so that it acts like pendulum free to move in only one plane.

infiltration Water on surface soaking into ground.

inosilicate Mineral with crystal structure containing silicon-oxygen tetrahedra in single or double chains.

intensity Measure of effects of earthquake waves on human beings, structures, and Earth's surface at particular place. Contrast with *magnitude,* which is measure of total energy released by an earthquake.

intermediate belt Subdivision of *zone of aeration.* Belt that lies between *belt of soil moisture* and *capillary fringe.*

intermediate lava Lava composed of 60 to 65 percent silica.

intrusive rock Rock solidified from mass of magma that invaded earth's crust but did not reach surface.

ion Electrically unbalanced form of an atom or group of atoms, produced by gain or loss of electrons.

ionic bond Bond in which ions are held together by electrical attraction of opposite charges.

ionic radius Average distance from center to outermost electron of an ion. Commonly expressed in *angstroms.*

island-arc deeps Arcuate trenches bordering some continents; some reach depths of 9,000 m or more below sea surface. Also called *trenches.*

isoclinal folding Beds on both *limbs* nearly parallel whether fold upright, *overturned,* or *recumbent.*

isograd Line connecting similar temperature-pressure values; line marking the boundary between two metamorphic facies.

isomorphism The characteristic of 2 or more crystalline substances to have similar chemical composition, axial ratios, and crystal forms. Such substances form an isomorphous series.

isoseismic line Line connecting all points on surface of earth where intensity of shaking from earthquake waves is same.

isostasy Ideal condition of balance that would be attained by earth materials of differing densities if gravity were the only force governing heights relative to each other.

isotope Alternative form of an element produced by variations in number of neutrons in nucleus.

jasper Granular, cryptocrystalline silica usually colored red by *hematite* inclusions.

joint Break in rock mass with no relative movement of rock on opposite sides of break.

joint system Combination of intersecting joint sets, often at approximately right angles.

kame Steep-sided hill of stratified glacial *drift.* Distinguished from *drumlin* by lack of unique shape and by stratification.

kame terrace Stratified glacial drift deposited between wasting glacier and adjacent valley wall. When ice melts this material stands as a terrace along valley wall.

kaolinite Clay mineral, hydrous aluminous silicate $Al_4Si_4O_{10}(OH)_8$. Structure consists of one sheet of silicon-oxygen tetrahedra that each share three oxygens to give ratio of $(Si_4O_{10})^{4-}$ linked with one sheet of aluminum and hydroxyl. Composition of pure kaolinite does not vary as for other clay minerals, *montmorillonite* and *illite,* in which ready addition or substitution of ions takes place.

karstic topography Irregular topography characterized by *sinkholes,* streamless valleys, and streams that disappear underground—all developed by action of surface and underground water in soluble rock such as limestone.

kettle Depression in ground surface formed by melting of a block of ice buried or partially buried by glacial *drift* either *outwash* or *till.*

kimberlite Porphyritic mafic pluton with abundant phenocrysts of altered olivine in fine-grained groundmass of calcite, olivine, and phlogopite, with accessory iron and titanium minerals; name derived from Kimberley, South Africa, where the rock contains diamonds.

kinetic energy Energy of movement. Amount possessed by an object or particle depends on mass and speed.

kyanite A silicate mineral characteristic of temperatures of middle-grade metamorphism. Al_2SiO_5 in bladed blue crystals is softer than a knife along the crystal. Its crystalline structure is based on independent tetrahedra. Compare with *andalusite,* which has same composition but has different crystal habit. Contrast with *sillimanite* which has same composition but different crystal habit and forms at highest metamorphic temperatures.

L Symbol for earthquake *surface waves* (from *large waves*).

laccolith *Concordant pluton* that has domed up strata into which it is intruded.

lag time On stream *hydrograph* time interval between center of mass of precipitation and center of mass of resulting flood.

laminar flow Mechanism by which fluid (such as water) moves slowly along a smooth channel or through a tube with smooth walls with fluid particles following straight-line paths parallel to channel or walls. Contrast with *turbulent flow.*

landslide General term for relatively rapid mass movement such as *slump, rock slide, debris slide, mudflow,* and *earthflow.*

lapilli *Pyroclastic debris* in pieces about walnut size.

large waves Earthquake *surface* waves.

latent heat of fusion Number of calories per unit volume that must be added to a material at melting point to complete process of melting. These calories do not raise temperature.

lateral moraine Ridge of *till* along edge of valley glacier. Composed largely of material fallen to glacier from valley walls.

laterite Tropical soil rich in hydroxides of aluminum and iron and formed under conditions of good drainage.

lava *Magma* poured out on surface of Earth or rock solidified from such magma.

leading continental margin That margin that occurs on the side of a continent that is riding on a plate in motion toward another plate; commonly this margin is very active tectonically.

left-lateral fault *Strike-slip fault* where ground opposite you appears to have moved left when you face it.

levee (natural) Bank of sand and silt built by river during floods, where suspended load deposited in greatest quantity close to river. Process of developing natural levees tends to raise river banks above level of surrounding *flood plains.* Break in natural levee sometimes called *crevasse.*

lignite Low-grade coal, with about 70 percent carbon. Intermediate between peat and bituminous coal.

limb One of two parts of *anticline* or *syncline,* on either side of axis.

limestone Sedimentary rock composed largely of mineral *calcite,* $CaCO_3$, formed by either organic or inorganic processes. Most limestones have clastic texture, but nonclastic, particularly crystalline, textures are common. Carbonate rocks, limestone and *dolomite,* constitute estimated 12 to 22 percent of sedimentary rocks exposed above sea level.

limonite Iron oxide with no fixed composition or atomic structure; a *mineraloid.* Always of secondary origin, not a true mineral. Is encountered as ordinary rust or coloring material of yellow clays and soils.

liquefaction Process of changing soil and unconsolidated sediments into water mixture immediately following earthquake; often results in foundation failure, with sliding of ground under building structures.

liquid State of matter that flows readily so that the mass assumes form of container but retains independent volume.

lithic sandstone See *graywacke.*

lithification Process by which unconsolidated rock-forming materials are converted into consolidated or coherent state.

lithosphere Rigid outer layer of Earth; includes *crust* and upper part of *mantle.* Relatively strong layer in contrast to underlying *asthenosphere.*

loess Unconsolidated, unstratified aggregation of small, angular mineral fragments, usually buff in color. Generally believed to be wind-deposited; characteristically able to stand on very steep to vertical slopes.

longitudinal dune Long ridge of sand oriented in general direction of wind movement. A small one is less than 3 m high and 60 m long. Very large ones are called *seif dunes.*

longitudinal wave *Push-pull wave.*

lopolith Tabular *concordant pluton* shaped like spoon bowl, with both roof and floor sagging downward.

mafic Containing magnesium ("ma") and iron ("fic"); ferromagnesian.

magma Naturally occurring silicate melt, which may contain suspended silicate crystals, dissolved gases, or both. These conditions may be met in general by a mixture containing as much as 65 percent crystals but no more than 11 percent dissolved gases.

magmatic segregation (magmatic differentiation) The process by which heavy, early crystallized minerals settle out from a *magma* and become concentrated.

magnetic declination Angle of divergence between geographic meridian and magnetic meridian. Measured in degrees east and west of geographic north.

magnetic inclination Angle that magnetic needle makes with surface of Earth. Also called *dip* of magnetic needle.

magnetic pole North magnetic pole is point on Earth's surface where north-seeking end of a magnetic needle free to move in space points directly down. At south magnetic pole the same needle points directly up. These poles are also known as *dip poles.*

magnetic reversal Shift of 180° in Earth's magnetic field such that north-seeking needle of magnetic compass would point south rather than to north magnetic pole.

magnetite A mineral; iron oxide, Fe_3O_4. Black; strongly magnetic. Important ore of iron.

magnetosphere Region 1,000 to 64,000 km above Earth, where magnetic field traps electrically charged particles from Sun and space. First believed to consist of two bands, Van Allen belts.

magnetostratigraphy Use of magnetized rocks to determine history of events in record of changes in Earth's magnetic field in past geologic ages.

magnitude Measure of total energy released by an earthquake. Contrast with *intensity,* which is measure of effects of earthquake waves at particular place.

manganese nodule A *nodule* found on the deep oceanic floors and composed largely of iron and manganese, with smaller amounts of such metals as cobalt, titanium, copper, and nickel.

mantle Intermediate zone of Earth. Surrounded by *crust,* it rests on *core* at depth of about 2,900 km.

marble Metamorphic rock of granular texture, with no rock cleavage, and composed of *calcite,* dolomite, or both.

maria Dark-toned "seas" of Moon. Mark Moon's topographically low areas.

marl Calcareous clay or intimate mixture of clay and particles of *calcite* or dolomite, usually shell fragments.

marsh gas *Methane,* CH_4; simplest paraffin hydrocarbon. Predominant component of *natural gas.*

mass A number that measures quantity of matter. It is obtained on Earth's surface by dividing weight of a body by acceleration due to gravity.

mass movement Surface movement of earth materials induced by gravity.

mass number Number of protons and neutrons in atomic nucleus.

mass unit One-twelfth mass of carbon atom. Approximately mass of hydrogen atom.

matter Anything that occupies space. Usually defined by describing its states and properties: solid, liquid, or gas; possesses mass, inertia, color, density, melting point, hardness, crystal form, mechanical strength, or chemical properties. Composed of *atoms.*

meander (1) Turn or sharp bend in stream's course. (2) To turn, or bend, sharply. Applied to stream courses in geological usage.

mechanical weathering Process by which rock is broken down into smaller and smaller fragments as result of energy developed by physical forces. Also known as *disintegration.*

medial moraine Ridge of *till* formed by junction of two *lateral moraines* when two valley glaciers join to form single ice stream.

mélange Heterogeneous mixture of rock materials. Mappable body of deformed rocks that may be several kilometers in length and consists of highly sheared clayey matrix, thoroughly mixed with angular native and exotic blocks of diverse origin and geologic ages.

Mercalli intensity scale Scale to evaluate intensity of earthquake shaking on basis of effects at given place.

metal Substance fusible and opaque, good conductor of electricity, and with characteristic luster. Examples: gold, silver, aluminum. Of the elements 77 are metals.

metallic bonding Special kind of bonding in atoms of metallic elements whereby outermost electrons are not shared or exchanged but are free to move around and connect to any atoms in solid. Relative freedom of movement of electrons accounts for high level of electrical conductivity in metals.

metalloid Element of some metallic and some nonmetallic characteristics. There are nine metalloids. See also Appendix B.

metamorphic facies Assemblage of minerals that reached equilibrium during metamorphism under specific range of temperature and pressure.

metamorphic rock "Changed-form rock"; any rock changed in texture or composition by heat, pressure, or chemically active fluids after original formation.

metamorphic zone Area subjected to *metamorphism* and characterized by certain metamorphic facies formed during process.

metamorphism A process whereby rocks undergo physical or chemical changes or both to achieve equilibrium with conditions other than those under which they were originally formed (weathering arbitrarily excluded from meaning). Agents of metamorphism are heat, pressure, and chemically active fluids.

metasomatism Process whereby rocks are altered when *volatiles* exchange ions with them.

meteor Transient celestial body that enters Earth's atmosphere with great speed, becoming incandescent from heat generated by air resistance.

meteoric water Groundwater derived primarily from precipitation.

meteorite Stony or metallic body fallen to Earth from outer space.

methane Simplest paraffin hydrocarbon, CH_4. Principal constituent of *natural gas.* Sometimes called *marsh gas.*

micas Group of silicate minerals characterized by perfect sheet or scale cleavage resulting from atomic pattern, in which silicon-oxygen tetrahedra linked in sheets. *Biotite* is ferromagnesian dark mica. *Muscovite* is potassic white mica.

microcontinent A comparatively small continental landmass. May be welded by plate motion to the margin of a larger continent.

microplates Small masses, usually of continental material, that behave like other larger lithospheric plates.

microseism "Small shaking." Specifically limited in technical usage to Earth waves generated by sources other than earthquakes and, most frequently, to waves with periods of from 1 to about 9 s, from sources associated with atmospheric storms.

Midatlantic Ridge The broad, generally north-south ridge beneath the Atlantic Ocen which, more or less, bisects it. The rift along its crests marks the diverging boundary between the American plates and the Eurasian and African plates.

migmatite Mixed rock produced by intimate interfingering of *magma* and invaded rock.

mineral Naturally occurring solid element or compound, exclusive of biologically formed carbon components. Has definite composition or range of composition and orderly internal atomic arrangement (crystalline structure), which gives unique physical and chemical properties, including tendency to assume certain geometrical forms known as *crystals.*

mineral deposit Occurrence of one or more minerals in such concentration and form as to make possible removal and processing for use at profit.

mineraloid Substance that does not yield definite chemical formula and shows no sign of crystallinity. Examples: *bauxite, limonite,* and *opal.*

miogeosyncline That part of a geosyncline in which volcanism is absent, generally located near *craton.*

Mohorovičić discontinuity (Moho) Base of crust marked by abrupt increases in velocities of Earth waves.

molecule Smallest unit of compound that displays properties of that compound.

monocline Double flexure connecting strata at one level with same strata at another level.

montmorillonite Clay mineral family, hydrous aluminous silicate with structural sandwich of one ionic sheet of aluminum and hydroxyl between two $(Si_4O_{10})^{4-}$ sheets. Sandwiches piled on each other with water between and with nothing but weak bonds to hold them together. As result, additional water can enter lattice readily, causing mineral to swell appreciably and further weakening attraction between structural sandwiches. Consequently a lump of montmorillonite in a bucket of water slumps rapidly into a loose, incoherent mass. Compare with other clay minerals, *kaolinite* and *illite.*

moon A natural satellite.

moraine General term applied to certain landforms composed of *till.*

mountain Any part of landmass projecting conspicuously above its surroundings.

mountain chain Series or group of connected mountains having well-defined trend or direction.

mountain structure Structure produced by deformation of rocks.

mud cracks Cracks caused by shrinkage of drying deposit of silt or clay under surface conditions.

mudflow A rapid movement of material downslope similar to a *debris flow* but containing less coarse material.

mudstone Fine-grained, detrital sedimentary rock made up of *silt-* and *clay-sized* particles. Distinguished from *shale* by lack of fissility.

muscovite "White mica." Nonferromagnesian rock-forming silicate mineral with tetrahedra arranged in sheets. Sometimes called *potassic mica.*

mylonite Fine-grained rock formed by grinding during intense folding or faulting associated with cataclastic metamorphism.

native state State in which an element occurs uncombined in nature. Usually applied to metals, as in "native copper," "native gold," etc.

natural gas Gaseous hydrocarbons that occur in rocks. Dominated by *methane.*

natural remanent magnetism Magnetism of rock. May or may not coincide with present magnetic field of Earth. Abbreviation, NRM.

natural resources Energy and materials made available by geological processes.

neck cutoff Breakthrough of a river across narrow neck separating two meanders, where downstream migration of one has been slowed and next meander upstream has overtaken it. Compare with *chute cutoff.*

negative charge Condition resulting from surplus of electrons.

nesosilicate Mineral with crystal structure containing silicon-oxygen tetrahedra arranged as isolated units.

neutron *Proton* and *electron* combined and behaving like fundamental particle of matter. Electrically neutral with mass of 1.00896 u. If isolated, may decay to form proton and electron.

névé Granular ice formed by recrystallization of snow or area containing same. Intermediate between snow and glacier ice. Sometimes called *firn.*

nodule Irregular, knobby-surfaced mineral body that differs in composition from rock in which formed. Silica in form of *chert* or *flint* is common component of nodules. They are commonly found in limestone and dolomite.

nonclastic texture Applied to sedimentary rocks in which rock-forming grains are interlocked. Most sedimentary rocks with nonclastic texture are crystalline.

nonconformity *Unconformity* separating younger rocks from distinctly different older highly metamorphosed or igneous rocks.

nondipole magnetic field Portion of earth's magnetic field remaining after dipole field and external field are removed.

nonferromagnesians Silicate minerals that do not contain iron or magnesium.

nonmetal Element that does not exhibit metallic luster, conductivity, or other features of metal. Of the elements 17 are nonmetals.

normal fault Fault in which *hanging wall* appears to have moved downward relative to *footwall;* opposite of *thrust fault.* Also called *gravity fault.*

normal-faulted mountain Mountain whose primary geologic structures are *normal faults.*

North Atlantic deep water Seawater in Arctic that sinks in North Atlantic and drifts southward as far as 60°S.

northeast trades Winds blowing from the northeast toward the equator in the Northern Hemisphere.

NRM See *natural remanent magnetism.*

nucleus *Protons* and *neutrons* constituting central part of an atom.

nuée ardente "Fiery cloud." French term applied to highly heated mass of gas-charged lava ejected from vent or pocket at volcano summit more or less horizontally onto an outer slope, down which it moves swiftly, however slight the incline.

obduction Process whereby part of the subducted plate and/or associated igneous rocks and deep-sea sediments are broken off and pushed up onto the overriding plate.

oblique slip fault Fault with components of relative displacement along both *strike* and *dip.*

obsidian Glassy equivalent of granite.

oil In geology refers to *petroleum.*

oil shale Shale containing such proportion of hydrocarbons as to be capable of yielding *petroleum* on slow distillation.

olivine Rock-forming ferromagnesian silicate mineral that crystallizes early from magma and weathers readily to Earth's surface. Crystal structure based on isolated $(SiO_4)^{4-}$ ions and positive ions of iron, magnesium, or both. General formula: $(Mg, Fe)_2SiO_4$.

oölites Spheroidal grains of sand size, usually composed of calcium carbonate, $CaCO_3$, and thought to have originated by inorganic precipitation. Some limestones largely made up of oölites.

ooze Deep-sea deposit consisting of 30 percent or more by volume of hard parts of very small, sometimes microscopic, organisms. If particular organism dominant, its name is used as modifier, as in *globigerina ooze,* or *radiolarian ooze.*

opal Amorphous silica, with varying amounts of water; a mineral gel.

open-pit mining Surface mining represented by sand and gravel pits, stone quarries, and copper mines of some western states.

ophitic Rock texture in which lath-shaped plagioclase crystals are enclosed wholly or in part in later-formed augite, as commonly occurs in diabase.

order of crystallization Chronological sequence in which crystallization of various minerals of an assemblage takes place.

ore deposit Metallic minerals in concentrations that can be worked at profit.

orogen A linear or arcuate belt along the junction between two plates where mountain building *(orogeny)* takes place.

orogeny Process of mountain building.

orthoclase Feldspar in which K+ is diagnostic positive ion; $K(AlSi_3O_8)$.

orthoquartzite Sandstone composed completely—or almost completely—of *quartz grains. Quartzose sandstone* is synonym.

outwash Material carried from a glacier by meltwater and laid down in stratified deposits.

outwash plain Flat or gently sloping surface underlain by outwash.

overbank deposits Sediments (usually clay, silt, and fine sand) deposited on flood plain by river overflowing banks.

overturned fold Fold with at least one *limb* rotated through more than 90°.

oxbow Abandoned *meander* caused by a *neck cutoff.*

oxbow lake Abandoned *meander* isolated from main stream channel by deposition and filled with water.

oxide mineral Mineral formed by direct union of an element with oxygen. Examples: ice, corundum, hematite, magnetite, cassiterite.

P Symbol for earthquake *primary waves.*

pahoehoe lava *Lava* whose surface is smooth and billowy, frequently molded into forms resembling huge rope coils. Characteristic of basic lavas.

paired terraces *Terraces* that face each other across stream at same elevation.

paleomagnetism Study of Earth's magnetic field as it has existed during geologic time.

paleosol Soil formed in past environment; often buried.

Pangaea Hypothetical continent from which all others are postulated to have originated through process of fragmentation and drifting.

parabolic dune *Dune* with long, scoop-shaped form that, when perfectly developed, exhibits parabolic shape in plan, with horns pointing upwind. Contrast *barchan,* in which horns point downwind. Characteristically covered with sparse vegetation; often found in coastal belts.

peat Partially reduced plant or wood material, containing approximately 60 percent carbon. An intermediate material in process of coal formation.

pebble size Volume greater than that of a sphere with diameter of 4 mm and less than that of a sphere of 64 mm.

pedalfer *Soil* characterized by accumulation of iron salts or iron and aluminum salts in *B* horizon. Varieties of pedalfers include red and yellow soils of southeastern United States and *podsols* of northeastern quarter of United States.

pedocal *Soil* characterized by accumulation of calcium carbonate in its profile. Characteristic of low rainfall. Varieties include black and chestnut soils of northern Plains states and red and gray desert soils of drier western states.

pedology Science that treats of *soils*—origin, character, and utilization.

pegmatite Small *pluton* of exceptionally coarse texture, with crystals up to tens of meters in length, commonly formed at margin of *batholith.* Nearly 90 percent of all pegmatites are simple pegmatites of *quartz, orthoclase,* and unimportant percentages of *micas;* others are extremely rare ferromagnesian pegmatites and complex pegmatites. Complex pegmatites have as major components *sialic* minerals of simple pegmatites but also contain a variety of rare minerals.

pegmatitic texture Said of the texture of an exceptionally coarsely crystalline igneous rock.

pelagic deposit Material formed in deep ocean and deposited there. Example: *ooze.*

pendulum Inertia member so suspended that, after displacement, restoring force will return it to starting position. If displaced and then released, oscillates, completing one to-and-fro swing in time called *period.*

perched water table Top of *zone of saturation* that bottoms on impermeable horizon above level of general *water table* in area. Is generally near surface and frequently supplies a hillside spring.

peridotite Coarse-grained igneous rock dominated by dark-colored minerals, consisting of about 75 percent *ferromagnesian silicates* and balance *plagioclase feldspars.*

periglacial Refers to conditions of near-glacial climate.

period For oscillating system, length of time required to complete one oscillation.

permafrost See *permanently frozen ground.*

permanently frozen ground Occurs in areas where mean annual temperature is zero degrees Celsius or below. Also called *permafrost.*

permeability For rock or earth material, ability to transmit fluids. Permeability equal to velocity of flow divided by hydraulic gradient.

petroleum An oily mixture of hydrocarbons extracted from subsurface Earth structures. Results from physical and chemical conversion of remains of animals and plants. A fuel in natural or refined state, yielding on distillation such products as gasoline, kerosene, naphtha.

phaneritic texture Individual grains large enough to be identified without the aid of a microscope; also called *granular texture.*

phase (1) Homogeneous, physically distinct portion of matter in physical-chemical system not homogeneous, as in three phases of ice, water, and aqueous vapor. (2) Group of seismic waves of one type.

phenocryst A crystal significantly larger than crystals of surrounding minerals.

phosphate rock Sedimentary rock containing calcium phosphate.

photosynthesis Process by which carbohydrates are compounded from carbon dioxide and water in presence of sunlight and chlorophyll.

phreatic explosions Volcanic explosions of steam, mud, or other material that is not incandescent; such eruptions are caused by heating and consequent expansion of ground water from an underlying igneous heat source.

phyllite Clayey metamorphic rock with rock cleavage intermediate between *slate* and *schist.* Commonly formed by the regional metamorphism of *shale* or *tuff.* Micas characteristically impart a pronounced sheen to rock cleavage surfaces. Has phyllitic cleavage.

phyllitic cleavage Rock cleavage in which flakes are produced barely visible to unaided eye. Coarser than *slaty* and finer than *schistose cleavage.*

phyllosilicate Mineral and crystal structure containing silicon-oxygen tetrahedra arranged as sheets.

physical geology Branch of *geology* that deals with nature and properties of material composing the Earth, distribution of materials throughout globe, processes by which they are formed, altered, transported, and distorted, and nature and development of landscape.

piedmont glacier Glacier formed by coalescence of *valley glaciers* and spreading over plains at foot of mountains from which valley glaciers came.

pirate stream (=capturing stream) A stream into which the headwaters of another stream have been diverted by capture.

placer A concentration of relatively heavy and resistant minerals in stream or beach deposits; two examples are some deposits of gold and of diamonds.

planet Natural satellite orbiting about Sun.

planetology Organized body of knowledge about planetary system.

plastic deformation Permanent change in shape or volume not involving failure by rupture and, once started, continuing without increase in deforming force.

plastic solid Solid that undergoes deformation continuously and indefinitely after stress applied to it passes a critical point.

plate Earth's *lithosphere,* varying in thickness from several tens of kilometers to as much as 100 km and including crust and part of upper *mantle* above *asthenosphere.*

plate tectonics Theory of worldwide dynamics involving movement and interactions of the many rigid plates of earth's *lithosphere.*

plateau basalt Basalt poured out from fissures in floods that tend to form great plateaus. Sometimes called *flood basalt.*

playa Flat-floored center of undrained desert basin.

playa lake Temporary lake formed in a playa.

pleochroic halo Minute, concentric-spherical zones of darkening or coloring that form around inclusions of radioactive minerals in *biotite, chlorite,* and a few other minerals. About 0.075 mm in diameter.

plume Pipelike convection cells thought to carry heat and *mantle* material from lower mantle up to *crust,* producing hot spots at surface.

plunge Acute angle that axis of folded rock mass makes with horizontal plane.

pluton A body of igneous rock formed beneath Earth surface by consolidation from magma. Sometimes extended to include bodies formed beneath surface by metasomatic replacement of older rock.

plutonic igneous rock   Rock formed by slow crystallization, which yields coarse texture. Once believed to be typical of crystallization at great depth, but not a necessary condition.

pluvial lake   Lake formed during a *pluvial period.*

pluvial period   Period of increased rainfall and decreased evaporation; prevailed in nonglaciated areas during time of ice advance elsewhere.

podsol   Ashy-gray or gray-brown soil of *pedalfer* group. Highly bleached soil, low in iron and lime, formed under moist and cool conditions.

point bars   Accumulations of sand and gravel deposited in slack waters on inside of bends of winding, or meandering, river.

polarity epoch   Interval of time during which Earth's magnetic field has been oriented dominantly in either normal or reverse direction. May be marked by shorter intervals of opposite sign, called *polarity events.*

polarity event   See *polarity epoch.*

polar wandering, or migration   Movement of position of magnetic pole during past time in relation to present position.

polymorphism   Existence of several different morphologic kinds occurring in species or mineral.

porosity   Percentage of open space or interstices in rock or other Earth material. Compare with *permeability.*

porphyritic   Textural term of igneous rocks in which large crystals, called *phenocrysts,* are set in finer groundmass, which may be crystalline or glassy or both.

porphyry   Igneous rock containing conspicuous *phenocrysts* in fine-grained or glassy *groundmass.*

positive charge   Condition resulting from deficiency of electrons.

potassic feldspar (=K-spar)   *Orthoclase,* $K(AlSi_3O_8)$.

potential energy   Stored energy waiting to be used. Energy that a piece of matter possesses because of position or because of arrangement of parts.

prairie soils   Transitional soils between *pedalfers* and *pedocals.*

Pratt hypothesis   Explains isostasy by assuming all portions of the crust have same total mass above certain elevation, called *level of compensation.* Higher sections would have proportionately lower density.

precipitation   Discharge of water, in rain, snow, hail, sleet, fog, or dew, on land or water surface. Also, process of separating mineral constituents from solution by evaporation *(halite, anhydrite)* or from *magma* to form igneous rocks.

predictor or precursor   Relating to earthquakes, refers to events immediately preceding actual shaking of ground. Includes changes in seismic velocities, groundwater levels, and tilt of ground surface.

pressure   Force per unit area.

prevailing westerlies   Winds blowing from the west and characteristic of the middle latitudes both north and south of the equator.

primary magmas   A magma originating below the Earth's crust; not contaminated by other magmatic material; sometimes called parental magma.

primary wave   Earthquake body wave that travels fastest and advances by *push-pull* mechanism. Also known as *longitudinal,* compressional, or *P wave.*

proton   Fundamental particle of matter with positive electrical charge of 1 unit (equal in amount, but opposite in effect, to the charge of *electron*) and mass of 1.00758 u.

proton-proton fusion   Rapidly moving *protons* in hot interior of stars collide and fuse to form atoms of helium from atoms of hydrogen in continuous buildup of higher elements.

pumice   Pieces of magma up to several centimeters across that have trapped bubbles of steam or other gases as they were thrown out in eruption. Sometimes they have sufficient buoyancy to float on water.

push-pull wave   Wave that advances by alternate compression and rarefaction of medium, causing particles in path to move forward and backward along direction of wave's advance. In connection with earth waves, also known as *primary wave,* compressional wave, *longitudinal wave,* or *P wave.*

pyrite   A sulfide mineral, iron sulfide, FeS_2.

pyroclastic debris   Fragments blown out by explosive volcanic eruptions and subsequently deposited on ground. Include *ash, cinders, lapilli,* blocks, bombs, and *pumice.*

pyroxene group   Ferromagnesian silicates with a single chain of silicon-oxygen tetrahedra. Common example: augite. Compare with *amphibole group* (example: hornblende), which has a double chain of tetrahedra.

pyrrhotite   A mineral, iron sulfide. So commonly associated with nickel minerals that it has been called "world's greatest nickel ore."

quartz   A silicate mineral, SiO_2, composed exclusively of silicon-oxygen tetrahedra, with all oxygens joined in a three-dimensional network. Crystal form is six-sided prism tapering at end, with prism faces striated transversely. An important rock-forming mineral.

quartzite   Metamorphic rock commonly formed by metamorphism of sandstone and composed of quartz. No rock cleavage. Breaks *through* sand grains in contrast to sandstone, which breaks *around* grains.

radial drainage   Arrangement of stream courses in which streams radiate outward in all directions from central zone.

radiant energy   Electromagnetic waves travelling as wave motion.

radioactivity   Spontaneous breakdown of atomic nucleus, with emission of radiant energy.

rain wash   Water from rain after it has fallen on ground and before concentrated in definite stream channels.

range   Elongated series of mountain peaks considered to be a part of one connected unit, such

as Appalachian Range or Sierra Nevada Range.

rank   Term used to designate extent to which *metamorphism* has advanced. Compare with *grade.* Rank is more commonly employed in designating stage of metamorphism of *coal.*

ray craters   Lunar craters marked by *rays.* Young on lunar time scale.

rays   Light-toned streaks that spread outward from such lunar craters as Tycho, Kepler, and Copernicus.

reaction series   See *Bowen's reaction series.*

recessional moraine   Ridge or belt of *till* marking period of moraine formation, probably in period of temporary stability or slight readvance, during general wastage of a glacier and recession of its front.

recorder   Part of a *seismograph* that makes record of ground motion.

rectangular stream pattern   Arrangement of stream courses in which tributaries flow into larger streams at angles approaching 90°.

recumbent fold   Fold with axial plane more or less horizontal.

reflection seismic prospecting   Uses reflected waves and places seismographs at distances only a fraction of depths investigated.

refraction seismic prospecting   Uses travel times of refracted waves and spreads seismographs over lines roughly four times depth being investigated.

refractory   Mineral or compound that resists action of heat and chemical reagents.

regional metamorphism   Metamorphism occurring over tens or scores of kilometers.

rejuvenation   Change in conditions of erosion that causes a stream to begin more active erosion and a new cycle.

relative time   Dating of events by place in chronologic order of occurrence rather than in years. Compare with *absolute time.*

relief   See Appendix D.

reverse fault   Fault in which *hanging wall* appears to have moved upward relative to *footwall;* contrast with *normal,* or *gravity fault.* Also called *thrust fault.*

rhyolite   Fine-grained igneous rock with composition of granite.

rift zone   System of fractures in Earth's crust. Often associated with lava extrusion.

right-lateral fault   *Strike-slip fault* in which ground opposite you appears to have moved right when you face it.

rigidity   Resistance to elastic *shear.*

rilles   Trenchlike depressions on Moon's surface. Some are straight walled, others sinuous.

ring dike   Arcuate (rarely circular) *dike* with steep *dip.*

ripple marks   Small waves produced in unconsolidated material by wind or water. See *ripple marks of oscillation.*

ripple marks of oscillation   *Ripple marks* formed by oscillating movement of water such as may be found along seacoast outside surf zone. Symmetrical, with sharp or slightly rounded ridges separated by more gently rounded troughs.

rock   Aggregate of minerals of one or more

kinds in varying proportions.

rock avalanche See *rock slide.*

rock cycle Concept of sequences through which earth materials may pass when subjected to geological processes.

rockfall Sudden fall of one or more large pieces of rock from a cliff face.

rock flour Finely divided rock material pulverized by glacier and carried by streams fed by melting ice.

rock-forming silicate minerals Minerals built around framework of silicon-oxygen tetrahedra: olivine, augite, hornblende, biotite, muscovite, orthoclase, albite, anorthite, quartz.

rock glacier Tongue of rock waste found in valleys of certain mountainous regions. Characteristically lobate and marked by series of arcuate, rounded ridges that give it aspect of having flowed as viscous mass.

rock melt Liquid solution of rock-forming mineral ions.

rock salt *Halite,* or common salt, NaCl.

rock slide Very rapid downslope movement of large masses of rock. Also called *rock avalanche.* Called *debris avalanche* if largely unconsolidated debris is involved.

Rossi-Forel scale Scale for rating earthquake intensities, devised in 1878.

runoff Water that flows off land.

rupture Breaking apart or state of being broken apart.

S Symbol for *secondary wave* of an earthquake; shear wave.

salt In geology this term usually refers to *halite,* or *rock salt,* NaCl, particularly in such combinations as salt water and *salt dome.*

saltation Mechanism by which a particle moves by jumping from one point to another.

salt dome Mass of NaCl generally of roughly cylindrical shape and with diameter of about 2 km near top. Such mass has pushed up through surrounding sediments into present position. Reservoir rocks above and alongside salt domes trap *oil* and *gas.*

sand Clastic particles of *sand size,* commonly but not always composed of mineral *quartz.*

sand size Volume greater than that of a sphere with diameter of 0.0625 mm and less than that of a sphere with diameter of 2 mm.

sandstone Detrital *sedimentary rock* formed by cementation of individual grains of sand size and commonly composed of mineral *quartz.* Sandstones constitute estimated 12 to 28 percent of sedimentary rocks.

sapropel Aquatic ooze or sludge rich in organic matter. Believed to be source material for *petroleum* and *natural gas.*

satellite crater *Crater* formed by impact of a fragment ejected during creation of a primary crater. Also called *secondary crater.*

scale See Appendix D.

schist Metamorphic rock dominated by fibrous or platy minerals. Has *schistose cleavage* and is product of *regional metamorphism.*

schistose cleavage Rock cleavage with grains and flakes clearly visible and cleavage surfaces rougher than in *slaty* or *phyllitic cleavage.*

sea-floor spreading Process by which ocean floors spread laterally from crests of main ocean ridges. As material moves laterally from ridge, new material replaces it along ridge crest by welling upward from mantle.

seamount Isolated steep-sloped peak rising from deep ocean floor but submerged beneath surface. Most have sharp peaks, but some have flat tops and are called *guyots,* or *tablemounts.* Seamounts are volcanic in origin.

secondary crater See *satellite crater.*

secondary magmas Magma formed by changes in composition of primary magma; often contaminated by primary magmas incorporating other crustal matter.

secondary wave Earthquake *body wave* slower than *primary wave. Shear* or *S wave.*

secular variation of magnetic field Change in inclination, declination, or intensity of Earth's magnetic field. Detectable only from long historical records.

sedimentary facies Accumulation of deposits that exhibits specific characteristics and grades laterally into other sedimentary accumulations that were formed at same time but exhibit different characteristics.

sedimentary rock Rock formed from accumulations of sediment, which may consist of rock fragments of various sizes, remains or products of animals or plants, products of chemical action or of evaporation, or mixtures of these. *Stratification* is the single most characteristic feature of sedimentary rocks, which cover about 75 percent of land area.

sedimentation Process by which mineral and organic matters are laid down.

seif dune Very large *longitudinal dune.* As high as 100 m and as long as 100 km.

seismic prospecting Method of determining nature and structure of buried rock formations by generating waves in ground (commonly by small explosive charges) and measuring length of time these waves require to travel different paths.

seismic sea wave Large wave in ocean generated at time of earthquake. Popularly but incorrectly known as *tidal wave.* Sometimes called *tsunami.*

seismicity General term for earthquake activity.

seismogram Record obtained on a *seismograph.*

seismograph Instrument for recording vibrations, most commonly employed for recording Earth vibrations during earthquakes.

seismology Scientific study of earthquakes and other Earth vibrations.

serpentine Silicate of magnesium common among metamorphic minerals. Occurs in two crystal habits: platy, known as *antigorite;* fibrous, known as *chrysotile,* an asbestos. "Serpentine" comes from mottled shades of green on massive varieties, suggestive of snake markings.

shale Fine-grained, detrital sedimentary rock made up of *silt-* and *clay-sized* particles. Contains clay minerals as well as particles of *quartz, feldspar, calcite, dolomite,* and other minerals. Distinguished from *mudstone* by breaking into small pieces more or less parallel to bedding.

shatter cone Conical fragment of rock ranging from centimeter to meter size. Striations radiate from apex to base. Associated with impact craters.

shear Change of shape without change of volume.

shear wave Wave that advances by shearing displacements (which change shape without changing volume) of medium. This causes particles in path to move from side to side or up and down at right angles to direction of wave's advance. Also called *secondary wave.*

sheet flow See *sheet wash.*

sheeting Joints essentially parallel to ground surface. More closely spaced near surface and become progressively farther apart with depth. Particularly well developed in granitic rocks, but sometimes in other massive rocks as well.

sheet wash Water accumulating on a slope in thin sheet of water. May begin to concentrate in *rills.* Also called *sheet flow.*

shield Nucleus commonly of Precambrian rocks around which a continent has grown. See *craton.*

shield volcano Volcano built up almost entirely of lava, with slopes seldom as great as 10° at summit and 2° at base. Examples: five volcanoes on island of Hawaii.

sial A term coined from chemical symbols for silicon and aluminum. Designates composite of rocks dominated by *granites, granodiorites,* and their allies and derivatives, which underlie continental areas of globe. Specific gravity about 2.7.

sialic rock Igneous rock composed predominantly of silicon and aluminum, from whose chemical symbols term is constructed. Average specific gravity, about 2.7.

siderite A mineral; iron carbonate, $FeCO_3$. An ore of iron.

silicate minerals Minerals with crystal structure containing *silicon-oxygen tetrahedra* arranged as isolated units *(nesosilicates),* single or double chains *(inosilicates),* sheets *(phyllosilicates),* or three-dimensional frameworks *(tectosilicates).*

silicon-oxygen tetrahedron Complex ion composed of silicon ion surrounded by four oxygen ions. Negative charge of 4 units, and represented by symbol $(SiO_4)^{4-}$. Diagnostic unit of silicate minerals, and makes up central building unit of nearly 90 percent of materials of earth's crust.

sill Tabular igneous intrusion that parallels the planar structure of the surrounding rocks.

sillimanite A silicate mineral, Al_2SiO_5, characteristic of highest metamorphic temperatures and pressures. Occurs in long slender crystals, brown, green, white. Crystalline structure based on independent *tetrahedra.* Contrast with *kyanite* and *andalusite,* which have same composition but different crystal habits and form at lower temperatures.

silt size Volume greater than that of a sphere with diameter of 0.0039 mm and less than that of a sphere with diameter of 0.0625 mm.

sima Term coined from silicon and magnesium. Designates worldwide shell of dark, heavy rocks. Sima is outermost rock layer under deep, principal ocean basins. Basaltic in composition.

simatic rock Igneous rock composed predominantly of ferromagnesian minerals.

sink See *sinkhole*.

sinkhole Depression in surface of ground caused by collapse of roof over solution cavern.

slate Fine-grained metamorphic rock with well-developed *slaty cleavage*. Formed by low-grade *regional metamorphism* of *shale*.

slaty cleavage Rock cleavage in which ease of breaking occurs along planes separated by microscopic distances.

slip face Steep face on lee side of a dune.

slope failure See *slump*.

slump Downward and outward movement of rock or unconsolidated material as unit or as series of units. Also called *slope failure*.

snowfield Stretch of perennial snow existing in area where winter snowfall exceeds amount of snow that melts away during summer.

snow line Lower limit of perennial snow.

soapstone See *talc*.

sodic feldspar *Albite*, Na(AlSi$_3$O$_8$).

soil Superficial material that forms at Earth's surface as result of organic and inorganic processes. Soil varies with climate, plant and animal life, time, slope of land, and parent material.

soil horizon Layer of soil approximately parallel to land surface with observable characteristics produced through operation of soil-building processes.

solar constant Average rate at which radiant energy received by Earth from Sun. Equal to little less than 2 cal/cm²-min on plane perpendicular to Sun's rays at outer edge of atmosphere, when Earth is at mean distance from Sun.

solar system Sun with group of celestial bodies held by its gravitational attraction and revolving around it.

sole mark Cast of sedimentary structures such as cracks, tracks, or grooves formed on lower surface or underside of sandstone bed, commonly revealed after original underlying sedimentary layer has weathered away.

solid Matter with definite shape and volume and some fundamental strength. May be crystalline, glassy, or amorphous.

solid solution Single crystalline phase that may vary in composition within specific limits.

solifluction Mass movement of soil affected by alternate freezing and thawing. Characteristic of saturated soils in high latitudes.

sorosilicates Mineral with crystal structure containing *silicon-oxygen tetrahedra* arranged as double units.

southeast trades Winds blowing from the southeast toward the equator in the Southern Hemisphere.

specific gravity Ratio between weight of given volume of material and weight of equal volume of water at 4°C.

specific heat Amount of heat necessary to raise temperature of 1 g of any material 1°C.

sphalerite A mineral; zinc sulfide, ZnS. Nearly always contains iron, (Zn,Fe)S. Principal ore of zinc. (Also known as *zinc blende* or *blackjack.*)

spheroidal weathering Spalling off of concentric shells from rock masses of various sizes as result of pressures built up during chemical weathering.

spit Sandy bar built by currents into a bay from a promontory.

spreading-center ridge Continuous, seismically active, ridge extending through North and South Atlantic, Indian, and South Pacific Oceans.

spring Place where *water table* crops out at ground surface and water flows out more or less continuously.

stack Small island that stands as isolated, steep-sided rock mass just off end of promontory. Has been isolated from land by erosion and weathering concentrated behind end of a headland.

stalactite Icicle-shaped accumulation of *dripstone* hanging from cave roof.

stalagmite Post of *dripstone* growing upward from cave floor.

star A heavenly body that seems to stay in same position relative to other heavenly bodies.

staurolite Silicate mineral characteristic of middle-grade *metamorphism*. Crystalline structure based on independent tetrahedra with iron and aluminum. Has unique crystal habit that makes it striking and easy to recognize: six-sided prisms intersecting at 90° to form cross or at 60° to form an X shape.

steady-state theory Theory that universe is developing by continuous creation of matter as newly formed galaxies replace those expanding out of sight, thus keeping mass density of universe constant. Compare with *big-bang theory.*

stishovite Quartz with high density of 4.3. Associated with impact craters.

stock Discordant *pluton* that increases in size downward, has no determinable floor, and shows area of surface exposure less than 100 km². Compare with *batholith.*

stoping Mechanism by which *batholiths* have moved into crust by breaking off and foundering of blocks of rock surrounding magma chamber.

strain Change of dimensions of matter in response to *stress:* commonly, unit strain, such as change in length per unit length (total lengthening divided by original length), change in width per unit width, change in volume per unit volume.

stratification Structure produced by deposition of sediments in layers or beds.

strategic mineral Mineral considered to be vital to the security of a nation, but that must be obtained from foreign sources: e.g., chromium-, titanium-, tin-, and manganese-bearing minerals for the United States.

stratigraphic trap Structure that traps *petroleum* or *natural gas* because of variation in permeability of reservoir rock or termination of inclined reservoir formation on up-dip side.

streak Color of fine powder of mineral; may be different from color of hand specimen. Usually determined by rubbing mineral on piece of unglazed porcelain (*hardness* about 7) known as a "streak plate," which is, of course, useless for minerals of greater hardness.

stream capture (piracy) The natural diversion of the head-waters of one stream into the channel of another stream having greater erosional activity.

steam order Hierarchy in which segments of a stream system are arranged.

stream terrace Surface representing remnants of stream's channel or flood plain when stream was flowing at higher level. Subsequent downward cutting by stream leaves remnants of old channel or *flood plain* standing as *terrace* above present stream level.

strength *Stress* at which rupture occurs or plastic deformation begins.

stress Force applied to material that tends to change its dimensions: commonly, unit stress or total force divided by the area over which applied. Contrast with *strain.*

striations (1) Scratches, or small channels, gouged by glacial action. Bedrock, pebbles, and boulders may show striations produced when rocks trapped by ice were ground against bedrock or other rocks. Striations along bedrock surface are oriented in direction of ice flow across that surface. (2) In minerals, parallel, threadlike lines, or narrow bands, on face of mineral. Reflect internal atomic arrangement.

strike Direction of line formed by intersection of a rock surface with a horizontal plane. Strike is always perpendicular to direction of *dip.*

strike-slip fault *Fault* in which movement is almost in direction of fault's *strike.*

strip mining Surface mining in which soil and rock covering sought-for commodity are moved to one side. Some coal mining is pursued in this manner.

Strombolian Volcanic activity is characterized by weak to moderate ejection of pasty blobs of lava. It is named for the Italian island, Stromboli.

structural relief Difference in elevation of parts of deformed stratigraphic horizon.

structure Attitudes of deformed masses of rock.

subduction Process of one lithospheric plate descending beneath another.

subduction zone Elongate region along which lithospheric plate descends relative to another lithospheric plate.

sublimation Process by which solid material passes into gaseous state without first becoming liquid.

subsurface water Water below ground surface. Also referred to as *underground water* and *subterranean water.*

subterranean water Water below ground surface. Also referred to as *underground water* and *subsurface water.*

subtropical high pressure zones Zones of atmospheric high pressure located about 30° north and 30° south of the equator.

sulfate mineral   Mineral formed by combination of complex ion $(SO_4)^{2-}$ with positive ion. Common example: gypsum, $CaSO_4 \cdot 2H_2O$.

sulfide mineral   Mineral formed by direct union of element with sulfur. Examples: argentite, chalcocite, galena, sphalerite, pyrite, and cinnabar.

superposition   Law by which, if series of sedimentary rocks has not been overturned, topmost layer is always youngest and lowermost always oldest.

surface wave   Wave that travels along free surface of medium. Earthquake surface waves sometimes represented by symbol L.

surge   Applied to glaciers; rapid and sometimes catastrophic advance of ice.

suspect terrane   See *tectonostratigraphic terrane.*

suspended water   Underground water held in *zone of aeration* by molecular attraction exerted on water by rock and earth materials and by attraction exerted by water particles on one another.

suspension   Process by which material is buoyed up in air or water and moved about without making contact with surface while in transit. Contrast with *traction.*

symmetrical fold   *Fold* in which axial plane is essentially vertical. Limbs dip at similar angles.

syncline   A configuration of folded, stratified rocks in which rocks dip downward from opposite directions to come together in a trough. Reverse of *anticline.*

tablemount   See *guyot.*

tabular   Shape with large area relative to thickness.

taconite   Unleached iron formation of Lake Superior District. Consists of chert with hematite, magnetite, siderite, and hydrous iron silicates. Ore of iron, averaging 25 percent iron, but natural leaching turns it into ore with 50 to 60 percent iron.

talc   Silicate of magnesium common among metamorphic minerals. Crystalline structure based on tetrahedra arranged in sheets; greasy and extremely soft. Sometimes known as *soapstone.*

talus   Slope established by accumulation of rock fragments at foot of cliff or ridge. Rock fragments that form talus may be rock waste, slide rock, or pieces broken by frost action. Actually, term "talus" widely used to mean rock debris itself.

tarn   Lake formed in bottom of *cirque* after glacier ice has disappeared.

tectonic change of sea level   Change in sea level produced by land movement.

tectosilicate   Mineral with crystal structure containing *silicon-oxygen tetrahedra* arranged in three-dimensional frameworks.

temporary base level   Nonpermanent *base level,* such as that formed by lake.

tension   Stretching stress that tends to increase volume of a material.

tephra   General term for explosively fragmented volcanic ejecta (in contrast to lava flows); ranges in size from ash size to lapilli (2 to 64 mm) to bombs and blocks (more than 64 mm).

terminal moraine   Ridge or belt of *till* marking farthest advance of a glacier. Sometimes called *end moraine.*

terminal velocity   Constant rate of fall eventually attained by grain or body when acceleration caused by influence of gravity is balanced by resistance of fluid through which grain falls.

terrace   Nearly level surface, relatively narrow, bordering a stream or body of water and terminating in a steep bank. Commonly term is modified to indicate origin, as in *stream terrace* and *wave-cut terrace.*

terrae   Light-toned highlands of Moon.

terrigenous deposit   Material derived from above sea level and deposited in deep ocean. Example: volcanic ash.

tetrahedron   A four-sided solid. Used commonly in describing silicate minerals as shortened reference to *silicon-oxygen tetrahedron.*

texture   General physical appearance of rock, as shown by size, shape, and arrangement of particles that make it up.

thermal gradient   In Earth, rate at which temperature increases with depth below surface.

thermal pollution   Increase in normal temperatures of natural waters caused by intervention of human activities.

thermal spring   Spring that brings warm or hot water to surface. Temperature usually 6.5°C or more above mean air temperature. Sometimes called *warm spring,* or *hot spring.*

thermoremanent magnetism   Magnetism acquired by igneous rock as it cools below Curie temperatures of magnetic minerals in it. Abbreviation, TRM.

thin section   Slice of rock ground so thin as to be translucent.

tholeiite   Groups of basalts primarily composed of *plagioclase* (approximately An$_{50}$), *pyroxene,* and iron oxides as *phenocrysts* in glassy groundmass of *quartz* and alkali *feldspar;* little or no *olivine* present.

thrust fault   Fault in which *hanging wall* appears to have moved upward relative to *footwall;* opposite of *gravity,* or *normal, fault.* Also called *reverse fault.*

thrust-faulted mountain   Mountain whose primary geologic structures are *thrust faults.*

tidal current   Water current generated by tide-producing forces of Sun and Moon.

tidal inlet   Waterway from open water into a lagoon.

tidal wave   Popular but incorrect designation for *tsunami.*

tide   Alternate rising and falling of surface of ocean, other bodies of water, or Earth itself in response to forces resulting from motion of Earth, Moon, and Sun relative to each other.

till   Unstratified, unsorted glacial *drift* deposited directly by glacier ice.

tillite   Rock formed by lithification of *till.*

time-distance graph   Graph of travel time against distance.

tombolo   Sand bar connecting an island to mainland or joining two islands.

topographic deserts   Deserts deficient in rainfall because they are either located far from oceans toward center of continents or cut off from rain-bearing winds by high mountains.

topography   Shape and physical features of land.

topset bed   Layer of sediment constituting surface of *delta.* Usually nearly horizontal and covers edges of inclined *foreset beds.*

tourmaline   Silicate mineral of boron and aluminum with sodium, calcium, fluorine, iron, lithium, or magnesium. Formed at high temperatures and pressures through agency of fluids carrying boron and fluorine. Particularly associated with *pegmatites.*

township and range   See Appendix D.

traction   Process of carrying material along bottom of a stream. Traction includes movement by saltation, rolling, or sliding.

trailing continental margin   The margin of a continent that occurs on the side opposite the margin moving toward another continental or oceanic plate (leading edge); commonly is less active tectonically, behaving in a passive manner. The margin of a continent moving away from the locus of spreading along a midoceanic ridge as, for example, the eastern coasts of North and South America.

transcurrent fault   See *strike-slip fault.*

transducer   Device that picks up relative motion between mass of seismograph and ground and converts this into form that can be recorded.

transform fault   Point at which strike-slip displacements stop and another structural feature, such as a ridge, develops.

transition element   Element in series in which inner shell is being filled with electrons after outer shell has been started. All transition elements metallic in free state.

transpiration   Process by which water vapor escapes from a living plant and enters atmosphere.

transverse dune   Dune formed in areas of scanty vegetation and in which sand has moved in ridge at right angles to wind. Exhibits gentle windward slope and steep leeward slope characteristic of other dunes.

transverse wave   See *shear,* or *shake, wave.*

trap rock   Popular synonym for *basalt* or *diabase.*

travel time   Total elapsed time for wave to travel from source to designated point.

travertine   Form of calcium carbonate, $CaCO_3$, which forms stalactites, stalagmites, and other deposits in limestone caves or incrustations around mouths of hot and cold calcareous springs. Sometimes known as *tufa,* or *dripstone.*

trellis stream pattern   Roughly rectilinear arrangement of stream courses in pattern reminiscent of garden trellis, developed in region where rocks of differing resistance to erosion have been folded, beveled, and uplifted.

trenches   See *island-arc deeps.*

TRM   See *thermoremanent magnetism.*

tropical deserts   Deserts lying between 5° to 30° north and south of equator.

truncated spur   Beveled end of divide between two tributary valleys where they join a main valley that has been glaciated. Glacier of main valley has worn off end of divide.

tsunami   Large wave in ocean generated at time of earthquake. Sometimes called *seismic sea wave*. Popularly but incorrectly known as *tidal wave*.

tufa   See *Travertine*.

tuff   Rock consolidated from volcanic ash.

tundra   Arctic treeless plain often developed on top of permanently frozen ground. Extensive tundra regions occur in parts of North America, Europe, and Asia.

turbidites   Sedimentary deposits settled out of turbid water carrying particles of widely varying grade size. Characteristically display *graded bedding*.

turbidity current   Current in which limited volume of turbid or muddy water moves relative to surrounding water because of current's greater density.

turbulent flow   Mechanism by which fluid (such as water) moves over or past a rough surface. Fluid not in contact with irregular boundary outruns that slowed by friction or deflected by uneven surface. Fluid particles move in series of eddies or whirls. Most stream flow is turbulent; turbulent flow is important in both erosion and transportation. Contrast with *laminar flow*.

twin (=twin crystal)   An intergrowth of 2 or more single crystals of the same mineral in a mathematically describable manner, the symmetry of the 2 parts may be reflected about a common plane, axis, or center.

twinning   The development of a twin crystal by growth, transformation, or gliding.

ultimate base level   Sea level, lowest possible *base level* for a stream.

unconformity   Buried erosion surface separating two rock masses, older exposed to erosion for long interval of time before deposition of younger. If older rocks were deformed and not horizontal at time of subsequent deposition, surface of separation is *angular unconformity*. If older rocks remained essentially horizontal during erosion, surface separating them from younger rocks is called *disconformity*. Unconformity that develops between massive igneous or metamorphic rocks exposed to erosion and then covered by sedimentary rocks is called *nonconformity*.

underground water   Water below ground surface. Also referred to as *subsurface water* and *subterranean water*.

uneven fracture   Mineral habit of breaking along rough, irregular surfaces.

uniformitarianism   Concept that present is key to past. This means that processes now operating to modify Earth's surface have also operated in geologic past, that there is uniformity of processes past and present.

unit cell   In crystalline structure of mineral, three-dimensional grouping of atoms arbitrarily selected so that mineral's structure is represented by periodic repetition of this unit in *space lattice*.

unpaired terrace   A *terrace* formed when an eroding stream, swinging back and forth across a valley, encounters resistant rock beneath unconsolidated alluvium and is deflected, leaving behind single terrace with no corresponding terrace on other side of stream.

valley glacier   Glacier confined to stream valley. Usually fed from a *cirque*. Sometimes called *Alpine glacier* or *mountain glacier*.

valley train   Gently sloping plain underlain by glacial outwash and confined by valley walls.

varve   Pair of thin sedimentary beds, one coarse, one fine. This couplet has been interpreted as representing a cycle of 1 year or interval of thaw followed by interval of freezing in lakes fringing a glacier.

ventifact   Pebble, cobble, or boulder that has had its shape or surface modified by wind-driven sand.

vertical exaggeration   See Appendix D.

vesicle   Small cavity in aphanitic or glassy igneous rock, formed by expansion of bubble of gas or steam during solidification of rock.

viscosity   An internal property of rock that offers resistance to flow. Ratio of deforming force to rate at which changes in shape are produced.

volatile components   Materials in magma, such as water, carbon dioxide, and certain acids, whose vapor pressures are high enough to cause them to become concentrated in any gaseous phase that forms.

volcanic ash   *Dust-sized pyroclastic* particle: volume equal to, or less than, that of sphere with diameter of 0.06 mm.

volcanic block   Angular mass of newly congealed *magma* blown out in eruption. Contrast with *volcanic bomb*.

volcanic bomb   Rounded mass of newly congealed *magma* blown out in eruption. Contrast with *volcanic block*.

volcanic breccia   Rock formed from relatively large pieces of congealed lava embedded in mass of *ash*.

volcanic dust   *Pyroclastic* detritus consisting of particles of *dust size*.

volcanic earthquakes   Earthquakes caused by movements of *magma* or explosions of gases during volcanic activity.

volcanic eruption   Explosive or quiet emission of *lava, pyroclastics,* or volcanic gases at Earth's surface, usually from volcano but rarely from fissures.

volcanic mountain   Mountain formed primarily by volcanic activity.

volcanic neck   Solidified material filling vent, or pipe, of dead volcano.

volcanic tremor   Continuous shaking of ground associated with certain phase of volcanic eruption.

volcano   Landform developed by accumulation of magmatic products near central vent.

vug   Small unfilled cavity in rock, usually lined with crystalline layer of different composition from surrounding rock.

Vulcanian   Describes a type of volcanic eruption characterized by the explosive ejection of fragments of new lava. The fragments are often incandescent when they leave the vent but either solid or too viscous to assume any appreciable degree of rounding during their flight through the air.

warm glacier   Reaches melting temperature throughout thickness during summer season.

warm spring   *Thermal spring* that brings warm water to surface. Temperature usually 6.5°C or more above mean air temperature.

warp   Large section of continent composed of horizontal strata gently bent upward or downward.

water gap   Gap cut through resistant ridge by superimposed or antecedent stream.

water table   Upper surface of *zone of saturation* for underground water. An irregular surface with slope or shape determined by quantity of groundwater and permeability of Earth materials. In general, highest beneath hills and lowest beneath valleys.

water witching   See *dowsing*.

wave   Configuration of matter that transmits energy from one point to another.

weathering   Response of materials once in equilibrium within Earth's crust to new conditions at or near contact with water, air, or living matter. See also *chemical weathering* and *mechanical weathering*.

wrinkle ridges   Ridges found on surfaces of lunar *maria* and flooded craters. May be caused by uplift due to volcanism or to compression.

xenolith   Rock fragment foreign to igneous rock in which it occurs. Commonly inclusion of country rock intruded by igneous rock.

yazoo-type river   Tributary unable to enter main stream because of *natural levees* along main stream. Flows along back-swamp zone parallel to main stream.

yield point   Maximum stress that solid can withstand without undergoing permanent deformation, either by plastic flow or by rupture.

zone of aeration   Zone immediately below ground surface, in which openings partially filled with air and partially with water trapped by molecular attraction. Subdivided into (1) *belt of soil moisture*, (2) *intermediate belt*, and (3) *capillary fringe*.

zone of saturation   Underground region within which all openings filled with water. Top of zone of saturation is called *water table*. Water contained within zone of saturation is called *groundwater*.

zones of regional metamorphism   High-grade, above 700°C; middle-grade, 400 to 700°C; low-grade, 150 to 400°C.

INDEX

St. Peter Sandstone, 115
Salinity, ocean, 355–57
Salsano, D. Domenico, 181
Salt:
    dissolution by water, 21
    formation of, 20
    rock (see Halite)
    in seawater, 355–56
Saltation, 282, 385–86
Salton Sea, California, 126
Salton trough, 126
Saltwater invasion, 320–21
San Andreas fault, 170–72, 174–76, 193–95, 202, 231,
        243–45, 365
San Bernardino Mountains, California, 231
Sand, 103
    as building material, 429, 430
    tar, 412, 413
    wind-deposited, 390–94
    wind-driven, 383–86, 394
Sand dunes, 390–94
Sandstone, 99, 101, 103–4, 106, 124, 149
Sandstorms, 384
San Fernando Valley, California, 315
San Gabriel Mountains, California, 231, 243
San Joaquin Valley, California, 321
San Juan Formation, New Mexico, 86
Santa Clara Valley, California, 321
Santa Monica Mountains, California, 263
Santa Monica pier, Los Angeles, California, 374
Sapphires, 431, 468–69
Satellite craters, 441
Saturation, zone of, 307, 315
Saturn, 437, 452–53
Scale, 484
Scanning electron microscope, 25
Scheelite, 427, 476–77
Schist, 124, 127–28
Schistocity, development of, 127
Scratchability of minerals, 26–27
Sea, 369
Sea arches, 372, 373
Sea caves, 373
Sea cliff, 371
Sea floor:
    continental margins, 360–65
    deposits on, 368
    oceanic basins, 366
    oceanic ridges, 366–67
    seamounts, 367
    trenches, 226, 227, 367
Sea-floor spreading, 9–12, 212, 224–25, 227–34
Sea level, 374–77
Seamounts, 367
Seawater, 354–56, 424, 481
Sea waves, seismic, 200–201
Secondary craters, 441
Secondary magmas, 46
Secondary rocks, 144
Secondary waves, 184, 185
Sectile tenacity, 465
Sedimentary facies, 114–115

Sedimentary parent material, 80
Sedimentary rocks, 5, 6, 9, 96–116
    bedding of, 107–8
    chemical, 97, 100, 103–7
    chemical composition of, 107
    classification of, 103–4
    color of, 114
    concretions in, 110
    correlation of, 137–38
    detrital, 97, 99, 100, 103–4
    diagenesis of, 102
    features of, 107–15
    formation of, 96–102
    fossils in (see Fossils)
    geodes in, 110
    lithification of, 102
    magnetic reversals and, 224–25
    mineral composition of, 99–100
    mud cracks preserved in, 109
    nodules in, 110
    of oceans, 367–68, 403, 405
    origin of material, 96–98
    radioactive dating of, 149, 151
    relative abundance of, 97, 106
    relative time determination and, 136–44
    ripple marks preserved in, 109
    in rock cycle, 8–9, 12, 13
    slope shape and, 265–67
    sole marks preserved in, 109–10
    texture of, 100–103
    transportation by wind, 383–86
    types of, 103–7
    (See also Sedimentation)
Sedimentation:
    absolute time and, 149–50
    defined, 98
    factors controlling, 97
    mineral deposits and, 423–24
    plate tectonics and, 115–116
Seif dunes, 394
Seismic exploration for fossil fuels, 407, 408
Seismic gap, 195
Seismic sea waves, 200–201
Seismic-sea-wave warning system (SSWWS), 200, 201
Seismic waves, 203, 229–31
Seismograms, 181, 182, 184
Seismograph networks, 187
Seismographs, 181–83
Seismology, 180–83
Seismometers, 181
Seismoscopes, 180, 181
Semiprecious stones, 431
Seneca, 272
Serpentine (chrysotile), 430, 463, 464, 476–77
Sessa, F., 3
Shackleton, N. J., 350
Shakespeare, William, 179
Shale, 106, 123, 124, 265
    oil, 409, 411–12
Shatter cones, 456
Shear, 159, 244
Shear strength, 159

The following numbers and exponents (and prefixes) are essentially an outline of the metric system:

$$1,000,000,000,000 = 10^{12} \text{ (tera-)}$$
$$1,000,000,000 = 10^{9} \text{ (giga-)}$$
$$1,000,000 = 10^{6} \text{ (mega-)}$$
$$1,000 = 10^{3} \text{ (kilo-)}$$
$$100 = 10^{2} \text{ (hecto-)}$$
$$10 = 10^{1} \text{ (deka-)}$$
$$1 = 10^{0} \text{ [unit]}$$
$$0.1 = 10^{-1} \text{ (deci-)}$$
$$0.01 = 10^{-2} \text{ (centi-)}$$
$$0.001 = 10^{-3} \text{ (milli-)}$$
$$0.000001 = 10^{-6} \text{ (micro-)}$$
$$0.000000001 = 10^{-9} \text{ (nano-)}$$

The fundamental unit of length is the meter, originally defined as one ten millionth of the distance from the equator to the North Pole, later defined as the distance between two marks inscribed on the standard meter bar in Paris, and most recently specified in terms of the wavelength of krypton. So:

$$1 \text{ km (kilometer)} = 10^{3} \text{ m (meter)}$$
$$1 \text{ cm (centimeter)} = 10^{-2} \text{ m}$$
$$1 \text{ mm (millimeter)} = 10^{-3} \text{ m}$$
$$1 \text{ } \mu\text{m (micrometer)} = 10^{-6} \text{ m}$$

The official definition of an inch is based on the length of a meter:

$$1 \text{ in (inch)} = 2.54 \times 10^{-2} \text{ m}$$

Temperature is measured in degrees Celsius, a scale in which the interval between the freezing and boiling points of water is divided into 100 degrees, with 0° representing the freezing point and 100° the boiling point. So:

$$\frac{5}{9}\,°\text{C (Celsius)} = 1\,°\text{F (Fahrenheit)} \quad \text{ or }$$

$$°\text{C} = \frac{(°\text{F} - 32°)}{1.8} \quad \text{ or } \quad °\text{F} = (°\text{C} \times 1.8) + 32°$$

Conversion of mass

	Grams, g	Kilograms, kg	Pounds, lb	Ounces, oz
g	1	1,000	453.6	28.35
kg	0.001	1	0.4536	2.835×10^{-2}
lb	2.205×10^{-3}	2.205	1	6.25×10^{-2}
oz	3.527×10^{-2}	35.27	16	1

Multiply units in the column heads by the figures in the table to convert to units at left. (For example, to convert pounds to kilograms, multiply the number of pounds by 0.4536).

Conversion of volume

	Cubic meters, m³	Cubic yards, yd³	Cubic centimeters, cm³	Cubic inches, in³	Cubic feet, ft³
m³	1	0.7646	1×10^{-6}	1.639×10^{-5}	2.832×10^{-2}
yd³	1.308	1	1.308×10^{-6}	2.143×10^{-5}	3.704×10^{-2}
cm³	1×10^{6}	7.646×10^{5}	1	16.39	2.832×10^{4}
in³	6.102×10^{4}	46,656	6.102×10^{-2}	1	1,728
ft³	35.31	27	3.531×10^{-5}	5.787×10^{-4}	1

Multiply units in the column heads by the figures in the table to convert to units at left. (For example, to convert cubic inches to cubic centimeters, multiply the number of cubic inches by 16.39).